Analysis and Partial Differential Equations

PURE AND APPLIED MATHEMATICS

A Program of Monographs, Textbooks, and Lecture Notes

LECTURE NOTES

IN PURE AND APPLIED MATHEMATICS

1. *N. Jacobson,* Exceptional Lie Algebras
2. *L.-Å. Lindahl and F. Poulsen,* Thin Sets in Harmonic Analysis
3. *I. Satake,* Classification Theory of Semi-Simple Algebraic Groups
4. *F. Hirzebruch, W. D. Newmann, and S. S. Koh,* Differentiable Manifolds and Quadratic Forms (out of print)
5. *I. Chavel,* Riemannian Symmetric Spaces of Rank One (out of print)
6. *R. B. Burckel,* Characterization of C(X) Among Its Subalgebras
7. *B. R. McDonald, A. R. Magid, and K. C. Smith,* Ring Theory: Proceedings of the Oklahoma Conference
8. *Y.-T. Siu,* Techniques of Extension on Analytic Objects
9. *S. R. Caradus, W. E. Pfaffenberger, and B. Yood,* Calkin Algebras and Algebras of Operators on Banach Spaces
10. *E. O. Roxin, P.-T. Liu, and R. L. Sternberg,* Differential Games and Control Theory
11. *M. Orzech and C. Small,* The Brauer Group of Commutative Rings
12. *S. Thomeier,* Topology and Its Applications
13. *J. M. Lopez and K. A. Ross,* Sidon Sets
14. *W. W. Comfort and S. Negrepontis,* Continuous Pseudometrics
15. *K. McKennon and J. M. Robertson,* Locally Convex Spaces
16. *M. Carmeli and S. Malin,* Representations of the Rotation and Lorentz Groups: An Introduction
17. *G. B. Seligman,* Rational Methods in Lie Algebras
18. *D. G. de Figueiredo,* Functional Analysis: Proceedings of the Brazilian Mathematical Society Symposium
19. *L. Cesari, R. Kannan, and J. D. Schuur,* Nonlinear Functional Analysis and Differential Equations: Proceedings of the Michigan State University Conference
20. *J. J. Schäffer,* Geometry of Spheres in Normed Spaces
21. *K. Yano and M. Kon,* Anti-Invariant Submanifolds
22. *W. V. Vasconcelos,* The Rings of Dimension Two
23. *R. E. Chandler,* Hausdorff Compactifications
24. *S. P. Franklin and B. V. S. Thomas,* Topology: Proceedings of the Memphis State University Conference
25. *S. K. Jain,* Ring Theory: Proceedings of the Ohio University Conference
26. *B. R. McDonald and R. A. Morris,* Ring Theory II: Proceedings of the Second Oklahoma Conference
27. *R. B. Mura and A. Rhemtulla,* Orderable Groups
28. *J. R. Graef,* Stability of Dynamical Systems: Theory and Applications
29. *H.-C. Wang,* Homogeneous Branch Algebras
30. *E. O. Roxin, P.-T. Liu, and R. L. Sternberg,* Differential Games and Control Theory II
31. *R. D. Porter,* Introduction to Fibre Bundles
32. *M. Altman,* Contractors and Contractor Directions Theory and Applications
33. *J. S. Golan,* Decomposition and Dimension in Module Categories
34. *G. Fairweather,* Finite Element Galerkin Methods for Differential Equations
35. *J. D. Sally,* Numbers of Generators of Ideals in Local Rings
36. *S. S. Miller,* Complex Analysis: Proceedings of the S.U.N.Y. Brockport Conference
37. *R. Gordon,* Representation Theory of Algebras: Proceedings of the Philadelphia Conference
38. *M. Goto and F. D. Grosshans,* Semisimple Lie Algebras
39. *A. I. Arruda, N. C. A. da Costa, and R. Chuaqui,* Mathematical Logic: Proceedings of the First Brazilian Conference

40. *F. Van Oystaeyen,* Ring Theory: Proceedings of the 1977 Antwerp Conference
41. *F. Van Oystaeyen and A. Verschoren,* Reflectors and Localization: Application to Sheaf Theory
42. *M. Satyanarayana,* Positively Ordered Semigroups
43. *D. L. Russell,* Mathematics of Finite-Dimensional Control Systems
44. *P.-T. Liu and E. Roxin,* Differential Games and Control Theory III: Proceedings of the Third Kingston Conference, Part A
45. *A. Geramita and J. Seberry,* Orthogonal Designs: Quadratic Forms and Hadamard Matrices
46. *J. Cigler, V. Losert, and P. Michor,* Banach Modules and Functors on Categories of Banach Spaces
47. *P.-T. Liu and J. G. Sutinen,* Control Theory in Mathematical Economics: Proceedings of the Third Kingston Conference, Part B
48. *C. Byrnes,* Partial Differential Equations and Geometry
49. *G. Klambauer,* Problems and Propositions in Analysis
50. *J. Knopfmacher,* Analytic Arithmetic of Algebraic Function Fields
51. *F. Van Oystaeyen,* Ring Theory: Proceedings of the 1978 Antwerp Conference
52. *B. Kedem,* Binary Time Series
53. *J. Barros-Neto and R. A. Artino,* Hypoelliptic Boundary-Value Problems
54. *R. L. Sternberg, A. J. Kalinowski, and J. S. Papadakis,* Nonlinear Partial Differential Equations in Engineering and Applied Science
55. *B. R. McDonald,* Ring Theory and Algebra III: Proceedings of the Third Oklahoma Conference
56. *J. S. Golan,* Structure Sheaves over a Noncommutative Ring
57. *T. V. Narayana, J. G. Williams, and R. M. Mathsen,* Combinatorics, Representation Theory and Statistical Methods in Groups: YOUNG DAY Proceedings
58. *T. A. Burton,* Modeling and Differential Equations in Biology
59. *K. H. Kim and F. W. Roush,* Introduction to Mathematical Consensus Theory
60. *J. Banas and K. Goebel,* Measures of Noncompactness in Banach Spaces
61. *O. A. Nielson,* Direct Integral Theory
62. *J. E. Smith, G. O. Kenny, and R. N. Ball,* Ordered Groups: Proceedings of the Boise State Conference
63. *J. Cronin,* Mathematics of Cell Electrophysiology
64. *J. W. Brewer,* Power Series Over Commutative Rings
65. *P. K. Kamthan and M. Gupta,* Sequence Spaces and Series
66. *T. G. McLaughlin,* Regressive Sets and the Theory of Isols
67. *T. L. Herdman, S. M. Rankin, III, and H. W. Stech,* Integral and Functional Differential Equations
68. *R. Draper,* Commutative Algebra: Analytic Methods
69. *W. G. McKay and J. Patera,* Tables of Dimensions, Indices, and Branching Rules for Representations of Simple Lie Algebras
70. *R. L. Devaney and Z. H. Nitecki,* Classical Mechanics and Dynamical Systems
71. *J. Van Geel,* Places and Valuations in Noncommutative Ring Theory
72. *C. Faith,* Injective Modules and Injective Quotient Rings
73. *A. Fiacco,* Mathematical Programming with Data Perturbations I
74. *P. Schultz, C. Praeger, and R. Sullivan,* Algebraic Structures and Applications Proceedings of the First Western Australian Conference on Algebra
75. *L. Bican, T. Kepka, and P. Nemec,* Rings, Modules, and Preradicals
76. *D. C. Kay and M. Breen,* Convexity and Related Combinatorial Geometry: Proceedings of the Second University of Oklahoma Conference
77. *P. Fletcher and W. F. Lindgren,* Quasi-Uniform Spaces
78. *C.-C. Yang,* Factorization Theory of Meromorphic Functions
79. *O. Taussky,* Ternary Quadratic Forms and Norms
80. *S. P. Singh and J. H. Burry,* Nonlinear Analysis and Applications
81. *K. B. Hannsgen, T. L. Herdman, H. W. Stech, and R. L. Wheeler,* Volterra and Functional Differential Equations

82. *N. L. Johnson, M. J. Kallaher, and C. T. Long,* Finite Geometries: Proceedings of a Conference in Honor of T. G. Ostrom
83. *G. I. Zapata,* Functional Analysis, Holomorphy, and Approximation Theory
84. *S. Greco and G. Valla,* Commutative Algebra: Proceedings of the Trento Conference
85. *A. V. Fiacco,* Mathematical Programming with Data Perturbations II
86. *J.-B. Hiriart-Urruty, W. Oettli, and J. Stoer,* Optimization: Theory and Algorithms
87. *A. Figa Talamanca and M. A. Picardello,* Harmonic Analysis on Free Groups
88. *M. Harada,* Factor Categories with Applications to Direct Decomposition of Modules
89. *V. I. Istrăţescu,* Strict Convexity and Complex Strict Convexity: Theory and Applications
90. *V. Lakshmikantham,* Trends in Theory and Practice of Nonlinear Differential Equations
91. *H. L. Manocha and J. B. Srivastava,* Algebra and Its Applications
92. *D. V. Chudnovsky and G. V. Chudnovsky,* Classical and Quantum Models and Arithmetic Problems
93. *J. W. Longley,* Least Squares Computations Using Orthogonalization Methods
94. *L. P. de Alcantara,* Mathematical Logic and Formal Systems
95. *C. E. Aull,* Rings of Continuous Functions
96. *R. Chuaqui,* Analysis, Geometry, and Probability
97. *L. Fuchs and L. Salce,* Modules Over Valuation Domains
98. *P. Fischer and W. R. Smith,* Chaos, Fractals, and Dynamics
99. *W. B. Powell and C. Tsinakis,* Ordered Algebraic Structures
100. *G. M. Rassias and T. M. Rassias,* Differential Geometry, Calculus of Variations, and Their Applications
101. *R.-E. Hoffmann and K. H. Hofmann,* Continuous Lattices and Their Applications
102. *J. H. Lightbourne, III, and S. M. Rankin, III,* Physical Mathematics and Nonlinear Partial Differential Equations
103. *C. A. Baker and L. M. Batten,* Finite Geometries
104. *J. W. Brewer, J. W. Bunce, and F. S. Van Vleck,* Linear Systems Over Commutative Rings
105. *C. McCrory and T. Shifrin,* Geometry and Topology: Manifolds, Varieties, and Knots
106. *D. W. Kueker, E. G. K. Lopez-Escobar, and C. H. Smith,* Mathematical Logic and Theoretical Computer Science
107. *B.-L. Lin and S. Simons,* Nonlinear and Convex Analysis: Proceedings in Honor of Ky Fan
108. *S. J. Lee,* Operator Methods for Optimal Control Problems
109. *V. Lakshmikantham,* Nonlinear Analysis and Applications
110. *S. F. McCormick,* Multigrid Methods: Theory, Applications, and Supercomputing
111. *M. C. Tangora,* Computers in Algebra
112. *D. V. Chudnovsky and G. V. Chudnovsky,* Search Theory: Some Recent Developments
113. *D. V. Chudnovsky and R. D. Jenks,* Computer Algebra
114. *M. C. Tangora,* Computers in Geometry and Topology
115. *P. Nelson, V. Faber, T. A. Manteuffel, D. L. Seth, and A. B. White, Jr.* Transport Theory, Invariant Imbedding, and Integral Equations: Proceedings in Honor of G. M. Wing's 65th Birthday
116. *P. Clément, S. Invernizzi, E. Mitidieri, and I. I. Vrabie,* Semigroup Theory and Applications
117. *J. Vinuesa,* Orthogonal Polynomials and Their Applications: Proceedings of the International Congress
118. *C. M. Dafermos, G. Ladas, and G. Papanicolaou,* Differential Equations: Proceedings of the EQUADIFF Conference
119. *E. O. Roxin,* Modern Optimal Control: A Conference in Honor of Solomon Lefschetz and Joseph P. Lasalle
120. *J. C. Díaz,* Mathematics for Large Scale Computing
121. *P. S. Milojević,* Nonlinear Functional Analysis
122. *C. Sadosky,* Analysis and Partial Differential Equations: A Collection of Papers Dedicated to Mischa Cotlar

Other Volumes in Preparation

Mischa Cotlar
Chicago, 1952

Analysis and Partial Differential Equations

A Collection of Papers Dedicated to Mischa Cotlar

edited by

Cora Sadosky

Howard University
Washington, D.C.

MARCEL DEKKER, INC. New York and Basel

Library of Congress Cataloging-in--Publication Data

Analysis and partial differential equations : a collection of papers
dedicated to Mischa Cotlar / edited by Cora Sadosky.
p. cm. -- (Lecture notes in pure and applied mathematics ;
122)
ISBN 0-8247-8302-6 (alk. paper)
1. Harmonic analysis. 2. Functions of complex variables.
3. Functional analysis. 4. Operator theory. 5. Differential
equations, Partial. 6. Cotlar, Mischa. I. Sadosky, Cora.
. II. Cotlar, Mischa. III. Series: Lecture notes in pure and
applied mathematics ; v. 122.
QA403.A56 1990
515--dc20 89-28898
CIP

This book is printed on acid-free paper

MARCEL DEKKER, INC.
270 Madison Avenue, New York, New York 10016

Current printing (last digit):
10 9 8 7 6 5 4 3 2 1

PRINTED IN THE UNITED STATES OF AMERICA

To

Mischa Cotlar

on the occasion of his seventy-fifth birthday

Preface

In May 1988, Mischa Cotlar turned seventy-five years of age. For a long time, his work and his life have been a source of inspiration for many colleagues, students, and friends throughout the world. On the occasion of his birthday, we wanted to express our appreciation to him, and so this book came about.

Cotlar himself has worked on several topics in analysis, mainly on harmonic analysis and operator theory. Both these fields being closely connected with the theory of partial differential equations, it appeared natural to group the contributed papers to form the first three parts of this volume, although some of them could fit under more than one heading.

Much of Cotlar's original work is not easily available, and many of his contributions are referred to only through secondary sources. Thus it seemed appropriate to include as a fourth part to this volume a descriptive account of Cotlar's work along with a complete bibliography of his mathematical books and papers. The material for this part was compiled with the very valuable help of John Horváth, whose collaboration and support I want to especially acknowledge.

Mischa Cotlar is an autodidact and a devoted vegetarian. His modesty is overwhelming, and his interest in philosophical questions is as deep as his commitment to the quest for peace and understanding among people. Together with his wife, he leads a very simple life, filled with deep friendships and very far from ordinary social conventions. Since most of what even his friends know about him has a legendary ring, it was to this point to include not only a biography of Cotlar but also the recollections of some of his oldest friends. We were fortunate to get the permission to print the speech of introduction of Cotlar as a new member of the National Academy of Exact Sciences of Argentina, delivered by Alberto P. Calderón on April 19, 1988. In addition we include the words of José Luis Massera and Manuel Sadosky from a meeting in honor of Cotlar at the University of Buenos Aires, under the auspices of the Secretariat of Science and Technology of Argentina, on April 18 and 19, 1988.

The scientific standing of the contributors to this volume, and the importance of their work dedicated to Mischa Cotlar, speak of the extraordinary qualities of this very modest man.

Cora Sadosky

Contents

Contributors

R. Arocena Universidad de la República, Montevideo, Uruguay

N. Asmar Department of Mathematics, University of Missouri–Columbia, Columbia, Missouri

J. Ball Department of Mathematics, Virginia Polytechnic Institute, Blacksburg, Virginia

C. A. Berenstein Mathematics Department, University of Maryland, College Park, Maryland

H. Berestycki Université de Paris VI, Paris, France

E. Berkson Department of Mathematics, University of Illinois, Urbana, Illinois

L. Caffarelli Institute for Advanced Study, School of Mathematics, Princeton, New Jersey

A. P. Calderón Instituto Argentino de Matemática, Buenos Aries, Argentina

D. -C. Chang Department of Mathematics, University of Maryland, College Park, Maryland

R. R. Coifman Department of Mathematics, Yale University, New Haven, Connecticut

G. Corach Instituto Argentino de Matemática, Buenos Aires, Argentina

B. Dahlberg Department of Mathematics, Washington University, St. Louis, Missouri

G. David Centre de Mathématiques, École Polytéchnique, Palaiseau, France

H. Donnelly Department of Mathematics, Purdue University, West Lafayette, Indiana

H. Dym Department of Theoretical Mathematics, The Weizmann Institute of Science, Rehovot, Israel

C. Fefferman Department of Mathematics, Princeton University, Princeton, New Jersey

R. Fefferman Department of Mathematics, University of Chicago, Chicago, Illinois

M. Feldman Department of Mathematics and Statistics, South Illinois University at Edwardsville, Edwardsville, Illinois
C. Foias Department of Mathematics, Indiana University, Bloomington, Indiana
D. J. G. Universidad de Buenos Aires, Buenos Aires, Argentina
J. García-Cuerva Facultad de Ciencias, Universidad Autonoma de Madrid, Madrid, Spain
A. E. Gatto Department of Mathematics, DePaul University, Chicago, Illinois
R. Gay Université de Bordeaux I, Talence, France
T. A. Gillespie Department of Mathematics, University of Edinburgh, Edinburgh, Scotland
I. Gohberg School of Mathematical Sciences, Tel Aviv University, Tel Aviv, Israel
S. Gokhale Western Illinois University, Macomb, Illinois
F. A. Grünbaum Department of Mathematics, University of California at Berkeley, Berkeley, California
H. Helson Department of Mathematics, University of California at Berkeley, Berkeley, California
D. A. Herrero Department of Mathematics, Arizona State University, Tempe, Arizona
J. Horváth Department of Mathematics, University of Maryland, College Park, Maryland
P. W. Jones Department of Mathematics, Yale University, New Haven, Connecticut
J.-P. Kahane Département de Mathématique, Université de Paris-Sud, Orsay, France
C. Kenig Department of Mathematics, University of Chicago, Chicago, Illinois
J. L. Massera Instituto de Matemática y Estadística, Universidad de la República, Montevideo, Uruguay
M. Milman Department of Mathematics, Florida Atlantic University, Boca Raton, Florida
A. Nagel Department of Mathematics, University of Wisconsin, Madison, Wisconsin
N. K. Nikolskii Steklov Mathematics Institute, Leningrad, U.S.S.R.
L. Nirenberg Courant Institute, New York University, New York, New York
J. Peetre University of Stockholm, Stockholm, Sweden
V. V. Peller Steklov Mathematics Institute, Leningrad, U.S.S.R.
G. Ponce Department of Mathematics, Pennsylvania State University, University Park, Pennsylvania
H. Porta Department of Mathematics, University of Illinois, Urbana, Illinois
L. Qing Academia Sinica, Beijing, China
L. Recht Departamento de Matemáticas, Universidad Simón Bolívar, Caracas, Venezuela
F. Ricci Istituto Politecnico di Torino, Torino, Italy
R. Rochberg Department of Mathematics, Washington University, St. Louis, Missouri
L. Rodman Department of Mathematics, College of William and Mary, Williamsburg, Virginia
C. Sadosky Department of Mathematics, Howard University, Washington D.C.
M. Sadosky Universidad de Buenos Aires, Buenos Aires, Argentina

Y. Sagher Department of Mathematics, University of Illinois at Chicago, Chicago, Illinois

N. Salinas Department of Mathematics, University of Kansas, Lawrence, Kansas

D. Sarason Department of Mathematics, University of California at Berkeley, Berkeley, California

T. Schonbek Department of Mathematics, Florida Atlantic University, Boca Raton, Florida

C. Segovia Instituto Argentino de Matemática, Buenos Aires, Argentina

S. Semmes Department of Mathematics, Rice University, Houston, Texas

F. Soria Facultad de Ciencias, Universidad Autónoma de Madrid, Madrid, Spain

E. M. Stein Department of Mathematics, Princeton University, Princeton, New Jersey

A. Tannenbaum Department of Electrical Engineering, University of Minnesota, Minneapolis, Minnesota

J. L. Torrea Facultad de Ciencias, Universidad Autónoma de Madrid, Madrid, Spain

S. Vági Department of Mathematics, DePaul University, Chicago, Illinois

A. L. Volberg Steklov Mathematics Institute, Leningrad, U.S.S.R.

G. Weiss Department of Mathematics, Washington University, St. Louis, Missouri

A. Yger Université de Bordeaux I, Talence, France

A. Založnik FAGG University of Ljublijana, Ljublijana, Yugoslavia

K. Zhou Department of Mathematics, University of Illinois at Chicago, Chicago, Illinois

Mischa Cotlar: A Biography

Mischa Cotlar was born in Sarney, Ukraine, in 1913, the youngest of the two sons of Ovsey Cotlar and Sara Medved, both of Jewish descent. His father, who administered a flour mill, was an avid chess player, was passionately interested in mathematics and in music, and had a considerable library. In 1928 the Cotlars emigrated to Uruguay, a country then known as "the Switzerland of the Americas." At the time, Mischa was 15 and had had only a year of formal education. However, his father had taught him mathematics and music, and Mischa was already a good pianist.

Upon their arrival to Montevideo, the Cotlars were very poor. The four lived crammed in a single room, and each of them had to work. The father sold newspapers at a downtown street corner, the oldest son became a tramway motorman, and Mischa played the piano from 4 p.m. to 4 a.m. in a harbor bar. The first thing Mr. Cotlar bought (in installments) was a piano for Mischa, which had to be squeezed in the already tight living space.

Once they were settled and their immediate survival assured, Mr. Cotlar turned his attention to chess and won the annual contest of the Sociedad Uruguaya de Ajedrez. For this he received much publicity, and the journalists were surprised and amused at realizing that Ovsey Cotlar was not the pseudonym of a known Uruguayan chess player but rather the real name of a Ukraine-born newspaper street vendor, who claimed to have a son gifted in mathematics.

The founder of the Uruguayan school of mathematics, Rafael Laguardia, was an avid chess player himself, and he read the interviews with the prize-winning newspaper vendor with interest. Professor Laguardia immediately went to visit Mr. Cotlar, who introduced Mischa to him.

Laguardia, a young professor of mathematics at the Engineering School of the Universidad de la Republica, had studied with Picard at La Sorbonne. Although he had specialized in the new integration theory, his main interest was promoting mathematics among young people. His house in Montevideo was the meeting place for young men interested in mathematics, and students gathered there almost daily. He had the best mathematical library in town. Mischa was invited to his seminar, introduced to Laguardia's students and friends, and given access to the library. At the time, Mischa was very much interested in number theory, and he had solved several open problems suggested in Grave's textbook, which he had read in Sarney. He showed his results to Laguardia, who was astonished: a boy that had completed only one year of elementary school had actually proved several theorems, and had done so with absolute rigor, although with strange notations and rudimentary techniques. In fact, some of the results had been already proved by professional mathematicians, something that Mischa did not know. Two years later Laguardia invited Mischa to teach a course on number theory at his seminar in the Engineering School of the Universidad de la República.

Mischa continued playing the piano at the harbor bar until 1931, when Laguardia got him a job as a pianist in a chamber music trio playing at the fancy British Hotel in Punta del Este, the exclusive seaside resort. There, Mischa played with Jaime Tomasow, one of the great Argentine virtuosi, who later reached international renown, and who would become one of his closest friends.

In 1934 Professor Julio Rey Pastor, the Spanish mathematician, then living in Buenos Aires, was invited to lecture in Montevideo on the new developments in mathematics. Those lectures became a landmark in the history of Uruguayan mathematics and induced Mischa to move to Argentina. In 1935, he left Montevideo for Buenos Aires, which was to become his hometown.

Buenos Aires changed Mischa's life in many ways. He carried a letter for Professor J. C. Vignaux, who immediately recommended him as a private tutor of mathematics, and who later introduced him to the theory of analytic functions. Tutoring substituted for piano playing as a source of income.

Mischa shared a small room with his friends, the violinist Tomasow and the young philosopher Jaime Resta. Soon after arrival, he met two students of mathematics, Cora Ratto and Manuel Sadosky, with whom Mischa was to develop multiple bonds. In 1937, he also met Yanny Frenkel, a Russian student of Rey Pastor, later to be one of the first doctoral graduates in mathematics from the Universidad de Buenos Aires. In 1938, Mischa and Yanny were married.

Those were years of great intellectual excitement and deep emotional involvement. Mischa corresponded with Professor Maurice Fréchet, who helped him publish his results. In 1935 he became deeply interested in Hinduism, Hindu philosophy, and Buddhism. In Buenos Aires Mischa met J. Krishnamurty, who was to be his lifelong teacher and friend. The arrival of Beppo Levi to Rosario, Argentina, was also very significant in his development, since they shared scientific and philosophical interests. Their interaction was very fruitful, and Levi published Mischa's results in the two journals he founded in Rosario, *Mathematicae Notae* and the *Revista del Instituto de Matemáticas de Rosario*. A few years later Mischa met Rodolfo Ricabarra, the young mathematician with whom he established one of the deepest and most important intellectual, philosophical, and mathematical relationships of his life.

But Mischa still had no academic job. In 1947 he was appointed research instructor at the Universidad Nacional de La Plata, but was fired six months later, when the authorities found out that he had no official diploma. After World War II, several prominent mathematicians, including Marshall Stone, George Birkhoff, A. Adrian Albert, and Antoni Zygmund, visited Argentina. These visits were of historic relevance. It was then that Zygmund met Alberto Calderón, a young engineer, and Birkhoff discovered Mischa. Calderón went to Chicago and Birkhoff recommended Mischa for a Guggenheim Fellowship, which he won in 1950.

Mischa and Yanny traveled to the United States in 1951. They spent six months at Yale University, where Mischa studied ergodic theory with S. Kakutani. In the meantime, Professor Stone had arranged for him to be accepted as a graduate student at the University of Chicago, in spite of his lack of formal education. Mischa went to Chicago and worked under Antoni Zygmund, writing a thesis, "On the Theory of the Hilbert Transform." He received his Ph.D. from the University of Chicago in 1953. Finally he had a diploma!

Yet Mischa and Yanny yearned for Argentina, and they returned. There were no jobs at the Universidad de Buenos Aires, except for card-carrying members of the ruling party. However, a new group was forming at the Universidad Nacional de Cuyo, inspired by the exiled Portuguese mathematician A. Monteiro. Mischa was invited as Director of the Mathematics Institute of the Departmento de Investigaciones Científicas, where most of the young and active Argentine mathematicians—without political exclusions—clustered. There Mischa wrote his four celebrated papers expanding the work done in his doctoral thesis. These were published in the Revista Matematica Cuyana.

In 1957 Cotlar was appointed Professor of Mathematics at the School of Sciences of the Universidad de Buenos Aires. There he shaped several generations of Argentine mathematicians and directed the doctoral dissertations of Rafael Panzone, Cora Ratto de Sadosky, Eduardo Ortiz, and Concepción Ballester. Although during that period Mischa published relatively few papers, he wrote two textbooks and three monographs. The monographs appeared in the series "Cursos y Seminarios de Matemática" from the School of Science of the Universidad de Buenos Aires, edited by Cora Ratto de Sadosky, and they contained most of his original results of the time. The textbooks, one on the first algebra course, written in collaboration with Cora Ratto de Sadosky, and the other on functional analysis, in collaboration with R. Cignoli, were published by EUDEBA, the University of Buenos Aires Press.

In July 1966, a military junta that deposed President Arturo Illia ordered the assault and vicious beating of the faculty and students of the School of Sciences of the Universidad de Buenos Aires. Cotlar resigned—along with more than four hundred other faculty members–and returned briefly to Montevideo. A year later he was appointed professor at Rutgers University (1967–1971), and he taught also at the Université de Nice (1969–70) and at the Universidad Central de Venezuela (1971). In 1972 Mischa and Yanny returned to Argentina, where Mischa joined the faculty of the Universidad Nacional de La Plata. During that stay in Argentina, he initiated a steady collaboration with Cora Sadosky, the daughter of his friends Manuel and Cora, with whom he had occasionally worked. In 1974, under a deteriorating political situation in Argentina, the Cotlars went back to Caracas, where Mischa created a research group in analysis, and directed the doctoral dissertation of Rodrigo Arocena.

Cotlar has lectured and traveled extensively. He has taught at Washington University (1958), the University of Chicago (1962), Darmouth College (1963), Universidad de la República at Montevideo (1966), Universidad de Ingeniería at Lima (1967), McGill University (1972), and Howard University (1982). In 1989. he gave one of the Toeplitz Lectures at the University of Tel Aviv. He received the Waissman

Prize of the Consejo Nacional de Investigaciones Científicas y Tecnicas of Argentina (1964), the Premio Nacional de Ciencias of Venezuela (1984), and the Prize of the Academia de Ciencias of Spain (1950). He is Honorary Professor of the Universidad de Buenos Aires and the Universidad Nacional de La Plata. Since 1987, he has been a member of the Academia Nacional de Ciencias Exactas, Físicas y Naturales of Argentina.

During the period 1956–1966 Cotlar was intensely involved in the international movement for the social responsibility of scientists. In 1961, he founded with Cora Ratto de Sadosky and Mario Bunge, the Sociedad Argentina por la Resposabilidad Social de la Ciencia. In 1962, he carried to the International Congress of Mathematicians, held in Stockholm, Bertrand Russell's letter urging the noncooperation of mathematicians with the military establishment. In 1964, again with Cora Ratto de Sadosky and with Francisco Bullrich, he coedited *Columna 10*, a newsletter devoted to the analysis of the impact of science and technology on international politics. He also was on the board of directors of the Fundación Alberto Einstein, which supported talented students from the School of Sciences of the Universidad de Buenos Aires. Many young people who later became distinguished Argentine scientists received Fundación Einstein fellowships.

Cotlar now works in Caracas, and, since the return of democracy in Argentina in 1984, he has taught periodically in Buenos Aires.

Mischa and Yanny, although both born in Russia, consider themselves Argentinians. With his stubborn devotion for his adopted country, Mischa's life fits well into the title of Arturo Mallea's classic novel *The Story of an Argentine Passion*. For the deep influence he has had in the development of Argentine and Venezuelan mathematicians, it is a pride for Latin Americans to claim him as their own.

D. J. G.

Introduction of Professor Mischa Cotlar to the National Academy of Exact Sciences of Argentina

Alberto P. Calderón

It is a particular pleasure for me to introduce to this Academy its new member, Dr. Mischa Cotlar, who is not only an old friend of mine—our friendship spans more than forty years—but also an admired colleague, as well as a musician, a pianist.

Dr. Cotlar was born in Sarney, Ukraine, in 1913, and emigrated to Uruguay in the late twenties. He lived in Montevideo several years, earning his living as a professional pianist. In 1939, already a mature mathematician, he moved to Buenos Aires, and started his public mathematical career with the presentation of a paper, "Théorie d'Anagènes," at an international congress in Bordeaux. This work was later published, in Spanish, in *Anales de la Sociedad Científica Argentina*. In Buenos Aires, Cotlar began a stage of intense and uninterrupted mathematical activity, participating in the meetings of the *Unión Matemática Argentina*, as well as in the seminar directed by Professor Rey Pastor, and publishing in local and foreign mathematical journals.

Mischa Cotlar was an autodidact. This is why he did not have diplomas of any kind. This lack created bureaucratic problems that blocked his access to teaching positions in our universities. His only two appointments during that period were as "Investigator" at the Universities of La Plata (1946) and Buenos Aires (1948). This obstacle was happily overcome for the benefit of Argentine mathematics in 1953.

It is interesting to recall how this happened. The famous American mathematician Marshall Harvey Stone, son of the also famous American jurist Harlann Fiske Stone, visited Argentina several times after 1943, and he met Cotlar. When appointed Chairman of the Department of Mathematics at the University of Chicago, to which he managed to attract several of the greatest luminaries of contemporary mathematics, Professor Stone urged Cotlar to apply for a fellowship and visit Chicago. Cotlar received the Guggenheim Fellowship and went to the United States.* The University of Chicago, with its characteristic flexibility, discarded all the encumbering formalities and rapidly gave him the Doctorate in Philosophy in Math-

*After being recommended to the Guggenheim Foundation by Professor George Birkhoff, Cotlar went first to Yale University, where he spent one semester and became acquainted with Professor Katutani and other mathematicians. It was thanks to Professor Stone's help that he had his fellowship renewed in order to attend the University of Chicago, where he could work toward a doctorate. [Editor]

ematics. This was one of the important services to Argentine mathematics done by that department of the university where, the presence of Argentine mathematicians has since been, but for brief periods, a constant feature. Cotlar returned to Argentina with his Ph.D. and the country derived the benefits of having him teaching at the University. In 1953 he was appointed Director of the Instituto de Matemática of the Departmento de Investigación (DIC), at the Universidad de Cuyo, until it was closed in 1956. That same year he was appointed Full Professor of Mathematics at the Universidad de La Plata, and in 1957 he joined the faculty of the Department of Mathematics of the School of Sciences of the Universidad de Buenos Aires, where he became Professor Plenario. In 1966 he resigned his position after the events that disrupted that school, and went to teach at Rutgers University (1967–1971) and universities of Nice (1969) and Central de Venezuela (1971) which he joined permanently in 1974 and where he continues with his mathematical activity today.

Dr. Cotlar's bibliography includes eight books and monographs and more than eighty articles published in scientific journals. He has been widely acknowledged and rewarded. He has received among others, the Premio Nacional de Ciencias, of the CONICIT (Consejo Nacional de Investigaciones Científicas y Técnicas) from Venezuela, the Premio Waissman of our CONICET and the Premio de la Academia de Ciencias of Madrid.

His contributions are, for the greatest part, in the area of Analysis, and touch a wide variety of chapters of this discipline, as lattice theory, the theory of semi-ordered groups, the theory of integration, ergodic theory, Banach algebras, the theory of normal families of functions, potential theory, Toeplitz kernel theory, and many more. Many of these studies were made in collaboration, a fact that underlines his generosity and the pure scientific interest that inspires his work. Among his collaborators are his wife, Yanny Frenkel, and Rodolfo Ricabarra, Cora Sadosky, Rodrigo Arocena, Eduardo Zarantonello, Beppo Levi, Rafael Panzone, and Juan Carlos Vignaux.

Dr. Cotlar's mathematical work has very singular characteristics. One is its insights, bringing to light the deep roots and motivations of theories and theorems. The other is the vision that uncovers links and unsuspected relations between subjects that apparently have no connection at all. It is for these characteristics, I believe, that his works have a very definite taste of philosophical essays.

Examples of this are the four consecutive papers that appeared in the *Revista Matemática Cuyana, volume 1 (1955), Fascículo 2.* In one of them, the result now known as Cotlar's Lemma shows the reason why, independently from Fourier Theory, the Hilbert Transform is bounded in L^2. In another of those papers, he gives the unified treatment of the Hilbert Transform and the Ergodic Theorem, which is known as the Principle of Transference in modern ergodic theory.

Finally, I want to call attention to the fact that some of Cotlar's results, published in journals of little accessibility and slow distribution, have been named after other mathematicians that discovered them independently but years after him. One such example is the theorem of representation of σ-Boole algebras as a σ-algebra of sets modulo an ideal of "null" sets, which today is known as the Loomis–Sikorski theorem.

Dr. Cotlar, we congratulate ourselves for your incorporation into this Academy. The institution not only is enriched by your presence, but is also strengthened by your prestige. Thank you, Dr. Cotlar.

Buenos Aires
April, 1988

Mischa in Montevideo

José Luis Massera

I have been asked to say a few words on this occasion for which we celebrate the 75th birthday of our dear and admired Mischa Cotlar, to recall shared experiences and events, and the feelings evoked by our old friendship.

I will start by his name. Mischa Cotlar is, really, just and simply Mischa. Mischa is Mischa for his friends and old acquaintances, but also for many others. Mischa may be translated into Spanish as "Miguelito," but, as often happens, in the translation important nuances are lost. Mischa is a very common and typical variation of a Russian name, which besides being a diminutive implies a strong load of affection. And, in the case of our Mischa, this last component has a lot of importance and it is deeply ingrained in the personality of its holder.

The history—my history—of my friendship with Mischa begins some 50 years ago and is closely tied to that of another great friend, Rafael Laguardia. Laguardia was the elder of our group of young Montevideans of the 1930s, avidly interested in mathematics. He was the oldest not only in age but also in background, since he had completed a degree in science at the Sorbonne, studying under some of the most distinguished professors of France.

Since my adolescence, I was strongly attracted by mathematics and has started to study—completely alone—subjects that were far beyond the high school requirements. I did that in the most crazy, absurd, extravagant way, but this is not the occasion to enter into details. In fact, I learned many things, which constituted an indescribably chaotic mass of knowledge.

In 1935 I realized that this was leading nowhere, and one day I visited Laguardia, without any sort of recommendation, in his very modest home, which was some three

---

Talk given at the session in honor of Professor Mischa Cotlar on the occasion of his seventy-fifth birthday at the University of Buenos Aires, Argentina, on April 18, 1988.

blocks from my own. I did not know Laguardia, but I knew who he was because a schoolmate had pointed him out to me in the quadrangle of the old Instituto Vázquez Acevedo (Montevideo's leading college preparatory school.). This visit started a very close friendship that lasted until his much lamented death a few years ago. I was in jail at the time and could not be with him then. A short time before dying, Laguardia presented me with an *Enciclopedia Larousse* which somehow passed through the jail censors, and which was, and still is, a source of remembrance of him. It was, for me and for many other prisoners, an open window to the world which all of us appreciated greatly. Laguardia introduced me to the small group of mathematical *aficionados*: Antonio Petracca, Fernando Fortezza, Carlos A. Infantozzi, Luis Castagnetto, and ... Mischa. Laguardia had met—as I also did, later on—a very cultured Russian immigrant who had reached Uruguay in the second decade of this century. His life was very modest, and he had a small stand, where he sold newspapers and magazines, in front of a hotel, on the Avenida 18 de Julio (Montevideo's main avenue). This gentleman had a teenage son who showed a great mathematical vocation, and he asked Laguardia to help the youngster in his studies. The son was Miguelito, Mischa, who joined our small group of enthusiasts. In this way our long-lasting friendship began, which has continued uninterrupted until now, in spite of all the accidents of life and the geographical constraints.

I want to say a few more words to describe the ambience in the small mathematical circle at Montevideo at the time of our encounter. For a rather long while, every weekend "Don" Julio Rey Pastor came to visit us from Buenos Aires and taught us on some of the new developments of mathematics—of which we knew nothing. Our mathematical relation with Rey Pastor was very fruitful, since we were able to discuss our mathematical problems freely in a one-to-one basis. In this way, our horizons were open to the then new mathematics: general topology, algebra, modern analysis. Don Julio had a very complex personality, and occasionally he was very difficult to deal with; but the overall balance of our relationship with him is, undoubtedly, enormously positive for the development of mathematics in Uruguay.

On the other hand, our small group established a close and relatively assiduous contact with the much larger and more mature group of the Argentine mathematicians. We participated in the main scientific meetings of the Unión Matemática Argentina, and published most of our first papers in its journal, the *Revista de la UMA*. I want to mention here González Domínguez, Babini, Zarantonello, among so many others, later joined by several expatriate mathematicians: Santaló, Pi Calleja, and Balanzat from Spain, Beppo Levi and Terracini from Italy, and Monteiro from Portugal. These distinguished mathematicians, in different moments and circumstances, had to emigrate from their own countries due to complex and bitter situations as well as to persecutions, and settled in Argentina and Uruguay. For us Uruguayans our scientific molding was, from the very beginning, really exceptional. We are much indebted to this truly internationalist environment, both rioplatense and European.

As I said before, Mischa was in the Uruguayan group. It was immediately evident that he not only was very talented but showed a great originality. I remember the impression made on me by his manuscript "Estructuras de Anágenos," which, under the title "Aritmética Abstracta," was published in the *Boletín de la Facultad de Ingeniería* of Montevideo in 1937. Later on, after many years of working in Argentina, came his doctorate from the University of Chicago and the many more years of fruitful investigations of the highest quality that he is still pursuing, in spite of the exiles and other obstacles. And, paradoxically, these exiles allowed him to renew his bonds with Uruguayan mathematicians and, in particular, with our colleague Rodrigo Arocena.

On many issues outside mathematics, Mischa and I have positions that are very different, ranging from dietary habits to philosophical viewpoints. But these divergences never encumbered our friendship: on the contrary, I would dare to say that this diversity has been a very rich ingredient in a human bond that is very intense and personal.

Yet Mischa has a fault, a very pronounced fault. He has always been, and always will be, too modest. What is worse, his is a modesty that is tremendously authentic and true. I can imagine his state of mind at this moment, in this session that was intended to be, and really is—in spite of Mischa's opinions—more than a well-deserved homage to his talent as a mathematician, an homage—why not say it?—to Mischa the man, the loving and lovable friend.

But Mischa should not be upset, because even if he dislikes the word homage, he knows that this is not a ceremonious, stiff, formal homage. It is an homage in which, irrespective of the high intellectual respect, predominates the deep, human affection felt for Mischa by all of us who have the pleasure of knowing him. And this sort of homage I think he will accept with pleasure, without having modesty making him uneasy.

I want to finish by expressing my friendship and appreciation to Yanny, his compañera for so many years. I think that she also deserves to be included in this friendly homage.

It is sad that Laguardia is not here with us today. I knew him very well, and I am sure that if he were alive he would have been as happy as I am now. As happy as we all are, young and old, in this reunion with Mischa, our dear friend.

My Friend Mischa Cotlar

Manuel Sadosky

After more than fifty years, I still see the young man recently arrived to Buenos Aires from Montevideo, shy, tall and husky, yet trying to look small, as not to be noticed. I see him entering the little classroom in the old building at Calle Perú, for more than a century home to the School of Sciences—which was not much more than an engineering college. The "irregular" courses of what was pompously called the Doctoral Program in Mathematics and Physics took place in a little classroom at the side of the stairs. The only paid professor was Dr. Julio Rey Pastor, but there was also Dr. Juan Carlos Vignaux, who taught for free. Needless to say, Rey Pastor and Vignaux were mortal enemies, and we, the students, had a hard time, since the examination panels were required by law to consist of three professors! That day we met, in 1935, Mischa had come to school looking for Professor Vignaux, to whom he had a letter of introduction and who has to become his protector. Since he had no high school diploma, Mischa could not register, but nevertheless he continued to come to class.

In Uruguay Mischa had been in contact with young Professor Rafael Laguardia and with José Luis Massera, then a student, who had a small but active study group in mathematics. Mischa did not know much of the contents of the elementary courses but was well versed in classical algebra and in number theory.

Cora Ratto—then my classmate and later my wife—and I immediately took a liking to the "Russian peasant," who was withdrawn and extremely humble. After a short while, Mischa told Cora that he missed playing the piano very much. Music allowed him to take refuge from the often dismal surrounding reality to enter a world of peace and harmony. Cora immediately invited him to play at her home and, after a long argument, convinced Mischa to accept the offer. Cora's mother delighted in Mischa's playing and always considered it a great personal vindication to have met a talented young man willing to use an instrument she had not succeeded in making important to her own children. Mischa, in his desire not to intrude, sat on the very edge of the piano stool, and we were always worried about the stability of the system and imagined the pianist rolling on the floor!

Little by little, Mischa carved his place in Buenos Aires. He earned his living tutoring students and even got a more stable situation teaching in a private academy that prepared candidates to enter the Naval School. Only once, around 1945, did he get a teaching assistantship at a university, the Universidad Nacional de La Plata, during the brief tenure of Ing. A. Martínez Civelli as Rector. We had succeeded in explaining to the authorities that Mischa was already the most creative mathematical researcher the country had. The assignment did not last long, since the political changes in 1946 brought a new Rector, who rushed to take him off the payroll.

One fortunate circumstance had a great impact on Mischa's life. In 1942 the eminent American mathematician George D. Birkhoff visited Buenos Aires and met with some young people interested in mathematical research. Well trained in the detection of talent, Birkhoff immediately spotted Mischa and made a note to recommend him for a Guggenheim Fellowship. But the process was not completed due to Birkhoff's death in 1944. Fortunately, his son Garrett, also a well-known mathematician, found his father's note among his papers and forwarded it to the Guggenheim Foundation. Mischa was able to travel to the United States in 1951 and he received a doctoral degree from the University of Chicago. He returned to Argentina with a title, yet his chances to get a university position were still bleak due to the fact that at the time it was necessary to give proof of allegiance to the official party, which neither Mischa nor his friends were ready to do.

By 1953, thanks to the "diplomatic" efforts of that great man, Antonio Monteiro, a Departamento de Investigaciones Científicas (DIC), was created at the Universidad Nacional de Cuyo. And the only requirements to join the DIC were professional! The appointment of Mischa as Director of DIC was a "bomb" in our small circle. The job of director or chairman entails bureaucratic chores; it is necessary to deal with files and budgets; but, worst of all, it requires wearing a tie for official ceremonies! Who could imagine Mischa in such a garb?

A group of intelligent, enthusiastic, and heterodox people gathered at DIC. They published a journal, the *Revista Matemática Cuyana*, to record the research produced there. Its contents received very favorable comments in the mathematical reviews. The results of the work at DIC fructified Argentine mathematics for years to come, since after it was closed in 1956, again for political reasons, its researchers went on to teach students who are now well-known mathematicians themselves. In fact, when the university situation changed dramatically in 1956, Mischa and his collaborators were asked to join university faculties all over the country.

Mischa was offered a full professorship at the School of Sciences in Buenos Aires. The dialogue he had with then Dean José Babini deserves to be recalled. Mischa objected that the salary offered by Babini was excessive. Startled by this unusual objection, and fearing a long and frustrating argument, Babini replied: "Mischa, don't you see it as unbecoming to a man of your philosophical views to discuss money matters?" Mischa was left speechless and joined our faculty, where he attracted very gifted young people to a very active analysis group.

I want to recall some nonacademic aspects of the period we shared with Mischa from 1956 to 1966 at the School of Sciences of the University of Buenos Aires. In July 1957 Mischa actively participated in one of Cora Ratto's most successful initiatives, the setting of the Fundación Alberto Einstein. This private institution provided financial and academic support to gifted students in the School of Sciences and allowed their full-time concentration on academic work.

Mischa was also very active in the Movement for Social Responsibility of Scientists, which gathered researchers from many parts of the country and had connections with similar movements in other countries. These organizations, which appeared after the atomic blasts in Hiroshima and Nagasaki, had among their objectives to alert the public to the distorted uses of scientific research.

In 1965, again with Cora Ratto and with Francisco Rullrich, he edited a little magazine called *Columna 10.* In its nine issues the magazine brought to the attention of Argentinians some key international themes of the time, such as the Vietnam War, the politics and economics of the nuclear arm race, and the problems of neocolonialism in Asia, Africa, and Latin America. To help shed light on world events the magazine made available in Spanish important articles by Bertrand Russell, Albert Camus, Celso Furtado, and others.

The military coup of 1966 ended a democratic and progressive period in Argentine universities. The military plotted the "Night of the Big Sticks" to serve as a warning to academics wanting a modern university. People were again forced to leave. Mischa, with a group of colleagues, first went to Peru, then to the United States, France, and Venezuela. He returned to Argentina at the end of 1972 and taught briefly at La Plata. In 1973 the popularly elected government did not redress the university situation, and the group that had been excluded was not invited back. The "democratic oasis" was very short-lived, and Mischa and Yanny once more packed their bags, this time to settle in Caracas, where Mischa is professor at the Universidad Central de Venezuela, and where he has formed a thriving group of young analysts.

I have omitted much from these recollections. For all of us who have the good fortune of being close to him, Mischa is, above all, a great friend of unfailing loyalty and devotion. His life has been enriched by many close friendships with remarkably diverse people. Many of his friends differ from Mischa in matters of philosophy and politics, but in our relationship we always sustained the right to noncoincidence. In all cases, we always placed social responsibility before any consideration of personal gain or fear.

I wish to conclude by saying that Mischa's lifelong companion, Yanny Frenkel, deeply identified with this exceptional man, wholly deserves sharing with him the celebration by his friends and colleagues.

Buenos Aires
April, 1988

Part I. Harmonic and Complex Analysis

1

Summability Methods for Transferring Fourier Multipliers and Transference of Maximal Inequalities

Naklé Asmar, Earl Berkson, and T. A. Gillespie

I. Introduction. Let $\mathbb{T}$ be the circle group in the complex plane $\mathbb{C}$, and let $\mathbb{Z}$ (respectively $\mathbb{R}$) be the additive group of integers (respectively, real numbers). Let $\mathbb{N}$ denote the set of positive integers, and put $\mathbb{Z}_+ = \mathbb{N} \cup \{0\}$. For $1 \leq p < \infty$ and G a locally compact abelian group with dual group $\hat{G}$, denote by $M_p(\hat{G})$ the space of Fourier multipliers for $L^p(G)$ with the usual multiplier norm $\|\cdot\|_{M_p(\hat{G})}$.

Let us call a compactly supported continuous function $\Lambda: \mathbb{R} \longrightarrow \mathbb{C}$ a *summability kernel for* L^p*-multipliers* if, for each $\phi \in M_p(\mathbb{Z})$, the function $\Psi(x) \equiv \sum_{n=-\infty}^{\infty} \phi(n)\Lambda(x-n)$ belongs to $M_p(\mathbb{R})$. (Notice that if the summability kernel Λ has support contained in $[-1,1]$, and $\Lambda(0) = 1$, then Ψ extends ϕ to $\mathbb{R}$.) In [9, Theorem (3.5)] it was shown that for each $\phi \in M_p(\mathbb{Z})$, the piecewise linear extension of ϕ with nodes at the integers belongs to $M_p(\mathbb{R})$, and in [6, Theorem C] it was shown that a canonical piecewise quadratic extension of ϕ belongs to $M_p(\mathbb{R})$ (see also [3] and [5] for abstract versions of piecewise quadratic multiplier extension). In terms of summability kernels for multipliers, these results in [9] and [6] state, respectively, that for $1 \leq p < \infty$, Δ and Δ^2 are summability kernels for L^p-multipliers, where $\Delta(x) \equiv \max\left\{1 - |x| , 0\right\}$. In Corollary (III.20) below, we generalize these results by introducing a broad class of functions serving as summability kernels for L^p-multipliers throughout the range $1 < p < \infty$. Our approach in Section III is based on suitable modifications of the methods developed in [6, §3], which exploit the dual space relationship of M_p-spaces to A_p-spaces [4, Theorem 1] by taking adjoints of suitable linear mappings between A_p-spaces.

In our main result, Theorem (IV.3) below, we present a maximal operator counterpart to the work of Section III. As described in (IV.14), Theorem (IV.3) readily implies the well-known fact [11] that the Carleson-Hunt theorem [8] transplants to $\mathbb{R}$.

In Section V we utilize the full precision of general transference bounds in order to develop Hilbert-transform based summability methods for converting $L^p(\mathbb{T})$-Fourier multipliers into $L^p(\mathbb{R})$-Fourier multipliers which are discontinuous functions on $\mathbb{R}$. Throughout all our considerations we have been motivated by the transference viewpoint in analysis initiated by R.R. Coifman and G. Weiss (see [2]).

It will be convenient to summarize here for later use some basic facts from [4] about A_p-spaces. Suppose that $1 < p < \infty$, and q is the index conjugate to p . Let G be a locally compact abelian group with dual group $\hat{G}$. Then $A_p(G)$ is the linear space, under pointwise operations, consisting of all functions $h: G \longrightarrow \mathbb{C}$ of the form

$$(I.1) \qquad h = \sum_{n=1}^{\infty} f_n * g_n ,$$

where $*$ denotes convolution over G , $\{ f_n \}_{n=1}^{\infty} \subseteq L^p(G)$, $\{ g_n \}_{n=1}^{\infty} \subseteq L^q(G)$, and

$$(I.2) \qquad \sum_{n=1}^{\infty} \| f_n \|_{L^p(G)} \| g_n \|_{L^q(G)} < +\infty .$$

For $h \in A_p(G)$, $\| h \|_{A_p(G)}$ is defined to be the infimum of all sums in (I.2) corresponding to representations of h having the form (I.1). The space $A_p(G)$ endowed with the norm $\| \cdot \|_{A_p(G)}$ is a Banach space. It is obvious from these definitions that each $h \in A_p(G)$ is a bounded continuous function on G such that

$$(I.3) \qquad \| h \|_u \le \| h \|_{A_p(G)} ,$$

where $\| h \|_u = \sup \left\{ |h(t)| : \ t \in G \right\}$. In fact, standard approximations in $L^p(G)$ and $L^q(G)$ by compactly supported continuous functions, together with (I.1) and (I.2), show that $A_p(G)$ is a subset of the space of all complex-valued continuous functions on G vanishing at infinity. Moreover, it is known [4, p. 500] that

(I.4) the algebra $\mathcal{A}(G)$, consisting of all functions on G which are Fourier transforms of functions in $L^1(\hat{G})$, is a subset of $A_p(G)$.

It is obvious that for $h \in A_p(G)$ with a representation as in (I.1) and (I.2), the series on the right of (I.1) converges to h in the Banach space $A_p(G)$, and

$$\sum_{n=1}^{\infty} \| f_n * g_n \|_{A_p(G)} < +\infty .$$

The central structural facts about $A_p(G)$ which we shall require are contained in the following proposition introduced in [4].

(I.5) ***Proposition*** ([4, Theorem 1]). *Suppose* $1 < p < \infty$. *There is a linear isometry of* $M_p(\hat{G})$ *onto the dual space of* $A_p(G)$. *If* $\phi \in M_p(\hat{G})$ *and* $T_\phi^{(p)}$ *denotes the multiplier transform on* $L^p(G)$ *defined by* ϕ, *then the corresponding functional* $\mathcal{K}_\phi$ *on* $A_p(G)$ *satisfies:*

$$\text{(I.6)} \qquad \mathcal{K}_\phi(h) = \sum_{n=1}^{\infty} \left[\left[T_\phi^{(p)} f_n\right] * g_n \right](0) ,$$

for every $h \in A_p(G)$ *and every representation for* h *satisfying* (I.1) *and* (I.2).

(I.7) <u>Remarks</u>. If $\phi \in M_p(\hat{G})$, then the restriction to

$\mathcal{A}(G)$ of the functional $\mathcal{K}_\phi$ in (I.6) takes the following form:

$$\mathcal{K}_\phi(f) = \int_{\hat{G}} \phi(\gamma)\Psi(\gamma)d\gamma ,$$

where $d\gamma$ is Haar measure on $\hat{G}$, $\Psi \in L^1(\hat{G})$, and $f(x) = \int_{\hat{G}} \Psi(\gamma)\gamma(x)d\gamma$ for all $x \in G$ ([4, proof of Theorem 1]).

Since we shall be applying the Coifman-Weiss General Transference Result [2, Theorem 2.4] to A_p-spaces, we quote its statement for ease of reference (see [1, Theorem (2.3)] for the generality stated here).

(I.8) ***Theorem*** (Coifman-Weiss General Transference Result). *Suppose that* G *is a locally compact abelian group with Haar measure* m, $(\mathcal{M},\mu)$ *is an arbitrary measure space,* $1 \le p < \infty$, *and* $u \longrightarrow R_u$ *is a strongly continuous representation of* G *in* X , *where* X *is a closed subspace of* $L^p(\mathcal{M},\mu)$. *Assume also that*

$$c \equiv \sup \left\{ \| R_u \| : \ u \in G \right\} < +\infty .$$

Let $k \in L^1(G,m)$, *and denote by* $N_p(k)$ *the norm of convolution by* k *on* $L^p(G,m)$. *Define*

$$H_k f = \int_G k(u)\left[R_{-u} f \right] dm(u) , \quad \textit{for} \quad f \in X ,$$

where the integral is taken as an X*-valued Bochner integral. Then*

$$\| H_k \| \le c^2 N_p(k) . \tag{I.9}$$

II. <u>A Useful Lemma</u>. In this section we combine some

standard techniques from Fourier analysis on $\mathbb{R}$ in the form of a lemma tailored to our subsequent purposes. For $f \in L^1(\mathbb{R})$ we denote the Fourier transform of f by $\hat{f}$. Thus $\hat{f}(y) \equiv \int_{\mathbb{R}} f(t)\, e^{-ity} dt$.

(II.1) ***Lemma***. *Let* $\mathcal{F}$ *be the class consisting of all* $J \in L^1(\mathbb{R})$ *such that* J *is continuous on* $\mathbb{R}$ *, and:*

(i) $\hat{J}$ *has compact support;*

(ii) $\hat{J}$ *is absolutely continuous on each compact interval of* $\mathbb{R}$ *;*

(iii) *the derivative* $\left[\hat{J}\right]'$ *of* $\hat{J}$ *belongs to* $L^2(\mathbb{R})$.

Then if $J \in \mathcal{F}$ *,*

$$\text{(II.2)} \qquad \alpha(J) \equiv \sup_{x \in \mathbb{R}} \sum_{m=-\infty}^{\infty} \left| J(x + 2\pi m) \right| < +\infty .$$

<u>Proof</u>. Notice that the sum in (II.2) is (2π)-periodic, and so $\alpha(J)$ is unchanged if the supremum in (II.2) is taken over all $x \in [-\pi,\pi]$. Let $N \in \mathbb{N}$ be such that $\operatorname{supp} \hat{J} \subseteq [-N,N]$, where $\operatorname{supp} \hat{J}$ denotes the support of $\hat{J}$. By Fourier inversion on $\mathbb{R}$, we have for $|x| \le \pi$, $m \in \mathbb{Z}$,

$$\text{(II.3)} \qquad J(x + 2\pi m) = (2\pi)^{-1}\int_{-N}^{N} \hat{J}(t)e^{itx}e^{2\pi itm}\, dt .$$

Thus by a change of the variable of integration,

$$\text{(II.4)} \qquad \pi N^{-1}J(x + 2\pi m) = (2\pi)^{-1}\int_{-\pi}^{\pi} \hat{J}(Nu/\pi)e^{ixNu/\pi}e^{i2mNu}du ,$$

$$\text{for } |x| \le \pi , \quad m \in \mathbb{Z} .$$

Let $j(u) = \hat{J}(Nu/\pi)$ for $u \in [-\pi,\pi]$. By our hypotheses in (II.1)-(ii),(iii), the function j is absolutely continuous on $[-\pi,\pi]$ and $j' \in L^2[-\pi,\pi]$. For $|x| \leq \pi$, let $W_x(u) = j(u)e^{ixNu/\pi}$, for $|u| \leq \pi$. Notice that $W_x(-\pi) = W_x(\pi) = 0$, W_x is absolutely continuous on $[-\pi,\pi]$, and $W_x' \in L^2[-\pi,\pi]$. A standard fact about absolute convergence of Fourier series [10, I.(6.2), p.32] now gives us for each $x \in [-\pi,\pi]$:

$$\sum_{n=-\infty}^{\infty} \left|\hat{W}_x(n)\right| \leq (2\pi)^{-1}\int_{-\pi}^{\pi} |W_x(t)|\, dt + \left[2\sum_{n=1}^{\infty} n^{-2}\right]^{1/2}\left[(2\pi)^{-1}\int_{-\pi}^{\pi} |W_x'(t)|^2\, dt\right]^{1/2} \leq \alpha_0 ,$$

where α_0 is a real constant depending only on J. The proof of Lemma (II.1) is now complete, since, by (II.4), $J(x + 2\pi m) = (N/\pi)\hat{W}_x(-2mN)$ for $x \in [-\pi,\pi]$, $m \in \mathbb{Z}$.

III. **Mappings of $A_p(\mathbb{R})$ into $A_p(\mathbb{T})$ and Summability Kernels**. Let J belong to the class $\mathcal{F}$ defined in Lemma (II.1). We shall show in Corollary (III.20) below that, for $1 < p < \infty$, $\hat{J}$ is a summability kernel for L^p-multipliers. In the particular case when J is taken to be $\kappa(t) \equiv (2\pi)^{-1}(t/2)^{-2}\sin^2(t/2)$, this result becomes the process for extension of Fourier multipliers by piecewise linearity [9, Theorem (3.5)]. When J is specialized to be $\kappa * \kappa$, this result becomes the piecewise quadratic extension process for Fourier multipliers in [6, Theorem C]. In outline, Section III follows the strategy introduced for the piecewise quadratic extension process in [6, §3] by setting up, appropriately for our context, a linear mapping U of $A_p(\mathbb{R})$ into $A_p(\mathbb{T})$, and then using Proposition (I.5) to treat the adjoint of U as a mapping of $M_p(\mathbb{Z})$ into $M_p(\mathbb{R})$.

To begin with, fix p in the range $1 < p < \infty$, let J_1 and J_2 belong to $\mathcal{F}$, and put $K = J_1 * J_2$. Notice that $K \in \mathcal{F}$. The real constant $\alpha(K)$ occurring in (II.2) is completely determined by J_1 and J_2, and has the property that

$$\text{(III.1)} \qquad \sum_{m=-\infty}^{\infty} | K(x + 2\pi m) | \le \alpha(K) , \quad \text{for all } x \in \mathbb{R} .$$

For $F \in A_p(\mathbb{R})$ define $UF: \mathbb{T} \longrightarrow \mathbb{C}$ by putting

$$\text{(III.2)} \qquad \left[UF\right](e^{ix}) = \sum_{m=-\infty}^{\infty} K(x + 2\pi m)\, F(x + 2\pi m) , \quad \text{for all } x \in \mathbb{R} .$$

In view of (III.1) and (I.3) applied to F, the series on the right of (III.2) is absolutely convergent for all $x \in \mathbb{R}$, and

$$\text{(III.3)} \qquad \| UF \|_u \le \alpha(K) \| F \|_u \le \alpha(K) \| F \|_{A_p(\mathbb{R})} .$$

In fact, as the following theorem shows, U is a linear mapping of $A_p(\mathbb{R})$ into $A_p(\mathbb{T})$.

(III.4) ***Theorem.*** *Suppose that* $1 < p < \infty$. *Then, with the above notation, we have that* U *is a bounded linear mapping of* $A_p(\mathbb{R})$ *into* $A_p(\mathbb{T})$ *satisfying*

(III.5) $$\| U \| \le 2\pi\alpha_1\alpha_2 ,$$

where α_1 *and* α_2 *are the real constants corresponding to* J_1 *and* J_2 *, respectively, in* (II.2).

<u>Proof</u>. Let q be the index conjugate to p . We claim that it suffices for the proof of the theorem to show that if $f \in L^p(\mathbb{R})$ and $g \in L^q(\mathbb{R})$, then $U(f * g) \in A_p(\mathbb{T})$, and

(III.6) $$\left\| U(f * g) \right\|_{A_p(\mathbb{T})} \le 2\pi\alpha_1\alpha_2 \left\| f \right\|_{L^p(\mathbb{R})} \left\| g \right\|_{L^q(\mathbb{R})} .$$

To establish this claim, assume that (III.6) holds, and let $h \in A_p(\mathbb{R})$ have the representation (I.1). By linearity of U together with (III.6), we have for $M,N \in \mathbb{N}$ with $N \le M$,

(III.7) $$\left\| U\left[\sum_{n=N}^{M} f_n * g_n \right] \right\|_{A_p(\mathbb{T})} \le 2\pi\alpha_1\alpha_2 \sum_{n=N}^{M} \| f_n \|_p \| g_n \|_q .$$

Thus, upon putting $h_N = \sum_{n=1}^{N} f_n * g_n$, we have that $\{ U(h_N) \}_{N=1}^{\infty}$ is Cauchy in $A_p(\mathbb{T})$. Let $H \in A_p(\mathbb{T})$ and $\left\| U(h_N) - H \right\|_{A_p(\mathbb{T})} \longrightarrow 0$ as $N \longrightarrow +\infty$. In particular, $\{ U(h_N) \}$ tends to H uniformly on $\mathbb{T}$. However, by (III.3), $\{ U(h_N) \}$ tends to Uh uniformly on $\mathbb{T}$. Combining these facts, we see that $Uh = H \in A_p(\mathbb{T})$, and, by virtue of (III.7) with $N = 1$ and M arbitrary, we obtain (III.5). This establishes the claim.

By standard approximations in Lebesgue spaces of $\mathbb{R}$, the general definition of the A_p-norm, and similar reasoning to the above, it is easy to see that we need only prove (III.6) under the assumption, which we now adopt, that f and g belong to $\mathcal{D}$, where here and henceforth $\mathcal{D}$ denotes the class of all complex-valued, infinitely differentiable functions on $\mathbb{R}$ having compact support. Put $F = f * g$. The proof of the theorem will be completed with the aid of a device originated in [7] for treating pointwise products of convolutions such as presently occur on the right of (III.2). For $t \in \mathbb{R}$, define $\Phi_t: \mathbb{T} \longrightarrow \mathbb{C}$ and $\Psi_t: \mathbb{T} \longrightarrow \mathbb{C}$ by putting:

$$\text{(III.8)} \qquad \Phi_t(e^{ix}) = \sum_{m=-\infty}^{\infty} f(x + 2\pi m)\, J_1(x + t + 2\pi m), \quad \text{for } x \in \mathbb{R};$$

$$\text{(III.9)} \qquad \Psi_t(e^{ix}) = \sum_{m=-\infty}^{\infty} g(x + 2\pi m)\, J_2(x - t + 2\pi m), \quad \text{for } x \in \mathbb{R}.$$

Since f and g are compactly supported, we see, in particular, that Φ_t and Ψ_t are continuous functions on $\mathbb{T}$ for each $t \in \mathbb{R}$. Straightforward calculations with integrals using Fubini's theorem and the compact supports of f and g can be performed on (III.8) and (III.9) to show that

$$\text{(III.10)} \qquad \left[UF\right](z) = 2\pi \int_{\mathbb{R}} \left[\Phi_t * \Psi_t\right](z)\, dt, \quad \text{for all } z \in \mathbb{T},$$

where the convolution sign on the right designates

convolution over $\mathbb{T}$ with respect to normalized Haar measure.

In order to convert the pointwise expressions in (III.10) into estimates concerning the norm in $A_p(\mathbb{T})$, we next consider the mapping $\theta : \mathbb{R} \longrightarrow A_p(\mathbb{T})$ defined by $\theta(t) = \Phi_t * \Psi_t$. It is clear from (III.8) and (III.9) that θ is continuous with respect to $\|\cdot\|_{A_p(\mathbb{T})}$. With the aid of Fubini's theorem it is easy to see from (III.8) that

(III.11)

$$\int_{\mathbb{R}} \| \Phi_t \|^p_{L^p(\mathbb{T})} \, dt$$

$$= (2\pi)^{-1} \int_{-\pi}^{\pi} \int_{\mathbb{R}} \left| \sum_{m=-\infty}^{\infty} f(x + t + 2\pi m) J_1(x + 2\pi m) \right|^p dt \, dx.$$

If we define the representation R of $\mathbb{Z}$ in $L^p(\mathbb{R})$ by taking R_m to be translation by $(2\pi m)$ on $L^p(\mathbb{R})$, for each $m \in \mathbb{Z}$, and put $f_x(t) = f(x + t)$ for $x \in \mathbb{R}$, $t \in \mathbb{R}$, then we can rewrite (III.11) in the form

$$\int_{\mathbb{R}} \| \Phi_t \|^p_{L^p(\mathbb{T})} \, dt$$

$$= (2\pi)^{-1} \int_{-\pi}^{\pi} \int_{\mathbb{R}} \left| \sum_{m=-\infty}^{\infty} \left[R_{-m} f_x\right](t) \, J_1(x - 2\pi m) \right|^p dt \, dx.$$

An application of the Coifman-Weiss General Transference Result (Theorem (I.8)) to the inner integral over $\mathbb{R}$ now shows that

$$\text{(III.12)} \qquad \int_{\mathbb{R}} \| \Phi_t \|^p_{L^p(\mathbb{T})} \, dt \le \alpha_1^p \, \| f \|^p_{L^p(\mathbb{R})} .$$

Similar reasoning shows that

$$\text{(III.13)} \qquad \int_{\mathbb{R}} \| \Psi_t \|^q_{L^q(\mathbb{T})} \, dt \le \alpha_2^q \| g \|^q_{L^q(\mathbb{R})} .$$

From (III.12), (III.13), and Hölder's inequality,

$$\text{(III.14)} \qquad \int_{\mathbb{R}} \| \theta(t) \|_{A_p(\mathbb{T})} \, dt$$

$$\le \int_{\mathbb{R}} \| \Phi_t \|_{L^p(\mathbb{T})} \| \Psi_t \|_{L^q(\mathbb{T})} \, dt$$

$$\le \alpha_1 \alpha_2 \| f \|_{L^p(\mathbb{R})} \| g \|_{L^q(\mathbb{R})} .$$

In particular, the $A_p(\mathbb{T})$-valued Bochner integral $h = \int_{\mathbb{R}} \theta(t) \, dt$ exists. Since by (I.3) point evaluations are continuous linear functionals on A_p-spaces, we see from (III.10) that $UF = 2\pi h \in A_p(\mathbb{T})$. Application of (III.14) to h now gives (III.6) and completes the proof of the theorem.

Remarks. By virtue of Lemma (II.1), a variant of Theorem (III.4) could be inferred directly from the abstract result in [3, Theorem 4]. We have given a detailed proof of Theorem (III.4) since the main steps will be needed again when we consider maximal multiplier transforms in Section IV. While the above demonstration of Theorem (III.4) benefits from the simplicity afforded by the transference viewpoint, the full sharpness of the general transference estimate (I.9) will not become indispensable until we come to the relaxed hypotheses of Section V, where estimates by

$\ell^1(\mathbb{Z})$-norms no longer completely suffice (see the opening discussion for Section V and the proof of Lemma (V.10)).

(III.15) *Theorem. Let* J_1 *and* J_2 *belong to the class* $\mathcal{F}$ *defined in* Lemma (II.1), *and put* $K = J_1 * J_2$. *Fix* p *in the range* $1 < p < \infty$. *Then for each* $\phi \in M_p(\mathbb{Z})$, *the function* $Q_\phi : \mathbb{R} \longrightarrow \mathbb{C}$, *defined by*

$$Q_\phi(x) \equiv \sum_{n=-\infty}^{\infty} \phi(n)\hat{K}(x-n) , \tag{III.16}$$

belongs to $M_p(\mathbb{R})$, *is continuous on* $\mathbb{R}$, *and satisfies*

$$\| Q_\phi \|_{M_p(\mathbb{R})} \le (2\pi)^2 \alpha_1 \alpha_2 \| \phi \|_{M_p(\mathbb{Z})} , \tag{III.17}$$

where α_1 *and* α_2 *are the real constants in* (II.2) *corresponding to* J_1 and J_2 , *respectively.*

<u>Proof</u>. We notice at the outset that since $\hat{K}$ has compact support, the sum on the right of (III.16) has only finitely many non-zero terms at each $x \in \mathbb{R}$. In fact, if we choose $M \in \mathbb{N}$ so that $\operatorname{supp} \hat{K} \subseteq [-M,M]$, then for each $n_0 \in \mathbb{Z}$,

$$Q_\phi(x) = \sum_{n=n_0-M+1}^{n_0+M} \phi(n)\, \hat{K}(x-n) \quad \text{for all } x \in [n_0, n_0+1].$$

In particular, Q_ϕ is bounded and continuous on $\mathbb{R}$, with $\| Q_\phi \|_u \le 2 \| \phi \|_u \| \hat{K} \|_u M$.

Standard approximations, together with Theorem (III.4), show that it suffices for the proof of the present theorem to establish that if $\phi : \mathbb{Z} \longrightarrow \mathbb{C}$ has finite support, $f \in \mathcal{D}$, and $g \in \mathcal{D}$, then, with the notation of (I.5) and

(III.2), and putting $\Psi(n) \equiv \phi(-n)$, we have:

$$\text{(III.18)} \qquad \mathcal{K}_{\Psi}\left[U(f * g)\right] = (2\pi)^{-2}\int_{\mathbb{R}} \hat{f}(-t)\hat{g}(-t)Q_{\phi}(t)\, dt .$$

For this purpose observe first that since $K \in \mathcal{S}$, it follows with the aid of Fourier inversion that $K \in C^{\infty}(\mathbb{R})$, the space of all infinitely differentiable, complex-valued functions on $\mathbb{R}$. Since we also have that $(f * g) \in \mathcal{D}$, it is clear from (III.2) that $U(f * g) \in C^{\infty}(\mathbb{T})$. In particular, by [10, Theorem I.(6.2)], $U(f * g)$ has an absolutely convergent Fourier series. For $n \in \mathbb{Z}$, let $e_n(z) \equiv z^n$ for all $z \in \mathbb{T}$. Clearly we now have

$U(f * g) = \sum_{n=-\infty}^{\infty} \left[U(f * g)\right]^{\wedge}(n)\ (e_n * e_n)$. It follows by (I.6) (or, alternatively, recourse to (I.7)) that

$$\text{(III.19)} \qquad \mathcal{K}_{\Psi}\left[U(f * g)\right] = \sum_{n=-\infty}^{\infty} \Psi(n) \left[U(f * g)\right]^{\wedge}(n) .$$

Standard elementary considerations with (III.2) show that for each $n \in \mathbb{Z}$,

$$\left[U(f * g)\right]^{\wedge}(n) = (2\pi)^{-1} \left[(f * g)\ K\right]^{\wedge}(n) .$$

Hence by Fourier inversion on $\mathbb{R}$ (or, alternatively, by Parseval's Formula), for each $n \in \mathbb{Z}$,

$$\left[U(f * g)\right]^{\wedge}(n) = (2\pi)^{-2} \int_{\mathbb{R}} \hat{f}(-t)\hat{g}(-t)\hat{K}(t + n)\, dt.$$

Using this in (III.19) gives (III.18), and completes the proof of the theorem.

It is desirable for structural simplicity as well as for applications to replace K in Theorem (III.15) by an arbitrary $J \in \mathcal{F}$. In fact, this can be accomplished as a consequence of Theorem (III.15).

(III.20) *Corollary. Let* $\mathcal{F}$ *be the class of functions defined in* Lemma (II.1), *let* $J \in \mathcal{F}$, *and fix* p *in the range* $1 < p < \infty$. *Then for each* $\phi \in M_p(\mathbb{Z})$, *the function* $Q_\phi : \mathbb{R} \longrightarrow \mathbb{C}$ *given by*

$$Q_\phi(x) \equiv \sum_{n=-\infty}^{\infty} \phi(n)\hat{J}(x - n) \tag{III.21}$$

belongs to $M_p(\mathbb{R})$, *is continuous on* $\mathbb{R}$, *and satisfies*

$$\| Q_\phi \|_{M_p(\mathbb{R})} \leq \Omega_J \| \phi \|_{M_p(\mathbb{Z})} , \tag{III.22}$$

where Ω_J *is a real constant depending only on* J .

<u>Proof</u>. Let $\{ V_\lambda \}_{\lambda > 0}$ be the de la Vallée-Poussin kernel for $\mathbb{R}$,

$$V_\lambda(x) \equiv 2\mathcal{k}_{2\lambda}(x) - \mathcal{k}_\lambda(x) ,$$

where $\{ \mathcal{k}_\lambda \}_{\lambda > 0}$ is Fejér's kernel for $\mathbb{R}$ given by,

$$\mathcal{k}_\lambda(x) \equiv \lambda(2\pi)^{-1}(\lambda x/2)^{-2}\sin^2(\lambda x/2) .$$

In particular, for each $\lambda > 0$, $V_\lambda \in \mathcal{F}$, and $\hat{V}_\lambda = 1$ on $[-\lambda,\lambda]$. Choose $\lambda_0 > 0$ so that $[-\lambda_0,\lambda_0]$ contains the support of $\hat{J}$. Put $J_1 = J$ and $J_2 = V_{\lambda_0}$. Then we see by taking Fourier transforms that $J = J_1 * J_2$, and hence application of Theorem (III.15) completes the proof of the Corollary.

IV. **<u>Summability Kernels and Transference of Bounds for Maximal Multiplier Transforms</u>**. We shall show that if, in the notation of Corollary (III.20), J is a fixed function belonging to $\mathcal{F}$, and p is a fixed index satisfying $1 < p < \infty$, then for each sequence $\{ \phi_k \} \subseteq M_p(\mathbb{Z})$, the strong type (p,p) norm of the maximal multiplier transform on $L^p(\mathbb{T})$ defined by the sequence $\{ \phi_k \}$ can be transferred to control the strong type (p,p) norm of the maximal multiplier transform on $L^p(\mathbb{R})$ defined by the sequence $\left\{ Q_{\phi_k} \right\}$. In order to achieve this, our main result, we shall require the following variant of [11, §1], which can be established by adapting the reasoning in [11].

(IV.1) ***Linearization Lemma for Maximal Multiplier Transforms.*** *Let* $\hat{G}$ *be the dual group of a locally compact abelian group* G. *Suppose that* $1 < p < \infty$, q *is the index conjugate to* p, *and* $N \in \mathbb{N}$. *Let* ψ_j, $j = 1,2,\cdots,N$, *belong to* $M_p(\hat{G})$, and *let* $\mathcal{M}$ *be the maximal operator on* $L^p(G)$ *corresponding to their multiplier transforms* $T_{\psi_j}^{(p)}$, $j = 1,2,\cdots,N$, *defined on* $L^p(G)$. *Then the strong type* (p,p) *norm* $\| \mathcal{M} \|_p$ *of* $\mathcal{M}$ *is the least non-negative real number* c *such that:*

$$\left\| \sum_{j=1}^{N} T_{\psi_j}^{(q)} f_j \right\|_{L^q(G)} \leq c \left\| \sum_{j=1}^{N} |f_j| \right\|_{L^q(G)},$$

for all $f_1, f_2, \cdots, f_N \in L^q(G)$.

The proof of the main result will be facilitated by the following lemma.

(IV.2) ***Lemma.*** *Let* $\hat{G}$ *be the dual group of a locally* *compact abelian group* G . *Suppose that* $1 \le p < \infty$, $\nu \in \mathbb{N}$, *and* $\{ m_j \}_{j=1}^{\nu} \subseteq M_p(\hat{G})$. *Let* $\mathcal{M}$ *be the maximal operator on* $L^p(G)$ *corresponding to the multiplier transforms of* $\{ m_j \}_{j=1}^{\nu}$. *For* $1 \le j \le \nu$, *let* $\{ m_{j,n} \}_{n=1}^{\infty}$ *be a sequence in* $M_p(\hat{G})$ *such that*:

(i) *for* $1 \le j \le \nu$, $m_{j,n} \longrightarrow m_j$ *as* $n \longrightarrow +\infty$ *pointwise on* $\hat{G}$;

(ii) $\sup \left\{ |m_{j,n}(\gamma)| : \ 1 \le j \le \nu \ , n \in \mathbb{N} \ , \gamma \in \hat{G} \right\} < +\infty$;

(iii) $\sup_{n \in \mathbb{N}} \| \mathcal{M}_n \|_p \le C$,

where C *is a real constant, and* $\mathcal{M}_n$ *denotes the maximal operator (with* (p,p) *norm* $\| \mathcal{M}_n \|_p$ *) corresponding to the multiplier transforms* $\left\{ T_{m_{j,n}}^{(p)} \right\}_{j=1}^{\nu}$ *on* $L^p(G)$. *Then,*

$$\| \mathcal{M} \|_p \le C .$$

<u>Proof</u>. The proof follows from standard reasoning which we sketch. If $f \in L^1(G)$ and $\hat{f}$ is compactly supported, then by dominated convergence on $\hat{G}$, and Fourier inversion, we infer that for each j , $T_{m_{j,n}}^{(p)} f \longrightarrow T_{m_j}^{(p)} f$ a.e. on G as $n \longrightarrow +\infty$. Hence $\mathcal{M}_n f \longrightarrow \mathcal{M} f$ as $n \longrightarrow +\infty$ a.e. on G . Fatou's lemma now shows that $\| \mathcal{M} f \|_p \le C \| f \|_p$. The proof of the lemma is completed by utilizing the density of such f in $L^p(G)$.

The stage is now set for our main result, which is expressed in the following theorem.

(IV.3) ***Theorem***. *Let* J *belong to the class* $\mathcal{F}$ *defined in* Lemma (II.1), *and fix* p *in the range* $1 < p < \infty$. *Let*

$\{ \phi_j \}$ be a finite or infinite sequence contained in $M_p(\mathbb{Z})$, and for each j, let $Q_j \in M_p(\mathbb{R})$ be the function corresponding to ϕ_j in accordance with (III.21). If the maximal operator $\mathcal{M}$ on $L^p(\mathbb{T})$ corresponding to the sequence $\left\{ T^{(p)}_{\phi_j} \right\}$ is of strong type (p,p), then the maximal operator $\mathcal{M}^{\#}$ on $L^p(\mathbb{R})$ corresponding to the sequence $\left\{ T^{(p)}_{Q_j} \right\}$ is of strong type (p,p), and the (p,p) norms of $\mathcal{M}$ and $\mathcal{M}^{\#}$ satisfy

$$\| \mathcal{M}^{\#} \|_p \le \eta_J \| \mathcal{M} \|_p ,$$

where η_J is a non-negative real constant depending only on J.

<u>Proof</u>. As in the proof of Corollary (III.20), we put $J_1 = J$, and take $J_2 \in \mathcal{F}$ to be a de la Vallée-Poussin kernel of large enough order so that $J = J_1 * J_2$. We shall show that the conclusion of the theorem is valid with $\eta_J = (2\pi)^2\alpha_1\alpha_2$, where α_1 and α_2 are the constants corresponding to J_1 and J_2, respectively, in (II.2).

By monotone convergence, we can assume without loss of generality that $\{ \phi_j \}$ is a finite sequence, say $\{ \phi_j \}_{j=1}^{\nu}$. If we put $\phi_{j,n} = \hat{\tau}_n \phi_j$, where the sequence $\{ \tau_n \}_{n=0}^{\infty}$ is Fejér's kernel for $\mathbb{T}$, then it is easy to see from Lemmas (IV.1) and (IV.2) that we can further assume without loss of generality that each ϕ_j is a finitely supported sequence on $\mathbb{Z}$.

With these reductions for the proof, let

$g, f_1, f_2, \cdots, f_\nu \in \mathcal{D}$. For $j = 1,2,\cdots,\nu$, let $\Psi_j \in M_p(\mathbb{Z})$ be defined by $\Psi_j(k) = \phi_j(-k)$ for all $k \in \mathbb{Z}$, and put $\tilde{f}_j(x) = f_j(-x)$ for all $x \in \mathbb{R}$. Notice that by Lemma (IV.1), the (p,p) norm of the maximal operator on $L^p(\mathbb{T})$ defined by $\left\{ T_{\Psi_j}^{(p)} \right\}_{j=1}^{\nu}$ coincides with the (p,p) norm of the maximal operator on $L^p(\mathbb{T})$ defined by $\left\{ T_{\phi_j}^{(p)} \right\}_{j=1}^{\nu}$.

Let $U: A_p(\mathbb{R}) \longrightarrow A_p(\mathbb{T})$ be the mapping defined in (III.2), and considered in Theorem (III.4), corresponding presently to $K = J_1 * J_2 = J$. For $1 \le j \le \nu$, we apply Plancherel's theorem to (III.18), which was established in the proof of Theorem (III.15), to get:

$$\text{(IV.4)} \qquad \mathcal{K}_{\Psi_j}\left[U(f_j * g)\right] = (2\pi)^{-1}\int_{\mathbb{R}}\left[T_{Q_j}^{(p)} \tilde{f}_j\right](x)\, g(x)\, dx ,$$

where $\mathcal{K}_{\Psi_j}$ is the linear functional on $A_p(\mathbb{T})$ corresponding to Ψ_j in accordance with Proposition (I.5). For $t \in \mathbb{R}$, we let $\Phi_{t,j}: \mathbb{T} \longrightarrow \mathbb{C}$ be defined for f_j in accordance with (III.8), and we let $\Psi_t: \mathbb{T} \longrightarrow \mathbb{C}$ be defined for g in accordance with (III.9). With this set-up we shall utilize the reasoning in the proof of Theorem (III.4) to complete the demonstration of the present theorem with the aid of Lemma (IV.1). For this purpose, we begin by observing that, as the penultimate step in the proof of Theorem (III.4) shows, we have for $1 \le j \le \nu$:

$$\text{(IV.5)} \qquad U(f_j * g) = 2\pi\int_{\mathbb{R}} \Phi_{t,j} * \Psi_t \, dt ,$$

where the integral on the right is an $A_p(\mathbb{T})$-valued Bochner integral. From (IV.4) and (IV.5) we infer that:

(IV.6)

$$(2\pi)^{-2} \left| \sum_{j=1}^{\nu} \int_{\mathbb{R}} \left[T_{Q_j}^{(p)} \tilde{f}_j \right] g \, dx \right|$$

$$= \left| \sum_{j=1}^{\nu} \int_{\mathbb{R}} \left[\left[T_{\Psi_j}^{(p)} \Phi_{t,j} \right] * \Psi_t \right] (1) \, dt \right|$$

$$\leq \int_{\mathbb{R}} \left\| \sum_{j=1}^{\nu} T_{\Psi_j}^{(q)} \Phi_{t,j} \right\|_{L^q(\mathbb{T})} \left\| \Psi_t \right\|_{L^p(\mathbb{T})} dt \ ,$$

where q is the index conjugate to p. The reasoning used to establish (III.13) shows that:

$$\text{(IV.7)} \qquad \int_{\mathbb{R}} \left\| \Psi_t \right\|_{L^p(\mathbb{T})}^p dt \leq \alpha_2^p \left\| g \right\|_{L^p(\mathbb{R})}^p .$$

Using (IV.7) in (IV.6), we get by Hölder's inequality:

(IV.8)

$$(2\pi)^{-2} \left| \sum_{j=1}^{\nu} \int_{\mathbb{R}} \left[T_{Q_j}^{(p)} \tilde{f}_j \right] g \, dx \right|$$

$$\leq \alpha_2 \left\| g \right\|_{L^p(\mathbb{R})} \left[\int_{\mathbb{R}} \left\| \sum_{j=1}^{\nu} T_{\Psi_j}^{(q)} \Phi_{t,j} \right\|_{L^q(\mathbb{T})}^q dt \right]^{1/q} .$$

Applying Lemma (IV.1) to the integrand in the majorant of (IV.8), we see that

(IV.9)

$$(2\pi)^{-2}\left|\sum_{j=1}^{\nu}\int_{\mathbb{R}}\left[T_{Q_j}^{(p)}\tilde{f}_j\right]g\,dx\right|$$

$$\leq \alpha_2\|g\|_{L^p(\mathbb{R})}\,\|\mathcal{M}\|_p\left[\int_{\mathbb{R}}\left\|\sum_{j=1}^{\nu}\left|\Phi_{t,j}\right|\right\|_{L^q(\mathbb{T})}^q dt\right]^{1/q}.$$

Put

$$\text{(IV.10)}\qquad Z(x)=\sum_{j=1}^{\nu}|f_j(x)|,\quad\text{for}\quad x\in\mathbb{R}.$$

It is easy to see from the definition of $\Phi_{t,j}$ that for $x \in \mathbb{R}$, $t \in \mathbb{R}$, we have:

$$\sum_{j=1}^{\nu}\left|\Phi_{t,j}(e^{ix})\right| \leq \sum_{m=-\infty}^{\infty} Z(x+2\pi m)\left|J_1(x+t+2\pi m)\right|.$$

Hence

$$\int_{\mathbb{R}}\left\|\sum_{j=1}^{\nu}\left|\Phi_{t,j}\right|\right\|_{L^q(\mathbb{T})}^q dt$$

$$\leq (2\pi)^{-1}\int_{\mathbb{R}}\int_{-\pi}^{\pi}\left[\sum_{m=-\infty}^{\infty} Z(x+2\pi m)\left|J_1(x+t+2\pi m)\right|\right]^q dx\,dt$$

$$= (2\pi)^{-1}\int_{\mathbb{R}}\int_{-\pi}^{\pi}\left[\sum_{m=-\infty}^{\infty} Z(x-t+2\pi m)\left|J_1(x+2\pi m)\right|\right]^q dx\,dt$$

$$= (2\pi)^{-1}\int_{-\pi}^{\pi}\int_{\mathbb{R}}\left[\sum_{m=-\infty}^{\infty} Z(x+t-2\pi m)\left|J_1(x-2\pi m)\right|\right]^q dt\,dx.$$

A transference argument on the last iterated integral similar to the argument used to establish (III.12) now shows that:

$$\text{(IV.11)}\qquad \left[\int_{\mathbb{R}} \left\| \sum_{j=1}^{\nu} \left|\Phi_{t,j}\right| \right\|_{L^q(\mathbb{T})}^{q} dt\right]^{1/q} \le \alpha_1 \| Z \|_{L^q(\mathbb{R})} .$$

Using (IV.11) in (IV.9), and recalling the definition of Z in (IV.10), we see that for $g, f_1, f_2, \cdots, f_\nu \in \mathcal{D}$,

(IV.12)

$$(2\pi)^{-2} \left| \sum_{j=1}^{\nu} \int_{\mathbb{R}} \left[T_{Q_j}^{(q)} f_j\right] g \, dx \right|$$

$$\le \alpha_1 \alpha_2 \| g \|_{L^p(\mathbb{R})} \| \mathcal{M} \|_p \left\| \sum_{j=1}^{\nu} \left| f_j \right| \right\|_{L^q(\mathbb{R})} .$$

Standard approximations in $L^p(\mathbb{R})$ and $L^q(\mathbb{R})$, and a further application of Lemma (IV.1) complete the proof of Theorem (IV.3).

(IV.13) <u>Remarks</u>. (i) A comparison of the proof of Corollary (III.20) with the opening comments in the proof of Theorem (IV.3) shows that for a given $J \in \mathcal{F}$, the corresponding constants Ω_J and η_J used in proving Corollary (III.20) and Theorem (IV.3), respectively, can be taken to be the same. (ii) For $\beta > 0$ and $g: \mathbb{R} \longrightarrow \mathbb{C}$, let $\left[\delta_\beta g\right](x) = g(\beta x)$ for all $x \in \mathbb{R}$. It is elementary and straightforward to see that if $1 < p < \infty$, $\left\{ C_j \right\}$ is a sequence of bounded linear mappings of $L^p(\mathbb{R})$ into $L^p(\mathbb{R})$ with corresponding maximal operator $\mathcal{C}$ of strong type

(p,p), then the maximal operator $\mathscr{C}^{(\beta)}$ on $L^p(\mathbb{R})$ corresponding to the sequence $\{ \delta_{1/\beta} C_j \delta_\beta \}$ equals $\delta_{1/\beta} \mathscr{C} \delta_\beta$, and hence is also of strong type (p,p) with $\| \mathscr{C}^{(\beta)} \|_p = \| \mathscr{C} \|_p$. In the particular case when $\{ C_j \} = \{ T_{m_j}^{(p)} \}$ for some sequence $\{ m_j \} \subseteq M_p(\mathbb{R})$, then $\{ \delta_{1/\beta} C_j \delta_\beta \} = \{ T_{\delta_\beta(m_j)}^{(p)} \}$. Consequently, under the hypotheses, and in the notation, of Theorem (IV.3), we see that for each positive real number β, the maximal operator $[\mathscr{M}^{\#}]^{(\beta)}$ corresponding to $\{ T_{\delta_\beta(Q_j)}^{(p)} \}$ is of strong type (p,p) with $\| [\mathscr{M}^{\#}]^{(\beta)} \|_p \le \eta_J \| \mathscr{M} \|_p$.

(iii) By special reasoning one can show that Corollary (III.20) and Theorem (IV.3) remain valid for the case $p = 1$, with $2\pi\alpha(J)$ replacing Ω_J and η_J, respectively. We omit the details for expository reasons.

(IV.14) <u>**An application of Theorem (IV.3)**</u>. Fix p in the range $1 < p < \infty$. For each positive integer j, let $\omega_j \in M_p(\mathbb{Z})$ be the characteristic function defined on $\mathbb{Z}$ of $\{ k \in \mathbb{Z} : |k| \le j \}$. The corresponding multiplier transform s_j on $L^p(\mathbb{T})$ is the $j^{\underline{th}}$ partial sum operator for Fourier series. The celebrated Carleson-Hunt Theorem [8] asserts that the maximal operator $\mathscr{A}$ on $L^p(\mathbb{T})$ defined by the sequence $\{ s_j \}_{j=1}^{\infty}$ is of strong type (p,p). Let S_j be the $j^{\underline{th}}$ partial sum operator on

$L^p(\mathbb{R})$ -- that is, S_j is the multiplier transform on $L^p(\mathbb{R})$ corresponding to $\chi_j \in M_p(\mathbb{R})$, where χ_j denotes the characteristic function defined on $\mathbb{R}$ of the interval $[-j,j]$. We shall now see that Theorem (IV.3) readily implies the well-known fact [11] that the maximal operator $\mathcal{S}$ defined by the sequence $\left\{ S_j \right\}_{j=1}^{\infty}$ on $L^p(\mathbb{R})$ inherits strong type (p,p) boundedness from $\mathfrak{s}$. To accomplish this application of Theorem (IV.3) we specialize $J \in \mathcal{F}$ to be $\kappa(t) \equiv (2\pi)^{-1}(t/2)^{-2}\sin^2(t/2)$. Hence $\hat{J}(x) = \max\left\{1 - |x|, 0\right\}$ for all $x \in \mathbb{R}$. For each $j \in \mathbb{N}$, the function $Q_j \in M_p(\mathbb{R})$ corresponding to $\omega_j \in M_p(\mathbb{Z})$ as in Theorem (IV.3) has the form: $Q_j(x) = 1$ if $|x| \le j$; $Q_j(x) = 0$ if $|x| \ge j + 1$; Q_j is linear on $[-j-1,-j]$ and on $[j,j+1]$. By virtue of Theorem (IV.3) and Remark (IV.13)-(ii), it follows from the Carleson-Hunt Theorem for Fourier series that for each $n \in \mathbb{N}$, the maximal operator $\left[\mathcal{M}^{\#}\right]^{(n)}$ on $L^p(\mathbb{R})$ corresponding to $\left\{ T^{(p)}_{\delta_n(Q_j)} \right\}_{j=1}^{\infty}$ is of strong type (p,p), with

$$\text{(IV.15)} \qquad \left\| \left[\mathcal{M}^{\#}\right]^{(n)} \right\|_p \le \eta_\kappa \left\| \mathfrak{s} \right\|_p .$$

Let $m_{j,n} = \delta_n(Q_{jn})$, for $n \in \mathbb{N}$, $j \in \mathbb{N}$. Thus $m_{j,n}$ is linear on $[j,j+(1/n)]$, and on $[-j-(1/n),-j]$, while $m_{j,n}(x) = 1$ if $|x| \le j$, and $m_{j,n}(x) = 0$ if $|x| \ge j + (1/n)$. Temporarily fix a positive integer j_0, and let $\mathcal{S}_{j_0}$ be the maximal operator corresponding to the finite sequence of partial sum operators $\left\{ S_j \right\}_{j=1}^{j_0}$ on $L^p(\mathbb{R})$. It follows from (IV.15) and Lemma (IV.2) applied to $\left\{ X_j \right\}_{j=1}^{j_0}$ and $\left\{ m_{j,n} \right\}_{1 \le j \le j_0,\ n \in \mathbb{N}}$ that:

$$\| \mathscr{S}_{j_0} \|_p \le \eta_\kappa \| \mathfrak{s} \|_p , \quad \text{for all } j_0 \in \mathbb{N} .$$

It follows from this by monotone convergence that the maximal operator $\mathscr{S}$ corresponding to the sequence $\left\{ S_j \right\}_{j=1}^{\infty}$ of partial sum operators on $L^p(\mathbb{R})$ is of strong type (p,p) with

$$\| \mathscr{S} \|_p \le \eta_\kappa \| \mathfrak{s} \|_p , \tag{IV.16}$$

where η_κ is the constant in Theorem (IV.3) corresponding to the particular choice $J = \kappa$ in this application.

In closing Section IV, we comment on the size of η_κ. In fact, elementary considerations show that we can choose a value of η_κ not exceeding 5. This can be seen as follows. In the notation of the proof of Theorem (IV.3), we have $\eta_\kappa = (2\pi)^2 \alpha_1 \alpha_2$, where, in the present circumstances, α_1 (respectively, α_2) is the constant in (II.2) corresponding to κ (respectively, corresponding to V_1, the de la Vallée-Poussin kernel for $\mathbb{R}$ of order 1). As is well-known, simple considerations with Fourier transforms show that the periodic function $\sum_{m=-\infty}^{\infty} \kappa(x + 2\pi m)$ is identically $(2\pi)^{-1}$ on $\mathbb{R}$. Since $\kappa \ge 0$, we have $\alpha_1 = (2\pi)^{-1}$. From $V_1(x) \equiv 4\kappa(2x) - \kappa(x)$, we see that

$\alpha_2 \leq 5(2\pi)^{-1}$.

V. **The Hilbert Transform and Summability Methods for Fourier Multipliers**. The crux of the Coifman-Weiss General Transference Result is that under appropriate circumstances convolution norms of L^1-kernels can be transferred to L^p-subspaces acted on by groups (see (I.9)). This provides much sharper bounds for transferred convolution operators than the crude and obvious estimates obtained from the L^1-norms of the convolution kernels. Although in the preceding sections (notably in (III.12), (III.13), and (IV.11)), our structural viewpoint has been fostered by the transference of convolution operators, we have only needed to transfer along with such operators their cruder bounds involving L^1-norms of kernels. This state of affairs is reflected in the roles played by α_1 and α_2 in Theorems (III.4) and (III.15), and is attributable to the absolute summability which Lemma (II.1) provides for the context of Section III.

In this, our final section, we shall relax the hypotheses of Section III, and, in effect, replace J in Corollary (III.20) by its Hilbert transform. This will associate with $L^p(\mathbb{T})$-multipliers discontinuous functions on $\mathbb{R}$ which are $L^p(\mathbb{R})$-multipliers (see Theorem (V.1), Corollary (V.13), and the remarks in (V.15)). As (V.9) shows, estimates by ℓ^1-norms of convolution kernels will no longer suffice for this task, which will require us to take full advantage of the sharpness inherent in (I.9). Moreover, in contrast to the considerations in Section III, the relaxed hypotheses of the current section do not lead us to mappings defined *globally* on $A_p(\mathbb{R})$ or the adjoints of such mappings (see the discussion at the outset of the proof for Lemma (V.10)).

(V.1) ***Theorem***. *Let* $\mathcal{F}$ *be the class of functions defined in* Lemma (II.1), *let* $J \in \mathcal{F}$, *and suppose that* $1 < p < \infty$. *For each* $t \in \mathbb{R}$, *denote the signum of* t *by* $\operatorname{sgn}(t)$. *Then for each* $\phi \in M_p(\mathbb{Z})$, *the function* $\mathcal{T}_\phi : \mathbb{R} \longrightarrow \mathbb{C}$ *defined by*

$$\text{(V.2)} \qquad \mathcal{T}_\phi \equiv \sum_{m=-\infty}^{\infty} \phi(m)\, \hat{J}(x - m)\, \operatorname{sgn}(x - m)$$

belongs to $M_p(\mathbb{R})$, *and*

$$\text{(V.3)} \qquad \left\| \mathcal{T}_\phi \right\|_{M_p(\mathbb{R})} \le \sigma_{p,J} \left\| \phi \right\|_{M_p(\mathbb{Z})} ,$$

where $\sigma_{p,J}$ *is a real constant depending only on* p *and* J.

The proof of Theorem (V.1) will be accomplished with the aid of Lemmas (V.6) and (V.10) below, the latter being obtained by suitably generalizing the reasoning in Section III. Throughout this section, the index p and the function J in the hypotheses of Theorem (V.1) will be fixed, and we also fix a positive real number ρ such that the support of $\hat{J}$ is contained in the interval $[-\rho,\rho]$. Following the notation in the proof of Corollary (III.20), we let V_ρ denote the de la Vallée-Poussin kernel for $\mathbb{R}$ of order ρ . Denote by $\tilde{V}_\rho$ the Hilbert transform of V_ρ . In particular, $\tilde{V}_\rho$ belongs to $L^s(\mathbb{R})$ for $1 < s < \infty$, and its $L^2(\mathbb{R})$-Fourier transform $\left[\tilde{V}_\rho\right]^\wedge$ is given by $\left[\tilde{V}_\rho\right]^\wedge(t) \equiv -i\left[\operatorname{sgn}(t)\right]\hat{V}_\rho(t)$. Hence by Fourier inversion we can express $\tilde{V}_\rho$ by

(V.4) $$\tilde{V}_\rho(y) = (2\pi i)^{-1}\int_{\mathbb{R}}\left[\operatorname{sgn}(t)\right]\hat{V}_\rho(t)\, e^{ity}\, dt ,$$

for all $y \in \mathbb{R}$.

Thus $\tilde{V}_\rho$ is a bounded continuous function on $\mathbb{R}$. Put

(V.5) $$J_1 = \tilde{V}_\rho , \quad J_2 = J , \quad \text{and} \quad K = J_1 * J_2 .$$

In particular, K is a bounded function on $\mathbb{R}$ belonging to $L^2(\mathbb{R})$, and by Fourier inversion $K \in C^\infty(\mathbb{R})$.

(V.6) ***Lemma***. *For each* $x \in [-\pi,\pi]$, *the sequence* $\left\{ \tilde{V}_\rho(x + 2\pi m) \right\}_{m=-\infty}^{\infty}$ *defines a bounded convolution operator on* $\ell^p(\mathbb{Z})$. *For each* $n \in \mathbb{N}$, *let* ω_n *be the characteristic function defined on* $\mathbb{Z}$ *of* $\left\{ m \in \mathbb{Z}: \ |m| \le n \right\}$, *and let* $N_{p,x,\rho,n}$ *denote the norm of the convolution operator on* $\ell^p(\mathbb{Z})$ *defined by the sequence* $\left\{ \omega_n(m)\, \tilde{V}_\rho(x + 2\pi m) \right\}_{m=-\infty}^{\infty}$. *Then*

$$\mathcal{N}_{p,\rho} \equiv \sup\left\{ N_{p,x,\rho,n}: \ x \in [-\pi,\pi] , \ n \in \mathbb{N} \right\} < +\infty .$$

<u>Proof of Lemma</u> (V.6). Easy direct calculations with (V.4) show that:

$$\text{(V.7)} \qquad \pi\, \tilde{V}_\rho(y) = y^{-1} - \rho^{-1} y^{-2} \left\{ \sin(2\rho y) - \sin(\rho y) \right\}, \quad \text{for } y \in \mathbb{R} \setminus \{0\}.$$

If $x \in [-\pi,\pi]$, and $m \in \mathbb{Z} \setminus \{0\}$, then $| x + 2\pi m | \geq 2\pi | m | - \pi = 2\pi \left[| m | - 2^{-1} \right] > 0$. Hence from (V.7) we have:

$$\begin{aligned} \text{(V.8)} \qquad & \pi\, \tilde{V}_\rho(x + 2\pi m) \\ &= (x+2\pi m)^{-1} - (2\pi m)^{-1} + (2\pi m)^{-1} \\ &\quad - \rho^{-1}(x+2\pi m)^{-2} \left\{ \sin 2\rho(x+2\pi m) - \sin \rho(x+2\pi m) \right\} \\ &= (2\pi m)^{-1} - x(2\pi m)^{-1}(x+2\pi m)^{-1} \\ &\quad - \rho^{-1}(x+2\pi m)^{-2} \left\{ \sin 2\rho(x+2\pi m) - \sin \rho(x+2\pi m) \right\}. \end{aligned}$$

Let $\hbar$ be the discrete Hilbert kernel. Thus, $\hbar(n) = n^{-1}$ for $n \in \mathbb{Z} \setminus \{0\}$, $\hbar(0) = 0$. Let ω_0 denote the characteristic function defined on $\mathbb{Z}$ of the singleton set $\{0\}$. It follows from (V.8) that for $x \in [-\pi,\pi]$,

$$\text{(V.9)} \qquad \left\{ \tilde{V}_\rho(x + 2\pi m) \right\}_{m=-\infty}^{\infty} = \tilde{V}_\rho(x)\omega_0 + 2^{-1}\pi^{-2}\hbar + b_{x,\rho},$$

where $b_{x,\rho} \in \ell^1(\mathbb{Z})$, and,

$$\sup \left\{ \| b_{x,\rho} \|_{\ell^1(\mathbb{Z})} : x \in [-\pi,\pi] \right\} < +\infty .$$

Since the discrete Hilbert transform is bounded on $\ell^p(\mathbb{Z})$, while $\tilde{V}_\rho$ is bounded on $\mathbb{R}$, and the truncates of the discrete Hilbert kernel define a uniformly bounded sequence of convolution operators on $\ell^p(\mathbb{Z})$, the conclusions of Lemma (V.6) follow from (V.9).

The next lemma provides the analogue of Theorem (III.4)

appropriate to the weaker hypotheses of the present set-up.

(V.10) ***Lemma***. *Let* $f \in \mathcal{D}$, $g \in \mathcal{D}$, *and put* $F = f * g$. *Define* $UF: \mathbb{T} \longrightarrow \mathbb{C}$ *by* (III.2), *where, in the present circumstances,* $K \in L^2(\mathbb{R}) \cap C^\infty(\mathbb{R})$ *is the bounded function on* $\mathbb{R}$ *defined in* (V.5). *Then* $UF \in A_p(\mathbb{T})$, *and*

$$\text{(V.11)} \qquad \| UF \|_{A_p(\mathbb{T})} \le 2\pi \, \mathcal{N}_{p,\rho} \, \alpha(J) \, \| f \|_{L^p(\mathbb{R})} \| g \|_{L^q(\mathbb{R})} ,$$

where $\mathcal{N}_{p,\rho}$ *is as described in* Lemma (V.6), *and* $\alpha(J)$ *is as in* (II.2).

Proof of Lemma (V.10). Before proceeding to the details of the lemma's demonstration, we make some clarifying observations about the differences between the current setting and the context of Theorem (III.4). Notice that since F has compact support, the definition of UF in Lemma (V.10) is meaningful. However, since we can no longer assert that the sequence $\left\{ K(x + 2\pi m) \right\}_{m=-\infty}^{\infty}$ is absolutely summable for all $x \in \mathbb{R}$ (consider, for instance, the special case when $J(x) \equiv (2\pi)^{-1}(x/2)^{-2}\sin^2(x/2)$ and ρ is taken to be 1), no attempt is being made in this section to define U on all of $A_p(\mathbb{R})$. As will be seen in due course, the current definition of U will suffice for purposes of establishing Theorem (V.1).

With J_1 and J_2 as in (V.5), we define Φ_t and Ψ_t , for $t \in \mathbb{R}$, by (III.8) and (III.9). As previously, for each $t \in \mathbb{R}$, Φ_t and Ψ_t are continuous functions on

$\mathbb{T}$. Moreover, since $\tilde{V}_\rho \in L^p(\mathbb{R})$, it is clear that Φ_t moves continuously in $L^p(\mathbb{T})$ as t varies in $\mathbb{R}$. Similarly, Ψ_t moves continuously in $L^q(\mathbb{T})$, where q is the index conjugate to p . Moreover, with due regard for the weaker properties currently enjoyed by J_1 , it is not difficult to see that (III.10) remains valid. The inequality (III.13) remains valid just as before, and so does the equation immediately preceding (III.12). However, special care is now needed in applying the Coifman-Weiss General Transference Result (I.8) to the latter equation, because, in the present setting, as (V.9) shows, for $x \in [-\pi,\pi]$, the sequence $\left\{ J_1(x - 2\pi m) \right\}_{m=-\infty}^{\infty}$ does not belong to $\ell^1(\mathbb{Z})$. For each $x \in [-\pi,\pi]$, and $n \in \mathbb{N}$, it follows from the General Transference Result and Lemma (V.6) that:

$$\int_{\mathbb{R}} \left| \sum_{m=-n}^{n} \left[R_{-m} f_x\right](t)\ J_1(x - 2\pi m) \right|^p dt$$

$$\leq \mathcal{N}^p_{p,\rho} \left\| f \right\|^p_{L^p(\mathbb{R})} .$$

An application of Fatou's Lemma now shows that for $x \in [-\pi,\pi]$:

$$\int_{\mathbb{R}} \left| \sum_{m=-\infty}^{\infty} \left[R_{-m} f_x\right](t)\ J_1(x - 2\pi m) \right|^p dt$$

$$\leq \mathcal{N}^p_{p,\rho} \left\| f \right\|^p_{L^p(\mathbb{R})} .$$

Hence

(V.12) $$\int_{\mathbb{R}} \| \Phi_t \|^p_{L^p(\mathbb{T})} \, dt \le \mathcal{N}^p_{p,\rho} \| f \|^p_{L^p(\mathbb{R})} .$$

The proof of Lemma (V.10) can now be completed by reasoning analogous to that used to finish the proof of Theorem (III.4).

<u>Proof of Theorem</u> (V.1). The $L^2(\mathbb{R})$-Fourier transform $\hat{K}$ of the function K in (V.5) can be taken to satisfy

$$\hat{K}(t) = -i\left[\mathrm{sgn}(t)\right] \hat{V}_\rho(t) \hat{J}(t) = -i\left[\mathrm{sgn}(t)\right] \hat{J}(t) , \quad \text{for all } t \in \mathbb{R} .$$

With the aid of Lemma (V.10), the reasoning used in the demonstration of Theorem (III.15) (especially the reasoning used to establish (III.18) therein) adapts directly to the present context of K and its $L^2(\mathbb{R})$-Fourier transform $\hat{K}$, and thereby provides a proof of Theorem (V.1) with

$$\sigma_{p,J} = (2\pi)^2 \mathcal{N}_{p,\rho} \alpha(J) ,$$

where $\mathcal{N}_{p,\rho}$ is defined in Lemma (V.6), and $\alpha(J)$ is as described in (II.2).

(V.13) *Corollary. Let* $\Delta(x) = \max\left\{ 1 - |x| , 0 \right\}$ *for all* $x \in \mathbb{R}$, *and suppose* $1 < p < \infty$. *Then for each* $\phi \in M_p(\mathbb{Z})$, *the function* $\mathcal{U}_\phi : \mathbb{R} \to \mathbb{C}$ *defined by*

$$\mathcal{U}_\phi(x) \equiv \sum_{m=-\infty}^{\infty} \phi(m) \Delta(x - m) \,\mathrm{sgn}(x - m)$$

belongs to $M_p(\mathbb{R})$, *and*

(V.14) $$\| \mathcal{U}_\phi \|_{M_p(\mathbb{R})} \le (2\pi) \mathcal{N}_{p,1} \| \phi \|_{M_p(\mathbb{Z})} ,$$

where $\mathscr{N}_{p,1}$ *is the constant defined in* Lemma (V.6) *corresponding to the special case* $\rho = 1$.

Proof. Take J in Theorem (V.1) to be the function defined by $J(x) \equiv (2\pi)^{-1}(x/2)^{-2}\sin^2(x/2)$. Thus, $\hat{J} = \Delta$, ρ can be taken to be 1 , and, as noted at the end of Section IV, $\alpha(J) = (2\pi)^{-1}$. Reference to the value for $\sigma_{p,J}$ obtained at the end of the proof of Theorem (V.1) completes the demonstration of this corollary.

(V.15) **Remarks**. The function $\mathcal{U}_\phi$ defined in Corollary (V.13) has the following description. For each $n \in \mathbb{Z}$: $\mathcal{U}_\phi(n) = 0$, $\mathcal{U}_\phi$ is linear on the open interval $(n,n+1)$, and

$$\lim_{x \to n^+} \mathcal{U}_\phi(x) = \phi(n) ,$$

$$\lim_{x \to (n+1)^-} \mathcal{U}_\phi(x) = -\phi(n + 1) .$$

In particular, $\mathcal{U}_\phi$ is equal almost everywhere on $\mathbb{R}$ to a function continuous on $\mathbb{R}$ if and only if $\phi(n) = 0$ for all $n \in \mathbb{Z}$.

ACKNOWLEDGEMENTS. The work of the second author was supported by a grant from the National Science Foundation. The third author is grateful to the Science and Engineering Research Council for its financial support, and to the University of Illinois for its hospitality.

References

1. E. Berkson, T.A. Gillespie, and P.S. Muhly, *Generalized analyticity in* UMD *spaces*, Arkiv för Matematik, to appear.

2. R.R Coifman and G. Weiss, *Transference methods in analysis*, C.B.M.S. Regional Conference Series in Math., No. 31, Amer. Math. Soc., Providence, R.I., 1977 (reprinted 1986).

3. M. Cowling, *Extension of Fourier* L^p-L^q *multipliers*, Trans. Amer. Math. Soc., 213(1975), 1-33.

4. A. Figà-Talamanca, *Translation invariant operators in* L^p, Duke Math. J., 32(1965), 495-501.

5. A. Figà-Talamanca and G.I. Gaudry, *Extensions of multipliers*, Boll. Unione Mat. Italiana, 3(1970), 1003-1014.

6. A. Figà-Talamanca and G.I. Gaudry, *Multipliers of* L^p *which vanish at infinity*, J. Functional Analysis, 7(1971), 475-486.

7. C. Herz, *Remarques sur la note précédente de Varopoulos*, C.R. Acad. Sci. Paris, 260(1965), 6001-6004.

8. R.A. Hunt, *On the convergence of Fourier series*, D.T. Haimo, ed., "Orthogonal Expansions and Their Continuous Analogues", Southern Illinois Univ. Press, Carbondale, 1968, 235-255.

9. M. Jodeit, *Restrictions and extensions of Fourier multipliers*, Studia Math., 34(1970), 215-226.

10. Y. Katznelson, *An introduction to harmonic analysis*, 2nd edition, Dover, New York, 1976.

11. C. Kenig and P. Tomas, *Maximal operators defined by Fourier multipliers*, Studia Math., 68(1980), 79-83.

2

The Three Squares Theorem, A Local Version

C. A. Berenstein, R. Gay, and A. Yger

One of us had the good luck of starting his research career in Buenos Aires under the guidance of Mischa Cotlar. Every subject he taught was deceptively simple. Mischa seemed to have the gift of making the deepest mathematical thoughts look easy. Twenty-five years later he is still regaling us with beautiful work written with the same simplicity. The three of us take the opportunity to offer in return some simple remarks on the local properties of the Pompeiu transform. The results below can be understood either as covering lemmas (a subject very close to Cotlar's work) or, as we prefer to think of them, as part of the integral geometry problem of recovering a function from its averages and do this only using "local" knowledge of these averages. In this context, it fits into a line of research pursued by us for some time, in one guise it appears as a deconvolution problem [5], in another as an overdetermined inverse problem, [4], [6], etc. The fact that the Pompeiu transform is injective locally [1], [2] and even has a local inverse, has thrown a monkey wrench into the idea that it was an object completely tied to the global action of a group on a measure space or, more restrictedly, to symmetric spaces [6]. What is the "true" nature of this transform, that is, the most general (and ultimately simpler) conditions under which one can invert it, are not yet clear to us. We hope the present paper will help to stir interest in this kind of question.

We start by recalling a rather general formulation of the Pompeiu transform [6]. Let G be a continuous group acting transitively on a locally compact space X such that there is a Radon measure $\mu \geq 0$ invariant under G. Let $E_1, \ldots, E_N$ be a finite collection of compact sets of positive measure. The Pompeiu transform

$$P : C(X) \longrightarrow (C(G))^N$$

is defined by

$$Pf(g) = \left[\int_{gE_j} f d\mu\right]_{j=1}^{N}.$$

The family $\{E_1, \ldots, E_N\}$ is said to have the Pompeiu property if P is injective. Of course, it is very hard to say something interesting in such a general context. For that reason, it is natural to restrict consideration to some natural cases.

(1) $X = \mathbb{R}^n$, $d\mu$ = Lebesgue measure, $G = M(n)$ = motion group of $\mathbb{R}^n$, $N = 1$. Let Ω be an open bounded set, $E = \bar{\Omega}$, such that E^c is connected, and $\partial\Omega$ is Lipschitz then E has the Pompeiu property if and only if there are no positive eigenvalues α for the following overdetermined Neumann problem:

$$(\mathcal{N}) \qquad \begin{cases} \Delta u + \alpha u = 0 & \text{in } \Omega, \\ \dfrac{\partial u}{\partial n} = 0, \quad u = 1 & \text{on } \partial\Omega. \end{cases}$$

This was basically proved in [7]. One sees that Ω = ball, always fails to have the Pompeiu property. If Ω = ellipsoid $\neq$ ball, then the Pompeiu property holds. The same thing happens if $\partial\Omega$ fails to be real analytic [5].

(2) $X = G/K$ symmetric space of the non-compact type, rank 1. All the items in (1) still hold [4] with the natural definition for $d\mu$. Here, Δ is the Laplace Beltrami operator.

(3) The same as in (2) when rank > 1. We know nothing general except

to say that if there is a solution to $(\mathcal{N})$ then $\bar{\Omega}$ does not have the Pompeiu property.

(4) $X = G = \mathbb{R}^n$. A single set will never have the Pompeiu property since every $\alpha \in \mathbb{C}^n$, which is a zero of the Fourier transform $\hat{\chi}_E(\alpha) = 0$ leads to a function $f(x) = e^{i\alpha x}$ in the kernel of P. On the other hand, there are some families known to have the Pompeiu property. For instance:

(i) (Three Squares Theorem). Let $0 < a_1 < a_2 < \dots < a_{n+1}$ be linearly independent over $\mathbb{Q}$, then the family of hypercubes $E_j = [-a_j, a_j]^n$, $j = 1, \dots, n+1$, has the Pompeiu property.

This was originally proved in [3] via a theorem about spectral analysis for certain ideals in $\mathcal{E}'(\mathbb{R}^n)$. Later a simple proof based on the well known case of $n = 1$ was given in [10].

(ii) Let $0 < r_1 < r_2$ be such that $r_2/r_1 \notin Z_n = \{\xi/\eta : \xi > 0, \eta > 0, J_{n/2}(\xi) = J_{n/2}(\eta) = 0\}$ then the family of two balls $E_j = \bar{B}(0, r_j)$, $j = 1, 2$, has the Pompeiu property.

This was proved independently in [7], [13]. Generalizations of (i) as well as a general technique depending on the concept of slowly decreasing families is given in [5].

(5) $X = G$ = semisimple Lie group. Note that while in $X = \mathbb{R}^n$, if $f \in L^1(\mathbb{R}^n)$ (or $L^2(\mathbb{R}^n)$) then f cannot be in the kernel of P even for $N = 1$, this property fails in other groups. See [12] for some interesting examples, both positive and negative.

There are many open questions and generalizations replacing sets E_j by distributions, etc. An excellent and witty survey of some of these questions is [14].

Up to this point, the techniques generally employed to study the injectivity of P seemed to indicate that the Pompeiu property could be only a "global" property, requiring the transitive action of a Lie group or something of that sort. It came in some sense as a surprise that some of these theorems had a "local" version. Let us explain now briefly what is proved in [1], [2]. We coach the problem in the same general terms as we had at the beginning.

Let Y be an open subset of X, define subsets G_j of G as follows:

$$G_j := \{g \in G : gE_j \subseteq Y\}.$$

The local Pompeiu transform acts on $C(Y)$,

$$P : C(Y) \longrightarrow \bigoplus_{j=1}^{N} C(G_j),$$

defined by the same formula as before. For the case $X = \mathbb{R}^n$, $G = M(n)$, if we want to prove the injectivity of the local Pompeiu transform for certain kinds of sets, it is natural to consider balls of a fixed radius R, since then it will also be done for any set that can be obtained as the union of balls of radius R. Elsewhere we have proved the following:

<u>Proposition 1</u> [1]. Let $0 < r_1 < r_2$, $r_2/r_1 \notin Z_n$, $r_1 + r_2 < R$. Let $Y = B(0,R)$, $E_j = \bar{B}(0,r_j)$, $j = 1,2$. Then the local Pompeiu transform is injective in $B(0,R)$.

<u>Corollary</u>. Let K a closed convex set in $\mathbb{R}^n$, $r_2/r_1 \notin Z_n$. If $f \in C(K^c)$ is such that

$$\int_B f dx = 0$$

for every closed ball of radius r_1 and r_2 such that $B \cap K = \emptyset$, then $f = 0$ in K^c. (This is the equivalent to Helgason's Hole Theorem for the Radon transform.)

<u>Proposition 2</u> [2]. Let E have the Pompeiu property, $R > 0$ such that

$E \subseteq B(0,R/2)$, ∂E sufficiently smooth. Then the local Pompeiu transform corresponding to E is injective in $B(0,R)$. Actually one needs some smoothness of ∂E near a single extreme point of ∂E. Moreover, one has a corollary similar to the previous one, where B is replaced by E.

In the proof of Propositions 1 and 2 use very strongly that one is dealing with the motion group; one especially uses the orthogonal group to more or less reduce the problem to a one variable (radial) case though the situation is not so simple as in the global Pompeiu problem. We show next that the local version of the Three-Squares Theorem also holds.

Theorem 3. Let $0 < r_1 < r_2 < \dots < r_{n+1}$, $n+1$ numbers linearly independent over $\mathbb{Q}$ and let

$$R > r_1 + r_2 + \dots + r_{n+1}.$$

Then if $f \in C(]-R,R[^n)$ satisfies for every j, $1 \leq j \leq n+1$,

$$\int_{[-r_j,r_j]^n} f(x-t)dt = 0 \quad \text{if} \quad \max_{1 \leq i \leq n} |x_i| < R - r_j$$

then $f \equiv 0$.

Proof. As usual, we can assume $f \in C^\infty(]-R,R[^n)$. The proof proceeds by recurrence on the dimension. We start with the following.

Lemma 4. $0 < r_1 < r_2$ be such that $r_2/r_1 \notin \mathbb{Q}$ and $R > r_1 + r_2$. If f is a C^∞ function in $]-R,R[$ such that

$$\int_{-r_j}^{r_j} f(x-t)dt = 0, \quad |x| < R - r_j, \; j = 1,2$$

then $f \equiv 0$ in $]-R,R[$.

Proof of Lemma 4. Let $\chi_j = \chi_{[-r_j,r_j]}$ be the characteristic function of the interval $[-r_j, r_j]$. The hypothesis states that $\chi_j * f(x) = 0$ for $|x| < R - r_j$. The classical theory of mean periodicity in one variable [8], [9], [11] allows us to find a sequence $(P_k)_{k\geq 1}$ of exponential-polynomial solutions of the equation

$$\chi_1 * P_k = 0 \quad \text{in} \quad |x| < R - r_1$$

such that

$$\lim_{k\to\infty} P_k = f \quad \text{in} \quad \mathscr{E}(]-R, R[).$$

Since

$$\hat{\chi}_1(z) = \int_{-r_1}^{r_1} e^{-itz} dt = \frac{2 \sin r_1 z}{z},$$

the exponential polynomials P_k are finite sums of the form

$$P_k(x) = \sum_{j\in\mathbb{Z}^*} P_{k,j}\, e^{(ij\pi/r_1)x} \qquad (P_{k,j} \in \mathbb{C}).$$

Let us consider the compactly supported distributions T_ℓ, $\ell \in \mathbb{Z}^*$, defined by

$$\hat{T}_\ell(z) := \begin{cases} \dfrac{\hat{\chi}_1(z)}{z - \dfrac{\ell\pi}{r_1}} & z \neq \dfrac{\ell\pi}{r_1}, \\[2ex] \hat{\chi}_1'\left(\dfrac{\ell\pi}{r_1}\right) & z = \dfrac{\ell\pi}{r_1}. \end{cases}$$

Clearly $\hat{T}_\ell$ is a function in the space $\hat{\mathscr{E}}'(\mathbb{R})$ as a consequence of the Paley-Wiener theorem. From the exact growth conditions one can conclude that $\text{cv}(\text{supp}\, T_\ell) = [-r_1, r_1]$. Another way of seeing this is that the above definition implies that $T_\ell \neq 0$ and satisfies the differential equation

$$\left(\frac{1}{i}\frac{d}{dx} - \frac{\ell\pi}{r_1}\right)T_\ell = \left(\frac{\delta'}{i} - \frac{\ell\pi}{r_1}\delta\right)*T_\ell = \chi_1.$$

Therefore, in the set $\mathbb{R}\setminus[-r_1,r_1]$, T_ℓ is a function of the form $c_\pm e^{(i\ell\pi/r_1)x}$ ($c_\pm$ two possibly different constants in each half-ray). Since T_ℓ has compact support it follows that $c_+ = c_- = 0$. Being a solution of a differential equation shows that in fact

$$\operatorname{supp} T_\ell = [-r_1,r_1].$$

We can compute the convolution of f with T_ℓ with help of the exponential polynomials P_k:

$$T_\ell*f(x) = \lim_{k\to\infty} \sum_{j\in\mathbb{Z}^*} P_{k,j}\hat{T}_\ell\left(\frac{j\pi}{r_1}\right) e^{(ij\pi/r_1)x} = \lim_{k\to\infty} P_{k,\ell}\,\hat{\chi}_1'\left(\frac{\ell\pi}{r_1}\right) e^{(i\ell\pi/r_1)x},$$

where the limit can be taken in the sense of $\mathcal{E}(]-R+r_1,R-r_1[)$. Since f is also mean-periodic with respect to χ_2 we have that in $]-R+r_1+r_2,R-r_1-r_2[$

$$0 = T_\ell*\chi_2*f(x) = (\chi_2*T_\ell*f)(x)$$

$$= \lim_{k\to\infty} \chi_2*\left[\hat{\chi}_1'\left(\frac{\ell\pi}{r_1}\right)P_{k,\ell}\, e^{(i\ell\pi/r_1)x}\right]$$

$$= \lim_{k\to\infty} \hat{\chi}_2\left(\frac{\ell\pi}{r_1}\right)\cdot\hat{\chi}_1'\left(\frac{\ell\pi}{r_1}\right)\cdot P_{k,\ell}\, e^{(i\ell\pi/r_1)x}.$$

We know that $\hat{\chi}_1'\left(\frac{\ell\pi}{r_1}\right) \neq 0$ and it is easy to see that the $\mathbb{Q}$-linear independence of r_1,r_2 guarantees that $\hat{\chi}_2\left(\frac{\ell\pi}{r_1}\right) \neq 0$ also. Therefore we can conclude that for every $\ell \in \mathbb{Z}^*$,

$$\lim_{k\to\infty} P_{k,\ell} = 0.$$

Hence

$$T_\ell * f(x) = 0 \quad \text{for} \quad x \in]-R+r_1, R-r_1[.$$

In order to show that $f \equiv 0$ in $-]R,R[$ we use the following division formula.

<u>Lemma 5</u>. Every distribution $\sigma \in \mathcal{E}'(\mathbb{R})$ such that $\mathrm{cv}(\mathrm{supp}\,\sigma) = [-\rho,\rho]$ $(\rho \geq 0)$ can be written in the form

$$\sigma = \sigma_1 + \chi_1 * S,$$

where $\sigma_1, S \in \mathcal{E}'(\mathbb{R})$ are such that

$$\mathrm{supp}\,\sigma_1 \subseteq [-r_1, r_1]$$

and

$$\mathrm{supp}\,S \subseteq [-\max(r_1,\rho) + r_1,\ \max(r_1,\rho) - r_1].$$

<u>Proof of Lemma 5</u>. If we compare Fourier transforms in the decomposition we want to prove we see that $\hat{\sigma}_1$ interpolates the values of $\hat{\sigma}$ on the zero set of $\hat{\chi}_1$. It is therefore natural to define $\hat{\sigma}_1$ by the interpolation formula

$$\hat{\sigma}_1(z) = z^p \sum_{\ell \in \mathbb{Z}^*} \frac{\hat{\sigma}\left(\frac{\ell\pi}{r_1}\right)}{\hat{\chi}_1'\left(\frac{\ell\pi}{r_1}\right)} \frac{\hat{T}_\ell(z)}{\left(\frac{\ell\pi}{r_1}\right)^p},$$

where p is an integer ≥ 1 choosen sufficiently large so that the above series converges in $\hat{\mathcal{E}}'(\mathbb{R})$. It is then clear that

$$\mathrm{supp}\,\sigma_1 \subseteq [-r_1, r_1]$$

since all the T_ℓ have that support. Furthermore, there is an entire function h such that

$$\hat{\sigma} - \hat{\sigma}_1 = \hat{\chi}_1 \cdot h$$

since $\hat{\sigma} - \hat{\sigma}_1$ vanishes on every zero of $\hat{\chi}_1$, and those zeros are all simple. The function $\hat{\chi}_1$ is slowly decreasing, hence $h \in \hat{\mathcal{E}}'(\mathbb{R})$. (A direct proof is as follows: on the sides of squares centered at $\frac{\ell\pi}{r_1}$ and of side $\frac{\ell\pi}{r_1}$, $|\hat{\chi}_1(z)| \geq \frac{c}{|z|}$. From the Paley-Wiener theorem it follows that h has polynomial growth on the real axis. For $|\mathrm{Im}\, z| \geq \frac{\pi}{r_1}$ we have an even better lower bound for $|\hat{\chi}_1|$. Therefore h also satisfies the Paley-Wiener estimates and we conclude that $h \in \hat{\mathcal{E}}'(\mathbb{R})$. Let $S \in \mathcal{E}'(\mathbb{R})$ be such that $\hat{S} = h$. Then

$$\sigma - \sigma_1 = \chi_1 * S.$$

The Titchmarsh-Lions theorem on supports says that

$$\mathrm{cv}(\mathrm{supp}(\sigma - \sigma_1)) = [-r_1, r_1] + \mathrm{cv}(\mathrm{supp}\, S).$$

If $0 \leq \rho \leq r_1$, we have $\mathrm{cv}(\mathrm{supp}(\sigma - \sigma_1)) \subseteq [-r_1', r_1']$ for some $r_1' \in [0, r_1]$, where one of the endpoints $\pm r_1'$ is an element of $\mathrm{supp}(\sigma - \sigma_1)$. Therefore, $\mathrm{cv}(\mathrm{supp}\, S) = \{0\}$, $r_1 = r_1'$ and $\max(r_1, \rho) - r_1 = 0$. So the lemma is correct in this case.

If $r_1 < \rho$, then $\mathrm{cv}(\mathrm{supp}(\sigma - \sigma_1)) = [-\rho, \rho]$, and

$$\mathrm{cv}(\mathrm{supp}\, S) = [-\rho + r_1, \rho - r_1].$$

This proves completely Lemma 5. □

Let us return to the proof of Lemma 4. We let

$$\sigma = \sigma_\rho = \frac{1}{2}(\delta_\rho + \delta_{-\rho}),$$

which by Lemma 5 can be written

$$\sigma_\rho = \sigma_{1,\rho} + \chi_1 * S_\rho.$$

If $\chi_1 * f = 0$ in $]-R+r_1, R-r_1[$ and if $|x| < R - \max(r_1, \rho)$, then

(i) $\sigma_{1,\rho} * f(x)$ is well defined since $\mathrm{supp}(\sigma_{1,\rho}) \subseteq [-r_1, r_1]$;

(ii) $\chi_1 * f(x)$ is well defined since $\mathrm{supp}\, \chi_1 = [-r_1, r_1]$. We have $\chi_1 * f = 0$;

(iii) $\chi_1 * S_\rho * f(x)$ is defined in $|x| < \inf(R-r_1, R-\rho) = R - \max(r_1, \rho)$ since $\mathrm{supp}(\chi_1 * S_\rho) \subseteq [-\max(r_1,\rho), \max(r_1,\rho)]$. By commutativity of the convolution we also have $\chi_1 * S_\rho * f(x) = 0$ in this interval.

Therefore, if $|x| < R - \max(r_1, \rho)$ we have

$$\sigma_\rho * f(x) \equiv \sigma_{1,\rho} * f(x).$$

On the other hand we have

$$\sigma_{1,\rho} = (-1)^p \delta_0^{(p)} * \sum_{\ell \in \mathbb{Z}^*} \frac{\hat{\sigma}_p\left(\frac{\ell\pi}{r_1}\right)}{\hat{\chi}_1'\left(\frac{\ell\pi}{r_1}\right)} \frac{\hat{T}_\ell(z)}{\left(\frac{\ell\pi}{r_1}\right)^p},$$

the series converges in $\mathcal{E}'(\mathbb{R})$. Since

$$T_\ell * f(x) = 0 \quad \text{when} \quad |x| < R - \max(r_1, \rho)$$

we can conclude that

$$\sigma_{1,\rho} * f(x) = 0$$

in the same interval. Therefore, the same statement is true about $\sigma_\rho * f$.

If we let $\rho = 0$ we have $\sigma_\rho = \delta_0$, and we conclude

$$f(x) = \delta_0 * f(x) = 0 \quad \text{in} \quad |x| < R - r_1.$$

If we now let $0 < \rho \le r_1$, we have $r_1 + \rho \le r_1 + r_1 < r_1 + r_2 < R$. Hence if $|x| < R - r_1 = R - \max(r_1, \rho)$, we have

$$\sigma_\rho * f(x) = \frac{f(x+\rho)+f(x-\rho)}{2} = 0.$$

If $0 \le x < R - r_1$, then the inequality $x - \rho < R - r_1$ is clear. Moreover $x - \rho > -R + r_1$, since $-R + r_1 + \rho \le -R + r_2 < -R + r_1 + r_2 < 0$. Thus $x - \rho \in\ ]-R+r_1, R-r_1[$ and $f(x-\rho) = 0$. Therefore, we also have $f(x+\rho) = 0$. Since x is arbitrary in $[0, R-r_1[$ and ρ in $[0, r_1]$, we conclude that $f \equiv 0$ in $[0,R[$. The same reasoning shows that $f \equiv 0$ in $]-R,0]$. This concludes the proof of Lemma 4 and Theorem 3 in the case $n = 1$. □

To continue the proof of Theorem 3 by induction on the dimension n, let us write, as in [10],

$$\chi_{[-r_j,r_j]^n} = \mu_j * \tau_j \quad (1 \le j \le n+1)$$

$$\mu_j = \chi_{[-r_j,r_j]}(x_1) \otimes \delta(x_2) \otimes \ldots \otimes \delta(x_n)$$

$$\tau_j = \delta(x_1) \otimes \chi_{[-r_j,r_j]^{n-1}}(x_2, \ldots, x_n).$$

For $k \in \{1, \ldots, n+1\}$ let

$$\nu_k = \mu_1 * \ldots * [\mu_k] * \ldots * \mu_{n+1},$$

where [ ] represents a missing term.

Fix now two distinct indices i, k. The convolution $(\nu_k * f)(x)$ is defined for those points x such that its first coordinate x_1 satisfies $|x_1| < R - \sum_{j \ne k} r_j$, While the convolution $\tau_i * \nu_k * f(x)$ is defined for those $x = (x_1, \ldots, x_n)$ such that $|x_1| < R - \sum_{j \ne k} r_j$ and $|x_\ell| < R - r_i$ for $2 \le \ell \le n$. In fact,

$$\operatorname{supp} \nu_k = \left[-\sum_{j \ne k} r_j, \sum_{j \ne k} r_j\right] \times \{0\}$$

$$\operatorname{supp} \tau_i = \{0\}\times[-r_i, r_i]^{n-1}.$$

Furthermore, $\tau_i * \nu_k$ contains $\tau_i * \mu_i$ as a factor. But $\tau_i * \mu_i * f = \chi_{[-r_j, r_j]^n} * f = 0$ in $(]-R+r_i, R-r_i[)^n$. Therefore we have that for $i \neq k$

$$\tau_i * \nu_k * f(x) = 0 \quad \text{if} \quad \begin{cases} |x_1| < R - \sum_{j\neq k} r_j, \\ |x_\ell| < R - r_i, \quad 2 \le \ell \le n. \end{cases}$$

Consider the auxiliary functions g_k defined and C^∞ for $x' \in]-R, R[^{n-1}$,

$$g_k(x') = (\nu_k * f)(x_1, x')$$

whenever we fix a real number x_1 satisfying $|x_1| < R - \sum_{j\neq k} r_j$. Let $\tilde{\chi}_i = \chi_{[-r_j, r_j]^{n-1}}$, then if $|x_\ell| < R - r_i$ for $\ell \ge 2$ and $k \neq i$ we have

$$\begin{aligned} \tilde{\chi}_i * g_k(x') &= \tau_i * \nu_k * f(x_1, x') \\ &= (\sigma_1 * \ldots * [\sigma_i] * \ldots * [\sigma_k] * \ldots * \sigma_{n+1}) * (\sigma_i * \tau_i) * f(x_1, x') = 0. \end{aligned}$$

By the inductive hypothesis we can conclude that

$$g_k(x') = 0 \quad \text{if} \quad |x_\ell| < R, \quad \ell \ge 2.$$

Hence, if we fix $x' = (x_2, \ldots, x_\ell)$ with $|x_\ell| < R$, we conclude that for every k, $1 \le k \le n+1$,

$$\nu_k * f(x_1, x') = 0 \quad \text{for} \quad |x_1| < R - \sum_{j\neq k} r_j.$$

To continue the proof we need the following

<u>Lemma 6</u>. Let $0 < r_1 < \ldots < r_{n+1}$ be $n+1$ numbers which are $\mathbb{Q}$-linearly independent. Let $R > r_1 + \ldots + r_{n+1}$ and $h \in \mathcal{E}(]-R, R[)$ be such that for every k, $1 \le k \le n+1$, and $|t| < R - \sum_{j\neq k} r_j$, one has

$$(\chi_1 * \ldots * [\chi_i] * \ldots * \chi_{n+1} * h)(t) = 0.$$

Then $h \equiv 0$ in $]-R,R[$.

Proof of Lemma 6. The case $n = 1$ is Lemma 4. We proceed by induction and let $\tilde{h} = \chi_{n+1} * h$, which is defined and C^∞ in $]-R+r_{n+1}, R-r_{n+1}[$. For $1 \le k \le n$ and $|x| < (R-r_{n+1}) - \sum_{\substack{j \neq k \\ j \neq n+1}} r_j$ we have

$$\chi_1 * \ldots * [\chi_k] * \ldots * \chi_n * \tilde{h}(x) = 0.$$

Therefore, by the inductive hypothesis, $\tilde{h} \equiv 0$ in $|x| < R - r_{n+1}$. The same way we can prove that $\chi_j * h \equiv 0$ in $|x| < R - r_j$, for every j. By Lemma 4 we conclude that $h \equiv 0$ in $]-R,R[$. □

We can now finish the proof of Theorem 3. Simply, if x' is fixed in $]-R,R[^{n-1}$, we have that for every k

$$\nu_k = \chi_1 * \ldots * [\chi_k] * \ldots * \chi_{n+1} * \delta(x'),$$

and

$$\nu_k * f(x_1, x') = 0.$$

We can conclude from Lemma 6 that

$$x_1 \longmapsto f(x_1, x') = 0 \quad |x_1| < R.$$

In other words, $f \equiv 0$ in $]-R,R[^n$. □

We note that Theorem 3 provides a different proof of the Three-Squares Theorem, albeit more complicated!

Corollary 7. Let $0 < r_1 < \ldots < r_{n+1}$ be $\mathbb{Q}$-linearly independent. Assume Ω is an open set in $\mathbb{R}^n$ with the property that for every $x \in \Omega$ there is $y \in \Omega$ and $R_y > 0$ such that

$$x \in y +]-R_y, R_y[^n \subseteq \Omega$$

and

$$R_y > r_1 + \dots + r_{n+1}.$$

Then the only continuous function f in Ω that satisfies

$$\int_{x+[-r_j,r_j]^n} f(t)dt_1 \dots dt_n = 0$$

for every j, $1 \leq j \leq n+1$, and every $x \in \Omega$ such that $x + [-r_j, r_j]^n \subseteq \Omega$, is the function identically zero.

<u>Remark</u>. a) There are sets Ω with the property assumed in Corollary 7. One example is the complement of a closed convex set. Another example is a set whose complement Ω^c looks like

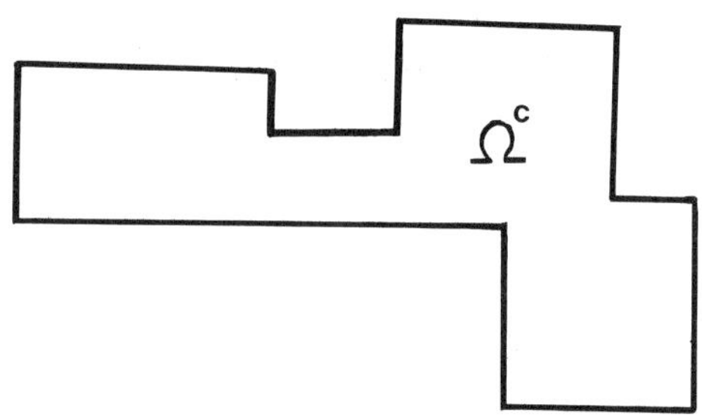

b) For $n = 2$, let f be a continuous function in an open set $\Omega \subseteq \mathbb{C}$ with the property of Corollary 7 for $Q_j = [-r_j, r_j] + i[-r_j, r_j]$, $j = 1,2,3$, such that $0 < r_1 < r_2 < r_3$ are $\mathbb{Q}$-linearly independent and such that for every j, $1 \leq j \leq 3$

$$\int_{\partial(z+Q_j)} f(\zeta)d\zeta = 0$$

for every $z \in \Omega$ such that $z + Q_j \subseteq \Omega$, then f is holomorphic in Ω.

Acknowledgements. We thank Professor Cora Sadosky for having invited us to participate in this volume dedicated to Mischa Cotlar.

C. Berenstein's work was partly supported by the National Science Foundation. He also thanks the Research Board of the University of Maryland for its support during the academic year 1988/89.

References

[1] C.A. Berenstein and R. Gay, A local version of the two-circles theorem, Israel J. Math. 55 (1986), 267-288.

[2] C.A. Berenstein and R. Gay, Le problème de Pompeiu local, to appear in J. Analyse Math.

[3] C.A. Berenstein and B.A. Taylor, The "three-squares" theorem for continuous functions, Arch. Rat. Mech. Anal. 63 (1977), 253-259.

[4] C.A. Berenstein and M. Shahshahani, Harmonic analysis and the Pompeiu problem, Amer. J. Math. 105 (1983), 1257-1229.

[5] C.A. Berenstein and A. Yger, Le problème de la déconvolution, J. Fcnal. Anal. 54 (1983), 113-160.

[6] C.A. Berenstein and L. Zalcman, The Pompeiu problem in symmetric spaces, Comment. Math. Helvetici 55 (1980), 593-621.

[7] L. Brown, B.M. Schreiber and B.A. Taylor, Spectral synthesis and the Pompeiu problem, Am. Inst. Fourier 23 (1973), 125-154.

[8] L. Hormander, The Analysis of Linear Partial Differential Operators, vol. 1 and 2, Springer-Verlag, 1983.

[9] J.P. Kahane, Sur quelques problèmes d'unicité et de prolongement, Ann. Inst. Fourier 5 (1953/54), 39-130.

[10] P.G. Laird, A reconsideration of the "three-squares" problem, Aequationes Mathematicae 21 (1980), 98-104.

[11] L. Schwartz, Théorie générale des fonctions moyénne périodiques, Ann. of Math. 48 (1947), 857-929.

[12] D.R. Scott and A. Sitaram, Some remarks on the Pompeiu problem for groups, Proc. Amer. Math. Soc., to appear.

[13] L. Zalcman, Analyticity and the Pompeiu problem, Arch. Rat. Mech. Anal. 47 (1972), 237-254.

[14] L. Zalcman, Offbeat integral geometry, Amer. Math. Monthly 87 (1980), 161-175.

[15] S. Williams, A partial solution of the Pompeiu problem, Math. Ann. 223 (1976), 183-190.

3

An Application of Ricci-Stein Theorem to Estimates of the C-R Equations

Der-Chen Chang

0. Introduction.

Let Ω be a bounded, open subset in $\mathbb{C}^{n+1}$ with smooth boundary studying the solution of the inhomogeneous Cauchy-Riemann equation $\bar{\partial}u = f$ on Ω with good bounds has been one of the main themes in the theory of complex analysis for many years. In studying the existence and regularity properties of the solution (Henkin or the Kohn solution) for the $\bar{\partial}$-equation, attention has mainly centered around the question of the integral representation of the kernel and subelliptic a priori estimates for the $\bar{\partial}$-Neumann problem. For strongly pesudoconvex domains, the solutions are quite complete, see Krantz [Kr], Lieb-Range [LR], Phong-Stein [PS2] and other related articles.

Do we have corresponding estimates in weakly pseudo-convex domains? In $\mathbb{C}^2$, the results of Chang-Nagel-Stein [CNS], Christ [Ch], Kohn-Fefferman [KF], Nagel-Rosay-Stein-Wainger [NRSW] shows the solution of the $\bar{\partial}$-equation which maps L^∞ to $\Lambda_{1/2m}$ and S_k^p to S_{k+1}^p where $2m$ is the "type" of the domain and S_k^p are the non-isotropic Sobolev spaces. In this paper, we compute the integral kernel for the Henkin solution $\mathbf{H}$ on domains

$$H_m = \{(z', z_{n+1}) \in \mathbb{C}^{n+1} : \operatorname{Im} z_{n+1} > \left[\sum_{j=1}^{n} |z_j|^2\right]^m, \quad m \in \mathbb{N}\}.$$

These are weakly pseudo-convex but finite type domains in the sense of Kohn [K]. The kernel K for $\mathbf{H}$ is an integral kernel with mixed type homogeneities, i.e., $\ker(H) = K = \sum_{k,\ell\geq 0} E_k F_\ell$ where

$$E_k(\lambda\cdot(z',w',t-s;\rho+\mu)) = E_k(\lambda z',\lambda w',\lambda(t-s);\lambda(\rho+\mu))$$
$$= \lambda^{-k}E_k(z',w',t-s;\rho+\mu), \quad \forall\ \lambda > 0;$$

$$F_\ell(\lambda\cdot(z',w',t-s;\rho+\mu)) = F_\ell(\lambda z',\lambda w',\lambda^{2m}(t-s);\lambda^{2m}(\rho+\mu))$$
$$= \lambda^{-\ell}F_\ell(z',w',t-s;\rho+\mu), \quad \forall\ \lambda > 0.$$

Thus the problem reduces to getting the sharp estimates for these kernels $E_k F_\ell$, especially in the case $k = 2n$, $\ell = 4m$.

To handle this critical case, we need to study the non-isotropic balls on ∂H_m and the corresponding singular Radon transforms (in section 2). Suppose $x = (w',s) \in \partial H_m$ and define the singular Radon transform:

$$R(f)(x) = \langle K(x,\cdot), f|_{\partial H_{m,x}}\rangle, \quad \text{where } f \in C_0^\infty(\partial H_m).$$

$\partial H_{m,x}$ is a smooth submanifold passing through x, which is defined by

$$\partial H_{m,x} = \{(x',t) \in \partial H_m : t-s-2m\mathrm{Im}\left[\sum_{j=1}^{n} |z_j|^2\right]^{m-1}\left[\sum_{j=1}^{n} \bar{z}_j w_j\right] = 0\}.$$

When we consider the L^2-estimate for R, the problem reduces to estimating the corresponding oscillatory integral:

$$T(f)(w',s) = \int_{\mathbb{C}^n} e^{2im\mathrm{Im}[|z'|^{2m-2}(\bar{z}'\cdot w')]} K_{\rho+\mu}(z',w',t-s)f_\rho(z',t)dz'.$$

The phase function will degenerate on those weakly pseudo-convex points (i.e., Re z_{n+1}-axis). Fortunately, this phase function is a polynomial in z' and w' variables which allows us to apply Ricci-Stein's theorem [RS] to get the result.

In section 1 of this paper, we construct a Cauchy-Fantappiè kernel which coincides with the Szegö kernel. But this Cauchy-Fantappiè kernel is just continuous (not C^1), hence we cannot apply the same method as Lieb-Range [LR] to construct the integral kernel for the Kohn solution. It is very interesting to find a "good" Cauchy-Fantappiè form in the future for these domains (even more general domains) which allows us to follow the Kerzman-Stein theory [KS] to construct an "explicit" asymptotic expansion for the Szegö kernel.

We also want to point out the Henkin solution maps S^p_k to S^p_{k+1} which tells us domains H_m are maximal hypoelliptic when $n \geq 2$ in the sense of Maire [M]. In Rothschild's paper [Ro], she proved nonexistence of optimal L^2-estimates for $\square_b$ on $\Omega = \{(z', z_{n+1}) \in \mathbb{C}^{n+1} : \operatorname{Im} z_{n+1} > |z_1|^4 + \ldots + |z_n|^4\}$; we still can get the optimal L^p estimates for $\bar{\partial}^* N$ for the case $n = 1$ (in section 3). To understand the interdependence between $\bar{\partial}^* N$ and $\square_b$ on these domains, we still need some work.

The author would like to thank his teacher and advisor Eli Stein for sharing his mathematical ideas generously. The author would also like to thank K. Diaz for helpful mathematical conversations.

1. The kernel of the Henkin solution for $\bar{\partial}$-equation on the domains $\{(z', z_{n+1}) \in \mathbb{C}^{n+1} : \operatorname{Im} z_{n+1} > \left[\sum_{1 \leq j \leq n} |z_j|^2\right]^m,\ m \in \mathbb{N}\}$.

Let $H_m = \{(z', z_{n+1}) \in \mathbb{C}^{n+1} : \operatorname{Im} z_{n+1} > \left[\sum_{1 \leq j \leq n} |z_j|^2\right]^m,\ m \in \mathbb{N}\}$, where $z' = (z_1, \ldots, z_n) \in \mathbb{C}^n$. The boundary of $H_1, \partial H_1$ is strongly pseudo-convex everywhere which is isomorphic to the Heisenberg gruop $\mathbb{H}^n$. For $m > 1$,

$\partial H_m = \{\mathrm{Im}\, z_{n+1} = \left[\sum_{1\le j\le n} |z_j|^2\right]^m\}$ is strongly pseudo-convex for $z_j \neq 0$, $j = 1,\ldots,n$, and weakly pseudo-convex of type $2m$ in the sense of Kohn [K] for $z_1 = \ldots = z_n = 0$. Thus the set $\mathrm{Re}\, z_{n+1} \approx \mathbb{R}$ forms a one real dimensional line of points in ∂H_m of type $2m$. In fact, these domains H_m are biholomorphically equivalent to the bounded domains $E_m = \{(w', w_{n+1}) \in \mathbb{C}^{n+1} : \left[\sum_{1\le j\le n} |w_j|^2\right]^m + |w_{n+1}|^2 < 1\}$ via the following map:

$$z_1 = \frac{w_1}{(1+w_{n+1})^{1/m}}, \ldots, z_n = \frac{w_n}{(1+w_{n+1})^{1/m}}; \quad \text{and} \quad z_{n+1} = \frac{i(1-w_{n+1})}{1+w_{n+1}}.$$

Let U be a boundary neighborhood of $0 \in \partial H_m$. Note that U is naturally parametrized by $z', t = \mathrm{Re}\, z_{n+1}$ and $\mathrm{Im}\, z_{n+1} = \rho + \left[\sum_{1\le j\le n} |w_j|^2\right]^m$, where $\rho = \mathrm{Im}\, z_{n+1} - \left[\sum_{1\le j\le n} |z_j|^2\right]^m$ is the "height" function defined on $\bar{H}_m$. Hence we have the following notation:

Points in U are denoted $z, w,$ etc.

$$z = (z', z_{n+1}) = (z', t+i[\left[\sum_{1\le j\le n} |z_j|^2\right]^m + \rho]) \approx (z', t; \rho)$$

where $z' \in \mathbb{C}^n$, $t = \mathrm{Re}\, z_{n+1} \in \mathbb{R}$ and $\rho = \mathrm{Im}\, z_{n+1} - \left[\sum_{1\le j\le n} |z_j|^2\right]^m \in \mathbb{R}^+$;

$$w = (w', w_{n+1}) = (w', s+i[\left[\sum_{1\le j\le n} |w_j|^2\right]^m + \rho]) \approx (w', s; \mu)$$

where $w' \in \mathbb{C}^n$, $s = \mathrm{Re}\, w_{n+1} \in \mathbb{R}$ and $\mu = \mathrm{Im}\, w_{n+1} - \left[\sum_{1\le j\le n} |w_j|^2\right]^m \in \mathbb{R}^+$.

The construction of the Henkin solution for the inhomogeneous Cauchy-Riemann equation is well known. Let us define an operator $T : C^1_{(0,1)}(\bar{U}) \to C_{(0,0)}(\bar{U})$ as follows:

$$\text{(1.1)} \qquad Tf(w) = \int_U f \wedge (\overline{\vartheta_z \mathcal{T}_0} + \partial_z \overline{\Gamma_0})$$

then $\overline{\partial_w}(Tf(w)) = f(w) + S_{-\infty}(f(w))$ on U if $f \in C^1_{(0,1)}(\bar{U})$ and $\bar{\partial} f = 0$. Here $S_{-\infty}$ is a smoothing operator. The proof can be found in Henkin-Leiterer [HL] and Range [R]. Now we need to compute the kernel $\overline{\vartheta_z \mathcal{T}_0} + \partial_z \overline{\Gamma_0}$ for our domain H_m. First of all, let us consider the complex tangent space to ∂H_m, which we denote $T^{\mathbb{C}}(\partial H_m)$, is spanned by the real and imaginary part of the holomorphic tangent vectors of

$$Z_j = \frac{\partial}{\partial z_j} + 2im\left[\sum_{i=1}^{n} |z_i|^2\right]^{m-1} \overline{z_j} \frac{\partial}{\partial z_{n+1}}, \quad j = 1,2,\ldots n.$$

Using the parametrized coordinates on U, then

$$Z_j = \frac{\partial}{\partial z_j} + im\left[\sum_{i=1}^{n} |z_i|^2\right]^{m-1} \overline{z_j} \frac{\partial}{\partial t}, \quad j = 1,2,\ldots n.$$

It is easy to see that, at $T^{\mathbb{C}}$, the weakly pseudo-convex points $T^{\mathbb{C}}(\partial H_m)$ is simply the z'-plane $(\approx \mathbb{C}^n)$. For other z, $T^{\mathbb{C}}(\partial H_m)$ is tilted slightly from the z'-plane. As $|z'|$ increases the tilt increases, but $T^{\mathbb{C}}(\partial H_m)$ is never perpendicular to the z'-plane. We also introduce another vector field

$$Z_{n+1} = i\sqrt{2} \frac{\partial}{\partial z_{n+1}} = \frac{i}{\sqrt{2}}\left(\frac{\partial}{\partial t} - i\frac{\partial}{\partial \rho}\right), \quad t = \frac{1}{\sqrt{2}}(\bar{Z}_{n+1} - Z_{n+1}).$$

Then $\{Z_1, \ldots, Z_n, Z_{n+1}\}$ forms a basis for the subbundle $T^{(1,0)}(U)$ of the complex tangent bundle $T^{\mathbb{C}}(U)$. Let $\{\omega_1, \ldots, \omega_n, \omega_{n+1}\}$ be the dual frame of $\{Z_1, \ldots, Z_n, Z_{n+1}\}$. In our case

$$\omega_{n+1} = \sqrt{2}\partial\rho \quad \text{and} \quad \omega_j = dz_j, \quad 1 \le j \le n.$$

Since the kernel Γ_0 depends on the geodesic distance ψ, so we may choose a "natural" Hermitian metric defined on $\bar{U} \cap \bar{H}_m$ which makes $\{\omega_1, \ldots, \omega_n, \omega_{n+1}\}$ is an orthonormal basis with respect to this metric, i.e., $ds^2 = \sum_{j=1}^{n+1} \omega_j \otimes \bar{\omega}_j$.

Then the function ρ will be the geodesic distance of a point $(z',t;\rho)$ to the boundary $V_0 \subset \partial H_m$. Under this assumption, we know that ψ can be approximated by:

$$\psi^2(z,w) \approx 2\sum_{j=1}^{n} |z_j - w_j|^2 + (t-s-2m\mathrm{Im}\left[\sum_{i=1}^{n} |z_i|^2\right]^{m-1} \cdot \left[\sum_{j=1}^{n} \bar{z}_j w_j\right])^2 + (\rho-\mu)^2.$$

(Note that ψ is not the geodesic distance between z and w with respect to the metric ds^2, but ψ^{-2n} can be treated as a parametrix for the complex Laplacian, see Phong-Stein [PS2].) So the principal part of the kernel $*\partial_z\bar{\Gamma}_0$ will be

$$*\partial_z\bar{\Gamma}_0(z,w) = c\cdot\frac{\sum_{j=1}^{n+1}(\bar{Z}_j\psi^2)\omega_1\wedge\ldots\wedge\omega_{n+1}\wedge\bar{\omega}_1\wedge\ldots\wedge\mathring{\bar{\omega}}_j\wedge\ldots\wedge\bar{\omega}_{n+1}}{\psi^{2n+2}(z,w)},$$

where $|Z_j\psi^2(z,w)| \approx \mathcal{E}_1$. Here $\mathcal{E}_j$ represents an integral kernel which has size $\|z-w\|^j$. From the theory of singular integrals, it is easy to see that this kernel is a bounded mapping from $L^p_k(U)$ to $L^p_{k+1}(U)$ for $k \in \mathbb{Z}^+$ and $1 < p < \infty$. (Note that $\partial_z\overline{\Gamma_0}$ defines a fractional integral, but when we differentiate the kernel $\partial_z\overline{\Gamma_0}$ which will produce a Calderón-Zygmund operator.)

Next, let us consider the kernel $*\overline{\vartheta_z\mathcal{T}_0} = (-1)\bar{\partial}_z\mathcal{A}_0(\mathcal{W},\mathcal{B})$ is the transition form between the generating forms $\mathcal{B}$ and $\mathcal{W}$ (see [R]). (Here the boundary geometry comes into play!) It is a straightforward computation to get $\varphi(z,w) = \langle\partial\gamma(z),z-w\rangle$. Here $\gamma(z) = \mathrm{Im}z_{n+1} - \left(\sum_{j=1}^{n}|z_j|^2\right)^m$ is the defining function for H_m.

But when we construct the kernel $\overline{\vartheta_z\mathcal{T}_0} + \partial_z\overline{\Gamma_0}$, we need to consider two new functions $\tilde{\psi}^2(z,w) = \psi^2(z,w) + 4\rho\mu$ and $\tilde{\varphi}(z,w) = \varphi(z,w) + \rho$ to substitute $\psi^2(z,w)$ and $\varphi(z,w)$ respectively. Note that $\tilde{\psi}^2(z,w) = \psi^2(z,w)$ and $\tilde{\varphi}(z,w) = \varphi(z,w)$ on the boundary ∂H_m and $\tilde{\psi}^2(z,w)$ and $\tilde{\varphi}(z,w)$ just vanish at $(z',t) = (w',s)$ on the boundary! From now on, we always consider $\tilde{\psi}^2(z,w)$ and $\tilde{\varphi}(z,w)$ in our kernel. (This is a key idea of Henkin and Romanov, see [HL]), and Krantz [Kr].) We drop the "~" for convenience. Hence we have

$$(1.2)\quad \psi^2(z,w) = 2\sum_{j=1}^{n} |z_j - w_j|^2 + (t-s-2m\mathrm{Im}\left[\sum_{i=1}^{n}|z_i|^2\right]^{m-1}\left[\sum_{j=1}^{n}\overline{z_j}w_j\right])^2 + (\rho+\mu)^2,$$

and

$$(1.3)\quad \varphi(z,w) = \frac{1}{2}\left[\left[\sum_{i=1}^{n}|w_i|^2\right]^m - \left[\sum_{i=1}^{n}|z_i|^2\right]^m\right] + m\left[\sum_{i=1}^{n}|z_i|^2\right]^{m-1}\cdot\sum_{j=1}^{n}\overline{z_j}(z_j-w_j)$$
$$+\frac{i}{2}(t-s) + \frac{1}{2}(\rho+\mu).$$

According to the definition of $\mathcal{A}_0(\mathcal{W},\mathcal{B})$:

$$\mathcal{A}_0(\mathcal{W},\mathcal{B}) = \frac{1}{(2\pi i)^{n+1}}\sum_{j=0}^{n-1} a_j \mathcal{W}\wedge\mathcal{B}\wedge(\overline{\partial}_z\mathcal{W})^j\wedge(\overline{\partial}_z\mathcal{B})^{n-1-j},$$

where

$$\mathcal{B} = \frac{\sum_{j=1}^{n+1}(Z_j\psi^2)\omega_j}{\psi^2} \quad\text{and}\quad \mathcal{W} = \frac{\sum_{j=1}^{n+1}\eta_j(z)dz_j}{\varphi} = \frac{\omega_{n+1}}{\varphi}.$$

Then

$$\mathcal{A}_0(\mathcal{W},\mathcal{B}) = \frac{1}{(2\pi i)^{n+1}}\sum_{j=0}^{n-1}\sum_{I} a_j\frac{(\overline{\partial}_z\omega_{n+1})^j\wedge(\overline{\partial}_z\upsilon_{i_2})\wedge\ldots\wedge(\overline{\partial}_z\upsilon_{i_{n-j}})}{\psi^{2(n-j)}(z,w)\varphi^{j+1}(z,w)},$$

where $\upsilon_i(z,w) = 2(\overline{z_i}-\overline{w_i})$, for $i = 1,2,\ldots,n+1$ and the summation is over all permutations $I = (i_1,\ldots,i_{n-j})$ of $(1,\ldots,n+1)$. Here we plug in the properties

$$\omega_{n+1} = \sqrt{2}\partial\rho = \sqrt{2}(m\left[\sum_{i=1}^{n}|z_i|^2\right]^{m-1}\sum_{j=1}^{n}\overline{z_j}dz_j + \frac{i}{2}dz_{n+1})$$

and

$$\overline{\partial}_z\omega_{n+1} = m\left[\sum_{i=1}^{n}|z_i|^2\right]^{m-1}\sum_{j=1}^{n}\overline{\omega_j}\wedge\omega_j + m(m-1)\left[\sum_{i=1}^{n}|z_i|^2\right]^{m-2}\sum_{j,k=1}^{n}\overline{z_j}z_k\overline{\omega_k}\wedge\omega_j$$

+ more smoothing terms.

Hence the kernel $*\overline{\vartheta_z\mathcal{T}_0} = (-1)\overline{\partial}_z\mathcal{A}_0(W,\mathcal{B})$ will be

$$(-1)\overline{\partial}_z\mathcal{A}_0(W,\mathcal{B}) = \sum_{j=0}^{n}\sum_{k=1}^{n+1}\delta_{jk}a_j\frac{(\overline{\partial}_z\omega_{n+1})^j(\bigwedge_{j\neq k}\overline{\omega_j})}{\psi^{2(n-j)}(z,w)\varphi^{j+1}(z,w)}$$

$$+\sum_{j=0}^{n}\sum_{k=1}^{n+1}C_{jk}\frac{(\overline{Z_k}\psi^2)\wedge(\overline{\partial}_z\omega_{n+1})^j(\bigwedge_{j\neq k}\overline{\omega_j})}{\psi^{2(n-j+1)}(z,w)\varphi^{j+1}(z,w)}$$

$$+\sum_{j=0}^{n}\sum_{k=1}^{n+1}D_{jk}\frac{(\overline{Z_k}\varphi)\wedge(\overline{\partial}_z\omega_{n+1})^j(\bigwedge_{j\neq k}\overline{\omega_j})}{\psi^{2(n-j)}(z,w)\varphi^{j+2}(z,w)}$$

where C_{jk}'s and D_{jk}'s are smooth, bounded functions on $\overline{U}$. This is a "mixed type" homogeneities kernel. We will discuss the regularity properties for such kernels in Section 2.

To simplify the computations, we just assume $n = 1$. In this case,

$$\mathcal{A}_0(W,\mathcal{B}) = \frac{1}{4\pi^2}W\wedge\mathcal{B} = \frac{1}{4\pi^2}\frac{\omega_2}{\varphi}\wedge\frac{(Z_1\psi^2)\omega_1+(Z_2\psi^2)\omega_2}{\psi^2} = \frac{1}{4\pi^2}\cdot\frac{(Z_1\psi^2)\omega_1\wedge\omega_2}{\varphi\psi^2}.$$

To compute $*\overline{\vartheta_z\mathcal{T}_0}$, we just need to consider

$$(-1)\overline{\partial}_z\mathcal{A}_0 = \frac{1}{4\pi^2}\left\{\overline{Z_1}\left[\frac{Z_1\psi^2}{\varphi\psi^2}\right]\overline{\omega_1}\wedge\omega_1\wedge\omega_2 + \overline{Z_2}\left[\frac{Z_1\psi^2}{\varphi\psi^2}\right]\overline{\omega_2}\wedge\omega_1\wedge\omega_2\right\}$$

$$= \frac{1}{4\pi^2}\left\{\frac{-\overline{Z_1}Z_1\psi^2}{\varphi\psi^2}+\frac{(\overline{Z_1}\varphi)(Z_1\psi^2)}{\varphi^2\psi^2}+\frac{(Z_1\psi^2)(Z_1\psi^2)}{\varphi\psi^4}\right\}\omega_1\wedge\omega_2\wedge\overline{\omega_1}$$

$$+\frac{1}{4\pi^2}\left\{\frac{-\overline{Z_2}Z_1\psi^2}{\varphi\psi^2}+\frac{(\overline{Z_2}\varphi)(Z_1\psi^2)}{\varphi^2\psi^2}+\frac{(Z_2\psi^2)(Z_1\psi^2)}{\varphi\psi^4}\right\}\omega_1\wedge\omega_2\wedge\overline{\omega_2}.$$

Now we can write down our first theorem:

Theorem 1. *The kernel for the operator* T *has the following form*

$$\sum_{k,\ell\geq 0} E_k F_\ell,$$

where $E_k = \frac{P}{\psi^a}$ *is a homogeneous kernel of degree* $-k$ *in the sense*

$$\delta\cdot(z',w',t-s;\rho+\mu)\longrightarrow(\delta z',\delta w',\delta(t-s);\delta(\rho+\mu)),\quad \delta>0,$$

and $F_\ell = \frac{Q}{\varphi^b}$ *is a homogeneous kernel of degree* $-\ell$ *in the sense*

$$\delta\cdot(z',w',t-s;\rho+\mu)\longrightarrow(\delta z',\delta w',\delta^{2m}(t-s);\delta^{2m}(\rho+\mu)),\quad \delta>0.$$

Here P,Q *are polynomials and* ψ,φ *are functions given by (1.2) and (1.3), respectively.*

Remarks. Here we have a quite different situation as the strongly pseudo-convex case. If we look at the kernel $H(z,w)$ and $H^*(z,w)$, the difference of $\varphi(z,w) = \langle\partial\gamma(z),z-w\rangle$ and its adjoint $\varphi^*(z,w)$:

$$\varphi(z,w)-\varphi^*(z,w) \approx O(|z-w|^{2m}) \quad \text{(No gain!)}.$$

Hence $H(z,w)-H^*(z,w)$ is still a type zero operator in the sense of Nagel-Rosay-Stein-Wainger [NRSW]. We cannot use this Cauchy-Fantappiè form $H(z,w)$ (at least not for this one) to "approximate" the Szegö kernel as the results of Kerzman-Stein [KS]. If we deal with strongly pseudo-convex case $(m = 1)$, we have

$$\varphi(z,w)-\varphi^*(z,w) \approx O(|z-w|^3) = O(|z-w|^{2\cdot 1+1}) \quad \text{(Gain one)}.$$

Hence we may construct a kernel to solve $\bar\partial$ which agrees with the Kohn solution $\bar\partial^*\mathbf{N}$ modulo a smoothing operator (see Lieb-Range [LR]). Since we have a lot of freedom to pick up a Cauchy-Fantappiè form, for domains we discussed here $\{(z',z_{n+1})\in\mathbb{C}^{n+1} : \operatorname{Im} z_{n+1} > \left[\sum_{1\le j\le n}|z_j|^2\right]^m\}$, we can actually find a Cauchy-Fantappiè form which is equal to the Szegö kernel:

Proposition. *Let*

$$g_j(z,w) = \left[\frac{i}{2}(\overline{z_{n+1}}-w_{n+1})\right]^{(m-1)/(n+1)m}\overline{z_j}, \quad \textit{for} \quad j = 1,2,\dots,n;$$

$$g_{n+1}(z,w) = \left[\frac{i}{2}(\overline{z_{n+1}}-w_{n+1})\right]^{(m-1)/(n+1)m}\cdot\left\{\frac{\left[\frac{i}{2}(\overline{z_{n+1}}-w_{n+1})\right]^{1/m}-\left[\sum_{i=1}^{n} z_i\overline{z_i}\right]}{z_{n+1}-w_{n+1}}\right\}.$$

Then the Cauchy-Fantappiè form

$$\Omega_0(g) = \frac{(-1)^{n(n+1)/2}n!}{(2\pi i)^{n+1}}\cdot\frac{\sum_{j=1}^{n} g_j(\bigwedge_{j\neq k}\overline{\partial}_z g_k)^n}{g^{n+1}}\wedge dz_1\wedge\dots\wedge dz_{n+1} = \text{Szegö kernel},$$

where $g(z,w) = \sum_{j=1}^{n+1} g_j(z,w)(z_j-w_j)$.

Proof. To simplify the computation, we just assume $n = 1$ (i.e., $\mathbb{C}^2$ case): In this case

$$g_1(z,w) = \left[\frac{i}{2}(\overline{z_2}-w_2)\right]^{(m-1)/2m}\overline{z_1},$$

and

$$g_2(z,w) = \left[\frac{i}{2}(\overline{z_2}-w_2)\right]^{(m-1)/2m}\cdot\left\{\frac{\left[\frac{i}{2}(\overline{z_2}-w_2)\right]^{1/m}-z_1\overline{z_1}}{z_2-w_2}\right\}$$

and

$$\Omega_0(g) = \frac{1}{4\pi^2}\cdot\frac{g_1\overline{\partial}_z g_2-g_2\overline{\partial}_z g_1}{g^2}\wedge dz_1\wedge dz_2$$

$$= \frac{1}{4\pi^2}\cdot\frac{\{\frac{i}{2m}\overline{z_1}d\overline{z_2}-[\frac{i}{2}(\overline{z_2}-w_2)]d\overline{z_1}\}\wedge dz_2\wedge dz_2}{(z_2-w_2)[\frac{i}{2}(\overline{z_2}-w_2)]^{(m-1)/m}\{[\frac{i}{2}(\overline{z_2}-w_2)]^{1/m}-w_1\overline{z_1}\}^2}$$

on the surface $\{(z_1\overline{z_1})^m+\frac{i}{2}(z_2-\overline{z_2}) = \gamma = 0\} = \partial H_m$. In this case, the unit normal is $\frac{\nabla\gamma}{\|\nabla\gamma\|}$ and we have

$$d\overline{z_2} \wedge dz_1 \wedge dz_2 = \frac{4}{\|\nabla\gamma\|} \cdot \frac{\partial\gamma}{\partial\overline{z_1}} d\sigma \quad \text{and} \quad d\overline{z_1} \wedge dz_1 \wedge dz_2 = \frac{-4}{\|\nabla\gamma\|} \cdot \frac{\partial\gamma}{\partial\overline{z_2}} d\sigma.$$

This gives, on ∂H_m

$\Omega_0(g) =$

$$= \frac{1}{4\pi^2} \left\{ \frac{1}{[\frac{i}{2}(\overline{z_2}-w_2)]^{(m-1)/m}([\frac{i}{2}(\overline{z_2}-w_2)]^{1/m}-w_1\overline{z_1})^2} \right\} \cdot \frac{\frac{i}{2m}\overline{z_1} \cdot m z_1^m \overline{z}_1^{m-1} + \frac{i}{2}(\overline{z_2}w_2)(-\frac{i}{2})}{(z_2-w_2)\|\nabla\gamma\|} \cdot 4d\sigma$$

$$= S(z,w) \cdot \frac{4 \cdot \frac{i}{2} \cdot \frac{i}{2} \cdot (\overline{z_2}-z_2-\overline{z_2}+w_2)}{(z_2-w_2)\|\nabla\gamma\|} \cdot d\sigma = S(z,w) \cdot \frac{d\sigma}{\|\nabla\gamma\|} = S(z,w)d(\text{Re } z_1)d(\text{Im } z_1)dt.$$

Instead of the standard surface measure $d\sigma$ here, we are using the measure $d\sigma/\|\nabla\gamma\|$ which equals the result of Greiner-Stein [GS].

Even though we have this result, unfortunately, this Cauchy-Fantappiè form is not C^1! (Note that g_{n+1} has continuous coefficient only.) It is very interesting to know how to find a "good" Cauchy-Fantappiè form such that we have a "smoothing" operator when we consider $H(z,w) - H^*(z,w)$.

2. Optimal L^p-estimates for the kernels with mixed type homogeneities on domains H_m

From the Theorem 1, we know the integral kernel for Henkin solution of $\bar\partial$-equation on domains H_m has the form

$$K(z',w',t-s;\rho+\mu) = \sum_{k,\,\ell\geq 0} E_k F_\ell.$$

It is easy to see that the relations between $k+\ell$ or $k+\frac{\ell}{2m}$ and the critical degree of integrability (2n+4m or 2n+2) determine the L^p boundedness of the operator Op(K). In this section, we will consider the critical case for Op(K) when $k = 2n$ and $\ell = 4m$ which arises from the differentiation of the kernel of the Henkin solution. We may use the ideas in [PS1] and [PS2] to handle this situation. Let us consider the corresponding singular Radon transform for our operator, i.e.,

$$R(f)(P) = \langle K(P,\cdot), f|_{\partial H_{m,p}}\rangle = \langle K(P,\cdot), f|_{M_P}\rangle, \quad \text{where } f \in C_0^\infty(\partial H_m). \tag{2.1}$$

Here we consider the underlying manifold ∂H_m as $M = \mathbb{C}^n \times \mathbb{R}^1$ and $M_P = \partial H_{m,(w's)}$ is a smooth submanifold passing through P, which is defined by

$$M_P = \{(z',t) \in \partial H_m : t-s-2m \operatorname{Im}\left[\left[\sum_{i=1}^n |z_i|^2\right]^{m-1}\left[\sum_{j=1}^n w_j\overline{z_j}\right]\right] = 0\}.$$

We also define

$$K_{w'}(z'-w',t;\rho+\mu) = K_{w',t,\rho+\mu}(z'-w') = K(w',z'-w',t;\rho+\mu)$$

and

$$f_{t,\rho}(z',s) = f(z',t+s+2m \operatorname{Im}|z'|^{2m-2}(\bar{z}'\cdot w');\rho).$$

Then we have

$$\begin{aligned}
&\operatorname{Op}(K)(f)(w',s;\mu)\\
&= \int_0^\infty\int_{-\infty}^\infty\int_{\mathbb{C}^n} K(w',w'-z',t-s-2m \operatorname{Im}|z'|^{2m-2}(\bar{z}'\cdot w');\rho+\mu)f(z',t;\rho)dz'dtd\rho\\
&= \int_0^\infty\int_{-\infty}^\infty\int_{\mathbb{C}^n} K_{w'}(z'-w',t;\rho+\mu)f(z',t+s+2m \operatorname{Im}|z'|^{2m-2}(\bar{z}'\cdot w');\rho)dz'dtd\rho\\
&= \int_0^\infty\int_{-\infty}^\infty\int_{\mathbb{C}^n} K_{w't,\rho+\mu}(z'-w')f_{t,p}(z',s+2m \operatorname{Im}|z'|^{2m-2}(\bar{z}'\cdot w'))dz'dtd\rho
\end{aligned}$$

$$= \int_0^{\infty}\int_{-\infty}^{\infty} (R_{t,\rho+\mu}f_{t,p})(w',s)dtd\rho.$$

Here $R_{t,\rho+\mu}(f_{t,\rho})(w',s)$ is the corresponding singular Radon transform defined in (2.1):

$$R_{t,\rho+\mu}(f_{t,\rho})(w',s) = \int_{\mathbb{C}^n} K_{w',t,\rho+\mu}(z'-w')f_{t,\rho}(z',s+2m\ \mathrm{Im}|z'|^{2m-2}(\bar{z}'\cdot w'))dz'.$$

First we consider the L^p boundedness for the operator $R_{t,\rho+\mu}$. Suppose

$$f(z',s) = \frac{1}{2\pi}\int_{-\infty}^{\infty} e^{i\lambda s}\hat{f}(z',\lambda)d\lambda$$

where $\hat{f}(z',\lambda)$ is the Fourier transform of $f(z',s)$ in the s-variable, then

$$R_{t,\rho+\mu}(f_{t,\rho})(w',s)$$

$$= \frac{1}{2\pi}\int_{\mathbb{C}^n}\int_{-\infty}^{\infty} e^{i\lambda s}e^{2mi\lambda \mathrm{Im}|z'|^{2m-2}(\bar{z}'\cdot w')}K_{w',t,\rho+\mu}(z'-w')\hat{f}_{t,\rho}(z',\lambda)d\lambda dz'$$

$$= \frac{1}{2\pi}\int_{-\infty}^{\infty} e^{i\lambda s}\int_{\mathbb{C}^n} \varphi(|z'-w'|^{2m}\lambda)\cdot e^{2mi\lambda \mathrm{Im}|z'|^{2m-2}(\bar{z}'\cdot w')}\cdot$$

$$\cdot K_{w',t,\rho+\mu}(z'-w')\hat{f}_{t,\rho}(z',\lambda)dz'd\lambda$$

$$+ \frac{1}{2\pi}\int_{-\infty}^{\infty} e^{i\lambda s}\int_{\mathbb{C}^n} \{1-\varphi(|z'-w'|^{2m}\lambda)\}\cdot e^{2mi\lambda \mathrm{Im}|z'|^{2m-2}(\bar{z}'\cdot w')}\cdot$$

$$\cdot K_{w',t,\rho+\mu}(z'-w')\hat{f}_{t,\rho}(z',\lambda)dz'd\lambda'$$

$$= I_1 + I_2,$$

where φ is a fixed smooth function on $[0,\infty)$ which is $\equiv 1$ near the origin, and has compact support contained in $|t| \le 1$.

It is easy to see the integrals in I_1 and I_2 for the λ-variable are oscillatory integrals in $\mathbb{C}^n$. We want to apply Plancherel formula and the results in [RS] to get the L^2 estimates for $(R_{t,\rho+\mu}(f_{t,\rho}))(w',s)$. In order to prove the L^2 estimate for $(R_{t,\rho+\mu}(f_{t,\rho}))(w',s)$ we need to look at the kernel $K_{w',t,\rho+\mu}(z'-w')$ more carefully. In our case:

$$E_k(z',w',t-s;\rho+\mu) \approx (2|z'-w'|^2+(t-s-2m\ \mathrm{Im}|z'|^{2m-2}(\bar{z}'\cdot w'))^2+(\rho+\mu)^2)^{-k/2}$$
$$= \psi^{-k}(z',w',t-s;\rho+\mu) = \psi^{-2n}(z',w',t-s;\rho+\mu)$$

and

$$F_\ell(z',w',t-2;\rho+\mu) \approx \{\frac{1}{2}|w'|^{2m}+\frac{2m-1}{2}|z'|^{2m}-m|z'|^{2m-2}(\bar{z}'\cdot w')$$
$$+\frac{i}{2}(t-s)+\frac{1}{2}(\rho+\mu))^{-\ell/2m}$$
$$= \varphi^{-\ell/2m}(z',w',t-s;\rho+\mu) = \varphi^{-2}(z',w',t-s;\rho+\mu)$$

with $k = 2n$, $\ell = 4m$. Now

$$\varphi(z',w',t-s;\rho+\mu) = \mathrm{Re}\ \varphi(z',w',t-s;\rho+\mu) + i\ \mathrm{Im}\ \varphi(z',w',t-s;\rho+\mu)$$
$$= \{\frac{1}{2}|w'|^{2m}+\frac{2m-1}{2}|z'|^{2m}-m\ \mathrm{Re}|z'|^{2m-2}(\bar{z}'\cdot w')+\frac{1}{2}(\rho+\mu)\}$$
$$+\frac{i}{2}\{t-s-2m\ \mathrm{Im}|z'|^{2m-2}(\bar{z}'\cdot w')\}.$$

Since we restrict ourselves to the hypersurfaces $\partial H_{m,(w',s)}$, i.e.,

$$t-s-2m\ \mathrm{Im}|z'|^{2m-2}(\bar{z}'\cdot w') = \text{constant},$$

so the imaginary part of φ essentially gives us $t-s$. The real part of φ is a nonnegative function only depends on z', w', and $\rho+\mu$. Now $\frac{1}{2}(\rho+\mu) \in \mathbb{R}^+$ is obvious. We also have

$$\frac{1}{2}|w'|^{2m}+\frac{2m-1}{2}|z'|^{2m}-m\ \mathrm{Re}|z'|^{2m-2}(\bar{z}'\cdot w')$$

$$= \frac{1}{2}[|w'|^2 - |z'|^2]^2\left\{\sum_{q=0}^{m-2}|z'|^{2q}\left[\sum_{p=0}^{m-2-q}|w'|^{2(m-2-q-p)}|z'|^{2p}\right]\right\}$$

$$+\frac{m}{2}|z'|^{2m-2}\cdot|z'-w'|^2 \geq 0.$$

It is easy to see that there exist two universal constants c_1, c_2 (depend on $\bar{U}$) such that $c_1|z'-w'|^{2m} + \frac{1}{2}(\rho+\mu) \leq \text{Re}\ \varphi(z,w) \leq c_2|z'-w'|^{2m} + \frac{1}{2}(\rho+\mu)$. Hence we know the size of the kernel $K_{w',t,\rho+\mu}(z'-w')$ satisfies the properties:

$$(2.3.1) \qquad |K_{w',t,\rho+\mu}(z'-w')| \leq \text{constant}\cdot((\rho+\mu)+|t-s|)^{-2}\cdot|z'-w'|^{-2n}.$$

Next we need to estimate the size of the kernel $\nabla_{z'}K_{w',t,\rho+\mu}$ and $\nabla_{w'}K_{w',t,\rho+\mu}$. Before we do that, let us prove the following lemma:

Lemma.

$$(2.4) \qquad \text{Re}\ \varphi(z',w',t-s;\rho+\mu) \geq \text{constant}(|z'|^{2m-2}+|w'|^{2m-2})|z'-w'|^2.$$

Proof. As we have seen

$$\text{Re}\ \varphi(z',w',t-s;\rho+\mu) = \frac{1}{2}|w'|^{2m} + \frac{2m-1}{2}|z'|^{2m-2}(\bar{z}'\cdot w'+z'\cdot\bar{w}').$$

We want to prove $\text{Re}\ \varphi \geq \text{constant}(|z'|^{2m-2}+|w'|^{2m-2})|z'-w'|^2$. If $|z'| \geq 100|w'|$ or $|w'| \geq 100|z'|$, the result of (2.4) is obvious by homogeneity argument. Now we just assume $n = 1$ and $z_1 = \alpha w_1$, $\alpha \in \mathbb{C}$ and $\frac{1}{100} < |\alpha| < 100$. Then we must show

$$(2.5) \qquad \frac{1}{2} + \frac{2m-1}{2}|\alpha|^{2m} - \frac{m}{2}|\alpha|^{2m-2}(\alpha+\bar{\alpha}) \geq c|\alpha-1|^2.$$

The left hand side of (2.5) has a real zero 1. Suppose it has a complex zero α, then

$$\frac{1}{2} + \frac{2m-1}{2}|\alpha|^{2m} = \frac{m}{2}|\alpha|^{2m-2}(\alpha+\bar{\alpha}) \leq m|\alpha|^{2m-1}.$$

On the other hand, let $f(t) = \frac{1}{2}+\frac{2m-1}{2}t^{2m} - mt^{2m-1}$. Then

$$f(t) \geq 0 \quad \text{when} \quad \frac{1}{100} < t < 100,$$

and $f(t) = 0$ only when $t = 1$. Since

$$f'(t) = (2m-1)mt^{2m-1} - m(2m-1)t^{2m-2} = 0 \ \text{ if and only if } \ t = 1 \ \text{ or } \ t = 0.$$

Hence $\frac{1}{2}+\frac{2m-1}{2}|\alpha|^{2m} - m|\alpha|^{2m-1} = 0$ only with $|\alpha| = 1$. Also we have $\frac{1}{2}+\frac{2m-1}{2}|\alpha|^{2m} \geq m|\alpha|^{2m-1}$. So if α is a zero of the left-hand side of (2.5), then $\frac{1}{2}(\alpha+\bar{\alpha}) = |\alpha| = 1$, i.e., $\alpha = 1$. We also know that

$$\frac{1}{2}+\frac{2m-1}{2}|\alpha|^{2m} - \frac{m}{2}|\alpha|^{2m-2}(\alpha+\bar{\alpha}) > 0$$

when $|\alpha|$ away from 1. So we just need to worry about when α is close to 1, $\alpha \in \mathbb{C}$, i.e., $\alpha = re^{i\theta}$, $r \approx 1$ and $\theta \approx 0$. In this situation, we need to prove:

$$\frac{1}{2}+\frac{2m-1}{2}r^{2m} - mr^{2m-1}\cos\theta \geq \text{constant}[(r\cos\theta-1)^2 + \sin^2\theta].$$

Now

$$\begin{aligned}\frac{1}{2}+\frac{2m-1}{2}r^{2m} - mr^{2m-1}\cos\theta &= \frac{1}{2}(1+(2m-1)r^{2m}-2mr^{2m-1}+2mr^{2m-1}(1-\cos\theta))\\ &= \frac{1}{2}(r-1)^2((2m-1)r^{2m-2}+(2m-2)r^{2m-3}+\dots\\ &\quad + 2r^{2m-2-(2m-3)}+1) + mr^{2m-1}(1-\cos\theta)\\ &\geq c(r-1)^2 + c\theta^2 \geq \frac{c}{100}[(r-1-r(1-\cos\theta))^2+\sin^2\theta].\end{aligned}$$

This completes the proof of (2.4). Now we estimate the terms $\nabla_{z'}K_{w',t,\rho+\mu}$ and $\nabla_{w'}K_{w',t,\rho+\mu}$. Suppose $j = 1$, then

$$\begin{aligned}\frac{\partial}{\partial z_1}\text{Re}\,\varphi &= \frac{2m-1}{2}\cdot m\cdot|z'|^{2m-2}\bar{z}_1 - \frac{m}{2}(m-1)|z'|^{2m-4}\bar{z}_1(\bar{z}'\cdot w'+z'\cdot\bar{w}') - \frac{m}{2}|z'|^{2m-2}\bar{w}_1\\ &= \frac{m^2}{2}|z'|^{2m-2}(\bar{z}_1-\bar{w}_1) + \frac{m(m-1)}{2}|z'|^{2m-4}\bar{z}_1^2(z_1-w_1).\end{aligned}$$

This gives us

$$\left|\frac{\partial}{\partial z_j}\mathrm{Re}\ \varphi\right| \le \text{constant}\ |z'|^{2m-2}\cdot|z_j-w_j|$$

and

$$(2.6)\qquad \left|\frac{\frac{\partial}{\partial z_j}\mathrm{Re}\ \varphi}{\mathrm{Re}\ \varphi}\right| \le \frac{\text{constant}}{|z'-w'|}.$$

On the other hand

$$\frac{\partial}{\partial w_1}\mathrm{Re}\ \varphi = \frac{m}{2}|w'|^{2m-2}\,\bar{w}_j - \frac{m}{2}|z'|^{2m-2}\,\bar{z}_j.$$

Hence we have

$$\left|\frac{\partial}{\partial w_j}\mathrm{Re}\ \varphi\right| \le \text{constant}\ (|w'|^{2m-2}+|z'|^{2m-2})|z_j-w_j|$$

and

$$(2.7)\qquad \left|\frac{\frac{\partial}{\partial w_j}\mathrm{Re}\ \varphi}{\mathrm{Re}\ \varphi}\right| \le \frac{\text{constant}}{|z'-w'|}.$$

Using (2.6) and (2.7), we can prove

$$(2.3.2)\qquad |\nabla_{z'}K_{w',t,\rho+\mu}(z'-w')| + |\nabla_{w'}K_{w',t,\rho+\mu}(z'-w')|$$
$$\le \text{constant}\cdot((\rho+\mu)+|t-s|)^{2}\cdot|z'-w'|^{-2n-1}.$$

Since the critical case $k = 2n$, $\ell = 4m$ is derived from the differentiation of the kernel of the Henkin solution, i.e., $K_{w',t,\rho+\mu}(z'-w') = X_jL_{w',t,\rho+\mu}(z'-w')$ with $X_j = \Xi_j = \frac{\partial}{\partial w_j} + im|w'|^{2m-2}\bar{w}_j\frac{\partial}{\partial s}$ or $X_j = \bar{\Xi}_j = \frac{\partial}{\partial \bar{w}_j} - im|w'|^{2m-2}w_j\frac{\partial}{\partial s}$ for $j = 1,2,\ldots,n$ are vector fields depending on (w',s)-variables. Just use the same techniques, it is easy to see that the kernel $L_{w',t,\rho+\mu}(z'-w')$ also satisfies the properties:

$$(2.8.1)\qquad |L_{w',t,\rho+\mu}(z'-w')| \le \text{constant}\cdot((\rho+\mu)+|t-s|)^{-2}\cdot|z'-w'|^{-2n+1};$$

(2.8.2) $$|\nabla_{z'}L_{w',t,\rho+\mu}(z'-w')| + |\nabla_{z'}L_{w',t,\rho+\mu}(z'-w')|$$

$$\leq \text{constant}\cdot((\rho+\mu)+|t-s|)^{-2}\cdot|z'-w'|^{-2n}.$$

Now let us introduce the following definition:

Definition. *An operator* Op(K) *is defined by a mixed type homogeneities kernel, i.e.,*

$$K = \frac{\mathcal{E}_j(\bar{\varphi}^c)}{\varphi^a\psi^b}$$

is called an admissible operator of type λ *if* $a - c \geq 0$ *and* $j - b + 2m(c-a) + 2n + 4m = \lambda \geq 0$. *Suppose* $j - b = -2n$ *and* $c - a = -2$, *i.e.,* $\lambda = 0$, *we call* Op(K) *is type* 0 *if* K *satisfies the conditions in David-Journé's theorem [DJ] (see below). When* $a - c < 0$, *we call* Op(K) *is an admissible operator of type* λ *if* $j + c - a - b + 2n + 2 = \lambda \geq 0$. *In the case where* $j + c - a - b + 2n + 2 = 0$, *we call* Op(K) *is of type* 0 *if* K *satisfies the conditions in David-Journé's theorem.*

Remark. The assumption on K for the case $j - b = -2n$ and $c - a = -2$ guarantees that K defines a bounded operator on the submanifold $M_p \approx \mathbb{C}^n$. In fact, when we consider the model domains H_m here, we can assume K satisfies the following more precise conditions:

(i) Size condition:

$$|K(z,w)| \approx |K^*(z,w)| \leq \text{constant}\cdot|z'-w'|^{-2n};$$

$$|\nabla_z K(z,w)| + |\nabla_w K(z,w)| \approx |\nabla_z K^*(z,w)| + |\nabla_w K^*(z,w)| \leq \text{constant}\cdot|z'-w'|^{-2n-1}.$$

(ii) Mean-value zero condition:

$$\int_{a<|z'-w'|<b} K(z',w',t-s;\rho+\mu)dz' = 0, \quad \forall\ 0 < a \leq b < \infty,$$

$$\text{and}\quad \forall\ (t-s;\rho+\mu) \in \mathbb{R}\times\mathbb{R}$$

$$\int_{a<|z'-w'|<b} K(z',w',t-s;\rho+\mu)dz' = 0, \quad \forall\ 0 < a \le b < \infty,$$
$$\text{and} \quad \forall\ (t-s;\rho+\mu) \in \mathbb{R}\times\mathbb{R}$$

Here $K^*(z,w)$ is the kernel for the adjoint operator.

As we have seen at the beginning of this section, our kernel for the singular Radon transform satisfies the size condition (i). We also can check the kernel $K_{w',t,\rho+\mu}(z'-w')$ satisfies the mean-value zero property (ii) by using the fact

$$c_1|z'-w'|^{2m}+\tfrac{1}{2}(\rho+\mu) \le \operatorname{Re}\varphi(z,w) \le c_2|z'-w'|^{2m}+\tfrac{1}{2}(\rho+\mu)$$

for two universal constants c_1 and c_2. We will not go through all the details here. The proof can be found in [C].

Proposition 2.9. *The operator* $T_{t,\rho+\mu}$ *defined by*

$$T_{t,\rho+\mu}(f_{t,\rho})(w') = \int_{\mathbb{C}^n} K_{w',t,\rho+\mu}(z'-w')f_{t,\rho}(z')dz'$$

is a bounded mapping from $L^2(\mathbb{C}^n)$ *to itself.*

Proof. Once we know the fact that $K_{w',t,\rho+\mu}(z'-w')$ comes from the differentiation of the Henkin kernel, it is a straightforward computation to check the kernel $K_{w',t,\rho+\mu}(z'-w')$ satisfies the conditions of David-Journé's theorem:

(i) $T_{t,\rho+\mu}(1) \in BMO$

(ii) $T^*_{t,\rho+\mu}(1) \in BMO$

(iii) $|\langle T_{t,\rho+\mu}\varphi,\psi\rangle| \le c\cdot\delta^{-2n}$ where φ and ψ are "bump functions" supported on the Euclidean ball $B(w',\delta)$ with $\sup_{w'}|\nabla_{z'}\varphi(z')| \le c\cdot\delta^{-1}$.

We omit the details.

Now we go back to the estimations of I_1 and I_2. Let us prove the L^2-boundedness for I_1 first. The term I_1 localized the problem. In fact, we just need to assume the phase function in the oscillatory integral is smooth (without any decay). Define

$$K^0_{w',t,\rho+\mu}(z'-w') = \varphi(|z'-w'|^{2m}\lambda)K_{w',t,\rho+\mu}(z'-w')$$

and

$$T_0(f)(w') = \text{P.V.}\int_{\mathbb{C}^n} e^{2mi\lambda \operatorname{Im}|z'|^{2m-2}(\bar{z}'\cdot w')}K^0_{w',t,\rho+\mu}(z'-w')\hat{f}_{t,\rho}(z',\lambda)dz'.$$

Note that the support of the kernel $K^0_{w',t,\rho+\mu}(z'-w')$ contained in the set $\{z' \in \mathbb{C}^n : |z'-w'| < \lambda^{-1/2m}\}$ so we first prove

$$(2.10)\qquad \int_{|w'|\le\lambda^{-1/2m}} |T_0(\hat{f}_{t,\rho})(w')|^2 dw' \le C\int_{|z'|\le\lambda^{-1/2m}} |(\hat{f}_{t,\rho})(z')|^2 dz'.$$

After change variables $z_j = \lambda^{-1/2m}\tilde{z}_j$, $w_j = \lambda^{-1/2m}\tilde{w}_j$, $j = 1,2,\ldots,n$, then

$$T_0(\hat{f}_{t,\rho})(w') = \text{P.V.}\int_{\mathbb{C}^n} e^{2mi\lambda \operatorname{Im}|z'|^{2m-2}(\bar{z}'\cdot w')}K^0_{w',t,\rho+\mu}(\tilde{z}'-\tilde{w}')\hat{f}_{t,\rho}(\tilde{z}')d\tilde{z}'$$

and the support of the kernel $K^0_{w',t,\rho+\mu}(\tilde{z}'-\tilde{w}')$ contained in the unit ball $\{\tilde{z}' \in \mathbb{C}^n, |\tilde{z}'-\tilde{w}'| < 1\}$. In order to prove (2.10) reduces to prove

$$(2.10)'\qquad \int_{|\tilde{w}'|\le 1} |T_0(\hat{f}_{t,\rho})(\tilde{w}')|^2 d\tilde{w}' \le C\int_{|\tilde{z}'|\le 2} |(\hat{f}_{t,\rho})(\tilde{z}')|^2 d\tilde{z}'$$

where C is independent of λ but depends on n and m. Consider $|\tilde{w}'| \le 1$ then

$$\left|\int_{|\tilde{z}'|\le 2} e^{2mi\lambda \operatorname{Im}|\tilde{z}'|^{2m-2}(\tilde{z}'\cdot\tilde{w}')}K^0_{\tilde{w}',t,\rho+\mu}(\tilde{z}'-\tilde{w}')\hat{f}_{t,\rho}(\tilde{z}')d\tilde{z}'\right|$$

$$\le \left|\int_{|\tilde{z}'|\le 2} K^0_{\tilde{w}',t,\rho+\mu}(\tilde{z}'-\tilde{w}')\hat{f}_{t,\rho}(\tilde{z}')d\tilde{z}'\right|$$

$$+ C_1 \int_{|\tilde{z}'|\le 2} |\tilde{z}'-\tilde{w}'| \cdot |K^0_{\tilde{w}',t,\rho+\mu}(\tilde{z}'-\tilde{w}')| \cdot |\hat{f}_{t,\rho}(\tilde{z}')| d\tilde{z}'$$

$$= A_1 + A_2.$$

Here we know that $2mi \ \mathrm{Im}|\tilde{z}'|^{2m-2}(\bar{\tilde{z}}'\cdot\tilde{z}') \equiv 0$. Since

$$|K^0_{\tilde{w}',t,\rho+\mu}(\tilde{z}'-\tilde{w}')| \le \frac{\text{constant}}{((\rho+\mu)+|t-s|)^2} \cdot |\tilde{z}'-\tilde{w}'|^{-2n}$$

and $K^0_{\tilde{w}',t,\rho+\mu}(\tilde{z}'-\tilde{w}')$ satisfies the conditions of the David-Journé theorem, so we can prove the L^2-boundedness for A_1 by Proposition 2.9. The kernel for A_2 belongs to L^1, so we also have the L^2-boundedness for A_2. Now we want to move the ball of center at the origin $\tilde{z}' = 0'$ and $\tilde{w}' = 0'$ to balls of center at arbitrary points ζ, i.e., we want to prove

$$(2.10)'' \quad \int_{|\tilde{w}-\zeta|\le 1} |T_0(\hat{f}_{t,\rho})(\tilde{w}')|^2 d\tilde{w}' \le C_2 \int_{|\tilde{z}-\zeta|\le 2} |\hat{f}_{t,\rho}(\tilde{z}')|^2 d\tilde{z}', \quad \forall\, \zeta \in \mathbb{C}^n,$$

with C_2 independent of ζ. Now we translate $\tilde{z}'$ by ζ and $\tilde{w}'$ by ζ. Even though T_0 is not a translation invariant operator, but we will get another operator

$$\tilde{T}_0 = e^{-2mi\,\mathrm{Im}|\tilde{z}'-\zeta|^{2m-2}(\tilde{z}'\cdot\bar{\zeta}+\zeta\cdot\bar{\tilde{w}}')} T_0.$$

Hence we have

$$\int_{\mathbb{C}^n} \chi_1(\tilde{w}'-\zeta)|T_0(\hat{f}_{t,\rho})(\tilde{w}')|^2 d\tilde{w}' \le C_{n,m} \int_{\mathbb{C}^n} \chi_2(\tilde{z}'-\zeta)|\hat{f}_{t,\rho}(\tilde{z}')|^2 d\tilde{z}'$$

where χ_1 and χ_2 are characteristic functions of balls radii equal to 1 and 2, respectively. Then

$$\int_{\mathbb{C}^n}\int_{\mathbb{C}^n} \chi_1(\tilde{w}'-\zeta)|T_0(\hat{f}_{t,\rho})(\tilde{w}')|^2 d\tilde{w}' \le C_{n,m} \int_{\mathbb{C}^n}\int_{\mathbb{C}^n} \chi_2(\tilde{z}'-\zeta)|\hat{f}_{t,\rho}(\tilde{z}')|^2 d\tilde{z}'$$

which implies

$$\|T_0(\hat{f}_{t,\rho})\|_{L^2} \le C_{n,m}\|\hat{f}_{t,\rho}\|_{L^2}.$$

To prove the desired result for I_2, let us embed the operator $R_{t,\rho+\mu}$ in an analytic family T_z, $z = \alpha + i\beta$, so that $T_0 = R_{t,\rho+\mu}$. We can prove the L^2-boundedness for T_z when $\alpha > 0$ but while α is negative, the situation becomes more akin to the standard singular integral operator. Suppose $\mathrm{Re}(z) < 0$ and sufficiently small, we define

$$T_z(\hat{f}_{t,\rho})(w') = \int_{\mathbb{C}^n} \{1-\varphi(|z'-w'|^{2m}\lambda)\} \frac{|z'-w'|^{-2mz}}{|\lambda|^z}\cdot e^{2mi\lambda \mathrm{Im}|z'|^{2m-2}(\bar{z}'\cdot w')}\cdot$$

$$\cdot K_{w',t,\rho+\mu}(z'-w')\hat{f}_{t,\rho}(z')dz'$$

$$= |\lambda|^{-z}\int_{\mathbb{C}^n} \{1-\varphi(|z'-w'|^{2m}\lambda)\}\cdot e^{2mi\lambda \mathrm{Im}|z'|^{2m-2}(\bar{z}'\cdot w')}\cdot$$

$$\cdot K_{w',t,\rho+\mu}(z'-w')\hat{f}_{t,\rho}(z')dz'.$$

Let the operator $T_{z,\lambda}$ defined as follows

$$(2.12)\quad T_{z,\lambda}(f)(w') = \int_{\mathbb{C}^n} e^{2mi\lambda \mathrm{Im}|z'|^{2m-2}(\bar{z}'\cdot w')}\{1-\varphi(|z'-w'|^{2m}\lambda)\}|z'-w'|^{-2mz}\cdot$$

$$\cdot K_{w',t,\rho+\mu}(z'-w')f(z')dz'.$$

We also rescale the operator $\{\tilde{T}_{z,\lambda}(f)\}(w') = \{T_{z,\lambda}(f)\}\left(\frac{w'}{\lambda^{1/2m}}\right)$ and function $\tilde{f}(z') = \left(\frac{z'}{\lambda^{1/2m}}\right)$, then we have $\|\tilde{f}\|_{L^2} = |\lambda|^{n/2m}\cdot\|f\|_{L^2}$. Since the oscillatory integral in (2.12) has a polynomial phase function, we can apply a result of Ricci and Stein [RS] to obtain

$$|\lambda|^{n/2m}\|T_{z,\lambda}(f)\|_{L^2} = \|\tilde{T}_{z,\lambda}(f)\|_{L^2} = \|\,|\lambda|^z\cdot T_{z,1}(\tilde{f})\|_{L^2} = |\lambda|^z\cdot\|T_{z,1}(\tilde{f})\|_{L^2}$$

$$\leq C\cdot|\lambda|^{z}\|\tilde{f}\|_{L^2} = C\cdot|\lambda|^{z+(n/2m)}\cdot\|f\|_{L^2},$$

whenever $0 < -\mathrm{Re}(z) < \alpha_{2m}\cdot\left(\frac{1}{2}-|\frac{1}{2}-\frac{1}{2}|\right) = \alpha_{2m}/2$ for some $\alpha_{2m} > 0$ (see [RS], p. 192). Now we may use Plancherel formula to prove the uniform L^2-boundedness for the singular Radon transforms whenever $-\alpha_{2m}/2 < \mathrm{Re}(z) < 0$:

$$\iint|(R_{t,\rho+\mu}f_{t,\rho})(w',s)|^2dw'ds = \frac{1}{4\pi^2}\iint\left|\int e^{i\lambda s}|\lambda|^{-z}\{T_{z,\lambda}(\hat{f}_{t,\rho})(\cdot,\lambda)\}(w')d\lambda\right|^2dw'ds$$

$$= \int_{\mathbb{C}^n}\left(\int_{\mathbb{R}^1}|\lambda|^{-2z}\cdot|\{T_{z,\lambda}(\hat{f}_{t,\rho})(\cdot,\lambda)\}(w')d\lambda|^2dw'\right.$$

$$= \int_{\mathbb{R}^1}|\lambda|^{-2z}\cdot\|T_{z,\lambda}(\hat{f}_{t,\rho})(\cdot,\lambda)\|^2_{L^2}d\lambda$$

$$\leq \int_{\mathbb{R}^1}|\lambda|^{-2z}\cdot\|T_{z,\lambda}\|^2_{L^2\to L^2}\cdot\|\hat{f}_{t,\rho}(\cdot,\lambda)\|^2_{L^2(\mathbb{C}^n)}d\lambda$$

$$\leq A^2\int_{\mathbb{R}^1}|\lambda|^{-2z}\cdot|\lambda|^{2z}\cdot\|(\hat{f}_{t,\rho})(\cdot,\lambda)\|^2_{L^2(\mathbb{C}^n)}d\lambda$$

$$= A^2\|\hat{f}_{t,\rho}\|^2_{L^2(\mathbb{C}^n)}.$$

Now we turn to the cases when $0 < \mathrm{Re}(z)$. We will prove the L^p-boundedness for the family of operators T_z. Since the kernel $|z'-w'|^{-2mz}\cdot K_{w',t,\rho+\mu}(z'-w')$ is integrable near ∞, so we just need to use Young's inequality and Plancherel formula to get the L^2-boundedness for the family T_z. We can write down the result immediately:

<u>Lemma 2.13</u>. *Suppose* $\mathrm{Re}(z) > 0$ *and* K *is an operator satisfying the conditions of the David-Journé theorem, then the operator*

$$T_{z,\lambda}(f)(w') = \int_{\mathbb{C}^n} e^{2mi\lambda \operatorname{Im}|z'|^{2m-2}(\bar{z}'\cdot w')}\{1-\varphi(|z'-w'|^{2m}\lambda)\}|z'-w'|^{-2mz}\cdot$$

$$\cdot K_{w',t,\rho+\mu}(z'-w')f(z')dz'$$

can be extended to be a bounded operator on $L^2(\mathbb{C}^n)$ *to itself. The bound of the operator depends on* $s-t$ *and* $\rho+\mu$ *but is independent of the parameter* z.

Next we want to prove T_z is weak type $(1,1)$ with operator norm independent of the parameter z. Suppose $z'-w' = \theta'$ and $t - s - 2m\operatorname{Im}|z'|^{2m-2}(\bar{z}'\cdot w') = \sigma$, we shall temporarily write

$$K^z_{\rho+\mu}(x,y) = \Psi^z_{\rho+\mu}(x;\theta',\sigma)$$

where $[\theta',\sigma] = [(\theta_1,\ldots,\theta_n),\sigma]$ are the coordinates of y with respect to x. If we define a pseudo-metric $\tilde{d}(x,y) < \delta \Longleftrightarrow$ "$|\theta'| < \delta$, and $|\sigma| < |z'|^{2m-2}\delta^2 + \delta^{2m}$," and a family of non-isotropic balls $\tilde{B}(x;\delta) = \{y : \tilde{d}(x,y) < \delta\}$ (see Nagel-Stein-Wainger [NSW]), then we can prove the following lemma:

<u>Lemma 2.14</u>. *Let* $\alpha = \operatorname{Re}(z) > 0$. *Then*

i) *for* $|\sigma| \le |\theta'|^2$,

(a) $|\Psi^z_{\rho+\mu}(x;\theta',\sigma)| \le \text{constant}\cdot|\theta'|^{-2n-2m\alpha}\cdot|\sigma|^{-1+\alpha}$,

(b) $\left|\dfrac{\partial\Psi^z_{\rho+\mu}(x;\theta',\sigma)}{\partial\theta_j}\right| \le \text{constant}\cdot|\theta'|^{-2n-2m\alpha-1}\cdot|\sigma|^{-1+\alpha}$, *for* $j = 1,2,\ldots,n$;

(c) $\left|\dfrac{\partial\Psi^z_{\rho+\mu}(x;\theta',\sigma)}{\partial\sigma}\right| \le \text{constant}\cdot|\theta'|^{-2n-2m\alpha}\cdot|\sigma|^{-1+\alpha-1}$.

ii) *for* $|\sigma| > |\theta'|^{2m}$

(a) $|\Psi^z_{\rho+\mu}(x;\theta',\sigma)| \le \text{constant}\cdot|\sigma|^{(-n/m)-1}$

(b) $\left|\dfrac{\partial\Psi^{z}_{\rho+\mu}(x;\theta',\sigma)}{\partial\theta_j}\right| \le \text{constant}\cdot|\sigma|^{(-n/m)-2-(1/2m)}$, *for* $j = 1,2,\ldots,n$;

(c) $\left|\dfrac{\partial\Psi^{z}_{\rho+\mu}(x;\theta',\sigma)}{\partial\sigma}\right| \le \text{constant}\cdot|\sigma|^{-(n/m)-1-1}$.

Proof. First we can look at

$$\Phi_z(v) = \frac{1}{2\pi}\int_0^{\infty} e^{iv\lambda}|\lambda|^{-z}d\lambda = \frac{1}{2\pi}\int_0^{\infty}\{1-\varphi(\lambda)\}e^{-iv\lambda}\cdot\lambda^{-z}d\lambda, \quad 0 < \mathrm{Re}(z) < 1$$

satisfies the following properties:

i) $\Phi_z(v) = C_z\cdot|v|^{-1+\mathrm{Re}z}$ + smooth function of v, and

$\Phi'_z(v) = C_z\cdot|v|^{-2+\mathrm{Re}z}$;

ii) $\Phi_z(v)$ is rapidly decreasing at $+\infty$, together with all its derivatives;

iii) $C_1 = \pi^{z-(1/2)}\cdot\dfrac{\Gamma(\frac{1}{2}-\frac{z}{2})}{\Gamma(\frac{z}{2})} \approx \frac{z}{2}$, as $\mathrm{Re}(z)\to 0$.

Plugging in all these three properties to the integral representation of the kernel $\Psi^{z}_{\rho+\mu}(x;\theta',\sigma)$, we may prove this lemma by direct computation.

Now we may follow the same method in [PS1], (with respect to different dilation structure $\delta\cdot(z',t;\rho)\to(\delta z',\delta^{2m}t;\delta^{2m}\rho)$, to get the following:

Lemma 2.15 *(Calderón-Zygmund condition). Assume* $K^{z}_{\rho+\mu}(x;y) = \Psi^{z}_{\rho+\mu}(x,\theta',\sigma)$ *satisfies the conclusions of Lemma 2.14. Then*

$$\int_{\tilde{d}(x,y_1)\ge c'\tilde{d}(y_1,y_2)} |K^{z}_{\rho+\mu}(x;y_1) - K^{z}_{\rho+\mu}(x;y_2)|dx \le A$$

where $c' \gg 1$.

Combine all the results above, we can prove the operator T_z is bounded on L^2 for $\mathrm{Re}(z) > 0$ with bounds growing at most polynomially in $\mathrm{Im}(z)$.

By Lemma 2.15, we can also invoke the theory of singular integrals to prove T_z is weak type (1,1). The conclusion is that T_z is bounded on L^p, for $1 < p \le 2$, (when $\mathrm{Re}(z) > 0$, again with bounds growing at most polynomially in $\mathrm{Im}(z)$. On the other hand, T_z is bounded on L^2 for $-\alpha_{2m}/2 < \mathrm{Re}(z) < 0$ with bounds independent of z. By the Stein-Weiss interpolation theorem (see Stein-Weiss [SW], chapter 3), we know that $R_{t,\rho+\mu} = T_0$ is bounded on L^p, $1 < p \le 2$. The case $2 \le p < \infty$ is handled by the similar estimates for the adjoint operator $(R_{t,\rho+\mu})^* = R_{s,\rho+\mu}$ where $R_{s,\rho+\mu}$ is defined by

$$R_{s,\rho+\mu}(g_{s,\mu})(z',t) = \int_{\Omega^*_{(z',t)}} K_{z',s,\rho+\mu}(w'-z')g_{s,\mu}(w')dw'.$$

Here $\Omega^*_{(z',t)} = \{(w',s) \in \partial H_m : s-t-2m\,\mathrm{Im}|w'|^{2m-2}(z'\cdot\bar{w}') = 0\}$. Now we can prove the following proposition:

<u>Proposition 2.16</u>. *Assume* $m \ge 1$. *Then the operator*

$$R_{t,\rho+\mu}(f_{t,\rho})(w',s) = \frac{1}{2\pi}\int_{\mathbb{C}^n} K_{w',t,\rho+\mu}(z'-w')f_{t,\rho}(z',s+2m\,\mathrm{Im}|z'|^{2m-2}(\bar{z}'\cdot w')dz'$$

initially defined for $f \in C_0^\infty(\partial H_m)$ *extends to a bounded operator on* $L^p(\partial H_m)$ *to itself*, $1 < p < \infty$; *i.e.*,

$$\|R_{t,\rho+\mu}(f_{t,\rho})(w',s)\|_{L^p(\partial H_m)} \le \frac{A_p}{((\rho+\mu)^2+|t-s|^2)}\cdot\|f_{t,\rho}\|_{L^p(\partial H_m)}.$$

Now we can prove our main theorem:

<u>Theorem 2</u>. *Let* $K(z,w) \in C^\infty(U)$ *be a kernel of the type* E_kF_ℓ *and consider the Hilbert integral operator* Op(K) *associated to* K *by the formula*

$$[\mathrm{Op}(K)](f)(w',s;\mu) = \int_0^\infty\int_{\partial H_m} K(z',w',t-s;\rho+\mu)f(z',t;\rho)dV(z')dtd\rho.$$

Then

i) Op(K) *can be extended as a bounded operator from* $L^p(\Omega_1)$ *to* $L^p(\Omega_2)$ $(1 \le p \le \infty)$ *for any compact subsets* $\Omega_1, \Omega_2 \subset \bar{H}_m$ *when either*

$$k + \ell < 2n + 4m; \; k < 2n,$$

or

$$k + \frac{\ell}{2m} < 2n + 2; \; \ell < 4m.$$

ii) Op(K) *can be extended as a bounded operator from* $L^p(H_m)$ *to* $L^p(H_m)$ *when either*

$$k + \ell = 2n + 4m; \; k < 2n$$

or

$$k + \frac{\ell}{2m} = 2n + 2; \; \ell < 4m.$$

iii) *Assume that* $k = 2n$, $\ell = 4m$, *and the operator* $T_{t,\rho+\mu}$ *defined by*

$$T_{t,\rho+\mu}(f_{t,\rho})(w') = \int_{\mathbb{C}^n} K_{w',t,\rho+\mu}(z'-w')f_{t,\rho}(z')dz'$$

is a bounded mapping from $L^2(\mathbb{C}^n)$ *to itself for all* $(t-s;\rho+\mu) \in \mathbb{R}^1 \times \mathbb{R}^+$. *Then* Op(K) *extends as a bounded operator on* $L^p(U)$ *to itself,* $1 < p < \infty$.

<u>Proof</u>. i) First observe that

$$\frac{1}{\{2|z'-w'|^2+(t-s)^2+(\rho+\mu)^2\}^{k/2}\{|z'-w'|+|t-s|^{1/2m}+(\rho+\mu)^{1/2m}\}^{\ell}} \tag{2.17}$$

$$\le \frac{\text{constant}}{|z'-w'|^k \cdot \{|z'-w'|+|t-s|^{1/2m}+(\rho+\mu)^{1/2m}\}^{\ell}},$$

and

$$\frac{1}{\{2|z'-w'|^2+(t-s)^2+(\rho+\mu)^2\}^{k/2}\{|z'-w'|+|t-s|^{1/2m}+(\rho+\mu)^{1/2m}\}^{\ell}} \tag{2.18}$$

$$\leq \frac{\text{constant}}{\{2|z'-w'|^2+(t-s)^2+(\rho+\mu)^2\}^{k/2}\cdot\{|t-s|^{1/2m}+(\rho+\mu)^{1/2m}\}^{\ell}},$$

and that the kernels on the right hand sides of (2.17) and (2.18) are respectively non-isotropically and isotropically homogeneous of degree $-(k+\ell)$ and $-(k+\ell/2m)$. When $k < 2n$ the function $|z'-w'|^{-k}$ is locally integrable. The same remark holds for $\ell < 4m$ and the function $\{|t-s|^{1/2m}+(\rho+\mu)^{1/2m}\}^{\ell}$. It follows that K is locally integrable for $k+\ell < 2n+4m$, $k < 2n$, and for $k+\frac{\ell}{2m} < 2n+2$, $\ell < 4m$.

ii) For the case $k+\frac{\ell}{2m} = 2n+2$, $\ell < 4m$, we know the kernel satisfies

(a) $K(\lambda z',\lambda w',\lambda(t-s);\lambda(\rho+\mu)) = \lambda^{-2n-2}K(z',w',t-s;\rho+\mu)$, and

(b) $\int_{\mathbb{R}^1}\int_{\mathbb{C}^n} |K(z',w',t-s;1)|dV(z')dt < \infty$.

Hence we may apply a theorem in [PS2] about the Hilbert integral operator to get the result. For the case $k+\ell = 2n+4m$, $k < 2n$, we know the kernel satisfies

(a′) $K(\lambda z',\lambda w',\lambda^{1/2m}(t-s);\lambda^{1/2m}(\rho+\mu)) = \lambda^{-2n-4m}K(z',w',t-s;\rho+\mu)$

and (b). Once again we may apply the same theorem due to Phong and Stein to get the L^p-boundedness.

iii) To prove the L^p-boundedness for the operator $Op(K)$, we need to check:

$$\int_0^\infty\int_{\partial H_m} |Op(K)(f)(w',s;\mu)|^p dV(w')dsd\mu.$$

First we fix μ, then

$$\|Op(K)(f)(\cdot,\cdot;\mu)\|_{L^p(\partial H_m)} = \left\|\int_0^\infty\int_{-\infty}^\infty (R_{t,\rho+\mu}f_{t,\rho})(w',s)dtd\rho\right\|_{L^p(\partial H_m)}.$$

Use Minkowski integral inequality, we may have

$$\le \int_0^\infty\int_{-\infty}^\infty \|(R_{t,\rho+\mu}f_{t,\rho})(w',s)\|_{L^p(\partial H_m)} dtd\rho.$$

By proposition 2.16 for the singular Radon transforms we can bet

$$\le A\int_0^\infty\int_{-\infty}^\infty \frac{\|f_{t,\rho}\|_{L^p(\partial H_m)}}{(t-s)^2+(\rho+\mu)^2} dtd\rho.$$

Note that

$$\|f_{t,\rho}\|_{L^p(\partial H_m)} = \|f(z',t;\rho)\|_{L^p(\partial H_m)} = \|f_{0,\rho}\|_{L^p(\partial H_m)}.$$

Hence we have

$$A\int_0^\infty\int_{-\infty}^\infty \frac{\|f_{t,\rho}\|_{L^p(\partial H_m)}}{(t-s)^2+(\rho+\mu)^2} dtd\rho = A\int_0^\infty\int_{-\infty}^\infty \frac{\|f_{0,\rho}\|_{L^p(\partial H_m)}}{t^2+(\rho+\mu)^2} dtd\rho = \tilde{A}\int_0^\infty \frac{\|f_{0,\rho}\|_{L^p(\partial H_m)}}{\rho+\mu} d\rho.$$

Now we may use the classical Hilbert integral theory (see [PS2]) to get

$$\|Op(K)(f)(w',s;\mu)\|^p_{L^p(H_m)} = \int_0^\infty \|Op(K)(f)(w',s;\mu)\|^p_{L^p(H_m)} d\mu$$

$$= \int_0^\infty\int_{\partial H_m} |Op(K)(f)(w',s;\mu)|^p dV(w')dsd\mu$$

$$\le A^p\int_0^\infty\left[\int_0^\infty \frac{\|f_{0,\rho}\|_{L^p(\partial H_m)}}{\rho+\mu} d\rho\right]^p d\mu \le \tilde{A}^p\|f\|^p_{L^p(H_m)}.$$

From the proof of Theorem 2 (part (i)), it is easy to get the following corollary:

Corollary 2.19. *Suppose the operator* Op(K) *defined by the kernel* $g_1(w',s;\mu)K(z',w',t-s,\rho+\mu)g_2(z',t;\rho) = g_1\cdot E_k F_\ell \cdot g_2$ *where* g_1, g_2 *are two cut-off functions defined on* $\overline{H}_m$. *Then we consider the boundedness of the operator* Op(K) *in two different cases*:

Case 1. $\ell \geq 4m,\ k < 2n,\ k+\ell < 2n+4m$; *we have the following*:

i) $\mathrm{Op}(K) = L^p(U) \to L^q(U),\ q^{-1} = p^{-1} - (2n+4m-k-\ell)/(2n+4m),\ 1 < p < \frac{2n+4m}{2n+4m-k-\ell}$;

ii) $\left|\int_U\int \exp\left[\alpha|\mathrm{Op}(K)(f)/\|f\|_{L^{(2n+4m)/(2n+4m-k-\ell)}}|^{(2n+4m)/(k+\ell)}\right] dw'dsd\mu\right| \leq$

$C < \infty$, α *is a sufficiently small constant*;

iii) $\mathrm{Op}(K) : L^1(U) \to L^{\frac{2n+4m}{k+\ell}-\varepsilon}(U),\ \forall\ \varepsilon > 0$;

iv) $\mathrm{Op}(K) : L^{\frac{2n+4m}{2n+4m-k-\ell}}(\log^+ L)^{\frac{2n+4m}{2n+4m-k-\ell}-1}(U) \to L^{\frac{2n+4m}{k+\ell}}(U)$;

v) $\mathrm{Op}(K) : L^{\frac{2n+4m}{2n+4m-k-\ell}}(\log^+ L)^{\frac{2n+4m}{2n+4m-k-\ell}-1+\varepsilon}(U) \to L^\infty(U),\ \ \forall\ \varepsilon > 0$.

Case 2. $\ell < 4m,\ k + \frac{\ell}{2m} < 2n+2$; *we have the following*:

i′) $\mathrm{Op}(K) : L^p(U) \to L^q(U),\ q^{-1} = p^{-1} - \frac{4nm+4m-2km-\ell}{4nm+4m},\ \ 1 < p < \frac{4nm+4m}{4nm+4m-2km-\ell}$;

ii′) $\left|\int_U\int \exp\left[\alpha|\mathrm{Op}(K)(f)/\|f\|_{L^{(4nm+4m)/(4nm+4m-2km-\ell)}}|^{(4nm+4m)/(2km+\ell)}\right]\cdot\right.$

$\left.\cdot dw'dsd\mu\right| \leq c < \infty$, α *is a sufficiently small constant*;

iii′) $\mathrm{Op}(K) : L^1(U) \to L^{\frac{4nm+4m}{2km+\ell}-\varepsilon}(U),\ \forall\ \varepsilon > 0$;

iv′) $\mathrm{Op}(K) : L^{\frac{4nm+4m}{4nm+4m-2km-\ell}}(\log^+ L)^{\frac{4nm+4m}{4nm+4m-2km-\ell}-1}(U) \to L^{\frac{4nm+4m}{2km+\ell}}(U)$;

v′) $\mathrm{Op}(K) : L^{\frac{4nm+4m}{4nm+4m-2km-\ell}}(\log^+ L)^{\frac{4nm+4m}{4nm+4m-2km-\ell}-1+\varepsilon}(U) \to L^\infty(U),\ \ \forall\ \varepsilon > 0$.

Here U *is a boundary neighborhood of any point* $p \in \partial H_m \subset \mathbb{C}^{n+1}$.

3. Optimal L^p-estimates for the Henkin solutions of $\bar{\partial}$-equations on domains H_m.

In this section we apply Theorem 2 to discuss the optimal L^p-estimates of the kernel **H** which we obtained in section 1. To simplify the computation, we just assume $n = 1$ and $m = 2$. It is easy to see from Theorem 1 the integral kernel $\ker(\mathbf{H}) = (K_1, K_2)$ for **H** as the following form (in terms of E_k, F_ℓ):

$$K_1 = C_1E_2F_4 + C_2E_1F_8 + C_3E_3F_0,$$

$$K_2 = C_1'E_2F_4 + C_2'E_0F_8.$$

We can get the following result immediately by applying Theorem 2 (i):

Proposition 3.1. *Suppose* U *is a boundary neighborhood of a point* $p \in \partial H_m \subset \mathbb{C}^{n+1}$, *then* **H** *extends to a bounded operator from* $L^p(U)$ *to itself*, $1 \le p \le \infty$.

The importance of these operators K_1 and K_2 lies in their smoothing properties. They are not only bounded on $L^p(U)$, $1 < p < \infty$, but also gain 1 in "good" direction. By the direct inspection of the kernel K_1 and K_2, we may apply Theorem 2 to get the following result:

Theorem 3. *Let* U *be a boundary neighborhood of any point* $p \in \partial H_m \subset \mathbb{C}^{n+1}$, $f \in C^\infty_{(0,1),0}(U)$ *with* $\bar{\partial} f = 0$. *Then the Henkin solutions* **H** *of the Cauchy-Riemann equation* $\bar{\partial} u = f$ *(i.e.,* $\bar{\partial}\mathbf{H}f = f + S_{-\infty}f$*) satisfies*

i) $\mathbf{H} : S^p_k(U) \longrightarrow S^p_{k+1}(U)$, $k \in \mathbb{Z}^+$, $1 < p < \infty$;

ii) $\bar{Z}_{n+1}\mathbf{H} : L^p_k(U) \longrightarrow L^p_k(U)$, $k \in \mathbb{Z}^+$, $1 < p < \infty$.

The computation is long and tedious, the proof can be found in [C]. We will not repeat it again.

We can also apply Corollary 2.19 to obtain the following corollary:

Corollary 3.2. i) $\mathbf{H} : L^p(U) \rightarrow L^q(U)$, $\frac{1}{q} = \frac{1}{p} - \frac{1}{2n+4m}$, *if* $1 < p < 2n+4m$;

ii) $\left| \int_U\int \exp\left(\left| \frac{\alpha \mathbf{H}(f)}{\|f\|_{L^{2n+4m}}} \right|^{(2n+4m/(2n+4m-1)} \right) dw' ds d\mu \right| \leq c < \infty$, α *is a sufficiently small constant*;

iii) $\mathbf{H} : L^1(U) \rightarrow L^{\frac{2n+4m}{2n+4m-1}-\varepsilon}(U)$, $\forall\ \varepsilon > 0$;

iv) $\mathbf{H} : L^{2n+4m}(\log^+ L)^{2n+4m-1}(U) \rightarrow L^{\frac{2n+4m}{2n+4m-1}-\varepsilon}(U)$;

v) $\mathbf{H} : L^{2n+4m}(\log^+ L)^{2n+4m-1+\varepsilon}(U) \rightarrow L^{\infty}(U)$, $\forall\ \varepsilon > 0$.

On the other hand, we know the regularity properties of the Szegö kernel and Bergman kernel on finite type domains in $\mathbb{C}^2$ (see [NRSW]), it is easy to obtain regularity properties for the Kohn solution for the Cauchy-Riemann equation by combining the results of Theorem 3, Corollary 3.2 and [NRSW].

References

[C] D.C. Chang, Optimal L^p estimates for the solutions of $\bar{\partial}$-equations on a family of weakly pseudo-convex domains, MSRI report 01021-88.

[CNS] D.C. Chang, A. Nagel, E.M. Stein, Estimates for the $\bar{\partial}$-Neumann problem for pseudo-convex domains in $\mathbb{C}^2$ of finite type, Proc. Natl. Acad. Sci. USA, 85 (1988), 8771-8774.

[Ch] M. Christ, Regularity properties of the $\bar{\partial}_b$ equation on weakly pseudo-convex CR manifolds of dimension three, J. of the AMS, 1 (1988), 587-646.

[DJ] G. David, J.-L. Journé, A boundedness criterion for generalized Calderón-Zygmund operators, Ann. of Math. 120 (1984), 371-397.

[FK] C.L. Fefferman, J.J. Kohn, Hölder estimates on domains of complex dimension two and on three dimensional CR manifolds, Adv. in Math., 69 (1988), 223-303.

[GS] P.C. Greiner, E.M. Stein, On the solvability of some differential operators of type $\square_b$, Proceedings of International Conferences, Cortona, Italy 1976-1977, Sc. Norm. Sup., Pisa (1978), 106-165.

[HL] G. Henkin, J. Leiterer, Theory of functions on complex manifolds, Birkhäuser Verlag, Basel-Boston-Stuttgart, 1984.

[KS] N. Kerzman, E.M. Stein, The Szegö kernel in terms of Cauchy-Fantappiè kernels, Duke Math. J., 45 (1978), 197-224.

[K] J.J. Kohn, Boundary behavior of $\bar{\partial}$ on weakly pseudo-convex manifolds of dimension two, J. Diff. Geometry, 6 (1972), 523-542.

[Kr] S.G. Krantz, Optimal Lipschitz and L^p-regularity for the equation $\bar{\partial}u = f$ on strongly pseudo-convex domains, Math. Annalen, 219 (1976), 213-233.

[LR] I. Lieb, R.M. Range, Integral representations and estimates in the theory of the $\bar{\partial}$-Neumann problem, Ann. of Math. 123 (1986), 265-301.

[M] H.-M. Maire, Régularité optimale des solutions de systèmes différentiels et du Laplacien associé; applications au $\square_b$, Math. annalen, 258 (1981), 55-63.

[NRSW] A. Nagel, J.-P. Rosay, E.M. Stein, S. Wainger, Estimates for the Bergman and Szegö kernels in certain weakly pseudo-convex domains, to appear in Ann. of Math.

[NSW] A. Nagel, E.M. Stein, S. Wainger, Balls and metrics defined by vector fields I: Basic properties, Acta Math., 155 (1985), 103-147.

[PS1] D.H. Phong, E.M. Stein, Hilbert integrals, singular integrals and Radon transforms I, Acta Math. 157 (1986), 99-157.

[PS2] D.H. Phong, E.M. Stein, Hilbert integrals, singular integrals and Radon transforms II, Inventiones Math. 86 (1986), 75-113.

[R] R.M. Range, Holomorphic functions and integral representations in several complex variables, Springer-Verlag, New York-Berlin-Heidelberg-Tokyo, 1986.

[RS] F. Ricci, E.M. Stein, Harmonic analysis on nilpotent groups and singular integrals I: Oscillatory integrals, J. of Functional Analysis, 73 (1987), 179-194.

[Ro] L.P. Rothschild, Nonexistence of optimal L^2 estimates for the boundary Laplacian operator on certain weakly pseudo-convex domains, Comm. in PDE 5 (1980), 897-912.

[SW] E.M. Stein, G. Weiss, Fourier analysis on Euclidean spaces, Princeton Univ. Press, Princeton, NJ, 1971.

4

Real-Analytic Operator-Valued Functions Defined in BMO

R. R. Coifman and S. Semmes

There are a number of operators arising in analysis that depend nonlinearly on a functional parameter, and for which this dependence is real-analytic when the parameter ranges through a suitably chosen function space. It sometimes happens that BMO is the "best" choice for this space. We shall investigate in this paper certain cases in which BMO is a natural candidate for this best choice.

The specific cases we treat are variations of problems considered in [**CMM**] and [**KM**], involving perturbations of $\frac{d^2}{dx^2}$ on $\mathbf{R}$. We shall also encounter the general problem of knowing which multilinear operators (or nonlinear functions) have the property that when you conjugate them by a change of variables $h : \mathbf{R} \to \mathbf{R}$, you get a real-analytic function of $\log h' \in$ BMO.

Before descending into the details, let us review some definitions and well-known examples.

Let BMO_r and BMO_c denote the spaces of real- and complex-valued BMO functions on $\mathbf{R}$. Set $N_r(\delta) = \{u \in \text{BMO}_r : \|u\|_* < \delta\}$, and define $N_c(\delta)$ similarly.

Let E be some Banach space, and let λ be an E-valued function defined on some neighborhood of 0 in BMO_r. We say that λ is real-analytic at 0 (sometimes we'll be sloppy and simply say "real-analytic") if it has a convergent power series in a neighborhood of 0. This means that there is a $\delta > 0$ so that

$$\lambda(b) = \Sigma\lambda_k(b) \text{ for } b \in N_r(\delta), \tag{0.1}$$

where each λ_k is a homogeneous polynomial of degree k, i.e., there is a k-linear function $\tilde{\lambda}_k(b_1, \cdots, b_k)$ defined for $b_i \in$ BMO_r and taking values in E such that $\lambda_k(b) = \tilde{\lambda}_k(b, \cdots, b)$. We also require that

$$\|\lambda_k(b)\|_E \le C\delta^{-k}\|b\|_*^k \tag{0.2}$$

for some $C > 0$, so that the series (0.1) converges. (See [**HP**] for general results concerning real-analytic functions between Banach spaces.)

A simple example is given by the weighted Hilbert transform. Let H denote the Hilbert transform, and take E to be $B(L^2)$, the space of bounded linear operators on L^2. Then $\lambda(b) = e^b He^{-b}$ is real-analytic in a neighborhood of the origin.

It is easy to see what λ_k is in this case: $\lambda_k(b) \in B(L^2)$ is given by

$$\lambda_k(b)(f)(x) = PV \int_{-\infty}^{\infty} \frac{1}{k!} \frac{(b(x) - b(y))^k}{x - y} f(y)dy.$$

Both authors supported by the NSF. The second author is also supported by the Sloan Foundation.

That λ_k satisfies (0.2) was proved in [**CRW**]. Let us sketch how this goes, because we shall be taking for granted similar arguments throughout this paper.

We know that $\|\lambda(b)\| \leq C$ if $b \in N_r(\delta)$ for δ small enough. There is an obvious extension of $\lambda(b)$ to $N_c(\delta)$, and the same bound still holds trivially. It is clear that λ is holomorphic in b, at least in some suitable weak sense. At any rate, we certainly have

$$\lambda_k(b) = \frac{1}{2\pi}\int_0^{2\pi} \lambda(e^{i\theta}b)e^{-ik\theta}\,d\theta.$$

Hence $\|\lambda_k(b)\| \leq C$ if $b \in N_c(\delta)$, which implies (0.2).

The general pattern of this argument is commonly used: to show that λ is real-analytic on $N_r(\delta)$, you show that its analytic continuation to $N_c(\delta)$ is bounded, and use this to control the power series of λ.

A more subtle example is given by

$$C(\beta)f(x) = P.V.\int_{-\infty}^{\infty} \frac{e^{\beta(y)}}{\int_y^x e^{\beta(t)}dt} f(y)dy.$$

When β is real, $C(\beta) = V_h H V_h^{-1}$, where $V_h f = f \circ h$, $h' = e^{\beta}$. When β is purely imaginary, $C(\beta)$ gives the Cauchy integral operator on a curve whose arclength parameterization $z(t)$ satisfies $z' = e^{\beta}$. (If $\|\beta\|_*$ is small, this will be a chord-arc curve with small constant.) When β is complex-valued, $C(\beta)$ is the Cauchy integral operator on a curve conjugated by a change of variables.

It is well-known that there exist $\delta, C > 0$ so that $\|C(\beta)\| \leq C$ if $\beta \in N_c(\delta)$, and $C(\beta)$ is represented by a convergent power series on $N_c(\delta)$. (See [**CM**].) Just as for the weighted Hilbert transform, the power series for $C(\beta)$ can be obtained explicitly from the kernel, and (0.2) is obtained from the bound on $C(\beta)$ for $\beta \in N_c(\delta)$.

Let us now discuss the examples that will be the focus of this paper.

Let $a \in L^\infty(\mathbf{R})$ be accretive, i.e., $Re\ a \geq \eta$ for some $\eta > 0$. Let $D = \frac{1}{i}\frac{d}{dx}$. There is a natural definition of $(DaD)^{\frac{1}{2}}$, where we let a denote also the corresponding multiplication operator. It is proved in [**CMM**] that $(DaD)^{\frac{1}{2}}D^{-1}$ defines a bounded operator on L^2 for all accretive $a \in L^\infty(\mathbf{R})$.

Is there a natural variant of this operator-valued function which is defined and bounded on a domain in BMO_c instead of L^∞? Consider K_α defined formally by

$$K_\alpha = (De^{2\alpha}D)^{\frac{1}{2}}(e^{\alpha}D)^{-1}. \tag{0.3}$$

Is K_α defined and bounded for $\alpha \in N_c(\delta)$ for some $\delta > 0$? Notice that K_α is not changed if we add a constant to α.

At this point it is not even clear how to define K_α when α is complex, if $e^{2\alpha}$ is not accretive. We can avoid this technical problem by reformulating the question as follows: if we restrict ourselves to α real, is K_α real-analytic at 0 in BMO_r? In this case we can use spectral theory to define K_α, and it is even unitary.

Following Kenig and Meyer [**KM**], we can consider a more general version of this question. Define $K_{\alpha,\beta}$ by

$$K_{\alpha,\beta} = (e^{2\alpha} D e^{2\beta} D)^{\frac{1}{2}} (e^{\alpha+\beta} D)^{-1}. \tag{0.4}$$

When $e^{2\alpha}$ and $e^{2\beta}$ are bounded and accretive, Kenig and Meyer showed that $K_{\alpha,\beta}$ can be defined and gives a bounded operator on L^2, and that $K_{\alpha,\beta}$ reduces to the Cauchy integral on Lipschitz graphs when $\alpha = \beta$.

Can we define $K_{\alpha,\beta}$ when $\alpha, \beta \in N_c(\delta)$, for δ small enough, and will it be bounded? As before, it is better to reformulate this by asking if $K_{\alpha,\beta}$ is real-analytic at $(0,0)$ in $\mathrm{BMO}_r \times \mathrm{BMO}_r$. In the case where $\alpha = \beta$, this is true, because it reduces to $C(\beta)$.

THEOREM 1. *$K_{\alpha,\beta}$ is not real-analytic at $(0,0)$ in $BMO_r \times BMO_r$.*

Let us rephrase the question in a more convenient way. If h is a homeomorphism of $\mathbf{R}$ to itself, then

$$V_h D V_h^{-1} = \frac{1}{h'} D, \tag{0.5}$$

as one can easily check using the chain rule. If $h' = e^{-2\alpha}$, then

$$K_{\alpha,\beta} = V_h (D e^{2\tilde{\beta} - 2\tilde{\alpha}} D)^{\frac{1}{2}} (e^{\tilde{\beta}-\tilde{\alpha}} D)^{-1} V_h^{-1}$$

where $\tilde{\beta} = \beta \circ h^{-1}, \tilde{\alpha} = \alpha \circ h^{-1}$. In view of this, Theorem 1 reduces to the following.

THEOREM 1′. *Set $J(b,f) = (D(1-b)^{-1} D)^{\frac{1}{2}} D^{-1} f$ and define $\lambda(\alpha, b) \in B(L^2)$ by*

$$\lambda(\alpha, b) f = V_h J(V_h^{-1} b, V_h^{-1} f), \quad h' = e^{\alpha}.$$

Then $\lambda(\alpha, b)$ is not real-analytic at $(0,0)$ in $BMO_r \times L^\infty$.

We shall prove Theorem 1′ by arguing as follows. We know from **[CMM]** that $J(b,f)$ has a nice power series expansion $J(b,f) = \sum_{n=0}^{\infty} J_n(b,f)$ which converges when $\|b\|_\infty < 1$, where each $J_n(b,f)$ is linear in f and is a homogeneous polynomial of degree n in b. We shall prove that if $J_{n,\alpha}(b,f) = V_h J_n(V_h^{-1} b, V_h^{-1} f), h' = e^{\alpha}$, has a convergent power series in α on $N_r(\delta)$, then $\delta \leq C(\log n)^{-1}$. This implies Theorem 1: if $K_{\alpha,\beta}$ were real analytic, then we could take $\delta > 0$ to be independent of n.

There are, however, some positive results.

THEOREM 2. *For each n, $J_{n,\alpha}$ is real-analytic at 0, with radius of convergence at least $(C \log n)^{-1}$.*

THEOREM 3. *$K_{\alpha,-\alpha}$ is real-analytic at 0 in BMO_r.*

In this last case it turns out that $(a e^{\alpha} D e^{-\alpha} b D)^{\frac{1}{2}} D^{-1}$ is real-analytic about $(a, \alpha, b) = (1,0,1)$ in $L^\infty \times$ BMO $\times L^\infty$, in contrast to Theorem 1′.

Theorems 1′ and 2 raise the following basic question. Let $F(a)$ be an operator-valued function. When is it true that $V_h F(V_h^{-1} a) V_h^{-1}, h' = e^{\alpha}$, is real-analytic in $\alpha, \alpha \in \mathrm{BMO}_r$?

In order for $V_h F(V_h^{-1}a)V_h^{-1}$ to be a well-defined function of α, it must be unchanged if we replace h by $ch + d, c > 0, d \in \mathbf{R}$. Thus $F(a)f$ must commute with simultaneous translations and dilations in a and f. If $F(a)$ is constant (i.e., does not depend on a), then it must be a linear combination of the identity and the Hilbert transform.

Suppose $F(a)$ is not constant, but that it is real-analytic at $a = 0$. Hence $F(a) = \Sigma F_k(a)$ for a small, where each $F_k(a)$ is an operator-valued homogeneous polynomial of degree k. One approach to dealing with $V_h F(V_h^{-1}a)V_h^{-1}$ is to first understand $V_h F_k(V_h^{-1}a)V_h^{-1}$ for each k.

This leads us to the following problem. Let $T(a_1, \cdots, a_k, f)$ be a multilinear operator that commutes with simultaneous translations and dilations in the a_i's and f. This implies that $T(a_1, \cdots, a_k, f)$ is given by

$$T(a_1, \cdots, a_k, f)(x) = \int_{\mathbf{R}} \int_{\mathbf{R}^k} K(x - y_1, \cdots, x - y_k, x - z)a_1(y_1) \cdots a_k(y_k)f(z)dydz$$

for some distribution kernel K which is homogeneous of degree $-(k+1)$. Can we characterize the distribution kernels K such that $V_h T(V_h^{-1}a_1, \cdots, V_h^{-1}a_k, V_h^{-1}f), h' = e^{\alpha}$, is real-analytic in α?

In proving Theorem 1′ we shall obtain some necessary conditions on K for this to hold. Theorem 2 gives a family of examples of such kernels, and its proof shows that there are many more, in sharp contrast to the $k = 0$ case.

It is not clear whether there should be a nice characterization of these kernels. Even in the bilinear case it is not clear exactly what kinds of singularities can arise, although we can say a lot about them. Two simple examples the reader should think of are $T(a, f) = aHf$ and $H(aHf)$. Both are real-analytic under conjugations.

The proof of Theorem j is given in Section j, $j = 1, 2, 3$.

1.

From the bottom half of p. 384 in [**CMM**], we have that

$$J_n(b, f) = \int_0^{\infty} Q_t(bP_t)^n f \frac{dt}{t}, \tag{1.1}$$

where: $J(b, f) = \sum_n J_n(b, f)$, $J(b, f)$ as in Theorem 1′; we have let b also denote the corresponding multiplication operator; $P_t f = k_t \star f, k_t(x) = \frac{1}{2}\frac{1}{t}e^{-|x|/t}$, and $Q_t f = h_t \star f, h_t(x) = i\ \mathrm{sgn}(x)k_t(x)$; and we have dropped an inessential multiplicative constant.

Hence

$$J_h(b, f)(x) = \int_{\mathbf{R}} \int_{\mathbf{R}^n} j_n(x - y_1, \cdots, x - y_n, x - z)b(y_1) \cdots b(y_n)f(z)dydz, \tag{1.2}$$

with

$$\begin{aligned} j_n(y_1, \cdots, y_n, z) &= \int_0^\infty h_t(y_1)k_t(y_1 - y_2)\cdots k_t(y_{n-1} - y_n)k_t(y_n - z)\frac{dt}{t} \\ &= \operatorname{sgn}(y_1)\int_0^\infty (\frac{1}{t})^{n+1}\exp\{-(|y_1| + |y_1 - y_2| + \cdots + |y_{n-1} - y_n| + |y_n - z|)\}\frac{dt}{t} \\ &= \operatorname{sgn}(y_1)\cdot(|y_1| + |y_1 - y_2| + \cdots + |y_{n-1} - y_n| + |y_n - z|)^{-n-1}. \end{aligned}$$

In each of the last two equalities we have dropped multiplicative constants.

Let $\tilde{J}_n(b_1, \cdots, b_n, f)$ be the $(n+1)$-linear operator which is symmetric in the b's and satisfies $\tilde{J}_n(b, \cdots, b, f) = J_n(b, f)$. Define $\tilde{J}_{n,\alpha}(b_1, \cdots, b_n, f)$ by

$$\tilde{J}_{n,\alpha}(b_1, \cdots, b_n, f) = V_h\tilde{J}_n(V_h^{-1}b_1, \cdots, V_h^{-1}b_n, V_h^{-1}f),$$

with $h' = e^\alpha$. Standard polarization arguments imply that if $J_{n,\alpha}$ has a convergent power series expansion in α on $N_r(\delta)$, then so does $\tilde{J}_{n,\alpha}$.

We should perhaps be a little more precise here. We are viewing $\tilde{J}_{n,\alpha}$ as a function in α that takes values in E_n, the Banach space of bounded $(n+1)$-linear maps from $L^\infty \times \cdots \times L^\infty \times L^2$ into L^2. We are saying that if $\lambda(\alpha, b)$ - defined in Theorem 1′ - is real analytic at $(0, 0)$, then for each n, $\tilde{J}_{n,\alpha}$ is real-analytic at $\alpha = 0$, with a radius of convergence that is bounded from below, independently of n. (In fact, $\tilde{J}_{n,\alpha}$ is really just the nth order differential of $\lambda(\alpha, b)$ in b evaluated at $b = 0$.) Our goal is to show that the radius of convergence of $\tilde{J}_{n,\alpha}$ must actually be $\leq c(\log n)^{-1}$.

The kernel $\tilde{j}_n$ of $\tilde{J}_n$ is given by symmetrizing j_n: if we set

$$\tilde{j}_n(y_1, \cdots, y_n, z) = \frac{1}{n!}\sum_{\pi\in\text{Sym}(n)} j_n(y_{\pi(1)}, \cdots, y_{\pi(n)}, z), \tag{1.3}$$

where Sym(n) denotes the group of all permutations on n letters, then $\tilde{J}_n(b_1, \cdots, b_n, f)$ is given in terms of $\tilde{j}_n$ by a formula like (1.2). From this and definitions we get

$$\begin{aligned} &\tilde{J}_{n,\alpha}(b_1, \cdots, b_n, f)(x) \\ &= \int_{\mathbf{R}}\int_{\mathbf{R}^n} \tilde{j}_n(h(x) - h(y_1)h(x) - h(y_n), h(x) - h(z))h'(y_1)\cdots h'(y_n)h'(z) \\ &\qquad \times b_1(y_1)\cdots b_n(y_n)f(z)dydz. \end{aligned} \tag{1.4}$$

The idea behind our upper bound on the radius of convergence of $\tilde{J}_{n,\alpha}$ is as follows. If $\tilde{J}_{n,\alpha}$ is real-analytic in α, then we can analytically continue $\tilde{J}_{n,\alpha}$ to α complex. The explicit form of $\tilde{j}_n$ will allow us to compute the kernel of this analytic continuation. We shall see that if the radius of convergence of $\tilde{J}_{n,\alpha}$ is too large, then the kernel of its analytic continuation has singularities that it shouldn't have. A technical problem in making this precise is that the kernel is a priori only a distribution. The following lemma takes care of that issue.

LEMMA 1.5. *Suppose that $n \geq 10$ and that $\tilde{J}_{n,\alpha}$ has a convergent power series representation for $\alpha \in N_r(\delta)$ in the sense of (0,1), (0,2), with $E = E_n$ defined above. Set $N_r'(\delta) = \{\alpha \in N_r(\delta)\colon \alpha$ is constant on $[-\frac{2}{n}, \frac{4}{n}]$ and on $[1,2]\}$. Let $x^0, y_1^0, \cdots, y_n^0$, and z^0 be any set of pairwise distinct points on $\mathbf{R}$ such that $x^0, y_i^0 \in [-\frac{1}{n}, \frac{3}{n}], i = 1, \cdots, n$, and $z^0 \in [\frac{4}{3}, \frac{5}{3}]$. Then the real-valued function*

$$f(\alpha) = \tilde{j}_n(h(x^0) - h(y_1^0), \cdots, h(x^0) - h(y_n^0), h(x^0) - h(z^0))e^{n\alpha(0)+\alpha(1)}, \; h' = e^{\alpha},$$

has a convergent power series representation on $N_r'(\delta/2)$.

The proof of this is fairly easy. Write

$$BL_n(\alpha, b_1, \cdots, b_n, f) = \tilde{J}_{n,\alpha}(b_1, \cdots, b_n, f)$$

and define $L_{n,\beta}$ by

$$L_{n,\beta}(\alpha, b_1, \cdots, b_n, f) = V_k L_n(V_k^{-1}\alpha, V_k^{-1}b_1, \cdots, V_k^{-1}b, V_k^{-1}f),$$

where $k' = e^{\beta}$. Definition chasing gives

$$L_{n,\beta}(\alpha, b_1, \cdots, b_n, f) = \tilde{J}_{n,\alpha+\beta}(b_1, \cdots, b_n, f).$$

Hence $L_{n,\beta}(\alpha, -)$ is an E_n-valued function of (α, β) that has a convergent power expansion on $(\alpha, \beta) \in N_r(\delta/2) \times N_r(\delta/2)$. In particular, if we differentiate $L_{n,\beta}(\alpha, -)$ finitely many times in β and evaluate at $\beta = 0$, the resulting function of α has a convergent power series representation on $N_r(\delta/2)$.

We are going to use this differentiation trick for certain special β's. Given small numbers $r, s_1, \cdots, s_n$, and t, we can find $\beta = \beta(r, s, t) \in \mathrm{BMO}_r$ and $k : \mathbf{R} \to \mathbf{R}$ such that β depends real-analytically on (r, s, t), $\beta(0,0,0) = 0$, $k' = e^{\beta}$, $k(y) = y + r$ for y near x^0, $k(y) = y + s_i$ for y near y_i, and $k(y) = y + z$ for y near z^0. (How near is near is allowed to depend on everything, especially n, δ, x^0, y_i^0, and z^0.)

If $\alpha \in N_r'(\delta/2)$ – so that α is constant near x^0, y_i^0, and z^0 – and if $b_1, \cdots, b_n$ and f are supported near $y_1^0, \cdots, y_n^0$ and z^0, respectively, and if $\beta = \beta(r, s, t)$ and k are as above, then

(1.6)

$$\begin{aligned} &L_{n,\beta}(\alpha, b_1, \cdots, b_n, f)(x) \\ &= \int_{\mathbf{R}}\int_{\mathbf{R}^n} \tilde{j}_n(h(x+r) - h(y_1+s_1), \cdots, h(x+r) - h(x+s_n), h(x+r) - h(z+t)) \\ &\qquad \times e^{n\alpha(0)+\alpha(1)} b(y_1)\cdots b(y_n)f(z)\,dy\,dz \end{aligned}$$

for x near x^0. Here $h' = e^{\alpha}$.

Let $d_n(\alpha, r, s, t)$ denote the distribution

$$\tilde{j}_n(h(x+r)-h(y_1+s_1),\cdots,h(x+r)-h(z+t))e^{n\alpha(0)+\alpha(1)}$$

defined for $(x,y_1,\cdots,y_n,z)$ near $(x^0,y_1^0,\cdots,y_n^0,z^0)$ in $\mathbf{R}^{n+2}$. At this stage the proof of Lemma 1.5 follows from words. We can view $d_n(\alpha,r,s,t)$ as a real-analytic function of α,r,s,t with values in the space of distribution kernels of operators in E_n restricted to a small neighborhood of (x^0,y^0,z^0) in $\mathbf{R}^{n+2}$. Equivalently, we can view $d(\alpha,r,s,t)$ as a real-analytic map of $\alpha\in N_r'(\delta/2)$ into the space of real-analytic functions of (r,s,t) near $(0,0,0)$ into distributions. For fixed α, the real-analyticity of $d_n(\alpha,r,s,t)$ as a distribution-valued function of (r,s,t) and the definition of $d(\alpha,r,s,t)$ imply that $d(\alpha,0,0,0)$ is actually real-analytic near (x^0,y^0,z^0) (as a function of (x,y,z)) and not just a distribution. Lemma 1.5 follows from all these things.

Let us use Lemma 1.5 to obtain an upper bound for the radius of convergence for $\tilde{J}_{n,\alpha}$. Let δ and everything else be as in the lemma. The lemma implies that $f(\alpha)$ analytically continues to $N_c'(\delta/2)$. Using the explicit form of $\tilde{j}_n$, we can get an upper bound on δ.

We compute $f(\alpha)$ first for $\alpha\in N_r'(\delta/2)$. We may as well take $\alpha(1)=0$. From (1.3) we get $f(\alpha)=$

$$\frac{1}{n!}\sum_{\pi\in\mathrm{Sym}(n)}\mathrm{sgn}(h(x^0)-h(h^0_{\pi(1)}))\Big(\sum_{i=1}^{n}|h(y^0_{\pi(i)})-h(y^0_{\pi(i-1)})|+|h(y^0_{\pi(n)})-h(z^0)|\Big)^{-n-1}.$$

For notational convenience we have set $y_0^0=x^0$ and $\pi(0)=0$. Because of our assumptions on α we get that $f(\alpha)$ equals

$$\frac{1}{n!}\sum_{\pi\in\mathrm{Sym}(n)}\mathrm{sgn}(x^0-y^0_{\pi(1)})\Big(e^{\alpha(0)}\sum_{i=1}^{n}|y^0_{\pi(i)}-y^0_{\pi(i-1)}|+z^0-1-e^{\alpha(0)}y^0_{\pi(n)}+\int_0^1 e^{\alpha}\Big)^{-n-1}.$$

The formula for the analytic continuation of $f(\alpha)$ to $N_c'(\delta/2)$ is of course the same.

We can choose $\alpha\in N_c'(\delta/2)$ so that $\int_0^1 e^{\alpha}=1$ and $\alpha(0)\in\mathbf{C}$ is anything we want, subject only to the constraint $|\alpha(0)|\le c\delta\log n$, where c is an absolute constant. If $c\delta\log n\ge 1$, we can take $\alpha(0)=-1$. This leads to a contradiction as follows.

Under these restrictions on α we have

$$f(\alpha)=\frac{1}{n!}\sum_{\pi\in\mathrm{Sym}(n)}\mathrm{sgn}(x^0-y^0_{\pi(1)})\Big(-\sum_{i=1}^{n}|y^0_{\pi(i)}-y^0_{\pi(i-1)}|+z^0+y^0_{\pi(n)}\Big)^{-n-1}. \tag{1.7}$$

We can choose x^0 and y_i^0 as in Lemma 1.5 so that $x^0<y_i^0$ for $i=1,\cdots,n$ and also

$$-\sum_{i=1}^{n}|y_i^0-y_{i-1}^0|+y_n^0=-\frac{3}{2}.$$

This produces a singularity at $z^0 = \frac{3}{2}$ in the term in (1.7) coming from $\pi =$ identity. We must check that we don't destroy the singularity when we sum over π. If we fix x^0, y_i^0, and α as above, and view $f(\alpha)$ as a function of z^0, we get

$$\frac{1}{n!}\sum_{\pi\in\mathrm{Sym}(n)} (-1)\cdot(z^0 - a_\pi)^{-n-1},$$

where $a_\pi \in \mathbf{R}$ for each $\pi \in \mathrm{Sym}(n)$. Clearly the singularities cannot cancel out.

Thus if $c\delta \log n \geq 1$, then $f(\alpha)$ does not analytically continue to $N_c'(\delta/2)$. Theorem 1′ now follows from Lemma 1.5.

In the introduction we raised the issue of finding necessary conditions on a multilinear operator T so that when we conjugate it by $V_h, h' = e^\alpha$, we get a real-analytic function of $\alpha \in \mathrm{BMO}_r$. Notice that Lemma 1.5 applies in this more general context, and its proof implies that the kernel $K(y_1, \cdots, y_n, z)$ of T must be real-analytic on the set where $y_1, \cdots, y_n$, and z are pairwise distinct and none of them are zero.

We can also get some information about what happens near the singular set. Suppose $n = 1$, so that T is a bilinear operator with kernel $K(y, z)$. As before, the kernel of $V_h T(V_h^{-1}a, V_h^{-1}f)$ is

$$K(h(x) - h(y), h(x) - h(z))h'(y)h'(z).$$

Suppose that $z < y < x$, and y is closer to x than to z. Because K is homogeneous of degree -2, we can rewrite this as

$$K\left(\frac{h(x) - h(y)}{h(x) - h(z)}, 1\right)h'(y)h'(z)(h(x) - h(z))^{-2}.$$

Let $g(u) = K(e^u, 1), u \in \mathbf{R}$. In order that $T_\alpha - T_\alpha(a, f) = V_h T(V_h^{-1}a, V_h^{-1}f), h' = e^\alpha$ – be real-analytic in α, it is necessary that $g(u)$ have a holomorphic extension to $\{w = u + iv : u \leq -N, |v| \leq \eta|u|\}$ for some $N, \eta > 0$ (depending on the radius of convergence of T_α). This is proved using a minor variation of the proof of Lemma 1.5.

It is not clear however exactly which g's arise from T's as above. One can show that g can't be too badly behaved as $u \to -\infty$, and the arguments in the next section show that the g's which are well-behaved are allowed.

2.

The idea behind the proof of Theorem 2 is roughly the reverse of its counterpart for Theorem 1′. We represent $J_{n,\alpha}$ in terms of integration against a kernel, and we show that this has a suitable analytic continuation to α complex. We obtain bounds on the operator norm of this analytic continuation using the $T(b)$ theorem. At that point we are essentially finished.

Let us commit a small abuse of notation by letting $J_n(b_1, \cdots, b_n, f)$ denote the multilinear operator with kernel j_n. Thus $J_n(b_1, \cdots, b_n, f)$ is given by a formula like (1.2), but with the b's on the right side not all the same. By definition $J_n(b, f) = J_n(b_1 \cdots b_n, f)$, but $J_n(b_1, \cdots, b_n, f)$ is not the same as $\tilde{J}_n$ because it is not symmetric in the b's. As always, we set

$$J_{n,\alpha}(b_1, \cdots, b_n, f) = V_h J_n(V_h^{-1}b_1, \cdots, V_n^{-1}b_n, V_n^{-1}f),$$

$h' = e^{\alpha}$. We must show that this is real-analytic in α.

By definitions,

(2.1)

$$J_{n,\alpha}(b_1, \cdots, b_n, f)(x) =$$

$$\int_{\mathbf{R}}\int_{\mathbf{R}^n} \operatorname{sgn}(x - y_1) K_{n,\alpha}(x, y_1, \cdots, y_n, z) h'(y_1) \cdots h'(y_n) h'(z) b_1(y_1) \cdots b_n(y_n) \times f(z) dy dz,$$

with $K_{n,\alpha}(x, y_1, \cdots, y_n, z) = \left(\sum_{i=1}^{n+1}(h(y_i) - h(y_{i-1}))\operatorname{sgn}(y_i - y_{i-1})\right)^{-n-1}$. We have set $y_0 = x$ and $y_{n+1} = z$ in this last formula for notational convenience. The z-integral in (2.1) should be viewed as a principal value integral with singularity at $z = x$.

This formula for $J_{n,\alpha}$ still makes sense at least formally if α is complex. We have to check that no singularities were created if $\|\alpha\|_\star$ is small enough.

Given $x, y_1, \cdots, y_n$, and z, let $I = [a, b]$ denote the smallest interval that contains these points. Set $h(I) = h(b) - h(a)$. If $\|\alpha\|_\star \log n$ is small enough, then

$$|\sum_{i=1}^{n+1}(h(y_i) - h(y_{i-1}))\operatorname{sgn}(y_i - y_{i-1})| \geq \frac{1}{2}|h(I)|. \tag{2.2}$$

Let us sketch the proof. If $|y_i - y_{i-1}| \leq \frac{1}{100n}|I|$, then this term does not have a significant contribution. When the reverse inequality holds,

$$\frac{h(y_i) - h(y_{i-1})}{y_i - y_{i-1}} \quad \text{and} \quad \frac{h(I)}{|I|}$$

must be close to each other, because $\|\alpha\|_\star \log n$ is small. This prevents significant cancellation from occurring in the left side of (2.2).

Thus, even if α is complex, as long as $\|\alpha\|_\star \log n$ is small enough, $K_{n,\alpha}$ is well-behaved. In particular, if α is purely imaginary,

$$\begin{aligned} |K_{n,\alpha}| &\leq C|I|^{-n-1} \\ |\nabla K_{n,\alpha}| &\leq C|I|^{-n-2}, \end{aligned} \tag{2.3}$$

where I is as above and $\nabla K_{n,\alpha}$ denotes the gradient in all variables. These are the natural analogues of the standard estimates for a kernel of a Calderón-Zygmund operator, but in the multilinear setting.

Let us show that $J_{n,\alpha}$ maps $L^\infty \times \cdots \times L^\infty \times L^2$ into L^2. Assume first that α is purely imaginary. The estimates for $J_{n,\alpha}$ are automatically the same as those for $J'_{n,\alpha}$, where

$$\begin{aligned} &J'_{n,\alpha}(b_1, \cdots, b_n, f)(x) \\ &= \int_{\mathbf{R}}\int_{\mathbf{R}^n} \operatorname{sgn}(x - y_1) K_{n,\alpha}(x, y_1, \cdots, y_n, z) b_1(y_1) \cdots b_n(y_n) f(z) dy dz. \end{aligned}$$

(We've simply dropped the factors of h'.)

We shall obtain our L^2-estimate for $J'_{n,\alpha}$ using the $T(b)$ theorem [**DJS**] and a standard induction argument on n. A simple proof of the $T(b)$ theorem is given in [**CJS**]. Note, however, that we shall use the theorem when $b = e^\alpha$, α imaginary, $\|\alpha\|_\star$ small, which is much easier than the general case. Indeed, the primary technical difficulty in [**DJS**] goes away in this case.

Assume that we have the desired estimate on $J'_{n-1,\alpha}$ and let us prove it for $J'_{n,\alpha}$. (Note that the $n = 0$ case reduces to the Hilbert transform conjugated by a change of variables.)

Let $b_1, \cdots, b_n \in L^\infty$ be given, and let T denote the linear operator defined by $T(f) = J'_{n,\alpha}(b_1, \cdots, b_n, f)$. Thus $T(f)$ is given by

$$Tf(x) = PV \int_{-\infty}^{\infty} K(x,y) f(y) dy,$$

where $|K(x,y)| \leq C|x-y|^{-1}, |\nabla K(x,y)| \leq C|x-y|^{-2}$. This follows from (2.3) and the definition of $J'_{n,\alpha}$.

Let us check that T satisfies the assumptions for the $T(b)$ theorem with $b = e^\alpha$. Using Cauchy's theorem one can check that $T^t(e^\alpha) = 0$ and $T(e^\alpha) = J'_{n-1,\alpha}(b_1, \cdots, b_{n-1}, b_n)$. (These formulae can be viewed as analytic continuations of their counterparts in the $\alpha = 0$ case.) It is not difficult to check the weak boundedness property for T, i.e., that

$$| < Te^\alpha \theta_1, e^\alpha \theta_2 > | \leq C|I|^{-3} \|\nabla \theta_1\|_\infty \|\nabla \theta_2\|_\infty$$

whenever θ_1 and θ_2 are supported in I. Here $< f, g >= \int f(x) g(x) dx$. (The proof of this uses the fact that

$$\int_{\mathbf{R}} \operatorname{sgn}(x - y_1) K_{n,\alpha}(x, y_1, \cdots, y_n, z) e^{\alpha(x)} dx = 0$$

as long as $y_1, \cdots, y_n$ and z do not all coincide. This formula is obtained from viewing the integral as a contour integral and using complex variables.)

To conclude that T is bounded on L^2 using the $T(b)$ theorem we only need to know that $J'_{n-1,\alpha}(b_1, \cdots, b_{n-1}, b_n)$ lies in BMO. This follows from our induction hypothesis, our estimates on the kernel of $J'_{n-1,\alpha}$, and standard reasoning.

The L^2 estimate for $J_{n,\alpha}$ when α is complex can be reduced to the imaginary case as follows. Given α, $\|\alpha\|_\star$ small, we can write $h = h_0 \circ h_1$, where $h' = e^\alpha, \alpha_0 = \log h'_0$ is purely imaginary, $\alpha_1 = \log h'_1$ is real, and both have small BMO norm. By definitions

$$J_{n,\alpha}(b_1, \cdots, b_n, f) = V_{h_1} J_{n,\alpha_0}(V_{h_1}^{-1} b_1, \cdots, V_{h_1}^{-1} b_n, V_{h_1}^{-1} f). \tag{2.4}$$

Set $a_i = V_{h_1}^{-1} b_i$. Define U_{h_1} by $U_{h_1}(f) = (f \circ h_1)(h'_1)^{\frac{1}{2}}$, so that U_{h_1} is unitary. We can rewrite (2.4) as

$$U_{h_1} W J_{n,\alpha_0}(a_1, \cdots, a_n, W^{-1} U_{h_1}^{-1} f), \tag{2.5}$$

where W denotes the operator of multiplication by $(h'_1)^{\frac{1}{2}} \circ h_1^{-1}$.

The linear operator $f \mapsto J_{n,\alpha_0}(a_1, \cdots, a_n, f)$ is a Calderón-Zygmund operator composed with multiplication by h_0', which is unimodular. If we conjugate this operator by W, we still get a bounded operator on L^2. Because U_{h_1} is unitary on L^2, we conclude that (2.5) defines a bounded operator on L^2.

Thus we have proved that $J_{n,\alpha}$ maps $L^\infty \times \cdots \times L^\infty \times L^2$ if $\|\alpha\|_\star \log n$ is small enough, and with a uniform bound on the norm. The real-analyticity of $J_{n,\alpha}$ is proved in the same way as the analogous results for the weighted Hilbert transform and $C(\beta)$, as described In the introduction.

The arguments in this section can be applied to other multilinear operators besides J_n. The explicit form for the kernel of J_n is convenient, but not really needed; all we have to have is enough control on the kernel so that the analytic continuation to α complex can be made, and so that the kernel remains reasonably well-behaved under this continuation. Of course, the kernel must also satisfy a cancellation condition, but that is automatic if the kernel represents a bounded multilinear operator. Indeed, this condition forces the Fourier transform of the kernel to be bounded, which in turn ensures that the kernel has integral zero over the unit sphere in $\mathbf{R}^{n+1}$.

Although the methods of this section are fairly general, we know that they do not give the complete picture, since bilinear operators like $B(a, f) = aHf$ or $H(aHf)$ have perfectly good analytic continuations, but they cannot be handled as above because their kernels are too singular. Perhaps the complete picture is generated in some reasonable way by simple examples like these and the kernels that can be treated as above.

3.

The basic plan of the proof of Theorem 3 is similar to that of Theorem 2. We shall show that $K_{\alpha,-\alpha}$ admits an analytic continuation to small, complex α that is uniformly bounded. The bounds will be obtained by a variant of the $T(b)$ theorem.

The difficulty lies in controlling the kernel of $K_{\alpha,-\alpha}$. The problem is that the standard methods for dealing with the resolvent don't seem to work here.

For various reasons of convenience we write $L_\alpha = (e^\alpha De^{-\alpha}D)^{-\frac{1}{2}}(e^\alpha De^{-\alpha})$, which is the same as $K_{\frac{1}{2}\alpha,-\frac{1}{2}\alpha}$. L_α is given by the integral

$$L_\alpha = \int_0^\infty (t^2 e^\alpha De^{-\alpha}D + I)^{-1} e^\alpha tDe^{-\alpha} \frac{dt}{t}. \tag{3.1}$$

We view (3.1) as the definition of L_α for α complex. We shall show that (3.1) makes sense and defines a bounded operator on L^2 when $\|\alpha\|_\star$ is small enough.

To control the resolvent $R_t = (t^2 e^\alpha De^{-\alpha}D + I)^{-1}$ it is useful to split tD into two parts, one with high frequencies, the other with low frequencies. Given $t > 0$, we choose operators X_t and Y_t such that $X_tY_t = tD$,

$$(X_tf)^\wedge(\xi) = \chi(t\xi)\hat{f}(\xi), (Y_tf)^\wedge(\xi) = \eta(t\xi)\hat{f}(\xi),$$

$\chi(\xi) = \xi$ if $|\xi| \geq 1$, $\chi(\xi)$ is smooth and never vanishes. Thus $\chi(\xi)$ takes complex values on $[-1, 1]$. Set $Z_t = X_t^{-1}$.

Thus we have

$$R_t = (t^2 De^{-\alpha} D + e^{-\alpha})^{-1} e^{-\alpha} \tag{3.2}$$
$$= Z_t^\star (Y_t e^{-\alpha} Y_t^\star + Z_t e^{-\alpha} Z_t^\star)^{-1} Z_t e^{-\alpha}.$$

This is a useful formula, because Y_t and Z_t are much friendlier than tD.

Let us first show that $A_t = (Y_t e^{-\alpha} Y_t^\star + Z_t e^{-\alpha} Z_t^\star)^{-1}$ is well-defined. For this we compute as follows:

(3.3)
$$A_t = A_t e^{-\alpha} e^{\alpha} = (e^{\alpha} Y_t e^{-\alpha} Y_t^\star + e^{\alpha} Z_t e^{-\alpha} Z_t^\star)^{-1} e^{\alpha}$$
$$= (Y_t^\star Y_t^\star + Z_t Z_t^\star + [e^{\alpha}, Y_t] e^{-\alpha} Y_t^\star + [e^{\alpha}, Z_t] e^{-\alpha} Z_t^\star)^{-1} e^{\alpha}.$$

Set $M_t = Y_t Y_t^\star + Z_t Z_t^\star, B_t = [e^{\alpha}, Y_t] e^{-\alpha} Y_t^\star$, and $C_t = [e^{\alpha}, Z_t] e^{-\alpha} Z_t^\star$. Then M_t, M_t^{-1}, B_t, and C_t are all bounded on L^2, and the norms of B_t and C_t are dominated by $\|\alpha\|_\star$. Hence,

$$A_t e^{-\alpha} = (I + M_t^{-1}(B_t + C_t))^{-1} M_t^{-1} \tag{3.4}$$

is bounded on L^2 if $\|\alpha\|_\star$ is small enough.

We want to get control on the kernel of R_t, and in particular show that it is well localized. This is not hard to do, using (3.2) and (3.4). Let β denote the real part of α, and set $\beta_t(x) = (2t)^{-1} \int_{x-t}^{x+t} \beta$.

LEMMA 3.5. *Given $j > 0$, there is a $\delta(j) > 0$ so that if $\alpha \in N_c(\delta(j))$, then*

$$R_t(e^{\alpha} f)(x) = \int_{-\infty}^{\infty} K_t(x,y) f(y) dy, \quad f \in L^2,$$

where $K_t(x,y)$ is a continuous function of x and y and

$$|K_t(x,y)| \le C e^{\beta_t(y)} \frac{t^{j-1}}{(|x-y|+t)^j}.$$

By definition, $R_t(e^{\alpha} f) = Z_t^\star A_t Z_t(f)$. To prove the lemma, we must first show that A_t is well-localized. Let j and t be fixed, and let us say that an operator $T \in B(L^2)$ is localized with constant C if for each interval $I, |I| = t$, and each f supported in I we have

$$\left(\int_{I+\ell} |Tf(x)|^2 dx\right)^{\frac{1}{2}} \le C(1+|\ell|)^{-j} \|f\|_2 \tag{3.6}$$

for all ℓ. Let us show that $A_t e^{-\alpha}$ satisfies (3.6) if $\|\alpha\|_\star$ is small enough.

It is easy to check that M_t^{-1}, B_t, and C_t each satisfy (3.6), and that for B_t and C_t we can take the constant to be $C(j)\|\alpha\|_\star$. It can also be checked that if T_1 and T_2 are localized with constants C_1 and C_2, then $T_1 T_2$ is localized with constant $C(j) C_1 C_2$. The desired estimate for $A_t e^{-\alpha}$ now follows from these facts by expanding $A_t e^{-\alpha}$ into a Neumann series using (3.4).

Lemma 3.5 now follows easily from the fact that $A_t e^{-\alpha}$ satisfies (3.6), and also the nice behavior of Z_t, namely that $Z_t f = f \star \zeta_t, \zeta_t(x) = \frac{1}{t}\zeta(\frac{x}{t}), \zeta(x)$ is smooth except at 0 where it has a jump discontinuity and $\zeta(x)$ decays rapidly at ∞.

We could again use an operational approach to obtain smoothness estimates on $K_t(x, y)$, but we are better off using the differential equations that $K_t(x, y)$ satisfies by dint of being the kernel of $R_t e^{\alpha}$; to wit,

$$(t^2 e^{\alpha} D_x e^{-\alpha} D_x + 1) K_t(x, y) = 0 \tag{3.7}$$

and

$$(t^2 e^{\alpha} D_y e^{-\alpha} D_y + 1) K_t(x, y) = 0 \tag{3.8}$$

when $x \neq y$, because

$$(t^2 e^{\alpha} D e^{-\alpha} D + I) R_t = R_t (t^2 e^{\alpha} D e^{-\alpha} D + I) = I.$$

Let us use Lemma 3.5, (3.7), and (3.8) to get control on the kernel of L_α. Let $H_t(x, y)$ denote the kernel of the integrand in (3.1), so that

$$H_t(x, y) = -e^{-\alpha(y)} t D_y K_t(x, y). \tag{3.9}$$

Notice that we do not pick up Dirac masses in $D_y K_t(x, y)$ at $x = y$ because $K_t(x, y)$ is continuous there.

LEMMA 3.10. *Under the same hypotheses as in Lemma 3.5,*

$$|H_t(x, y)| \leq C \frac{t^{j-1}}{(|x - y| + t)^j} \tag{3.11}$$

$$|t D_y H_t(x. y)| \leq C e^{\beta_t(y) - \beta(y)} \frac{t^{j-1}}{(|x - y| + t)^j}, \quad x \neq y \tag{3.12}$$

$$|t D_x H_t(x, y)| \leq C e^{\beta(x) - \beta_t(x)} \frac{t^{j-1}}{(|x - y| + t)^j}, \quad x \neq y. \tag{3.13}$$

The second inequality follows from (3.9), (3.8), and Lemma (3.5). Let us check (3.11). The idea is to play off (3.12) against (3.9) and Lemma 3.5. Let y be given. Assume that $x \notin [y, y + t]$; otherwise, replace $y + t$ by $y - t$ in the argument below. From (3.12) we have that

$$|H_t(x, u) - H_t(x, y)| \leq C \frac{t^{j-1}}{(|x - y| + t)^j} \quad \text{if} \quad u \in [y, y + t].$$

Using this, (3.9), and Lemma 3.5 we get

$$\begin{aligned}
C e^{\beta_t(y)} \frac{t^{j-1}}{(|x - y| + t)^j} &\geq |K_t(x, y + t) - K_t(x, y)| \\
&= \left| \int_y^{y+t} \frac{1}{t} e^{\alpha(u)} H_t(x, u) du \right| \\
&\geq \left| \int_y^{y+t} \frac{1}{t} e^{\alpha(u)} H_t(x, y) du \right| - \int_y^{y+t} C e^{\beta(u)} \frac{t^{j-2}}{(|x - y| + t)^j} du \\
&\geq |H_t(x, y)| \left| \frac{1}{t} \int_y^{y+t} e^{\alpha(u)} du \right| - C e^{\beta_t(y)} \frac{t^{j-1}}{(|x - y| + t)^j}.
\end{aligned}$$

From this we get (3.11).

The proof of (3.13) is similar to the proof of (3.11). From (3.7) and (3.9), or just the definition of $H_t(x,y)$, we have

$$t^2 e^{\alpha} D_x e^{-\alpha} D_x H_t(x,y) = -H_t(x,y) \quad \text{when} \quad x \neq y. \tag{3.14}$$

Thus we have control on $H_t(x,y)$ and $D_x e^{-\alpha} D_x H_t(x,y)$, and the desired estimate for $e^{-\alpha(x)} D_x H_t(x,y)$ is obtained by playing these two off each other, as before. This proves Lemma 3.10.

Note that (3.12) and (3.13) imply in particular that

$$\begin{aligned} &|H_t(x,y) - H_t(x,y')| \le C\left(\frac{|y-y'|}{t}\right)^{\frac{1}{2}} \frac{t^{j-1}}{(|x-y|+t)^j}, \quad x \notin [y,y'], \\ &|H_t(x,y) - H_t(x',y)| \le C\left(\frac{|x-x'|}{t}\right)^{\frac{1}{2}} \frac{t^{j-1}}{(|x-y|+t)^j}, \quad y \notin [x,x'], \end{aligned} \tag{3.15}$$

as long as $|x-x'| \le t, |y-y'| \le t$. Thus if $L_\alpha(x,y) = \int_0^\infty H_t(x,y)\frac{dt}{t}$ for $x \neq y$, then

$$\begin{aligned} &|L_\alpha(x,y)| \le C|x-y|^{-1} \\ &|L_\alpha(x,y) - L_\alpha(x,y)| \le C\frac{|y-y'|^{\frac{1}{2}}}{|x-y|^{\frac{3}{2}}}, \quad |y-y'| \le \frac{1}{2}|x-y|, \\ &|L_\alpha(x,y) - L_\alpha(x',y)| \le C\frac{|x-x'|^{\frac{1}{2}}}{|x-y|^{\frac{3}{2}}}, \quad |x-x'| \le \frac{1}{2}|x-y|. \end{aligned} \tag{3.16}$$

This follows from (3.11) and (3.15), with $j = 2$.

We want to show that L_α is bounded on L^2 using the $T(b)$ theorem. We shall apply the $T(b)$ theorem with $b_1 = e^\alpha$ and $b_2 = e^{-\alpha}$. When α is not purely imaginary, this does not fit exactly into the usual statement of the $T(b)$ theorem, and we'll address that issue shortly. Let's first check that L_α satisfies the hypotheses of the $T(b)$ theorem.

The kernel $L_\alpha(x,y)$ of L_α satisfies the standard estimates by (3.16). The weak boundedness property relative to b_1, b_2 also holds for L_α; i.e., given θ_1, θ_2 supported in I,

$$|\langle L_\alpha(b_1\theta_1), b_2\theta_2\rangle| \le C|I|^{-3}\|\nabla\theta_1\|_\infty\|\nabla\theta_2\|_\infty,$$

where $\langle f,g\rangle = \int f(x)g(x)dx$. This follows from (3.1), (3.11), and the fact that $\int_{-\infty}^\infty H_t(x,y)b_1(y)dy = 0$, which is itself a consequence of (3.9). This also gives $L_\alpha(b_1) = 0$.

We need to compute $L_\alpha^t(b_2)$. Let θ be a compactly supported Lipschitz function such that $\int \theta b_1 = 0$. Then

$$\begin{aligned} \langle L_\alpha^t(b_2), b_1\theta\rangle &= \langle b_2, L_\alpha(b_1\theta)\rangle \\ &= \int_{\mathbf{R}}\int_0^\infty\int_{\mathbf{R}} H_t(x,y)b_1(y)\theta(y)dy\frac{dt}{t}b_2(x)dx \\ &= \int_0^\infty\int_{\mathbf{R}}\int_{\mathbf{R}} H_t(x,y)b_1(y)\theta(y)dyb_2(x)dx\frac{dt}{t} \\ &= \int_0^\infty\int_{\mathbf{R}}\int_{\mathbf{R}} K_t(x,y)t\theta'(y)dyb_2(x)dx\frac{dt}{t}. \end{aligned} \tag{3.17}$$

(Our various estimates on $H_t(x,y)$ and $K_t(x,y)$ and our assumptions on θ can be used to show that all these integrals converge, and to justify the use of Fubini's theorem.) Because

$$(t^2 e^{\alpha} D_x e^{-\alpha} D_x + I)R_t = I,$$

we have

$$t^2 D_x e^{-\alpha} D_x K_t(x,y) = -e^{-\alpha(x)} K_t(x,y) + \delta_y(x), \tag{3.18}$$

where $\delta_y(x)$ denotes the Dirac mass at y. This equation should be interpreted in the sense of distributions: $D_x K_t(x,y)$ lies in L^2, and $D_x e^{-\alpha} D_x K_t(x,y)$ denotes the distributional derivative of $e^{-\alpha(x)} D_x K_t(x,y)$. From (3.18) we get

$$\int_{\mathbf{R}} \int_{\mathbf{R}} e^{-\alpha(x)} K_t(x,y) t\theta'(y) dx dy = \int_{\mathbf{R}} t\theta'(y) dy = 0.$$

Inserting this into (3.17) gives $L^t_\alpha(b_2) = 0$.

We would like to invoke the $T(b)$ theorem to conclude that L_α is bounded on L^2. Our situation does not quite fit into [**DJS**], because e^α is not bounded. It does satisfy a pseudo-accretivity condition: if P_t is any reasonable approximation to the identity on $\mathbf{R}$, then $|P_t(e^\alpha)| \approx P_t(e^\beta)$, where $\beta = Re\ \alpha$ as before. Thus e^α could be said to be pseudo-accretive relative to the weight e^β, rather than to 1. Similarly, $e^{-\alpha}$ is pseudo-accretive relative to $e^{-\beta}$.

Arguments completely analogous to those in [**DJS**] allow us to conclude that L_α is bounded on L^2 in this case. The proof is in fact much easier than the general result in [**DJS**], because the assumption $\|\alpha\|_*$ small makes life much easier than in the general pseudo-accretive case.

Of course, it is even easier to adapt the proof of the $T(b)$ theorem given in [**CJS**] to this case.

Thus the L^2-operator norm of L_α is bounded when $\alpha \in N_c(\delta)$ for δ small enough. As usual, the real-analyticity of L_α can be obtained from this and the computation of a formal power series for L_α.

As we indicated In the introduction, Theorem 3 can be extended to show that $(ae^\alpha De^{-\alpha} bD)^{\frac{1}{2}} D^{-1}$ is real-analytic at $(a, \alpha, b) = (1,0,1)$ in $L^\infty \times$ BMO $\times L^\infty$. The proof is essentially the same as the one above.

References

[CJS] R. Coifman, P. Jones, and S. Semmes, *Two elementary proofs of the L^2 boundedness of Cauchy integrals on Lipschitz curves*, to appear.

[CM] R. Coifman and Y. Meyer, *Au delá des opérateurs pseudo-differentiels*, Astérisque **57**(1978), Société Mathématique de France.

[CMM] R. Coifman, A. McIntosh, and Y. Meyer, *L'intégrale de Cauchy définit un opérateur borné sur L^2 pour les courbes lipschitziennes*, Ann. Math. **116** (1982), 361–387.

[CRW] R. Coifman, R. Rochberg, and G. Weiss, *Factorization theorems for Hardy spaces in several complex variables*, Ann. Math. **103** (1976), 611–635.

[DJS] G. David, J.L. Journé and S. Semmes, *Opérateurs de Calderón-Zygmund, fonctions para-accretive, et interpolation*, Revista Mat. Iberoamericana **1, No. 4** (1985), 1–56.

[HP] E. Hille and R. Phillips, "Functional Analysis and Semigroups," Amer. Math. Soc. Colloquium Publications, 1957.

[KM] C. Kenig and Y. Meyer, *Kato's square roots of accretive operators and Cauchy kernels on Lipschitz curves are the same*, Proceedings of the conference at El Escorial, Spain (**1983**), pages 123-143 (North-Holland Math. Stud. **111**), in "Recent Progress in Fourier Analysis."

5

Strong A_∞ Weights, Sobolev Inequalities and Quasiconformal Mappings

Guy David and Stephen Semmes

Introduction.

An A_∞ weight ω on $\mathbf{R}^n$ is called strongly A_∞ if its measure distance is equivalent in size to the associated geodesic distance. The first is defined by

$$\delta(x,y) = \left\{ \int_{B_{x,y}} \omega(u)du \right\}^{1/n}, \tag{0.1}$$

where $B_{x,y}$ denotes the ball with diameter $|x-y|$ that contains x and y. The geodesic distance associated to ω is defined to be the length of the shortest curve joining a given pair of points, where "length" is defined in terms of $\delta(x,y)$. This is the same as the geodesic distance associated to the Riemannian metric on $\mathbf{R}^n$ obtained by multiplying the standard metric by $\omega^{2/n}$ (so that the associated volume element is just $\omega(u)du$), except for a multiplicative constant. More precise definitions are given below.

In the first section we discuss some situations in which the strong-A_∞ condition arises. For instance, the Jacobian of a quasiconformal homeomorphism on $\mathbf{R}^n$ is always strongly A_∞, by an argument of Gehring. We do not know if any reasonable converse to this statement holds. A necessary condition for a weight ω to be the Jacobian of a q.c. mapping is that certain Sobolev and Poincaré inequalities hold, to wit, the inequalities with respect to ω that can be reduced to the standard case using the q.c. map as a change of variables. In Sections 2 and 3 we show that these inequalities always hold for a strongly A_∞ weight.

An attractive feature of our approach is the geometrical information it gives. We shall show that for proving these Sovolev and Poincaré inequalities relative to a strongly A_∞ weight there is a version of the classical argument that controls $(g(x)-g(y))$ in terms of ∇g by connecting x to y by a nice family of curves, integrating ∇g on each curve, and averaging the result over the whole family. The main point is to find this nice family of curves. If the weight were the Jacobian of a q.c. map, this family could be obtained from the standard situation by a change of variables.

1. Strong A_∞–weights.

Recall that a positive, locally integrable function $\omega(x)$ on $\mathbf{R}^n$ is called an A_∞ weight if for each $\epsilon > 0$ there is a $\delta > 0$ so that if $Q \subseteq \mathbf{R}^n$ is any cube and $E \subseteq Q$ satisfies $|E| \leq \delta|Q|$, then $\omega(E) \leq \epsilon\omega(Q)$. Here $|E|$ denotes the Lebesque measure of E, and $\omega(E) = \int_E \omega(x)dx$. The main point in this condition is that δ does not depend on Q; this is a uniform absolute continuity condition. See [**GR**] or [**J**] for basic properties of A_∞ weights.

Let us be more precise about the definition of strong-A_∞ indicated in the introduction. Given an arc $\gamma(t), 0 \leq t \leq 1$, we would like to define its ω–length to be

$$\int_0^1 \omega(\gamma(t))^{1/n} \mid d\gamma(t) \mid,$$

The second author is partially supported by the NSF and the Sloan Foundation

where $|d\gamma(t)|$ denotes arclength measure on γ. This would be fine if ω were continuous, but to avoid technical problems it is better to take a slightly different tack.

Let us define the ω-length of $\gamma(t)$ by

$$\omega\text{-length of } \gamma = \limsup\{\sum \delta(\gamma(t_{i+1}), \gamma(t_i))\}, \tag{1.1}$$

where $\{t_i\}$ is any partition of $[0,1]$, and $\delta(\cdot,\cdot)$ is as in (0.1), and the lim sup is taken as the partition becomes finer and finer. This agrees with the above when ω is continuous, except for a multiplicative constant. More generally, we could define the "length" of a set using a standard Hausdorff measure construction.

We define the geodesic distance $d_\omega(x,y) = d(x,y)$ associated to ω to be the infimum of the lengths of all curves that join x to y. If ω is an A_∞ weight, it is not hard to check that

$$d(x,y) \leq C\delta(x,y). \tag{1.2}$$

This is proved by taking the average ω-length of a suitable family of curves joining x to y, and using the fact that ω satisfies a reverse Hölder inequality to control this average.

We say that ω is strongly A_∞ if the reverse inequality holds, i.e.,

$$\delta(x,y) \leq Cd(x,y) \qquad \text{for all } \ x,y \in \mathbf{R}^n. \tag{1.3}$$

This is equivalent to requiring that

$$\omega(B)^{1/n} \leq C(\omega\text{-length of } \gamma) \tag{1.4}$$

for any curve γ contained in the ball B with diam γ = diam B.

The relation between strong-A_∞ and the other A_p classes is as follows. Every A_1 weight is strongly A_∞, but for each $p > 1$ there is an A_p weight which is not strongly A_∞. Indeed, $|x_1|^\alpha$ is not strongly-A_∞ for any $\alpha > 0$, as one can see by applying (1.4) with γ contained in the x_2-axis. Conversely, for arbitrarily large p there is a strongly A_∞ weight that is not A_p, namely $|x|^\alpha$ for α large enough.

Note that a doubling measure that satisfies (1.4) is automatically an A_∞ weight. Indeed, Gehring proved that (1.4) implies that ω satisfies a reverse Hölder inequality.

Let us briefly indicate why the Jacobian ω of a q.c. map h on $\mathbf{R}^n$ is always strongly A_∞. In this case we have

$$\omega(B_{x,y}) \approx |h(x) - h(y)|^n$$

from a standard distortion theorem and a change of variables, where $B_{x,y}$ is as in the introduction. On the other hand, if γ joins x to y, then the ω-length of γ dominates $|h(x) - h(y)|$, because $\omega^{1/n} \approx |\nabla h|$.

There are other situations where strong-A_∞ weights arise. Suppose that ρ is a quasisymmetric embedding of $\mathbf{R}^n$ into $\mathbf{R}^m, n < m$, and that its image $\Gamma = \rho(\mathbf{R}^n)$ satisfies

$$C^{-1}R^n \leq H_n(\Gamma \cap B(y,R)) \leq CR^n, \tag{1.5}$$

where $B(y,R)$ is the ball with center y and radius $R, y \in \Gamma$, and H_n denotes n–dimensional Hausdorff measure. An argument like Gehring's for the $m = n$ case can again be used to show that the Jacobian is strongly A_∞.

Such an embedding ρ always exists if $n = 2$ and if Γ satisfies (1.5), certain a priori smoothness assumptions, and also has the property that for each $y \in \Gamma$, and $R > 0$ there is a topological disk D such that $B(y,R) \cap \Gamma \subseteq D \subseteq B(y, CR) \cap \Gamma$. This is proved in [**Se**]. These hypotheses hold in particular for the so-called "chord–arc surfaces with small constant" which are two–dimensional.

A more general example is given by the ω–regular mappings considered in [**D**]. Given $\omega \in A_\infty$, we say that $\rho : \mathbf{R}^n \to \mathbf{R}^m$ is ω–regular if

$$(1.6) \qquad |\rho(x) - \rho(y)| \leq C d_\omega(x,y) \qquad \text{for all } \; x, y \in \mathbf{R}^n, \text{ and}$$

$$(1.7) \qquad \omega(\rho^{-1}(B(z,R))) \leq CR^n \qquad \text{for all } \; z \in \mathbf{R}^m, R > 0.$$

One can show that (1.6) is equivalent to the assumption that $\nabla\rho \in L^1_{\text{loc}}$ and $|\nabla\rho| \leq \omega^{1/n}$ a.e.

PROPOSITION 1.8. *If $\omega \in A_\infty$ and if an ω–regular map exists, then ω is strongly A_∞.*

It would be very interesting to know if the converse is true.

Using the usual rescaling arguments we are reduced to showing that if γ is an arc with diameter $\geq \frac{1}{2}$ which is contained in the unit ball, and if $\omega(B(0,1)) = 1$, then $\eta =$ (the ω–length of γ) is not too small.

Suppose to the contrary that η is very small. From (1.6) we have that diam $(\rho(\gamma)) \leq C\eta$. Given $A > 0$, if we let T denote the tube $\{z \in \mathbf{R}^n : d_\omega(z,y) \leq A\eta \text{ for some } y \in \gamma\}$, then diam $\rho(T) \leq C(A+1)\eta$, so that $\omega(T) \leq C(1 + A^n)\eta^n$ by (1.7). We want to get a lower bound on $\omega(T)$ that contradicts this if η is too small.

For each $\beta < \frac{1}{2}$ we can find more that $N_\beta = (3\beta)^{-1}$ points x_i on γ which are at least (Euclidian) distance β apart. For each i let B_i be the ball centered at x_i with $\omega(B_i) = \eta^n$. From (1.2) we get $d(y, x_i) \leq C\delta(y, x_i) \leq C\eta$ if $y \in B_i$. Thus $\cup B_i \subseteq T$ if A is reasonably large.

If η is small enough (how small depending only on β), then the A_∞ condition on ω forces the B_i's to have radius $< \frac{1}{2}\beta$, so that they are disjoint. In this case we have

$$\omega(T) \geq \omega(\cup B_i) \geq \sum \omega(B_i) = \eta^n \, N_\beta.$$

Our earlier estimate gives $\omega(T) \leq C\eta^n$, with C independent of η. This is a contradiction if β is small enough. Thus η couldn't be too small, which proves the proposition.

Let $\mathcal{W}, \mathcal{W}_1$, and $\mathcal{W}_2$ denote the class of strongly A_∞ weights, the subclass of weights for which there exists an ω–regular mapping, and the subclass for which there is a q.c. mapping whose Jacobian is equivalent in size to ω. Thus $\mathcal{W} \supseteq \mathcal{W}_1 \supseteq \mathcal{W}_2$, and we do not know if any of the inclusions are strict. Note that if $\mathcal{W}_1 = \mathcal{W}_2$, then for any ω–regular mapping ρ the image $\rho(\mathbf{R}^n)$ can be reparameterized by a 1–regular mapping.

We do not even know the answer to any of these questions in their "small constant" versions. This is closely related to the question of whether chord–arc surfaces with small constant admit bilipschitz parameterizations, or just quasisymmetric ones when the dimension of the surface is larger than 2.

One way to try to distinguish between these classes of weights would be to show that they have different properties vis–à–vis Sobolev estimates, for instance. If ω is comparable to the Jacobian of a q.c. map on $\mathbf{R}^n$, then a change of variables allows one to obtain the Sobolev inequality

$$(1.9)\qquad \Big(\int_{\mathbf{R}^n} | g(x) |^q \, \omega(x)dx\Big)^{1/q} \leq C\Big(\int_{\mathbf{R}^n} (\omega(x)^{-1/n} \mid \nabla g(x) \mid)^p \omega(x)dx\Big)^{1/p},$$

$1 \leq p < n, \frac{1}{q} = \frac{1}{p} - \frac{1}{n}$, and also the Poincaré inequality

$$(1.10)\qquad \omega(B)^{-\frac{(n+1)}{n}} \int_B \int_B | g(x) - g(y) | \, \omega(x)\omega(y)dxdy \leq C \int_{2B} (\omega(x)^{-1/n} \mid \nabla g(x) \mid) \, \omega(x)dx.$$

for, say, C^1 functions with compact support, where C does not depend on g or B, of course.

It turns out that (1.9) and (1.10) do not allow us to distinguish between $\mathcal{W}, \mathcal{W}_1$, and $\mathcal{W}_2$.

THEOREM. *If ω is a strong-A_∞ weight, then (1.9) and (1.10) hold.*

A remarkable feature of this result is that the right sides of (1.9) and (1.10) can be finite without ∇g being locally integrable. This is reminiscent of the fact that if f is a holomorphic function on the unit disk, or a Riesz system on the ball, then $\int_{\partial B} |f|\omega$ can control f on the interior for certain ω's even when $f \notin L^1(\partial B)$. This is a consequence of the subharmonicity of $|f|^p$ for some $p < 1$.

Thus we are still left with the problem of finding analytical invariants of weights that might distinguish between $\mathcal{W}, \mathcal{W}_1$, and $\mathcal{W}_2$. This is related to the problem of finding analytical invariants of surfaces—e.g., chord–arc surfaces with small constants—that detect whether they are bilipschitz images of $\mathbf{R}^n$.

Coifman raised the following question. Given a function h, is there some way of generating an almost-smallest strongly-A_∞ weight bigger than h? For A_1 weights, you could take $((h^p)^*)^{1/p}$, $p > 1$.

The proof of the theorem is given in the next two sections. We first make some reductions that are fairly standard, and then we carry out the main step. The two sections can be read in any order.

2. Reductions.

We can assume that ω is smooth, using the obvious limiting argument.

We shall deduce (1.9) and (1.10) from

$$(2.1) \qquad | g(x) - g(y) | \leq C \int_{B_{x,y}} \{\omega(B_{x,u})^{-\frac{(n-1)}{n}} + \omega(B_{y,u})^{-\frac{(n-1)}{n}}\} \mid \nabla g(u) \mid \omega(u)^{\frac{(n-1)}{n}} du.$$

When $w \equiv 1$, this reduces to the familiar Riesz potential.

It is not hard to derive (1.10) from (2.1);

$$\int_B \int_B | g(x) - g(y) | \omega(x)\omega(y)dxdy$$

$$\leq C \iiint_{\substack{x\in B\\ y\in B\\ u\in B_{x,y}}} \{\omega(B_{x,u})^{-\frac{(n-1)}{n}} + \omega(B_{y,u})^{-\frac{(n-1)}{n}}\} \mid \nabla g(u) \mid \omega(u)^{\frac{n-1}{n}} \omega(x)\omega(y)dxdydu$$

$$\leq C \int_{2B} \left\{ \int_B \int_B \omega(B_{x,u})^{-\frac{(n-1)}{n}} \omega(x)\omega(y)dxdy \right\} \mid \nabla g(u) \mid \omega(u)^{\frac{n-1}{n}} du.$$

For each u we have

$$\int_B \omega(B_{x,u})^{-\frac{(n-1)}{n}} \omega(x)dx \leq C \sum_{k\geq 0} \int_{\substack{x\in B\\ \omega(B_{x,u})\approx 2^{-k}\omega(B)}} 2^{k(n-1)/n}\omega(B)^{-\frac{(n-1)}{n}} \omega(x)dx$$

$$\leq C\omega(B)^{-\frac{(n-1)}{n}} \sum_{k\geq 0} 2^{k(n-1)/n}\, 2^{-k}\omega(B) = C\omega(B)^{1/n}.$$

Substituting this into the above gives (1.10).

Consider now (1.9) when $p > 1$. Define an "ω–Riesz potential of order 1" by

$$R_\omega f(x) = \int_{\mathbf{R}^n} \omega(B_{x,u})^{-\frac{(n-1)}{n}} f(u)\omega(u)du.$$

Just as in the standard $\omega \equiv 1$ case we have

$$(2.2) \qquad \left(\int | R_\omega f |^q \omega\right)^{1/q} \leq C(p)\left(\int | f |^p \omega\right)^{1/p},$$

$1 < p < n, \frac{1}{q} = \frac{1}{p} - \frac{1}{n}$, whenever ω is a doubling measure. For strong A_∞ weights we also have

$$(2.3) \qquad | g(x) | \leq CR_\omega(\omega^{-\frac{1}{n}} \mid \nabla g \mid)(x)$$

if g is compactly supported, because of (2.1). Thus (1.9) follows from (2.2) and (2.3) when $p > 1$.

Let us prove (1.9) when $p = 1$. We shall derive it from an isoperimetric inequality, which will itself be obtained from (1.10). The isoperimetric inequality is

$$\int_\Omega \omega(x)dx \leq C\Big(\int_{\partial\Omega} \omega(y)^{\frac{n-1}{n}} dH_{n-1}(y)\Big)^{\frac{n}{n-1}}, \tag{2.4}$$

where Ω is any bounded open set, and $dH_{n-1}(y)$ denotes $(n-1)$–dimensional Hausdorff measure. One should think of the right hand side as being

$$\Big(\int \mid \nabla\chi_\Omega \mid \omega^{\frac{n-1}{n}}\Big)^{\frac{n}{n-1}}.$$

Let us prove (1.9) when $p = 1$ using (2.4).

We may assume that $g \geq 0$; otherwise, replace g by $|g|$. We can write

$$g(x) = \int_0^\infty \chi_{R_t}(x)dx,$$

where $R_t = \{x : g(x) > t\}$. The co–area formula (see [**Gi**], p.20) gives

$$\mid \nabla g \mid = \int_0^\infty \mid \nabla\chi_{R_t} \mid dt,$$

where both sides are viewed as measures on $\mathbf{R}^n$. This allows us to reduce (1.9) for $p = 1$ to the case where $g = \chi_{R_t}$. In this case, (1.9) is the same as (2.4).

Now we want to obtain (2.4) from (1.8). The idea is to use a covering argument to reduce to the easy case where Ω is contained in a ball B such that $\omega(\Omega \cap B)$ and $\omega(\Omega^c \cap B)$ are both comparable to $\omega(B)$. In this case we can apply (1.10) to $g = \chi_\Omega$ to get (2.4).

Given any bounded open set Ω, we can certainly cover Ω by balls B such that

$$\omega(\frac{1}{2}B \cap \Omega) \geq C^{-1}\omega(B), \ \omega(\frac{1}{2}B \cap \Omega^c) \geq C^{-1}\omega(B). \tag{2.5}$$

A covering lemma of Vitali type (see [**St**], p.9) says that we can find a sequence of disjoint balls B_i satisfying (2.5) and also

$$\omega(\Omega) \leq C \sum_i \omega(B_i).$$

Applying (1.10) with $g = \chi_\Omega$ and $B = \frac{1}{2}B_i$ we get that

$$\omega(B_i)^{-\frac{(n+1)}{n}}\omega(\frac{1}{2}B_i \cap \Omega)\omega(\frac{1}{2}B_i \cap \Omega^c) \leq C \int\limits_{\partial\Omega\cap B_i} \omega(y)^{\frac{n-1}{n}} dH_{n-1}(y).$$

From (2.5) we now get

$$\begin{aligned}\omega(\Omega) &\leq C\sum_i \omega(B_i) \leq C\sum_i \Big\{ \int_{\partial\Omega\cap B_i} \omega(y)^{\frac{n-1}{n}} dH_{n-1}(y)\Big\}^{\frac{n}{n-1}} \\ &\leq C\Big\{ \sum_i \int_{\partial\Omega\cap B_i} \omega(y)^{\frac{n-1}{n}} dH_{n-1}(y)\Big\}^{\frac{n}{n-1}} \\ &\leq C\Big\{ \int_{\partial\Omega} \omega(y)^{\frac{n-1}{n}} dH_{n-1}(y)\Big\}^{\frac{n}{n-1}},\end{aligned}$$

which is (2.4).

Thus we are reduced to proving (2.1), which we do in the next section.

In dimension 2 we can obtain the isoperimetric inequality (2.4) directly from the definition of strong-A_∞ weights. It suffices to prove (2.4) when Ω is connected and simply connected. Let B be the smallest ball containing Ω. The ω-length of $\partial\Omega$ controls $\omega(B)^{\frac{1}{2}}$, because of the strong-A_∞ condition, and because $\partial\Omega$ is connected. This implies (2.4), since $\omega(\Omega) \leq \omega(B)$. Note that (1.9) can be derived from (2.4) even when $p > 1$, using well-known symmetrization arguments.

The preceeding argument seems to have no counterpart in higher dimensions. Even in dimension 2, the arguments of the next section seem to be more satisfactory and informative.

3. The Proof of (2.1).

As indicated in the introduction, we shall prove (2.1) by finding a nice family of curves joining x to y, integrating ∇g along these curves, and then averaging the result. We start with the 2-dimensional case, where the notations are simpler. We can take $x = (-1,0)$ and $y = (1,0)$.

If $\omega \equiv 1$, a good family of paths is obtained by connecting x to $(0,t)$, and then $(0,t)$ to y, both by line segments, for each $t \in (-1,1)$. This is not good enough for a general ω. Roughly speaking, we need to have more control on the ω-length of our curves. If ω were the Jacobian of a q.c. mapping $h, h(x) = x,\ h(y) = y$, then we could apply h^{-1} to the simple family of curves above to get a good curve family for ω.

Let D denote the square with vertices $(-1,0),(0,1),(1,0)$, and $(0,-1)$. We want to find a family of paths $\gamma_t, 0 \leq t \leq 1$, that lie in D, join x to y, and such that if μ_t denotes arclength measure on γ_t—viewed as a measure on $\mathbf{R}^2$—then

$$\int_0^1 \mu_t dt \leq C\{\omega(B_{x,u})^{-\frac{1}{2}} + \omega(B_{y,u})^{-\frac{1}{2}}\}\omega(u)^{\frac{1}{2}} du, \tag{3.1}$$

as an inequality between measures. If we can find such γ_t's, then (2.1) follows at once from

$$| g(x) - g(y) | \leq \int_0^1 \int_{\gamma_t} | \nabla g | \, d\mu_t dt$$

and using (3.1).

It is helpful to transform the problem of building the γ_t's into a problem about building functions on D. Suppose, for simplicity, that each γ_t is a simple curve, and let Ω_t denote the set of points beneath it. Define f on D by

$$f = \int_0^1 \chi_{\Omega_t}\, dt.$$

Let L^+ and L^- denote the pieces of ∂D that lie above and below the real axis, so that $f = 0$ on L^+ and $f = 1$ on L^-. Also, $|\nabla f|$ is a measure that satisfies

$$|\nabla f| \leq \int_0^1 |\nabla \chi_{\Omega_t}|\, dt = \int_0^1 \mu_t dt \leq C\{\omega(B_{x,u})^{-\frac{1}{2}} + \omega(B_{y,u})^{-\frac{1}{2}}\}\omega(u)^{\frac{1}{2}} du.$$

The converse of this is more interesting for us. Suppose we have a function f defined on D, $0 \leq f \leq 1$, $f = 0$ on L^+, $f = 1$ on L^-, such that

$$|\nabla f| \leq C\{\omega(B_{x,u})^{-\frac{1}{2}} + \omega(B_{y,u})^{-\frac{1}{2}}\}\omega(u)^{\frac{1}{2}} du. \tag{3.2}$$

Set $\Omega_t = \{w \in D \setminus \{x, y\} : f(w) > t\}$. The co–area formula ([**Gi**], p.20) gives

$$|\nabla f| = \int_0^1 |\nabla \chi_{\Omega_t}|\, dt, \tag{3.3}$$

as an equality between measures.

Because f is continuous except at x and $y, \partial\Omega_t$ must connect x to y for each t. (If this were false—if x and y lie in different components of $\partial\Omega_t$—then you could find a path connecting L^+ to L^- that does not intersect $\partial\Omega_t$. This is impossible because f is continuous.) For almost all $t, \partial\Omega_t$ is rectifiable, and we can pick our path γ_t inside Ω_t. The paths γ_t satisfy (3.1) because of (3.2) and (3.3).

Thus, modulo some technical details, we need only find f. We shall in fact give a slightly different, cleaner proof instead, which avoids these technicalities, but still requires such an f.

To build f we use the geodesic distance $d(u, v)$ associated to ω. Let $\delta > 0$ be small, to be chosen soon. Let R_+ denote the region above the angle whose vertex is $(0, 1 + \delta)$ and whose boundary rays eminating from $(0, 1 + \delta)$ pass through x and y. Thus D sits "below" R_+. Let R_- denote the region obtained by flipping R_+ across the real axis, so that $(0, 1 + \delta)$ is replaced by $(0, -1 - \delta)$, and D lies "above" R_-. Set

$$d_+(u) = \inf_{v \in R_+} d(u, v), d_-(u) = \inf_{v \in R_-} d(u, v).$$

Because ω is strongly $A_\infty, d_\pm(u)$ is comparable to $\omega(B)^{\frac{1}{2}}$, where B is the largest ball centered at u and disjoint from $R_\pm$.

If δ is small enough then

$$d_+(u) \leq \frac{1}{10} d_-(u) \text{ on } L^+,\ d_-(u) \leq \frac{1}{10} d_+(u) \text{ on } L^-.$$

Let ϕ be a smooth function on $\mathbf{R}$ such that $0 \leq \phi \leq 1$, $\phi(s) = 0$ if $s \leq \frac{1}{10}$, and $\phi(s) = 1$ if $s \geq 10$. Thus

$$f(u) = \phi\Big(d_+(u)/d_-(u)\Big)$$

satisfies $0 \leq f \leq 1$, $f = 0$ on L^+, $f = 1$ on L^-, and also (3.2). The latter follows from the fact that $| \nabla d_\pm | \leq C\omega^{\frac{1}{2}}$, by definition, and also that $d_+(u)$ and $d_-(u)$ are both comparable to

$$\min(\omega(B_{x,u})^{\frac{1}{2}},\ \omega(B_{y,u})^{\frac{1}{2}})$$

on D.

Let us now deduce (2.1) from the existence of f, without going through the construction of the curves γ_t. For $\epsilon > 0$ small, set

$$D_\epsilon = \{u = (u_1, u_2) \in D : -1 + \epsilon \leq u_1 \leq 1 - \epsilon\},$$

i.e., D_ϵ is just D with two bits cut off near x and y. From Stokes' theorem we get

$$\int_{\partial D_\epsilon} g\ df = \int_{D_\epsilon} dg \wedge df. \tag{3.4}$$

Because f satisfies (3.2), and $D_\epsilon \subseteq D \subseteq B_{x,y}$, we have

$$\left|\int_{D_\epsilon} dg \wedge df\right| \leq C \int_{B_{x,y}} \{\omega(B_{x,u})^{-\frac{1}{2}} + \omega(B_{y,u})^{-\frac{1}{2}}\} \mid \nabla g(u) \mid \omega(u)^{\frac{1}{2}} du.$$

Also, df is zero on $\partial D_\epsilon \cap (L^+ \cup L^-)$, because f is constant there. Denote by S_x and S_y the two remaining segments in ∂D_ϵ, equipped with the orientations inherited from ∂D_ϵ. Then

$$\int_{S_x} df = 1 \qquad \text{and} \qquad \int_{S_y} df = -1.$$

This and our a priori smoothness assumptions allow us to conclude that

$$\lim_{\epsilon \to 0+} \int_{\partial D_\epsilon} g\ df = g(x) - g(y).$$

This finishes the proof of (2.1) in the 2–dimensional case.

The proof in higher dimensions is almost the same, but with a few more notations. Given $x, y \in \mathbf{R}^n$, let $[x, y]$ denote the line segment that joins them, and let H denote the hyperplane through $\frac{x+y}{2}$ that is orthogonal to $[x, y]$. Let Q be an $(n-1)$ dimensional cube centered at $\frac{x+y}{2}$ with sidelength $|x - y|$. Let $F_i^\pm, 1 \leq i \leq n - 1$, be the $2(n-1)$ faces of Q, with F_i^+ opposite to F_i^-. Let D be the convex hull of $Q \cup \{x, y\}$. When $n = 3, D$ is an octohedran, and in general it has $4(n-1)$ faces. For each i, let $L_{i,x}^+$ denote the convex hull of $F_i^+ \cup \{x\}$, and similarly for $L_{i,y}^+$. When $n=3$ they are two triangular faces of D, and in general they are pyramidal faces having the same base. Set $L_i^+ = L_{i,x}^+ \cup L_{i,y}^+$, and define L_i^- similarly.

Using the geodesic distance function as before, we can find functions f_i defined on $D \setminus \{x,y\}$ such that $0 \le f_i \le 1$, $f_i = 0$ on L_i^+, $f_i = 1$ on L_i^-, and

$$| \nabla f_i(u)| \le C\{\omega(B_{x,u})^{-\frac{1}{n}} + \omega(B_{y,u})^{-\frac{1}{n}}\}\ \omega(u)^{\frac{1}{n}}. \tag{3.5}$$

Set $D_\epsilon = D \cap \{u \in \mathbf{R}^n : \epsilon \mid y-x \mid^2 \le \, < u-x, y-x > \, \le (1-\epsilon) \mid y-x \mid^2\}$, so that D_ϵ is just D minus little neighborhoods near x and y. From Stokes' theorem we have

$$\int_{\partial D_\epsilon} g\ df_1 \wedge \cdots \wedge df_{n-1} = \int_{D_\epsilon} dg \wedge df_1 \wedge \cdots \wedge df_{n-1}. \tag{3.6}$$

The right hand side is controlled by the right hand side of (2.1), because of (3.5). The contributions to the left side coming from $\partial D_\epsilon \cap L_{i,y}^\pm$ or $\partial D_\epsilon \cap L_{i,x}^\pm$ are zero, because f_i is constant there. What remains on the left is

$$\int_{D\cap H_x} g\ df_1 \wedge \cdots \wedge df_{n-1} + \int_{D\cap H_y} g\ df_1 \wedge \cdots \wedge df_{n-1},$$

where $H_x = \{u :< u-x, y-x >= \epsilon \mid y-x \mid^2\}$ and $H_y = \{u :< u-x, y-x >= (1-\epsilon) \mid y-x \mid^2\}$ are the hyperplanes used to cut D_ϵ out of D. We equip $D_\epsilon \cap H_x$ and $D_\epsilon \cap H_y$ with the orientations inherited from ∂D_ϵ. Some easy calculus gives

$$\int_{D\cap H_x} df_1 \wedge \cdots \wedge df_{n-1} = - \int_{D\cap H_y} df_1 \wedge \cdots \wedge df_{n-1} = \pm 1,$$

where the sign depends on how the faces $F_i^\pm$ were numbered. As before, this and our a priori smoothness assumptions imply

$$\lim_{\epsilon \to 0+} \int_{\partial D_\epsilon} g\ df_1 \wedge \cdots \wedge df_{n-1} = \pm(g(x) - g(y)).$$

From here we get (2.1), and the theorem is proved.

As in the 2–dimensional case, we could have used the f_i's to find a nice family of curves connecting x to y. Let us sketch the argument.

Let G be the $\mathbf{R}^{n-1}$–valued function defined on $D\setminus\{x,y\}$ by $G(w) = (f_1(w), \cdots, f_{n-1}(w))$. Let $u \in (0,1)^{n-1}$ be given. We want to first check that $G^{-1}(u)$ connects H_x to H_y for each choice of $\epsilon > 0$ (and therefore connects x to y).

If not, we could find a continuous function f_n defined on $G^{-1}(u) \cup H_x \cup H_y$ that is 0 on H_x, 1 on H_y, never takes the value $\frac{1}{2}$, and f satisfies $0 \le f_n \le 1$. We can of course extend f_n continuously to D_ϵ. We use elementary degree theory (see [**N**]) to get a contradiction. To wit, the map $\tilde{G} = (f_1, \cdots, f_n)$ from D_ϵ to $[0,1]^n$ has degree ± 1 on $(0,1)^n$, the sign depending on our choices. This implies that $\tilde{G}$ takes the value $(u_1, \cdots, u_{n-1}, \frac{1}{2})$, a contradiction.

Thus $G^{-1}(u)$ always connects x to y. Theorem 3.2.15 of **[F]** tells us that $G^{-1}(u)$ is rectifiable for almost all u. It remains to show that if α_u is the arclength measure on $G^{-1}(u)$, viewed as a measure on $\mathbf{R}^n$, then

$$\int \alpha_u du \leq C\{\omega(B_{x,z})^{-\frac{(n-1)}{n}} + \omega(B_{y,z})^{-\frac{n-1}{n}}\}\omega(z)^{\frac{n-1}{n}} dz, \tag{3.7}$$

as an inequality between measures. Indeed, Theorem 3.2.12 of **[F]** (the co–area formula) allows us to control the left side by $df_1 \wedge \cdots \wedge df_{n-1}$, which is itself controlled by the right side of (3.7).

Although this approach is technically more obnoxious, it is more intuitive geometrically. It is also nice to know that we can always build this family of curves for any strongly A_∞ weight, and not just the weights that are comparable to the Jacobian of a q.c. mapping.

References

[D] G. David, *Opérateurs d'intégrale singulière sur les surfaces régulières*, Ann. Scient. Ec. Norm. Sup., **21** (1988), 225–258.

[F] H. Federer, "Geometric Measure Theory," Grundlehren der Mathematischen Wissenschaften 153, Springer-Verlag, 1969.

[Ge] F. Gehring, *The L^p-integrability of the partial derivatives of a quasiconformal mapping*, Acta Math. **130** (1973), 265–277.

[Gi] Giusti, *Minimal surfaces and functions of bounded variation*, Monographs in Math. **80**. Birkhäuser, (1984).

[GR] J. Garcia-Cuerva and J.L. Rubio de Francia, *Weighted norm inequalities and related topics*, North Holland, Amsterdam 1985.

[J] J.L. Journé, *Calderón-Zygmund operators, Pseudo-differential operators and the Cauchy integral of Calderón*, Lecture Notes in Mathematics **994**, Springer-Verlag, 1983.

[N] L. Nirenberg, *Topics in Nonlinear Functional Analysis*, Lecture notes from the Courant Institute of Mathematical Sciences, New York, N.Y. (1974).

[Se] S. Semmes, *Chord-arc surfaces with small constants I and II*, to appear in Advances in Math.

[St] E. Stein, *Singular integrals and differentiability properties of functions*, Princeton University Press, 1970.

Centre de Mathématiques, Ecole Polytechnique, 91128 Palaiseau Cedex, France
Dept. of Math, Rice University, Houston, Texas

6

Some Remarks on Harmonic Analysis and Elliptic Partial Differential Equations

Robert Fefferman

In this article, we shall discuss the interplay of two topics. One of these is the theory of Muckenhoupt weights, which plays a central role in Harmonic Analysis on Euclidean spaces. The other is the study of the harmonic measure associated to an elliptic operator with non-smooth coefficients.

The nature of this harmonic measure is important because it determines the solvability of the Dirichlet problem for these operators when the given boundary data belongs to an L^p class.

Our results announced here are joint work with Carlos Kenig and Jill Pipher. Our purpose in this paper is to provide the reader with the motivation behind the theorems and an understanding of the relationship between the Harmonic analysis and the differential equations. Detailed proofs will appear elsewhere [**1**].

Let us begin with the theory of Muckenhoupt weights. The reader will recall that $w(x) \geq 0$ defined and locally integrable ond $\mathbb{R}^n$ is said to be a Muckenhoupt weight if and only if it belongs to one of the A^p classes, defined as follows:

$$w \in A^p(\mathbb{R}^n), \quad 1 < p < \infty, \quad \text{iff} \quad \Big(\frac{1}{|Q|}\int_Q w(x)\,dx\Big)\Big(\frac{1}{|Q|}\int_Q w(x)^{-\frac{1}{p-1}}\,dx\Big)^{p-1} \leq C$$

for all cubes $Q \subseteq \mathbb{R}^n$. Recall also that when $p = \infty$, $A^\infty(\mathbb{R}^n) = \bigcup\limits_{1<p<\infty} A^p(\mathbb{R}^n)$. The importance of these classes is partially due to the theorem (see[**2**] and [**3**]) that for $1 < p < \infty$,

$$\|T(f)\|_{L^p(w\,dx)} \leq C\|f\|_{L^p(w\,dx)}$$

holds when T is either a classical Calderón - Zygmund Singular Integral on $\mathbb{R}^n$, or the Maximal Operator of Hardy - Littlewood if and only if $w \in A^p(\mathbb{R}^n)$. Our concern here is centered around the intimate connection of these spaces A^p with the space BMO($\mathbb{R}^n$) of John and Nirenberg [**4**]. The space BMO($\mathbb{R}^n$) is defined by

$$\varphi \in \text{BMO}(\mathbb{R}^n) \quad \text{if and only if} \quad \frac{1}{|Q|}\int_Q |\varphi(x) - \varphi_Q|\, dx \leq C$$

for every cube $Q \subseteq \mathbb{R}^n$, where $\varphi_Q = \frac{1}{|Q|}\int_Q \varphi(x)\, dx$. The reader familiar with the basic classical theory of weights and BMO will recognize the following very well known facts:

1. $w \in A^\infty(\mathbb{R}^n)$ implies $\log w \in \text{BMO}(\mathbb{R}^n)$. The converse is obviously false, since $\frac{1}{|x|}$ is not a measure on $\mathbb{R}^1$, yet $\log\frac{1}{|x|} \in \text{BMO}(\mathbb{R}^1)$.
2. $\varphi \in \text{BMO}(\mathbb{R}^n)$ if and only if $\varphi = f_0 + \sum_{j=1}^n R_j f_j$ where the f_j, $j = 1 \ldots, n$ are in $L^\infty(\mathbb{R}^n)$ and R_j are the Riesz transforms. This is the Decomposition Theorem of C. Fefferman and E. M. Stein [**5**].
3. On $\mathbb{R}^1$, $w \in A^2(\mathbb{R}^1)$ if and only if $w = e^{f_1 + H(f_2)}$ for some pair of bounded functions f_1 and f_2 with
$$\|f_2\|_{L^\infty} < \pi/2.$$
This is the Helson-Szego Theorem [**6**].
4. If $\psi \in C_c^\infty(\mathbb{R}^n)$, $\int \psi\, dx = 0$, and ψ is suitably non-trivial, then a function $\varphi(x)$ on $\mathbb{R}^n$ belongs to BMO if and only if
$$\iint_{\substack{x\in Q \\ 0<t<\text{side length }(Q)}} |\varphi * \psi_t(x)|^2 \frac{dx\,dt}{t} \leq C|Q| \qquad \text{for all cubes } Q \subseteq \mathbb{R}^n,$$
i.e.,$\varphi \in$ BMO if and only if $|\varphi * \psi_t(x)|^2 \frac{dx\,dt}{t}$ is a Carleson measure on $\mathbb{R}^{n+1}_+$ (see C. Fefferman [**7**]).

Our result is a characterization of the class $A^\infty(\mathbb{R}^n)$, which is motivated by the four facts mentioned above from harmonic analysis.

We know from (1) and (4) above that if we are trying to determine whether a given locally integrable positive function $f(x)$ belongs to $A^\infty(\mathbb{R}^n)$, it is not enough to have $\log f \in$ BMO, i.e.

$$\text{(a)} \qquad |(\log f) * \psi_t(x)|^2 \frac{dx\,dt}{t}$$

to be a Carleson measure on $\mathbb{R}^{n+1}_+$, (ψ is an appropriate Schwartz function such that $\int \psi = 0$). By making two small changes in (a) we shall derive a condition which does in fact characterize A^∞.

Suppose, as we may, that in (a), ψ is a bump function of mean zero which arises by differentiation from another bump function $\varphi(x) \geq 0$, i.e., assume that $\psi_t(x) = t\frac{\partial \varphi_t}{\partial t}(x)$ where $\varphi \in C_0^\infty(\mathbb{R}^n)$, $\varphi \geq 0$, then (a) can be written

$$\text{(b)} \qquad |t\frac{\partial}{\partial t}(\log f * \varphi_t)|^2 \frac{dxdt}{t} \quad \text{is Carleson.}$$

The first change we make is to commute the process of extending f to $\mathbb{R}^{n+1}_+$ by regularizing with the process of taking $\log f$. In other words we substitute in (a)

$$\log(f * \varphi_t) \quad \text{for} \quad (\log f) * \varphi_t.$$

This produces the new (and inequivalent) condition

$$\text{(c)} \qquad |t\frac{\partial}{\partial t}\log(f * \varphi_t)|^2 \frac{dxdt}{t} = \frac{|f * \psi_t(x)||^2}{[f * \varphi_t(x)]^2}\frac{dxdt}{t}$$

is a Carleson measure. Easy examples show that this new condition (c) is still not enough to characterize $A^\infty(\mathbb{R}^n)$. However, if we make the second change that $f(x)dx$ be a doubling measure on $\mathbb{R}^n$, then we do have:

THEOREM. *Let $f(x) \geq 0$ be a locally integrable function on $\mathbb{R}^n$. Then $f \in A^\infty(\mathbb{R}^n)$ if and only if*

(i) *$f(x)\,dx$ is a doubling measure*

(ii) *$\frac{|f*\psi_t(x)|^2}{[f*\varphi_t(x)]^2}\frac{dxdt}{t}$ is a Carleson measure on $\mathbb{R}^{n+1}_+$.*

In this theorem, φ and ψ are suitable functions in $\mathcal{S}$ with $\varphi \geq 0$ and $\int \psi = 0$.

This characterization of $A^\infty(\mathbb{R}^n)$ plays the role for this space that the theorem of Helson-Szego plays for $A^2(\mathbb{R}^1)$. Here, the interest in this theorem comes from a conjecture of Dahlberg in elliptic equations, which we shall now describe.

Suppose B denotes the unit ball in $\mathbb{R}^n$. Let $A(x)$, $x \in B$ denote a symmetric matrix whose entries $a_{ij}(x)$ are all in $L^\infty(B)$ and such that $A(x)$ is uniformly positive definite, i.e., there exits $c > 0$ such that

$$\langle A(x)\xi, \xi\rangle \geq c|\xi|^2 \qquad \text{for all } \xi \in \mathbb{R}^n, x \in B.$$

Then the elliptic equation

$$L(u) = \operatorname{div}(A(x)\nabla u) = 0$$

has been studied extensively in the literature. In particular we should mention the results of Moser [**8**] on interior regularity for solutions of such equations and those of Littman-Stampacchia-Weinberger [**9**] proving that the classical Dirichlet problem in B is always solvable for an elliptic operator of this type. It follows from this solvability and the maximum principle that a harmonic measure ω exists for the operator L. This means that there is a unique probability measure ω on ∂B such that if f is continuous on ∂B and $L(u) = 0$ on B while $u(x) = f(x)$ on ∂B, then

$$u(0) = \int_{\partial B} f\, d\omega.$$

The nature of the measure $d\omega$ on ∂B is of great importance. For instance, if we define a function $k(x) \geq 0$ on ∂B to belong to the class B^p if and only if

$$\left(\frac{1}{\sigma(\Delta)}\int_\Delta k^p\, d\sigma\right)^{1/p} \leq C\left(\frac{1}{\sigma(\Delta)}\int_\Delta k\, d\sigma\right)$$

for every surface ball $\Delta \subseteq \partial B$, (here $d\sigma$ is the surface measure on ∂B) then the solvability of the Dirichlet problem

$$\begin{cases} L(u) = 0 & \text{in } B \\ u(x) = f(x) & \text{on } \partial B \end{cases}$$

for all boundary data $f \in L^p(d\sigma)$ is essentially equivalent to $\omega \in B^{p'}(d\sigma)$ where p' is the dual exponent to p. Also, if u is a non-negative solution of $L(u) = 0$ in B, then according to a result of Caffarelli, Fabes, Mortella, and Salsa [**10**] u has radial limits except on a set of ω measure 0. The question of existence a.e. of radial limits with respect to surface measure is reduced to proving the absolute continuity of $d\sigma$ with respect to $d\omega$. This will be proven if, for instance, one can show that $\omega \in A^\infty(d\sigma)$.

Unfortunately, (see [**11**])there exist operators $L(u) = \operatorname{div}(A\nabla u)$ as above whose harmonic measures ω are mutually singular with respect to $d\sigma$. It is therefore of interest to obtain criteria which guarantee the good behavior of the harmonic measure associated with L. This is a topic which has a rich history in its own right, but we shall be content here to state one result in this direction.

To do this, we recall that a non-negative measure μ on B is called a Carleson measure with vanishing trace, written $\mu \in \mathcal{C}_0$ provided

$$\frac{\mu[S(\Delta(x_0;r))]}{\sigma(\Delta(x_0;r))} \to 0 \quad \text{as } r \to 0 \text{ uniformly in } x_0 \in \partial B.$$

(Here $S(\Delta(x_0;r)) = B(x_0;r) \cap B$, the Carleson region associated to the surface ball $\Delta(x_0;r)$.) Then we have

THEOREM OF DAHLBERG. **[12]** *Suppose $L_0(u) = \operatorname{div}(A_0 \nabla u)$ and $L_1(u) = \operatorname{div}(A_1 \nabla u)$ are two elliptic operators as above. For $z \in B$, set*

$$a(z) = \sup_{x \in B(z;\frac{\delta(z)}{2})} \|A_0(x) - A_1(x)\|,$$

where $\delta(z) = 1 - |z|$. Suppose that the measure $a^2 \frac{dz}{\delta} \in \mathcal{C}_0$. Then $\omega_{L_0} \in B^p(d\sigma)$ implies that $\omega_{L_1} \in B^p(d\sigma)$ for every $1 < p < \infty$.

In addition to this theorem, Dahlberg made the following

CONJECTURE. *Suppose L_0 and L_1 are elliptic operators, and define $a(z)$, $z \in B$ as above. Assume that $a^2 \frac{dz}{\delta}$ is a Carleson measure in B (not necessarily of vanishing trace). Then if $\omega_{L_0} \in A^\infty(d\sigma)$ this implies that $\omega_{L_1} \in A^\infty(d\sigma)$.*

In **[1]** we prove this conjecture. This is accomplished by reducing it to the following special case:

THEOREM (S). *Suppose L_0 and L_1 are elliptic operators and define $S(a)(x) = \int_0^1 a^2((1-t)x)\frac{dt}{t}$, for each $x \in \partial B$. Assume that $S(a) \in L^\infty(\partial B)$, and that $\omega_{L_0} \in A^\infty(d\sigma)$. Then $\omega_{L_1} \in A^\infty(d\sigma)$.*

This is proven in **[13]**. The key idea in the proof of Theorem (S) in **[13]** is to make use of the entirely distinct properties of the Banach space BMO as opposed to those of the spaces L^p. (For example, we make use of the John-Nirenberg inequality in a crucial way.) BMO arises in Theorem (S) whereas L^p spaces arise in Dahlberg's Theorem, because Theorem (S) relates to A^∞ which in turn is related to BMO. The B^p condition is, of course, defined in terms of L^p estimates.

Finally, let us discuss how our characterization of A^∞ relates to the Dahlberg conjecture.

There is a class of fundamental examples of elliptic operators with bounded measureable coefficients which provide the link. This class arises by considering quasi-conformal mappings, Φ, from $\mathbf{R}^2_+$ to itself, and observing that if u is harmonic in $\mathbf{R}^2_+$ then $u \circ \Phi$ is a solution to an elliptic operator L with bounded measurable coefficients which can be easily computed in terms of Φ.

According to a result of Beurling and Ahlfors [**14**], given a positive measure f on $\mathbf{R}^1$ which satisfies a doubling condition (i.e., $f(\tilde{I}) \leq Cf(I)$, where $\tilde{I}$ denotes the concentric double of the interval $I \subseteq \mathbf{R}^1$ there exists a quasiconformal map $\Phi_f : \mathbf{R}^2_+ \to \mathbf{R}^2_+$ such that $\Phi_f(x,0) = (f(x),0)$. Then if we consider the operator L_{Φ_f}, with coefficients $A(x,t),(x,t) \in \mathbf{R}^2_+$, and define

$$a(x,t) = \sup_{(y,s)\in B((x,t);t/2)} \|A(y,s) - I\|$$

then it turns out that we essentially have

$$a(x,t) \leq \frac{|f * \psi_t(x)|}{f * \varphi_t(x)}$$

for the appropriate φ, ψ such that $\varphi \geq 0$ and $\int \psi = 0$. Also the harmonic measure associated to the operator L_{Φ_f} is essentially f. The one-dimensional version of the Dahlberg conjecture, applied to the operator L_{Φ_f} and the Laplacian would say that if the measure

$$\frac{|f * \psi_t(x)|^2}{[f * \varphi_t(x)]^2} \frac{dxdt}{t}$$

is Carleson then $f \in A^\infty(\mathbf{R}^1)$.

Conversely, given any A^∞ measure f on $\mathbf{R}^1$, the operator L_{Φ_f} has f as its harmonic measure (up to multiplication by a function bounded above and below by positive constants) and has the disagreement between its coefficients and those of the Laplacian, as measured by the function a satisfying

$$a^2 \frac{dz}{\delta} \in \mathcal{C}.$$

This means that our characterization of A^∞ in terms of Carleson measures implies that the Dahlblerg conjecture is sharp. Thus, under the hypotheses of the conjecture aplied to $L_0 = \Delta$ and L_1, all A^∞ weights occur as harmonic measure ω_{L_1}.

Conversely, it can be shown that if we allow perturbations for which

$$a^2 \frac{dz}{\delta} \notin \mathcal{C},$$

then there is such a perturbation L, of Δ, whose harmonic measure $\omega_L \notin A^\infty$. This also uses the characterization of A^∞ weights to construct the counterexample. It is therefore quite clear that Dahlberg's conjecture is not only correct, but extremely sharp as well.

BIBLIOGRAPHY

[1] R. Fefferman, J. Pipher, and C. Kenig, To appear.

[2] R. Hunt, B. Muckenhoupt, and R. Wheeden, *Weighted Norm Inequalities for the Conjugate Function and Hilbert Transform*, Trans. AMS **176** (1973), 1026–1029.

[3] R. Coifman and C. Fefferman, *Weighted Norm Inequalities for Maximal Functions and Singular Integrals*, Studia Math. **51**(1974), 241–250.

[4] F. John and L. Nirenberg, *On Functions of Bounded Mean Oscillation*, Comm. Pure and Appl. Math. **14** (1961), 415–426.

[5] C. Fefferman and E. M. Stein, H^p Spaces of Several Variables, Acta Math. **129** (1972), 137–193.

[6] H. Helson and G. Szegö *A Problem in Prediction Theory*, Ann. Math. Pura Appl. **51** (1960), 107–138.

[7] C. Fefferman, *Characterization of Bounded Mean Oscillation*, Bulletin of the AMS **77** (1971), 587–588.

[8] J. Moser, *On Harnack's Theorem of Elliptic Differential Equations*, Comm. Pure and Appl. Math. **14** (1961), 577–591.

[9] W. Littman, G. Stampacchia and H. Weinberger, *Regular Points for Elliptic Equations with Discontinuous Coefficients*, Ann. della Sc. N. Sup. d Pisa **17** (1963), 45–79.

[10] L. Caffarelli, E. Fabes, S. Mortella, and S. Salsa, *Boundary Behavior of Non-Negative Solutions of Elliptic Operators in Divergence Form*, Indiana U. Math. Journal **30** (1981), 621–640.

[11] L. Caffarelli, E. Fabes, and C. Kenig, *Completely Singular Elliptic Harmonic Measures*, Indiana U. Math. Jour. **30** (1981), 917–924.

[12] B. Dahlberg, *On the Absolute Continuity of Elliptic Measures*, American Journal of Math. **108** (1986), 1119–1138.

[13] R. Fefferman, *A Criterion for the Absolute Continuity of the Harmonic Measure Associated with an Elliptic Operator*, to appear in the Journal of the AMS, 1989.

[14] L. Ahlfors and A. Beurling, *The Boundary Correspondence Under Quasiconformal Mappings*, Acta Math. **96** (1956), 125–142.

7

Singular Value Estimates for Commutators and Hankel Operators on the Unit Ball and the Heisenberg Group

Mark Feldman and Richard Rochberg[1]

I. INTRODUCTION AND SUMMARY.

We begin with an informal presentation of our main results. Definitions and precise statements are in the following sections.

Let L^2 be the Lebesgue space of $\sum_n$, the unit sphere in $\mathbb{C}^{n+1}$, and let H^2 be the associated Hardy space. We always assume $n > 0$; some of the results must be modified when $n = 0$.

Let P be the orthogonal projection (= the Cauchy-Szego projection) from L^2 to H^2 and let Q be the orthogonal projection from L^2 to $\overline{H}_0^2$, the space of complex conjugates of functions in H^2 having mean zero. For a function b defined on $\sum_n$ we denote by M_b the operator of pointwise multiplication by b. We will assume b is the boundary value of a conjugate analytic function. We will be interested in the question of boundedness and membership in the Schatten-von Neumann ideals for three operators acting on L^2: the commutator

$$C_b = [M_b, P], \tag{1.1}$$

[1]Supported in part by NSF Grant DSM 8701271

the big Hankel operator

$$H_b = (I - P)M_bP = C_bP, \tag{1.2}$$

and the small (or reduced) Hankel operator

$$h_b = QM_bP = QH_b = QC_bP. \tag{1.3}$$

(In the traditional case, $n = 0$, $H_b = h_b$ and is called a Hankel operator. In other contexts, such as operators on the Bergman space, operators of the form (1.2) and (1.3) have both been called Hankel operators, resulting in some mild confusion.) Note that both H_b and h_b can be regarded as operators whose domain is the Hardy space.

For $n = 0$ it is known that C_b is bounded if and only if b is in BMO, and that for $0 < p < \infty$, C_b is in the Schatten-von Neumann ideal S_p if an only if b is in the Besov space B^p [RS1]. For the analogous operators on $L^2(\mathbb{R}^n)$, $n > 1$, it was shown by Janson and Wolff [JW] that there is an interesting dichotomy: for $p > n$, certain commutators (such as C_b) are in S_p if and only if the associated function, b, is in a particular Besov space; however, for $p \leq n$ the commutator is in S_p if and only if b is constant (and thus the commutator vanishes). A similar Janson-Wolff phenomenon has been demonstrated by Arazy, Fisher, and Peetre [AFP] for the big Hankel operator acting on the Bergman space of the disk.

One of our goals here was to find out if there is a similar dichotomy for the analogous class of operators on the Hardy spaces of the spheres. For C_b our result is

<u>THEOREM</u> <u>A</u>:

(i) C_b *is bounded iff b is in BMO.*

(ii) For $p > 2n + 2$, C_b *is in* S_p *if and only if b is in the Besov space* B^p.

(iii) For $p \leq 2n + 2$, C_b *is in* S_p *if and only if* $C_b = 0$.

At first glance, this is a bit of a surprise. The topological dimension of $\sum_n$ is 2n + 1. However, there is a natural (for this problem) metric on $\sum_n$ with the property that a ball of radius R contains, roughly, $\epsilon^{-(2n+2)}$ balls of radius ϵR. Thus the "homogenous dimension" of $\sum_n$ for this problem is 2n + 2. That turns out to be decisive for separating cases (ii) and (iii) of the theorem.

For the Hankel operators we have

<u>THEOREM</u> <u>B</u>: *Theorem A holds with* C_b *replaced by* H_b.

<u>THEOREM</u> <u>C</u>:

(i) h_b *is bounded iff b is in BMO.*

(ii) For $1 \leq p < \infty$, h_b *is in* S_p *if and only if b is in* B^p.

Thus, as is true for operators on the Bergman space of the disk, there is no Janson-Wolff phenomenon for the small Hankel operator. This is more than an analogy; we will see that Theorem C implies the corresponding result for operators on the Bergman space.

We won't always prove the theorems in the setting of $\sum_n$. Using a Cayley transform we show that the operators in question are unitarily equivalent to operators on the Lebesgue and Hardy spaces associated to the Heisenberg group and we usually work in that context. Combining this unitary equivalence and the various implications we get proofs of all three theorems in both settings.

In Theorem A, which we prove on the Heisenberg group, we actually prove a more general result involving commutators with Calderon-Zygmund kernels and nonanalytic symbols. The techniques used are extensions of those of [RS2] in which nearly weakly orthogonal (NWO) sequences are used to study operators acting on $L^2(\mathbf{R}^n)$. In fact, one of our goals here is to show that use of NWO sequences can be adapted to the geometry intrinsic to the Heisenberg group.

Sufficiency in part (ii) of Theorem C follows from a decomposition theorem of Coifman-Rochberg and the theory of NWO sequences. Necessity is proved using reproducing kernel techniques in ways similar to their use in [R1] to obtain the result with n = 0.

Because H_b is intermediate in size between C_b and h_b, parts (i) and (ii) of Theorem B follow from Theorems A and C. To prove part (iii) of Theorem B, we follow the lead of [AFP]. We first show that if there is any non-zero h_b in S_{2n+2} then h_ζ (Here $(\zeta, z) = (\zeta, z_1, \ldots, z_n)$ are coordinates in $\mathbf{C}^{n+1}$.) would have to be in S_{2n+2}, also. We then verify that h_ζ is not in S_{2n+2} by computing its action on the monomials.

In addition to the theorems on Schatten-von Neumann ideal criteria we describe the (group) Fourier transforms of the Hankel operators acting on the Hardy space of the Heisenberg group. The form of these transforms is similar to the classical case. We also find an interesting relation between these operators and the Hankel operators on the Fock space studied in [JPR]. (The Fock spaces arise in the study of the Fourier transform of functions on the Heisenberg group.)

Some of our results concerning boundedness and Schatten-von Neumann criteria for Hankel operators have been obtained by others. See [B] and [Z] in the references.

Here are the contents.

Section 2 contains material on the Cayley transform, the Heisenberg group and the various function spaces we will be using. In Section 3 we introduce the idea of nearly weakly orthonormal sequences in L^2 of the Heisenberg group. We describe the Fourier transforms in Section 4. Section 5 contains the proof of (a result a bit more general than) Theorem A for the Heisenberg group. In Section 6 we prove part (iii) of Theorem B. In Section 7 we prove Theorem C. In Section 8 we derive results for the small Hankel operator on the Bergman spaces of the ball B_n, and explain the relation of Theorem C to the results of [AFP]. In the last section we note some open problems.

II. PRELIMINARIES.

A. The Heisenberg Group.

The *generalized upper half space* in $\mathbb{C}^{n+1}$, $n \geq 1$, is the set

$$D_n = \{(\zeta,z) \ \varepsilon \ \mathbb{C} \times \mathbb{C}^n\text{: } \operatorname{Im} \zeta - |z|^2 > 0\}.$$

By making the change of coordinates

$$(\zeta,z) \to (\operatorname{Im} \zeta - |z|^2, \operatorname{Re} \zeta, z) = (\tau,t,z)$$

we can identify D_n with $(0,\infty) \times H_n$, where H_n is the boundary (i.e., $H_n = \{(\zeta,z)\text{: } \operatorname{Im} \zeta - |z|^2 = 0\}$) and can be identified with $\mathbb{R}^{2n+1}$ (as a manifold). H_n acts simply and transitively on D_n by $(t,z)\ ((\eta,w)) = (t + \eta + 2iw{\cdot}z + i|z|, z + w)$ (here $w{\cdot}z = \sum_{i=1}^{n} w_i\ \bar{z}_i$) . These maps are biholomorphic automorphisms of D_n and the induced map on H_n,

$$(t,x,y) \cdot (t',x',y') = (t,z) \cdot (t',z') = (t + t' + 2 \operatorname{Im} z{\cdot}z', z + z')$$

makes H_n into the *Heisenberg group*. For our purposes there is nothing special about the case $n = 1$ so we fix n at this value for the remainder of the paper, though at times we display an n to clarify the general formulation of a statement.

The multiplicative group $\mathbf{R}^+$ acts on $H(=H_1)$ through dilations $(a(t,z) = (a^2t, az))$ and $|(t,z)| = (t^2 + |z|^4)^{1/4}$ is a *homogeneous norm*, i.e., a continuous map from H to $\mathbf{R}^+$ which is C^∞ on $H\backslash\{0\}$ and satisfying $|g^{-1}| = |g|$, $|g| = 0$ iff $g = 0$ and $|a(t,z)| = a|(t,z)|$. A translation invariant metric is given by $d(g,h) = |gh^{-1}|$. Haar measure on H is dtdxdy, i.e., Lebesque measure on $\mathbf{R}^3$. Basis vectors for the left invariant vector fields on H are $X = \frac{\partial}{\partial x} - 2y\frac{\partial}{\partial t}$, $Y = \frac{\partial}{\partial y} + 2x\frac{\partial}{\partial t}$, and $T = \frac{\partial}{\partial t}$. For a multi-index $\alpha = (\alpha_t,\alpha_x,\alpha_y)$ we write $h^\alpha = t^{\alpha_t}\, x^{\alpha_x}\, y^{\alpha_y}$ and define the *homogeneous degree* of α as $|\alpha| = 2\alpha_t + \alpha_x + \alpha_y$. We also denote by H^α the vector field $T^{\alpha_t}\, X^{\alpha_x}\, Y^{\alpha_y}$ acting in the h variable. If $E \subseteq H$ is measurable, denote the (Haar) measure of E by $|E|$. Then $|rE| = r^4|E|$ and $4(=2n + 2)$ is the *homogeneous dimension* of H (cf. [FS], [CW]).

Fix $A > 1$. Let $\mathcal{B}$ be the collection of balls of radius $A^j(5)^{1/4}$, $j \in \mathbb{Z}$, centered at the points $A^j(k,l,m)$, $(k,l,m) \in \mathbb{Z}^3 \subseteq H$. For each j the collection of balls in $\mathcal{B}$ of radius $A^j(5)^{1/4}$ covers H and has the property that no point is in more than c(A) of them. There is also a constant k so that if $B \in \mathcal{B}$ and rad(B) = $A^j(5)^{1/4}$ then there are at most k balls in $\mathcal{B}$ of radius $A^{j-1}(5)^{1/4}$ that meet B (see [F]). We set (σ_B, g_B) = (rad(B), center (B)).

B. Function Spaces.

The *Hardy* space $H^2(D)(=H^2(D_1))$ is the collection of holomorphic functions on D satisfying $\|f\|^2_{H^2} = \sup_{\tau>0} \int_H |f(\tau,h)|^2\, dh < \infty$. We identify functions in $H^2(D)$ with their boundary values, and denote this space by $H^2(H)$. Let P be the orthogonal projection of $L^2(H)$ onto $H^2(H)$. P(f) is the boundary limit of S(f) where S is given by $S(f)(\tau,g) = \int_H f(h)\, S_{(\tau,g)}(h)\, dh$ and $S_{(\tau,t,z)}(s,w) = c((s - t)i + \tau + |w - z|^2)^{-(n+1)}$ is the Szego kernel (or $S_{(\zeta,z)}(\eta,w) = c_n(i(\zeta - \bar{\eta}) - 2z\cdot w)^{-(n+1)}$ in complex coordinates). P(f) is equal to the singular integral operator $\lim_{\epsilon\to 0} \int_{|h|>\epsilon} f(h)\, S_{(0,g)}(h)\, dh$. (See [KV]).

In order to discuss the group Fourier transform for H we need to introduce the *Fock* spaces F_λ. These are defined for $\lambda > 0$ ($\lambda < 0$) as the collection of holomorphic (antiholomorphic) functions on $\mathbb{C}$ satisfying

$$||f||^2_{F_\lambda} = \int_{\mathbb{C}} |f(t)|^2\, e^{-4\pi|\lambda||z|^2} |\lambda| dz < \infty.$$

F_λ is a closed subspace of $L^2\left(\mathbb{C},\, |\lambda|\, e^{4\pi|\lambda||z|^2} dz\right)$ with reproducing kernel $E_W(z) = e^{4\pi\lambda w\bar{z}}$ for $\lambda > 0$ and $E_W(z) = e^{-4\pi\lambda z\bar{w}}$ for $\lambda < 0$. We denote the orthogonal projection of $L^2(\mathbb{C},\, |\lambda|\, e^{4\pi|\lambda||z|^2} dz)$ onto F_λ by P_λ.

Though the group Fourier transform of a function in $L^2(H)$ is an operator valued function on $\mathbb{R}\backslash\{0\}$ taking values in $\mathrm{Hom}(F_\lambda,F_\lambda)$, for functions in the Hardy space the operators are of the form $\frac{1}{\sqrt{2\lambda}} \otimes h_\lambda$ and the Fourier transform can be identified with the vector valued function h_λ. To be precise, let $(\mathcal{F}f)(\lambda)$ be the (operator–valued) group Fourier transform of $f \in L^1(H) \cap L^2(H)$ and let $\hat{f}(\lambda,z) =$

$\int_{\mathbb{R}} f(t,z)\ e^{-2\pi i\lambda t} dt.$

<u>THEOREM</u> <u>(2.1)</u>: *A function $f \varepsilon L^1(H) \cap L^2(H)$ belongs to $H^2(H)$ iff $(Ff)(\lambda)$ is zero for a.e. $\lambda < 0$ and for a.e. $\lambda > 0$ $\big(((Ff)(\lambda))g\big)(z) = \lambda^{-1/2}\hat{f}(\lambda,z)\ e^{2\pi\lambda|z|^2} g(0)$.*

The Fourier transform of P also has a simple form.

<u>THEOREM</u> <u>(2.2)</u>: *Let $f \varepsilon L^1(H) \cap L^2(H)$. Then $\big((F(Pf)(\lambda))g\big)(t) = \lambda^{-1/2}P_\lambda\big(\hat{f}(\lambda,z)\ e^{2\pi\lambda|z|^2}\big)\ g(0)$.*

The first two theorems are essentially theorems (8.3) and (8.2) of [OV] specialized to the Fock spaces. From now on we will identify $(Ff)(\lambda)$ with the function $\lambda^{-1/2}\hat{f}(\lambda,z)\ e^{2\pi\lambda|z|^2}$. The group Plancherel formula restricted to $H^2(H)$ then becomes

$$||f||^2_{H^2} = c \int_0^\infty ||(Ff)(\lambda)||^2_{F_\lambda} \lambda\ d\lambda \tag{2.3}$$

and the inversion formula becomes

$$f(t,z) = c \int_0^\infty \Big(\int_{\mathbb{C}} (Ff)(\lambda,w)\ e^{4\pi\lambda z\overline{w}}\ e^{-4\pi\lambda|w|^2} \lambda\ dz\Big)\ \lambda^{-1/2} e^{-2\pi\lambda|z|^2} e^{2\pi i\lambda t}\ \lambda\ d\lambda \tag{2.4}$$

For $\beta > 0,\ 0 < p < \infty$ the weighted *Bergman* space $A_p^\beta(D_n)$ is the set of functions holomorphic on D satisfying $\int_0^\infty \int_{H_n} \tau^{\beta p}\, |f(\tau,u)|^p\, du\frac{d\tau}{\tau} = ||f||_{\beta,p}^p < \infty.$

$A_2^{1/2}(D_n)$ is, of course, the subspace of $L^2(D_n)$ consisting of holomorphic functions and thus is a space with a reproducing kernel, the Bergman kernel, which is given by $B_{(\tau,h)}(\sigma,g) = C_n\big((s - t)i + (\tau + \sigma) + |w - z|^2\big)^{-(n+2)}$ $\qquad (= S_{(\tau,h)}(\sigma,g)^{\frac{n+1}{n+2}})$.

Thus,

$$\text{(2.5)} \qquad ||f||_{\beta,p}^p = c\int_0^\infty \int_{H_n} B_{(\tau,g)}^{\frac{1-\beta p}{n+2}}(\tau,g)\, |f(\tau,g)|^p\, dud\tau$$

$$= c\int_{D_n} B_{(\zeta,z)}^{\frac{1-\beta p}{n+2}}(\zeta,z)\, |f(\zeta,z)|^p\, dz.$$

Similarly, the space of square integrable holomorphic functions defined on B_n has a Bergman kernel which we also write as $B_{(\zeta,z)}(\eta,w) = c_n(1 - \zeta\bar{\eta} - z\cdot w)^{-(n+2)}$ and we can define the weighted Bergman spaces $A_p^\beta(B_n)$ by using the obvious analagous definition; i.e., by using the right hand side of (2.5) with D_n replaced by B_n.

The following theorem of Coifman-Rochberg will be important. (Here we will assume that the constant A appearing in the definition of the covering $\mathcal{B}$ is sufficiently small.)

<u>THEOREM (2.6)</u>: *(Coifman–Rochberg [CR]) Let $0 < p < \infty$, $\beta > 0$, and* $\theta > \left(\frac{n+1+\beta p}{n+2}\right) max(-1, p-2)$. *Then*

(a) if $f \in A_p^\beta$ there exists $\lambda \in \ell^p$ such that

$$f(\tau,h) = \sum_B \lambda_B \, \sigma_B^{2(\theta+1)(\beta+\frac{n+1}{p})} B_{(\sigma_B^2, g_B)}^{\frac{\theta+2}{n+2}(\beta+\frac{n+1}{p})}(\tau,h) \tag{2.7}$$

$$\left(= \sum_B \lambda_B \, |B|^{\frac{\theta+1}{n+1}(\beta+\frac{n+1}{p})} B_B^{\frac{\theta+2}{n+2}(\beta+\frac{n+1}{p})}(\tau,h)\right)$$

and

$$\|\lambda\|_p \le c\, \|f\|_{\beta,p}.$$

(b) if $\lambda \in \ell^p$ then f defined by (2.7) is in A_p^β and $\|f\|_{\beta,p} \le c\, \|\lambda\|_p$.

A similar cover yields the same result for B_n, where B is replaced by $\varphi^{-1}(B)$ and φ is the Cayley transform described in (2.D) below.

For a function defined on B_n we set the radial derivative $R = \sum z_i \frac{\partial}{\partial z_i}$ and the operator $D = I + \frac{1}{n+1} R$. For m an integer and f analytic,

$$D^m f(z) = \langle f, S_z^{\frac{n+1+m}{n+1}} \rangle \tag{2.8}$$

(see page 104 of [Ru]).

Similarly, we define $D^m f(\tau,h) = c < f, S^{\,n+1+m}_{(\tau,h)}{}^{n+1} >$ for f analytic on D_n and use the same formula to extend the definitions to any $m \varepsilon \mathbb{R}$. Note that for functions on D_n, $Df(\tau,h) = \frac{\partial}{\partial\tau} f(\tau,h)$.

COROLLARY (2.9): *Let $0 < p < \infty$, $0 < \beta$ and $r, s \geq 0$. Then $D^s f \,\varepsilon\, A^\beta_p(D_n)(A^\beta_p(B_n))$ if and only if $D^{s+r} \,\varepsilon\, A^{\beta+r}_p(D_n)\ (A^{\beta+r}_p(B_n))$.*

Proof: Use the decomposition in Theorem 2.6.

For $b \,\varepsilon\, L^1_{loc}(H)$ and $B \varepsilon \mathcal{B}$ let $m_B(b) = |B|^{-1} \int_B b$ and $osc(b,B,r,k) = |B|^{-1}\left(\int_{kB} |b - m_B(b)|^r\right)^{1/r}$. Define the *mean oscillation space* $MO^p_{r,k} = \{b \,\varepsilon\, L^1_{loc}(H)$: $\{osc(b,B,r,k)\}_{B \varepsilon \mathcal{B}} \,\varepsilon\, \ell^p\}$. It is easy to see that MO^p_r $(= MO^p_{p,k})$ is independent of k (cf. [RS2]). As is well known MO^∞_r is independent of r, $1 \leq r < \infty$ and is the space of functions of *bounded mean oscillation*, BMO.

THEOREM (2.10): *The following are equivalent:*

(a) $f \,\varepsilon\, MO^p_r$ $\forall r$, $1 \leq r < \infty$.

(b) $f \,\varepsilon\, MO^p_1$.

(c) *There is a C^1 function $F(\tau,u)$ on D with $\lim_{\tau\to 0} F(\tau,u) = f(u)$ a.e. and so that*

$$F^* \,\varepsilon\, L^p(D, dr \frac{dt}{t^6}) \text{ where } F^*(t,v) = \sup_{\substack{u \varepsilon B(v,t) \\ t/2 < \tau < t}} \{\tau \,|(\frac{\partial}{\partial\tau} F(\tau,u), XF(\tau,u), YF(\tau,u))|\}$$

(Here $B(v,t)$ is the ball of radius t centered at v).

Furthermore, if f is analytic, then the above are equivalent (for $p > 2n + 2$) to

(d) $D^{n+1} f \varepsilon A_p^{(n+1)(1-\frac{1}{p})}$.

The first three equivalences are proved as in [RS2] and we won't repeat the details. That (a) implies (d) is a consequence of Proposition 3.18 in [F]. The analyticity of f and Corollary 2.9 show that (d) implies (c).

From (c), MO^p is the Besov space B^p when $p > 2n + 2$. For $p \leq 2n + 2$, MO^p is trivial; to get a non-trivial space, higher order moments must be subtracted. (See [RS2] or [RT] for the case $n = 0$ or [F] for $n > 0$.) For analytic f, B^p is defined by (d) for $p > 1$ and by $D^{n+2} f \varepsilon A_1^1$ for $p = 1$.

C. Singular Values.

Let A be a bounded linear operator on a Hilbert space. The *singular numbers* of A are

$$s_n = \inf \{||A - K|| : \text{rank } K \leq n\}. \tag{2.11}$$

A is in the *Schatten–von Neumann* class S_p, $0 < p < \infty$, if $||A||_p^p = \sum s_n^p < \infty$. Each S_p is a two sided ideal, i.e.

$$\|XAY\|_p \leq \|X\| \, \|A\|_p \, \|Y\|, \tag{2.12}$$

whenever X, Y are bounded operators. For a more detailed treatement of S_p see [S] or [GK].

The following will be useful:

$$\|A\|_p^p \geq \sum |\langle A\varphi_i, \varphi_i\rangle|^p \tag{2.13}$$

whenever $\{\varphi_i\}$ is an orthonormal system and $1 \leq p < \infty$.

D. The Cayley Transform.

The generalized upper half space is biholomorphically equivalent to the unit ball and this fact allows us to transfer results between the settings of B_n and D_n. The equivalence can be given by the Cayley transform from B_n to D_n defined by $\varphi(\zeta,z) = \frac{i}{1-\zeta}(1 + \zeta,z)$.

We will define the small Hankel operator on $L^2(H_n)$ by $h_b = \overline{P}\, M_b P$.

<u>LEMMA</u> <u>(2.14)</u>: *Let* $\varphi\colon B_n \to D_n$ *be the Cayley transform and let* $U_\varphi f = \left(\frac{\partial\varphi}{\partial\zeta}\right)^{1/2} \prod_{i=1} \frac{\partial\varphi}{\partial z_i}\, f\circ\varphi$. *Then*

(i) $U_\varphi : H^2(D_n) \to H^2(B_n)$ *is unitary and* $U_\varphi^{-1} = U_{\varphi^{-1}}$,

(ii) $C_{b\circ\varphi} = U_\varphi\, C_b\, U_\varphi^{-1}$

(iii) $H_{b\circ\varphi} = U_\varphi\, H_b\, U_\varphi^{-1}$

(iv) $h_{b\circ\varphi} = U_\varphi\, h_b\, U_\varphi^{-1} - P_1$

(Here P_1 is the projection onto the subspace of $H^2(\sum_n)$ spanned by 1.).

<u>Proof</u>: (i) is simply a change of variables, combined with the fact that φ is biholomorphic and the observation that $|\frac{\partial\varphi}{\partial\zeta}|\ |\prod_{i=1}^{n} \frac{\partial\varphi}{\partial\zeta_i}|^2$ is the Jacobian of φ. (This can be computed directly – φ is conformal and a unit normal is mapped to $|\frac{\partial\varphi}{\partial\zeta}|$ times a unit normal – or see [K].). The proofs of (ii), (iii) and (iv) are all similar. To prove (iv), for example, let f ε H_2 and g ε $\overline{H}_0^2$.

Then

$$<h_{b\circ\varphi}\, U_\varphi f,\, g>_{B_n} = <Q(\overline{b\circ\varphi}\, \left(\frac{\partial\varphi}{\partial\zeta}\right)^{1/2} \prod_{i=1}^{n} \frac{\partial\varphi}{\partial z_i}\ P(f\circ\varphi)),\, g>_{B_n}$$

$$= <\overline{b\circ\varphi}\ \left(\frac{\partial\varphi}{\partial\zeta}\right)^{1/2} \prod_{i=1}^{n} \frac{\partial\varphi}{\partial z_i}\ P(f\circ\varphi),\, Qg>_{B_n}$$

$$= <U_\varphi(\overline{b}\ Pf),\, \overline{P}g>_{B_n} = <\overline{b}\ Pf,\, U_\varphi^{-1}\ \overline{P}g>_{D_n}$$

$$= <\overline{b}\ Pf,\, \overline{P}(U_\varphi^{-1}\ \overline{P}g)>_{D_n} = <\overline{P}\ \overline{b}\ Pf,\, U_\varphi^{-1}\ \overline{P}g>_{D_n}$$

$$= <U_\varphi\ h_b\ f,\, \overline{P}g>_{B_n} - 0 = <U_\varphi\ h_b\ f,\, \overline{P}g> - <U_\varphi\ f,\, P_1\ g>.$$

III. NEARLY WEAKLY ORTHONORMAL SEQUENCES.

A sequence of functions $\{e_B\}_{B \varepsilon \mathcal{B}}$ is said to be *nearly weakly orthonormal* (NWO) if $f \to f^*$ is bounded on $L^2(H)$ where

$$f^*(h) = \sup_{\substack{h \varepsilon B \\ B \varepsilon \mathcal{B}}} \left\{ |B|^{-1/2} \, |< f, e_B >| \right\}.$$

An important example is $\{\sigma_B^{n+1} \, S_{(\sigma_B^2, g_B)}\}$. That this is an NWO sequence is clear since f^* is dominated by the classical maximal function in D_n (See Theorem 6.3.1 of [Ru] or [KV].). Other examples are provided by the following lemma (which we use in section 7).

<u>LEMMA (3.1)</u>: *For* $a \geq \frac{1}{2n+2}$, *the sequence* $\{\sigma_B^{(1+2a)(n+1)} \, S_B^{1+a}\}$ *is nearly weakly orthonormal.*

<u>Proof</u>: First recall that $D^{1/2}$ is a bounded map from H^2 to $A_2^{1/2}$. (This can be proved just as in the Euclidean case by taking Fourier tranforms.) Then from the definition of NWO

$$f^*(h)^2 \leq \sum_{\substack{h \varepsilon B \\ B \varepsilon \mathcal{B}}} |B|^{-1} \, |< f, e_B >|^2$$

and

$$||f^*||_2^2 \leq c \sum_{B \varepsilon \mathcal{B}} |< f, e_B >|^2 .$$

From this and Corollary 2.9,

$$||f^*||_2^2 \leq c \sum \sigma_B^{(2n+2)(1+2a)} |< f, S_B^{1+a} >|^2$$

$$= c \sum \sigma^{(2n+2)+4a(n+1)} |D^{a(n+1)-1/2} D^{1/2} f|^2$$

$$\leq c ||D^{a(n+1)-1/2} D^{1/2} f||_{a(n+1),2} \leq c||D^{1/2} f||_{1/2,2} \leq c ||Pf||_2 \leq c ||f||_2.$$

(We have actually proved that the sequences in the lemma are weakly orthonormal. A sequence is said to be weakly orthonormal if $\sum |<f,e_B>|^2 \leq c||f||_2^2$ whenever $f \varepsilon L^2$. The first two inequalities in the proof show that any weakly orthonormal sequence is NWO. See [RS1] and [RS2].)

For a more complete exposition of NWO sequences and examples of their use, see the introduction to [RS2].

We will use NWO sequences to study Schatten-von Neumann ideal membership. In order to do this we need to extend some of the results of [RS2] from the Euclidean setting to H. We want to know that:

(3.2)

(i) if an operator $A = \sum_{B \epsilon \mathcal{B}} \lambda_B < \cdot\, , \varphi_B > \psi_B$ with $\{\varphi_B\}$ and $\{\psi_B\}$ NWO and $\{\lambda_B\} \;\varepsilon\; \ell^p$, then $\|A\|_p \le c_p \|\lambda\|_p$, $0 < p < \infty$.

(ii) if A is a bounded operator and $\{\varphi_B\}$ and $\{\psi_B\}$ are NWO then

$$\sum |< A\varphi_B, \psi_B >|^p \le c_p\, \|A\|^p_{S^p} \text{ for } 1 < p < \infty.$$

The proofs in [RS1] and [RS2] of these two facts generalize directly to the Heisenberg setting except for Lemma 4.1 of [RS1]. The proof there uses the fact that on $\mathbf{R}^n$ the geometry is simple enough so that dyadic cubes can be used in place of $\mathcal{B}$. We now give the needed variation of the lemma.

For a sequence $\{\lambda_B\}_{B \varepsilon \mathcal{B}}$ define a new sequence $\{M(\{\lambda_R\})(B)\}_{B \varepsilon \mathcal{B}}$ by $M(\lambda)(B) = \sum_i \sum_{A_i(B)} |\lambda_R| \frac{|R|}{|B|}$ where $A_o(B) = \{B\}$ and for $i = 1, 2, 3, \ldots,$

$$A_i(B) = \{B' \;\varepsilon\; \mathcal{B} : \text{rad}\,(B') = \frac{\text{rad}(B)}{2^i} \text{ and } B' \cap P \neq \emptyset \text{ for some } P \;\varepsilon\; A_{i-1}\}.$$

We define the discrete Carleson measure norm of a sequence $\{\lambda_B\}$ to be

$$\sup_{B \varepsilon \mathcal{B}} \frac{1}{|B|} \int_{\hat{B}} \left(\sum |\lambda_B|\, |B|\, \delta_{(\sigma^2_B, g_B)}\right) = \|\lambda_B\|_{cmd}.$$

Here $\hat{B}$ is the tent over B, i.e. $\hat{B} = \{(\tau,h) : (\text{ball of radius } \tau \text{ centered at } h) \subseteq B\}$. Since $\bigcup_{i=0}^{\infty} A_i(B)$ is contained in a ball $B' \;\varepsilon\; \mathcal{B}$ of radius no more than 2^5 it is easy to see that

$||M(\lambda)||_\infty \sim ||\sum |\lambda_B|\, |B|\, \delta_{(\sigma_B^2, g_B)}||_{cmd}$. Let $M(\lambda)^*(n)$ be the nonincreasing rearrangement of $M(\lambda)(B)$ and label the B's so that $M(\lambda)^*(n) = M(\lambda)(B_n)$. Let $\lambda^k(B_j) = 0$ for $j = 0, 1, \ldots, k-1$; and $\lambda^k(B_j) = \lambda_B$ for $j \geq k$.

<u>LEMMA</u> <u>(3.2)</u>: *$|| M(\lambda^k)(B) ||_\infty \leq c\, M(\lambda)^*(k)$.*

<u>Proof</u>: Let $A_i'(B) = A_i(B) \setminus \{B_0, \ldots, B_{k-1}\}$ and $L = \sup \{i : A_i(B) \neq A_i'(B)\} \leq k-1$ (or one if $A_0 = A_0'$). Let UA_i' denote the union of the sets in A_i'. Then since if $R \in A_i'(B)$ and $R \cap UA_k' \neq \emptyset$ for some $0 \leq k \leq i-1$ then $R \in UA_L(R_L)$ for some R_L:

$$M(\lambda^k)(B) = \sum_{i=0}^{\infty} \sum_{A_i'(B)} |\lambda_R|\, |R|\, |B|^{-1}$$

$$\leq |B|^{-1} \Bigg(\sum_{R_0 \in A_0'(B)} |R_0| \Big(\sum_{i=0}^{\infty} \sum_{R \in A_i(R_0)} |\lambda_R|\, |R|\, |R_0|^{-1} \Big)$$

$$+ \sum_{\substack{R_1 \in A_1'(B) \\ R_1 \cap UA_0'(B) = \emptyset}} |R_1| \Big(\sum_{i=0}^{\infty} \sum_{R \in A_i(R_i)} |\lambda_R|\, |R|\, |R_0|^{-1} \Big)$$

$$+ \cdots + \sum_{\substack{R_{L+1} \in A_{L+1}'(B) \\ R_{L+1} \cap U(A_0'(B) \cup \cdots \cup A_L'(B)) = \emptyset}} |R_{L+1}| \cdot \Big(\sum_{i=0}^{\infty} \sum_{R \in A_i(R_{L+1})} |\lambda_R|\, |R|\, |R_{L+1}|^{-1} \Big) \Bigg)$$

$$= |B|^{-1} \Big(\sum_{R_0 \in A_0'(B)} |R_0|\, M(\lambda^k)(R_0) + \cdots + \sum_{\substack{R_{L+1} \in A_{L+1}'(B) \\ R_{L+1} \cap U(A_0'(B) \cup \cdots \cup A_L'(B)) = \emptyset}} |R_{L+1}|\, M(\lambda^k)(R_{L+1}) \Big)$$

$$= |B|^{-1} \sum_{i=0}^{L+1} \sum_{R_i \varepsilon A_i'(B)} |R_i|\, M(\lambda^k)(R_i) \leq (\max M(\lambda^k)\ (R_i)).$$

$$R_i \cap U(A_0' \cap \cdot \cdot \cdot \cap A_i') = \emptyset$$

The last inequality follows from the fact that for each i, $\left\{\frac{1}{2}\cdot R_i\right\}_{R_i \epsilon A_i}$ is a disjoint collection and $\bigcup_{R_i \varepsilon A_i} R_i \subseteq 2\ B$. (Hence, $\sum_{A_i} |R_i| = 4 \sum_{A_i} |\frac{1}{2}\cdot R_i| \subseteq 16\ |B|$.)

We close this section with two remarks: (i) there are other possible variations of the lemma. For example, we could have chosen B to be a collection of disjoint sets, mimicking more closely the idea of dyadic cubes. (ii) A version of Lemma 3.2 is valid in a space of homogeneous type in the sense of Coifman and Weiss [CW]. We will use this extension on $\sum_n$ where $\varphi^{-1}(B)$ replaces B. (Recall that φ is the Cayley transform.).

IV. FOURIER TRANSFORM.

For simplicity, in this section, we change viewpoint a bit and consider b analytic and h_b as a map from $(H^2)^*$ to (H^2), and the bilinear form $h_b(f,g) = \langle \overline{h_{\overline{b}} f}, g \rangle = \langle b, f\, g \rangle_{H^2}$. $\left(\text{Here } (H^2)^* = R\ \overline{H^2} \text{ where } (Rf)\ (t,z\ ) = f\ (-t,\ z).\right)$ Let $h_d^{\alpha+\beta}(f,g) = \langle d, fg \rangle_{F_{\alpha+\beta}}$ be the Hankel form on $F_\lambda \otimes F_S$ with symbol d. We prove two propositions which extend results for classical Hankel operators and forms. The first is a Heisenberg version of $\langle h_b\ f, g \rangle =$

$< b, f\, g > = \int_0^\infty \int_0^\infty \hat{b}(\lambda + s)\, \hat{f}(\lambda)\, \hat{g}(s)\, d\lambda ds$ and the second is the analog of

$$\left(h_b(f)\right)^{\hat{}}(\lambda) = \int_0^\infty \hat{b}(\lambda + s)\, \hat{f}(s)\, ds\, .$$

<u>THEOREM</u> <u>(4.1)</u>: $h_b(f,g) = \int_0^\infty \int_0^\infty h^{\lambda + s}_{(Fb)(\lambda+s)} \left((Ff)(\lambda),\, (Fg)(s)\right) \lambda\, d\lambda s ds.$

<u>Proof</u>: The proof follows from the Plancherel Theorem. Specifically,

$$h_b(f,g) = < bf,\, g > \; = c \int_0^\infty \int_{\mathbb{C}} [bf]^{\hat{}}(\lambda,z)\, \bar{\hat{g}}(\lambda,z)\, dz d\lambda$$

$$= c \int_0^\infty \int_{\mathbb{C}} \left(\int_0^\infty \hat{b}(\lambda + s,z)\, \bar{\hat{f}}(s,z)\, ds\right) \bar{\hat{g}}(\lambda,z)\, dz d\lambda$$

$$= c \int_0^\infty \int_0^\infty \left(\int_{\mathbb{C}} (\lambda + s)^{-1}\, \hat{b}(\lambda + s,z) e^{2\pi(\lambda+s)|z|^2} s^{-1}\, \bar{\hat{f}}(s,z)\, e^{2\pi s|z|^2} \lambda^{-1}\right.$$

$$\times \quad \left. \bar{\hat{g}}(\lambda,z) e^{2\pi\lambda|z|^2} (\lambda+s) e^{-4\pi(\lambda+s)|z|^2}\, dz\right) s ds \lambda d\lambda$$

$$= c \int_0^\infty \int_0^\infty \int_{\mathbb{C}} (Fb)(\lambda + s,z)\, \overline{(Ff)}\, (s,z)\, \overline{(Fg)}\, (\lambda,z)\, (\lambda + s)\, e^{-4\pi(\lambda+s)|z|^2}\, dz(t) s ds \lambda d\lambda$$

which is the desired identity.

<u>THEOREM</u> <u>(4.2)</u>:

Let h_b be the (reduced) Hankel operator with symbol b. Then

$$\left(F(h_b(f))\right)(s,w) = c \int_0^\infty \int_{\mathbb{C}} (Fb)\left(\lambda + s,\, \frac{\lambda z + sw}{\lambda + s}\right) \overline{(Ff)}(\lambda,z)\, \lambda\, e^{-4\pi\lambda|z|^2}\, dz \lambda d\lambda.$$

<u>Proof</u>: Let $\{\psi_j\}$ be an orthonormal basis for $H^2(H)$. By Plancherel

$$\left(\mathcal{F}(h_b(f))\right)(s,w) = \sum_j < h_b f, \hat{\psi}_j > \hat{\psi}_j\,(s,w).$$

We show that

$< h_b\, f, \psi_j > =$

$$c\int_0^\infty \int_{\mathbf{C}} \left(\int_0^\infty \int_{\mathbf{C}} (\mathcal{F}b)\left(\lambda + s, \frac{\lambda z + sw}{\lambda + s}\right)\overline{(\mathcal{F}f)}(\lambda,z)\, \lambda\, e^{-4\pi\lambda|z|^2} dz\lambda d\lambda\right) \overline{(\mathcal{F}\psi_j)}\,(s,w)\, dwds$$

and thus conclude that the quantity in parenthesis is $\left(F(h_b(f))\right)(s,w)$. Using the reproducing kernel for F_λ and Theorem 4.1,

$< h_b\, f, \psi_j > = h_b\,(f,\psi_j)$

$$= \int_0^\infty \int_0^\infty \int_{\mathbf{C}} (\mathcal{F}b)(\lambda+s,z)\, \overline{(\mathcal{F}f)}(\lambda,z)\, \overline{(\mathcal{F}\psi_j)}(s,z)(\lambda+s)e^{-4\pi(\lambda+s)|z|^2} dz\lambda d\lambda sds$$

$$= \int_0^\infty \int_0^\infty \int_{\mathbf{C}} (\mathcal{F}b)(\lambda + s,z)\, \overline{(\mathcal{F}f)}(\lambda,z) \left(\int_{\mathbf{C}} \overline{(\mathcal{F}\psi_j)}\,(s,w)\, e^{4\pi s\bar{z}w} e^{-4\pi s|w|^2} sdw\right)$$

$$\times \quad (\lambda + s)\, e^{-4\pi(\lambda + s)|z|^2} dz\lambda d\lambda sds$$

$$= \int_0^\infty \int_{\mathbf{C}} \left(\int_0^\infty \int_{\mathbf{C}} (\mathcal{F}b)(\lambda+s,z)\overline{(\mathcal{F}f)}(\lambda,z)e^{4\pi s\bar{z}w}(\lambda+s)\lambda e^{-4\pi(\lambda+s)|z|^2} dzd\lambda\right)$$

$$\times \quad \overline{(F\psi_j)}(s,w)e^{-4\pi s|w|^2} sdw.$$

Again using the reproducing kernel (this time to evaluate the integral in brackets) we get

$$\int_0^\infty \int_{\mathbb{C}} (\mathcal{F}b)(\lambda + s,z)\, \overline{(\mathcal{F}f)}(\lambda,z)\, e^{4\pi(s\,\bar{z}w-(\lambda+s)|z|^2)} (\lambda + s)\, dz\lambda d\lambda$$

$$= \int_0^\infty \int_{\mathbb{C}} \Big[\int_{\mathbb{C}} \overline{(\mathcal{F}f)}(\lambda,u)\, e^{4\pi\lambda u\bar{z}}\, e^{-4\pi\lambda|u|^2} \lambda du \Big] (\mathcal{F}b)(\lambda + s,z)$$

$$\times \quad e^{4\pi(s\,\bar{z}w-(\lambda+s)|z|^2)} (\lambda + s)\; dz\lambda d\lambda$$

$$= \int_0^\infty \int_{\mathbb{C}} \Big[\int_{\mathbb{C}} (\mathcal{F}b)(\lambda + s,z)\, e^{4\pi(\lambda u\bar{z} + s\bar{z}w - (\lambda+s)|z|^2)}$$

$$\times \quad (\lambda + s)\, dz \Big] \overline{(\mathcal{F}f)}(\lambda,u)\, \lambda\, e^{-4\pi\lambda|u|^2} dwd\lambda$$

$$= \int_0^\infty \int_{\mathbb{C}} (\mathcal{F}b)\Big(\lambda + s, \frac{\lambda u + sw}{\lambda + s}\Big) \overline{(\mathcal{F}f)}(\lambda,u)\, e^{-4\pi\lambda|u|^2} \lambda du\lambda d\lambda.$$

The bilinear form $h_{b(\lambda+s)}^{\lambda+s}$ is studied in [JPR] (c.f. Theorem 7.6). We conclude with the remark that Proposition (4.1) is also true for the big Hankel operator H_b where $h_d^{\alpha+\beta}$ is replaced by the Hankel form $H_d^{\alpha+\beta}$ on $L^2(\mathbb{C}, e^{-4\pi|\alpha||z|^2}dz) \times L^2(\mathbb{C}, e^{-4\pi|\beta||z|^2}dz)$. The proof (though easy) involves more facts from non-commutative Fourier analysis and we omit it.

V. PROOF OF THEOREM A.

As stated in the introduction, we will prove a slightly more general version of parts (ii) and (iii) in the setting of the Heisenberg group, H_1.

The proof that $b \varepsilon$ BMO implies that C_b is bounded can be found in [J]. The proof there is in the setting of $\mathbb{R}^n$, but all the arguments and tools used are available in the setting of a space of homogeneous type. We discuss the converse at the end of this section. For the proof of part (iii) we actually show that if C_b is in S_b then b is in MO^p ($p \geq 1$). Part (iii) then follows from the fact that MO^p contains only the constant functions for $p \leq 4$. This can be seen by arguing as in Section 4 of [RS2] and making use of the fact that the homogeneous dimension of $\sum$ (or H) is four.

We now turn to the proof of (ii). The techniques of [RS2] (specifically, Section 2) apply to our setting with minor modifications except for a lemma (Lemma 2.9 of [RS2]). We will briefly outline the general approach and then prove the needed version of the lemma in detail.

Suppose b is in B^p. We will study the commutator, C_b, of M_b with a Calderon-Zygmund operator with kernel K(h,g). We will assume K(h,g) satisfies

$$(5.1) \qquad |H^\alpha G^\beta K(h,g)| \leq C(\alpha,\beta) \cdot |hg^{-1}|^{-4-|\alpha|-|\beta|}.$$

(Here H^α and G^β are differential operators in the variables h and g respectively.) An induction argument shows that $H^\alpha G^\beta S(h,g) = P(h,g)\ S(h,g)^{(1 + \frac{|\alpha| + |\beta|}{2})}$

where P is a polynomial of homogeneous degree less than or equal to $|\alpha| + |\beta|$. Hence, S satisfies (5.1). We will decompose $\left(C_b f\right)(h) = \int (b(h) - b(g))\ (h,g)\ f(g)$ dg as a sum of the form $\sum_{\alpha \varepsilon A} \sum_{B \varepsilon \mathcal{B}} \lambda_{B,\alpha}\ \gamma_B < f, G_{B,\alpha} > F_{B,\alpha}$ where A is $\mathbb{Z}^6 \times$ (a finite set), $\{G_{B,\alpha}\}_{B \epsilon \mathcal{B}}$ and $\{F_{B,\alpha}\}_{B \epsilon \mathcal{B}}$ are NWO for each α and $|\lambda_{B,\alpha}| \leq C(k)(1 + |\alpha|)^{-k}$ for every k with constants C(k) independent of B and , most importantly, $\gamma_B \leq c\ \mathrm{osc}(b,B)$. We then apply the theory of NWO sequences, including Lemma (1.16) of [RS2] about sums of operators, to obtain that $C_b \varepsilon S^p$. We obtain the decomposition of C_b by first writing the kernel as

$$(5.2) \qquad \sum_{\overline{B}\ \varepsilon\ \overline{\mathcal{B}}} (b(h) - b(g))\ K(h,g)\ \psi_{\overline{B}}\ (h,g)$$

where $\overline{\mathcal{B}}$ is a Whitney decomposition of $H \times H \backslash \Delta$ consisting of sets of the form $B^1 \times B^2$ where $|B^1| = |B^2|$ and $B^1, B^2\ \varepsilon\ \mathcal{B}$ and $\{\psi_{\overline{B}}\}$ is a smooth partition of unity subordinate to the cover. We then decompose the functions $K_{\overline{B}}\ (h,g) = K(h,g)\psi_{\overline{B}}(h,g)$ as

$$(5.3) \qquad \sum_{j \varepsilon \mathbb{Z}^6} \lambda_{B,i}\ |B^1|^{-1/2}\ e_{B^1,j}\ (h)\ |B^2|^{-1/2}\ f_{B^2,j}(g).$$

A rearrangment of (5.3) then gives us (5.2). We now prove the lemma.

<u>LEMMA (5.4)</u>: *Suppose $K(h,g)$ satisfies (5.1) and $\overline{B}$ is a Whitney decomposition of $(H \times H)\setminus\Delta$ as above. Then for $\overline{B} \in \overline{B}$ $(\overline{B} = B^1 \times B^2)$ and $|j| = |(j_1, j_2, j_3)| + |(j_4, j_5, j_6)| = 0, 1, 2, \ldots$ there exist functions $e_{\overline{B},j}(h)$ and $f_{\overline{B},j}(g)$ with supp $e_{\overline{B},j} \subset B^1$ and supp $f_{\overline{B},j} \subset B^2$, $\|e_{\overline{B},j}\|_\infty \leq 1$, $\|f_{\overline{B},j}\|_\infty \leq 1$ and such that (5.3) holds with $|\lambda_{\overline{B},j}| \leq c(k)\,(1 + |j|)^{-k}$ for*
$k = 0, 1, 2, \ldots$.

<u>Proof</u>: Let $B^1 = B(h_0,r)$ and $B^2 = B(g_0,r)$. Choose $\epsilon < 1$ so that $B(h_0,r(1 + \epsilon)) \times B(g_0,r(1 + \epsilon))$ so that misses the diagonal. Let $\psi(h,g)$ be smooth with supp $\psi \subseteq B(0, (1 + \epsilon)r) \times B(0, (1 + \epsilon)r)$, $0 \leq \psi \leq 1$, $\psi \equiv 1$ on $B(0, (1 + \frac{\epsilon}{2})\, r) \times B(0, (1 + \frac{\epsilon}{2})\, r)$ and $|H^\alpha G^\beta \psi| \leq C(\alpha,\beta)$. Set $\widetilde{K}_{\overline{B}}(h,g) = \psi(h,g)\, K_{\overline{B}}(h\, h_0^{-1}, g\, g_0^{-1})$. Since $(t,x,y) \in B(0, (1 + \epsilon)\, r)$ implies that $(t,x,y) \in [-(1 + \epsilon)^2\, r^2, (1 + \epsilon)^2\, r^2] \times [-(1 + \epsilon)\, r, (1 + \epsilon)\, r] \times [-(1 + \epsilon)\, r, (1 + \epsilon)\, r]$,
$\widetilde{K}_{\overline{B}}$ can be expanded in a multiple Fourier series, $\widetilde{K}_{\overline{B}}(h,g)$
$= c\sum_j \widetilde{\lambda}_{\overline{B},j}\, r^{-2}\, \widetilde{e}_{\overline{B},j}(h) r^{-2}\, \widetilde{f}_{\overline{B},j}(g)$ where
$\widetilde{e}_{\overline{B},j}(h) = \widetilde{e}_{\overline{B},j}(t,x,y) = e^{i(j_1 r^{-2}(1+\varepsilon)^{-2} t^2 + j_2 r^{-1}(1+\varepsilon)^{-1} x \ + j_3 r^{-1}(1+\varepsilon)^{-1} y)}$. Thus,

$K_{\overline{B}}(h,g) = c\sum_j \lambda_{\overline{B},j}\, |B^1|^{-1/2}\, e_{\overline{B},j}(h)\, \chi_{B^1}(h)\, |B^2|^{-1/2}\, f_{\overline{B},j}(g)\, \chi_{B^2}(g)$. Note that after Heisenberg translation $\widetilde{e}\,\widetilde{f}$ is no longer a product of exponentials but it is still a product of a function of h and a function of g, which is exactly what we need.

To estimate $|\lambda_{\overline{B},j}|$ we integrate by parts (see [FS]). Let $\left(\frac{\partial}{\partial g}\right)^{\alpha} = \frac{\partial^{\alpha_t}}{\partial t^{\alpha_t}} \frac{\partial^{\alpha_x}}{\partial x^{\alpha_x}} \frac{\partial^{\alpha_y}}{\partial y^{\alpha_y}}$.

$$|\lambda_{\overline{B},j}| = c\, |r^{-4} \int_{B_0 \times B_0} K_{\overline{B}}^{\,2}(h,g)\, r^{|\alpha+|\beta|}\, (1+\varepsilon)^{|\alpha|+|\beta|}$$

$$\times \left(\frac{\partial}{\partial g}\right)^{\alpha}\left(\frac{\partial}{\partial h}\right)^{\beta} j_1^{-\alpha_t}\, j_2^{-\alpha_x}\, j_3^{-\alpha_y}\, j_4^{-\beta_t}\, j_5^{-\beta_x}\, j_6^{-\beta_y}\quad e_{\overline{B},j}(h)\; f_{\overline{B},j}(g)\, dhdg|$$

$$\leq \quad c\,(1+\varepsilon)^{|\alpha|+|\beta|}\, r^{4+|\alpha|+|\beta|}\quad |(j_1,j_2,j_3)|^{-|\alpha|}\quad |(j_4,j_5,j_6)|^{-|\beta|}$$

$$\times \sup_{(h,g)} |\left(\frac{\partial}{\partial g}\right)^{\alpha} \left(\frac{\partial}{\partial h}\right)^{\beta} K_{\overline{B}}^{\,2}(g,h)|$$

$$= \quad c\,(1+\varepsilon)^{|\alpha|+|\beta|}\, r^{4+|\alpha|+|\beta|}\quad |(j_1,j_2,j_3)|^{-|\alpha|}\quad |(j_4,j_5,j_6)|^{-|\beta|} \times$$

$$\sup_{(h,g)} |T^{\alpha_t}(X+2yT)^{\alpha_x}(Y-2xT)^{\alpha_y}T^{-\beta_t}(X'+2yT')^{\beta_x}(Y'-2xT')^{\beta_y}K((t,x,y),(t',x',y'))|$$

$$\leq \quad c\,(1+\varepsilon)^{|\alpha|+|\beta|}\quad |(j_1,j_2,j_3)|^{-|\alpha|}\quad |(j_4,j_5,j_6)|^{-|\beta|}\, C(\alpha,\beta).$$

For the converses of (i) and (ii) we need only assume K is smooth away from the origin, homogeneous of (Heisenberg) degree -4, and doesn't vanish identically. Clearly, S satisfies these conditions.

Again, the techniques of [RS2] (this time Section 3) apply now that we have Lemma 5.4, so we just recall that approach. The idea is to select sequences $\{e_B\}$ and $\{f_B\}$ so that the numbers $|< C_b e_B, f_B >|$ control, for example, the ℓ^p norm of the mean oscillation numbers of b. This is done by applying Lemma 5.4 to $K^{-1}(h,g)\chi_{B^1 \times B^2}$ and summing the operators with these kernels weighted by a $\ell^{p'}$ sequence, λ. Then, $\|b\|_{MO^p} \leq \text{trace}\,(C_b h) \leq c\, \|C_b\|_{S_p}\, \|\lambda\|_{p'}$. The estimates used on $|\langle C_b e_b, f_b \rangle|$ also show that C_b is bounded when $b \in \text{BMO}$.

This completes Theorem A for the Heisenberg group. Theorem A for the sphere follows from unitary equivalance and Theorem C. (Recall that we only consider conjugate analytic functions on the sphere.)

VI. THE PROOF OF THEOREM B(iii).

We return to the setting of $\sum_n$ (Unitary equivalence proves the theorem for H_n.). Suppose H_b is a non-zero Hankel operator in some S_p. By our standing assumption, b is conjugate analytic. We now regard b as extended to B_{n+1}, the unit ball, as a conjugate analytic function. Since H_b is not the zero operator, b is not the constant function. Hence there is a point (ζ', z') in B_{n+1} so that either

$$(6.1) \qquad \left(\frac{\partial}{\partial \bar{\zeta}}\, b\right)(\zeta') \neq 0 \qquad \text{or} \qquad \left(\frac{\partial}{\partial \bar{z}_j}\, b\right)(z'_j) \neq 0 \qquad \text{for some } j.$$

By an easy extension of Lemma 2.10, for any conformal automorphism U of B_{n+1}, H_b is unitarily equivalent to $H_{b \circ U}$. Since (6.1) holds we can choose U so that $\tilde{b} = b \circ U$ satisfies

$$(6.2) \qquad \left(\frac{\partial}{\partial \bar{\zeta}}\, \tilde{b}\right)(0) = c \neq 0$$

Let H_θ be the Hankel operator with symbol $e^{-i\theta}\,\widetilde{b}\,(e^{i\theta}\zeta,z)$. H_θ will also be in S_p as will

$$H = \frac{1}{2\pi}\int_0^{2\pi} H_\theta \, d\theta$$

By direct computation (using, for instance, the associated bilinear form) H is the Hankel operator with symbol $b(\zeta,z) = c\bar{z}$ with the constant c from (6.2). Hence, $H_{\bar{\zeta}}$ is in S_p. (This pattern of analysis is from [AFP].)

If R is in S_p for some $p \geq 2$ and e_n is any orthonormal set then, by (2.9) applied to R^*R,

(6.3) $$\sum \|Re_n\|^p < \infty$$

We will now do the explicit computation to see what (6.3) says when $R = H_{\bar{\zeta}}$ and the e_n are the monomials. We will write monomials as $\zeta^\alpha\, z^\beta$ where $\alpha \in \mathbb{Z}^+$ and $\beta \in (\mathbb{Z}^+)^n$. We will write H for $H_{\bar{\zeta}}$.

$$H(\zeta^\alpha\, z^\beta) = (I - P)\,(\bar{\zeta}\,\zeta^\alpha\, z^\beta) = \bar{\zeta}\,\zeta^\alpha\, z^\beta - P(\bar{\zeta}\,\zeta^\alpha\, z^\beta).$$

To compute $P(\bar{\zeta}\,\zeta^\alpha\, w^\beta)$, it suffices to sum the projections of $\bar{\zeta}\,\zeta^\alpha\, z^\beta$ onto span of each of the normalized analytic monomials. However, the projection of $\bar{\zeta}\,\zeta^\zeta z^\beta$ onto the span of $\zeta^\lambda z^\gamma$ will be zero unless $\alpha - 1 = \lambda$, $\beta = \gamma$. (This is seen be writing $\zeta = r\,e^{i\theta}$ and computing the appropriate inner product by first doing the θ integral; or doing the analogous computation with one of the components of (ζ, z).) Thus

(6.4) $$H(\zeta^\alpha z^\beta) = \bar{\zeta}\, \zeta^\alpha z^\beta - k\, \zeta^{\alpha-1} z^\beta$$

where the constant k is determined by

(6.5) $$k = < \bar{\zeta}\, \zeta^\alpha z^\beta, \zeta^{\alpha-1} z^\beta > / \; ||\zeta^{\alpha-1} z^\beta||^2$$

Here the norm and inner product are on $L^2(\Sigma_n)$.

Set

$$d_{2\alpha,2\beta} = \int_{\Sigma_n} |\zeta|^{2\alpha}|z|^{2\beta}\, dr.$$

Thus,

(6.6) $$k = d_{2\alpha,2\beta} \,/\, d_{2\alpha-2,2\beta}$$

From (6.4)

$$||H(\zeta^\alpha z^\beta)||^2 = d_{2\alpha+2,\beta} + k^2 d_{2\alpha-2,2\beta} - 2k d_{2\alpha,2\beta}.$$

By (6.6) this is

$$||H(\zeta^\alpha z^\beta)||^2 = d_{2\alpha+2,2\beta} - d^2_{2\alpha,2\beta} \,/\, d_{2\alpha-2,2\beta}.$$

Hence,

$$\frac{||H(\zeta^\alpha z^\beta)||^2}{||\zeta^\alpha z^\beta||^2} = \frac{d_{2\alpha+2,2\beta}}{d_{2\alpha,2\beta}} - \frac{d_{2\alpha,2\beta}}{d_{2\alpha-2,2\beta}}$$

We now simplify this using the fact that

$$d_{2h,2k} = \frac{n!\,h!\,k!}{(n + h + |k| - 1)!}$$

([Ru] pg 16). So

$$\frac{||H\,(\zeta^{\alpha}\,z^{\beta})||^2}{||\zeta^{\alpha}\,z^{\beta}||^2} = \frac{n + |\beta|}{(n + 1 + \alpha + |\beta|)\,(n + \alpha + |\beta|)}$$

Thus, letting $e_{\alpha,\beta} = \zeta^{\alpha}\,z^{\beta}/\,||\zeta^{\alpha}\,z^{\beta}||$ we have

(6.7) $$||H(e_{\alpha,\beta})|| \sim \frac{(n + |\beta|\,)^{1/2}}{n + \alpha + |\beta|}$$

Hence, by (6.3) we have

(6.8) $$S = \sum_{\alpha \varepsilon \mathbb{Z}^{+}} \sum_{\beta \varepsilon (\mathbb{Z}^{+})^{n}} \left(\frac{(n + |\beta|)^{1/2}}{n + \alpha + |\beta|}\right)^{p} < \infty$$

Let $j = |\beta|$. Elementary counting shows that there is a constant C so that the number of β with $|\beta| = j$ is at least $(cj)^{n-1}$. Thus

$$S \geq c \sum_{\alpha \varepsilon \mathbb{Z}^{+}} \sum_{j \varepsilon \mathbb{Z}^{+}} \frac{(n + j)^{p/2}}{(n + \alpha + j)^{p}} \cdot j^{n-1}.$$

Interchange order of summation and estimating the sum by an integral we find

$$S \geq c \sum_{j} \frac{j^{n-1}}{(n + j)^{\frac{p}{2}-1}}.$$

By (6.7) $S < \infty$. Hence we must have $\frac{p}{2} - n > 1$; which is the required conclusion.

VII. PROOF OF THEOREM C.

Part (i) is Theorem VII of [CRW] (The atomic decomposition for $H^1(H_n)$ can be found in [D].), so we prove part (ii). The following arguments work for H_n and $\sum_n$ (modulo the rank one operator that subtracts the mean from an $\overline{H}^2$ function).

Suppose $h_b \varepsilon S_p$. First, consider $p > 1$. By 3.1 (ii), 2.8 and the discussion at the beginning of Section 3,

$$\|h_b\|_{S_p}^p \geq c\sum |<\overline{P}\ b\ P(\sigma^{n+1} S_B), \sigma^{n+1} \bar{S}_B>|^p \geq c\sum |<b, \sigma^{2n+2} \bar{S}_B^2>|^p$$

$$= c\sum \sigma^{(2n+2)p} |(D^{n+1} b)(\sigma_B^2, h_B)|^p$$

$$= c\sum \sigma^{(2n+4)+2(\beta p-1)} |(D^{n+1} b)(\sigma_B^2, h_B)|^p$$

where $\beta = (n+1)(1-\frac{1}{p})$. Arguing as in the proof of Theorem 1 of [R1] the right hand side is a Riemann sum for $\|D^{n+1} b\|^p_{(n+1)(1-\frac{1}{p})p}$ and thus $D^{n+1} b\ \varepsilon$ $A_p^{(n+1)(1-\frac{1}{p})}$ as required.

Now consider p=1 and suppose $h_b\ \varepsilon\ S_1$. By applying Theorem 2.6 to $A_2^{1/2}$ we have that $\{\sigma_B^{(\theta+2)(n+1)} B_B^{\frac{\theta+2}{2}}(\tau,h)\}$ is the image of an orthonormal basis for $A_2^{1/2}$ under a bounded map. Since $D^{-1/2}$ is bounded from $A_2^{1/2}$ to H^2, $\{\sigma_B^{(\theta+1)(n+1)} D^{-1/2} B_B^{\frac{\theta+2}{2}}(\tau,h)\}$ is the image of an orthonormal basis for H^2.

Arguing as for p > 1 and using (2.12) and (2.13),

$||h_b||_{S_1} \geq c \sum \sigma_B^{(\theta+1)(2n+4)} |< D^{(\theta+1)(n+2)} b >|$, and, therefore, $D^{(n+2)}\, b \varepsilon\, A_1^{\theta(n+2)+1}$.

By Corollary 2.9 D^{n+1} b ε A_1^0.

In the other direction, suppose $D^{n+1}(b)\ \varepsilon\ A_p^{(n+1)(1-\frac{1}{p})}$, $(p \geq 1)$. We will use the Coifman-Rochberg decomposition to write h_b as a sum of operators of a fixed finite rank. (In the case n = 0 one actually gets rank one.) By Theorem 2.6,

$$(D^{n+1}\ b)\ (\tau,h) = \Sigma\ \sigma_B^{2(\theta+1)(n+1)}\ B_B^{(\theta+2)(\frac{n+1}{n+2})},$$

and thus,

$$\begin{aligned} b(\tau,h) &= \Sigma\ \sigma_B^{2(\theta+1)(n+1)}\ \bar{B}_B^{(\theta+1)(\frac{n+1}{n+2})}(\tau,h) + \text{constant} \\ &= \Sigma\ \sigma_B^{2(\theta+1)(n+1)}\ \bar{S}_B^{(\theta+1)}(\tau,h) + \text{constant}, \end{aligned}$$

and

$$b(h) = \lim_{\tau\to 0} b(\tau,h) = \Sigma \sigma_B^{2(\theta+1)(n+1)}\ \bar{S}_B^{(\theta+1)}(h) + \text{constant}.$$

Therefore,

$$(h_b f)\ (h) = \lim_{\tau\to 0} < \bar{P}\ b\ P\ f,\ \bar{S}_{(\tau,h)} > = \lim_{\tau\to 0}\ < b\ P\ f,\ \bar{S}_{(\tau,h)} >$$

$$= \lim_{\tau\to 0} \Sigma\ \lambda_B\ \sigma_B^{2(\theta+1)(+1)}\ < \bar{S}_B^{(\theta+1)}\ Pf,\ \bar{S}_{(\tau,h)} >$$

$$= \sum \lambda_B \, \sigma_B^{2(\theta+1)(n+1)} \lim_{\tau\to 0} < S_{(\tau,h)} \, Pf, S_B^{(\theta+1)} >$$

$$= \sum \lambda_B \, \sigma_B^{2(\theta+1)(n+1)} \lim_{\tau\to 0} D^{\theta(n+1)} (S_{(\tau,h)} \, Pf)(\sigma_B^2, g_B)$$

$$= \sum \lambda_B \, \sigma_B^{2(\theta+1)(n+1)} \lim_{\tau\to 0} \sum_{\ell=0}^{\theta(n+1)} \binom{\theta(n+1)}{\ell} \left(D^{\ell}(Pf)\right)(\sigma_B^2, g_B)$$

$$\times \left(D^{\theta(+1)-\ell} \, S_{(\tau,h)}\right) (\sigma_B^2, g_B)$$

$$= \sum \lambda_B \, \sigma_B^{2(\theta+1)(n+1)} \lim_{\tau\to 0} \sum_{\ell=0}^{\theta(n+1)} \binom{\theta(n+1)}{\ell} \left(D^{\ell}(Pf)\right) (\sigma_B^2, g_B)$$

$$\times \left(S^{1+\theta-\frac{\ell}{n+1}}_{(\tau,h)}\right) (\sigma_B^2, g_B)$$

$$= \sum \lambda_B \, \sigma_B^{2(\theta+1)(n+1)} \lim_{\tau\to 0} \sum_{\ell=0}^{\theta(n+1)} \binom{\theta(n+1)}{\ell} < Pf, S_B^{1+\frac{\ell}{n+1}} > \bar{S}_B^{1+\theta-\frac{\ell}{n+1}} (\tau,h)$$

$$= \sum_{\ell=0}^{\theta(n+1)} \binom{\theta(n+1)}{\ell} \sum \lambda_B < Pf, \sigma_B^{(n+1)+2\ell} \, \bar{S}_B^{1+\frac{\ell}{n+1}} > \sigma_B^{(n+1)+2\theta(n+1)-2\ell}$$

$$\times \bar{S}_B^{1+\theta-\frac{\ell}{n+1}} (h) .$$

By Lemma (3.1), $\sigma_B^{(1+2a)(n+1)} \, S_B^{1+a}$ is NWO for $a \geq \frac{1}{2n+2}$. Thus, we have written h_b as a finite sum of operators in S_p (Recall from Section 3 that the case a=0 is NWO.).

VIII. HANKEL OPERATORS ON B_n.

The following three observations can be found on page 626 of [CRW] (Also, see Section 7.2 of [Ru].) :

(1) If $f \varepsilon L^p(\sum_n)$ and doesn't depend on ζ then

$$\int_{\Sigma_n} |f|^p \, d\sigma = c_n \int_{B_n} |f(1-|z|^2,z)|^p \, dz.$$

(2) If $f \varepsilon L^p (B_n, (1 - |(\zeta,z)|^2)^k d\zeta dz)$ and doesn't depend on ζ then

$$\int_{B_n} |f(\zeta,z)|^p (1 - |(\zeta,z)|^2)^k \, d\zeta dz = c_{k,n} \int_{B_{n-1}} f(1 - |z|^2,z)|^p (1 - (\zeta,z)|^2)^{k+1} \, dz.$$

(3) If $f, g \varepsilon H^2 (\sum_n)$ then $f_0(z) = \frac{1}{2\pi}\int_0^{2\pi} f(e^{i\theta}\zeta,z) \, d\theta$ is in $H^2(B_n)$ with $||f_0||_2 \leq ||f||_2$ and $(fg)_0 = f_0 g_0$.

It is also easy to see that

(4) If $S_{(\zeta,z)}(\eta,w)$ is the Cauchy-Szego kernel for $H^2(\sum_n)$, then

$$\left(< f_0, (S_{(\zeta,z)})_0 > \Sigma_n\right)_0 = c_n < f_0, B_Z > B_n.$$

Let $\widetilde{P}$ be the Bergman projection and $\widetilde{Q}$ the projection onto the $\overline{A}_2^{1/2}$ functions with mean zero.

From the above $QM_b\ Pf = \widetilde{Q}^2 M_{b_0}\ \widetilde{P}\ f_0$. In other words, for functions in $H^2(\Sigma_n)$ depending only on z, $h_b\ f = c_n\ \widetilde{h}_{b_0}\ f_0$ where $\widetilde{h}_d$ is the small Hankel operator acting on $A_2^{1/2}\ (B_n)$. Therefore these observations and Theorem C combine to prove:

<u>THEOREM</u> <u>8.1</u>: *For $p > 1$, $\widetilde{h}_b\ \varepsilon\ S_p$ if and only if $D^n b\ \varepsilon\ A_p^{n(\frac{1}{p}-1)+\frac{1}{p}}$. $\widetilde{h}_b\ \varepsilon\ S_1$ if and only if $D^{n+1}b\ \varepsilon\ A_1^2$.*

For n = 1 this is Theorems 9.3, 9.4 and Corollary 9.5 of [AFP].

IX. FINAL REMARKS AND SOME OPEN PROBLEMS.

We first note that Theorem C and Lemma 2.13 show that compostion with the Cayley tranform is a linear isometry from the Sobolev space of holomorphic functions $WA^p_{(n+1)(p-1),n+1}\ (D_n)$ onto $WA^p_{\beta_p,n+1}(B_n)$ for $p > 1$, and from $WA^1_{0,n+2}$ (D_n) onto $WA^1_{0,n+2}(B_n)$. (See [BB] for more on $W^p_{q,s}\ (B_n)$.) A natural question then is exactly when is the Cayley transform an isometry?

We were able to obtain as a corollary results for the small Hankel operator acting on the Bergman space of the ball in one less dimension. Can this be done for the large Hankel operator?

REFERENCES

[AFP] J. Arazy, S. Fisher and J. Peetre, Hankel operators on weighted Bergman spaces, to appear, Amer. J. Math.

[BB] F. Beatrous and J. Burbea, Sobolev spaces of holomorphic functions in the ball, to appear in Dissertationes Matematica.

[B] J. Burbea, Trace ideal criteria for Hankel operators in the ball, in preparation.

[CR] R. Coifman and R. Rochberg, Representation theorems for holomorphic and harmonic functions in L^p, Astérisque, 77(1980), pp. 12-66.

[CRW] R.R. Coifman, R. Rochberg, and Guido Weiss, Factorization theorems for Hardy Spaces in Several Variables, Ann. Math. **103**(1976), pp. 611-635.

[CW] ______ and G. Weiss, Extensions of Hardy spaces and their use in analysis, Bull. Amer. Math. Soc., **83**(1977), pp. 569-645.

[D] J. Dorronsorro, Weighted Hardy spaces on Hermitian Hyperbolic Space, dissertation, Washington Univ., St. Louis, Mo., 1981.

[F] M. Feldman, Mean Oscillation spaces on the Heisenberg group and the atomic decomposition of Bergman spaces, manuscript.

[FS] G. Folland and E.M. Stein, *Hardy Spaces on Homogeneous Groups*, Math. Notes, 628 (1982), Princeton Univ. Press.

[GK] I. Gohberg and M.G. Krein, *Introduction to the Theory of Linear Non-Self adjoint Operators*, 18(1969), AMS Translations.

[J] S. Janson, Mean oscillation and commutators of singular integral operators, Ark. Mat. 16 (1978), pp. 263-270.

[JW] S. Janson and T. Wolff, Schatten classes and commutators of singular integral operators, Ark. Mat. 20(1982), pp. 301-310.

[K] A. Koranyi, The Poisson integral for generalized half-planes and bounded symmetric domains, Ann. Math. 82(1965), pp. 332-350.

[KV] A. Koranyi and S. Vagi, Singular integrals in homogeneous spaces and some problems of classical analysis, Ann. Scuala Normale Superiore Pisa, 25(1977), pp. 575-648.

[OV] D. Ogden and S. Vagi, Harmonic analysis of a nilpotent group and function theory on Siegel domains of type II, Adv. in Math., 33(1979), pp. 31-92.

[RT] F. Ricci and M. Taibleson, Boundary values of harmonic functions in mixed norm spaces and their atomic structure, Ann. Scuola Norm. Sup. Pisa, C1, Sci. Ser. IV 10 (1983), pp. 1-54.

[R1] R. Rochberg, Trace ideal criteria for Hankel operators and commutators, Indiana U. Math. J., **31**(1982) pp. 913-925.

[R2] ______, Interpolation by functions in Bergman spaces, Mich. Math. Journal, 29 (1982), pp. 229-236.

[RS1] ______ and S. Semmes, A decomposition theorem for functions in BMO and applications, J. Functional Anal., 67(1986), pp. 228-263.

[RS2] _______________, Nearly weakly orthonormal sequences, singular value estimates, and Calderon-Zygmund operators, to appear in Journal of Functional Analysis.

[Ru] W. Rudin, *Function Theory in the Unit Ball of* $\mathbb{C}^n$, (1980), Springer-Verlag.

[S] B. Simon, *Trace Ideals and Their Applications*, (1979) Cambridge University Press.

[Z] G. Zhang, Hankel operators on Hardy spaces and Schatten classes, preprint.

8

General Endpoint Results in Extrapolation

José García-Cuerva

The purpose of this note is to extend to a general setting some of the results obtained in [7] for the A_p classes , in the same way as the extrapolation theorem of José-Luis Rubio de Francia (see [9] or [5],chapterIV) was extended by B.Jawerth[8].

We shall be working on a fixed σ -finite measure space X,with measure simply denoted by dx.For f measurable on X, we shall write:$\|f\|_p = (\int_X |f(x)|^p dx)^{1/p}$, for $1 \leq p < \infty$, whereas $\|f\|_\infty$ will stand for the essential supremum of $|f(x)|$ over X. We shall denote by L^p the usual Lebesgue space formed by those f (actually, equivalent classes modulo functions vanishing almost everywhere) for which $\|f\|_p < \infty$.

For u a non-negative measurable function and $1 \leq p < \infty$, $L^p(u)$ will be the space of those measurable f for which:

$$\|f\|_{p,u} = (\int_X |f(x)|^p u(x) dx)^{1/p} < \infty$$

and uL^∞ will be the space of those f such that $|f|u^{-1} \in L^\infty$. $L^p(u)$ is a Banach space with norm given by $\|f\|_{p,u}$. Likewise uL^∞ is a Banach space with norm $\|f\|_{\infty,u} = \||f|u^{-1}\|_\infty$

We shall be concerned with operators or mappings S taking some normed space $dom(S)$ of measurable functions into functions which are at least integrable over sets of finite measure. We shall say that such an operator S is **subadditive** if:

1) $|S(f+g)(x)| \leq |S(f)(x)| + |S(g)(x)|$ a.e. for every $f, g \in dom(S)$

We shall say that S is **sublinear** if it is subadditive and, besides:

2) $|S(\lambda f)(x)| = |\lambda||S(f)(x)|$ a.e. for every number λ and every $f \in dom(S)$

Finally for a sublinear operator S , we shall say that S is a **positive sublinear** operator if:

3) $|f| \leq g \in dom(S)$ implies $f \in dom(S)$ and $|S(f)(x)| \leq S(g)(x)$ a.e.

In particular $g \geq 0$ implies $S(g) \geq 0$

We shall say that a mapping T is bounded in some Banach space B of functions if $B \subset dom(T)$ and there is a constant C such that ,for every $f \in B$, $T(f)$ is also in B and $\|T(f)\|_B \leq C\|f\|_B$.
The infimum of all the constants C for which the inequality holds will be called the norm of T as an operator in B

Here is our basic result:

THEOREM 1. Let S be a positive sublinear operator and let T be a mapping satisfying the following condition:

Every time that S is bounded in vL^∞ for some $v \geq 0$, the mapping T is also bounded in vL^∞, with norm depending only on that of S

Let $1 \leq p < \infty$ and $w(x) \geq 0$. Suppose that S is bounded in $L^p(w)$. Then T is also bounded in $L^p(w)$, with norm depending only on that of S.

PROOF: To say that S is bounded in vL^∞ is equivalent to saying that $S(v)(x) \leq Cv(x)$ for a.e.x. Indeed,this condition is clearly necessary ,as one sees by applying S to $v \in vL^\infty$, and, conversely, if it holds, we have for $f \in L^\infty$:

$$|S(vf)|v^{-1} \leq S(v\|f\|_\infty)v^{-1} = \|f\|_\infty S(v)v^{-1} \leq C\|f\|_\infty$$

that is,writing $vf = g$:$\|S(g)\|_{\infty,v} \leq C\|g\|_{\infty,v}$

The fact that S is bounded in $L^p(w)$ means that:

$$\int_X |S(f)(x)w(x)^{1/p}|^p dx = \int_X |S(f)(x)|^p w(x) dx \leq$$

$$\leq C \int_X |f(x)|^p w(x) dx = C \int_X |f(x)w(x)^{1/p}|^p dx$$

If we make $fw^{1/p} = g$ or $f = w^{-1/p}g$, we have:

$$\int_X |S(w^{-1/p}g)w^{1/p}|^p \leq C \int_X |g|^p$$

In other words,the operator S^*, given by $S^*(f) = S(w^{-1/p}g)w^{1/p}$ is bounded in L^p.Now S^* is also a positive sublinear operator and we can apply to it the iterative process of José-Luis Rubio de Francia (see [5], chapter IV, section 5), in the following way:

Given $h \geq 0$, $h \in L^p$, we form

$$H = \sum_{j=0}^{\infty} (2\|S^*\|)^{-j} S^{*j}(h)$$

where S^{*j} is the result of iterating S^* j times and $\|S^*\|$ is the norm of S^* as an operator in L^p ,which is, of course, the same as the norm of S as an operator in $L^p(w)$.

We then have: $h \leq H$, $\|H\|_p \leq 2\|h\|_p$ and:

$$S^*(H) \leq \sum_{j=0}^{\infty} (2\|S^*\|)^{-j} S^{*j+1}(h) =$$

$$= 2\|S^*\| \sum_{j=1}^{\infty} (2\|S^*\|)^{-j} S^{*j}(h) \leq 2\|S^*\|H = CH$$

Note that C is twice the norm of S as an operator in $L^p(w)$.

It follows that $S^*(H) = S(w^{-1/p}H)w^{1/p} \leq CH$.If we call $v = w^{-1/p}H$, we have: $S(v) \leq Cv$, so that S is bounded in vL^∞.Consequently, T is also bounded in vL^∞

We shall need the following simple observation, taken from [7]: Given $\phi \in L^1$ with $\|\phi\|_1 = 1$, there exists $\psi \in L^1, \psi > 0$, with $\|\psi\|_1 \leq 2$, such that $\|\phi/\psi\|_\infty = 1$. We simply take $\psi(x) = |\phi(x)|$ if $\phi(x) \neq 0$ and $\psi(x) = \phi_0(x)$ if $\phi(x) = 0$, where $\phi_0(x) > 0$ is a fixed integrable function with $\|\phi_0\|_1 = 1$.Clearly, a condition necessary and sufficient for the existence of such a ϕ_0 is that the space X be σ -finite, as we are assuming.

Now, given $f \in L^p(w)$, we set $\phi = (\int_X |f|^p w)^{-1}|f|^p w$.Obviously $\|\phi\|_1 = 1$ and we can find $\psi > 0$ with $\|\psi\|_1 \leq 2$ and $\|\phi/\psi\|_\infty = 1$. Equivalently:

$$\||f|^p w/\psi\|_\infty = \int_X |f|^p w, \text{ or } \|f\|_{p,w} = \|\psi^{-1/p} f w^{1/p}\|_\infty$$

Let $h = \psi^{1/p}$. Then $\|h\|_p^p = \|\psi\|_1 \leq 2$

To $h \in L^p$, we associate H as above, so that T is bounded in $w^{-1/p}HL^\infty$, with norm depending only on the norm of S as an operator in $L^p(w)$, and not on h. Then:

$$\begin{gathered}\|f\|_{p,w} = \|fw^{1/p}h^{-1}\|_\infty \geq \|fw^{1/p}H^{-1}\|_\infty = \|f(w^{-1/p}H)^{-1}\|_\infty \geq \\ \geq C\|T(f)(w^{-1/p}H)^{-1}\|_\infty \geq C\|T(f)w^{1/p}H^{-1}\|_\infty(\int_X H(x)^p dx)^{1/p} \geq \\ \geq C(\int_X |T(f)(x)|^p w(x)H(x)^{-p}H(x)^p dx)^{1/p} = C\|T(f)\|_{p,w}\end{gathered}$$

Since C is independent of f and it only depends upon the norm of S as an operator in $L^p(w)$, we get that T is bounded in $L^p(w)$ with norm depending only on that of S.#

At this point one could be tempted to extend theorem 1 to pairs of weights, as it is done in [6] with Jawerth's theorem. However, that extension is a triviality. Indeed, suppose that T maps uL^∞ boundedly into vL^∞ every time that S does, i.e.: every time that $S(u) \leq Cv$. Then, for any f, we can take $u = |f|$ and $v = S(|f|)$, and we have $\|T(g)v^{-1}\|_\infty \leq C\|gu^{-1}\|_\infty$. In particular $\|T(f)/S(|f|)\|_\infty \leq C$, so that : $|T(f)| \leq CS(|f|)$. Of course, T is bounded in any space where S is bounded, but this is not very exciting.

Let us introduce some notation, which will seem reasonable after looking at some typical examples to be given below. For any mapping T and $1 < p < \infty$, we shall denote by $\mathcal{A}_p(T)$ the class of those weights $w \geq 0$, such that T is bounded in $L^p(w)$. Likewise, we shall denote by $\mathcal{A}_1(T)$ the class of those weights $w \geq 0$, such that T is bounded in wL^∞. Naturally, if T is sublinear and positive, $w \in A_1(T)$ if and only if $T(w) \leq Cw$ a.e.

With this notation, theorem 1 can be reformulated in the following way:

Let S be a positive sublinear operator and T a mapping such that $\mathcal{A}_1(S) \subset \mathcal{A}_1(T)$. Then, for every $1 < p < \infty$: $\mathcal{A}_p(S) \subset \mathcal{A}_p(T)$.

We shall devote the rest of the paper to give some examples,applications and further results.

1)Let $S = M$ be the Hardy-Littlewood maximal operator in $\mathbf{R}^n$ or even in a space of homogeneous type (see [2]), then $\mathcal{A}_1(M) = A_1$ and $\mathcal{A}_p(M) = A_p$, the classes of weights introduced by Muckenhoupt (see [5], chapter IV). In this case we get the theorem in [7], namely:

If T is bounded in wL^∞ for every $w \in A_1$, then T is bounded in $L^p(w)$ for every $w \in A_p$ and every $1 < p < \infty$.

2)For some $1 < r < \infty$, let $S = M_r$, where $M_r(f)(x) = (M(|f|^r)(x))^{1/r}$. It follows from Jensen's inequality that $M_r \geq M$. Now, for every $p > r$, $\mathcal{A}_p(M_r) = A_{p/r}$ and $\mathcal{A}_1(M_r) = \{v \geq 0 : v^r \in A_1\}$, so that we get:

If T is bounded in wL^∞ for every $w \geq 0$ such that $w^r \in A_1$, then T is bounded in $L^p(w)$ for every $w \in A_{p/r}$ and every $r < p < \infty$.

3)There are few operators to which theorem 1 can be immediately applied. Many interesting operators (singular integrals, for instance) do not preserve L^∞; they map L^∞ into $B.M.O.$. However, the sharp maximal function $f \mapsto f^\sharp$ introduced by C.Fefferman and E.M.Stein in [3], allows us to go back to L^∞. Let us define the relevant terms. We shall do it in $\mathbf{R}^n$, but it is clear how to extend everything to a space of homogeneous type:

First of all, for $f \in L^1_{\text{loc}}(\mathbf{R}^n)$, we define

$$f^\sharp(x) = \sup \frac{1}{|Q|} \int_Q |f(y) - f_Q| dy$$

where f_Q denotes the average of f over Q, that is:$f_Q = \frac{1}{|Q|}\int_Q f(y)dy$, and the sup is taken over all the cubes(with sides parallel to the coordinate axes) Q containing x.

Now $B.M.O.$ is the space of those functions f (modulo constants) such that $f^\sharp \in L^\infty$. $B.M.O.$ is a Banach space with the norm

$$\|f\|_{B.M.O.} = \|f^\sharp\|_\infty = \sup_Q \frac{1}{|Q|} \int_Q |f(y) - f_Q| dy$$

For $w \geq 0$, we can define $B.M.O.(w)$ as we did in [4]: It is the space of those f, modulo constants, for which

$$\|f\|_{B.M.O.(w)} = \sup_Q \frac{1}{w(Q)} \int_Q |f(y) - f_Q| dy < \infty$$

where $w(Q) = \int_Q w(x)dx$

We shall be mainly concerned with the case $w \in A_1$. In that case $f \in B.M.O.(w)$ implies $f^\sharp \in wL^\infty$. Indeed:

$$\frac{1}{|Q|} \int_Q |f(y) - f_Q| dy = \frac{w(Q)}{|Q|} \frac{1}{w(Q)} \int_Q |f(y) - f_Q| dy \leq$$

$$\leq C \frac{w(Q)}{|Q|} \leq CM(w)(x) \leq Cw(x)$$

for any $x \in Q$. Thus we get: $f^\sharp(x) \leq C\|f\|_{B.M.O.(w)}w(x)$ or, in other words: $\|f^\sharp\|_{\infty,w} \leq C\|f\|_{B.M.O.(w)}$ where C is the A_1-constant for w.

Actually, as noted in [1], the converse is also true, but we shall not need this. We just need the fact that for $w \in A_1$, the sharp maximal function is bounded from $B.M.O.(w)$ to wL^∞, with norm depending only on the A_1-constant for w

Now we can extend theorem 1 in the following way:

THEOREM 2. Let S be a positive sublinear operator in $\mathbf{R}^n$, such that $S \geq M$, the Hardy-Littlewood operator. Suppose that the mapping T satisfies the following property: For every $w \in \mathcal{A}_1(S)$, T is bounded from wL^∞ to $B.M.O.(w)$ with a norm depending only on the norm of S as an operator in wL^∞. Then, for every $1 < p < \infty$: $\mathcal{A}_p(S) \subset \mathcal{A}_p(T)$, provided we know, a priori, that for every $w \in \mathcal{A}_p(S)$ and every $f \in L^p(w)$, $T(f) \in L^p(w)$

In any case , if $1 < p < \infty$ and $w \in \mathcal{A}_p(S)$, the inequality:

$$\left(\int_{\mathbf{R}^n} |T(f)(x)|^p w(x)dx\right)^{1/p} \leq C\left(\int_{\mathbf{R}^n} |f(x)|^p w(x)dx\right)^{1/p}$$

with C independent of f, holds whenever the left hand side is finite.

PROOF: We shall apply theorem 1 to the operator $T^\sharp$ defined by $T^\sharp(f)(x) = (T(f))^\sharp(x)$. Every time that S is bounded in vL^∞,that is,if $v \in \mathcal{A}_1(S)$, we know that $T : vL^\infty \longmapsto B.M.O.(v)$ is bounded.Besides, since $S \geq M$, $\mathcal{A}_1(S) \subset A_1$, so that the sharp maximal function is bounded from $B.M.O.(v)$ to vL^∞. The composition $T^\sharp$ is, thus, bounded in vL^∞. Not only that, but also, the norm of $T^\sharp$ depends only on that of S. It follows from theorem 1, that, for every $1 < p < \infty$ and every $w \in \mathcal{A}_p(S)$, $T^\sharp$ is bounded in $L^p(w)$. To complete the proof, we just need to observe that $\mathcal{A}_p(S) \subset A_p \subset A_\infty$, and for $w \in A_\infty$ we have the inequality:

$$\int_{\mathbf{R}^n} |M(f)(x)|^p w(x)dx \leq C \int_{\mathbf{R}^n} |f^\sharp(x)|^p w(x)dx$$

(see [5], chapter IV), valid if the left hand side is finite.

We finally get:

$$\int_{\mathbf{R}^n} |T(f)(x)|^p w(x)dx \leq C \int_{\mathbf{R}^n} |(T(f))^\sharp(x)|^p w(x)dx \leq C \int_{\mathbf{R}^n} |f(x)|^p w(x)dx$$

as we wanted to prove #

The need to guarantee a priori that $T(f) \in L^p(w)$ for every $w \in \mathcal{A}_p(S)$ and every $f \in L^p(w)$ is an unsatisfactory aspect of this theorem. Since we are dealing with abstract operators, we cannot do better. However, in practice, it is generally possible to find a dense class of functions for which we have $T(f) \in L^p(w)$

In some of the following examples we have to make the same proviso, although we shall not write it explicitly all the time.

By using theorem 2 with $S = M$ or $S = M_r$, we can improve the results contained in 1) and 2) above, obtaining:

a)**If T is bounded from wL^∞ to $B.M.O.(w)$ for every $w \in A_1$, then T is bounded in $L^p(w)$ for every $w \in A_p$ and every $1 < p < \infty$.** (This result is in [7])

b)**If T is bounded from wL^∞ to $B.M.O.(w)$ for every $w \geq 0$ such that $w^r \in A_1$, where $1 < r < \infty$ is fixed, then T is bounded in $L^p(w)$ for every $w \in A_{p/r}$ and every $r < p < \infty$**

4)Given a function $0 < \nu < \infty$ a.e. in $\mathbf{R}^n$, we define the operator $\mathcal{M}_\nu$ by

$$\mathcal{M}_\nu(f) = M(f) + M(f/\nu)\nu$$

Then $\mathcal{A}_1(\mathcal{M}_\nu) = \{w \geq 0 : w \in A_1 \text{ and } w/\nu \in A_1\}$

This class is denoted by $A_1^{(\nu)}$ in [10]. Also $\mathcal{A}_p(\mathcal{M}_\nu)$ will be the class of weights w such that both M and $M(f/\nu)\nu$ are bounded in $L^p(w)$, that is :

$$\mathcal{A}_p(\mathcal{M}_\nu) = \{w \geq 0 : w \in A_p \text{ and } \nu^p w \in A_p\}$$

This class is denoted by $A_p^{(\nu)}$ in [10].

Theorem 2 applies in this situation, giving us the following result, contained in [10]:

If T is bounded from wL^∞ to $B.M.O.(w)$ for every $w \in A_1^{(\nu)}$, then T is bounded in $L^p(w)$ for every $w \in A_p^{(\nu)}$ and every $1 < p < \infty$

5)With ν as above and a fixed number $\lambda > 1$, we consider

$$S(f) = M(|f|^\lambda)^{1/\lambda} + M((|f|/\nu)^\lambda)^{1/\lambda}\nu$$

Clearly $S \geq M$. In this case:

$$\mathcal{A}_1(S) = \{w \geq 0 : w^\lambda \in A_1 \text{ and } w^\lambda \nu^{-\lambda} \in A_1\}$$

This class is denoted by $A_{1,\lambda}^{(\nu)}$ in [10].

$$\mathcal{A}_p(S) = \{w \geq 0 : w \in A_{p/\lambda} \text{ and } \nu^p w \in A_{p/\lambda}\}$$

provided $p > \lambda$. This class is denoted by $A_{p,\lambda}^{(\nu)}$ in [10].

By applying theorem 2, we obtain the following extrapolation theorem of [10]:

If T is bounded from wL^∞ to $B.M.O.(w)$ for every $w \in A_{1,\lambda}^{(\nu)}$, then T is bounded in $L^p(w)$ for every $w \in A_{p,\lambda}^{(\nu)}$ and every $\lambda < p < \infty$.

In [10] this theorem is used to prove a commutator theorem for singular integrals with kernels satisfying an L^s-Dini condition. It is given here as an illustration of the varied results one can obtain from the general principles by selecting the operator S appropriately

6)Let us look for a result dual to theorem 2. First of all, wL^∞ is the dual space of $L^1(w)$, in other words, there is a one to one correspondence (preserving norms) between continuous linear functionals Λ over $L^1(w)$ and functions $g \in wL^\infty$. This correspondence is given by

$$\Lambda(f) = \int f(x)g(x)dx$$

for every $f \in L^1(w)$

Likewise, $B.M.O.(w)$ is the dual space of the weighted Hardy space $H^1(w)$. These spaces were studied in [4] for the line, and in [11] for spaces of homogeneous type. In our case $w \in A_1$, and the situation is particularly simple.

There are at least two ways to define $H^1(w)$. The first one uses the "grand" maximal function $f \mapsto f^*$ (see [4] or [5], chapter III). $H^1(w)$ is the space of those functions f for which:

$$\|f\|_{H^1(w)} = \|f^*\|_{1,w} < \infty$$

The second one uses the notion of atom. For $w \in A_1$, an atom with respect to w, or w-atom, is a function a, supported on a cube Q, and satisfying:

(i) $$\int_{\mathbf{R}^n} a(x)dx = 0$$

(ii) $$|a(x)| \leq w(Q)^{-1} \qquad \text{for a. e. } x$$

The atomic space $H^1_{\mathrm{at}}(w)$ is defined as the class of those f which can be written as

$$f = \sum_{j=1}^{\infty} \lambda_j a_j$$

where the a_j's are w-atoms and $\sum_{j=1}^{\infty} |\lambda_j| < \infty$. The norm is given by the infimum of all the quantities $\sum_{j=1}^{\infty} |\lambda_j|$ for all the possible ways to write f as a sum of w-atoms

Since $w \in A_1$, it is well known that the two spaces coincide with equivalence of norms. It follows from the atomic description that the dual of $H^1(w)$ is $B.M.O.(w)$.

Indeed if a is a w-atom and $g \in B.M.O.(w)$:

$$|\int_{\mathbf{R}^n} ag| = |\int_Q a(x)(g(x) - g_Q)dx| \leq$$
$$\leq \|a\|_\infty \int_Q |g(x) - g_Q|dx \leq \frac{1}{w(Q)} \int_Q |g(x) - g_Q|dx \leq \|g\|_{B.M.O.(w)}$$

We conclude (as in [4] or [5]) that the pairing $\Lambda(f) = \int fgdx$ (with the appropriate interpretation of the integral) provides a one to one correspondence between linear functionals Λ over $H^1(w)$ and functions $g \in B.M.O.(w)$ with equivalence of norms.

We can obtain the following result, dual to theorem 2:

THEOREM 3. Let S be a positive sublinear operator in $\mathbf{R}^n$, such that $S \geq M$, and satisfying the following additional property:

(*) For every $1 < p < \infty$, $w \in \mathcal{A}_p(S)$ if and only if $w^{1-p'} \in \mathcal{A}_{p'}(S)$, where p' is the exponent conjugate to p, so that $(1/p) + (1/p') = 1$

Suppose that T is a linear operator which for every $w \in \mathcal{A}_1(S)$,maps $H^1(w)$ to $L^1(w)$ with a norm depending only on the best constant in the inequality $S(w) \leq Cw$. Then, for every $1 < p < \infty$, $\mathcal{A}_p(S) \subset \mathcal{A}_p(T)$

PROOF. We shall look at the dual operator T^*. For every $w \in \mathcal{A}_1(S)$, we have $T : H^1(w) \longmapsto L^1(w)$ and, consequently: $T^* : wL^\infty \longmapsto B.M.O.(w)$. We can apply theorem 2 to T^* ,concluding that $\mathcal{A}_{p'}(S) \subset \mathcal{A}_{p'}(T^*)$.

Now we just have to realize that the spaces $L^p(v)$ and$L^{p'}(v^{1-p'})$ are in duality with the usual pairing$(f,g) \longmapsto \int fg$,so that : $v \in \mathcal{A}_p(T)$ if and only if$v^{1-p'} \in \mathcal{A}_{p'}(T^*)$.Thus

$$w \in \mathcal{A}_p(S) \Longrightarrow w^{1-p'} \in \mathcal{A}_{p'}(S) \subset \mathcal{A}_{p'}(T^*) \Longrightarrow w \in \mathcal{A}_p(T) \qquad \#$$

Actually theorem 3 works as well for T non linear. The restriction to T linear was due to our method of proof by duality. Let us indicate how to prove theorem 3 for a general T.

Let $w \in \mathcal{A}_p(S) \subset A_p$. Then, given $f \in L^p(w)$, $\|T(f)\|_{p,w} = \int |T(f)|g$ for some $g \geq 0$, $g \in L^{p'}(w^{1-p'})$ with $\|g\|_{p',w^{1-p'}} = 1$.

The hypothesis (*) implies that $w^{1-p'} \in \mathcal{A}_{p'}(S)$. To g we can associate

$$G = \sum_{j=0}^{\infty}(2\|S\|)^j S^j g$$

,where $\|S\|$ is the norm of S as a bounded operator in $L^{p'}(w^{1-p'})$. Then we have $g \leq G$, $\|G\|_{p',w^{1-p'}} \leq 2$ and $S(G) \leq 2\|S\|G$, so that $G \in \mathcal{A}_1(S)$.

It follows that $T : H^1(G) \longmapsto L^1(G)$.

Now we shall use the first characterization we gave of weighted $H^1_,$ the one involving the "grand" maximal function f^*. Note that $f^* \leq CM(f)$. We have:

$$\|T(f)\|_{p,w} = \int |T(f)|g \leq \int |T(f)|G \leq$$

$$\leq C\int |f^*|G \leq C(\int |f^*|^p w)^{1/p}(\int G^{p'}w^{1-p'})^{1/p'} \leq C\|f\|_{p,w}$$

since $w \in A_p$. This ends the proof #

Observe that the condition (*) is also present in Jawerth's theorem ([8] theorem 6.7) every time one wants to extrapolate from below, whereas no condition is needed to extrapolate from above. In that sense, theorem 2 is the most satisfactory general result.

REFERENCES.

[1] COIFMAN,R.; DAVID,G.; MEYER,Y. and SEMMES,S.
"ω-Calderón-Zygmund operators" in Fourier Analysis and Partial Differential Equations. Proceedings of El Escorial. June 1987. To appear in Springer Lecture Notes.

[2] COIFMAN,R. and WEISS,GUIDO.
"Extensions of Hardy spaces and their use in Analysis" Bull. Amer. Math. Soc. 83(1977)569-645.

[3] FEFFERMAN,C. and STEIN,E.M.
"H^p spaces of several variables" Acta Math. 129(1972)137-193.

[4] GARCIA-CUERVA,J.
"Weigted H^p spaces" Dissert. Math. 162, Warsaw 1979.

[5] GARCIA-CUERVA,J. and RUBIO DE FRANCIA,J.L.
"Weighted Norm Inequalities and Related Topics" Math. Studies 116, North-Holland 1985.

[6] HERNANDEZ,E.
"Factorization and extrapolation of pairs of weights" preprint.

[7] HARBOURE,E.; MACIAS,R.A. and SEGOVIA,C.
"Extrapolation results for classes of weights" Amer. Jour. of Math. 110(1988)383-397.

[8] JAWERTH,B.
"Weighted inequalities for maximal operators:linearization,localization and factorization" Amer. Jour. of Math. 108(1986)361-414.

[9] RUBIO DE FRANCIA,J.L.
"Factorization and extrapolation of weights" Bull. Amer. Math. Soc. 7(1982)393-396.

[10] SEGOVIA,C. and TORREA,J.L.
"Extrapolation for pairs of related weights" this volume.

[11] STROMBERG,J.O. and TORCHINSKY,A.
"Weights, Sharp maximal functions and Hardy spaces" Bull. Amer. Math. Soc. 3(1980)1053-1056.

9

Fractional Integrals on Spaces of Homogeneous Type

A. Eduardo Gatto and Stephen Vági

INTRODUCTION

The purpose of this paper is to extend the classical results about fractional integrals to the context of spaces of homogeneous type. It turns out that they are valid, and that many of the classical proofs can be adapted if some mild "natural" conditions are imposed on the space. Depending on the result they are: normality, existence of an order, and a cancellation property. The case of a space of finite measure and the case of a space whose measure has atoms are included.

In IV. it is shown that the theory applies to the parabolic spaces, in particular to Euclidean spaces, homogeneous groups,and $\mathbb{Z}^n$.

The authors thank Professor Carlos Kenig for stimulating their interest in this subject.

Supported in part by a grant from the Faculty Research and Development Fund of the College of Liberal Arts and Sciences of DePaul University.

I. DEFINITIONS, NOTATIONS AND STATEMENT OF RESULTS

Let X be a non empty set. A non negative function $\delta(x,y)$ on XxX is a quasi–distance if it satisfies the following conditions:

for every x and y in X $\delta(x,y)=0$ if and only if $x=y$, (I,1)

for every x and y in X $\delta(x,y)=\delta(y,x)$, (I,2)

there exists a positive constant κ such that

$$\delta(x,z)\leq\kappa(\delta(x,y)+\delta(y,z)) \qquad (I,3)$$

for every x,y and z in X.

If X has more than one element then κ cannot be less than 1. This trivial case is excluded from consideration in this paper. If $\kappa=1$, then δ is a metric.If ρ is a metric on X and $\lambda>1$ then $\delta = \rho^{\lambda}$ is a quasi–distance which is not a metric.

The subsets $\{(x,y) : \delta(x,y)<\epsilon, \epsilon>0\}$ of XxX define a base of a metrizable uniform structure on X. The balls $B_r(x)= \{y: \delta(y,x) <r, r>0\}$ form a base of neighborhoods for the topology induced by the uniform structure. The "open" balls $B_r(x)$ are not necessarily open in the topology (however see the remarks following the definition of order below).

The non empty set X equipped with the quasi–distance δ is a space of homogeneous type (X,δ,μ) if there exists a positive measure μ defined on a σ–algebra of subsets of X which contains the open sets of X and the balls $B_r(x)$ with the following property: there exists a positive constant A such that for every x in X and every $r>0$

$$0<\mu(B_{2r}(x)) \leq A\mu(B_r(x)). \qquad (I,4)$$

A measure satisfying these conditions is said to be doubling; any such measure is totally σ–finite by (I,4). It was proved by Macías and Segovia (1979b) that the set of points of X which have positive measure is countable, and that for every such point there exists an r>0 such that $B_r(x)=\{x\}$. In view of this fact one can define for every x in X a number $r_x \geq 0$ as follows

$$r_x=0 \text{ if } \mu(\{x\})=0$$

$$r_x=\sup\{r>0 : B_r(x)=\{x\}\}, \text{ if } \mu\{x\}\neq 0. \tag{I,5}$$

Note that $B_{r_x}(x)=\{x\}$ if $r_x>0$. Similarly, for every x in X one can define a number R_x as follows

$$R_x=+\infty \text{ if } \mu(X) = \infty$$

$$R_x= \inf\{r>0 : B_r(x)=X\}. \tag{I,6}$$

A space of homogeneous type (X,δ,μ) is said to be <u>normal</u> if there are positive constants A_1 and A_2 such that for all x in X

$$A_1 r \leq \mu(B_r(x)) \text{ if } 0 < r < R_x \tag{I,7}$$

and

$$\mu(B_r(x)) \leq A_2 r \text{ if } r \geq r_x. \tag{I,8}$$

A space of homogeneous type (X,δ,μ) is said to be of <u>order γ</u>, $0<\gamma\leq 1$ if there exists a positive constant M such that for every x, y and z in X,

$$|\delta(x,z)-\delta(y,z)| \leq M\delta(x,y)^{\gamma} (\max\{\delta(x,z), \delta(y,z)\})^{1-\gamma}. \tag{I,9}$$

<u>Remark I.1.</u> If (X,δ,μ) is of order γ, then it follows readily that the "open" balls $B_r(x)$ are, in fact, open in the topology of X, and that consequently μ is a totally σ – finite Borel measure.

<u>Remark I.2.</u> If δ is a metric then (I,9) holds with κ =M=1. If ρ is a metric and $\delta = \rho^{\lambda}$ with $\lambda>1$, then it can be seen readily that δ satisfies (I,9) with $\gamma=1/\lambda$.

The above definitions of normality and order can be seen without difficulty to be equivalent to those given by Macías and Segovia (1979c).

The following property of spaces of homogeneous type is important in the theory of fractional integrals. It extends a result obtained in Gatto, Gutiérrez and Wheeden (1983).

A normal space of homogeneous type (X,δ,μ) which is of order γ is said to have the cancellation property of order $1-\alpha$ $(0<\alpha<\gamma)$ if the integral of the function of y

$$\delta(x_1,y)^{\alpha-1}-\delta(x_2,y)^{\alpha-1} \tag{I,10}$$

vanishes on $X-\{x_1,x_2\}$ for every choice of x_1 and x_2. The fact that (I,10) is integrable is the result of Corollary II.1.

If (X,δ,μ) is a space of homogeneous type then for $\beta>0$ the space $\mathrm{Lip}(\beta)$ is defined as the set of functions φ defined on X which satisfy

$$|\varphi(x)-\varphi(y)| \leq C\ \delta(x,y)^{\beta} \tag{I,11}$$

for all x and y with a constant C independent of x and y. The greatest lower bound of these constants is denoted by $\|\varphi\|_{\mathrm{Lip}(\beta)}$. If the space is of order γ then $\mathrm{Lip}(\beta)$ for $0<\beta\leq\gamma$ has non constant elements. Note that $\|\varphi\|_{\mathrm{Lip}\,(\beta)}=0$ if and only if φ is constant, therefore if functions which differ by a constant are identified one obtains a Banach space which will be denoted by $\underline{\mathrm{Lip}}(\beta)$.

If $0<q<\infty$ then $L^q=L^q(X,\delta,\mu)$ has its usual meaning and $\|f\|_{L^q}=\|f\|_q=(\int_X |f|^q d\mu)^{1/q}$. For a measurable function f the distribution function λ_f is defined for $t>0$ by $\lambda_f(t)=\mu(\{x:|f(x)|>t\})$, and f belongs to weak L^p $(=wL^p)$ for

$p\geq 1$ if $[f]_p^p=\sup\{t^p\lambda_f(t):t>0\}$ is finite. Chebyshev's inequality states that $[f]_p\leq\|f\|_p$.

A measurable function ψ defined on X which is integrable on bounded subsets of X is said to be of <u>bounded mean oscillation</u> (BMO) if there is a positive constant C such that for every ball B contained in X

$$\frac{1}{\mu(B)}\int_B |\psi(x)-m_B(\psi)|\,d\mu(x) \leq C, \tag{I,12}$$

where $m_B(\psi)$ in (I,12) is the average of ψ on B:

$$m_B(\psi)=\frac{1}{\mu(B)}\int_B\psi(x)d\mu(x) \tag{I,13}$$

The greatest lower bound of the constants in (I,12) will be denoted by $\|\psi\|_*$, it is a norm if functions differing by a constant are identified. The normed linear space thus obtained is complete in its norm and is denoted by BMO.

If $0<p\leq 1$ then a <u>p–atom</u> is a bounded measurable function a supported in a ball B such that

$$\|a\|_\infty\leq\mu(B)^{-1/p} \tag{I,14}$$

and

$$\int_X a(x)d\mu(x) =0 \tag{I,15}$$

If $\mu(X)<\infty$ then μ will be assumed normalized: $\mu(X)=1$. The spaces H^p are defined as follows. A measurable function f on a space of homogeneous type (X,δ,μ) is said to belong to H^1 if it is integrable and if there exists a sequence $\{a_i\}$ of 1-atoms and a sequence of numbers $\{\lambda_i\}$ with $\sum|\lambda_i|<\infty$ such that

$$f=\sum_{i=1}^{\infty}\lambda_i a_i\ , \tag{I,16}$$

convergence being in L^1. One defines $\|f\|_{H^1}$ by

$$\|f\|_{H^1} = \inf\left(\sum_{i=1}^{\infty} |\lambda_i|\right), \tag{I,17}$$

where the infimum is taken over all possible representations of the form (I,16). In the following definition (X,δ,μ) is a normal space of order γ, $0<\gamma \leq 1$. Let $(1+\gamma)^{-1}<p<1$. If ψ is an element of $\mathrm{Lip}(\frac{1}{p}-1)$ and a is a p–atom then

$$<a,\psi> = \int_X a(x)\,\psi(x)\,d\mu(x)$$

defines a bounded linear functional on $\underline{\mathrm{Lip}}(\frac{1}{p}-1)$, because from the definition of a p–atom it follows immediately that

$$|<a,\psi>| \leq C\|\psi\|_{\mathrm{Lip}(\frac{1}{p}-1)}, \tag{I,18}$$

where C is a constant independent of a and ψ. If $\{a_i\}$ is a sequence of p–atoms and $\{\lambda_i\}$ a sequence of numbers such that $\sum|\lambda_i|^p<\infty$ then the series $\sum\lambda_i<a,\psi>$ converges absolutely for every ψ in $\mathrm{Lip}(\frac{1}{p}-1)$ because by (I,18)

$$\left|\sum_{i=1}^{\infty}\lambda_i<a_i,\psi>\right| \leq C\sum_{i=1}^{\infty}|\lambda_i|\,\|\psi\|_{\mathrm{Lip}(\frac{1}{p}-1)} \leq$$

$$C\left(\sum_{i=1}^{\infty}|\lambda_i|^p\right)^{1/p}\|\psi\|_{\mathrm{Lip}(\frac{1}{p}-1)}.$$

This shows that $\sum\lambda_i a_i$ converges in the norm of the dual of $\underline{\mathrm{Lip}}(\frac{1}{p}-1)$ to a bounded linear functional, say f: i.e.

$$<f,\psi> = \sum_{i=1}^{\infty}\lambda_i<a_i,\psi>. \tag{I,19}$$

The norm of f as an element of the dual of $\underline{\mathrm{Lip}}(\frac{1}{p}-1)$ is bounded by $C(\sum|\lambda_i|^p)^{1/p}$. The space H^p is defined as the space of continuous linear functionals on $\underline{\mathrm{Lip}}(\frac{1}{p}-1)$ which can be represented in the form (I,17) where the a_i's are p–atoms, and the λ_i's are numbers such that $\sum_{i=1}^{\infty}|\lambda_i|^p<\infty$. One defines $\|f\|_{H^p}$ by

$$\|f\|_{H^p} = \inf \left(\sum_{i=1}^{\infty} |\lambda_i|^p\right)^{1/p}, \tag{I,20}$$

where the infimum is taken over all possible representations of the form (I,17). The expression in (I,20) is not a norm (except when p=1), but its pth power is a metric relative to which H^p is complete. Note: If $\mu(x)<\infty$ the space H^p defined above is the "homogeneous" one which does not contain the constants.

Now let $0<\alpha<1$, the fractional integral of order α is defined by

$$(I_\alpha f)(x) = \int_{X-\{x\}} \frac{f(y)d\mu(y)}{\delta(x,y)^{1-\alpha}}. \tag{I,21}$$

In some cases – Theorems 2, 3, 4, and 5 below – it is necessary to use a modified version of (I,21) when $\mu(X)=\infty$ in order to circumvent convergence problems at infinity. Let z be a point of X chosen once and for all and let ψ_z be the function which is identically 1 on X or the characteristic function of the complement of the ball $B_1(z)$. The modified version of (I,21) is defined as follows:

$$\tilde{I}_\alpha = I_\alpha \quad \text{if } \mu(X)<\infty$$

$$(\tilde{I}_\alpha f)(x) = \int_{X-\{x,z\}} \left[\frac{1}{\delta(x,y)^{1-\alpha}} - \frac{\psi_z(y)}{\delta(z,y)^{1-\alpha}}\right] f(y)d\mu(y) \quad \text{if } \mu(X)=\infty \tag{I,22}$$

In all cases in which $\tilde{I}_\alpha$ occurs it turns out that $\tilde{I}_\alpha f$ belongs to BMO or a Lipschitz space. These spaces don't distinguish between functions which differ by a constant, and therefore (the equivalence class of) $\tilde{I}_\alpha f$ does not depend on the choice of z. Both (I,21) and (I,22) are meaningful for continuous functions of bounded support, and it will be seen that they are well defined on appropriate larger classes of functions.

Notational Convention. If there is no risk of confusion then the integral $\int_{X-\{x\}}$ in (I,19) or (I,20) will be written as $\int'$, and similarly integrals of the form $\int_{B_r(x)-\{x\}}$ will be written as $\int'_{B_r(x)}$ or $\int'_{\delta(x,y)<r}$.

Now the main results of the paper can be stated.

Theorem I.1. Let $0<\alpha<1$, and let (X,δ,μ) be a space of homogeneous type which is normal.

(i) If f belongs to L^p, $1<p<1/\alpha$ then $I_\alpha f(x)$ converges absolutely for almost every x in X, and for q given by $1/q=1/p-\alpha$ one has

$$\|I_\alpha a\|_q \le C \|f\|_p . \tag{I,23}$$

(ii) If f belongs to L^1 then $I_\alpha f(x)$ converges absolutely for almost every x in X, and for $\lambda > 0$

$$\mu(\{x: |I_\alpha f(x)| > \lambda\}) \le \left(\frac{C}{\lambda}\|f\|_1\right)^{1/(1-\alpha)} \tag{I,24}$$

In both inequalities the constants C are independent of f.

Theorem I.2. Let $0<\alpha<1$, and let (X,δ,μ) be a space of homogeneous type which is normal and of order γ, $0<\gamma\le 1$. Let $\tilde{I}_\alpha$ be defined by (I,22) with $\psi_z=1-\chi_{B_1(z)}$. If f belongs to $wL^{1/\alpha}$, then $\tilde{I}_\alpha f(x)$ converges absolutely for a.e. x in X and

$$\|\tilde{I}_\alpha f\|_* \le C\, [f]_{1/\alpha} \tag{I,25}$$

with a constant C independent of f.

Theorem I.3. Let (X,δ,μ) be a space of homogeneous type which is normal and of order γ, $0<\gamma\leq 1$, and let $0<\alpha<1$, $0 < \alpha - 1/p < \gamma$. Let $\tilde{I}_\alpha$ be defined by (I,22) with $\psi_z \equiv 1$. If f belongs to wL^p, then $\tilde{I}_\alpha f(x)$ converges absolutely for every x, and

$$\|\tilde{I}_\alpha f\|_{Lip(\alpha-1/p)} \leq C\, [f]_p \tag{I,26}$$

with a constant C independent of f.

Theorem I.4. Let (X,δ,μ) be a space of homogeneous type which is normal and of order γ, $0<\gamma\leq 1$, and which has the cancellation property of order $1-\alpha$, $0<\alpha<\gamma$. Let $\tilde{I}_\alpha$ be defined by (I,22) with $\psi_z\equiv 1$. If f belongs to BMO, then $\tilde{I}_\alpha f(x)$ converges absolutely for every x, and

$$\|\tilde{I}_\alpha f\|_{Lip(\alpha)} \leq C\, \|f\|_* \tag{I,27}$$

with a constant C independent of f.

Theorem I.5. Let (X,δ,μ) be a space of homogeneous type which is normal and of order γ, $0<\gamma\leq 1$, and which has the cancellation property of order $1-\alpha$, $0<\alpha<\gamma$. Let $\tilde{I}_\alpha$ be defined by (I,22) with $\psi_z\equiv 1$. If f belongs to $Lip(\beta)$, with $\alpha+\beta<\gamma$, then $\tilde{I}_\alpha f(x)$ converges absolutely for every for every x, and

$$\|\tilde{I}_\alpha f\|_{Lip(\alpha+\beta)} \leq C\, \|f\|_{Lip(\beta)} \tag{I,28}$$

with a constant C independent of f.

Theorem I.6. Let (X,δ,μ) be a space of homogeneous type which is normal and of order γ, $0<\gamma\leq 1$. Let $(1+\gamma)^{-1}<p\leq 1$, and $0<\alpha<\gamma$.

(i) If $0<1/q=1/p-\alpha<1$ and a is a p–atom, then $I_\alpha a$ belongs to

L^q, there is a constant C independent of a such that

$$\|I_\alpha a\|_q \leq C, \tag{I,29}$$

and I_α extends to a continuous linear map of H^p to L^q.

(ii) If (X,δ,μ) has the cancellation property of order $1-\alpha$, $1\leq 1/q=1/p-\alpha$, and a is a p–atom, then $I_\alpha a$ belongs to H^q, there is a constant C independent of a such that

$$\|I_\alpha a\|_{H^q} \leq C, \tag{I,30}$$

and I_α extends to a continuous linear map of H^p to H^q.

The following two theorems establish the link between the results stated so far and some of the principal examples discussed in Section IV.

Theorem I.7. Let (X,δ,μ) be a space of homogeneous type such that $\delta=\rho^\lambda$ where ρ is a metric, $\lambda\geq 1$, and $\mu(B_r(x)) = \omega r$, for all $r>0$, where ω is a constant independent of x and r. Then (X,δ,μ) has the cancellation property of order $1-\alpha$, $0<\alpha<1/\lambda$.

The definitions of the concepts which occur in Theorem I.8 are recalled in section IV. If G is a homogeneous group the gauge of G defines a quasi distance on G, with this quasi–distance and a Haar measure G is a space of homogeneous type. A suitable power of the quasi distance makes G into a space of homogeneous type which is normal, and of order γ, for some γ, $0<\gamma\leq 1$.

Theorem I.8. Let G be a homogeneous group which is normal and of order γ, $0<\gamma\leq 1$. If $0<\alpha<\gamma$ then G has the cancellation property of order $1-\alpha$.

II. PRELIMINARY LEMMAS

Lemma II.1. Let (X,δ,μ) be a space of homogeneous type which is normal. Let x be a point of X, and let $r>0$. Then there exist positive constants B_1 and B_2 such that

$$\text{(i)} \quad \int_{\delta(x,y)<r}' \delta(x,y)^{\alpha-1} d\mu(x) \leq B_1 r^{\alpha}, \text{ for all } \alpha>0, \tag{II,1}$$

$$\text{(ii)} \quad \int_{\delta(x,y)\geq r} \delta(x,y)^{\alpha-1} d\mu(y) \leq B_2 r^{\alpha}, \text{ for all } \alpha<0. \tag{II,2}$$

Proof: A proof of (i) will be given; the proof of (ii) is similar and can be found in Macías and Segovia (1979c). If $r \leq r_x$ then the integral is zero and there is nothing to prove. Let then $r_x<r$ if $\mu(X)=\infty$, and $r_x<r\leq R_x$ if $\mu(X)<\infty$. For non negative integers k write $r_k=2^{-k}r$. Let first $\alpha>1$.

$$\int_{\delta(x,y)<r}' \delta(x,y)^{\alpha-1} d\mu(y) = \int_{r_x\leq\delta(x,y)<r} \delta(x,y)^{\alpha-1} d\mu(y) = \sum_{k=0}^{K} \int_{r_{k+1}\leq\delta(x,y)<r_k} \delta(x,y)^{\alpha-1} d\mu(y),$$

where K is the smallest positive integer such that $r_{k+1}\leq r_x$ if $r_x>0$, and $K=+\infty$ if $r_x=0$. The sum above is majorized by

$$\sum_{k=0}^{K} r_k^{\alpha-1} \int_{\delta(x,y)<r_k} d\mu(y) \leq A_2 \left(\sum_{k=0}^{\infty} 2^{-k\alpha} \right) r^{\alpha} .$$

Finally, if $R_x<r<\infty$ write the integral (II,1) as the sum of the integrals over $B_{R_x}(x)-\{x\}$ and over the set $\{y:\delta(x,y)=R_x\}$. The first of these integrals is majorized by const.r^{α}, by what has already been proved. The second integral is majorized by const.r^{α} because of normality. The argument for $0<\alpha<1$ is similar.

Lemma II.2. Let (X,δ,μ) be a space of homogeneous type which is of order γ, $0<\gamma\leq 1$. Then there exists a positive constant B_3 such that for $\delta(x,y) \geq \delta(x,z)$

$$|\delta(x,y)-\delta(z,y)| \leq B_3 \delta(x,z)^{\gamma} \delta(x,y)^{1-\gamma} \tag{II,3}$$

Proof. Let $\gamma<1$. By (I,9) it will suffice to prove that $\max\{\delta(x,y),\delta(x,y)\}\leq 2\kappa\delta(x,y)$. If this maximum is equal to $\delta(x,y)$ there is nothing to prove since $\kappa>1$. Let then $\delta(z,y) \leq \kappa(\delta(x,z)+\delta(x,y)) \leq 2\kappa\delta(x,y)$. By the "triangle inequality" (I,3) one has to $\delta(z,y)\leq\kappa(\delta(x,z)+\delta(x,y))\leq 2\kappa\delta(x,y)$. The lemma follows with $B_3=(2\kappa)^{1-\gamma}M$, where M is the constant in (I,9). If $\gamma=1$ there is nothing to prove.

Lemma II.3. Let (X,δ,μ) be a space of homogeneous type which is of order γ, $0<\gamma\leq 1$, and let $0<\alpha<1$. Then there exists a positive constant B_4 such that

$$|\delta(x,y)^{\alpha-1}-\delta(z,y)^{\alpha-1}| \leq B_4 \delta(x,z)^{\gamma} \delta(x,y)^{\alpha-\gamma-1} \tag{II,4}$$

for all x, y, z in X such that $\delta(x,y)\geq 2\kappa\delta(x,z)$.

Proof. Consider first the case $\delta(x,y)\leq\delta(z,y)$. The left hand side of (II,4) is equal to

$$(1-\alpha)\int_{\delta(x,y)}^{\delta(z,y)} t^{\alpha-2}dt.$$

Since $\alpha-2$ is negative the integral is majorized by

$$(1-\alpha)\delta(x,y)^{\alpha-2}(\delta(z,y)-\delta(x,y))\leq(1-\alpha)B_3\delta(x,z)^{\gamma}\delta(x,y)^{\alpha-\gamma-1},$$

where the last inequality is a consequence of Lemma II.2. Now let $\delta(x,y)>\delta(z,y)$. Arguing as in the case $\delta(x,y)\leq\delta(z,y)$ one sees that the left hand side of (II,4) is majorized by

$$(1-\alpha)B_3\delta(z,y)^{\alpha-2}\delta(x,z)^{\gamma}\delta(x,y)^{1-\gamma} \tag{II,5}$$

The inequality (I,3) and the hypothesis $2\kappa\delta(x,z)\leq\delta(x,y)$ yield $\delta(z,y)\geq\delta(x,y)/2\kappa$. Replacing in (II,5) $\delta(z,y)$ by the smaller number $\delta(x,y)/2\kappa$ one obtains (II,4), this time with constant $(1-\alpha)B_3(2\kappa)^{2-\alpha}$. Since $\kappa\geq1$ the value of B_4 is $(1-\alpha)B_3(2\kappa)^{2-\alpha}$.

<u>Corollary II.1</u>. Let (X,δ,μ) be a space of homogeneous type which is normal and which is (i) of finite measure, or (ii) of order γ, $0<\gamma\leq1$. Then the function

$$\delta(x_1,y)^{\alpha-1}-\delta(x_2,y)^{\alpha-1}$$

is absolutely integrable on $X-\{x_1,x_2\}$ as a function of y, for every choice of x_1 and x_2 for $0<\alpha<1$ in case (ii) and $0 < \alpha < \gamma$ in case (ii).

<u>Proof.</u> If $x_1=x_2$ there is nothing to prove. Let therefore, $x_1\neq x_2$, and let $r\geq\delta(x_1,x_2)$, let B be the ball $B_{2\kappa r}(x_1)$ and $\tilde{B}=B_{(2\kappa^2+\kappa)r}(x_2)$. Note that $B\subset\tilde{B}$. Write

$$\int_{X-\{x_1,x_2\}}(\delta(x_1,y)^{\alpha-1}-\delta(x_2,y)^{\alpha-1})d\mu(y) = I_1+I_2, \tag{II,6}$$

where I_1 is the integral on the ball B I_2 the integral on the complement B^C of B.

$$|I_1|\leq\int_B{}' (x_1,y)^{\alpha-1}d\mu(y) + \int_B{}' \delta(x_2,y)^{\alpha-1}d\mu(y), \tag{II,7}$$

and both integrals are finite by Lemma II.1. If the space has finite measure by taking r sufficiently large one has $I_2=0$, and the proposition is proved. If the space has infinite measure one how has to estimate I_2. Note that if y belongs to the complement of B then $\delta(x_1,y)\geq 2\kappa\delta(x_1,x_2)$ and one can estimate the integrand of I_2 by Lemma II.3. to obtain

$$|I_2|\leq B_4\delta(x_1,x_2)^{\gamma}\int_{B^c}\delta(x_1,y)^{\alpha-\gamma-1}d\mu(y), \tag{II,8}$$

the exponent is less than -1, therefore the integral is finite by Lemma II.1.(ii). The inequalities (II,7) and (II,8) show that the integral in (II,6) is absolutely convergent.

Lemma II.4. Let (X,δ,μ) be a space of homogeneous type which is normal. Let φ be an element of BMO. For $a>1$, $r>0$ and non negative integers j define $r_j=a^j r$. For z in X define $m_j=m_{B_{r_j}(z)}(\varphi)$. Then there exists a positive constant B_5 independent of φ and j such that

$$|m_j-m_o|<B_5\|\varphi\|_*j. \tag{II,9}$$

This is part of Lemma 1 in Macías and Segovia (1979a) where it is stated in a slightly different form, and under the assumption that μ has no atoms, and that the space has infinite measure. The modifications needed to include these cases are minimal, and the proof will be omitted.

The following result is well known in the context of $\mathbb{R}^n$. Since it does not seem to be in the literature for spaces of homogeneous type a proof will be given.

Lemma II.5. Let (X,δ,μ) be a space of homogeneous type which is normal. Let φ belong to BMO, let x be a point of X, $r>0$, and let $B=B_r(x)$. Then

(i) For $0<\epsilon\leq 1$ $$\int'_{\delta(x,y)<r} \frac{|\varphi(y)-m_B(\varphi)|}{\delta(x,y)^{1-\epsilon}} d\mu(y) \leq B_6\, r^{\epsilon}\|\varphi\|_* \tag{II,10}$$

(ii) For $0<\epsilon$ $$\int_{\delta(x,y)\geq r} \frac{|\varphi(y)-m_B(\varphi)|}{\delta(x,y)^{1+\epsilon}} d\mu(y) \leq B_7\, r^{-\epsilon}\|\varphi\|_*. \tag{II,11}$$

The constants B_6 and B_7 do not depend on φ and r.

Proof of Lemma II.5. If $r\leq r_x$ then the domain of integration in (II,10) is empty and the inequality is vacuously true. Let, then, $r>r_x$. If $\epsilon=1$ then, recalling that the space is normal (II,10) is simply the restatement of the fact that φ belong to BMO. Let, then $0<\epsilon<1$, and let $1<p<1/(1-\epsilon)$. By Hölder's inequality the left hand side of (II,10) is majorized by

$$\left[\int'_{\delta(x,y)<r} \frac{d\mu(y)}{\delta(x,y)^{(1-\epsilon)p}}\right]^{1/p} \left[\int_B |\varphi(y)-m_B(\varphi)|^{p'} d\mu(y)\right]^{1/p'} \tag{II,12}$$

Since $(1-\epsilon)p<1$ Lemma II.1. (i) applies, and the first factor of (II.12) is majorized by $B_1^{1/p} r^{(\epsilon-1)+1/p}$. To estimate the second factor one needs the following inequality satisfied by BMO functions

$$\left[\frac{1}{\mu(B)} \int_B |\varphi(y)-m_B(\varphi)|^q d\mu(y)\right]^{1/q} \leq C_q \|\varphi\|_*,\ 1<q<\infty \tag{II,13}$$

The constant C_q is independent of φ,B; it only depends on q. This inequality is a result of Coifman and Weiss(1977) which extends to spaces of homogeneous type earlier results of John and Nirenberg (1961). Using (II,13) and normality the second factor of (II,12) is majorized by $A_2^{1/p'} C_{p'} r^{1/p'}$. Combining the estimates of the two factors of (II,12) one obtains (II,10) with $B_6=B_1^{1/p} A_2^{1/p'} C_{p'}$.

To prove (II,11) note first that if the space has finite measure and $r>R_x$ then the integral is zero, and there is nothing to prove. Consequently one can assume $r\leq R_x$. One can also assume that $r\geq r_x$ because if $r<r_x$ then the integral over the set $\{y: r\leq\delta(x,y)<r_x\}$ is zero and $r^{-\epsilon}>r_x^{-\epsilon}$. Use the notation of Lemma II.4 with a=2, and write $B_j=B_{r_j}(x)$.

$$\int_{\delta(x,y)\geq r} \frac{|\varphi(y)-m_o|}{\delta(x,y)^{1+\epsilon}}d\mu(y) \leq \sum_{j=0}^{\infty} \int_{B_{j+1}-B_j} \frac{|\varphi(y)-m_o|}{r_j^{1+\epsilon}}d\mu(y) \leq \tag{II,14}$$

$$2A\sum_{j=2}^{\infty} (2^j r)^{-\epsilon}\frac{1}{\mu(B_j)}\int_{B_{j+1}} |\varphi(y)-m_o|d\mu(y).$$

Note that if the space is of finite measure only finitely many integrals are different from zero. In the last inequality normality was used to get $\mu(B_{j+1})/r_j\leq A_2 2r_j/r_j=2A_2$. Now $|\varphi(y)-m_o| \leq |\varphi(y)-m_{j+1}| + |m_{j+1}-m_o|$, and by Lemma II.4. $|m_{j+1}-m_o| \leq B_5(j+1)\|\varphi\|_*$. Therefore the last integral in (II,14) is majorized by $\|\varphi\|_*+(j+1)B_5\|\varphi\|_*$ and the left hand side of (II,14) is majorized by

$$r^{-\epsilon}2A_2\sum_{j=0}^{\infty} 2^{-j\epsilon}(1+(j+1)B_5)\|\varphi\|_*.$$

This finishes the proof of (II,14) and of the lemma.

III. PROOFS OF THEOREMS I.1 TO I.6

Proof of Theorem I.1. Both parts of the theorem are easy consequences of an inequality of Hedberg (1972) which here will be shown to hold in normal spaces of homogeneous type. If f is a measurable function on X which is integrable on bounded sets then the Hardy–Littlewood maximal function of f,Mf is defined by

$$(Mf)(x)=\sup_{r>0} \frac{1}{\mu(B_r(x))} \int_{B_r(x)} |f(y)|\,d\mu(y).$$

Let $0<\alpha<1$, $1\leq p<1/\alpha$, and f an element of L^p. The f satisfies the following inequality.

$$|I_\alpha f(x)| \leq \|f\|_p^{p\alpha}(Mf(x))^{1-p\alpha}. \tag{III,1}$$

Theorem I.1. follows from (III,1) by raising both sides of that inequality to the qth power $(1/q=1/p-\alpha)$ and using the fact that the Hardy–Littlewood maximal operator is of weak type (1,1), and of strong type (p,p) if $p>1$. Note also that (III,1) shows that $I_\alpha f(x)$ converges absolutely for almost every x.

To prove Hedberg's inequality fix x in X. It can be assumed that Mf(x) is finite, since otherwise there is nothing to prove. Let $r>0$, $r_j=2^{-j}r$ for non negative integers j, and let $B_j=B_{r_j}(x)$. Suppose first that $r>r_x$,

$$I_\alpha f(x) = \left[\int'_{\delta(x,y)<r} + \int_{\delta(x,y)\geq r} \right] \frac{f(y)d\mu(y)}{\delta(x,y)^{1-\alpha}} = I_1+I_2.$$

For I_1 one has

$$|I_1| \leq \sum_{j=0}^{J} \int_{B_j-B_{j+1}} \frac{|f(y)|d\mu(y)}{\delta(x,y)^{1-\alpha}} . \tag{III,2}$$

Here J is the smallest integer j for which $2^{-(j+1)}\leq r_x$ if $r_x>0$ and $J=+\infty$ if $r_x=0$. The right hand side is further majorized by

$$\sum_{j=0}^{J} \int_{B_j} \frac{|f(y)|\,d\mu(y)}{r_{j+1}^{1-\alpha}} \leq 2A_2 \sum_{j=0}^{\infty} (2^{-j-1}r)^{\alpha} \frac{1}{\mu(B_j)} \int_{B_j} |f(y)|\,d\mu(y) \leq$$

$$\leq 2A_2 \sum_{j=0}^{\infty} 2^{-(j+1)\alpha} Mf(x) r^{\alpha}. \tag{III,3}$$

If $r < r_x$ then $I_1=0$, and the above estimate holds a fortiori. To estimate I_2 apply Hölder's inequality:

$$|I_2| \leq \|f\|_p \left[\int_{\delta(x,y)\geq r} \frac{d\mu(y)}{\delta(x,y)^{(1-\alpha)p'}} \right]^{1/p'}. \tag{III,4}$$

Note that $(1-\alpha)p'>1$, and that therefore one can estimate the integral in (III,4) by Lemma II.1. (ii). Introducing this into (III,4) one finally has

$$|I_2| \leq \|f\|_p B_2^{1/p'} r^{\alpha-1/p}. \tag{III,5}$$

Combining the estimates of I_1 and I_2 one obtains

$$|I_\alpha f(x)| \leq C(r^{\alpha} Mf(x) + \|f\|_p r^{\alpha-1/p}), \tag{III,6}$$

where C is the larger of the constants in (III,3) and (III,5). Now choosing $r=(\|f\|_p/Mf(x))^p$ in (III,6) one obtains (III,1).

In the proof of Theorems I.2 and I.3 the following estimate is needed.

<u>Lemma III.1</u> Let (X,δ,μ) be a normal space, B^c the complement of the ball of center x_o and radius r, f in wL^p, $1<p<\infty$ and ϵ such that $\epsilon p'>1$. Then

$$\int_{B^c} |f(y)| \frac{1}{\delta(x_o,y)^\epsilon} d\mu(y) \le C\, r^{1-\epsilon-\frac{1}{p}} [f]_p, \tag{III,7}$$

where c is independent of r, x_o and f.

Proof of Lemma III.1. Let a be a positive number to be determined later, let $f^a(x) = f(x)$ if $|f(x)|>a$, and $f^a(x) = 0$ otherwise. Set $f_a = f-f^a$, and write

$$J = \int_{B^c} |f^a(y)|\, \delta(x_o,y)^{-\epsilon} d\mu(y) + \int_{B^c} |f_a(y)|\, \delta(x_o y)^{-\epsilon} d\mu(y) = J_1+J_2$$

clearly,

$$J_1 \le r^{-\epsilon} \|f^a\|_1 \,,$$

and

$$\|f^a\|_1 \le \lambda_f(a) + \int_a^\infty \lambda_f(t)dt \le a^{1-p}[f]_p^p + [f]_p^p \int_a^\infty t^{-p}dt = C_1 a^{1-p}[f]_p^p.$$

Together these estimates give

$$J_1 \le C_1 r^{-\epsilon} a^{1-p}[f]_p^p.$$

To estimate J_2 find $s > p$ such that $s'\epsilon>1$; this is possible since $\epsilon p' > 1$. Apply Hölder's inequality with exponents s,s', and then use Lemma II.1.(ii) to obtain

$$J_2 \le C_2\, r^{-\epsilon+\frac{1}{s'}} \|f_a\|_s.$$

This inequality in view of

$$\|f_a\|_s^s \le s[f]_p^p \int_0^a t^{s-p-1} dt = \frac{s}{s-p} a^{s-p} [f]_p^p$$

yields

$$J_2 \leq \frac{s}{s-p}(C_2 r^{-\epsilon+\frac{1}{s'}} a^{1-\frac{p}{s}} [f]_p^{\frac{p}{s}};$$

combining this inequality with the estimate of J above

$$J \leq C_1 r^{-\epsilon} a^{1-p} [f]_p^p + \frac{s}{s-p} C_2 r^{-\epsilon+\frac{1}{s'}} a^{1-\frac{p}{s}} [f]_p^{\frac{p}{s}}.$$

To conclude the proof of the lemma set $a = r^{-\frac{1}{p}} [f]_p$ in the last estimate.

Proof of Theorem I.2. Let B be a ball of center x_0 and radius $r>0$, and let $\tilde{B} = B_{2\kappa r}(x_o)$. Consider first the case $\mu(X) = \infty$, and define

$$a_B = \int_X \left(\frac{1 - \chi_{\tilde{B}}(y)}{\delta(x_o, y)^{1-\alpha}} - \frac{\psi_z(y)}{\delta(z, y)^{1-\alpha}} \right) f(y) d\mu(y).$$

and

$$I(x) = \int_{X-\{x\}} \left(\frac{1}{\delta(x,y)^{1-\alpha}} - \frac{1-\chi_{\tilde{B}}(y)}{\delta(x_o,y)^{1-\alpha}} \right) f(y)\, d\mu(y).$$

It will be shown that a_B is well defined and that

$$\int_B |I(x)|\, d\mu(x) \leq C\, \mu(B)[f]_{1/\alpha}. \tag{III,8}$$

with a constant C independent of B and f. The theorem will follow from (III,8) because

$$\tilde{I}_\alpha f(x) = I(x) + a_B,$$

and therefore it will be finite almost everyhere on B and

$$\int_B |\tilde{I}_\alpha f(x) - a_B| \, d\mu(x) \le C \, [f]_{1/\alpha}$$

which implies $\|\tilde{I}_\alpha f\|_{BMO} \le C \, [f]_{1/\alpha}$ by a well known property of BMO.

In order to see that a_B is well defined note that f is locally in L^1 and the bracket bounded, on the other hand the bracket is $O(\delta(x_o,y)^{\alpha-1-\gamma})$ for large y and therefore by Lemma III.1. the integral converges at infinity. Next, note that

$$I(x) = \int_{\tilde{B}-\{x\}} \frac{f(y) \, d\mu(y)}{\delta(x,y)^{1-\alpha}} + \int_{\tilde{B}^c} \left(\frac{1}{\delta(x,y)^{1-\alpha}} - \frac{1}{\delta(x_o,y)^{1-\alpha}}\right) f(y) d\mu(y)$$

$$= I_1(x) + I_2(x).$$

The proof of (III,8) is immediate from the following two inequalities:

$$\int_B |I_1| \, d\mu \le C_1 \, \mu(B) \, [f]_{1/\alpha}, \tag{III,9}$$

and

$$|I_2(x)| \le C_2 \, [f]_{1/\alpha} \quad \text{for } x \epsilon B, \tag{III,10}$$

both constants C_1 and C_2 being independent of B and f. To prove (III,9), by Fubini's Theorem and Lemma II,1 (i) the left hand side of (III,9) is less than or equal to

$$\int_{\tilde{B}} |f(y)| \left(\int_{\delta(x,y) < (2\kappa^2+\kappa)r} \frac{1}{\delta(x,y)^{1-\alpha}} \, d\mu(x)\right) d\mu(y) \le$$

$$\|f\,\chi_{\tilde{B}}\|_1\ C\, r^{\alpha}.$$

Now observe that for $\epsilon=[f]_{1/\alpha}/\mu(\tilde{B})^{\alpha}$ we have

$$\|f\,\chi_{\tilde{B}}\|_1 \le \int_0^{\epsilon} \mu(\tilde{B})\, dt + \int_{\epsilon}^{\infty} t^{-\frac{1}{\alpha}} [f]_{1/\alpha}^{1/\alpha} dt \le C\,[f]_{1/\alpha}\, \mu(\tilde{B})^{1-\alpha}$$

This concludes the proof of (III,9).

Now, by Lemma (III,3)

$$|I_2(x)| \le B_3 r^{\gamma} \int_{\tilde{B}^c} |f(y)| \frac{1}{\delta(x_o,y)^{1+\gamma-\alpha}}\, d\mu(y).$$

Since $(1+\gamma-\alpha)p' > 1$ applying Lemma III.1. (ii) with $\epsilon = 1+\gamma-\alpha$ one has (III,10).

In the case $\mu(X)<\infty$, $\tilde{I}_\alpha = I_\alpha$ and define

$$a_B = \int_X \frac{1-\chi_{\tilde{B}^c}(y)}{\delta(x_o,y)^{1-\alpha}} f(y)d\mu(y)$$

$$I(x) = \int_{X-\{x\}} \left[\frac{1}{\delta(x,y)^{1-\alpha}} - \frac{1-\chi_{\tilde{B}}(y)}{\delta(x_o,y)^{1-\alpha}}\right] f(y)\, d\mu(y).$$

With these changes the same proof goes through.

The proofs of Theorems I.3., I.4. and I.5. share traits which are similar, and it will be convenient to introduce some notation which will remain in force in all three proofs. Notice first that in all three cases $\tilde{I}_\alpha f(x)$ is defined with $\psi_z \equiv 1$. Next if $0<\alpha<\gamma$ and x_1, x_2 are in X define – for suitable f–Δ by

$$\Delta = \int_{X-\{x_1,x_2\}} \left(\frac{1}{\delta(x_1,y)^{1-\alpha}} - \frac{1}{\delta(x_2,y)^{1-\alpha}}\right) f(y)d\mu(y).$$

If $x_1 \neq x_2$, let $r = \delta(x_1,x_2)$, $B = B_{2\kappa r}(x_1)$, and

$$\Delta = \Delta_B + \Delta_{B^c},$$

where Δ_B and Δ_{B^c} are the integrals over B and B^c, respectively. Finally let $\tilde{B} = B_{(2\kappa^2+\kappa)r}(x_2)$ and note that $B \subset \tilde{B}$.

Proof of Theorem I.3 Note that $\tilde{I}_\alpha f(x)$ is Δ with $x_1=x$, $x_2=z$ and assuming $\tilde{I}_\alpha f(x)$ finite, also

$$\tilde{I}_\alpha f(x_1) - \tilde{I}_\alpha f(x_2) = \Delta.$$

It will be proved that

$$|\Delta| \leq C\ \delta(x_1,x_2)^{\alpha-\gamma}[f]_{p'} \tag{III,11}$$

with c independent of x_1, x_2 and f. In view of what was said above, all the assertions of the theorem follow from (III,11).

Clearly one has

$$|\Delta_B| \leq \int_B{}' \frac{|f(y)|d\mu(y)}{\delta(x_1,y)^{1-\alpha}} + \int_{\tilde{B}}{}' \frac{|f(y)|d\mu(y)}{\delta(x_2,y)^{1-\alpha}}. \tag{III,12}$$

If $\frac{1}{\alpha} < s < p$ then $f\chi_B$ and $f\chi_{\tilde{B}}$ belong to L^s. The estimates of both integrals are carried out exactly the same way, so it suffices to do the first one.

Note that for $\epsilon = [f]_p/\mu(B)^{1/p}$

$$\|f\chi_B\|_s^s = s\int_0^{\epsilon} t^{s-1}\mu(B)dt + s\int_{\epsilon}^{\infty} t^{s-1}\, t^{-p}\, [f]_p^p\, dt$$

$$\leq \epsilon^s\mu(B) + s\,\frac{\epsilon^{s-p}}{p-s}\,[f]_p^p$$

$$\leq C\,\mu(B)^{1-\frac{s}{p}}\,[f]_p^s.$$

Now applying Hölder's Inequality with $\frac{1}{s} + \frac{1}{s'} = 1$, the integral in question does not exceed

$$C\,\mu(B)^{\frac{1}{s}-\frac{1}{p}}\,[f]_p\,\left(\int_B \frac{d\mu(y)}{\delta(x_1,y)^{(1-\alpha)s'}}\right)^{1/s'}.$$

Since $s'(1-\alpha) > 1$, and the space is normal, Lemma II.1(i) applies and the last expression is majorized by

$$C\,r^{\alpha-\frac{1}{p}}\,[f]_p = C\,\delta(x_1,x_2)^{\alpha-\frac{1}{p}}\,[f]_p$$

To estimate Δ_{B^c} use Lemma II.3. and Lemma III.1 to obtain

$$|\Delta_{B^c}| \leq C\,\delta(x_1,x_2)^{\alpha-\frac{1}{p}}\,[f]_p.$$

This concludes the proof of the theorem.

Proof of Theorem I.4. (See the paragraph preceding the proof of Theorem I.3. for the notation.) The basic inequality to be proved is

$$|\Delta| \leq C\,\delta(x_1,x_2)^{\alpha}\,\|f\|_* \qquad \text{(III,13)}$$

where C is a constant independent of x_1,x_2 and f. Since X has the cancellation property of order $1-\alpha$ Δ does not change if f is replaced by $f-m_B(f)$ in the integrand. This will be done, and consequently also Δ_B and Δ_{B^c} will have $f-m_B(f)$ in their integrands.

$$|\Delta_B| \le \int_B' \frac{|f(y)-m_B(f)|}{\delta(x_1,y)^{1-\alpha}}\,d\mu(y) +$$

$$\int_{\tilde{B}}' \frac{|f(y)-m_B(f)|}{\delta(x_2,y)^{1-\alpha}}\,d\mu(y) = I_B + I_{\tilde{B}}. \tag{III,14}$$

By Lemma II.5.(i)

$$I_B \le C_1\delta(x_1,x_2)^{\alpha}\|f\|_*, \tag{III,15}$$

where the constant is independent of x_1,x_2 and f. By Lemma II.1.(i) one has

$$I_{\tilde{B}} \le |m_B(f)-m_{\tilde{B}}(f)|\,B_1(2\kappa^2+\kappa)r^{\alpha} + \int_{\tilde{B}} \frac{|f(y)-m_{\tilde{B}}(f)|}{\delta(x_2,y)^{1-\alpha}}\,d\mu(y), \tag{III,16}$$

and

$$|m_B(f)-m_{\tilde{B}}(f)| \le \frac{1}{\mu(B)}\int_{\tilde{B}} |f(y)-m_{\tilde{B}}(f)|\,d\mu(y) \le$$

$$\frac{\mu(\tilde{B})}{\mu(B)}\frac{1}{\mu(\tilde{B})}\int_{\tilde{B}} |f(y)-m_{\tilde{B}}(f)|\,d\mu(y) \le$$

$$\frac{A_2(2\kappa^2+\kappa)r}{r/A_1}\,\|f\|_* = C_2\|f\|_*\ . \tag{III,17}$$

In the last inequality normality was used and the fact that $r=\delta(x_1,x_2)$ implies that $r_{x_2} < r \leq R_{x_2}$. The integral in (III,16) does not exceed $C_3\delta(x_1,x_2)^{\alpha}\|f\|_*$ by Lemma II.5.(i); combining this estimate with (III,17) and (III,16) one obtains $I_B^{\sim} \leq C_4\delta(x_1,x_2)^{\alpha}\|f\|_*$. This inequality together with (III,31) and (III,14) yields

$$|\Delta_{B^c}| \leq B_4 r^{\gamma} \int_{B^c} \frac{|f(y)-m_B(f)|}{\delta(x_1,y)^{1-\alpha+\gamma}}\, d\mu(y) \leq$$

$$\leq C_5 r^{\alpha}\|f\|_*, \tag{III,19}$$

where to obtain the last inequality Lemma II.5.(ii) with $\epsilon=\gamma-\alpha>0$ was applied. The estimate (III.13) now follows from (III,18) and (III,19). The assertions of the theorem all follow from (III,13) by the same argument as in the proof of Theorem I.3.

Proof of Theorem I.5. (See the paragraph preceding the proof of Theorem I.3. for the notation.) The basic inequality to be proved in this case is

$$|\Delta| \leq C\, \delta(x_1,x_2)^{\alpha+\beta}\|f\|_{Lip(\beta)}, \tag{III,20}$$

C being a constant independent of x_1,x_2 and f. Since X has the cancellation property of order $1-\alpha$ Δ does not change if a constant is added to f. Add the constant $-f(x_1)$ to f, then Δ_B and Δ_{B^c} also will have $f(y)-f(x_1)$ in their integrands.

$$|\Delta_B| \leq \int_B \frac{|f(y)-f(x_1)|}{\delta(x_1,y)^{1-\alpha}}\, d\mu(y) +$$

$$|f(x_2)-f(x_1)| \int_B \frac{d\mu(y)}{\delta(x_2,y)^{1-\alpha}} +$$

$$\int_{\tilde{B}} \frac{|f(y)-f(x_2)|}{\delta(x_2,y)^{1-\alpha}}\, d\mu(y) \leq$$

$$\|f\|_{Lip(\beta)}\Big(\int_B{}' \delta(x_1,y)^{\beta+\alpha-1} d\mu(y) + r^{\beta} \int_{\tilde{B}}{}' \delta(x_2,y)^{\alpha-1} d\mu(y) +$$

$$\int_{\tilde{B}} \delta(x_2,y)^{\beta+\alpha-1} d\mu(y)\Big).$$

Each of these three integrals can be estimated by Lemma II.1.(i) and one obtains

$$|\Delta_B| \leq C_1 r^{\alpha+\beta} \|f\|_{Lip(\beta)}, \tag{III,21}$$

and C_1 only depends on B_1 and κ. To estimate Δ_{B^c} one uses Lemma II.3 which can be applied since for y in B^c $\delta(y, x_1) \geq 2\kappa r$ $\delta(y,x_1) \geq 2\kappa r$.

$$|\Delta_{B^c}| \leq B_4 r^{\gamma} \int_{B^c} \delta(x_1,y)^{\beta+\alpha-\gamma-1} d\mu(y) \|f\|_{Lip(\beta)}.$$

Since $\beta+\alpha-\gamma < 0$ one can use Lemma II.1.(ii) to conclude that

$$|\Delta_{B^c}| \leq B_4 B_2 r^{\gamma} (2\kappa r)^{\alpha+\beta-\gamma} \|f\|_{Lip(\beta)} =$$

$$< C_2 r^{\alpha+\beta} \|f\|_{Lip(\beta)}. \tag{III,22}$$

The inequalities (III,21) and (III,22) together prove (III,20), and the assertions of the theorem follow from this inequality by the same reasoning as in Theorem I.3.

The proof of Theorem I.6. will depend on molecules. The definition of the molecules used in this theorem is given below. The main fact about molecules is their atomic decomposition. This decomposition can be obtained under the present hypotheses following the method of Macías and Segovia (1979a). Since the necessary modifications are slight it will be sufficient to indicate them instead of reproducing the whole proof.

Let (X,δ,μ) be a space of homogeneous type which is normal and of order γ, $0<\gamma\le 1$. Let $\epsilon>0$, $(1+\gamma)^{-1}<s\le 1$, and $0<\alpha<\gamma$. A measurable function M or X is an <u>(s,ε)–molecule</u> if there exist a point x_0 in X and positive numbers r and L such that

$$|M(x)|\le L\, r^{-\frac{1}{s}} \text{ for all } x \tag{III,23}$$

$$|M(x)|\,\delta(x,x_0)^{\frac{1}{s}+\epsilon}\le Lr^{\epsilon} \text{ for all } x \tag{III,24}$$

$$\int_X M\, d\,\mu = 0 \tag{III,25}$$

Note that the first two conditions imply that M is in $L^1(X)$, hence it is legitimate to impose condition (III.25). This definition differs from that of Macías and Segovia only in that L^∞ estimates instead of L^2 estimates are used in (III,23) and (III,24).

The basic property of an (s,ϵ)–molecule is that it is an element of H^s and its norm is bounded by a constant that depends only on L and ϵ, but not on x_0 or r. This is the content of Lemma III.2. whose proof follows closely the proofs of Theorems 2 and 3 of Macías and Segovia (1979a). The main modifications will be indicated.

<u>Lemma III.2</u>. Let M be an (s,ϵ)–molecule. Then for φ in $\mathrm{Lip}(\frac{1}{s}-1)$ if $s<1$ and φ in BMO if $s=1$ $M\varphi$ belongs to $L^1(X)$, and

$$l_M(\overline{\varphi}) = \langle M, \varphi \rangle = \int_X M\varphi d\mu$$

defines a continuous linear functional on $\underline{\text{Lip}}(\frac{1}{s} - 1)$ and BMO, respectively. There exists a constant B which depends only on L and ϵ, a sequence $\{a_i\}_{i=0}^{\infty}$ of s–atoms and a sequence of numbers $\{\lambda_i\}_{i=0}^{\infty}$ such that $\sum_{i=0}^{\infty} |\lambda_i|^s \leq B$, and such that the functional $l_M(\overline{\varphi})$ is given by the absolutely convergent series.

$$l_M(\overline{\varphi}) = \sum_{i=0}^{\infty} \lambda_i \, l_{a_i}(\overline{\varphi}) \tag{III,27}$$

for φ in $\underline{\text{Lip}}(\frac{1}{s} - 1)$ if s<1, and for φ in BMO if s=1.

Observations on the proof of this lemma.

Observation 1. If x_o is a point of positive measure or if $\mu(X)=1$. then the number r in (III,23) and (III,24) can be assumed to satisfy

$$r_{x_o} \leq r < R_{x_o} . \tag{III,27}$$

In fact, if (III,23) and (III,24) are true with $r < r_{x_o}$ then obviously (III,24) is also true for r_{x_o}. From (III.24) – with r_{x_o} – and $\delta(x,x_o) \geq r_{x_o}$ it follows that

$$|M(x)| \leq \frac{L\, r_{x_o}^{\epsilon}}{\delta(x,x_o)^{\frac{1}{s}+\epsilon}} \leq \frac{L\, r_{x_o}^{\epsilon}}{r_{x_o}^{\frac{1}{s}+\epsilon}} = L r_{x_o}^{-\frac{1}{s}} .$$

On the other hand if $\delta(x,x_o) < r_{x_o}$ then $x=x_o$ and since $\int M \, d\,\mu = 0$ one has

$$|M(x_o)|\ \mu(\{x_o\}) \leq \int_{B_r^c(x_o)} |M|\,d\mu \leq L\,B_2\,r_{x_o}^{\frac{1}{s}+1}$$

by (III,24) and Lemma II.1.(ii). By normality $\mu(\{x_o\}) = \mu(B_{r_{x_o}}(x_o) \geq A_1 r_{x_o}$, hence

$$|M(x_o)| \leq \frac{L B_2}{A_1} r_{x_o}^{-\frac{1}{s}}.$$

Now assume that (III,23), and (III,24) hold for $r \geq R_{x_o}$. Clearly, $R_{x_o}^{-\frac{1}{s}} \geq r^{-\frac{1}{s}}$, hence (III,23) holds for R_{x_o}. On the other hand this inequality implies

$$|M(x)|\,\delta(x,x_o)^{\frac{1}{s}} \leq |M(x)|\,R_{x_o}^{\frac{1}{s}} \leq L,$$

and hence,

$$|M(x)|\,\delta(x,x_o)^{\frac{1}{s}+\epsilon} \leq L\,R_{x_o}^{\epsilon}.$$

Observation 2. The proof of the lemma proceeds exactly as in Macías and Segovia (1979a) with one exception: their definition of Lipschitz spaces has to be replaced by the one given in Section I of this paper.

Proof of Theorem I.6 Let a be a p–atom with support in $B=B_r(x_o)$. Notice that if $r \leq r_{x_o}$ then $\int a\,d\mu = 0$ forces a to be identically zero, hence one can assume that $r > r_{x_o}$. If $\mu(X) = 1$ it is easy to show that there is a constant $R > \sup\{R_x : x\epsilon X\}$ such that $\mu(B_r(x)) \geq A_1' r$ for all x in X, and $r \leq R$, where A_1' is a constant slightly smaller then A_1 in (I,7). Therefore if $\mu(X) = 1$ the radius of the support of a can be assumed to be less than or equal R, since if $r > R$ then $\|a\|_\infty \leq \mu(B_r(x_o))^{-1/p} \leq \mu(B_R(x_o))^{-1/p}$, and $B_R(x_o) = X$ contains the support of a.

Inequalities (I,29) and (I,30) will be proved first, the extension of I_α to all of H^p will be carried out later. The proof depends on the following estimate.

$$|I_\alpha a(x)| \leq \begin{cases} C \quad \mu(B)^{-\frac{1}{q}} \text{ if } \delta(x,x_o)<2\kappa r, \\ C\, r^{\gamma+1-\frac{1}{p}}\, \delta(x,x_o)^{\alpha-\gamma-1} \quad \text{if } \delta(x,x_o)\geq 2\kappa r. \end{cases} \qquad \text{(III,28)}$$

The constant C is independent of a, x, and r.

Let $\tilde{B} = B_{2\kappa r}(x_o)$ and $\tilde{\tilde{B}}(x)=B_{(2\kappa^2+\kappa)r}(x)$. Notice the $\tilde{B} \subset \tilde{\tilde{B}}(x)$, for $x\epsilon\tilde{B}$. Pick x in $\tilde{B}$. By Lemma II.1.(i) and normality

$$|I_\alpha a(x)| \leq \int_B' \frac{|a(y)|d\mu(y)}{\delta(x,y)^{1-\alpha}} \leq \int_{\tilde{\tilde{B}}(x)}' \frac{|a(y)|}{\delta(x,y)^{1-\alpha}} d\mu(y)$$

$$\leq \mu(B)^{-\frac{1}{p}} B_1(2\kappa^2+\kappa)^\alpha r^\alpha \leq B_1 \left(\frac{2\kappa^2+\kappa}{A_1}\right)^\alpha \mu(B)^{1/q}.$$

The first inequality is proved. Now let x be in $\tilde{B}^c$. Using the condition $\int a d\mu = 0$ one has

$$I_\alpha a(x) = \int_B' \left(\frac{1}{\delta(x,y)^{1-\alpha}} - \frac{1}{\delta(x,x_o)^{1-\alpha}}\right) a(y)d\mu(y).$$

Since $\delta(y,x_o)<r$ one has $2\kappa\delta(y,x_o) < 2\kappa r \leq \delta(x,x_o)$ and the bracket in the integral can be estimated by Lemma II.3.

$$|I_\alpha(x)| \leq B_4\mu(B)^{1/p}\, \delta(x,x_o)^{\alpha-\gamma-1} \int_B \delta(x,x_o)^\gamma d\mu(y) \leq$$

$$B_4 B_1 A_1^{-\frac{1}{p}}\, r^{\gamma+1-\frac{1}{p}}\, \delta(x,x_o)^{\alpha-\gamma-1}$$

where Lemma II.1.(i) and normality were used in the second inequality. The proof of (III,28) is complete. The proof of (I,29) now is immediate from (III,28):

$$\|I_\alpha a\|_q^q \le C^q \int_{\tilde{B}} \mu(\tilde{B})^{-\frac{1}{p}} d\mu + C^q r^{q(\gamma+1-\frac{1}{p})} \int_{\tilde{B}^c} \delta(x,x_o)^{(\alpha-\gamma-1)q} d\mu(x)$$

$$\le C^q + C^q B_2 (2\kappa)^{(\alpha-\gamma-1)q+1} = C$$

where in the second inequality Lemma II.1.(ii) was used to estimate the second integral.

If a≡1 then $I_\alpha a(x) = \int_X \frac{d\mu(y)}{\delta(x,y)^{1-\alpha}} \le B_1 R^\alpha$, by Lemma II.1.(ii), and R is independent of a.

Now let $q \le 1$. Let $\epsilon = \gamma+1-\frac{1}{p}$; this number is positive. Notice that $1+\gamma-\alpha = \frac{1}{q} + \epsilon$. To prove that $I_\alpha a$ is a (q,ϵ) molecule for every p–atom a, let first $x \epsilon \tilde{B}$. The first inequality (III,28) and normality give,

$$|I_\alpha a(x)| \le C\, A_2^{-\frac{1}{q}} r^{-\frac{1}{q}}$$

If $\delta(x,x_o) \ge 2\kappa r$ then the second inequality (III,28) gives

$$|I_\alpha a(x)| < C r^\epsilon (2\kappa r)^{-\frac{1}{q}-\epsilon},$$

and this proves (III,23). In view of the second inequality (III,28) one only has to prove (III,24) for x in $\tilde{B}$. If $\delta(x,x_o) < 2\kappa r$ then by (III,23)

$$|I_\alpha a(x)|\ \delta(x,x_o)^{\frac{1}{q}+\epsilon} \le C A_2^{-\frac{1}{q}} r^{-\frac{1}{q}} (2\kappa r)^{-\frac{1}{q}+\epsilon}$$

$$= CA_2^{-\frac{1}{q}} (2\kappa)^{\frac{1}{q} + \epsilon} r^{\epsilon}.$$

Now let L be the largest of the constants $CA_2^{-\frac{1}{q}} (2\kappa)^{\frac{1}{q}+\epsilon}$, $CA_2^{-\frac{1}{q}}$, $C(2\kappa)^{-\frac{1}{q}-\epsilon}$.

To prove (III,25) note first that $I_\alpha a$ is integrable because it is a molecule. Let $\lambda(x)$ be the characteristic function of $\{x_o\}^c$, then

$$\int_X I a\, d\mu = \int_X \Big(\int_X{}' \frac{a(y)d\mu(y)}{\delta(x,y)^{1-\alpha}} \Big)\, d\mu(x)$$

$$= \int_X \Big(\int_X{}' a(y) \Big(\frac{1}{\delta(x,y)^{1-\alpha}} - \frac{\lambda(x)}{\delta(x,x_o)^{1-\alpha}}\Big)\, d\mu(y) \Big) d\mu(x)$$

$$= \int_X a(y) \Big(\int_X \Big(\frac{1}{\delta(x,y)^{1-\alpha}} - \frac{\lambda(x)}{\delta(x,x_o)^{1-\alpha}}\Big)\, d\mu(x) \Big)\, d\mu(y)$$

$$= \int_X a(y) \Big(\int_{X-\{y,x_o\}} \Big(\frac{1}{\delta(x,y)^{1-\alpha}} - \frac{1}{\delta(x,x_o)^{1-\alpha}}\Big)\, d\mu(x) \Big)\, d\mu(y)$$

The inner integrand is zero by the cancellation property. This concludes the proof that $I_\alpha a$ is a(q,ϵ) molecule. The constant L in the definition of the molecule $I_\alpha a$ is independent of a, therefore by Lemma III.2 one has $\|I_\alpha a\| \le B$.

To prove that I_α can be extended to all of H^p it is necessary to consider the different possible combinations of p and q. These are,

1. $p = 1,\ q = \frac{1}{1-\alpha}$.
2. $p < 1,\ q > 1$.
3. $p = \frac{1}{1+\alpha},\ q = 1$.
4. $p < 1,\ q < 1$.

Consider the case 2 and $\mu(X) = \infty$. Let f be in H^p and $f = \sum_{i=1}^{\infty} \lambda_i a_i$

be a decomposition of f into p–atoms. Then by (I,29)

$$\sum_{i=1}^{n} \|\lambda_i I_\alpha(a_i)\|_q \leq C \sum_{i=1}^{n} |\lambda_i| \leq C\left(\sum_{i=1}^{n} |\lambda_i|^p\right)^{\frac{1}{p}},$$

i.e the series $\sum_{i=1}^{\infty} \lambda_i I_\alpha(a_i)$ converges in L^q norm to a limit g. This is the element of L^q that one wants to identify as $I_\alpha f$, however the fact that the set of atoms is not a linearly independent set requires that one show that g is independent of the particular decomposition of f. To do this let φ be a bounded measurable function supported in a bounded set. Notice that $q' > \frac{1}{\alpha}$, and that consequently by Theorem I.3.

$$\tilde{I}_\alpha: L^{q'} \to \mathrm{Lip}(\alpha - \tfrac{1}{q'}) = \mathrm{Lip}\,(\tfrac{1}{p} - 1).$$

For u and v in X define

$$K(u,v) = \begin{cases} \dfrac{1}{\delta(u,v)^{1-\alpha}} & \text{if } u \neq v \\ 0 \text{ if } u = v \end{cases}$$

Notice that K(u,v) = K(v,u). Now fix z in X, and let x be in X. Since a is an atom $\int a\,d\mu = 0$, and hence

$$\int_X I_\alpha(a_i)\varphi\, d\mu = \int_X \left(\int_X (K(x,y) - K(x,z))\, a(y)\,d\mu(y)\right)\varphi(x)\,d\mu(x)$$

$$= \int_X a(y) \left(\int_X (K(x,y) - K(x,z))\, \varphi(x)\right)d\mu(x))\,d\mu(y)$$

$$= \int_X a_i\, \tilde{I}_\alpha \varphi\, d\mu$$

The change in the order of integration is justified because it is readily checked that the integrand of the double integral is in L^1 (XxX). By the above equality

$$\int_X g\,\varphi d\mu = \int_X \Big(\sum_{i=1}^{\infty} \lambda_i I_\alpha(a_i)\Big)\varphi d\mu$$

$$= \sum_{i=1}^{\infty} \lambda_i \int_X a_i \tilde{I}_\alpha(\varphi) d\mu$$

$$= \sum_{i=1}^{\infty} \lambda_i l_{a_i}(\overline{I_\alpha(\varphi)}) = l_f(\overline{I_\alpha \varphi})$$

Since the bounded and boundedly supported functions are dense in $L^{q'}$ this shows that g does only depend of f, not on the decomposition of f into $\sum \lambda_i a_i$. Having shown this one can write $g=I_\alpha f$ and one has by (I,29)

$$\|I_\alpha f\|_q \le \sum_{i=1}^{\infty} |\lambda_i|\, \|I_\alpha(a_i)\| \le C \sum_{i=1} |\lambda_i| \le C\Big(\sum_{i=1}^{\infty} |\lambda_i|^p\Big)^{1/p}$$

Since this inequality hold for every decomposition of f into atoms one finally has

$$\|I_\alpha f\|_q \le C\|f\|_{H^p}.$$

The case $\mu(X)=1$ is simpler. The proofs in the other cases follow exactly the same pattern, and are left to the reader. In case 1. one has to use Theorem I.2., in case 3. Theorem I.4., and in case 4. Theorem I.5..

IV. PROOFS OF THEOREMS I.7 AND I.8, EXAMPLES

The following Lemma is needed in the proof of Theorem I.7.

Lemma IV.1. Let (X,δ,μ) be a space of homogeneous type such that $\mu(B_r(x)) = \omega r$ for all $r>0$, where ω is independent of x and r, and $\alpha>0$. Then

$$\int_{\delta(x,y)<r} \delta(x,y)^{\alpha-1} d\mu(y) = \frac{\omega}{\alpha} r^{\alpha}, \text{ for all } r>0. \tag{IV,1}$$

Proof The case $0<\alpha<1$ will be proved. The case $\alpha > 1$ is obtained in a similar way and if $\alpha=1$ there is nothing to prove. Let I be the left hand side of (IV.1), and $a>1$, then the following inequalities hold:

$$\frac{a-1}{a^{\alpha}-1} a^{\alpha-1} \omega r^{\alpha} \leq I \leq \frac{a-1}{a^{\alpha}-1} \omega r^{\alpha} \tag{IV,2}$$

In order to prove (IV.2), let $r_k = a^{-k} r$, $k = 0,1,2,\ldots$. Since

$$I = \sum_{k=0}^{\infty} \int_{r_{k+1} \leq \delta(x,y)<r_k} \delta(x,y)^{\alpha-1} d\mu(y),$$

and $0 < \alpha < 1$ it follows that

$$\omega \sum_{k=0}^{\infty} r_k^{\alpha-1} (r_k - r_{k+1}) \leq I \leq \omega \sum_{k=0}^{\infty} r_{k+1}^{\alpha-1} (r_k - r_{k+1}). \tag{IV,3}$$

Substituting r_k by its definition and computing the sums one obtains (IV,2). Finally, since both sides of (IV,2) tend to $\frac{\omega}{\alpha} r^{\alpha}$ when a tends to 1, the equality (IV,1) is obtained.

Proof of Theorem I.7 Observe that (X,δ,μ) has order $\frac{1}{\lambda}$ and since $\alpha < \frac{1}{\lambda}$, by Corollary II.1. the function

$$\frac{1}{\delta(x_1,y)^{1-\alpha}} - \frac{1}{\delta(x_2,y)^{1-\alpha}}$$

is absolutely integrable on X. Therefore, in order to show that the integral of this function is zero it is enough to show that

$$\lim_{R\to+\infty}\int_{\delta(x_1,y)<R}\left[\frac{1}{\delta(x_1,y)^{1-\alpha}}-\frac{1}{\delta(x_2,y)^{1-\alpha}}\right]d\mu(y)=0. \qquad \text{(IV,4)}$$

Rewrite the integrand in terms of ρ, and consider the integral over $\rho(x_1,y)< r$, where $r = R^{1/\lambda}$ and let

$$I = \int_{\rho(x_1,y)<r}\left(\frac{1}{\rho(x_1,y)^{\lambda(1-\alpha)}}-\frac{1}{\rho(x_2,y)^{\lambda(1-\alpha)}}\right)d\mu(y)$$

$$= \int_{\rho(x_1,y)<r}\frac{1}{\rho(x_1,y)^{\lambda(1-\alpha)}}\,d\mu(y) - \int_{\rho(x_1,y)<r}\frac{1}{\rho(x_2,y)^{\lambda(1-\alpha)}}\,d\mu(y)\,, \qquad \text{(IV,5)}$$

where the last equality follows from the fact both functions are locally integrable by Lemma II.1 (ii).

If $d=\rho(x_1,x_2)$ one has $B_{r-d}(x_2) \subset B_r(x_1) \subset B_{r+d}(x_2)$ where all balls are taken with respect to the metric ρ. Therefore, there is s, $r-d<s<r+d$ such that

$$\int_{\rho(x_1,y)<r}\frac{1}{\rho(x_2,y)^{\lambda(1-\alpha)}}d\mu(y)=\frac{\omega}{\alpha}s^{\lambda\alpha} = \int_{\rho(x_1,y)<s}\frac{1}{\rho(x_1,y)^{\lambda(1-\alpha)}}\,d\mu(y) \qquad \text{(IV,6)}$$

where the integrals are computed using the fact that $\rho=\delta^{1/\lambda}$ and Lemma IV.1. Using the expressions (IV,6) in (IV,5) one obtains

$$|I| \le \int_{r-d\le\rho(x,y)\le r+d}\frac{1}{\rho(x_1,y)^{\lambda(1-\alpha)}}\,d\mu(y)$$

$$\le (r-d)^{-\lambda(1-\alpha)}\omega\left[(r+d)^{\lambda}-(r-d)^{\lambda}\right] = 0\,(r^{\lambda\alpha-1}).$$

Finally $r^{\lambda\alpha-1}$ tends to 0 as r tends to infinity because $\lambda\alpha < 1$.

Example IV.1. The parabolic spaces associated to $\mathbb{R}^n$.

A detailed description of the theory of these spaces is in Calderón and Torchinsky (1975).

Let $X = \mathbb{R}^n$ and consider a multiplicative group A_t of linear transformations of $\mathbb{R}^n$ with the following property

$$t^{\alpha}|x| \leq |A_t x| \leq t^{\beta}|x|$$

for all $t \geq 1$, $x \epsilon \mathbb{R}^n$ and $|x|$ the Euclidean norm of x. For $x \neq 0$ let $\rho(x)$ be the unique value of t such that

$$|A_t^{-1} x|=1,$$

and for $x = 0$ let $\rho(0) = 0$. Then $\rho(x)$ satisfies the following properties

i) $\rho(x) = 0 <=> x = 0$

ii) $\rho(x) \leq 1 <=> |x| \leq 1$

iii) $\rho(A_t x) = t\rho(x)$

iv) $\rho(x_1 + x_2) \leq \rho(x_1) + \rho(x_2)$

Define $d(x,y) = \rho(x-y)$, the parabolic metric associated with A_t. Note that when $A_t = tI$ the parabolic metric is the euclidean metric. Let $B_r = \{y:\rho(y)<r\}$. Note that B_1, $= \{y:|y|<1\}$, and $B_r = A_r(B_1)$. Let P be the infinitesimal generator of A_t and λ the trace of P. Then $\det(A_t) = t^{\lambda}$ and

$$m(B_r) = r^{\lambda} m(B_1)$$

where $m(B_r)$ denotes the Lebesgue measure of B_r. Therefore (X,d,m) where m is the Lebesgue measure is a space of homogeneous type that can be normalized in the following way,

$$\delta(x,y) = d(x,y)^{\lambda}.$$

Now (X,δ,m) is a normal space of order $\frac{1}{\lambda}$, whose quasi distance is the λ–th power of a distance and such that $m(B_r) = \omega r$, for all $r>0$, where $\omega=m(B_1)$. Finally, in view of Theorem I.7., (X,δ,m) also has the cancellation property of order $1-\alpha$, $\alpha<1/\lambda$.

Example IV.2. Homogeneous Groups

General references for this example are Folland and Stein (1982), Korány and Vági (1971) and Korány et al. (1970).

A connected, simply connected nilpotent Lie group G with identity element e is called a homogeneous group if there exists a (multiplicatively written) one–parameter group $\{a(t)\}_{t>0}$ of automorphisms of G such that

(i) $\lim_{t\to 0} a(t)x=e$ for every x in G, and (IV,7)

(ii) the induced automorphisms $a_*(t)$ of the Lie algebra are diagonalizable.

The condition (ii) implies that the eigenvalues (of the infinitesimal generator) of a_* are positive, let q be their sum. Let dx denote a fixed Haar measure on G then

$$d(a(t)x) = t^q dx \tag{IV,8}$$

for all $t > 0$. (The Haar measure will also be denoted by μ, if convenient).

Given any real s, a function f defined on G–{e} is said to be homogeneous of degree s if for all $t > 0$, and all $x \in G-\{e\}$

$$f(a(t)x) = t^s f(x)$$

If $s > 0$ and f is continuous then f can be extended continuously to all of G by setting f(e)=0. On every homogeneous group there exist strictly positive functions on G–{e} which are continuously differentiable, homogeneous of degree 1 and whose values at x and x^{-1} are equal for all x in G–{e}. Let $x \longrightarrow |x|$ be such a function. It is a gauge as defined in Korány–Vági (1971), i.e. there is a constant $\kappa \geq 1$ such that

$$|xy| \leq \kappa(|x| + |y|)$$

for all x, in G and the sets $B_r(e) \doteq \{x: |x|<r, r>0\}$ form a basis of neighborhoods at e. By (IV,8)

$$\mu(B_r(e)) = \omega r^q. \tag{IV,9}$$

The smallest eigenvalue of the infinitesimal generator of $a_*(t)$ is denoted by Λ. If f is homogeneous of degree s and continuously differentiable in G–{e} then there exist positive constants M and N such that

$$\left.\begin{array}{l} |f(xy)-f(y)| \\ |f(yx)-f(y)| \end{array}\right\} M|x|^{\Lambda}|y|^{s-\Lambda} \tag{IV,10}$$

whenever $N|x| \leq |y|$.

It is immediate that a gauge defines a left in variant quasi–distance δ on G by $\delta(x,y) = |x^{-1}y|$. By (IV,9) μ has the doubling property, hence (G,δ,μ) is a space of homogeneous type. This space can be normalized by introducing

$|x|_1=|x|^q$ as a new gauge (and reparametrizing a(t) by $t \longrightarrow t^{1/q}$ to make $|x|_1$ homogeneous of degree 1). Using (IV,10) one can see that G with the new quasi–distance has order $\gamma = \Lambda/q \leq 1$. From here on it is assumed that the gauge has been so normalized, i.e. that G with $\delta(x,y) = |x^{-1}y|$ is a space of homogeneous type which is normal, and of order γ, $0<\gamma\leq 1$. Theorem I.8. asserts the (G,δ,μ) also has the cancellation property of order $1-\alpha$, $0<\alpha<\gamma$.

The following formula for integration in "polar coordinates" is needed from Folland and Stein (1982). Let $S = \{x: |x|=1\}$, then there exists a unique Radon measure σ on S such that for every f in $L^1(G)$

$$\int_G f(x)dx = \int_0^\infty \left(\int_S f(a(t)\, y)d\sigma(y)\right)dt. \tag{IV,11}$$

Proof of Theorem I.8. Since the integrand is left–invariant it is enough to prove the theorem for $x_1=x$ and $x_2=e$. By Corollary II.1. the function $y \longrightarrow |x^{-1}y|^{\alpha-1}-|y|^{\alpha-1}$ is absolutely integrable, therefore it suffices to show that

$$I_R= \int_{|y|<R} \left(|x^{-1}y|^{\alpha-1}-|y|^{\alpha-1}\right)dy \rightarrow 0, \text{ as } R \longrightarrow \infty.$$

Introducing $x^{-1}y$ as a new variable in $\int_{|y|<R} |x^{-1}y|^{\alpha-1}dy$ one obtains

$$I_R = \int_{|xy|<R} |y|^{\alpha-1}d\mu - \int_{|y|<R} |y|^{\alpha-1}d\mu$$

The integrals cancel in $B_R(e) \cap B_R(x^{-1})$ and

$$I_R = \int_{\substack{|xy|<R\\ |y|\geq R}} |y|^{\alpha-1}d\mu - \int_{\substack{|y|<R\\ |xy|\geq R}} |y|^{\alpha-1}d\mu = I_2 - I_1$$

Let $R>\kappa+\kappa|x|$ then $B_1(e) \subset B_R(x^{-1})$. If $|\dot{y}|=1$ define $t_{\dot{y}} = \inf\{t:|xa(t)\dot{y}|=R\}$. By continuity $|xa(t_{\dot{y}})\dot{y}| = R$. Let χ denote the characteristic function of $B_R(e) - B_R(x^{-1})$ by (IV,11)

$$I_1 = \int_S \left(\int_0^{\infty} \chi(a(t))\dot{y})t^{\alpha-1}dt\right)d\sigma(\dot{y})$$

$$\leq \int_{S'} \left(\int_{t_{\dot{y}}}^{R} t^{\alpha-1}dt\right)d\sigma(\dot{y})$$

$$\leq \int_{S'} (R-t_{\dot{y}})t_{\dot{y}}^{\alpha-1}d\sigma(\dot{y}),$$

where S' denotes the subset of S where $t_{\dot{y}} \leq R$. Note that $R=|xa(t_{\dot{y}})\dot{y}| \leq \kappa\,|x|+\kappa t_{\dot{y}}$, and

$$1< \frac{R-\kappa|x|}{\kappa} \leq t_{\dot{y}}.$$

Therefore $t_{\dot{y}} \longrightarrow \infty$ as $R \longrightarrow \infty$, and one can take R so large that $t_{\dot{y}} > |x|$. Then by Lemma II.2.

$$R-t_{\dot{y}} = |\chi a(t_{\dot{y}})\dot{y}| - |a(t_{\dot{y}})\dot{y}| \leq B_3|x|^{\gamma}\, t_{\dot{y}}^{1-\gamma}$$

and,

$$I_1 \leq B_3|x|^{\gamma} \int_{S'} t_{\dot{y}}^{\alpha-\gamma} d\sigma(\dot{y}).$$

Since $\alpha-\gamma < 0$, and $t_{\dot{y}} \geq \frac{R-\kappa|x|}{\kappa}$ the Lebesgue bounded convergence theorem shows that $I_1 \longrightarrow 0$ as $R \longrightarrow \infty$. The proof that $I_2 \longrightarrow 0$ is very similar.

Example IV.3.

Let $X = \mathbb{Z}^m$ where $\mathbb{Z}$ denotes the group of integers, $\delta(x,y) =$

$\|x-y\|_\infty^m$, where $\|x\|_\infty = \max\{|x_i|;\ 1 \le i \le m\}$, and μ is the counting measure. Then (X,δ,μ) is a space of homogeneous type which is normal and of order $\frac{1}{m}$ and which has the cancellation property of order $1-\alpha$ for $0<\alpha<\frac{1}{m}$. It is easy to see that the measure of any ball of radius $r>0$ is $(2\lceil r^{\frac{1}{m}}\rceil-1)^m$ where $\lceil k\rceil = \inf\{\ n: n \ge k, n \in \mathbb{Z}\}$. Note that $r_x = 1$ for all x and that

$$\mu(B_r(x)) \le 3^m r \text{ for } r \ge r_x ,$$

and

$$r \le \mu(B_r(x) \text{ for all } r > 0 .$$

Therefore (X,δ,μ) is a normal and μ is doubling. (X,δ,μ) is of order $\frac{1}{m}$ because δ is the m–th power of a metric. It will be shown next that for $0<\alpha<\frac{1}{m}$ (X,δ,μ) has the cancellation property of order $1-\alpha$.

It can be assumed that $x_1=0$ and write $x_2=x$.

$$\frac{1}{\delta(0,y)^{1-\alpha}} - \frac{1}{\delta(x,y)^{1-\alpha}}$$

is in L^1 as a function of y, it is enough to show that

$$\lim_{R\to+\infty} \int'_{\delta(0,y)<R} \left(\frac{1}{\delta(0,y)^{1-\alpha}} - \frac{1}{\delta(x,y)^{1-\alpha}}\right)d\mu(y) = 0.$$

Rewrite δ in terms of $\|\ \ \|_\infty$, replace $\int'$ by $\sum'$, $R^{1/m}$ by r and assume $r > \|x\|_\infty$. Let

$$I_r = \sum_{\|y\|_\infty<r}' \left(\frac{1}{\|y\|_\infty^{m(1-\alpha)}} - \frac{1}{\|x-y\|_\infty^{m(1-\alpha)}}\right)$$

$$= \sum_{\|y\|_\infty < r}' \frac{1}{\|y\|_\infty^{m(1-\alpha)}} - \sum_{\|x-y\|_\infty < r}' \frac{1}{\|y\|_\infty^{m(1-\alpha)}}.$$

Therefore

$$|I_r| \leq \sum_{r - \|x\|_\infty \leq \|y\|_\infty < r + \|x\|_\infty} \frac{1}{\|y\|_\infty^{m(1-\alpha)}}$$

$$\leq (r - \|x\|_\infty)^{m(\alpha-1)} \sum_{r - \|x\|_\infty \leq \|y\|_\infty \leq r + \|x\|_\infty} 1 .$$

The last sum can be computed using the formula for the measure of a ball above;

$$(2\lceil (r + \|x\|_\infty)\rceil - 1)^m - (2\lceil r - \|x\|_\infty \rceil - 1)^m \leq$$

$$(2 (r + \|x\|_\infty + 1)^m - (2 (r - \|x\|_\infty) - 1)^m = O(r^{m-1}).$$

Finally $|I_r| = O(r^{\alpha m-1})$ and since $\alpha < \frac{1}{m}$, $\lim_{r \to +\infty} I_r = 0$.

REFERENCES

Aimar, H.S. (1983). Operadores Integrales Sigulares y Aproximaciones de la Identidad en Espacios de Tipo Homogéneo. Thesis, University of Buenos Aires.

Calderón, A.P. and Torchinsky, A. (1975). Parabolic Maximal Functions Associated with Distributions. Advances in Math., 16, pp. 1–63.

Coifman, R.R. and Weiss, G. (1971). Analyse Harmonique sur certains Espaces Homogènes, Lecture Notes in Math., vol 242, Springer–Verlag Berlin and New York.

Coifman, R.R. and Weiss, G. (1977). Extensions of Hardy Spaces and their use in Analysis, Bull. of the Am. Math Soc., 83, pp. 569–645

Folland, G.B. and Stein, E.M. (1982). Hardy Spaces on Homogeneous Groups, Mathematical Notes vol 28, Princeton University Press, Princeton.

Gatto, A.E., Guitiérrez, C.E. and Wheeden, R.L. (1983). On Weighted Fractional Integrals, Conference on Harmonic Analysis in Honor of Antoni Zygmund, vol 1 pp. 124–137, Wadsworth International Group, Belmont, CA.

Hedberg, L.I. (1972). On certain Convolution Inequalities, Proceedings of the Am. Math. Soc., 36, pp. 505–510.

John, F. and Nirenberg, L. (1961). On Functions of Bounded Mean Oscillation, Comm. Pure Appl. Math; 14, pp. 415–426.

Koräny, A. and Vági, S. (1971). Singular Integrals on Homogeneous Spaces and some Problems of Classical Analysis, Annali della Scuola Normale Superiore Pisa, 25, pp. 575–648.

Korány, A., Vági, S. and Welland G.V. (1970), Remarks on the Cauchy Integral and the Conjugate Function in Generalized Half–Planes, Journal of Math and Mech., 19, pp. 1069–1081.

Macías, R.A. and Segovia, C. (1979a). Singular Integrals on Generalized Lipschitz and Hardy Spaces, Studia Math.., 65, pp. 55–75.

Macías, R.A., and Segovia, C. (1979b). Lipschitz Functions on Spaces of Homogeneous Type, Advances in Math., 33, pp. 257–270.

Macías, R.A., and Segovia, C. (1979c). A Decomposition into Atoms of Distributions on Spaces of Homogeneous Type, Advances in Math., 33, pp. 271–309.

Reimann, H.M., and Rychener, Th. (1975). Funktionen beschränketer mittlerer Oszillation, Lecture Notes in Math., vol 487, Springer–Verlag, Berlin, New York.

Taibleson, M.H. and Weiss, G. (1980). The Molecular Characterization of Hardy Spaces, Asterisque, vol 77.

Zygmund, A. (1969). Trigonometric Series 2nd ed.., Cambridge University Press, New York.

10

Large Analytic Functions, II

Henry Helson

This note adds some information about the Smirnov class $\mathfrak{A}$ of analytic functions in the unit disc to the results presented in [1]. In the first section we give a characterization of the functions in $\mathfrak{A}$ that are real, that is, functions of $\mathfrak{A}_r$ in the notation of [1]. The second section contains an application to the characterization of strongly outer functions. References to work of E. Hayashi and T. Nakazi on this subject were given in [1]. The third section describes a locally convex topology on $\mathfrak{A}$, different from a metric topology investigated by Shapiro and Shields [3].

1. Large Real Analytic Functions

The space $\mathfrak{A}$ is the set of all quotients f/g, where f, g are in H^∞ on the circle, and g is an outer function. These are called *large analytic functions*. Within $\mathfrak{A}$ the class of real functions is denoted by $\mathfrak{A}_r$.

Theorem 1. *The functions of $\mathfrak{A}_r$ are exactly the quotients $i(p+q)/(p-q)$, where p and q are inner and $p-q$ is outer.*

For any complex number z of modulus 1 the quotient $i(1+z)/(1-z)$ is real. Let p and q be inner, with $p-q$ outer. Then $i(p+q)/(p-q)$ is a real function, and it belongs to $\mathfrak{A}_r$. This is half the assertion of the theorem.

Now let f, g belong to H^∞, with g outer. If the quotient f/g is real then $f+ig$ and $f-ig$ have the same modulus. Therefore there is an outer function h such that $f+ig = ph$, $f-ig = qh$ for inner functions p, q. Thus $2f = (p+q)h$, $2ig = (p-q)h$. Since g is outer, $p-q$ must be outer. Dividing gives $f/g = i(p+q)/(p-q)$. This proves the other half of the theorem.

Note that f/g is constant if and only if p/q is constant. Furthermore, $p-q$ is constant only if p and q are both constant, unless $p = q$.

The inner functions p, q are unique up to constant factors in the representation of Theorem 1. To see this, let $\phi = i(p+q)/(p-q)$ with $p-q$ outer. Then $\phi+i = 2ip/(p-q)$, so that p is exactly the inner factor of $\phi+i$. Similarly, q is the inner factor of $\phi-i$. It is also obvious that the pair (p, q) is determined up to a single constant factor.

2. Strong Outer Functions

A function k in H^1 is *strongly outer* if the only functions f in H^1 such that f/k is positive are scalar multiples of k. (If k has norm 1, it has this property if and only if it is an exposed point of the unit ball of H^1.) Every strongly outer function is outer, and therefore has a square root in H^2. If g is in H^2, then $k = g^2$ is strongly outer if and only if g has this related property: for f in H^2, f/g is real only if f is a scalar multiple of g.

Theorem 2. *Let g be in H^2. Then g^2 is strongly outer in H^1 if and only if $g/(p-q)$ is not in H^2 for any inner functions p, q except when $p-q$ is constant.*

If $h = 2ig/(p-q)$ is in H^2 for some p, q such that $p-q$ is not outer, then g is not outer either. If h is in H^2 with $p-q$ outer and non-constant, then the pair of equations

$$2f = (p+q)h, \quad 2ig = (p-q)h \tag{1}$$

defines f in H^2 such that f/g is real and non-constant. In either case g^2 is not strongly outer. Conversely, if g is not strongly outer and f/g is real, then as in the last section we can find h to satisfy (1) and thus $g/(p-q)$ is in H^2 for some inner functions p, q such that $p-q$ is not constant.

It is known that if g^2 is strongly outer in H^1, then

$$\int |g|^2/|p-1|^2 \, d\sigma = \infty \qquad (d\sigma = dx/2\pi) \tag{2}$$

for every non-constant inner function p. D. Sarason offers evidence to support the conjecture that when g is outer in H^2, this condition is sufficient for g^2 to be strongly outer [2]. Now p-1 is outer for every non-constant inner p; therefore (2) is formally weaker than the condition of Theorem 2, and the question is whether the conditions are in fact the same. That is, does (2) (for every non-constant inner p) imply that

$$\int |g|^2/|p-q|^2 \, d\sigma = \infty \tag{3}$$

for every pair of inner functions p, q such that $p-q$ is outer and non-constant?

The question which outer functions can be represented in the form $p-q$ does not seem to be easy.

3. A topology for $\mathfrak{A}$

Shapiro and Shields [3] defined a metric on the Nevanlinna class N:

$$d(f, g) = \int \log (1+|f-g|)\, d\sigma. \tag{4}$$

They were mainly interested in showing that this metric has pathological properties, but they mentioned that its restriction to $\mathfrak{A}$ makes $\mathfrak{A}$ into a topological vector space that is complete but not locally convex. We propose another topology that will be locally convex.

Denote by $\mathfrak{W}$ the class of non-negative functions w such that w and $\log w$ are summable. For each such weight function form the Lebesgue space L^2_w based on the measure $w d\sigma$. In this space let $\mathfrak{F}^2_w$ be the closure of analytic trigonometric polynomials. It was observed in [1] that $\mathfrak{A}$ is precisely the union of all $\mathfrak{F}^2_w$ as w ranges over $\mathfrak{W}$. If $w \leq w'$ then $\mathfrak{F}^2_{w'}$ is continuously embedded in $\mathfrak{F}^2_w$. Therefore [4] we can form the inductive limit of the spaces $\mathfrak{F}^2_w$. This is a locally convex topology.

An algebraically-linear functional on $\mathfrak{A}$ is continuous if and only if its restriction to each space $\mathfrak{F}^2_w$ is continuous.

By Szegö's theorem, each Taylor coefficient is a continuous functional on the spaces $\mathfrak{F}^2_w$, and therefore on $\mathfrak{A}$. It follows that the new topology separates points.

Let F be a continuous linear functional on $\mathfrak{A}$. Then F is a continuous functional on each space $\mathfrak{F}^2_w$. One of these spaces is H^2 (obtained when w is a constant function). Therefore F determines a function f in H^2, and the formula

$$F(g) = \sum \overline{a_n(f)}\, a_n(g) \tag{5}$$

holds for g in H^2. ($a_n(f)$ and $a_n(g)$ are Fourier coefficients.) Since trigonometric polynomials are dense in every space $\mathfrak{F}^2_w$, this function f determines F.

It is not obvious which functions f in H^2 determine functionals on $\mathfrak{A}$. John McCarthy has observed that if f is analytic on a disc of radius greater than 1, then the series (5) converges and defines a linear functional on $\mathfrak{A}$. Here are some more conditions.

Theorem 3. *Each of these conditions is necessary and sufficient for f in H^2 to define a continuous linear functional in $\mathfrak{A}$:*

(a) *for each function $W = 1/w$, w in $\mathfrak{W}$, there is a function k in H^1_0 (the* *summable analytic functions with mean value* 0) *such that*

(6) $$\int |f + \bar{k}|^2 W \, d\sigma < \infty$$

(b) *for each* h *in* H^∞, f *is in the range of the Toeplitz operator* $T_{\bar{h}}$

(c) *for every* h *in* H^2, *the Taylor coefficients of* f *can be represented*

(7) $$\hat{f}(n) = (k, S^n h) \qquad (n \geq 0)$$

with some k *in* H^2. *Here* S *is multiplication by* exp *ix in* H^2, *and the parentheses denote the inner product in* H^2.

The proofs are not difficult, and we omit them.

References

[1] H. Helson, Large analytic functions, to appear in Proceedings of a conference held in Timisoara in June, 1988.

[2] D. Sarason, Exposed points in H^1, to appear.

[3] J. H. Shapiro and A. L. Shields, Unusual topological properties of the Nevanlinna class, Amer. J. Math. 97(1976), 915-936.

[4] A. Wilansky, *Modern Methods in Topological Vector Spaces*, McGraw-Hill, 1978.

11

Reiteration Phenomena in the Complex Method of Interpolation

Peter W. Jones

§1. Introduction.

The purpose of this note is to present constructive methods for solving various reiteration problems in Calderón's complex method of interpolation. (See [1] for definitions and basic results.) The two basic types of problems with which we will concern outselves are Calderón's reiteration theorem and Wolff's converse to it [2]. We now recall those results.

Let A_0 and A_1 be a compatible couple of Banach spaces and form the spaces $A_\theta = (A_0, A_1)_\theta$, $0 < \theta < 1$.

Theorem A. (Calderón's reiteration theorem)

Suppose $0 \leq \theta_1 < \theta_2 \leq 1$ and $0 < \psi < 1$. Then

$$(A_{\theta_1}, A_{\theta_2})_\psi = A_\theta$$

where $\theta = (1-\psi)\theta_1 + \psi\theta_2$.

Theorem B. (Wolff's theorem)

Let A_1, A_2, A_3, A_4 be Banach spaces and assume $A_1 \cap A_4$ is a dense subspace of both A_2 and A_3. Suppose $(A_2, A_4)_\theta = A_3$, $(A_1, A_3)_\phi = A_2$, $0 < \theta, \phi < 1$. Then

$$(A_1, A_4)_\xi = A_2, \ (A_1, A_4)_\psi = A_3$$

where

$$\xi = \frac{\phi\theta}{1-\phi+\phi\theta}, \ \psi = \frac{\theta}{1-\phi+\phi\theta}.$$

Proofs of both of these results have until now depended on applications of Calderón's so-called duality theorem:

Theorem C.

$$((A_0, A_1)_\theta)^* = (A_0^*, A_1^*)^\theta, \; 0 < \theta < 1.$$

Here A^* denotes the dual of A, and $(\cdot,\cdot)^\theta$ is Calderón's upper method of complex interpolation.

The idea of this paper is to give proofs of theorems A and B which are based only on elementary $\bar{\partial}$ methods plus the three lines phenomenon. We shall not discuss how these methods can be extended to cover more general cases (e.g. infinitely many Banach spaces and/or multiply connected domains). An interesting problem which we leave open is how to extend our methods to handle quasi-Banach spaces. (We shall indicate later where problems arise.)

§2. Theorem A.

To avoid certain (rather easily fixed) technical problems, we will work on the "unit annulus" $\mathcal{Q} = \{z : 1 \leq |z| \leq e\}$ rather than on the unit strip $\mathcal{S} = \{z : 0 \leq x \leq 1\}$. Let A_0, A_1 be a (compatible) Banach couple and let $\mathcal{F}(A_0, A_1)$ be the family of all functions taking values in $A_0 + A_1$, being holomorphic on the interior of $\mathcal{Q}$, and being continuous on $\mathcal{Q}$, taking values in A_0 when $|z| = 1$, and taking values in A_1 when $|z| = e$. For $f \in \mathcal{F}(A_0, A_1)$ set

$$\|f\|_{\mathcal{F}(A_0,A_1)} = \max\Big(\sup_{|z|=1} \|f\|_{A_0}, \sup_{|z|=e} \|f\|_{A_1} \Big).$$

We then define $A_\theta = (A_0, A_1)_\theta$ to be the space of all values $f(e^\theta)$ where $f \in \mathcal{F}(A_0, A_1)$ with norm

$$\|x\|_{A_\theta} = \inf\{\|f\|_{\mathcal{F}(A_0,\ A_1)} : f(e^\theta) = x\}.$$

It is well known (and an easy exercise) that the space A_θ above is the same as the one obtained by Calderón's original construction. The same procedure can, of course, be done on any annulus, and we will not belabor the reader with the obvious method of labeling the spaces so constructed. For $0 \le \theta_1 < \theta_2 \le 1$ we denote by $\mathfrak{a}_{\theta_1,\theta_2}$ the annulus $\{e^{\theta_1} \le |z| \le e^{\theta_2}\}$ and $\mathcal{F}(A_{\theta_1},\ A_{\theta_2})$ the obvious holomorphic functions on $\mathfrak{a}_{\theta_1,\theta_2}$ with boundary values in either A_{θ_1} or A_{θ_2}.

We now give the proof of theorem A. To avoid notational complexities we only treat the sample case $\theta_1 = {}^1/3$, $\theta_2 = {}^2/3$, $\psi = {}^1/2$, $\theta = {}^1/2$. The proof in the general case is the same. It is clear from the definition of A_θ that $A_{1/2} \subset (A_{1/3},\ A_{2/3})_{1/2}$, so we need only show the reverse inclusion. Pick $a \in A_{1/3}$ with norm $\|a\|_{(A_{1/3},\ A_{2/3})_{1/2}} < 1$ and select $f_0 \in \mathcal{F}(A_{1/3},\ A_{2/3})$ with norm $\|f_0\|_{\mathcal{F}(A_{1/3},\ A_{2/3})} \le 1$. We would like to extend f_0 to be holomorphic on all of $\mathfrak{a}$, but of course this is not in general possible. We do the next best thing by defining f_0 to be zero on $\mathfrak{a}\backslash\mathfrak{a}_{1/3,2/3}$. Then f_0 is holomorphic on $\mathfrak{a}$ except for jumps on $\partial\mathfrak{a}_{1/3,2/3}$. To remedy this we solve a $\bar\partial$ problem:

$$\bar\partial g = \bar\partial f_0 \quad \text{on } \mathfrak{a} \tag{2.1}$$

and

$$g(e^{1/2}) = 0. \tag{2.2}$$

Then if g has the correct boundary behavior on $\mathcal{Q}$, $f \equiv f_0 - g \in \mathcal{F}(A_0, A_1)$ and $f(e^{1/2}) = f_0(e^{1/2})$.

To construct g we first construct some auxiliary functions. Fix ζ, $|\zeta| = e^{1/3}$, and pick $G_\zeta(z) \in \mathcal{F}(A_0, A_1)$ satisfying

$$\|G_\zeta\|_{\mathcal{F}(A_0,\ A_1)} \le C \tag{2.3}$$

and

$$G_\zeta(\zeta) = f_0(\zeta). \tag{2.4}$$

This is possible by the definition of the complex method of interpolation. Multiplying G_ζ if necessary by a scalar valued holomorphic function we may assume that G_ζ satisfies the additional property

$$G_\zeta(e^{1/2}) = 0. \tag{2.5}$$

For $|\zeta| = e^{2/3}$ we also select functions G_ζ satisfying (2.3)-(2.5). It is not hard to see that for $|\zeta| = e^{1/3}$ or $e^{2/3}$, G_ζ may be picked continuously in ζ.

Now set

$$g(z) = \frac{1}{2\pi} \int_{|\zeta| = e^{1/3}} \frac{\frac{\zeta}{|\zeta|} G_\zeta(z)}{z-\zeta}\, ds(\zeta) - \frac{1}{2\pi} \int_{|\zeta|=e^{2/3}} \frac{\frac{\zeta}{|\zeta|} G_\zeta(z)}{z-\zeta}\, ds(\zeta),$$

where d_s denotes arclength. Then since $\bar{\partial} f_0 = \frac{1}{2} \frac{\zeta}{|\zeta|} f_0(\zeta) ds$ on

$|\varsigma| = e^{1/3}$ and $\overline{\partial} f_0 = \frac{-1}{2} \frac{\varsigma}{|\varsigma|} f_0(\varsigma) ds$ on $|\varsigma| = e^{2/3}$, one verifies without too much difficulty that (2.4) implies $\overline{\partial} f_0 - \overline{\partial} g$ (i.e. (2.1) holds) and that $f_0 - g$ is a holomorphic function on $\mathfrak{Q}$. By (2.3), $f_0 - g$ is continuous on $\mathfrak{Q}$ and

$$\|f_0 - g\|_{\mathcal{F}(A_0, A_1)} \le C. \tag{2.6}$$

Furthermore, condition (2.5) shows that $g(e^{1/2}) = 0$. Setting $f = f_0 - g$ we thus have $\|f\|_{\mathcal{F}(A_0, A_1)} \le C$ and $f(e^{1/2}) = a$. The theorem is proven.

It should be noted that Calderón's proof of theorem A shows that $(A_{\theta_1}, A_{\theta_2})_{\psi} = A_\theta$ with <u>equality</u> of norms, whereas because of (2.6) we have only proved <u>equivalence</u>. A more serious difficulty arises when one attempts to extend this argument to the case of quasi-Banach spaces. To deduce (2.6) one must verify e.g. that for $|z| = 1$, $\|g(z)\|_{A_0} \le C$. To do this one writes

$$\|g(z)\|_{A_0} \le \frac{1}{2\pi} \int_{|\varsigma|=e^{1/3}} \frac{\|G_\varsigma(z)_{A_0}\|}{|z-\varsigma|} ds(\varsigma)$$

$$+ \frac{1}{2\pi} \int_{|\varsigma|=e^{2/3}} \frac{\|G_\varsigma(z)\|_{A_0}}{|z-\varsigma|} ds(\varsigma)$$

$$\le C.$$

For quasi-Banach spaces this reasoning is no longer valid because failure of the triangle inequality means that one is not allowed to pass the $\|\cdot\|_{A_0}$ sign under the integral.

§3. Theorem B.

Our proof of theorem B is merely obtained by copying Wolff's proof in [2] for the real method, and using the $\bar{\partial}$ ideas of section 2 whenever necessary. To ease notational problems we only treat the case where $\theta=\phi={}^1/2$ (so that $\xi = {}^1/3$, $\upsilon = {}^2/3$). As noted in [2], the only difficulty is to prove that $A_2 \subset (A_1, A_4)_{1/3}$, $A_3 \subset (A_1, A_4)_{2/3}$. We shall only prove the first of these assertions, the proof of the other being the same. While we could prove $A_2 \subset (A_1, A_4)_{1/3}$ directly, we will instead employ an artiface to eliminate some minor technical problems. What we first prove is that $(A_2, A_3)_{1/2} = (A_1, A_4)_{1/2}$. Then the desired result follows from two applications of theorem A because

$$\begin{aligned} A_2 &= (A_1, A_3)_{1/2} \\ &= (A_1, A_5)_{2/3} \\ &= (A_1, A_4)_{1/3}. \end{aligned}$$

Here $A_5 \equiv (A_2, A_3)_{1/2}$. As noted in [2], we need only show $A_5 \subset (A_1, A_4)_{1/2}$.

Let $\mathfrak{a}_1 = \{1 \le |z| \in e\}$. We wish to paste together interpolation between A_1 and A_3 (on $\mathfrak{a}_1$) and A_2 and A_4 (on $\mathfrak{a}_2$) to get the desired interpolation between A_1 and A_4 on $\mathfrak{a}$. Pick $a_5 \in A_5$ with norm

$$\|a_5\|_{(A_2,A_3)_{1/2}} < 1$$

and select $f_0 \in \mathcal{F}(A_1, A_3)$ such that $f_0(e^{1/2}) = a_5$ and

$$\|f_0\|_{\mathcal{F}(A_1, A_3)} \le 1. \tag{3.1}$$

Extend f_0 to all of $\mathfrak{a}$ by defining $f_0 \equiv 0$ on $\mathfrak{a}\backslash\mathfrak{a}_1$. Then

$$\bar{\partial} f_0 = \frac{-1}{2} \frac{\zeta}{|\zeta|} f_0(\zeta) ds(\zeta) \tag{3.2}$$

where ds is arclength on $\{|\zeta| = e^{2/3}\}$. The idea of this section is to prove

Lemma 3.1. <u>Suppose</u> $b(\zeta)$ <u>is a continuous function on</u> $\{|\zeta| = e^{2/3}\}$ <u>taking values in</u> A_3 <u>and suppose</u> $\|b(\zeta)\|_{A_3} \le 1$ <u>for all</u> ζ <u>there. Then there is an analytic function</u> F <u>on</u> $\mathcal{Q}\backslash\{|\zeta| = e^{2/3}\}$ <u>such that</u> $F(e^{1/2}) = 0$ <u>and</u>

$$\|b(\zeta) - \bar{\partial}F(\zeta)\|_{A_3} \le {}^1/2, \quad |\zeta| = e^{2/3}.$$

<u>Furthermore</u> F <u>is continuous off</u> $\{|\zeta| = e^{2/3}\}$ <u>and</u>

$$\sup_{|z|=1} \|F(z)\|_{A_1}, \quad \sup_{|z|=e} \|F(z)_{A_4}\| \le C.$$

An easy iteration argument then shows that there is F analytic on $\mathcal{Q}\backslash\{|\zeta| = e^{2/3}\}$ such that $F(e^{1/2}) = 0$, $\bar{\partial}F = \bar{\partial}f_0$, $f_0 - F \in \mathcal{F}(A_1, A_4)$ and (by (3.1), (3.2)), $\|f_0 - F\|_{\mathcal{F}(A_1, A_4)} \le C_0$. This is an exercise which will be left to the reader. It will follow from the special construction of F.) That proves theorem B because $(f_0 - F)(e^{1/2}) = a_5$.

<u>Proof of Lemma 3.1</u>. For each $|\zeta| = e^{2/3}$ pick $\hat{G}_\zeta(z) \in \mathcal{F}(A_2, A_4)$ (so that $\hat{G}_\zeta$ is analytic on $\mathcal{Q}_2$) such that

$$\hat{G}_\zeta(\zeta) = b(\zeta),$$

$$\hat{G}_\zeta(e^{1/2}) = 0,$$

$$\|\hat{G}_\zeta\|_{\mathcal{F}(A_2,\ A_4)} \le C.$$

As in section 2, the existence of $\hat{G}_\zeta$ follows from the definition of the complex method of interpolation. Now let M be a large integer to be fixed later and define

$$G_\zeta(z) = z^M . \frac{\bar{\zeta}^M}{|\zeta|^{2M}} . \hat{G}_\zeta(z).$$

Then from the properties of $\hat{G}_\zeta$ it follows that

$$G_\zeta(\zeta) = b(\zeta), \tag{3.3}$$

$$G_\zeta(e^{1/2}) = 0, \tag{3.4}$$

$$\|G_\zeta\|_{\mathcal{F}(A_2,\ A_4)} \le C\ e^{M/3}, \tag{3.5}$$

$$\|G_\zeta(z)\|_{A_2} \le C\ e^{-M/3},\ |\ | = e^{1/3}. \tag{3.6}$$

As in section 2, we may choose G_ζ continuously in ζ. We define

$$f_1(z) = \frac{1}{\pi} \int\limits_{|\zeta|=e^{2/3}} \frac{\frac{\zeta}{|\zeta|} G_\zeta(z)}{z-\zeta}\, ds(\zeta)$$

so that by (3.3),

$$\bar{\partial} f_1 = b ds \quad \text{on} \quad \mathcal{Q}_2.$$

We define $f_1 \equiv 0$ on $\mathcal{Q}\backslash\mathcal{Q}_2$ so that on $\mathcal{Q}_1$,

$$\bar{\partial} f_1 = \frac{1}{2} \frac{\zeta}{|\zeta|} f_1(\zeta) ds(\zeta),$$

where ds is arclength on $\{|\zeta| = e^{1/3}\}$. Furthermore, by the definition of

f_1 and (3.6),

(3.7) $$\|f_1(\zeta)\|_{A_2} \le Ce^{-M/3}, \quad |\zeta| = e^{1/3}.$$

For each $|\zeta| = e^{1/3}$ we pick as before a function $\hat{H}_\zeta(z) \in \mathcal{F}(A_1, A_3)$, $\hat{H}_\zeta(z)$ analytic on $\mathcal{Q}_1$ so that $\hat{H}_\zeta(\zeta) = \frac{1}{2}\frac{\zeta}{|\zeta|} f_1(\zeta)$, $\hat{H}_\zeta(e^{1/2}) = 0$, $\|\hat{H}_\zeta\|_{\mathcal{F}(A_1, A_3)} \le C$. We then set

$$H_\zeta(\zeta) = z^{-M} \cdot \frac{|\zeta|^{2M}}{\zeta^{-M}} \hat{H}_\zeta(z)$$

and note that

(3.8) $$H_\zeta(\zeta) = \frac{1}{2}\frac{\zeta}{|\zeta|} f_1(\zeta),$$

(3.9) $$H_\zeta(e^{1/2}) = 0,$$

(3.10) $$\|H_\zeta\|_{\mathcal{F}(A_1, A_3)} \le Ce^{M/3},$$

(3.11) $$\|H_\zeta(z)\|_{A_3} \le Ce^{-M/3}.$$

If we define

$$f_2(z) = \frac{1}{\pi} \int\limits_{|\zeta|=e^{1/3}} \frac{H_\zeta(z)}{z-\zeta}\, ds(\zeta)$$

we then see that by (3.8),

$$\bar{\partial} f_2 = \bar{\partial} f_1 \quad \text{on } \mathcal{Q}_1,$$

while by (3.7) and (3.11),

$$\|f_2(\zeta)\|_{A_3} \le Ce^{-2M/3}, \quad |\zeta| = e^{2/3}.$$

Setting $F = f_1 - f_2$ we have

$\bar{\partial}F = (b(\varsigma) + \frac{1}{2}\frac{\varsigma}{|\varsigma|} f_2(\varsigma))ds(\varsigma)$ on Ω, where ds is arclength on $\{|\varsigma| = e^{2/3}\}$. If M is large enough our last estimate on f_2 yields $\|b(\varsigma) - \bar{\partial}F(\varsigma)\|_{A_3} \le 1/2$, $|\varsigma| = e^{2/3}$, as was required. By (3.4) and (3.9), $F(e^{1/2}) = 0$, while by (3.5) and (3.10), F has the correct boundary behavior. The reader should notice that estimate (3.11) is not really needed; estimate (3.6) is all that is required. We have merely added (3.11) to make the argument symmetric. As noted earlier, we leave to the reader the verification of the fact that the function F has the correct Banach space properties to ensure that iteration of lemma 3.1 gives us a rigorous proof of theorem B.

References

[1] J. Bergh and J. Löfström, Interpolation spaces, Springer-Verlag, 1976.

[2] T. Wolff, "A note on interpolation spaces", Springer Lecture Notes No. 908 (1982), 199-204.

12

Produits de Riesz et Séries de Dirichlet

Jean-Pierre Kahane

Les produits de Riesz

$$\rho(t) = \Pi(1 + a_p p^{it} + \bar{a}_p p^{-it}) \tag{1}$$

où p parcourt un ensemble fini de nombres premiers, et $|a_p| \leq \frac{1}{2}$, sont un moyen commode pour obtenir des théorèmes Ω pour les séries de Dirichlet $\sum_1^\infty c_n n^{-s}$ $(s = \sigma + it)$. Rappelons que $\varphi(t) = \Omega(\psi(t))$ signifie

$$\psi(t) \geq 0 \quad et \quad \limsup_{t \to \infty} \frac{|\varphi(t)|}{\psi(t)} > 0.$$

L'état de la question pour $\varsigma(\sigma + it)$ lorsque $\frac{1}{2} \leq \sigma \leq 1$ se trouve dans la dernière édition du livre de Titschmarch (1986). La méthode des produits de Riesz, appliquée à $\varsigma(\sigma + it)$, donne mieux que Titschmarch $(\frac{1}{2} \leq \sigma < 1)$, aussi bien que Levinson (1972) $(\frac{1}{2} \leq \sigma < 1)$, et moins bien que Montgomery (1977) $(\frac{1}{2} < \sigma < 1)$ et Balasubramanian-Ramachandra (1977) $(\sigma = \frac{1}{2})$. Appliquée à des séries de Dirichlet plus générales, elle donne d'assez bons résultats, comme le montrent les produits d'Euler aléatoires, pour lesquels les estimations 0 et Ω ne sont pas trop éloignées.

En développant, les produits (1) s'écrivent

$$\rho(t) = \sum a_r r^{it} \tag{2}$$

où r parcourt un ensemble fini de rationnels. Notons qu'on a

$$\rho(t) \geq 0 \ , \ a_1 = 1. \tag{3}$$

La méthode consiste à exploiter les propriétés (2) et (3). Désignons par X la borne supérieure des numérateurs des r figurant dans (2), écrits sous forme irréductible :

$$X = \sup\{m \in \mathcal{N} \ : \ \forall a_r \neq 0 \ \ \frac{m}{r} \in \mathcal{Z}\}. \tag{4}$$

Dans le cas (1), on a simplement

$$X = \prod p. \tag{5}$$

Nous supposons la série de Dirichlet $\sum_1^\infty c_n \, n^{-s}$ convergente pour $\sigma > \sigma_0$, et nous posons

$$f(s) = \sum_1^\infty c_n \, n^{-s} \quad (s = \sigma + it \ , \ \sigma > \sigma_0). \tag{6}$$

Pour $\sigma > \sigma_0$ fixé, on sait que la série converge uniformément sur tout intervalle borné, et plus précisément que

$$\frac{f(\sigma+it)}{1+\mid t \mid} = \lim_{N\to\infty} \text{unif} \left(\frac{1}{1+\mid t \mid} \sum_1^N c_n \, n^{-\sigma - it} \right). \tag{7}$$

Désormais, σ est fixé, $\sigma > \sigma_0$. Désignons par $\omega(t)$ une fonction majorant $\mid f(\sigma+it) \mid$, avec les conditions supplémentaires écrites ci-dessous :

$$\left. \begin{array}{r} \mid f(\sigma+it) \mid \leq \omega(t) \\ \omega(-t) = \omega(t) \\ \omega(t) \uparrow (t > 0) \\ \dfrac{\omega(t)}{t} \downarrow (t < 0) \end{array} \right\}. \tag{8}$$

La méthode des produits de Riesz est résumée dans le théorème suivant.

Théorème. *Sous les conditions (2), (3), (4), (6), (8), on a*

$$\omega(X) \geq \frac{1}{50} \mid \sum a_n \, c_n \, n^{-\sigma} \mid . \tag{9}$$

Preuve. Soit $\Delta(x)$ une fonction paire, ≥ 0, de type positif, de classe C^3, portée par l'intervalle $[-\frac{1}{2}, \frac{1}{2}]$, et $\hat{\Delta}(t)$ sa transformée de Fourier :

$$\hat{\Delta}(t) = \frac{1}{2\pi} \int_{\mathcal{R}} e^{itx} \, \Delta(x) dx \quad (\geq 0)$$

$$\Delta(x) = \int_{\mathcal{R}} e^{-itx} \hat{\Delta}(t) dt \qquad (\geq 0).$$

Posons

$$A = \int_{\mathcal{R}} X^{-1}\hat{\Delta}(X^{-1}t)f(\sigma+it)\rho(t)dt.$$

En vertu de (7) et de l'hypothèse de classe C^3 et du support compact sur $\Delta(x)$, qui entraîne $\hat{\Delta}(t) = 0(|\, t\,|^{-3})$ $(t \to \infty)$, on a

$$A = \sum_{n=1}^{\infty} c_n n^{-\sigma} B_n$$

avec

$$\begin{aligned} B_n &= \int_{\mathcal{R}} X^{-1}\hat{\Delta}(X^{-1}t)n^{-it}\rho(t)dt \\ &= \sum a_r \Delta(X \log \frac{n}{r}). \end{aligned}$$

Comme, pour $r \neq n$, on a

$$\log\frac{n}{r} \notin \left[\log\left(1-\frac{1}{X}\right), \log\left(1+\frac{1}{X}\right)\right],$$

et que $\Delta(x)$ est nulle hors de $[-\frac{1}{2}, \frac{1}{2}]$, on a $B_n = c_n$, donc

$$A = \sum a_n c_n n^{-\sigma}. \tag{10}$$

D'autre part, suivant (8), on a

$$|\, A\,| \leq \int_{\mathcal{R}} X^{-1}\hat{\Delta}(X^{-1}t)\omega(t)\rho(t)dt$$

avec

$$\begin{aligned} &\omega(t) \leq \omega(X) \quad pour\ 0 \leq t \leq X \\ &\frac{\omega(t)}{t} \geq \frac{\omega(X)}{X} \quad pour\ t \geq X \end{aligned}$$

donc en tous cas

$$\omega(t) \leq \omega(X)(1 + X^{-2}t^2). \tag{11}$$

Posons

$$\begin{aligned} \hat{\Delta}_1(t) &= (1+t^2)\hat{\Delta}(t) \\ \Delta_1(x) &= \int_{\mathcal{R}} e^{-itx}\hat{\Delta}(t)(1+t^2)dt \ (= \Delta(x) - \Delta''(x)). \end{aligned}$$

Ainsi

$$\begin{aligned} |\, A\,| &\leq \omega(X) \int \hat{\Delta}_1(X^{-1}t)\rho(t)dt \\ &= \omega(X) \sum a_r \Delta_1(-X \log r) \\ &= \omega(X) a_1 \Delta_1(0) = \omega(X)\Delta_1(0) \end{aligned}$$

et la comparaison à (10) donne (9) quand on choisit Δ proportionnelle à $\chi * \chi * \chi * \chi$, où χ est la fonction indicatrice de $[-\frac{1}{8}, \frac{1}{8}]$. CQFD

Remarques

1. La constante $\frac{1}{50}$ dans (9) est loin d'être optimale. Le choix ci-dessus donne $\frac{1}{49}$, et on obtient mieux en élargissant le support de Δ.

2. Les conditions sur $\omega(t)$ n'interviennent que sous la forme (11), et, moyennant des changements évidents et une autre constante dans (9), on pourrait remplacer (11) par une condition moins forte du type

$$\omega(t) \leq C\omega(X)(1 + X^{-N}t^N). \tag{12}$$

Cependant, en pratique, les conditions données en (8) sont faciles à appliquer et donnent des théorèmes Ω, comme on va le voir.

Application : Théorèmes Ω pour les fonctions (6).

1. Supposons que la série de Dirichlet dans (6) soit de la forme

$$\sum_p c_p p^{-s} \qquad (p \text{ premier}, \quad | c_p | = 1). \tag{13}$$

Choisissons $\rho(t)$ comme en (1), avec $| a_p | = \frac{1}{2}$ et $a_p c_p \geq 0$. Alors (9) s'écrit

$$\omega(p_1 p_2 \cdots p_\nu) \geq \frac{1}{100} \sum_{j=1}^{\nu} p_j^{-\sigma} \tag{14}$$

(p_j) étant la suite croissante des nombres premiers, et ν un entier arbitraire. (14) s'écrit aussi

$$\omega\left(\exp(\int_1^y \log x \, d\pi(x))\right) \geq \frac{1}{100} \int_1^y x^{-\sigma} d\pi(x)$$

donc, compte tenu du théorème des nombres premiers ($\pi(x) \sim \ell i x$)

$$\omega\left(e^{y(1+\epsilon)}\right) \geq \frac{1-\epsilon}{100(1-\sigma)} \frac{y^{1-\sigma}}{\log y}$$

pour tout $\epsilon > 0$, donc

$$\omega(t) \geq C_\sigma \frac{(\log t)^{1-\sigma}}{\log \log t}.$$

En particulier, le choix

$$\omega(-t) = \omega(t) = \frac{1}{2} C_\sigma \frac{(\log t)^{1-\sigma}}{\log \log t} \qquad (t \geq T > e^e)$$

$$\omega(t) = \omega(T) \quad (-T \leq t \leq T)$$

contredit (8). Comme les trois dernières lignes de (8) sont vérifiées, c'est la première ligne qui est contredite.

Corollaire 1. *Si la série (6) est de la forme (13),*

$$f(\sigma + it) = \Omega\left(\frac{(\log t)^{1-\sigma}}{\log \log t}\right). \tag{15}$$

2. Supposons que les coefficients dans (6) soient multiplicatifs ($(m,n) = 1 \Rightarrow c_{mn} = c_m c_n$) et de module 1. Choisissons $\rho(t)$ comme ci-dessus. Alors (9) s'écrit

$$\omega(p_1 p_2 \cdots p_\nu) \geq \frac{1}{50} \prod_{j=1}^{\nu} (1 + \frac{1}{2} p_j^{-\sigma})$$

soit

$$\omega\left(exp(\int_1^y \log x \, d\pi(x))\right) \geq c \exp\left(\frac{1}{2}\int_1^y x^{-\sigma} d\pi(x)\right)$$

d'où

$$\omega(t) \geq C_{\sigma,\epsilon} exp\left(\frac{1-\epsilon}{2(1-\sigma)} \frac{(\log t)^{1-\sigma}}{\log \log t}\right)$$

et on conclut comme précédemment.

Corollaire 2. *Si les coefficients dans (6) sont multiplicatifs et de module 1, on a*

$$f(\sigma + it) = \Omega\left(exp(\frac{1-\epsilon}{2(1-\sigma)} \frac{(\log t)^{1-\sigma}}{\log \log t}\right) \tag{16}$$

pour tout $\epsilon > 0$.

3. Supposons seulement $| c_n |= 1$ pour tout n, et $\sigma_0 < \sigma < \frac{1}{2}$. Choisissons X premier, et

$$\rho(t) = \left| \sum_{p \le X} \bar{c}_p p^{-1/2} p^{-it} \right|^2 \left(\sum_{p \le X} \frac{1}{p} \right)^{-1} . \tag{17}$$

Ainsi (2), (3), (4) sont vérifiées, et (9) s'écrit

$$\omega(X) \ge (\sum_{p \le X} p^{-\sigma - 1/2})(\sum_{p \le X} \frac{1}{p})^{-1}.$$

On conclut :

Corollaire 3. *Si les coefficients dans (6) sont de module 1, et $\sigma_0 < \sigma < \frac{1}{2}$, on a*

$$f(\sigma + it) = \Omega \left(\frac{t^{1/2-\sigma}}{(log\, t)(log\, log\, t)} \right) . \tag{18}$$

On voit que, pour $\sigma < \frac{1}{2}$, (18) est meilleur que (16).

Remarque sur les corollaires 1 et 2.
Le corollaire 1 peut s'appliquer à

$$\sum \epsilon_p p^{-s} \quad (p \text{ premier}) \tag{19}$$

où $\epsilon_p = \pm 1$. Si on considère les ϵ_p comme aléatoires, indépendants, prenant avec même probabilité les valeurs $+1$ et -1, la série (19) est presque sûrement convergente pour $\sigma > \frac{1}{2}$. Désignons sa somme par $F(s)$ ($= F(s, \omega)$, où ω appartient à ℓ espace de probabilité).

Proposition. *Pour $\frac{1}{2} < \sigma < 1$*

$$F(\sigma + it) = O \left(\frac{(log\, t)^{1-\sigma}}{(log\, log\, t)^{\sigma}} \right) \quad (t \to \infty) \ p.\ s. \tag{20}$$

Preuve. Posons $\ell = (\log t)(\log \log t)$, $\ell_0 = \ell$, $\ell_j = 2^j \ell_{j-1}$,

$$F(s) = \sum_{p \le \ell} \epsilon_p p^{-s} + \sum_{j=0}^{\infty} \sum_{\ell_j < p \le \ell_{j+1}} \epsilon_p p^{-s}.$$

On a

$$\left|\sum_{p\leq \ell} \epsilon_p p^{-s}\right| \leq \sum_{p\leq \ell} p^{-\sigma} \simeq \frac{\ell^{1-\sigma}}{(1-\sigma)\log \ell}$$

quels que soient les ϵ_p. On a

$$\left|\sum_{\ell_j<p\leq \ell_{j+1}} \epsilon_p p^{-s}\right| \leq C\left(\sum_{\ell_j<p\leq \ell_{j+1}} p^{-2\sigma}\right)^{1/2} (\log(t\ell_{j+1}))^{1/2}$$

presque-sûrement à partir d'un $j = j(\omega)$ assez grand, lorsque C est une constante convenable (inégalités de Salem-Zygmund). En ajoutant, le premier et le second termes sont prépondérants, et on obtient (20).

Il convient de comparer (20) à (15).

Considérons maintenant le produit d'Euler

$$\prod (1+\epsilon_p p^{-s})^{-1} \quad (p \text{ premier}) \tag{21}$$

qui est également presque sûrement convergent pour $\sigma > \frac{1}{2}$. Désignons sa valeur par $Z(s)$ (= $Z(s,\omega)$).

Corollaire de la proposition. *Pour* $\frac{1}{2} < \sigma < 1$ *il existe p. s.* $C_\sigma < \infty$ *tel que*

$$Z(\sigma+it) = O\left(exp\left(C_\sigma \frac{(log\, t)^{1-\sigma}}{(log\, log\, t)^{\sigma}}\right)\right) \quad (t\to\infty). \tag{22}$$

Remarquons que la série de Dirichlet de $Z(s)$ vérifie l'hypothèse du corollaire 2 du théorème. Il convient donc de comparer (22) à (16).

Pour des informations sur les produits d'Euler aléatoires on peut consulter Queffelec (1980).

Si on choisit

$$f(s) = \zeta(s)(1-2^{1-s}) = \sum_{1}^{\infty} (-1)^n n^{-s}$$

le corollaire 2 est encore valable. Le meilleur estimé du type Ω connu, pour $-\frac{1}{2} < \sigma < 1$, va exactement dans le sens opposé à (22)

$$f(\sigma + it) = \Omega \left(\exp \frac{(\log t)^{1-\sigma}}{20(\sigma - \frac{1}{2})(\log \log \sigma)^{\sigma}} \right)$$

(Montgomery [3]).

Références bibliographiques

[1] **R. Balasubramanian et K. Ramachandra** On the frequency of Titchmarsh's phenomenon for $\zeta(s)$ III. *Proc. Indian Acad. Sci. 86 A (1977), 156-165.*

[2] **N. Levinson** Ω−theorems for the Riemann zeta-function. *Acta Arith. 20 (1972), 319-332.*

[3] **H. L. Montgomery** Extreme values of the Riemann zeta-function. *Comment. Math. Helv. 52 (1977), 511-518.*

[4] **H. Queffelec** Propriétés presque sûres et quasi-sûres des séries de Dirichlet et des produits d'Euler. *Canad. J. Math. 32 (1980), 531-558.*

[5] **E. C. Titchmarsh** The theory of the Riemann Zeta-function. *2nd ed., revised by D. R. Heath-Brown, Oxford 1986.*

13

A Note on Commutation Properties for Interpolation Spaces

Mario Milman and Tomas Schonbek

0. Introduction. We study certain commutativity properties of the classical interpolation methods. In particular we consider commutation with intersections (§ 1, § 2) and commutation between real and complex methods (§ 3, § 4). In § 1 we extend an old result of Peetre [14] concerning commutation of real interpolation and intersections. Our results in this section were motivated by some problems proposed in [6], and provide a new and complete approach to them. (Independently, but much before us, a solution had been given in [11]). In § 2 we use the Lions–Karadzov commutation relation between the real and complex methods in order to extend the results of the previous section to complex interpolation spaces. In § 3 we give new conditions under which commutation relations of Grisvard type, between the complex and real methods, hold. In the last section, § 4 , we briefly indicate how our results can be used in the study of commutators of singular integrals and in particular we provide a new approach to part of the work by Janson and Wolff [8].

Finally, it is a real pleasure to contribute to this volume dedicated to our dear teacher and friend Mischa Cotlar. Professor Cotlar has been a source of inspiration for many generations of Latin American mathematicians.

*1. **Commutation between real interpolation and intersections.*** Our initial motivation here was a problem raised in [6] : Is $(L^p,L^r \cap L^s)_{\theta,q} = (L^p,L^r)_{\theta,q} \cap (L^p,L^s)_{\theta,q}$? The easy answer is yes and in fact the following result holds (cf. [11]).

(1.1) Proposition. *Let* X, Y_0, Y_1 *be Banach lattices , then* $(X,Y_0)_{\theta,q} \cap (X,Y_1)_{\theta,q} = (X,Y_0 \cap Y_1)_{\theta,q}$, $0 < \theta < 1$, $0 < q \leq \infty$.

It has been known for a long time that (1.1) does not hold, in general, for arbitrary Banach spaces. In fact, Triebel [16] gives a counterexample where all the spaces involved are

Hilbert spaces.

We shall now formulate a general result on commutation between real interpolation and intersections that extends earlier work by Peetre **[14]**.

(1.2) Theorem. *Let $X_0, X_1, \ldots X_n$ be quasi-Banach spaces continuously embedded in some topological vector space. Assume that for each $S \subset \{1,\ldots,n\}$, $t > 0$, $a \in \bigcap_{\nu \in S} (X_0 + X_1)$ there exists an element $e(t,a;S) \in \bigcap_{\nu \in S} X_\nu$ such that*

i. $e(t,a,\phi) = a \quad (\bigcap_{\nu\in\phi} (X_0 + X_\nu) = X_0 + X_1 + \ldots + X_n)$;

ii. there exists $c > 0$ such that for $\nu \in S \subset \{1,\ldots,n\}$, $t > 0$ we have

$$\text{(1.3)} \quad t\|e(t,a;S)\|_\nu \leq c \sum_{\mu\in S} K(t,a,X_0,X_\mu)$$

$$\text{(1.4)} \quad \|e(t,a;S\backslash\{\nu\}) - e(t,a;S)\|_0 \leq c \sum_{\mu\in S} K(t,a,X_0,X_\mu) .$$

Then, for $0 < \theta < 1$, $0 < q \leq \infty$,

$$(X_0, \bigcap_{\nu=1}^{n} X_\nu)_{\theta,q} = \bigcap_{\nu=1}^{n} (X_0,X_\nu)_{\theta,q} .$$

Proof. Let $Y = X_1 \cap X_2 \ldots \cap X_n$. It suffices to prove that there exist α, $\beta > 0$ such that

$$\text{(1.5)} \quad \alpha \sum_{\nu=1}^{n} K(t,a,X_0,X_\nu) \leq K(t,a,X_0,Y) \leq \beta \sum_{\nu=1}^{n} K(t,a,X_0,X_\nu)$$

with α, β independent of a, t. The first inequality in **(1.5)** is trivial. To prove the second one, let $S_0 = \phi$; $S_\nu = \{1,\ldots\nu\}$ for $\nu = 1,\ldots n$. Let $t > 0$, $a \in \bigcap_{\nu=1}^{n} (X_0 + X_\nu)$; let $a_1 = e(t,a;S_n)$, $a = a - a_1$. Then $a_0 = \sum_{\nu=1}^{n} (e(t,a;S_{\nu-1}) - e(t,a;S_\nu))$ and, by **(1.4)**,

$$\|a_0\|_0 \leq c \sum_{\nu=1}^{n} \sum_{\mu=1}^{\nu} K(t,a,X_0,X_\mu) \leq nc \sum_{\nu=1}^{n} K(t,a,X_0,X_\nu).$$

By **(1.3)**, $a_1 \in Y$ and

$$t\|a_1\|_Y = t \max_{1\leq\nu\leq n} \|a_1\|_\nu \leq ct \sum_{\nu=1}^{n} K(t,a,X_0,X_\nu)$$

Since $K(t,a,X_0,Y) \leq \|a_0\|_0 + t\,\|a_1\|_Y$, the second inequality in **(1.5)** follows. The theorem is proved.

The hypotheses of **(1.2)** are easily seen to hold in the *quasi–linearizable* case discussed by Peetre in **[14]** (Chapter X, Theorem 4). Moreover, **(1.2)** also applies to **(1.1)** while it is well known that the pair (L^p,L^r) is not quasi–linearizable and therefore Peetre's theorem does not apply in this case. For completeness' sake we provide a proof of **(1.1)** in the following formulation.

(1.1)' *Let $X_0,\ldots,X_n$ be quasi–Banach lattices of measurable functions on some measure space. Then,* $(X_0, \bigcap_{\nu=1}^{n} X_\nu)_{\theta,q} = \bigcap_{\nu=1}^{n} (X_0,X_\nu)_{\theta,q}$, $0 < \theta < 1$, $0 < q \leq \infty$.

Proof. Let $c > 1$. For $f \in (X_0 + X_\nu)_+$, $1 \leq v \leq n$, $t > 0$, we let $f_\nu(t) \in (X_\nu)_+$, $f_{0,\nu}(t) \in (X_0)_+$ be such that $f = f_\nu(t) + f_{0,\nu}(t)$ and

(1.6) $\|f_\nu(t)\|_0 + t\|f_\nu(t)\|_\nu \leq c\, K(t,f,X_0,X_\nu)$.

If $\phi \neq S \subset \{1,\ldots n\}$, $t > 0$, $f \in \bigcap_{\nu\in S} (X_0 + X_\nu)$ we set $e(t,f;S) = \min_{\nu\in S} f_\nu(t)$ (where min denotes the lattice minimum); we set $e(t,f;\phi) = f$ for f in $X_0 + \ldots + X_n$. Inequality **(1.3)** for $a = f \geq 0$ is then a trivial consequence of **(1.6)**. To see that **(1.4)** also holds for $a = f \geq 0$, let $\nu \in S \subset \{1,\ldots,n\}$ and set $g = e(t,f;S\backslash\{\nu\})$. Then,

$e(t,f;S\backslash\{\nu\}) - e(t,f;S) = (g - f_\nu(t))^+ = \max(g - f_\nu(t),0)$. If $S\backslash\{\nu\} = \phi$, then $g = f$, $(g - f_\nu(t))^+ = f_{0,\nu}(t)^+ = f_{0,\nu}(t)$; if $S\backslash\{\nu\} \neq \phi$, let $\mu \in S\backslash\{\nu\}$. Then,

$$(g - f_\nu(t))^+ \leq (f_\mu(t) - f_\nu(t))^+ = (f_{0,\nu}(t) - f_{0,\mu}(t))^+ \leq |f_{0,\nu}(t) - f_{0,\mu}(t)|.$$

In either case,

$$\|e(t,f;S\backslash\{\nu\}) - e(t,f,S)\|_0 = \|(g - f_\nu(t))^+\|_0 \leq c \sum_{\mu\in S} K(t,f,X_0,X_\mu)$$

follows immediately from **(1.6)**. It is now easy to extend the definition of e(t,f;S) to not necessarily positive elements f in such a way that the hypothesis of **(1.2)** are satisfied. Alternatively, the proof of **(1.2)** shows that **(1.3)** holds for $a = f \geq 0$. Since the K functional for pairs of lattices has the property that $K(t,g) = K(t,|g|)$ we conclude that **(1.3)** holds in general and the result follows.

Using a form of the reiteration theorem (cf. **[11]**) we can state a version of **(1.1)** valid for Banach spaces due to Maligranda **[11]**.

(1.7) Corollary. *Let $\bar{A}$ be a Banach pair, $0 < \theta_i < 1$, $1 \leq q_i \leq \infty$, $i=0, 1, 2, 3$. Then*

$$(\bar{A}_{\theta_o,q_o}, \bar{A}_{\theta_1,q_1} \cap \bar{A}_{\theta_2,q_2})_{\theta_3,q_3} = (\bar{A}_{\theta_o,q_o}, \bar{A}_{\theta_1,q_1})_{\theta_3,q_3} \cap (\bar{A}_{\theta_o,q_o}, \bar{A}_{\theta_2,q_2})_{\theta_3,q_3}.$$

Proof. For a lattice X of measurable functions on $(0,\infty)$ and $\bar{A}$ a Banach pair let $\bar{A}_X = \{a/a \in \Sigma(\bar{A})$ such that $K(t,a,\bar{A}) \in X$, and $\|a\|_{\bar{A}_X} = \|K(.,a,\bar{A})\|_X\}$. For example the choice $X = L^q(t^{-\theta q-1})$ yields $\bar{A}_X = \bar{A}_{\theta,q}$. Then, if X, Y and Z are Banach lattices of measurable functions on $(0,\infty)$ we have (cf. [13])

$$(1.8) \quad (\bar{A}_X,\bar{A}_Y)_Z = \bar{A}_{(X,Y)_Z}$$

(We are oversimplifying somewhat here, it is required that the lattices satisfy some mild restrictions which are always satisfied by $L^q(t^{-\theta q-1})$ or their intersections. We shall refer to these "good lattices" as "*parameter spaces.*)" Let $X_i = L^{q_i}(t^{-\theta q_i-1})$, $i = 0, ..3$. Then,

$$\begin{aligned}(\bar{A}_{\theta_o,q_o}, \bar{A}_{\theta_1,q_1} \cap \bar{A}_{\theta_2,q_2})_{\theta_3,q_3} &= (\bar{A}_{X_o}, \bar{A}_{X_1} \cap X_2)_{X_3} \\ &= \bar{A}_{(X_o,X_1 \cap X_2)_{X_3}} \quad \text{(by (1.8))} \\ &= \bar{A}_{(X_o,X_1)_{X_3} \cap (X_o,X_2)_{X_3}} \quad \text{(by (1.1))} \\ &= \bar{A}_{(X_o,X_1)_{X_3}} \cap \bar{A}_{(X_o,X_2)_{X_3}} \\ &=(\bar{A}_{\theta_o,q_o}, \bar{A}_{\theta_1,q_1})_{\theta_3,q_3} \cap (\bar{A}_{\theta_o,q_o}, \bar{A}_{\theta_2,q_2})_{\theta_3,q_3}.\end{aligned}$$

We conclude this section by indicating a different approach to the results presented in [6].

(1.9) Corollary. *Let X and Z be lattices and let $Y = (X,Z)_{\xi,r}$ for some $0 < \xi < 1$, $0 < r \leq \infty$. Then, for $0 < \theta < 1$, $0 < q \leq \infty$,*

$$(X \cap Y, X \cap Z)_{\theta,q} = X \cap (Y,Z)_{\theta,q}.$$

Proof. Let $\gamma = (1 - \theta)\xi + \theta$. Then, by **(1.1)** and reiteration,

$$\begin{aligned}(X \cap Y,X \cap Z)_{\theta,q} &= (X \cap (X,Z)_{\xi,r},X \cap Z)_{\theta,q} = ((X,X \cap Z)_{\xi,r},X \cap Z)_{\theta,q} \\ &= (X,X \cap Z)_{\gamma,q} = X \cap (X,Z)_{\gamma,q} = X \cap ((X,Z)_{\xi,r},Z)_{\theta,q} \\ &= X \cap (Y,Z)_{\theta,q}.\end{aligned}$$

Let $X = L(p_1,q_1)$, $Z = L(p_3,q_3)$, $p_1 < p_3$, $1/p_2 = (1-\xi)/p_1 + \xi/p_3$; then $Y = (X,Z)_{\xi,r} = L(p_2,q_2)$ and we obtain theorem 3 of [6]:

$$(L(p_1,q_1) \cap L(p_2,q_2), L(p_1,q_1) \cap L(p_3,q_3))_{\theta,q} = L(p_1,q_1) \cap (L(p_2,q_2) \cap L(p_3,q_3))_{\theta,q}$$

if $p_1 < p_3$.

*2. **Commutativity between real and complex interpolation.*** Our point of departure in this section is a well–known result of Lions [10], later extended by Karodzov (cf. [9]), relating the real and complex methods of interpolation. Let $\overline{A}$ be a Banach pair, then

$$(2.1)\quad [(A_0,A_1)_{\theta_0,q_0},(A_0,A_1)_{\theta_1,q_1}]_\alpha = (A_0,A_1)_{\theta,q}$$

where $0 < \theta_i$, $\alpha < 1$, $1 \leq p_i \leq \infty$ (but the case $p_1 = p_2 = \infty$ is excluded), $\theta = (1-\alpha)\,\theta_0 + \alpha\,\theta_1$ and $1/q = (1-\alpha)/q_0 + \alpha/q_1$.

To state a more general version, recall the following construction due to Calderón. Let X_0, X_1 be lattices, then we define for $0 < \theta < 1$,

$$[[X_0,X_1]]_\theta = X_0^{\ 1-\theta} X_1^{\ \theta} = \{x\colon |x| \leq |x_0|^{1-\theta}\, |x_1|^{\theta} \text{ for } x_i \in X_i,\ i = 0,1\}$$
$$\|x\|_\theta = \inf\{\|x_0\|_0^{\ 1-\theta}\, \|x_1\|_1^{\ \theta} : |x| = |x_0|^{1-\theta}\, |x_1|^{\theta}\}$$

Then, if the spaces X_i, i = 0, 1, are parameter spaces, we have

$$(2.2)\quad [(A_0,A_1)_{X_0},\ (A_0,A_1)_{X_1}]_\theta = (A_0,A_1)_{[[X_0,X_1]]_\theta}.$$

We present now a version of (1.7) for complex interpolation.

(2.3) Theorem. $[\overline{A}_{\theta_0,q_0}, \overline{A}_{\theta_1,q_1} \cap \overline{A}_{\theta_2,q_2}]_{\theta_3} = [\overline{A}_{\theta_0,q_0}, \overline{A}_{\theta_1,q_1}]_{\theta_3} \cap [\overline{A}_{\theta_0,q_0}, \overline{A}_{\theta_2,q_2}]_{\theta_3}$.

Proof. Observe that if X, Y, Z are lattices, then

$$(2.4)\quad [[X, Y \cap Z]]_\theta = [[X,Y]]_\theta \cap [[X,Z]]_\theta.$$

(The easy proof is left to the reader.) Let $X_i = L^{q_i}(t^{-\theta_i q_i - 1})$, i = 0,..3. Then,

$$\begin{aligned}[\overline{A}_{\theta_0,q_0}, \overline{A}_{\theta_1,q_1} \cap \overline{A}_{\theta_2,q_2}]_{\theta_3} &= [\overline{A}_{X_0}, \overline{A}_{X_1 \cap X_2}]_{\theta_3} \\ &= \overline{A}_{[[X_0,X_1 \cap X_2]]_{\theta_3}} \quad \text{(by (2.2))} \\ &= \overline{A}_{[[X_0,X_1]]_{\theta_3} \cap [[X_0,X_1]]_{\theta_3}} \quad \text{(by (2.4))}\end{aligned}$$

$$= \overline{A}_{[[X_0,X_1]]_{\theta_3}} \cap \overline{A}_{[[X_0,X_1]]_{\theta_3}} \quad \text{(by definition)}$$

$$= [\overline{A}_{\theta_0,q_0}, \overline{A}_{\theta_1,q_1}]_{\theta_3} \cap [\overline{A}_{\theta_0,q_0}, \overline{A}_{\theta_2,q_2}]_{\theta_3} \quad \text{(by (2.2))}.$$

Using the methods of [4] it is easy to extend the above results to quasi–Banach spaces but we shall not pursue this point any further here.

3. Commutation theorems of Grisvard. In [7] Grisvard has given general conditions under which it is possible to commute not only between real and complex interpolation functors but also the order of the factors. Grisvard gives very nice applications of these results and obtains among other things (the now classical) interpolation theorems for Besov and Sobolev spaces. In particular the results of Grisvard give conditions under which

$$\text{(3.1)} \quad [(A_0,B_0)_{\theta_0,q_0},(A_1,B_1)_{\theta_1,q_1}]_\alpha = ([A_0,A_1]_\alpha,[B_0,B_1]_\alpha)_{\theta,q}$$

where as usual $0 < \theta_i, \alpha < 1$, $1 \le q_i \le \infty$, $\theta = (1-\alpha)\theta_0 + \alpha\theta_1$, $1/q = (1-\alpha)/q_0 + \alpha/q_1$. Notice that, in particular, if $A_0 = A_1$ and $B_0 = B_1$ we are in the situation described in (2.2).

We consider now an application of the commutativity results for intersections to derive extreme forms of Grisvard type formulae. It is easier to formulate the results using the J method of interpolation. In this formulation the spaces $\overline{A}_{X,J}$, where X is a parameter space, are defined as follows.

$$\overline{A}_{X,J} = \{x/\ x = \int_0^\infty u(s)\, ds/s, \text{ convergence in } A_0 + A_1, \text{ with } \|u(s)\|_0 \in X,\ s\|u(s)\|_1 \in X\}$$

with the usual range norm. The classical J spaces are obtained letting $X = L^q(s^{-\theta q-1}) \cong L^q_{-\theta}$.

Let I denote the vector valued integral $I(u) = \int_0^\infty u(s)\, ds/s$, then a typical assumption in Grisvard's results is that this operation is onto for suitable spaces.

We shall assume that $B_0 = B_1 = B$, say. Moreover we assume that

$$\text{(3.2)} \quad [L^1_0(A_0), L^p_{-\theta_0}(A_1) \cap L^p_{1-\theta_0}(B)]_\theta = [L^1_0(A_0), L^p_{-\theta_0}(A_1)]_\theta \cap [L^1_0(A_0), L^p_{1-\theta_0}(B)]_\theta .$$

Then,

(3.3) Theorem. *$([A_0,A_1]_\theta,[A_0,B]_\theta)_{\theta_0,r} \subset [A_0,(A_1,B)_{\theta_0,p}]_\theta$, where $0 < \theta, \theta_0 < 1$, $1 \le p < \infty$, $1/r = (1-\theta) + \theta/p$.*

Proof. We consider the operator I. Trivially, $I : L^1_0(A_0) \to A_0$ and, by definition $I : L^p_{-\theta_0}(A_1) \cap L^p_{1-\theta_0}(B) \to (A_1,B)_{\theta_0,p}$. By interpolation,

$I : [L^1_0(A_0),L^p_{-\theta_0}(A_1) \cap L^p_{1-\theta_0}(B)]_\theta \to [A_0,(A_1,B)_{\theta_0,p}]_\theta$; i.e., in view of (3.2),

$I : [L^1_0(A_0),L^p_{-\theta_0}(A_1)]_\theta \cap [L^1_0(A_0),L^p_{1-\theta_0}(B)]_\theta \to [A_0,(A_1,B)_{\theta_0,p}]_\theta.$

But by the vector valued interpolation theorems of Calderón [2]

(3.4) $[L^1_0(A_0),L^p_{-\theta_0}(A_1)]_\theta \cap [L^1_0(A_0),L^p_{1-\theta_0}(B)]_\theta = L^r_{-\theta\theta_0}([A_0,A_1]_\theta) \cap L^r_{\theta(1-\theta_0)}([A_0,B]_\theta)$

and I maps the space in the right hand side of (3.4) <u>onto</u> $([A_0,A_1]_\theta,[A_0,B]_\theta)_{\theta_0,r}$, by the definition of the $\overline{X}_{\theta,r,J}$ spaces (cf. [7] p.147). The theorem follows.

(3.5) Remark. If we replace in **(3.2)** and **(3.4)** $[..]_\theta$ by $(...)_{\theta,r}$ then by an appropriate real interpolation theorem for vector valued L^p spaces we obtain

(3.6) $((A_0,A_1)_{\theta,r},(A_0,B)_{\theta,r})_{\theta_0,r} \subset (A_0,(A_1,B)_{\theta_0,p})_{\theta,r}$.

(3.7) Remark. This result should be compared with an end point version of **(2.1)**. In fact the proof of **(2.1)** in [1] (with indices $\theta_0 = 0$, $q_0 = 1$) shows that $(A,[A,B]_\theta)_{\theta_0,r} \subset [A,(A,B)_{\theta_0,p}]_\theta$ with the indices as in **(3.3)** (cf. also [5]). In [5] a sharper end point form of **(2.1)** is given in some special cases. Namely, let B be a Banach lattice of measurable functions satisfying other mild restrictions, then for all θ, $\theta_0 \in (0,1)$, p_0, $p \in [1,\infty]$, $1/r = (1-\theta)/p_0 + \theta/p$, and $p < \infty$, we have

(3.8) $[L^{p_0},(L^{p_0},B)_{\theta_0,p}]_\theta = (L^{p_0},B)_{\theta_0\theta,r}$.

To compare this result with our general formulae, let us write **(3.8)** in a more suggestive form. In fact we observe that, by reiteration, $(L^{p_0},B)_{\theta_0\theta,r} = (L^{p_0},[L^{p_0},B]_{\theta_0})_{\theta,r}$. Therefore **(3.8)** states that

(3.8)' $[L^{p_o},(L^{p_o},B)_{\theta_o,p}]_\theta = (L^{p_o},[L^{p_o},B]_{\theta_o})_{\theta,r}$.

If we let $A_o = A_1 = L^{p_o}$, and apply **(3.3)**, we obtain the somewhat weaker result $(L^{p_o},[L^{p_o},A]_\theta)_{\theta_o,r} \subset [L^{p_o},(L^{p_o},A)_{\theta_o,p}]_\theta$, $1/r = (1-\theta) + \theta/p$. It is likely that the methods of [5] could be used to obtain sharp versions of **(3.3)** under suitable extra assumptions.

More generally, to compare our results with those of Grisvard, we formulate a general version of **(3.3)**, which can be obtained with the same proof.

(3.9) Theorem. Let A_o, A_1, B_o, B_1 be compatible Banach spaces and let $0 < \theta_i, \alpha < 1$, $1 \leq q_i \leq 1$, $i = 0, 1$, $\mu = (1-\alpha)\theta_o + \alpha\theta_1$, $1/q = (1-\alpha)/q_o + \alpha/q_1$. Suppose that

(3.2)' $[L^{q_o}{}_{-\theta_o}(A_o) \cap L^{q_o}{}_{1-\theta_o}(A_1), L^{q_1}{}_{-\theta_1}(B_o) \cap L^{q_1}{}_{1-\theta_1}(B_1)]_\alpha = L^q{}_{-\mu}([A_o,B_o]_\alpha) \cap L^q{}_{1-\mu}([A_1,B_1]_\alpha)$.

Then, .

(3.10) $([A_o,B_o]_\alpha, [A_1,B_1]_\alpha)_{\mu,q} \subset [(A_o,A_1)_{\theta_o,q_o},(B_o,B_1)_{\theta_1,q_1}]_\alpha$

Using these methods it is also possible to deal with quasi–Banach spaces.

(3.11) Remark. Let S_α (resp S_μ) denote the space to the left (resp. to the right) in (3.2)', and observe that we always have $S_\alpha \subset S_\mu$. Then, if I defines a surjection I: $S_\alpha \to [(A_o,A_1)_{\theta_o,q_o},(B_o,B_1)_{\theta_1,q_1}]_\alpha$, we see that

$$([A_o,B_o]_\alpha,[A_1,B_1]_\alpha)_{\mu,q} \supset [(A_o,A_1)_{\theta_o,q_o},(B_o,B_1)_{\theta_1,q_1}]_\alpha .$$

Compare with [7] (theorem 2.1).

4. An application to integral operators. We briefly indicate a connection with the work of Russo [15], and Janson and Wolff [8] on conditions for certain bounded compact operators to belong to the Schatten classes S^p. More general results will be given elsewhere.

We consider first a form of the classical Hausdorff–Young inequality, valid for integral operators, given by Russo [15].

Let G: MxM $\to \mathbb{C}$ be a kernel, where (M,μ) is some measure space. Let

$L^p(L^q)^{symm} = L^p_x(L^q_y) \cap L^p_y(L^q_x)$, where the subindex indicates the variable with respect to which we take the L^s norm.

(4.1) Theorem. *Let $G \in L^p(L^{p'})^{symm}$ with $2 \le p \le \infty$, and $1/p + 1/p' = 1$. Let R_G be the integral operator defined by $R_G\varphi(x) = \int G(x,y)\varphi(y)\, d\mu(y)$ for $\varphi \in L^2$; then $R_G \in S^p$.*

Proof. The cases p = 2 (Hilbert –Schmidt operators) and p = ∞ are classical. Therefore, the operator $T(G) = R_G$ satisfies $T : (L^2(L^2)^{symm}, L^\infty(L^1)^{symm})_{\theta,p} \to (S^2,S^\infty)_{\theta,p} = S^p$, with $1/p = (1-\theta)/2$. Now, since $L^2(L^2)^{symm} = L^2(MxM) \cong L^2$, we can use **(1.1)** and theorem (2.3.1) (i) of **[12]** to obtain $(L^2, L^\infty_x(L^1_y) \cap L^1_y(L^\infty_x))_{\theta,p} \supset L^p(L^{p'})^{symm}$. Alternatively, we can use complex interpolation and **(2.4)**.

In their paper Janson and Wolff require a weak type version of Russo's result.

(4.2) Theorem. *Let $G \in L^p(L^{p',\infty})^{symm}$, $2 < p < \infty$, $1/p + 1/p' = 1$. Then, $R_G \in S^{p,\infty}$.*

Proof. It is enough to prove that $L^p(L^{p',\infty})^{symm} \subset (L^2, L^\infty(L^1)^{symm})_{\theta,\infty}$. Now, we have $(L^2, L^\infty(L^1)^{symm})_{\theta,\infty} = (L^2_x(L^2_y), L^\infty_x(L^1_y))_{\theta,\infty} \cap (L^2_y(L^2_x), L^\infty_y(L^1_x))_{\theta,\infty}$. At this point we use theorem (2.3.1) (i) of **[12]** to conclude that $(L^2_x(L^2_y), L^\infty_x(L^1_y))_{\theta,\infty} \cap (L^2_y(L^2_x), L^\infty_y(L^1_x))_{\theta,\infty} \supset L^p(L^{p',\infty})^{symm}$.

The proof in **[8]** is also simple and direct. In fact, their approach may be relevant when trying to implement the "second order" or commutator inequalities associated with interpolation methods (cf. **[3]** and the references therein).

References

[1] J. Bergh and J. Löfström, *Interpolation spaces. An introduction*, Grundlehren Math. Wiss., vol 223, Springer, Berlin/New York, 1976.

[2] A.P. Calderón, *Intermediate spaces and interpolation, the complex method*, Studia Math. **24** (1964), 113–190.

[3] M. Cwikel, B. Jawerth, M. Milman *and* R. Rochberg, *Differential estimates and commutators in interpolation theory*, Proc. Conf. Modern Analysis, to appear.

[4] M. Cwikel, M. Milman and Y. Sagher, *Complex interpolation of some quasi –Banach spaces*, J. Funct. Anal. **65** (1986), 339–347.

[5] M. Cwikel and Y. Sagher, *Relations between real and complex interpolation spaces*, Indiana Math. J. **36** (1987), 905–912.

[6] S.J. Dilworth, *Interpolation of intersections of L^p spaces*, Arch. Math. **50** (1988), 51–55.

[7] P. Grisvard, *Commutativite de deux foncteurs d'interpolation et applications*, J. Math. Pures et Appl. **45** (1966), 207–290.

[8] S. Janson and T. Wolff, *Schatten classes and commutators of singular integrals*, Univ. of Chicago, 1981.

[9] G.E. Karadzov, *On the commutativity of two interpolation functors*, Dokl. Akad. Nauk SSSR **223** (1975), 875–878.

[10] J.L. Lions, *Une propriete de stabilite pour espaces d'interpolation; applications*, C. R. Acad. Sci. Paris **256** (1963), 855–857.

[11] L. Maligranda, *On commutativity of interpolation with intersection* , Suppl. Rend. Circ. Mat. di Palermo **II**(1985), 113–118.

[12] M. Milman, *Complex interpolation and geometry of Banach spaces*, Ann. Mat. Pura Appl. **86** (1984), 317–328.

[13] P. Nilsson, *Reiteration theorems for real interpolation and approximation spaces*, Ann. Mat. Pura App. **32** (1982), 291–330.

[14] J. Peetre, *A theory of interpolation of normed spaces*, Notas de Matematica, Univ. of Brasilia, 1963

[15] B. Russo, *On the Hausdorff–Young theorem for integral operators*, Pac. J. Math. **68** (1977), 241–253.

[16] H. Triebel, *Eine Bemerkung zur nicht–kommutativen Interpolation*, Math. Nach. 69 (1975), 57–60.

14

Harmonic Analysis and Fundamental Solutions on Nilpotent Lie Groups

A. Nagel, F. Ricci, and E. M. Stein

§1. Introduction.

The purpose of this paper is to study the harmonic analysis of a broad class of (hypo-elliptic) partial differential operators on arbitrary (simply-connected) nilpotent Lie groups. We shall find the asymptotic development of their fundamental solutions, both at the origin and at infinity, and study the corresponding "Riesz transforms" and the analogues of the Hardy-Littlewood-Sobolev theorems on "fractional integration".

Background.

If we take the harmonic analysis of the usual Laplacian $\triangle$ on $\mathbb{R}^n$ as our model, then there are two ways of arriving at the basic operators in question. One in an "intrinsic" way which has as its starting point the Riesz transforms, $\frac{\partial}{\partial x_j}(-\triangle)^{-\frac{1}{2}}$, $\frac{\partial^2}{\partial x_i \partial x_j}\triangle^{-1}$, the fractional integration operators $(-\triangle)^{-\frac{\alpha}{2}}$, etc. The second approach, an "extrinsic" one, proceed by studying convolution operators $f \to f * K$ where K are Calderón-Zygmund kernels, or kernels displaying a more general homogeneity. For $\mathbb{R}^n$ these two approaches lead to essentially the same classes of operators. The situation is, broadly speaking, the same when $\mathbb{R}^n$ is replaced by a nilpotent group which carries automorphic dilations and Δ is replaced by a corresponding homogeneous operator. However it changes markedly in the general situation; then the operators that arise in these two classes are essentially different, although each class of operators has an independent interest. The study of singular integrals (the extrinsic approach) for general nilpotent groups was carried out in [RiS1] and [RiS2]. Here we deal with the questions that arise from the intrinsic approach. Our analysis proceeds by determining the asymptotics of the fundamental solutions of the partial differential operators in question, and then passes to Riesz transforms, fractional integration, etc.

The framework in which this is done consists in assigning to each suitable partial differential operator $\mathcal{L}$ on an arbitrary nilpotent Lie group G, two other nilpotent Lie groups G_0 and G_∞ which arise from G as appropriate contractions. These two groups have the additional feature that they are homogeneous (*i.e.* have automorphic dilations). What is important is that one can show that G_0 controls the local behavior of the fundamental solution of $\mathcal{L}$, and more precisely the first term of its asymptotic expansion is exactly the fundamental solution for a corresponding operator in G_0. Similarly, G_∞ plays an analogous role at infinity.

The operators $\mathcal{L}$ we deal with are the following. We let $X_1, \dots, X_k$ be any set of left-invariant vector fields on G which generate its Lie algebra $\mathfrak{g}$. We let $\mathcal{L} = P(X_1, \dots, X_k)$ be a (non-commutative) polynomial in $X_1, \dots, X_k$, which is "universally hypoelliptic". The precise definitions are in §5 below. Here we content ourselves with citing some examples of such P: first $X_k + \sum_{j=1}^{k-1} X_j^2$; and more generally $X_k + \sum_{j=1}^{k-1} (-1)^{b_j} X_j^{2b_j}$, where b_j are positive integers; the results also hold for systems of such operators $\{\mathcal{L}_j\}$, the simplest being $\mathcal{L}_j = X_j$, for $j = 1, \dots, k$.

Organization of the Paper.

This paper is organized as follows. Sections §2 and §3 are devoted to the construction and study of the groups G_0 and G_∞. One first constructs a family of Lie algebra structures $\mathfrak{g}_t$, $t > 0$ defined on the vector space of $\mathfrak{g}$ with the property that $\lim_{t \to 0} \mathfrak{g}_t = \mathfrak{g}_0$ exists, and where the contraction procedure can also be described as follows. Let $\mathcal{F}$ be an appropriate "free" nilpotent Lie algebra, and $\mathfrak{i}$ and ideal so that $\mathcal{F}/\mathfrak{i}$ is isomorphic with $\mathfrak{g}$. Then if we set $\mathfrak{i}_t$ to be $D_t^{-1}(\mathfrak{i})$, where D_t are the natural dilations on $\mathcal{F}$, one shows that $\lim \mathfrak{i}_t = \mathfrak{i}_0$ exists and $\mathcal{F}/\mathfrak{i}_t$ and $\mathcal{F}/\mathfrak{i}_0$ are isomorphic with $\mathfrak{g}_t$ and $\mathfrak{g}_0$ respectively. We can take G_0 to be the group whose Lie algebra is $\mathfrak{g}_0$. The group G_∞ is constructed analogously.

In section 4 we show that the multiplicative structure of G_0 approximates that of G near the origin; similarly the left-invariant vector fields of G_0 approximate those of G, and also the appropriate balls of G_0 approximate those of G locally. There are analogous approximations at infinity, with the group G_∞ replacing G_0 throughout.

In sections 5 and 6 we come to the asymptotic development of the fundamental solution of $\mathcal{L}$. Let us describe the main ideas in the simpler case that the degree of $\mathcal{L}$ is smaller that the homogeneous dimension of G_0. Together with $\mathcal{L}$ there is a similar operator $\tilde{\mathcal{L}}$ on the free nilpotent algebra $\mathcal{F}$. If $\tilde{K}$ is its fundamental solution, then one can show first that $K(x) = \int_{\mathfrak{i}} \tilde{K}(x+t)dt$ is a fundamental solution of $\mathcal{L}$. Next one has that simultaneously

$$K(x) = \int_{\mathfrak{i}_0} \tilde{K}(x+t+\psi(t))\,dt = \int_{\mathfrak{i}_\infty} \tilde{K}(x+t+\overline{\psi}(t))\,dt, \tag{1.1}$$

where ψ and $\overline{\psi}$ are respectively appropriate sub-homogeneous and super-homogeneous functions. Also one has that $K_0(x) = \int_{\mathfrak{i}_0} K(x+t)\,dt$ is a fundamental solution corresponding to G_0, so working with the first identity in (1.1) gives the asymptotic development of K near the origin, with leading term K_0. One argues similarly for the asymptotics at infinity. The general situation is more involved because the analogous integrals do not converge, and must be treated by a complicated subtraction procedure. Section 7 contains the derivations of the etimates for the Riesz transforms, and Sobolev-space inequalities. Given the properties of the fundamental solutions obtained above, thee proofs of these estimates can then be carried out using standard techinques.

Further remarks.

It should be noted that the results for L^p, L^q inequalities in the case of $\mathcal{L}_j = X_j$ are in Varopolous [V], and that paper can be considered as one of the starting points of the present work. Another source is the earlier study of (local) estimates for parametricies of operators which are suitable polynomials in vector fields satisfying the Hörmander condition, carried out in [NSW]. This study in turn has its roots in the ideas of "lifting" to the free nilpotent algebra, and the corresponding approximation of vector fields by those of a nilpotent group, which is the method of [RoS]. These ideas reappear here, although transformed and partly disguised. The gradations occuring in Sections 2 and 3 are developments of ideas appearing in Goodman [Go] and Hulanicki and Jenkins [Hu], but are used differently here.

Finally it should be noted that our asymptotics are based on the homogeneity properties of fundamental solutions in the special situation that arises when the groups have automorphic dilations. The appropriate properties of fundamental solutions in those cases goes back to results of Folland [F] and Geller [Ge].

§2. Construction of the local Lie algebra $\mathfrak{g}_0$.

Let $\mathfrak{g}$ be a nilpotent Lie algebra. Suppose we are given k elements $X_1, \dots, X_k$ in $\mathfrak{g}$ that generate $\mathfrak{g}$. We do not asume that the X_j's are linearly independent or even distinct. In addition, we assume that we assign to each X_j a positive integer a_j, which we shall refer to as the *degree* of X_j.

In terms of this data we shall construct a fimily of nilpotent Lie algebras $\mathfrak{g}_\delta$ for $\delta \geq 0$, where $\mathfrak{g}_0$ is a homogeneous nilpotent Lie algebra which incorporates the basic "local" structure of $\mathfrak{g}$. In order to do this we asign to each commutator $[X_{i_1}, [X_{i_2}, \cdots [X_{i_{l-1}}, X_{i_l}] \cdots]]$ the degree $a_{i_1} + a_{i_2} + \cdots + a_{i_l}$, and we denote by $\mathfrak{f}_j$ the subspace of $\mathfrak{g}$ of elements spanned by commutators of degrees $\leq j$. Then we have the increasing filtration

$$\mathfrak{f}_1 \subset \mathfrak{f}_2 \subset \cdots \subset \mathfrak{f}_n = \mathfrak{g}, \tag{2.1}$$

where n is the smallest integer so that any commutator of degree $\geq n+1$ vanishes. Clearly, $[\mathfrak{f}_i, \mathfrak{f}_j] \subset \mathfrak{f}_{i+j}$.

Now choose subspaces $\mathfrak{v}_j$ of $\mathfrak{g}$ so that

$$\mathfrak{v}_1 \oplus \cdots \oplus \mathfrak{v}_j = \mathfrak{f}_j, \qquad \mathfrak{v}_1 = \mathfrak{f}_1, \qquad \sum_{j=1}^{n} \oplus \mathfrak{v}_j = \mathfrak{g}. \tag{2.2}$$

Denote by p_j the projection on the jth component with respect to the direct sum decomposition $\sum \oplus \mathfrak{v}_j = \mathfrak{g}$. Now on the vector space $\mathfrak{g}$, define a new Lie algebra stucture $[\ ,\ ]_\delta$, for $\delta > 0$, by

$$[x, y]_\delta = (D_\delta^o)^{-1} \cdot [D_\delta^0(x), D_\delta^0(y)], \tag{2.3}$$

where D_δ^0 is defined by

$$D_\delta^0(x_j) = \delta^j\, x_j \qquad \text{if } x_j \in \mathfrak{v}_j,\ j = 1, \ldots, n. \tag{2.4}$$

Notice that (2.3) makes $\mathfrak{g}_\delta$ into a Lie algebra isomorphic to $\mathfrak{g}$. We also define the bracket $[\ ,\ ]_0$ by

$$[x_j, x_k]_0 = p_{j+k}[x_j, x_k] \qquad \text{if } x_j \in \mathfrak{v}_j \text{ and } x_k \in \mathfrak{v}_k. \tag{2.5}$$

PROPOSITION 1. *For any x, $y \in \mathfrak{g}$, $[x, y]_\delta \to [x, y]_0$ as $\delta \to 0$. Thus $\mathfrak{g}$ equipped with the bracket $[\ ,\ ]_0$ is a Lie algebra (which we denote by $\mathfrak{g}_0$). Moreover $\mathfrak{g}_0$ is homogeneous with respect to the dilations D_δ^0 given by equation (2.4)*

Remark: It is to be noted that while the $\mathfrak{g}_\delta$ are isomorphic to $\mathfrak{g}$ as Lie algebras when $\delta > 0$, this is not the case with $\mathfrak{g}_0$.

PROOF: If $x_j \in \mathfrak{v}_j$ and $x_k \in \mathfrak{v}_k$, then

$$[x_j, x_k] - p_{j+k}[x_j, x_k] \in \mathfrak{v}_1 \oplus \cdots \oplus \mathfrak{v}_{j+k-1} = \mathfrak{f}_{j+k-1}.$$

Thus

$$\begin{aligned}
[x_j, x_k]_\delta &= (D_\delta^0)^{-1} [D_\delta^0(x_j), D_\delta^0(x_k)] \\
&= (D_\delta^0)^{-1} [x_j, x_k]\, \delta^{j+k} \\
&= (D_\delta^0)^{-1} \big(\delta^{j+k}\, p_{j+k}[x_j, x_k] + \delta^{j+k}\, e_{j+k-1}\big),
\end{aligned}$$

with $e_{j+k-1} \in \mathfrak{v}_1 \oplus \cdots \oplus \mathfrak{v}_{j+k-1}$. Altogether then,

$$[x_j, x_k]_\delta = p_{j+k}\, [x_j, x_k] + O(\delta).$$

Hence as $\delta \to 0$, $[\ ,\ ]_\delta \to [\ ,\ ]_0$, proving that $[\ ,\ ]_0$ is a Lie bracket. That the Lie algebra $\mathfrak{g}_0$ is homogeneous follows immediately from equations (2.4) and (2.5).

Our construction above of $\mathfrak{g}_0$ made use of a fixed (non-unique) gradation (2.2) associated to the filtration (2.1). There is also a more intrinsic way of looking at the construction of $\mathfrak{g}_0$, which we now turn to.

Let $\mathcal{F}$ be the "free" nilpotent Lie algebra with k generators of degree n. That is $\mathcal{F}$ is the quotient of the free Lie algebra generated by $Y_1, Y_2, \ldots, Y_k$, divided by the ideal generated by all commutators of degree greater than n, where Y_1 has degree a_1, Y_2 has degree a_2, $\cdots$, and Y_k has degree a_k.

Let $\mathfrak{i}$ be the ideal in $\mathcal{F}$ corresponding to $\mathfrak{g}$. That is, the correspondence $Y_j \to X_j$, $j = 1, \ldots, k$ leads to an isomorphism $I : \mathcal{F}/\mathfrak{i}$ with $\mathfrak{g}$. Next on $\mathcal{F}$ we consider the dilations D_δ given by

$$D_\delta : Y_j \to \delta^{a_j}\, Y_j.$$

Because of the universal nature of $\mathcal{F}$, D_δ extends to an automorphism of $\mathcal{F}$. In this way we obtain the direct sum decomposition

$$\mathcal{F} = \sum_{j=1}^{n} \oplus \mathcal{F}_j$$

where $\mathcal{F}_j$ is the span of the commutators of degree j. We denote by π_j the corresponding projection of $\mathcal{F}$ on $\mathcal{F}_j$.

Next define $\mathfrak{i}_\delta$ by

$$\mathfrak{i}_\delta = D_\delta^{-1}(\mathfrak{i}) \qquad \text{for } \delta > 0. \tag{2.6}$$

Since $\mathfrak{i}$ is an ideal, so is $\mathfrak{i}_\delta$.

PROPOSITION 2. *As subspaces of $\mathcal{F}$, $\mathfrak{i}_\delta$ tend to a limit $\mathfrak{i}_0$ as $\delta \to 0$. The limiting subspace $\mathfrak{i}_0$ is an ideal. Moreover, $\mathcal{F}/\mathfrak{i}_0$ is isomorphic with $\mathfrak{g}_0$. Also*

$$\mathfrak{i}_0 = \sum_{j=1}^{n} \oplus \pi_j(\mathfrak{i} \cap (\mathcal{F}_1 \oplus \cdots \oplus \mathcal{F}_j)). \tag{2.7}$$

PROOF: We shall construct a particular basis of $\mathfrak{i}$ which will make the convergence of $\mathfrak{i}_\delta$ to $\mathfrak{i}_0$ clear. First construct any basis of $\mathfrak{i} \cap \mathcal{F}_1$, *i.e.* for the elements of $\mathfrak{i}$ which have degree 1. Next, construct a supplementary basis of $\mathfrak{i} \cap (\mathcal{F}_1 \oplus \mathcal{F}_2)$, supplementing the basis already constructed for $\mathfrak{i} \cap \mathcal{F}_1$. Continue in this way until a full basis of $\mathfrak{i} \cap (\mathcal{F}_1 \oplus \cdots \oplus \mathcal{F}_n) = \mathfrak{i}$ has been constructed.

Now what does it mean that $\mathfrak{i}_\delta \to \mathfrak{i}_0$ as $\delta \to 0$, as subspaces of $\mathcal{F}$? One requires: (a), for each $\delta > 0$, the $\mathfrak{i}_\delta$ and $\mathfrak{i}_0$ have the same dimension, and (b), for any $x_0 \in \mathfrak{i}_0$ there are $x_\delta \in \mathfrak{i}_\delta$ so that $x_\delta \to x_0$ as $\delta \to 0$. As far as the identity of dimensions of the $\mathfrak{i}_\delta$, $\delta > 0$ with $\mathfrak{i}$, this is obvious because of the linear isomorphism given by D_δ. Moreover, the dimension of $\mathfrak{i}$ and $\sum_{j=1}^{n} \oplus \pi_j(\mathfrak{i} \cap (\mathcal{F}_1 \oplus \cdots \oplus \mathcal{F}_j))$ are clearly the same because of the particular basis of $\mathfrak{i}$ which we have constructed.

Now let x_0 be a general element of $\pi_j(\mathfrak{i} \cap (\mathcal{F}_1 \oplus \cdots \oplus \mathcal{F}_j))$. Then $x_0 = \pi_j(y_1 + \cdots + y_j) = y_j$, where $y_l \in \mathcal{F}_l$, and $y_1 + \cdots + y_j \in \mathfrak{i}$. Now $D_\delta^{-1}(y_1 + \cdots + y_j) \in \mathfrak{i}$, so $\delta^{-1} y_1 + \cdots + \delta^{-j} y_j \in \mathfrak{i}_\delta$. Set

$$\begin{aligned} x_\delta &= y_j + \delta\, y_{j-1} + \cdots + \delta^{j-1} y_1 \\ &= x_0 + \delta\, y_{j-1} + \cdots + \delta^{j-1} y_1 = x_0 + O(\delta) \end{aligned}$$

Hence $x_\delta \to x_0$. Since every element in $\mathfrak{i}_0$ can be written as the sum of elements of the form x_0, the convergence of $\mathfrak{i}_\delta$ to $\mathfrak{i}_0$ is proved. The convergence of the ideals $\mathfrak{i}_\delta$ to $\mathfrak{i}_0$ also makes $\mathfrak{i}_0$ an ideal.

To set up the isomorphism between $\mathcal{F}/\mathfrak{i}_0$ and $\mathfrak{g}_0$ we need to make a choice of subspaces in $\mathcal{F}$, which choice corresponds to the choice of gradation (2.2) fixed earlier. We choose subspaces $\mathfrak{h}_j$ to be subspaces of $\mathcal{F}_j$ which are complementqary to $\pi_j(\mathfrak{i} \cap (\mathcal{F}_1 \oplus \cdots \oplus \mathcal{F}_j))$, *i.e.*

$$\mathcal{F}_j = \mathfrak{h}_j \oplus \pi_j(\mathfrak{i} \cap (\mathcal{F}_1 \oplus \cdots \oplus \mathcal{F}_j)),$$

and set

$$\mathfrak{h} = \sum_{j=1}^{n} \oplus \mathfrak{h}_j.$$

Clearly

$$\mathcal{F} = \mathfrak{h} \oplus \mathfrak{i}_0. \tag{2.8}$$

LEMMA.

$$\mathcal{F} = \mathfrak{h} \oplus \mathfrak{i}. \tag{2.9}$$

To prove this we shall show

(i) $I(\mathfrak{h}_j) \subset \mathfrak{f}_j$

(ii) $I(\pi_j(\mathfrak{i} \cap (\mathcal{F}_1 \oplus \cdots \oplus \mathcal{F}_j)) \subset \mathfrak{f}_{j-1}$

(iii) I is one to one on $\mathfrak{h}$.

Recall that I is the projection of $\mathcal{F}/\mathfrak{i}$ to $\mathfrak{g}$. Now (i) is obvious, because $\mathfrak{h}_j \subset \mathcal{F}_j$, and $I(\mathcal{F}_j)$ consists of elements of degree $\leq j$ in $\mathfrak{g}$, hence are in $\mathfrak{f}_j$. To prove (ii) suppopse $x_j \in \pi_j(\mathfrak{i} \cap (\mathcal{F}_1 \oplus \cdots \oplus \mathcal{F}_j))$. Then there are $x_1, \cdots, x_j$, with $x_l \in \mathcal{F}_l$ and $x_1 + \cdots + x_j \in \mathfrak{i}$. Hence $I(x_j) = -I(x_1 + \cdots + x_{j-1}) \subset \mathfrak{f}_{j-1}$.

To prove (iii) note that by (i) and (ii) since

$$\mathcal{F}_j = \mathfrak{h}_j \oplus \pi_j(\mathfrak{i} \cap (\mathcal{F}_1 \oplus \cdots \oplus \mathcal{F}_j)),$$

we get that $I(\mathcal{F}_j) \subset I(\mathfrak{h}_j) + \mathfrak{f}_{j-1}$, $j = 1, \cdots, n$ which gives inductively

$$I(\mathcal{F}_1 \oplus \cdots \oplus \mathcal{F}_j) = \mathfrak{f}_j \subset I(\mathfrak{h}_1 \oplus \cdots \oplus \mathfrak{h}_j), \quad j = 1, \cdots, n.$$

Hence $\mathfrak{g} = I(\mathcal{F}) = I(\mathfrak{h})$. But since $\mathfrak{h} \oplus \mathfrak{i}_0 = \mathcal{F}$, $\mathfrak{h}$ has the complementary dimension to $\mathfrak{i}$, and it therefore follows the $\mathcal{F} = \mathfrak{h} \oplus \mathfrak{i}$; this also proves (iii) and finishes the proof of the lemma.

We should also observe that once we have chosen the subspaces $\mathfrak{v}_j$ above, we can then determine the subspaces $\mathfrak{h}_j$ (as subspaces of $\mathcal{F}_j$ complementary to $\pi_j(\mathfrak{i} \cap (\mathcal{F}_1 \oplus \cdots \oplus \mathcal{F}_j))$ uniquely by requiring $I(\mathfrak{h}_j) = \mathfrak{v}_j$. The converse is also true. As a consequence of (2.9), since both $\mathcal{F}$ and $\mathfrak{h}$ are homogeneous, we have

$$\mathcal{F} = \mathfrak{h} \oplus \mathfrak{i}_\delta, \quad \delta > 0 \tag{2.10}$$

where $\mathfrak{i}_\delta = D_\delta^{-1}(\mathfrak{i})$.

Now we have that $D_\delta^0 : \mathfrak{g}_\delta \to \mathfrak{g}$ is a Lie algebra isomorphism, and that $I : \mathcal{F}/\mathfrak{i} \to \mathfrak{g}$ is the natural projection. It follows then that $D_\delta^{-1} \circ I \circ D_\delta^0 : \mathcal{F}/\mathfrak{i}_\delta \to \mathfrak{g}_\delta$ is an isomorphism of Lie algebras, for each $\delta > 0$. Since we know that $\mathfrak{g}_\delta \to \mathfrak{g}_0$ and $\mathfrak{i}_\delta \to \mathfrak{i}_0$, we get the desired isomorphism of $\mathcal{F}/\mathfrak{i}_0$ with $\mathfrak{g}_0$, concluding the proof of the proposition.

The parallel direct sum decompositions (2.8) and (2.9) have the following important consequence.

COROLLARY. *There is a linear mapping $\psi : \mathfrak{i}_o \to \mathcal{F}$ so that $\mathfrak{i} = \{x + \psi(x),\ x \in \mathfrak{i}_0\}$. Moreover ψ is strictly sub-homogeneous in the sense that if $x \in \mathcal{F}_j$, then*

$$\psi(x) \in \mathcal{F}_1 \oplus \cdots \oplus \mathcal{F}_{j-1}, \qquad j = 1, \ldots, n.$$

PROOF: The two decompositions $\mathcal{F} = \mathfrak{h} \oplus \mathfrak{i} = \mathfrak{h} \oplus \mathfrak{i}_0$ clearly induce a unique linear isomorphism ℓ of $\mathfrak{i}_0$ tp $\mathfrak{i}$. Now suppose $x \in \mathfrak{i}_0$, and $x \in \mathcal{F}_j$. Then $x \in \pi_j(\mathfrak{i} \cap (\mathcal{F}_1 \oplus \cdots \oplus \mathcal{F}_j))$ and hence there exist $x_1, \ldots, x_j$, with $x_k \in \mathcal{F}_k$, so that $x_1 + \cdots + x_j \in \mathfrak{i}$, and $x = x_j$. Thus

$$\ell(x) = x_j + (x_1 + x_2 + \cdots + x_{j-1}) = x_j + \psi(x_j) = x + \psi(x)$$

with $\psi(x) = x_1 + \cdots + x_{j-1}$ and the corollary is proved.

§3. Construction of the global Lie algebra $\mathfrak{g}_\infty$.

Having described the contraction of $\mathfrak{g}$ to the "local" algebra, we now describe the similar contraction of $\mathfrak{g}$ to the "global" algebra $\mathfrak{g}_\infty$. We deal with the same data: the given elements $X_1, \ldots, X_k$ of $\mathfrak{g}$ which generate $\mathfrak{g}$, and the corresponding degrees $a_1, a_2, \ldots, a_k$, assigned to them. We then consider the decreasing filtration

$$\mathfrak{g} = \bar{\mathfrak{f}}_1 \supset \bar{\mathfrak{f}}_2 \supset\supset \cdots \supset \bar{\mathfrak{f}}_{n+1} = \{0\} \tag{3.1}$$

where $\bar{\mathfrak{f}}_j$ is spanned by all commutators of degree $\geq j$.

Choose disjoint subspaces $\bar{\mathfrak{v}}_1, \bar{\mathfrak{v}}_2, \ldots, \bar{\mathfrak{v}}_n$, so that

$$\bar{\mathfrak{f}}_j = \bar{\mathfrak{v}}_j \oplus \cdots \oplus \bar{\mathfrak{v}}_n, \tag{3.2}$$

and let $\bar{p}_j$ be the corresponding projection on $\bar{\mathfrak{v}}_j$. On $\mathfrak{g}$ we define the Lie algebra structure $\mathfrak{g}_\infty$ with Lie bracket $[\ ,\ ]_\infty$, given by

$$[x_j . x_k]_\infty = \bar{p}_{j+k}\,[x_j, x_k] \qquad \text{if } x_j \in \bar{\mathfrak{v}}_j, x_k \in \bar{\mathfrak{v}}_k. \tag{3.3}$$

We also consider the dilation D_δ^∞ given by

$$D_\delta^\infty(x_j) = \delta^j\, x_j, \qquad \text{if } x_j \in \bar{\mathfrak{v}}_j. \tag{3.4}$$

We denote by $\bar{\mathfrak{g}}_\delta$ the Lie algebra image of $\mathfrak{g}$ under $(D_\delta^\infty)^{-1}$.

In analogy with Proposition 1 above one can show that $\overline{\mathfrak{g}}_\delta \to \mathfrak{g}_\infty$, as $\delta \to \infty$, making $\mathfrak{g}_\infty$ into a Lie algebra. Also one notes that $\mathfrak{g}_\infty$ is homogeneous with respect to the dilations D_δ^∞ given by (3.4).

Similarly, when one passes to the "free" algebra $\mathcal{F}$, and if we keep to the notation the $\mathfrak{i}_\delta = (D_\delta^\infty)^{-1}(\mathfrak{i})$, we can prove that the ideals $\mathfrak{i}_\delta$ converge to the ideal $\mathfrak{i}_\infty$, as $\delta \to \infty$. Moreover

$$\mathfrak{i}_\infty = \sum_{j=1}^{n} \oplus\, \pi_j(\mathfrak{i} \cap (\mathcal{F}_j \oplus \mathcal{F}_{j+1} \oplus \cdots \oplus \mathcal{F}_n)). \tag{3.5}$$

Also $\mathcal{F}/\mathfrak{i}_\infty$ is isomorphic to $\mathfrak{g}_\infty$. Next, if one chooses $\overline{\mathfrak{h}} = \sum_{j=1}^n \oplus \overline{\mathfrak{h}}_j$, so that $\overline{\mathfrak{h}}_j$ is a complementary subspace in $\mathcal{F}_j$ to $\pi_j(\mathfrak{i} \cap (\mathcal{F}_j \oplus \mathcal{F}_{j+1} \oplus \cdots \oplus \mathcal{F}_n))$, then of course $\mathcal{F} = \overline{\mathfrak{h}} \oplus \mathfrak{i}_\infty$, but also

$$\mathcal{F} = \overline{\mathfrak{h}} \oplus \mathfrak{i}. \tag{3.6}$$

Finally one can assert the following:

COROLLARY. *There exists a linear mapping $\overline{\psi} : \mathfrak{i}_\infty \to \mathcal{F}$ so that $\mathfrak{i} = \{x + \overline{\psi}(x), x \in \mathfrak{i}_\infty\}$. Moreover $\overline{\psi}$ is strictly super-homogeneous in the sense that if $x \in \mathcal{F}_j$, then $\overline{\psi}(x) \in \mathcal{F}_{j+1} \oplus \cdots \oplus \mathcal{F}_n$.*

The proofs of all these facts conserning $\mathfrak{g}_\infty$ are much the same as those for $\mathfrak{g}_0$ given in Section 2 above.

A final remark concerning the homogeneous dimensions of $\mathfrak{g}_0$ and $\mathfrak{g}_\infty$ will be useful below.

PROPOSITION 3. *The homogeneous dimension of $\mathfrak{g}_\infty$ is greater than or equal to the homogeneous dimension of $\mathfrak{g}_0$.*

PROOF: Recall that

$$\begin{aligned} hom.dim.(\mathfrak{g}_0) &= \sum_{k=1}^{n} k\, dim(\mathfrak{v}_k) = \sum_{k=1}^{n} k\,(dim\,(\mathfrak{f}_k) - dim\,(\mathfrak{f}_{k-1})) \\ &= n\, dim\,(\mathfrak{f}_n) - \sum_{k=1}^{n-1} dim\,(\mathfrak{f}_k). \end{aligned}$$

Similarly,

$$hom.dim.(\mathfrak{g}_\infty) = \sum_{k=1}^{n} k\,(dim\,(\overline{\mathfrak{f}}_k) - dim\,(\overline{\mathfrak{f}}_{k+1})) = \sum_{k=1}^{n} dim\,(\overline{\mathfrak{f}}_k).$$

However, $dim\,(\mathfrak{f}_k) + dim\,(\overline{\mathfrak{f}}_{k+1}) \geq dim\,(\mathfrak{g})$, since $\mathfrak{f}_k + \overline{\mathfrak{f}}_{k+1} = g$, and $\mathfrak{f}_n = \overline{\mathfrak{f}}_1 = \mathfrak{g}$. Therefore the result follows.

§4. Approximations.

We let G, G_0, and G_∞ denote the simply connected Lie groups whose Lie algebras are $\mathfrak{g}$, $\mathfrak{g}_0$, and $\mathfrak{g}_\infty$ respectively. Our purpose here is to show in what sense the multiplication structure of G is approximated near the identity by that of G_0, and near infinity by that of G_∞. Similarly we will see that the left-invariant vector fields of G will be suitably approximated locally by those coming from G_0, while at infinity by those coming from G_∞. Similar comparisons will also be made for a "norm function" defined on G, and the corresponding "balls".

We turn first to the multiplication structures. Let $\{X_{j,i}\}$ denote a basis of $\mathfrak{v}_j$, $1 \leq j \leq n$, and $1 \leq i \leq dim(\mathfrak{v}_j)$. Thus a general element in $\mathfrak{g}$ can be written as $x = \sum_{j,i} \xi_{j,i} X_{j,i}$. We shall use the $\xi = \{\xi_{j,i}\}$ as coordinates of $\mathfrak{g}$ (and hence also for $\mathfrak{g}_0$); also for G and G_0 via their identification with $\mathfrak{g}$ and $\mathfrak{g}_0$ by their respective exponential maps.

Suppose $x = \sum_{j,i} \xi_{j,i} X_{j,i}$, and $y = \sum_{j,i} \eta_{j,i} X_{j,i}$. Denote by $x \cdot y$ the product of x and y in the group G and $(x \cdot y)_{G_0}$ the product in the group G_0. Since the Lie algebras of $\mathfrak{g}$ and $\mathfrak{g}_0$ are nilpotent, it follows as is well known that

$$\begin{cases} x \cdot y = \sum_{j,i} Q_{j,i}(\xi, \eta) X_{j,i}, & \text{and} \\ (x \cdot y)_{G_0} = \sum_{j,i} Q^0_{j,i}(\xi, \eta) X_{j,i}, \end{cases}$$

where $Q_{j,i}(\xi, \eta)$ and $Q^0_{j,i}(\xi, \eta)$ are polynomials in ξ and η. Denote the dilations $D^0_\delta(\xi) = \xi'$, by $\xi'_{j,i} = \delta^j \xi_{j,i}$; similarly for $D^0_\delta(\eta)$.

PROPOSITION 4.

(a) *$Q^0_{j,i}(\xi, \eta)$ is a homogeneous polynomial (with respect to the dilations D^0_δ) of degree j.*

(b) *$Q_{j,i} = Q^0_{j,i} + R_{j,i}$ where $R_{j,i}$ is a sum of monomials each of degree $> j$.*

PROOF: If we recall the defining property (2.5), we see that when we set

$$\begin{cases} [x, y] = \sum_{j,i} B_{j,i}(\xi, \eta) X_{j,i}, & \text{and} \\ [x, y]_0 = \sum_{j,i} B^0_{j,i}(\xi, \eta) X_{j,i}, \end{cases}$$

the following assertions are straight foward. First $B_{j,i}(\xi, \eta)$ and $B^0_{j,i}(\xi, \eta)$ are bilinear forms. Second $B^0_{j,i}(\xi, \eta)$ is homogeneous of degree j. Finally $B_{j,i}(\xi, \eta) = B^0_{j,i}(\xi, \eta)$ modulo terms homogeneous of degrees $> j$.

Next we calculate the multiplication laws (4.1) from the Lie algebra laws (4.2) via the Campbell-Hausdorff formula. It is then clear by inspoection that our assertions (a) and (b) hold.

We now compare the left-invariant vector fields of G and G_0. The left-invariant vector field on G corresponding to $X_{j,i}$ is given by

$$\frac{d}{dt}(X_{j,i} f)(x \cdot \gamma(t))|_{t=0} = X^G_{j,i} f$$

where $\gamma(t) = exp_G(tX_{j,i})$. The corresponding vector field on G_0, $X_{j,i}^{G_0}$, is given by a similar formula with G replaced by G_0, and the group multiplication $x \cdot \gamma(t)$ replaced by that of G_0. We shall now express $X_{j,i}^{G}$ and $X_{j,i}^{G_0}$ in the coordinates ξ. We have by (4.1)

$$(X_{j,i}^{G} f)(\xi) = \frac{d}{dt} f(Q(\xi, \eta(t))|_{t=0}, \qquad \text{and}$$
$$(X_{j,i}^{G_0} f)(\xi) = \frac{d}{dt} f(Q^0(\xi, \eta(t))|_{t=0},$$

where $\eta(t)$ has coordinate t in the $(j,i)^{th}$ place, and coordinate zero elsewhere. It follows that

$$X_{j,i}^{G} = \sum_{\alpha,\beta} q_{j,i,\alpha.\beta}(\xi) \frac{\partial}{\partial \xi_{\alpha,\beta}}, \qquad \text{and}$$
$$X_{j,i}^{G_0} = \sum_{\alpha,\beta} q_{j,i,\alpha,\beta}^{0}(\xi) \frac{\partial}{\partial \xi_{\alpha,\beta}}, \qquad \text{where}$$
$$q_{j,i,\alpha,\beta}(\xi) = \frac{\partial}{\partial \eta_{j,i}} Q_{\alpha,\beta}(\xi, \eta)\,|_{\eta=0}$$

with a similar formula for q^0.

For any vector field with homogeneous polynomial coefficients, $\sum_{\alpha,\beta} q_{\alpha,\beta}(\xi) \frac{\partial}{\partial \xi_{\alpha,\beta}}$, define its *order* to be $\max_{\alpha,\beta}\{\text{degree } (\xi_{\alpha,\beta}) - \text{degree } (q_{\alpha,\beta})\}$.

PROPOSITION 5.

(a) $X_{j,i}^{G_0}$ *has order* j

(b) $X_{j,i}^{G} = X_{j,i}^{G_0}$ + *a sum of vector fields of order* $< j$

This follows immediately from (4.3) and Proposition 4 above.

Next we return to our original generating elements $X_1, X_2, \ldots, X_k$ of $\mathfrak{g}$ to which we have assigned degrees $a_1, a_2, \ldots, a_k$ respectively. We write $X_1^G, X_2^G, \ldots, X_k^G$ for the corresponding left-invariant vector fields on G.

PROPOSITION 6. *When* X_j^G *is expressed in the cannonical coordinates it is the sum of terms* $q_{\alpha,\beta}(\xi) \frac{\partial}{\partial \xi_{\alpha,\beta}}$, *each of order* $\leq a_j$.

In fact, we have that $X_j \in \mathfrak{f}_{a_j}$, thus $X_j \in \mathfrak{v}_1 \oplus \mathfrak{v}_2 \oplus \cdots \oplus \mathfrak{v}_{a_j}$. We now express X_j as a linear combination of the basis elements of the $\mathfrak{v}_l$, $1 \leq l \leq a_j$, and apply Proposition 5 to obtain the desired conclusion.

Next we point out that there is a complete analogy for the behavior of G at infinity, with the group G_∞ replacing G_0. One proceeds as follows. First one chooses a basis $\{\overline{X}_{j,i}\}$ of $\overline{\mathfrak{v}}_j$, and introduces coordinates $\{\overline{\xi}_{j,i}\}$ via $x = \sum_{j,i} \overline{\xi}_{j,i} \overline{X}_{j,i}$. We also consider the dilations D_δ^∞ defined by $D_\delta^\infty(\overline{\xi}_{j,i}) = \delta^j \overline{\xi}_{j,i}$. Next one expresses the multiplication laws on G and G_∞ respectively as

$$x \cdot y = \sum_{j,i} \overline{Q}_{j,i}(\overline{\xi}, \overline{\eta}) \overline{X}_{j,i},$$
$$(x \cdot y)_{G_\infty} = \sum_{j,i} \overline{Q}_{j,i}^{\infty}(\overline{\xi}, \overline{\eta}) \overline{X}_{j,i}.$$

It follows that $\overline{Q}_{j,i}^{\infty}$ is homogeneous of degree j and $\overline{Q}_{j,i} - \overline{Q}_{j,i}^{\infty}$ is a sum of terms each homogeneous of degree $< j$. Next $\overline{X}_{j,i}^{G_\infty}$ has order j exactly, in these coordinates, while $\overline{X}_{j,i}^{G} - \overline{X}_{j,i}^{G_\infty}$ is a sum of terms each having order $> j$. Finally X_j^G, when expressed in these cannonical coordinates is a sum of homogeneous terms each having order $\geq a_j$.

We shall now study an appropriate "norm function" on G, and the resulting family of balls. Beginning first with $\mathfrak{g}_0$ and its dilations D_δ^0 we construct a homogeneous norm function ρ_0 for it. That is, using the coordinates ξ above, we fix a function $\rho_0 : \mathfrak{g}_0 \to \mathbb{R}_+$ so that

(a) $\rho_0(\xi) > 0$ if $\xi \neq 0$;
(b) $\rho_0(D_\delta^0\,\xi) = \delta\,\rho_0(\xi)$;
(c) ρ_0 is C^∞ away from the origin;
(d) $\rho_0(-\xi) = \rho_0(\xi)$.

Similarly we fix a homogeneous norm function ρ_∞ on $\mathfrak{g}_\infty$. Then we fix a function $\rho : G \to \mathbb{R}$, so that ρ is smooth and positive away from the the group identity, and so that in a small neighborhood of the identity $\rho(x) = \rho_0(\xi)$ if $x = exp_G\left(\sum_{j,i} \xi_{j,i} X_{j,i}\right)$, while outside a conpact set $\rho(x) = \rho_\infty(\overline{\xi})$, if $x = exp_G\left(\sum_{j,i} \overline{\xi}_{j,i} \overline{X}_{j,i}\right)$, and so that $\rho(x) = \rho(x^{-1})$.

For each $x \in G$ and $r > 0$, we define $B_r(x)$, the "ball" centered at x of radius r to be $B_r(x) = \{y \in G \,|\, \rho(x^{-1}y) < r\}$.

PROPOSITION 7.

(i) *The collection $\{B_r(x)\}$ is left-invariant in the sense that $a \cdot B_r(x) = B_r(a \cdot x)$, for $a, x \in G$.*
(ii) *$|B_r(x)| \approx r^{Q_0}$ as $r \to 0$ and $|B_r(x)| \approx r^{Q_\infty}$ as $r \to \infty$.*
(iii) *There exists $c > 0$ so that if $B_{r_1}(x_1) \cap B_{r_2}(x_2) \neq \emptyset$ with $r_1 \geq r_2$, then $B_{c\,r_1}(x_1) \supset B_{r_2}(x_2)$.*

Note that (ii) implies the doubling property for the volumes of the balls, and (iii) the engulfing property, with together with (ii) is what is needed in the proof of a Vitali covering lemma.

PROOF OF THE PROPOSITION: First, (i) is obvious from the definition of $B_r(x)$. Next (ii) follows easily from (i) and the fact that the exponential mapping has Jacobian one, and the homogeneity of ρ_0 and ρ_∞ which imply that $|\{\rho_0(\xi) < r\}| = c_0\, r^{Q_0}$ and $|\{\rho_\infty(\overline{\xi}) < r\}| = c_\infty\, r^{Q_\infty}$. To prove (iii) we will show that

$$\rho(x \cdot y) \leq C_1\big(\rho(x) + \rho(y)\big) \quad \text{for } x, y \in G. \tag{4.4}$$

Let us assume that $\rho(x) \leq r$ and $\rho(y) \leq r$. When r is small we use ρ_0, *i.e.*, $\rho(x) = \rho_0(\xi)$ with $x = exp_G(\sum_{j,i} \xi_{j,i} X_{j,i})$, and $\rho(y) = \rho_0(\eta)$ with $y = exp_G(\sum_{j,i} \eta_{j,i} X_{j,i})$, while $\rho(x \cdot y) = \rho_0(Q(\xi, \eta))$ with $x \cdot y = exp_G(\sum_{j,i} Q_{j,i}(\xi, \eta) X_{j,i})$. However $\rho_0(\xi) \approx \sum_{j,i} |\xi_{j,i}|^{\frac{1}{j}}$, so $\rho_0(\xi) \leq r$ implies that $|\xi_{j,i}| \leq c\,r^j$. Similarly $|\eta_{j,i}| \leq c'\,r^j$, and hence by Proposition 4, $|Q_{j,i}(\xi, \eta)| \leq c\,r^j$ which proves that $\rho_0(Q(\xi, \eta)) \leq c'\,r$, and hence (4.4) in this case. This easily reduces the problem to the situation where $r \geq \overline{C}$ and $\rho(x \cdot y) \geq \overline{C}$, with $r = \max\big(\rho(x), \rho(y)\big)$, for a fixed large constant $\overline{C}$. Then we argue as we did when r

was small but we replace the representations for x, y, and $x \cdot y$ by $x = exp_G(\sum_{j,i} \overline{\xi}_{j,i} \overline{X}_{j,i})$, $y = exp_G(\sum_{j,i} \overline{\eta}_{j,i} \overline{X}_{j,i})$, and $x \cdot y = exp_G(\sum_{j,i} \overline{Q}_{j,i}(\overline{\xi}, \overline{\eta}) \overline{X}_{j,i})$, and we substitute $\rho_\infty(\overline{\xi})$ for $\rho_0(\xi)$.

Finally once (4.4) is proved we see that if $y \in B_{r_1}(x_1) \cap B_{r_2}(x_2)$, then $\rho(x_1^{-1} \cdot y) < r$ and $\rho(x_2^{-1} \cdot y) < r_2$ which implies that $\rho(x_1^{-1} \cdot x_2) < 2c_1 r$, since we also have $\rho(x^{-1}) = \rho(x)$. Then using (4.4) again we obtain that $B_{r_2}(x_2) \subset B_{c r_1}(x_1)$ with $c = 2c_1^2 + c_1$.

§5. Asymptotics for the fundamental solution.

We shall state here our first main theorem. Let $\mathcal{P} = \mathcal{P}(Y_1, \dots, Y_k)$ be a non-commutative polynomial in the indeterminates $Y_1, \dots, Y_k$. We shall asume that it is homogeneous and "universally hypoelliptic" in the following senses. First assigning degrees a_j to Y_j, we suppose that $\mathcal{P}$ is a linear combination of monomials of the form $Y_{i_1} \cdot Y_{i_2} \cdots Y_{i_l}$ with $a_{i_1} + a_{i_2} + \cdots + a_{i_l} = d$ with d fixed, and referred to as the degree of $\mathcal{P}$. We shall also assume that $\tilde{\mathcal{L}} = \mathcal{P}(Y_1, \dots, Y_k)$ is a hypoelliptic left-invariant partial differential operator on the simply connected Lie group whose Lie algebra is a free nilpotent Lie algebra $\mathcal{F}$, generated by $Y_1, \dots, Y_k$, but of arbitrarily high step.

Together with our Lie algebra $\mathfrak{g}$, and the elements $X_1, \dots, X_k$ in it which have been distinguished previously, (and to which we have assigned degrees $a_1, a_2, \dots, a_k$), we consider the partial differential operator $\mathcal{L}$ on G given by

$$\mathcal{L} = \mathcal{P}(X_1, \dots, X_k) \tag{5.1}$$

with X_j taken as left-invariant vector fields on G. As we shall see below it follows that $\mathcal{L}$ is hypoelliptic as a consequence of the fact that $\tilde{\mathcal{L}}$ is. We shall be concerned with the existence and further properties of a fundamental solution for $\mathcal{L}$. *i.e.* a distribution K on G so that $\mathcal{L}(K) = \delta_0$ where δ_0 is the Dirac delta function at the group identity. We prove that such a global fundamental solution K of $\mathcal{L}$ exists.

In order to formulate the properties of K that we are going to determine, we have to recall the results of Folland [F] and Geller [Ge] on fundamental solutions of homogeneous hypoelliptic operators on homogeneous groups. Therefore assume now that N is a homogeneous group with homogeneous dimension q, and let $\mathcal{L}$ be a homogeneous differential operator of degree d. Then

(i) if $d < q$, $\mathcal{L}$ has a unique fundamental solution that is homogeneous of degree $-q+d$;
(ii) if $d \geq q$, then $\mathcal{L}$ has a log-homogeneous fundamental solution $K(x)$ of degree $-q+d$, in the sense that

$$K(x) = h(x) + p(x) \log \rho(x)$$

where $h(x)$ is a homogeneous function of degree $-q+d$, $p(x)$ is a homogeneous polynomial of degree $-q+d$, and $\rho(x)$ is a smooth homogeneous norm on N. This log-homogeneous fundamental solution is unique modulo homogeneous polynomials of degree $-q+d$ (which can be added to the term $h(x)$ without altering the form of $K(x)$).

Parallel to the operator $\mathcal{L}$ on the group G, we consider the operators $\mathcal{L}_0$ on G_0 and $\mathcal{L}_\infty$ on G_∞ obained by a similar construction. Precisely, let $X_j' = p_{a_j}(X_j)$, $j = 1, \dots, k$,

$\overline{X}'_j = \overline{p}'_{a_j}(X_j)$. We set $\mathcal{L}_0 = \mathcal{P}(x'_1, \ldots, x'_k)$, $\mathcal{L}_\infty = \mathcal{P}(\overline{X}'_1, \ldots, \overline{X}'_k)$. Then $\mathcal{L}_0$ and $\mathcal{L}_\infty$ are hypoelliptic and furthemore they are homogeneous of degree d with respect to the dilations of the group on which they are defined. Then the results of Folland and Geller can be applied to exhibit homogeneous (or log-homogeneous) fundamental solutions $K^{(0)}$ and $K^{(\infty)}$ of $\mathcal{L}_0$ and $\mathcal{L}_\infty$ respectively.

In the theorem we are going to prove we see that $K^{(0)}$ and $K^{(\infty)}$ play the most relevant role in the asymptotic expansion of K respectively at the origin and at infinity. Since the statement of the theorem in full generality may be difficult to understand, we first explain its meaning distinguishing among several cases.

Case 1: $d < Q_0$. Then *a fortiori* $d < Q_\infty$ and we require that the fundamental solution K vanishes at infinity. This condition uniquely determines K, since on can show using [Ge] that two fundamental solutions of $\mathcal{L}$ differ by a polynomial. We prove that such a K exists and that it has an asymptotic expansion as $x \to \infty$

$$K(x) \sim \sum_{k=0}^{\infty} \overline{K}_m(x)$$

where $\overline{K}_0 = K^{(\infty)}$ and $\overline{K}_m$ is homogeneous of degree $-Q^\infty + d - m$ with respect to the dilations of G_∞. Furthermore as $x \to 0$

$$K(x) \sim \sum_{m=0}^{\infty} K_m(x)$$

where K_m is (log-)homogeneous of degree $-Q_0 + d + m$ with respect to the dilations of G_0, and K_0 is the homogeneous fundamental solution $K^{(0)}$ of $\mathcal{L}_0$.

Case 2: $Q_0 \le d < Q_\infty$. We still require that $K(x)$ vanishes at infinity. We obtain such a fundamental solution and it has an asymptotic expasnsion at infinity exactly as in the first case. As to the behavior of K at the origin, we can now say that there is a polynomial $A(x)$, of degree at most $d - Q_0 - 1$ (relative to the dilations of G_0), such that

$$K(x) - A(x) \sim \sum_{m=0}^{\infty} K_m(x)$$

where $K_0 = K^{(0)}$ is a log-homogeneous fundamental solution of $\mathcal{L}_0$ and K_m is log-homogeneous of degree $-Q_0 + d + m$.

Case 3: $d \ge Q_\infty$. We now require that $K(x) = O(\rho_\infty(x)^{d-Q_\infty} \log \rho_\infty(x))$ as $x \to \infty$. This condition only determines K modulo polynomials of degree at most $d - Q_\infty$ relative to the dilations of G_∞. The solution that we construct is such that

$$K(x) \sim \sum_{m=0}^{\infty} \overline{K}_m(x) \qquad \text{as } x \to \infty$$

where $\overline{K}_0 = K^{(\infty)}$ is a log-homogeneous fundamental solution of $\mathcal{L}_\infty$ and each $\overline{K}_m$ is (log-)homogeneous of degree $-Q_\infty + d - m$.

As $x \to 0$, there is again a polynomial $A(x)$ of Q_0-degree at most $-Q_0 + d - 1$ such that an asymptotic formula as is case 2 holds.

When we write an asymptotic expansion

$$K(x) \sim \sum_{m=0}^{\infty} K_m(x) \qquad \text{as } x \to 0, \tag{5.2}$$

we mean that K_m is (log-)homogeneous of degree $\mu + m$ with respect to the dilations of G_0, for some fixed μ, and that for every $N \geq 0$ and every differential monomial $P(x, D)$ of order n (in the sense defined immediately following (4.3))

$$\begin{aligned} &P(x,D)\big[K(x) - \sum_{m=0}^{N} K_m(x)\big] \\ &\quad = \begin{cases} O(\rho_0(x)^{\mu+N+1-n} & \text{if } \mu + N + 1 - n < 0 \\ O(\rho_0(x)^{\mu+N+1-n} \log \rho_0(x)) & \text{if } \mu + N + 1 - n \geq 0 \end{cases} \end{aligned} \tag{5.3}$$

as $x \to 0$.

The parallel notation

$$K(x) \sim \sum_{m=0}^{\infty} \overline{K}_m \qquad \text{as } x \to \infty$$

is defined similarly, replacing G_0 with G_∞, ρ_0 with ρ_∞, and assigning homogeneity $\mu - m$ to the term $\overline{K}_m$.. Consequently the exponents $\mu + N + 1 - n$ in (5.3) have to be replaced by $\mu - N - 1 - n$.

THEOREM 1. *Suppose $\mathcal{L}$ is defined by (5.1). Then $\mathcal{L}$ is hypoelliptic on G and it has a fundamental solution $K(x)$ such that*

(i) *$K(x) \to 0$ as $x \to \infty$ if $d < Q_\infty$.*

(ii) *$K(x) = O(\rho_\infty(x)^{d-Q_\infty} \log \rho_\infty(x))$ as $x \to \infty$ if $d \geq Q_\infty$.*

This fundamental solution is uniquely determined modulo a polynomial of G_∞ degree at most $d - Q_\infty$. Furthermore

(i)

$$K(x) \sim \sum_{m=0}^{\infty} \overline{K}_m(x) \qquad \text{as } x \to \infty$$

where $\overline{K}_0 = K^{(\infty)}$ is a (log-)homogeneous fundamental solution of $\mathcal{L}_\infty$;

(ii) *There is a polynomial $A(x)$ of G_0-degree at most $d - Q_0 - 1$ such that*

$$K(x) - A(x) \sim \sum_{m=0}^{\infty} K_m(x) \qquad \text{as } x \to 0$$

where $K_0 = K^{(0)}$ is a (log-)homogeneous fundamental solution of $\mathcal{L}_0$.

Before we go further we illustrate the main result with three simple examples.

Example (1) The group G is the three-dimensional Heisenberg group, whose Lie algebra is spanned by X, Y, Z, with $[X,Y] = Z$ the only non-zero commutation relation. Suppose $\mathcal{L} = X^2 + Y^2 + Z^2$. Then $K^{(0)}$ is the usual fundamental solution of the three-dimensional Laplacian $c_1||x||^{-1}$. At infinity we have $K^{(\infty)}(x) = c_2|x|^{-2}$ is the fundamental solution fo the sub-Laplacian $X^2 + Y^2$, where $|\cdot|$ is the norm function with respect to the usual automorphic dilations of G.

Example (2) The same group G, but now $\mathcal{L} = X^2 + Y^2 + Y$. Then $K^{(0)}(x) = c_2|x|^{-2}$, but $K^{(\infty)}$ is the fundamental solution of the Kolmogorov operator $X^2 + Y$ (which has a different homogeneity than $c_2|x|^{-2}$).

Example (3) In $\mathbb{R}^2$, $\mathcal{L} = \frac{\partial^4}{\partial x^4} - \frac{\partial^2}{\partial y^2} + \frac{\partial}{\partial x} + \frac{\partial}{\partial y}$, (*i.e.* $X_1 = \frac{\partial}{\partial x}$, $X_2 = \frac{\partial}{\partial y}$, $X_3 = \frac{\partial}{\partial x} + \frac{\partial}{\partial y}$. $\mathcal{L} = X_1^4 - X_2^2 + X_3$). Here the fundamental solutions at the origin and infinity have different axes diagonalizing their homogeneities.

We return to Theorem 1. As before, we can compare functions on G with functions on G_0 or G_∞ by means of the correspondences

$$exp_G(\sum_{j,i} \xi_{j,i} X_{j,i}) \leftrightarrow exp_{G_0}(\sum_{j,i} \xi_{j,i} X_{j,i})$$

and

$$exp_G(\sum_{j,i} \overline{\xi}_{j,i} \overline{X}_{j,i}) \leftrightarrow exp_{G_\infty}(\sum_{j,i} \overline{\xi}_{j,i} \overline{X}_{j,i}).$$

If we fix a Haar measure on G, these correspondences also determine uniquely Haar measure on G_0 and G_∞. If we also fix a Haar measure on $\tilde{G}$, we obtain Haar measure on the invariant subgroups I, I_0, I_∞, because $G \sim \tilde{G}/I$, $G_0 \sim \tilde{G}/I_0$, $G_\infty \sim \tilde{G}/I_\infty$.

At the Lie algebra level, let $\mathcal{F} = \mathfrak{i} \oplus \mathfrak{h} = \mathfrak{i}_0 \oplus \mathfrak{h}$, $\mathcal{F} = \mathfrak{i} \oplus \overline{\mathfrak{h}} = \mathfrak{i}_\infty \oplus \overline{\mathfrak{h}}$ be the decompositions of $\mathcal{F}$ constructed in Sections 2 and 3. Let $x \in \overline{\mathfrak{h}}$, $t \in \mathfrak{i}_\infty$, $y \in \mathfrak{h}$, $s \in \mathfrak{i}$. The Haar measures we have fixed on G and $\tilde{G}$ respectively correspond to fixed Lebesgue measures on $\mathfrak{h}$, $\overline{\mathfrak{h}}$, $\mathcal{F}$, and they are such that there is a Lebesgue measure ds on $\mathfrak{i}$ such that

$$ds\,dx = ds\,dy = \text{Lebesgue measure on } \mathcal{F}.$$

Suppose now we make the change of variables $s = t + \overline{\psi}(t)$, $t \in \mathfrak{i}_\infty$, and let dt be the measure on $\mathfrak{i}_\infty$ such that $dt\,dx =$ Lebesgue measure on $\mathcal{F} = \overline{\mathfrak{h}} \oplus \mathfrak{i}_\infty$.

We then have, for $f \in C_0(\mathcal{F})$

$$\int_{\mathfrak{i}\oplus\overline{\mathfrak{h}}} f(s+x)\,ds\,dx = \int_{\mathfrak{i}_\infty\oplus\overline{\mathfrak{h}}} f(t+\overline{\psi}(t)+x)\,dt\,dx$$

because the linear map $t + x \to t + \overline{\psi}(t) + x$ of $\mathcal{F}$ into itself is the identity plus a nilpotent map, due to the sub-homogeneity of $\overline{\psi}$.

Similarly

$$\int_{\mathfrak{i}\oplus\mathfrak{h}} f(s+y)\,ds\,dy = \int_{\mathfrak{i}_0\oplus\mathfrak{h}} f(u+\psi(u)+u)\,du\,dy.$$

From this it follows that

(i) Let $f \in L^1(\tilde{G})$ and define f^t on G by integrating f over the cosets of I with respect to the Haar measure constucted above. Then, keeping in mind that we can use both $\mathfrak{h}$ and $\overline{\mathfrak{h}}$ to parameterize the elements of G, we have

$$f^t(x) = \int_{\mathfrak{i}} f(x+s)\,ds = \int_{\mathfrak{i}_\infty} f(x+t+\psi(t))\,dt \quad (x \in \overline{\mathfrak{h}}),$$

$$f^t(y) = \int_{\mathfrak{i}} f(y+s)\,ds = \int_{\mathfrak{i}_0} f(y+u+\psi(u))\,du \quad (y \in \mathfrak{h});$$

(ii) Let $\Lambda : \mathfrak{i}_0 \oplus \mathfrak{h} \to \mathfrak{i}_\infty \oplus \overline{\mathfrak{h}}$ be the linear map defined by

$$\Lambda(u,y) = (t,u) \Leftrightarrow y+u+\psi(u) = x+t+\overline{\psi}(t).$$

Then

$$\frac{\partial\Lambda(w,y)}{\partial(u,y)} = 1.$$

We also observe that if $(t,x) = \Lambda(u,y)$, then $x = 0 \Leftrightarrow y = 0$. More than that,

$$\Lambda(u,y) = (\lambda_1(u,y), \lambda_2(y))$$

and

$$\lambda_1(u,0) = (I+\overline{\psi})^{-1}(I+\psi)u.$$

Before proving the theorem we need to establish the following lemma.

LEMMA. *Let N ba a graded group, L a homogeneous normal subgroup of N, $\mathfrak{n} = \mathfrak{l} \oplus \mathfrak{h}$ a decomposition of $\mathfrak{n}$ with $\mathfrak{h}$ a homogeneous subspace, identified with the Lie algebra of N/L. Let Q_1, Q_2 be the homogeneous dimensions of $\mathfrak{l}$ and $\mathfrak{h}$, $Q = Q_1 + Q_2$. Suppose $\tilde{K}(x+t)$, $(x \in \mathfrak{h}, t \in \mathfrak{l})$ is a smooth function on $N/\{0\}$ homogeneous of degree $-Q+k$, and let t^γ be a monomial on $\mathfrak{l}$ homogeneous of degree $|\gamma|$. Let also φ be a smooth function on $\mathfrak{l}$ such that $\varphi \equiv 0$ on a neighborhood of 0 and has compact support. Then as long as $k + |\gamma| > 0$, the integral*

$$K(x) = \int_{\mathfrak{l}} \varphi(t)\,t^\gamma \Big[\tilde{K}(x+t) - \sum_{|\alpha|<-Q_2+k+|\gamma|} \frac{1}{\alpha!} D_0^\alpha \tilde{K}(t) x^\alpha\Big]\,dt$$
$$+ \int_{\mathfrak{l}} (1-\varphi(t))\,t^\gamma \Big[\tilde{K}(x+t) - \sum_{|\alpha|\leq -Q_2+k+|\gamma|} \frac{1}{\alpha!} D^\alpha \tilde{K}(t)\,x^\alpha\Big]\,dt$$

is convergent and defines a (log-)homogeneous function on $\mathfrak{h}$ *of degree* $-Q_2 + k + |\gamma|$.

PROOF: The condition $k + |\gamma| > 0$ assures that each term separately in the first integral converges. For the second integral, we use Taylor's formula for graded groups [FS]. For x fixed and $|t|$ large,

$$
\begin{aligned}
|\tilde{K}(x+t) - & \sum_{|\alpha| \leq -Q_2+k+|\gamma|} \frac{1}{\alpha!} D^\alpha \tilde{K}(t)\, x^\alpha| \\
& \leq C \sup_{-Q_2+k+|\gamma|<|\beta|<N} \ \sup_{|y|<C|x|} |x|^{|\beta|}\, |D^\beta \tilde{k}(y+t)| \\
& \leq C|t|^{-Q+k+Q_2-k-|\gamma|-1} \\
& = C|t|^{-Q_1-|\gamma|-1}.
\end{aligned}
$$

If $k + |\gamma| < Q_2$, then

$$K(x) = \int_{\mathfrak{l}} t^\gamma \tilde{K}(x+t)\, dt$$

and this is a homogeneous function. If $k + |\gamma| \geq Q_2$, then

$$
\begin{aligned}
K(rx) - & r^{-Q_2+k+|\gamma|} K(x) = \\
& = r^{-Q_2+k+|\gamma|} \int [\varphi(rt) - \varphi(t)] \sum_{|\alpha|=-Q_2+k+|\gamma|} \frac{1}{\alpha!} D^\alpha \tilde{K}(t)\, x^\alpha\, dt \\
& = r^{-Q_2+k+|\gamma|} (\log r)\, P(x)
\end{aligned}
$$

where P is a polynomial homogeneous of degree $-Q_2 + k + |\gamma|$, by integrating in polar coordinates. Then if $x = |x|\, x'$,

$$K(x) - |x|^{-Q_2+k+|\gamma|} K(x') = |x|^{-Q_2+k+|\gamma|} (\log |x|)\, P(x')$$

i.e.

$$K(x) = |x|^{-Q_2+k+|\gamma|}\, K(x') + (\log |x|)\, P(x).$$

Q.E.D.

§6. Proof of Theorem 1.

Let $\tilde{G}$ be the homogeneous group associated to the free Lie algebra $\mathcal{F}$, where the degree of $\mathcal{F}$ is taken large enough so that the homogeneous dimension $\tilde{Q}$ of $\tilde{G}$ is strictly larger then d. Consider the operator $\mathcal{L} = \mathcal{P}(X_1, \ldots, X_k)$ on $\tilde{G}$, which is hypoelliptic by assumtion, and let K be its homogeneous fundamental solution, as described in the beginning of Section 5.

For $x \in \mathfrak{h}$, write

$$\begin{aligned} K(x) = &\int_{\mathfrak{i}_\infty} \varphi(t)\tilde{K}(x + t + \overline{\psi}(t))\, dt \\ &+ \int_{\mathfrak{i}_\infty} (1 - \varphi(t)\big[\tilde{K}(x + t + \overline{\psi}(t)) - \sum_{|\alpha| \leq d - Q_\infty} \frac{1}{\alpha!} D^\alpha \tilde{D}(t + \overline{\psi}(t))\, x^\alpha\big]\, dt \end{aligned} \tag{6.1}$$

where $\varphi(t)$ is a smooth function with compact support on $\mathfrak{i}_\infty$ such that $\varphi \equiv 1$ in a neighborhood of zero.

Step 1 The integrals in (6.1) converge and $K(x) \in L^1_{loc}$ and is smooth away from the origin.

First, $\tilde{K}$ is locally integrable on $\tilde{G}$, and the first integral conveges and gives a locally integrable function. The same holds for derivatives of K, for $x \neq 0$. We now prove that the second integral gives a smooth function on all of $\mathfrak{h}$. It follows from the sub-homogeneity of $\overline{\psi}$ that for $|t|$ large $|t + \overline{\psi}(t)| \sim |t|$. So if $|t| >> |x|$, we can use Taylor's formula for graded groups as in the lemma. The same argument can be applied to each derivative of $\tilde{K}$ showing that the second integral in (6.1) can be differentiated indefinitely.

Step 2 $K(x)$ is a fundamental solution of $\mathcal{L}$.

Let $f(t)$ be a C^∞-function with compact support on $\mathfrak{h}$, and set $\tilde{f}(x + t + \overline{\psi}(t)) = f(x)$. At the group level, if π is the quotient map from $\tilde{G}$ to G, we have defined $\tilde{f}$ as $f \circ \pi$. Then

$$\tilde{\mathcal{L}}^t \tilde{f} = (\mathcal{L}^t f) \circ \pi, \qquad i.e.,$$

$$(\tilde{\mathcal{L}}^t\, \tilde{f}(x + t + \overline{\psi}(t)) = (\mathcal{L}^t\, f)(x).$$

Let now η be a C^∞-function with compact support on $\tilde{G}$ such that $\eta \equiv 1$ in a neighborhood of zero, and define $\eta_R = \eta \circ D_R^{-1}$. Then

$$\tilde{\mathcal{L}}^t(\tilde{f}\,\eta_R) = (\tilde{\mathcal{L}}^t\, \tilde{f})\,\eta_R + \sum_\gamma (L_\gamma \tilde{f})(L'_\gamma\, \eta_R)$$

where L_γ and L'_γ are homogeneous differential operators, of degree $d(\gamma)$ and $d'(\gamma)$ respectively, such that $d'(\gamma) > 0$ and $d(\gamma) + d'(\gamma) = d$. We write

$$\int_{\mathfrak{h}} K(x)\,(\mathcal{L}^t f)(x)dx = \int\int \varphi(t)\,\tilde{K}(x+t+\overline{\psi}(t))\,(\tilde{\mathcal{L}}^t \tilde{f})(x+t+\overline{\psi}(t))\,dt\,dx$$
$$+\int\int (1-\varphi(t))\big[\tilde{K}(x+t+\overline{\psi}(t)) - \sum_{|\alpha|\le d-d_\infty} \frac{1}{\alpha!} D^\alpha \tilde{K}(t+\overline{psi}(t))\,x^\alpha\big]\times$$
$$\times\,(\tilde{\mathcal{L}}^t \tilde{f})(x+t+\overline{\psi}(t))\,dt\,dx$$

$$= \lim_{R\to\infty}\Big[\int\int \varphi(t)\eta_R(x+t+\overline{\psi}(t))\,\tilde{K}(x+t+\overline{\psi}(t))\,(\tilde{\mathcal{L}}^t \tilde{f})(x+t+\overline{\psi}(t))\,dt\,dx$$
$$+\int\int (1-\varphi(t))\,\eta_R(x+t+\overline{\psi}(t))\big[\tilde{K}(x+t+\overline{\psi}(t)) -$$
$$-\sum_{|\alpha|\le d-d_\infty} \frac{1}{\alpha!} D^\alpha \tilde{K}(t+\overline{psi}(t))\,x^\alpha\big]\,(\tilde{\mathcal{L}}^t \tilde{f})(x+t+\overline{\psi}(t))\,dt\,dx\Big]$$

$$= \lim_{R\to\infty}\Big[\int\int \eta_R(x+t+\overline{\psi}(t))\tilde{K}(x+t+\overline{\psi}(t))(\tilde{\mathcal{L}}^t \tilde{f})(x+t+\overline{\psi}(t))\,dt\,dx$$
$$-\sum_\alpha \frac{1}{\alpha!}\int\int (1-\varphi(t))D^\alpha\tilde{K}(t+\overline{\psi}(t))\,x^\alpha\times$$
$$\times\,\tilde{\mathcal{L}}^t \tilde{f}(x+t+\overline{\psi}(t))\eta_R(x+t+\overline{\psi}(t))\,dt\,dx\Big]. \tag{6.2}$$

The first integral is just the integral over $\tilde{G}$

$$\int_{\tilde{G}} \eta_r\,\tilde{K}(\tilde{\mathcal{L}}^t \tilde{f}) = \int_{\tilde{G}} \tilde{K}(\tilde{\mathcal{L}}^t(\eta_R \tilde{f}) - \sum_\gamma (L_\gamma \tilde{f})\,(L'_\gamma\,\eta_R))$$
$$= \eta_R(0)\tilde{f}(0) - \sum_\gamma \int_{\tilde{G}} \tilde{K}(L_\gamma \tilde{f})(L'_\gamma \eta_R)$$

because $\eta_R \tilde{f}$ has compact support. Observe that the first term equals $f(0)$. We will show that each of the other terms tends to zero with $R \to \infty$. We have

$$\int_G \tilde{K}(L_\gamma \tilde{f})(L'_\gamma \eta_R) = \pm \int_G \tilde{f}\,L^t_\gamma(\tilde{K}(L'_\gamma \eta_R))$$

because $d'(\gamma) > 0$, so that $L'_\gamma\,\eta_R$ is supported away from zero, where $\tilde{K}$ is smooth. Actually, $L'_\gamma\,\eta_R$ is supported between two balls of radius comparable with R, so that, since $d(\gamma) + d'(\gamma) = d$,

$$|\int_G \tilde{f}\,L^t_\gamma(\tilde{K}((L'_\gamma \eta_R))| \le C \int_{|z|\sim R} R^{d-\tilde{Q}-d}\,dz = C\,R^{-Q_\infty}.$$

We prove now that, for $|\alpha| \leq d - Q_\infty$, each integral

$$\int\int (1-\varphi(t))\, \eta_R(x+t+\overline{\psi}(t))\, D^\alpha \tilde{D}(t+\overline{\psi}(t))\, x^\alpha\, (\mathcal{L}^t f)(x)\, dt\, dx \tag{6.3}$$

tends to zero with R. Fix t in the support of $1-\varphi$, and consider the integral

$$\int \eta_R(x+t+\overline{\psi}(t))\, x^\alpha\, (\mathcal{L}^t f)(x)\, dx$$

as an integral on G. Call $\eta_R(x,t) = \eta_R(x+t+\overline{\psi}(t))$. Then

$$\int \eta_R(x,t)\, x^\alpha\, (\mathcal{L}^t f)(x)\, dx = \int \mathcal{L}(\eta_R(x,t)\, x^\alpha)\, f(x)\, dx.$$

We use now the homogeneous structure of G_∞. With respect to it, $\mathcal{L}$ is a sum of homogeneous operators, with degrees not smaller than d. Each of these operators involves therefore derivatives in the x-direction of degree at least d. Since the support of f is compact, we have, for $x \in supp\ f$, and $|\gamma| \geq d$

$$\frac{\partial^\gamma}{\partial x^\gamma}(\eta_R(x+t+\overline{\psi}(t))\, x^\alpha) = \sum_{\beta\leq\alpha} \binom{\gamma}{\beta} \frac{\partial^{\gamma-\beta}\eta_R}{\partial x^{\gamma-\beta}}(x+t+\overline{\psi}(t)) \frac{\partial^\beta}{\partial x^\beta} x^\alpha$$

and this is supported where $|x+t+\overline{\psi}(t)| \sim R$, because the hypothesis $|\alpha| \leq d - Q_\infty$ implies that every multiindex $\gamma - \beta$ is non-trivial. Therefore this expression is dominated by $R^{-|\gamma|+|\alpha|} \leq R^{-d+|\alpha|}$ and is supported on the set $\{x+t | x \in supp\ f, \quad |t| \sim R\}$, for R large.

Therefore the whole integral (6.2) is dominated by

$$\begin{aligned}\int_{x\in suppt\ f} \int_{|t|\sim R} |D^\alpha \tilde{K}(t+\overline{\psi}(t))|\, R^{-d+|\alpha|}\, dt\, dx &\leq C \int_{|t|\sim R} |t|^{-\tilde{Q}+d-|\alpha|} R^{-d+|\alpha|}\, dt \\ &\leq C\, R^{-Q_\infty}\end{aligned}$$

and tends to zero when $R \to \infty$.

Step 3 Asymptotics of $K(x)$ as $x \to \infty$.

The argument given above, modified by replacing $\overline{\psi}(t)$ with zero, shows that the function of $x \in \overline{\mathfrak{h}}$

$$\int_{\mathfrak{i}_\infty} \varphi(t)\tilde{K}(x+t)\, dt + \int_{\mathfrak{i}_\infty} (1-\varphi(t))\, [\tilde{K}(x+t) - \sum_{|\alpha|\leq d-Q_\infty} \frac{1}{\alpha!} D^\alpha \tilde{K}(t) x^\alpha]\, dt$$

is a fundamental solution of $\mathcal{L}_\infty$. However it is not (log-)homogeneous in general. We therefore use the kernel

$$K^{(\infty)}(x) = \int_{\mathfrak{i}_\infty} \varphi(t)\big[\tilde{K}(x+t) - \sum_{|\alpha|<d-Q_\infty} \frac{1}{\alpha!} D^\alpha \tilde{K}(t)\, x^\alpha\big]\, dt$$
$$+ \int_{\mathfrak{i}_\infty} (1-\varphi(t))\big[\tilde{K}(x+t) - \sum_{|\alpha|\le d-Q_\infty} \frac{1}{\alpha!} D^\alpha \tilde{K}(t)\, x^\alpha\big]\, dt.$$

By the lemma, this is (log-)homogeneous and it is still a fundamental solution of $\mathcal{L}_\infty$ because we have modified the previous kernel only by a polynomial of low degree.

We show that $K^{(\infty)}$ is the leading term in the asymptotic expansion of K at ∞. In order to do this we consider the difference

$$\begin{aligned} K(x) - K^{(\infty)}(x) = &\int_{\mathfrak{i}_\infty} \big[\tilde{K}(x+t+\overline{\psi}(t)) - \tilde{K}(x+t)\big]\, dt \\ &- \int \varphi(t) \sum_{|\alpha|\le d-Q_\infty} \frac{1}{\alpha!} D^\alpha \tilde{K}(t)\, x^\alpha\, dt \\ &+ \int (1-\varphi(t))\big[\tilde{K}(x+t+\overline{\psi}(t)) - \tilde{K}(x+t) \\ &\quad - \sum_{|\alpha|\le d-Q_\infty} \frac{1}{\alpha!} \big(D^\alpha \tilde{K}(t+\overline{\psi}(t)) - D^\alpha \tilde{K}(t)\big)\, x^\alpha\big)\, dt. \end{aligned} \tag{6.4}$$

In the first integral, since t is small and x is large, we obtain a term dominated by $|x|^{-\tilde{Q}+d}$, with derivatives which decay even faster. The second integral gives a polynomial of degree at most $d - Q_\infty - 1$. For the last integral, we consider the function M on $\overline{\mathfrak{h}} \times \mathcal{F}$ defined by

$$M(x,w) = \tilde{K}(x+t) - \sum_\alpha \frac{1}{\alpha!} D^\alpha \tilde{K}(w)\, x^\alpha$$

and estimate by the mean value theroem

$$|M(x,t+\overline{\psi}(t)) - M(x,t)| \le C \sum_{n)\beta)=1} |\overline{\psi}(t)^\beta| \sum_{|u|<C\,|\overline{\psi}(t)|} |D_w^\beta M(x,t+u)|$$

where $n(\beta)$ is the ordinary degree of β, and using the fact that for $|t|$ large, $|\overline{\psi}(t)|$ is very small compared with $|t|$.

Since

$$D_w^\beta M(x,w) = D^\beta \tilde{K}(x+w) - \sum_{|\alpha|\le d-Q_\infty} \frac{1}{\alpha!} D^{\beta+\alpha} \tilde{K}(w)\, x^\alpha,$$

we see that, by the graded Taylor formula, if $|t| > C\,|x|$,

$$|D_w^\beta M(x,t+u)| \le C \sum_{d-Q_\infty<|\gamma|\le d-Q_\infty+N} |x|^{|\gamma|}\, |t|^{-\tilde{Q}+d-|\beta|-|\gamma|},$$

whereas, if $|t| < C\,|x|$, just by homogeneity arguments, and since $|x+t+u| \sim |x|+|t| \sim |x|$,

$$|D_w^\beta M(x, t+u)| \le C\,|x|^{-\tilde{Q}+d-|\beta|}\log|x|.$$

Therefore the last integral in (6.4) is dominated by

$$\sum_{n(\beta)=1} |x|^{-\tilde{Q}+d-|\beta|}\log|x| \int_{|t|<C\,|x|} |\overline{\psi}(t)^\beta|\,dt$$
$$+ \sum_{\substack{n(\beta)=1 \\ d-Q_\infty<|\gamma|\le d-Q_\infty+N}} |x|^{|\gamma|} \int_{|t|>C\,|x|} |\overline{\psi}(t)^\beta|\,|t|^{-\tilde{Q}+d-|\beta|-|\gamma|}\,dt.$$

We now use the fact that $|\overline{\psi}(t)^\beta| \le C\,|t|^{|\beta|-1}$. Therefore the last integral in (6.4) is dominated by

$$\sum_{n(\beta)=1} |x|^{-\tilde{Q}+d-|\beta|}\log|x|\,|x|^{\tilde{Q}-Q_\infty+|\beta|-1} + \sum_{\beta,\gamma} |x|^{|\gamma|+|\beta|-1-Q_\infty+d-|\beta|-|\gamma|}$$
$$\le C\,|x|^{-Q_\infty+d-1}\log|x|.$$

Similar estimates hold when one considers $P(x,D)\big(K(x)-K^{(\infty)}(x)\big)$.

The construction of the other terms in the asymptotic expansion of K goes as follows: by expanding $\tilde{K}$ and its derivatives with respect to $\overline{\psi}(t)$, we have

$$\begin{aligned}
K(x) &= \int_{i_\infty} \tilde{K}(x+t+\overline{\psi}(t))\,dt \\
&\quad + \int_{i_\infty} (1-\varphi(t))\big[\tilde{K}(x+t+\overline{\psi}(t)) - \sum_{|\alpha|\le d-Q_\infty} \frac{1}{\alpha!} D^\alpha \tilde{K}(t+\overline{\psi}(t))\,x^\alpha\big]\,dt \\
&= \int_{i_\infty} \varphi(t)\big[\tilde{K}(x+t) + \sum_{1\le|\beta|\le N} \frac{1}{\beta!} D^\beta \tilde{K}(x+t)\overline{\psi}(t)^\beta\big]\,dt \\
&\quad + \int_{i_\infty} (1-\varphi(t))\big[\tilde{K}(x+t) - \sum_{|\alpha|\le d-Q_\infty} \frac{1}{\alpha!} D^\alpha \tilde{K}(t)\,x^\alpha + \\
&\quad \sum_{1\le|\beta|\le N} \frac{1}{\beta!}\big(D^\beta \tilde{K}(x+t) - \sum_{|\alpha|\le d-Q_\infty} \frac{1}{\alpha!} D^{\alpha+\beta}\tilde{K}(t)\,x^\alpha\big)\overline{\psi}(t)^\beta\big]\,dt \qquad (6.5) \\
&\quad + R_N(x) \\
&= K^{(\infty)}(x) + \sum_{1\le|\beta|\le N} \frac{1}{\beta!} \int_{i_\infty} \varphi(t)\,D^\beta\,\tilde{K}(x+t)\overline{\psi}(t)^\beta\,dt \\
&\quad \int_{i_\infty} (1-\varphi(t))\big[D^\beta\,\tilde{K}(x+t) - \sum_\alpha \frac{1}{\alpha!}\,D^{\alpha+\beta}\tilde{K}(t)\,x^\alpha\big]\,\overline{\psi}(t)^\beta\,dt \\
&\quad + \tilde{R}_N(x),
\end{aligned}$$

where $\tilde{R}_N$ is R_N plus a polynomial of degree $< d - Q_\infty$. We now consider each term in braces. Since $\overline{\psi}$ is superhomogeneous, $\overline{\psi}(t)^\beta$ is a sum of terms each of degree at most $|\beta| - n(\beta)$. Let t^γ be one of these terms, withn $|\gamma| \leq |\beta| - n(\beta)$. We apply the lemma to those terms for which the decay of $D^\beta \tilde{K}(x+t)$ is not sufficient to give integrability in t. This happens when $-\tilde{Q} + d - |\beta| + |\gamma| \geq -\tilde{Q} + Q_\infty$, *i.e.*, when$|\beta| - |\gamma| \leq d - Q_\infty$. By the lemma (setting $Q = \tilde{Q}, k = d - |\beta|$, so that $k + |\alpha| \geq Q_\infty > 0$) the function

$$\int_{\mathfrak{i}_\infty} \varphi(t)\, t^\gamma \big[D^\beta \tilde{K}((x+t) - \sum_{|\alpha|<-Q_\infty+d-|\beta|+|\gamma|} \frac{1}{\alpha!} D^{\alpha+\beta}\tilde{K}(t)\, x^\alpha\big]\, dt + \\ + \int_{\mathfrak{i}_\infty} (1-\varphi(t))\, t^\gamma \big[D^\beta \tilde{K}((x+t) - \sum_{|\alpha|<-Q_\infty+d-|\beta|+|\gamma|} \frac{1}{\alpha!} D^{\alpha+\beta}\tilde{K}(t)\, x^\alpha\big]\, dt \tag{6.6}$$

is log-homogeneous of degree $-Q_\infty + d - |\beta| + |\gamma| \leq -Q_\infty + d - n(\beta)$.

The terms present in (6.6) which are not in (6.5) (or *vice versa*) are the first integral, which is a low degree polynomial plus a function that decays like $|x|_{-\tilde{Q}-|\beta|}$, and terms in the second integral that again contribute a polynomial of low degree.

For those terms instead for which $|\beta| - |\gamma| > d - Q_\infty$, the second integral in (6.5) can be split into à sum of terms. We obtain terms like

$$\int_{\mathfrak{i}_\infty} (1-\varphi(t)) D^\beta \tilde{K}(x+t)\, t^\gamma\, dt = \int_{\mathfrak{i}_\infty} D^\beta \tilde{K}(x+t)\, t^\gamma\, dt + \text{rapidly decreasing}$$

which equals a homogenous function of degree $-Q_\infty + d - |\beta| + |\gamma|$ plus a rapidly decreasing function. The other terms are polynomials of degree not greater than $d - Q_\infty$.

The estimate for the remainder term R_N follows the same lines as the previous estimate for the difference $K - K^\infty$. Putting together these terms, we have found that for each N there is an expansion

$$K(x) = \sum_{m=0}^{N} K_{N,m}(x) + R_N$$

where $K_{N,m}$ have descending log-homogeneities. A simple argument shows that these terms $K_{N,m}$ do not depend on N.

Step 4 Asymptotics of $K(x)$ as $x \to 0$.

Observe that the last integral in (6.1) gives a smooth function of x. We therefore concentrate our attention on the first integral, *i.e.*

$$\int_{\mathfrak{i}_\infty} \varphi(t)\, \tilde{K}(x + t + \overline{\psi}(t))\, dt.$$

We switch to the coordinates y, u, with $y \in \mathfrak{h}$, $u \in \mathfrak{i}_0$, by means of the linear function Λ. We then replace x by $\lambda_2(y)$ and with a slight abuse of notation, we write $K(y)$ in place of

$K(\lambda_2(y))$. We have

$$K(y) = \int_{\mathfrak{i}_0} \varphi(\lambda_1(u,y))\,\tilde{K}(y+u+\psi(u))\,du + \text{smooth function}$$
$$= \int_{\mathfrak{i}_0} \varphi(u)\,\tilde{K}(y+u+\psi(u))\,du$$
$$+ \int_{\mathfrak{i}_0} \left[\varphi(\lambda_1(u,y)) - \varphi(u)\right] \tilde{K}(y+u+\psi(u))\,du + \text{smooth function.}$$

The last integral above is a smooth function of y. In fact if $|y|$ is small, $u \to \lambda_1(u,y)$ is a non-singular transformation of $\mathfrak{i}_0$ onte $\mathfrak{i}_\infty$, because when $y = 0$, $\lambda_1(u,0)$ is non-singular. Therefore the difference $\varphi(\lambda_1(u,y)) - \varphi(y) = 0$ for y and u small. Thus ultimately

$$K(x) = \int_{\mathfrak{i}_0} \varphi(u)\,\tilde{K}(y+u+\psi(u))\,du + \text{ smooth function.}$$

We then take a Taylor expansion of $\tilde{K}(y+u+\psi(u))$ with respect to $\psi(u)$ and obtain

$$K(y) = \int_{\mathfrak{i}_0} \varphi(u)\,\tilde{K}(y+u)\,du + \sum_{1\le|\beta|\le N} \frac{1}{\beta!} \int_{\mathfrak{i}_0} \varphi(u)\,D^\beta\,\tilde{K}(y+u)\,\psi(u)^\beta\,du$$
$$+ R_N(y) + \text{smooth function.}$$

If we compare $\int_{\mathfrak{i}_0} \varphi(u)\,\tilde{K}(y+u)\,du$ with the fundamental solution of $\mathcal{L}_0$

$$K^{(0)}(y) = \int_{\mathfrak{i}_0} \varphi(u)\big[\tilde{K}(y+u) - \sum_{|\alpha|<d-Q_0} \frac{1}{\alpha!} D^\alpha\,\tilde{K}(u)\,y^\alpha\big]\,du$$
$$+ \int_{\mathfrak{i}_0} (1-\varphi(u))\big]\big[\tilde{K}(y+u) - \sum_{|\alpha|\le d-Q_0} \frac{1}{\alpha!} D^\alpha\,\tilde{K}(u)\,y^\alpha\big]\,du$$

we see (as before) that

$$\int_{\mathfrak{i}_0} \varphi(u)\,\tilde{K}(y+u)\,du = K^{(0)}(y) + \text{ smooth funcion.}$$

Now take β such that $1 \le |\beta| \le N$. By the sub-homogeneity of ψ, $\psi(u)^\beta$ is a sum of terms u^γ with $|\gamma| \ge |\beta| + n(\beta)$. We observe that the total homogeneity of $D^\beta\,\tilde{K}(y+u)\,u^\gamma$ on $\tilde{G}$ is $-\tilde{Q}+d-|\beta|+|\gamma| \ge -\tilde{Q}+d+n(\beta)$. We can therefore construct the (log-)homogeneous function of degree $-\tilde{Q}+d-|\beta|+|\gamma|$

$$K_{\beta,\gamma}(y) = \int \varphi(u)\big[D^\beta\,\tilde{K}(y+u) - \sum_{|\alpha|<-\tilde{Q}+d-|\beta|+|\gamma|} \frac{1}{\alpha!} D^{\alpha+\beta}\,\tilde{K}(u)\,y^\alpha\big]u^\gamma\,du$$
$$+ \int (1-\varphi(u))\big[D^\beta\,\tilde{K}(y+u) - \sum_{|\alpha|<-\tilde{Q}+d-|\beta|+|\gamma|} \frac{1}{\alpha!} D^{\alpha+\beta}\,\tilde{K}(u)\,y^\alpha\big]u^\gamma\,du.$$

When we campare $K_{\beta,\gamma}(y)$ with the term $\int \varphi(u)\, D^\beta \tilde{K}(y+u)\, u^\gamma\, du$ present in the expansion of $K(y)$, we see that they differ by a smooth function. Therefore

$$K(y) = K^{(0)}(y) + \sum_{m=1}^{N} K_m(y) + R_N(y) + \text{ smooth function,}$$

where the K_m are (log-)homogeneous of ascending degrees.

In order to estimate the remainder $R_N(y)$ we use Taylor's formula:

$$\begin{aligned}|\tilde{K}(y+u+\psi(u)) - &\sum_{|\beta|\le N} \frac{1}{\beta!} D^\beta \tilde{K}(y+u)\, \psi(u)^\beta| \\ &\le C \sum_{N+1\le|\gamma|\le N+1} |\psi(u)|^{|\gamma|} \sup_{|z|\le C\,|\psi(u)|} |D^\gamma \tilde{K}(y+u+z)| \\ &\le C \sup_\gamma |u|^{(1+\epsilon)|\gamma|} (|y|+|u|)^{-\tilde{Q}+d-|\gamma|} \\ &\le C\,(|y|+|u|)^{-\tilde{Q}+d} |u|^{\epsilon(N+1)}.\end{aligned}$$

(The ϵ comes from the sub-homogeneity of ψ). Similarly, the derivatives $D^\alpha R_N(y)$ can be estimated by the Taylor expansion of $D^\alpha \tilde{K}(y+u+\psi(u))$. It follows that

$$|D^\alpha R_N(y)| \le C \int_{|u|<1} \frac{|u|^{(N+1)}}{(|y|+|u|)^{\tilde{Q}-d+|\alpha|}}\, du.$$

If N is large enough, we have that as many derivatives as required of R_N are bounded. Therefore for N large, R_N is a C^k-function at the origin, so that

$$R_N(y) = P_{k'}(y) + o(|y|^{k'})$$

where $P_{k'}$ is a polynomial of degree k'.

All this together gives an expansion of $K(y)$ as

$$K(y) = \sum_{m=0}^{k} K_m(y) + o(|y|^k).$$

The proof is therfore complete.

§7. Fractional integration and Riesz transforms.

We shall suppose we are given a system of l polynomials $\mathcal{P}_1, \mathcal{P}_2, \ldots, \mathcal{P}_l$, which are noncommutative polynomials in the indeterminates $Y_1, \ldots, Y_k$, each homogeneous of degree d (with Y_j having degree a_j). We also suppose that the system

$$\tilde{\mathcal{L}}_1 = \mathcal{P}_1(Y_1, \ldots, Y_k), \ldots, \tilde{\mathcal{L}}_l = \mathcal{P}_l(Y_1, \ldots, Y_k)$$

is universally hypoelliptic in the sense that if $\tilde{\mathcal{L}}_j(f) = g_j$ in $\tilde{G}$, $j = 1, \ldots, l$, and g_j are in C^∞, then $f \in C^\infty$. We then consider $\mathcal{L}_j = \mathcal{P}_j(X_1, \ldots, X_k)$.

THEOREM 2. *Let $\mathcal{R}(X_1, \ldots, X_k)$ be a monomial of degree $d' \leq d$. Then*

(a) *If $0 \leq d' < d$, we have the a priori inequality*

$$||\mathcal{R}(f)||_{L^q(G)} \leq C \sum_{j=1}^{l} ||\mathcal{L}_j(f)||_{L^p(G)} \tag{7.1}$$

for all $f \in C_0^\infty(G)$, as long as $1 < p, q < \infty$, and

$$\frac{d-d'}{Q_\infty} \leq \frac{1}{p} - \frac{1}{q} \leq \frac{d-d'}{Q_0}.$$

(b) *If $d' = d$, the the a priori inequality (7.1) holds with $q = p$, $1 < p < \infty$.*

PROOF: By a theorem of Helffer-Nourrigat [HeN] the operator $\sum_j \tilde{\mathcal{L}}_j^* \tilde{\mathcal{L}}_j$ is hypoelliptic and so we can invoke the previous results with $\mathcal{L} = \sum_j \mathcal{L}_j^* \mathcal{L}_j$ and $\tilde{\mathcal{L}} = \sum_j \tilde{\mathcal{L}}_j^* \tilde{\mathcal{L}}_j$. We thus have an appropriate fundamental solution K for $\mathcal{L}$ which satisfies $(\sum_j \mathcal{L}_j^* \mathcal{L}_j f, K) = f(0)$, for every $f \in C_0^\infty(G)$ where (,) is the usual L^2 inner product. It then follows from this that

$$f = \sum_{j=1}^{l} \mathcal{L}_j(f) * K_j, \quad \text{with } K_j = \overline{\mathcal{L}}_j(K). \tag{7.2}$$

Now write $K_j = K_j^I + K_j^{II}$, where $K_j^I = K_j$ for x near the origin, $K_j^{II} = K_j$ for large x, and K_j^I, K_j^{II} are smooth away from the origin. Write $A = \mathcal{R}(K_j^I)$, $B = \mathcal{R}(K_j^{II})$.

We consider first the case (a) of the theorem. Using the asymptotic expansion (part ii) of theorem 1 for K, and proposition 6 in section 4, we see that on $\mathfrak{g}_0$, the kernel A is dominated by $c\,|x|^{-Q_0+d-d'}$ (as long as $Q_0 > d-d'$), and moreover A has compact support. Thus A belongs to the weak type space $L^{r,\infty}$, with $r(Q_0 - d + d') \geq Q_0$. Therefore by the argument in [S], page 121, the mapping $f \to f * A$ is bounded from L^p to L^q, with $\frac{1}{q} = \frac{1}{p} - 1 + \frac{1}{r}$, which is $\frac{1}{p} - \frac{1}{q} \leq \frac{d-d'}{Q_0}$.

A similar argument shows that $f \to f * B$ is bounded for $\frac{d-d'}{Q_\infty} \leq \frac{1}{p} - \frac{1}{q}$, and together these prove part (a) of the theorem.

To prove part (b) we have to use the theory of singular integrals. Consider first the kernel A. By the asymptotic expansion and proposition 5 and 6, A is the sum of a kernel which is homogeneous of degree $-Q_0$ exactly and has mean zero on the unit sphere, plus an integrable function. So one may assume that near the orign A is itself homogeneous of degree $-Q_0$, and has vanishing mean value. In view of the approximate multiplication law (see proposition 4) one can also conclude that

$$|A(y^{-1}x) - A(x)| \leq C \frac{\delta}{|x|^{Q_0+1}} \tag{7.3}$$

whenever $\rho(y) < \delta$, and $|\rho(x)| \geq C\,\delta$, $\delta \leq 1$ where C is a sufficiently large constant. Hence for the kernel A one can show first that the operator $f \to f * A$ is bounded on L^2 by the almost orthogonality lemma, and then that it is bounded on L^p, $1 < p < \infty$ by the usual arguments of the Calderón-Zygmund theory, using the properties of the balls given in proposition 7. (Arguments of this kind are given in Goodman [Go]). A similar treatment works for the operators of the form $f \to f * B$ concluding the the theorem.

References

[F] Folland, G. B., *Sub-elliptic estimates and function spaces on nilpotent groups*, Arkiv fúr Math. **13** (1975), 161–207.

[FS] Folland, G. B., and Stein, E. M., "Hardy Spaces on Homogeneous Groups," Princeton Univ. Press, 1982.

[Ge] Geller, D., *Liouville's theorem for homogeneous groups*, Comm. in Partial Diff. Equ. **15** (1983), 1665–1677.

[Go] Goodman, R. W., "Nilpotent Lie Groups (Lecture Notes in Math. #562)," Springer, 1976.

[HeN] Helffer, B., and Nourrigat, J., *Characterization des operateurs hypoelliptiques homogènes invariants à gauche sur un groupe de Lie gradué*, Com. in Partial Diff. Equ. **3** (1978), 899–958.

[Hu] Hulanicki, A., and Jenkins, J., *Nilpotent Lie groups and summability of eigen-function expansions of Schrödinger operators*, Studia Math. **80** (1984), 235–244.

[NSW] Nagel, A., Stein, E. M., and Wainger, S., *Balls and metrics defined by vector fields*, Acta Math. **155** (1985), 103–147.

[RiS1] Ricci, F., and Stein, E. M., *Oscillatory singular integrals and harmonic analysis on nilpotent groups*, Proc. Nat. Acad. Sci. UI.S.A. **83** (1986), 1–3.

[RiS2] ——————, *Harmonic analysis on nilpotent groups and singular integrals*, Jour. Functional Analysis **78** (1988), 56–84.

[RoS] Rothschild, L., and Stein, E. M., *Hypoelliptic differential operators and nilpotent groups*, Acta Math. **137** (1976), 247–320.

[S] Stein, E. M., "Singular integrals and differentiablitiy properties of functions," Princeton Univ. Press, 1970.

[V] Varopolous, N.,, Jour. of Functional Analysis **66** (1986), 406–431.

15

Tangential and Approximate Free Interpolation

Nikolai K. Nikolskii and Alexander L. Volberg

I. INTRODUCTION

A. Preliminaries

The notion of free interpolation passed through several stages during the last three decades.· Starting about 100 years ago as a Hadamard theorem on the existence of an entire function f with arbitrary (prescribed) values $f(z_n) = a_n$ on an unbounded sequence of complex points $z_n \in \mathbb{C}$, $\lim_n |z_n| = \infty$, it became an important tool for different purposes (e.g., for exponential-polynomial description of solutions of convolution equations, for studies of bases of rational functions and exponentials, for discovering the multiplicative structure of holomorphic functions, and so on). For example, free interpolation is very useful in studying Hardy classes H^p, and especially, the algebra H^∞ of all bounded analytic functions on the unit disc $\mathbb{D} = \{z \in \mathbb{C}: |z| < 1\}$. For H^∞, the evolution of the free interpolation notion includes:

The well-known study of subsets $\sigma \subset \mathbb{D}$ for which the restrictions space $H^\infty \mid \sigma$ is an ideal space (i.e., $f \in H^\infty \mid \sigma$ and $|g| \leq |f|$ on σ implies $g \in H^\infty \mid \sigma$); such σ's turn out to be sequences $\sigma = \{\lambda_n\}_{n \geq 1}$ satisfying the Carleson condition

$$\inf_n |B_{\lambda_n}(\lambda_n)| > 0 \tag{C}$$

where $B_{\lambda_n} = \prod_{k \neq n} b_{\lambda_k}$, $b_\lambda = (\lambda - z)(1 = \bar{\lambda} z)^{-1} \cdot |\lambda|/\lambda$.

Multiple free interpolation with bounded multiplicities (under the Carleson condition).

General multiple free interpolation, and free interpolation with respect to an arbitrary sequence $\{\theta_n\}_{n \geq 1}$ of inner function where the problem is to find a function $f \in H^\infty$ with prescribed projections onto the corresponding z^*-invariant subspaces: $f - f_n \in \theta_n H^\infty$, $n \geq 1$ ($f_n, f_n \in H^\infty$, obeying the only boundedness condition $\sup_n \|f_n\| < \infty$, where $\|\cdot\|_\infty$ stands for the uniform norm over $\mathbb{D}$).

Finally, the most general free interpolation with respect to an arbitrary sequence $\{\theta_n\}_{n\geq 1}$ of H^∞ functions; this last problem can be described and solved as follows.

First, recall that a nonzero H^∞ function θ admits the canonical Nevanlinna–Smirnov factorization

$$\theta = \left(\prod_{\lambda\in\Lambda} b_\lambda^{k_\lambda}\right) \exp\left(-\int_{\mathbb{T}} \frac{\zeta+z}{\zeta-z}\, d\mu_\theta(\zeta)\right)$$

where $\Lambda = \Lambda(\theta) \subset \mathbb{D}$, k_λ are integers ≥ 1 with $\sum_{\lambda\in\Lambda} k_\lambda(1-|\lambda|) < \infty$, $b_\lambda = (\lambda - z)(1-\bar{\lambda}z)^{-1}\cdot|\lambda|/\lambda$ is a Moebius transform, $d\mu_\theta = \log|\theta|^{-1}dm + d\mu_s$; m stands for normalized Lebesgue measure on $\mathbb{T} = \{z: |z| = 1\}$, and μ_s is a singular measure (with respect to m), $\mu_s \geq 0$. If σ is a Borel subset of the closed unit disc $\operatorname{clos}\mathbb{D}$, let us define "the part of θ over σ" by the formula

$$\theta_\sigma = \left(\prod_{\lambda\in\Lambda\cap\sigma} b_\lambda^{k_\lambda}\right) \exp\left[-\int_{\sigma\cap\mathbb{T}} \frac{\zeta+z}{\zeta-z}\, d\mu_\theta(\zeta)\right]$$

Now, let $\{\theta_n\}_{n\geq 1}$ be a sequence of nonzero H^∞ functions normalized by the conditions $|\theta_n(z)| \leq 1$, $z \in \mathbb{D}$, $n = 1, 2, \ldots$. We say that $\{\theta_n\}$ is a *free interpolating sequence* (FI-sequence) if one (all) of the equivalent conditions (1.1)–(1.3) is (are) fulfilled. Hence, condition (1.4) serves as a description of FI-sequences.

Theorem A The following are equivalent.

(1.1) For every $c_n \in \mathbb{C}$, $\sup_n |c_n| < \infty$, there exist a function f, $f \in H^\infty$, and a constant $c > 0$ such that $f - c_n \in \theta_n H^\infty$, $\|(f - c_n)\theta_n^{-1}\|_\infty \leq c$, $n = 1, 2, \ldots$.

(1.2) For every $f_n \in H^\infty$, $\sup_n \|f_n\|_\infty < \infty$, there exists a function f, $f \in H^\infty$, with the property: $f - f_n \in \theta_n H^\infty$, $\sup_n \|(f - f_n)\theta_n^{-1}\|_\infty < \infty$.

(1.3) For every $f_n \in H^\infty$ with $\sup_n \operatorname{dist}(\bar{\theta}_n f_n, H^\infty) < \infty$, $\sup_n \|f_n\|_{L^\infty}(\Delta_n dm) < \infty$ (where $\Delta_n^2 \equiv 1 - |\theta_n|^2$) there exists $f \in H^\infty$ such that $f - f_n \in \theta_n H^\infty$, $\sup_n \|(f - f_n)\theta_n^{-1}\|_\infty < \infty$.

(1.4) There exist a function $\theta \in H^\infty$, $0 \not\equiv |\theta| \leq 1$, and a Borel partition $\{\sigma_n\}_{n\geq 1}$ of the closed disc $\operatorname{clos}\mathbb{D}$ ($\sigma_i \cap \sigma_j = \emptyset$, $i \neq j$, and $\cup\sigma_i = \operatorname{clos}\mathbb{D}$) such that $\varepsilon|\theta_{\sigma_n}(z)| \leq |\theta_n(z)| \leq |\theta_{\sigma_n}(z)|$ for all $z \in \mathbb{D}$, $n \geq 1$ and some $\varepsilon > 0$, and the inequality

$$|\theta(z)| \geq \delta \inf_n |\theta_{\sigma_n}(z)|, \qquad z \in \mathbb{D} \tag{CV}$$

holds with a positive δ.

It is worth mentioning that in applications θ_n are usually given at once in the form of a partition $\theta_n = \theta_{\sigma_n}$, $n \geq 1$; (CV) stands for L. Carleson and V. Vasyunin.

The equivalence 1.1 $\Leftrightarrow$ 1.4 was proved in Vasyunin (1976) [for the case of a partition of an inner function θ, i.e., for $|\theta| = 1$ a.e. on $\mathbb{T}$, see also Vasyunin (1978)]. 1.2 $\Leftrightarrow$ 1.4 was proved in Nikolskii (1978b) for inner θ [see also Nikolskii (1978a), (1986)] and for a partition of general $\theta \in H^\infty$ in Nikolskii (1987) and Nikolskii and Hruščëv (1987). The last two papers also contain the implications 1.3 $\Leftrightarrow$ 1.4 (as well as many comments and examples and some generalizations). It should be noted that the proof of the theorem is entirely elementary [especially in Nikolskii (1987) and Nikolskii and Hruščëv (1987)] but involves some Hilbert space operator theory. A purely function-theoretic proof of the equivalence 1.1 $\Leftrightarrow$ 1.4 (for a partition of a Blaschke product θ) is given by Carleson (1985).

In fact, in Nikolskii (1987) [see also Nikolskii and Hruščëv (1987)] it is proved that for a partition of an inner function satisfying the generalized Carleson condition (CV), all locally defined functions $f_n \in H^\infty(l_n)$, $l_n = \{z: |\theta_{\sigma_n}(z)| < \varepsilon\}$, $0 < \varepsilon < 1$ with the boundedness condition $\sup_n \sup\{|f(z)|: z \in l_n\} < \infty$ can be interpolated in the sense of Theorem A:

$$f - f_n \in \theta_{\sigma_n} H^\infty(l_n), \qquad n \geq 1; \sup_n \|(f - f_n)\theta_{\sigma_n}^{-1}\|_{H^\infty(l_n)} < \infty.$$

If θ reduces to a Blaschke product $\theta = \prod b_{\lambda_n}^{k_n}$ and if $\sigma_n = \{\lambda_n\}$, $n \geq 1$, this means the possibility of free interpolation by germs of H^∞ functions (with unbounded heights). In the singular inner function case ($k_\lambda \equiv 0$, $|\theta| = 1$ a.e. on $\mathbb{T}$) Theorem A leads to a construction of an H^∞ function with prescribed asymptotics along a system of radii: $f(r\zeta_n) = f_n(r\zeta_n) + O(\exp(-a_n(1-r)^{-1}))$ for $r \to 1$ ($|\zeta_n| = 1, n \geq 1$).

If θ is an outer function (i.e. $k_\lambda \equiv 0$, $\mu_s \equiv 0$) Theorem A reduces to the following proposition B.

Notation. Let w_n, $n = 1, 2, \ldots$ be measurable functions (weights) on $\mathbb{T}$, $0 \leq w_n \leq 1$ a.e. , and let

$$W_n(\zeta) = \inf\{w_k(\zeta): k \geq 1, k \neq n\} \tag{1.5}$$

for almost all $\zeta \in \mathbb{T}$ and all $n = 1, 2, \ldots$. Denote by f^{out} the outer function in $\mathbb{D}$ with $|f^{\text{out}}| = |f|$ a.e. on $\mathbb{T}$, and put $f^{\text{out}} \equiv 0$ if $\int_{\mathbb{T}} \log|f|\,dm = -\infty$; f^{inn} stands for the inner part of a nonzero H^∞ function f, i.e., $f^{\text{inn}} = f/f^{\text{out}}$.

Theorem B The following are equivalent.

(1.6) For every $c_n \in \mathbb{C}$, $\sup_n |c_n| < \infty$ there exists $f \in H^\infty$ and $c > 0$ such that $|f - c_n| \leq c w_n$ a.e. on $\mathbb{T}$, $n \geq 1$.

(1.7) For every $F_n \in H^\infty$, $\sup_n \|f_n\|_\infty < \infty$, there exists $f \in H^\infty$ such that $|f - f_n| \leq c w_n$ a.e. on $\mathbb{T}$, $n \geq 1$, where $c = c(\delta)\sup_n \|f_n\|_\infty$; $c(\delta)$ depends on δ of (1.8) only.

(1.8) There exists a positive δ for which

$$\sup_n |W_n^{\text{out}}(z)| \geq \delta, \qquad z \in \mathbb{D}$$

For the sake of completeness we mention some points in the proofs of Theorems A and B which are different from those given in Nikolskii (1987) and Nikolskii and Hruščëv (1987).

Proof of Theorem A. For a partition $\theta_n - \theta_{\sigma_n}$ of a nonzero H^∞ function θ a complete proof is contained in the papers mentioned above. To solve the general case from this partial one let us define a *sparse sequence* of H^∞ functions $\{\theta_n\}_{n\geq 1}$ by the following property: there exists a positive ε such that the level sets

$$L(\theta_n, \varepsilon) = \{z \in \operatorname{clos} \mathbb{D}: |\theta_n(z)| < \varepsilon\}$$

are pairwise disjoint. Of course, for $|z| = 1$ we mean θ_n equals the corresponding angular limit.

Lemma 1.9 Under condition (1.1) [or (1.6) in which case $[\theta_n = w_n^{\text{out}}, n \geq 1]$, one gets the following:

(1.10) There exists a constant, say α, such that $c \leq \alpha \sup_n |c_n|$.
(1.11) If $\int_{\mathbb{T}} \log w_k \, dm = -\infty$ for some $k \geq 1$, then $\inf_{n\neq k} w_n \geq \alpha^{-1}$ a.e. on $\mathbb{T}$, and hence no more than one of w_k^{out} is zero.
(1.12) $|\theta_n(z) + \theta_k(z)| \geq \alpha^{-1}$ for $z \in \mathbb{D}, n \neq k$.
(1.13) $\{\theta_n\}_{n\geq 1}$ is a sparse sequence of functions, namely,
(1.14) $L(\theta_n, 1/2\alpha) \cap L(\theta_k, 1/2\alpha) = \varnothing, \quad n \neq k$.
(1.15) If $\int_{\mathbb{T}} \log w_k \, dm > -\infty$ for all $k \geq 1$, then $\int_{\mathbb{T}} \log(\inf_{k\geq 1} w_k) \, dm > -\infty$.

Proof. (1.10) The existence of α is a standard consequence of the category theorem [for details see Nikolskii and Hruščëv (1987)].

(1.11) Let $c_k = 1, c_n = 0$ for $n \neq k$ and let f be an H^∞ function from (1.6). Then $|1 - f| \leq \alpha w_k$, $|f| \leq \alpha w_n$, $n \neq k$. This implies $f \equiv 1$, and hence $w_n \geq \alpha^{-1}, n \neq k$ a.e. on $\mathbb{T}$.

Remark. Since (1.12), (1.13), and (1.15) are obviously true under this last condition ($w_n \geq \alpha^{-1}, n \neq k$) we assume from now on that $w_k^{\text{out}} \not\equiv 0$ for every $k \geq 1$.

(1.12) Let $c_k = 1$, $c_n = 0$ for $n \neq k$ and let f be an H^∞ function from (1.1) or (1.6). Then $1 \leq |1 - f| + |f| \leq \alpha|\theta_k| + \alpha|\theta_n|$ on the disc $\mathbb{D}$ [for (1.6) we use the maximum principle: a.e. on $\mathbb{T}$ implies $|f| \leq \alpha|\theta_n| = |w_n^{\text{out}}|$ on $\mathbb{D}$].

(1.13) Obvious from (1.12).

(1.15) If $w_k \geq \alpha^{-1}$ a.e. on $\mathbb{T}$ for every $k \geq 1$, the assertion follows. If $\alpha \operatorname{ess\,inf}_{\mathbb{T}} w_k < 1$ for an integer k, consider a function $f \in H^\infty$ from the proof of (1.11) (above): $|1 - f| \leq \alpha w_k$ implies $f \not\equiv 0$, and hence $-\infty < \int_{\mathbb{T}} \log |f| \, dm \leq \int_{\mathbb{T}} \log(\alpha \inf_{n\neq k} w_n) \, dm$. Moreover, $\int_{\mathbb{T}} \log w_k \, dm > -\infty$ and the lemma follows. ∎

Corollary 1.16 For (1.1) or (1.6) (except trivial case 1.11) to be valid it is necessary that there exist an H^∞ function θ, $0 \not\equiv |\theta| \leq 1$ and a partition $\{\sigma_n\}_{n\geq 1}$ of $\operatorname{clos} \mathbb{D}$ such that $|\theta_{\sigma_n}| \asymp |\theta_n|$ [$\theta_n = w_n^{\text{out}}$ for (1.6)] in the sense of bilateral estimations on $\mathbb{D}$ with constants independent of n.

In fact, it is easy to see that (1.1) and 1.12 imply that the inner parts $\theta_k^{\text{inn}}, k = 1, 2, \ldots$, are pairwise coprime. Moreover, the product $\theta^{\text{inn}} \equiv \prod_{k\geq 1} \theta_k^{\text{inn}}$ converges (hint: as in Lemma 1.9 consider interpolating data $c_n = \delta_{nk}, \geq 1$). Since (1.14) we can consider an outer function θ^{out} with $|\theta^{\text{out}}| = |\theta_n|$ a.e. on $\mathbb{T} \cap L(\theta_n, 1/2\alpha)$, $n \geq 1$, and put $|\theta^{\text{out}}| = 1$ elsewhere. By (1.15), $\theta^{\text{out}} \not\equiv 0$. It is easy to see that the function $\theta = \theta^{\text{inn}}\theta^{\text{out}}$ and its parts θ_{σ_n} over $\sigma_n = L(\theta_n, 1/2\alpha)$ satisfy $(2\alpha)^{-1}|\theta_{\sigma_n}| \leq |\theta_n| \leq |\theta_{\sigma_n}|$ on $\mathbb{D}$, $n = 1, 2, \ldots$. ■

Corollary 1.16 reduces Theorem A to a partition sequence case $\{\theta_{\sigma_n}\}$. ■

Proof of Theorem B. By (1.11), $w_n \geq \alpha^{-1}$ a.e. on $\mathbb{T}$, $n \neq k$, if $\int_{\mathbb{T}} \log w2_k = -\infty$. In this case(1.6)–(1.8) are equivalent.

On the other hand, if $w_n^{\text{out}} \not\equiv 0$ for every $n = 1, 2, \ldots$ we use (1.13), (1.15), and the outer function $\theta = \theta^{\text{out}}$ from Corollary 1.16: $|\theta| = w_n$ a.e. on $\sigma_n \equiv L(w_n, 1/2\alpha)$, $n \geq 1$ and $|\theta| = 1$ elsewhere. Then $|w_n^{\text{out}}| \asymp |\theta_{\sigma_n}|$ [the same sense as in (1.16)] and $|\theta| \asymp \inf_n |w_n^{\text{out}}|$, $|\theta/\theta_{\sigma_n}| \asymp |W_n^{\text{out}}|$ and at last $\sup_n |W_n^{\text{out}}| \asymp |\theta|/\inf_n |\theta_{\sigma_n}|$. Now the theorem follows from Theorem A. ■

B. Aim

The aim of the chapter is twofold. First, we would like to give more details in a partial case of the general free interpolation problem (Section A) when

$$w_n = \varepsilon_n \chi_{\sigma_n} + \chi_{\mathbb{T}-\sigma_n}, \qquad n \geq 1. \tag{1.17}$$

Here χ_{σ_n} stands for the characteristic function of σ_n, $\{\sigma_n : n \geq 1\}$ is a sequence of pairwise disjoint measurable subsets of $\mathbb{T}$, and the reals ε_n are positive. In this case condition (1.7) means

(1.18) For every $f_n \in H^\infty$, $\sup_n \|f_n\|_\infty < \infty$, there exist a function f, $f \in H^\infty$, and a constant $c > 0$ such that

$$|f(\zeta) - f_n(\zeta)| \leq c\varepsilon_n \qquad \text{a.e. on } \sigma_n,\ n = 1, 2, \ldots. \tag{1.19}$$

Of course, we suppose $\lim_n \varepsilon_n = 0$.

Because (1.19) means that f is approximately (up to ε_n) equal to f_n on σ_n, we refer to this kind of relation between f (a solution) and $\{f_n\}$ (data) as *approximate free interpolation* (AFI). In Section III we examine the corresponding sparse and FI-sequences of sets (i.e., such ones for which w_n^{out} are sparse or FI) and, in particular, (1) establish a relationship of FI-sequences to the known "strong interpolating property" [see (3.36) below], and (2) prove that an arbitrary sequence of arcs σ_n, $\sigma_n \subset \mathbb{T}$ with $\sum_{k\geq 1} m\sigma_k < 1$ can be scattered throughout the unit circle $\mathbb{T}$ in such a way that AFI (1.18) holds with a choice of $\varepsilon \downarrow 0$.

The second aim of the chapter is to point out a relation between Theorem A and a Carleson theorem (Carleson, 1962) on a kind of local interpolation (see Theorem 2.1 below). We start with an arbitrary H^∞ function θ, $0 \not\equiv |\theta| \leq 1$, and with an arbitrary Borel partition $\{\sigma_n\}_{n\geq 1}$ of the closed unit disc $\operatorname{clos}\mathbb{D}$. Thus, the sequence $\{\theta_{\sigma_n}\}_{n\geq 1}$ should not be an FI-sequence. How do we reconstruct it to reach an FI-property? A universal suggestion consists in considering the level sets $L(\theta, \varepsilon)$,

$0 < \varepsilon < 1$, and their connected components. When σ_n are singletons ($\sigma_n \subset \operatorname{clos} \mathbb{D}$) or arcs ($\sigma_n \subset \mathbb{T}$), this rule leads to a grouping of σ_n into some unions $\sum_j = \cup \sigma_n$ which now enjoy the FI-property. We derive our theorem on grouping from Theorem A in a very elementary way not involving any kind of hard analysis [as the known "Carleson contours" originally used; see also Garnett (1981)].

In a very preliminary form a part of this chapter appeared in Nikolskii (1988).

II. GENERALIZING A CARLESON THEOREM

Corollary 1.16 shows that starting with some H^∞ functions ϑ_n, $0 \not\equiv |\vartheta_n| \leq 1$ and trying to define an FI-sequence θ_n, $n \geq 1$ by grouping ϑ_n "with sufficiently close spectra," $\theta_n = \prod_{k \in A_n} \vartheta_k$ we can think at once of partitions of a (fixed) H^∞ function θ. Let us begin with a description of partitions generated by level sets.

Let θ be an H^∞ function, $0 \not\equiv |\theta| \leq 1$, and $0 < \varepsilon < 1$. Denote by Ω_n, $n = 1, 2, \ldots$, the connected components of the open set $\mathbb{D} \cap L(\theta, \varepsilon)$, and by Ω_n^* the set of all $\zeta \in \mathbb{T}$ enjoying the following property: there exists a Stoltz angle

$$S(\zeta, \alpha, \varepsilon) = \{z \in \mathbb{D} \colon |\arg(\zeta - z)| < \alpha, 1 - |z| \leq \varepsilon\}, \qquad \alpha = \alpha_\zeta < \frac{\pi}{2}, \varepsilon = \varepsilon_\zeta$$

which is entirely contained in $\Omega_n \colon S(\zeta, \alpha, \varepsilon) \subset \Omega_n$. Further, let $\Omega_0 = \mathbb{D} \setminus L(\theta, \varepsilon)$, $\Omega_0^* = \mathbb{T} \setminus (\bigcup_{n \geq 1} \Omega_n^*)$, and finally

$$\sigma_n = \Omega_n \cup \Omega_n^*, \qquad n = 0, 1, 2, \ldots$$

By definition, $\{\sigma_n\}_{n \geq 0}$ is a Borel partition of the closed unit disc $\operatorname{clos} \mathbb{D}$. Moreover, if $\zeta \in \mathbb{T}$ and there exists an angular limit $\theta(\zeta)$, and if $|\theta(\zeta)| < \varepsilon$, then $\zeta \in \Omega_n^*$ for an integer $n \geq 1$. Some ζ's with $|\theta(\zeta)| \leq \varepsilon$ also belong to $\bigcup_{n \geq 1} \Omega_n^*$; all others belong to Ω_0^*. Further, it is known [e.g., see Garnett (1981)] that at μ_s-almost all (with respect to the singular measure μ_s from the Nevanlinna–Smirnov factorization of θ) points $\zeta \in \mathbb{T}$ there exists a zero angular limit $\theta(\zeta) = 0$. Hence, μ_s-almost all points of $\mathbb{T}$ belong to $\bigcup_{n \geq 1} \Omega_n^*$. It is also easy to see that the zero set $\Lambda(\theta) = \{\lambda \in \mathbb{D} \colon \theta(\lambda) = 0\}$ is contained in $\bigcup_{n \geq 1} \Omega_n$. So, we get a factorization

$$\theta = \prod_{n \geq 0} \theta_{\sigma_n}$$

in which θ_{σ_0} is outer and $\varepsilon \leq |\theta_{\sigma_0}(z)| \leq 1, z \in \mathbb{D}$, and for $n \geq 1$, θ_{σ_n} is described by its zero set $\Lambda(\theta_{\sigma_n}) = \sigma_n \cap \Lambda(\theta)$ and generating measure $\mu_{\theta_{\sigma_n}} = \mu_\theta \chi_{\sigma_n \cap \mathbb{T}}$. In particular, $|\theta_{\sigma_n}| \leq \varepsilon\, m$-a.e. on $\sigma_n \cap \mathbb{T}$ and $|\theta_{\sigma_n}| = 1\, m$-a.e. on $\mathbb{T} \setminus \sigma_n$.

We will refer to the partition described as the *level sets partition*.

Theorem 2.1 Let θ be an H^∞ function, $0 \not\equiv |\theta| \leq 1$, and $\{\sigma_n\}_{n \geq 0}$ be a level sets partition of θ. Then $\{\theta_{\sigma_n}\}_{n \geq 1}$ is an FI-sequence of functions.

Corollary 2.2 Let $\theta = \prod_{n\geq 1}\vartheta_n$, $\vartheta_n \in H^\infty$, and let the level sets $L(\vartheta_n, \varepsilon) \cap \mathbb{D}$ be connected for every ε, $0 < \varepsilon < 1$ (for example, this is the case if $\vartheta_n = b_{\lambda_n}^{k_n}$, or $\vartheta_n = \exp(-a_n(\zeta_n + z)(\zeta_n - z)^{-1})$, or $\vartheta_n = (\varepsilon_n \chi_{e_n} + \chi_{\mathbb{T}\setminus e_n})^{\text{out}}$ for an arc $e_n \subset \mathbb{T}$). Then every level sets partition θ_{σ_n}, $n \geq 1$, is equivalent to a grouping of ϑ_k: $|\theta_{\sigma_n}| \asymp |\prod_{k\in A_n}\vartheta_k|$, $n = 1, 2, \ldots$, where $\{A_n\}$ is a partition of the natural numbers.

In fact, every $\mathbb{D} \cap L(\vartheta_k, \varepsilon)$ is contained precisely in one component Ω_n of $\mathbb{D} \cap L(\theta, \varepsilon)$. ∎

Before proving Theorem 2.1 we establish an auxiliary result on outer functions on simply connected domains with nontrivial boundaries (i.e., conformally equivalent to the unit disc $\mathbb{D}$). Perhaps it will be useful for other purposes too. For a domain, say Ω, of the type mentioned, all notions have a meaning provided they have it for the unit disc $\mathbb{D}$. For example, one can conformally transplant (from $\mathbb{D}$) to Ω the notions (and properties) of the Nevanlinna class, outer and inner functions, singular inner functions, and so on. Now we prefer to deal with logarithms $\log|f|$ and speak about outer and singular harmonic functions on Ω. By the way, it is easy to see (and is well-known) that a nonnegative harmonic function u on Ω is outer if and only if there exists a sequence $\{u_n\}_{n\geq 1}$ of bounded harmonic functions such that $0 \leq u_n \uparrow u$ on Ω. Two useful corollaries follow from this description: (1) if u is outer on Ω and $\Omega' \subset \Omega$, $u \mid \Omega'$ is outer, too; and (2) if u is outer on Ω and ω is an analytic function on Ω' with $\omega(\Omega') \subset \Omega$, $u \circ \omega$ is outer, too. Moreover, from the Nevanlinna–Smirnov canonical factorization on $\mathbb{D}$ it follows that the property of being outer is local: if all restrictions $f \mid (\mathcal{O}_\zeta \cap \Omega)$, where ζ runs over the boundary $\partial\Omega$ and $\mathcal{O}_\zeta$ is an open neighborhood of ζ, are outer, so is f on Ω.

The next lemma will be useful in the following to check that some functions that occur are outer.

Lemma 2.3 Let Ω be a domain conformally equivalent to the disc $\mathbb{D}$ and $\mathcal{X}$, μ a measurable space, $\mu\mathcal{X} < \infty$. Further, u_x ($x \in \mathcal{X}$) is a nonnegative harmonic function on Ω, and the function $x \mapsto u_x(z)$ is measurable for every $z \in \Omega$ and is summable for at most one of them $\int_{\mathcal{X}} u_x(z_0)\, d\mu(x) < \infty$. Then the function u,

$$u(z) = \int_{\mathcal{X}} u_x(z)\, d\mu(x), \qquad z \in \Omega$$

is harmonic and outer provided all $u_x, x \in \mathcal{X}$, are the same.

Proof. By Harnack's inequality, the integral defining $u(z)$ converges uniformly on compact subsets of Ω. Hence u is harmonic. Without loss of generality we can deal with the unit disc case $\Omega = \mathbb{D}$. Then, if u_x is outer, there exists a function f_x on $\mathbb{T}$ such that $f_x \geq 0$, $f_x \in L^1(m)$, and $u_x(z) = f_x * P_z$, where P_z stands for the Poisson kernel, and $*$ is a convolution sign (on the group $\mathbb{T}$). It is easy to see that measurability of functions $x \mapsto f_x * P_z$, $z \in \mathbb{D}$, implies successively the measurability of the following ones: $x \mapsto f_x$ (in the weak $\sigma(L^1, C)$ sense); $x \mapsto |f_x - g| * P_z$ [for every $g \in L^1(m)$]; $x \mapsto f_x^{[n]} * P_z$, $n = 1, 2, \ldots$, where $f^{[n]}$ stands for the cutoff of a function: $f^{[n]} = \min(f, n)$.

Hence the functions

$$u_n(z) \equiv \int_{\mathcal{X}} f_s^{[n]} * P_z \, d\mu(x)$$

are harmonic and bounded and $0 \le u_n \uparrow u$. ■

Corollary 2.4 Let Ω, Ω' be domains conformally equivalent to the disc $\mathbb{D}$, let G be the Green's function of Ω' and μ be a finite Borel measure on $\Omega' \setminus \Omega$. Then the function u,

$$u(z) = \int_{\Omega' \setminus \Omega} G(z, \zeta) \, d\mu(\zeta), \qquad z \in \Omega$$

is outer.

In fact, it is enough to prove that $G(\cdot, \zeta)$ is outer on Ω provided $\zeta \in \Omega' \setminus \Omega$, or—after conformal transplantation—that $\log |(1 - \bar{\zeta} z)(\zeta - z)^{-1}|$ is outer on $\Omega'' \subset \mathbb{D}$ provided $\zeta \in \mathbb{D} \setminus \Omega''$. After one more conformal mapping, consider $\log |z|^{-1}$ on $\Omega''' \subset \mathbb{D}$, $0 \notin \Omega'''$, or simply the identity map z on Ω'''. It is equivalent to consider the Riemann mapping $\omega : \mathbb{D} \to \Omega'''$. But the latter is outer because $\omega \in H^p$, $1/\omega \in H^p$ for $0 < p < 1/2$, by a theorem of A. Baernstein (1976). ■

A particular case of Corollary 2.4 can be found in Nikolskii (1974, §2.1, Sec. 4).

Corollary 2.5 If B is a Blaschke product and Ω is an arbitrary simply connected domain, $\Omega \subset \mathbb{D}$ and if $B^{-1}(0) \cap \Omega = \varnothing$ then $B \mid \Omega$ is an outer function.

Corollary 2.6 Let e be a Borel subset of the circle $\mathbb{T}$, μ be a finite Borel measure on e, and Ω be a simply connected domain, $\Omega \subset \mathbb{D}$ satisfying the following condition: for every $\zeta \in e$ there exists a connected set $\gamma_\zeta \subset \mathbb{D} \setminus \Omega$ lying in a Stoltz angle $S(\zeta, \alpha, \varepsilon)$ and "ending" at ζ ($\zeta \in \operatorname{clos} \gamma_\zeta$). Then the function f,

$$f = \exp \left[- \int_e \frac{\zeta + z}{\zeta - z} \, d\mu(\zeta) \right],$$

is outer on Ω.

In fact, considering $\log |f|$ and applying Lemma 2.3 we reduce the problem to the Poisson kernel $(1 - |z|^2)|\zeta - z|^{-2}$, $\zeta \in e$ or to the simplest singular inner function $f_\zeta = \exp(-(\zeta + z)(\zeta - z)^{-1})$, $\zeta \in e$. This latter is holomorphic in the whole complex plane $\mathbb{C}$, except $z = \zeta$, and so is outer on a neighborhood of every point from $\partial\Omega \setminus \{\zeta\}$. As to ζ, it is easy to see (and is well known) that f_ζ is outer in every angle $\{z : 0 < \arg(\zeta - z) < \pi/2 + \alpha\}$ and $\{z : \alpha < \arg(\zeta - z) < \pi\}$ for $0 < \alpha < \pi/2$, and hence in every (connected component of a) neighborhood $\mathcal{O}_\zeta \cap \Omega$. Thus, f_ζ is outer on Ω, and the corollary follows. ■

Proof of Theorem 2.1. In view of Theorem A it is enough to check condition 1.4 (for $\theta_n = \theta_{\sigma_n}$).

For $z \in \mathbb{D} \setminus L(\theta, \varepsilon)$ the condition is obvious. To check it for $z \in L(\theta, \varepsilon)$ let us consider the restriction $\varphi = \theta_{\sigma_n}\theta^{-1} \mid \Omega_n$ to a component Ω_n. Note that on the inner part of the boundary, at almost all (with respect to m) points ζ of $\mathbb{T}$ which are angular limits of a sequence from Ω_n, the function φ has boundary values $\varphi(\zeta)$ and $|\varphi(\zeta)| = 1$ if $\zeta \in \Omega_n^*$, $|\varphi(\zeta)| \leq \varepsilon^{-1}$ if $\zeta \in \Omega_0^*$ (obviously, $\zeta \notin \Omega_k^*$, $k \neq n$, 0 for almost all such points). Moreover φ is an outer function on Ω_n. In fact, it is obvious for $\theta_{\sigma_n}^{\text{out}}/\theta^{\text{out}}$ and follows, respectively, from Corollary 2.5 and 2.6 for B_{σ_n}/B and $\theta_{\sigma_n}^s/\theta^s$ (i.e., for the Blaschke product part and for the singular inner part of θ).

Now, one can apply an elementary maximum principle (Lemma 2.7), which implies $|\theta_{\sigma_n}| \leq \varepsilon^{-1}|\theta|$ on Ω_n.

Condition 1.4 (and hence the theorem) follows. ∎

The maximum principle we just referred to is as follows [see Nikolskii and Hruščëv (1987) for a simple proof].

Lemma 2.7 Let Ω be a simply connected domain, $\Omega \subset \mathbb{D}$ and let φ be a function from the Nevanlinna–Smirnov class $N^+(\Omega)$ (= the set of all quotients f/g with $f, g \in H^\infty(\Omega)$ and with g outer) which is continuous on $\mathbb{D} \cap \partial\Omega$. If $|\varphi| \leq 1$ on $\mathbb{D} \cap \partial\Omega$ and if at almost all points ζ of $\mathbb{T}$ which are angular limits of Ω we have $\limsup |\varphi(z)| \leq 1$ ($z \to \zeta$ nontangentially, $z \in \Omega$), then $|\varphi| \leq 1$ on Ω.

III. APPROXIMATE FREE INTERPOLATION

A. Free Interpolating Arcs Systems

Now we turn to a partial case of the general free interpolation problem: to find necessary and sufficient conditions on an arcs system $\{\sigma_k\}_{k\geq 1}$, $\sigma_k \subset \mathbb{T}$ and a sequence of positive numbers $\{\varepsilon_k\}_{k>1}$, $\lim_k \varepsilon_k = 0$, to meet property (1.18), which means that for every $f_k \in H^\infty$, $\sup_k \|f_k\|_\infty < \infty$, there exists an H^∞ function f satisfying the inequalities

$$|f(\zeta) - f_k(\zeta)| \leq c\varepsilon_k \text{ a.e.} \qquad \zeta \in \sigma_k, k = 1, 2, \ldots.$$

In fact, c can be chosen such that $c = c(\delta)\sup_k \|f_k\|_\infty$ where $c(\delta)$ depends only on δ, from the generalized Carleson condition (CV) (see Section I.A). It is also easy to see that without loss of generality we can suppose $\sigma_k \cap \sigma_j = \varnothing, k \neq j$.

First, one has to rewrite slightly the generalized Carleson condition for the corresponding sequence of weights

$$w_n = \varepsilon_n \chi_{\sigma_n} + \chi_{\mathbb{T}\setminus\sigma_n}; \qquad n \geq 1. \tag{3.1}$$

Lemma 3.2 Let σ_n, $n = 1, 2, \ldots$, be pairwise disjoint Borel subsets of the unit circle $\mathbb{T}$ and w_n be weight sequence (3.1). The following are equivalent.

$$\{w_n^{\text{out}}\}_{n\geq 1} \text{ is an FI sequence.} \tag{3.3}$$

$$\inf_n \sum_{k,k\neq n} \omega(z,\sigma_k)\log\varepsilon_k^{-1} \leq \text{const}, \qquad z \in \mathbb{D}, \tag{3.4}$$

where $\omega(z,\sigma_k)$ stands for the harmonic measure of σ_k at a point $z \in \mathbb{D}$.

$$\sup_{z\in\mathbb{D}}\left(\sum_{k\geq 1}\omega(z,\sigma_k)\log\varepsilon_k^{-1} - \max_{k\geq 1}\omega(z,\sigma_k)\log\varepsilon_k^{-1}\right) < +\infty. \tag{3.5}$$

$$\min\left(\sum_{k\in K}\omega(z,\sigma_k)\log\varepsilon_k^{-1}, \sum_{k\in K'}\omega(z,\sigma_k)\log\varepsilon_k^{-1}\right) \leq \text{const} \tag{3.6}$$

with a constant independent of $z \in \mathbb{D}$ and on a subset $K \subset \mathbb{N}$.

Proof. The equivalence (3.3) $\Leftrightarrow$ (3.4) follows from Theorem B if we note that $W_n = \sum_{k,k\neq n}\varepsilon_k\chi_{\sigma_k} + \chi_n$, where χ_n is the characteristic function of $\mathbb{T}\setminus\bigcup_{k,k\neq n}\sigma_k$, and further that

$$\begin{aligned} |W_n^{\text{out}}(z)| &= \exp\left(-\sum_{k,k\neq n}\log\varepsilon_k^{-1}\int_{\sigma_k}\frac{1-|z|^2}{|\zeta - z|^2}\,dm(\zeta)\right) \\ &= \exp\left(-\sum_{k,k\neq n}\omega(z,\sigma_k)\,\log\varepsilon_k^{-1}\right). \end{aligned}$$

(3.4) $\Leftrightarrow$ (3.5) because for any sequence of real functions f_k we get

$$\inf_{n\geq 1}\sum_{k,k\neq n} f_k(z) = \sum_{k\geq 1} f_k(z) - \sup_{k\geq 1} f_k(z) \tag{3.7}$$

provided one side of this equality has a meaning.

The equivalence (3.5) $\Leftrightarrow$ (3.6) follows from elementary Lemma 3.8, a sharpened form of a lemma from Nikolskii (1987). ∎

Lemma 3.8 Let $p_k \in \mathbb{R}$, $p_k \geq 0$, and $p = \sum_k p_k$, $p_0 = \max_k p_k$. Then

$$\frac{1}{2}(p - p_0) \leq \sup_{k\subset\mathbb{N}}\min\left(\sum_{k\in K} p_k, \sum_{k\in K'} p_k\right) \leq p - p_0$$

(and both estimations are the best possible).

Proof. The right inequality is obvious because $p_0 = p_k$ for some k, and if, for example, $k \in K$ we have $\sum_{k \in K'} p_k \leq p - p_0$ and $\min(\sum_K, \sum_{K'}) \leq p - p_0$ for every K.

It is sufficient to check the left inequality for finite sequences $p_1, \ldots, p_n$ (with $K \subset [1, \ldots n]$, $K' = [1, \ldots n] \setminus K$). Let K_0, K_0' be an extreme partition, i.e.,

$$c \overset{\text{def}}{=} \min\left(\sum_{K_0} p_k, \sum_{K_0'} p_k\right) \geq \min\left(\sum_{K} p_k, \sum_{K'} p_k\right) \tag{3.9}$$

for every K. Say, $c = \sum_{K_0} p_k$. Since $K_0' = \varnothing$ (otherwise, $p_k = 0$ for all k with no more than one exception, and then all is obvious), let $\kappa \in K_0', p_\kappa > 0$. If $\sum_{K_0 \cup \kappa} p_k \leq \sum_{K_0' \setminus \kappa} p_k$, we get from (3.9) $c \geq c + p_\kappa$. Thus $\sum_{K_0 \cup \kappa} p_k > \sum_{K_0' \setminus \kappa} p_k$, and so $p - c - p_\kappa = \sum_{K'} p_k - p_\kappa = \sum_{K' \setminus \kappa} p_k \leq$ [by (3.9)] $\leq c$. This means $2c \geq p - p_\kappa \geq p - p_0$. ■

In the case when one and then all of (3.3)–(3.6) are fulfilled we refer to $(\sigma_n, \varepsilon_n)$ as the *approximate free interpolation system* (AFI). To get a notion of the geometric nature of AFI systems we consider in detail a specific case when all σ_n are intervals (arcs). For an interval (arc) σ of $\mathbb{T}$ and for real $R, R \geq 1$, let us define the disc $D(R\sigma)$ of radius $R|\sigma|$, the boundary of which meets σ at the endpoints of σ and which lies as is shown in Fig. 1.

Corollary 3.10 If $(\sigma_n, \varepsilon_n)$ is an AFI system, then

(3.11) $\sum_{n \geq 1} |\sigma_n| \log \varepsilon_n^{-1} < \infty$.

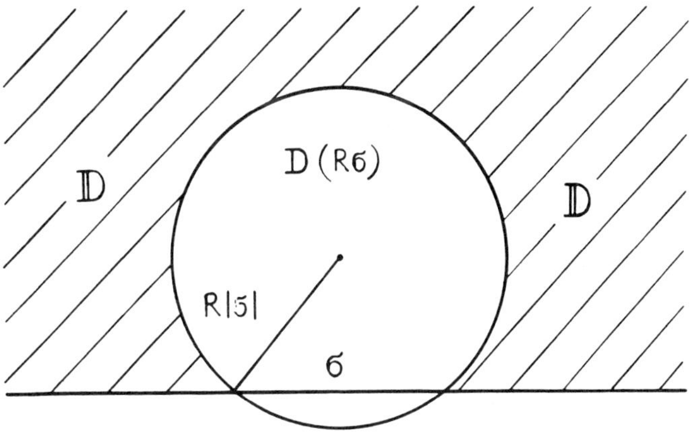

Fig. 1 The disc $D(R\sigma)$.

(3.12) $\{w_n^{\text{out}}\}_{n\geq 1}$ is a sparse sequence, i.e., there exists $A > 0$ such that

$$\{z \in \mathbb{D}: \omega(z, \sigma_n) > A/\log \varepsilon_n^{-1}\} \cap \left\{z \in \mathbb{D}: \omega(z, \sigma_k) > A/\log \varepsilon_k^{-1}\right\} = \varnothing, \qquad n \neq k$$

(3.13) If all σ_n are intervals (arcs), there exists $a > 0$ such that

$$D(a \log \varepsilon_n^{-1} \sigma_n) \cap D(a \log \varepsilon_k^{-1} \sigma_k) = \varnothing, \qquad n \neq k$$

In fact, (3.11) is obvious, (3.12) follows from Lemma 1.9 and (3.13) from (3.12) if we mention that

$$\text{(3.14)} \qquad \begin{aligned} &\omega(z, \sigma) < BR^{-1} && \text{for } z \in \mathbb{D} \setminus D(R\sigma) \\ &\omega(z, \sigma) > bR^{-1} && \text{for } z \in \mathbb{D} \cap D(R\sigma) \end{aligned}$$

for some absolute constants b and B. ■

Lemma 3.2 implies the following simple [but rough, see (3.57), (3.62)] geometric sufficient condition for a system of arcs to be an AFI system.

Corollary 3.15 Let σ_n be arcs of $\mathbb{T}$ and $\varepsilon_n > 0$ ($n = 1, 2, \ldots$). If the discs $D(L_n \log \varepsilon_n^{-1} \sigma_n)$ are pairwise disjoint

$$D(L_n \log \varepsilon_n^{-1} \sigma_n) \cap D(L_k \log \varepsilon_k^{-1} = \varnothing, \qquad n \neq k$$

and if $\sum_{n\geq 1} L_n^{-1} < \infty$, the conditions (3.3)–(3.6) are fulfilled [and so $(\sigma_n, \varepsilon_n)$ is an AFI system].

Indeed, from (3.14) we get $\omega(z, \sigma_k) \leq B/(L_k \log \varepsilon_k^{-1})$ for $z \in \mathbb{D} \setminus D(L_k \log \varepsilon_k^{-1} \sigma_k)$ and hence

$$\inf_n \sum_{k, k\neq n} \omega(z, \sigma_k) \log \varepsilon_k^{-1} \leq B \sum_{k\geq 1} L_k^{-1}, \qquad z \in \mathbb{D} \quad \blacksquare$$

To state a criterion for AFI we need a notation: $\lambda_k = (1 - |\sigma_k|)\zeta_k$, where ζ_k is the center of σ_k. *The Stoltz domain* $S_{\zeta,\alpha}$, $\zeta \in \mathbb{T}$, $0 < \alpha < \pi/2$, is the convex hull of $\{\zeta\}$ and the disc $\{z: |z| \leq \sin \alpha\}$.

Theorem 3.16 Let σ_n be arcs of $\mathbb{T}$ and $\varepsilon_n > 0$, $n = 1, 2, \ldots$. The following statements are equivalent.

(3.17) $\{(\sigma_n, \varepsilon_n)\}$ is an AFI system.

(3.18) The sequence of Blaschke factors $b_{\lambda_k}^{m_k}$, $m_k = [\log \varepsilon_k^{-1}]$. ($[x]$ is the greatest integer in x) is an FI-sequence and $\{(\sigma_n, \varepsilon_n)\}$ is sparse in the sense of (3.13).

(3.19) The sequence of inner functions θ_{ζ_k},

$$\theta_{\zeta_k}(z) = \exp(-|\sigma_k| \log \varepsilon_k^{-1}(\zeta_k + z)(\zeta_k - z)^{-1})$$

is an FI-sequence and $\{(\sigma_n, \varepsilon_n)\}$ is sparse in the sense of (3.13).

Proof. By Lemma 3.2, (3.17) is equivalent to (3.5). Taking into account condition (CV), it is easy to see that (3.18) is equivalent to the conjunction of (3.13) and

$$\text{(3.20)} \qquad \sup_{\mathbb{D}} \left(\sum_k |\sigma_k| \log \varepsilon_k^{-1} \frac{1 - |z|^2}{|z - \lambda_k|^2} - \max_k |\sigma_k| \log \varepsilon_k^{-1} \frac{1 - |z|^2}{|z - \lambda_k|^2} \right) < \infty.$$

Similarly, (3.19) is equivalent to the conjunction of (3.13) and

$$\text{(3.21)} \qquad \sup_{\mathbb{D}} \left(\sum_k |\sigma_k| \log \varepsilon_k^{-1} \frac{1 - |z|^2}{|z - \zeta_k|^2} - \max_k |\sigma_k| \log \varepsilon_k^{-1} \frac{1 - |z|^2}{|z - \zeta_k|^2} \right) < \infty$$

Now we prove that (3.17) $\Rightarrow$ (3.20). Let $n(z) = k$ if $z \in D(a \log \varepsilon_k^{-1} \sigma_k)$ and if such a k exists, and $n(z) = \infty$ otherwise. By virtue of (3.13), $n(z)$ is well defined. Taking (3.14) into account we see that (3.17) implies

$$\text{(3.22)} \qquad \sum_{k \neq n(z)} \log \varepsilon_k^{-1} \int_{\sigma_k} \frac{1 - |z|^2}{|z - \zeta|^2} dm(\zeta) \leq \text{const}, \qquad z \in \mathbb{D}$$

It is obvious that

$$\text{(3.23)} \quad |\sigma_k| \leq 2^{-1}(1 - |z|) \Longrightarrow a_2 \max_{\zeta \in \sigma_k} |z - \zeta| \leq |z - \lambda_k| \leq a_1 \min_{\zeta \in \sigma_k} |z - \zeta|.$$

There exists α, $0 < \alpha < \pi/2$, such that

$$\text{(3.24)} \qquad \lambda_k \in S_{\zeta,\alpha} \Longrightarrow \zeta \in \sigma_k.$$

Then for $z \in \mathbb{D}$ such that $2^{-1}(1 - |z|) \leq |\sigma_k|$, $k \neq n(z)$, $k \geq k_0$, we have $\lambda_k \notin S_{z/|z|,\alpha}$. In fact, if $\lambda_k \in S_{z/|z|,\alpha}$ one has $z/|z| \in \sigma_k$ by virtue of (3.24). Taking into account that $1 - |z| \leq 2|\sigma_k|$, we conclude that $\omega(z, \sigma_k) \geq \delta_0 > 0$, where δ_0 is an absolute constant. If k_0 is chosen in such a way that $B/\log \varepsilon_k^{-1} < \delta_0$, $k \geq k_0$, we see that

$z \in D(a \log \varepsilon_k^{-1} \sigma_k)$, which contradicts $k \neq n(z)$. It is obvious that

$$(3.25) \qquad 2^{-1}(1-|z|) \leq |\sigma_k|, \lambda_k \notin S_{z/|z|,\alpha} \Longrightarrow a_2 \max_{\zeta \in \sigma_k} |z-\zeta| \leq |z-\lambda_k|.$$

Now (3.22), (3.23) and (3.25) imply

$$\sum_{k \geq k_0, k \neq n(z)} |\sigma_k| \log \varepsilon_k^{-1} \frac{1-|z|^2}{|z-\lambda_k|^2} \leq a_2^{-1} c$$

and so (3.20) holds because k_0 is independent of z.

(3.17) $\Rightarrow$ (3.21). The proof of this implication proceeds along just the same lines.

(3.20), (3.13) $\Rightarrow$ (3.17). First note that the Harnack inequality and (3.14) imply the existence of $\tau = \tau(B, a) > 0$ such that

$$(3.26) \qquad D_k \stackrel{\text{def}}{=} \{z \colon |b_{\lambda_k}(z)|^{-1} \geq 1 + \tau(\log \varepsilon_k^{-1})^{-1}\} \subset D(a \log \varepsilon_k^{-1} \sigma_k).$$

Let $n'(z) = k$ if $z \in D_k$ and if such a k exists, and $n'(z) = \infty$ otherwise. By virtue of (3.13) D_k are pairwise disjoint and so the definition of $n'(z)$ is correct.

Now we will verify that (3.20) implies

$$(3.27) \qquad \sum_{k \neq n'(z)} |\sigma_k| \log \varepsilon_k^{-1} \frac{1-|z|^2}{|z-\lambda_k|^2} \leq \text{const}, \qquad z \in \mathbb{D}$$

Let $m(z) = k$ if $|\sigma_k| \log \varepsilon_k^{-1}(1-|z|^2)|z-\lambda_k|^{-2} = \max_{k \geq 1} |\sigma_k| \log \varepsilon_k^{-1}(1-|z|^2)|z-\lambda_k|^{-2}$. For $z \in \mathbb{D}$ such that $m(z) = n'(z)$ there is nothing to prove. So let $m(z) \neq n'(z)$. In this case (3.27) follows from the assertion

$$(3.28) \qquad z \in D_k \Longrightarrow |\sigma_k| \log \varepsilon_k^{-1}(1-|z|^2)|z-\lambda_k|^{-2} \leq \text{const}.$$

By the standard chain of inequalities we have (putting $\eta' = \min_k(1 + \tau/\log \varepsilon_k^{-1})^{-1}$) for $z \notin D_k$,

$$\begin{aligned} |\sigma_k| \log \varepsilon_k^{-1} \frac{1-|z|^2}{|z-\lambda_k|^2} &\leq \log \varepsilon_k^{-1} \frac{(1-|\lambda_k|^2)(1-|z|^2)}{|z-\lambda_k|^2} \\ &\leq C(\eta) \log \varepsilon_k^{-1} \log \left| \frac{1-\bar{\lambda}_k z}{z-\lambda_k} \right|^2 = C(\eta) \log |b_{\lambda_k}(z)|^{\log \varepsilon_k} \\ &\leq C(\eta) \log(1 + \tau/\log \varepsilon_k^{-1})^{\log \varepsilon_k^{-1}} \leq \text{const}, \end{aligned}$$

and this is (3.28). From (3.27) it immediately follows that

$$(3.29) \qquad \sum_{k \neq n(z)} |\sigma_k| \log \varepsilon_k^{-1}(1-|z|^2)|z-\lambda_k|^{-2} \leq \text{const.}$$

It remains to obtain (3.22), but unfortunately $\min_{\zeta \in \sigma_n} |z-\zeta| \cdot |z-\lambda_n|^{-1}$ can be very small even if $\lambda_k \notin S_{z/|z|,\alpha}$. But there exists β, $\alpha < \beta < \pi/2$, such that

$$(3.30) \qquad \lambda_k \notin S_{z/|z|,\beta} \Longrightarrow |z-\lambda_k| \leq a_1 \min_{\zeta \in \sigma_k} |z-\zeta|.$$

At the same time

$$(3.31) \qquad 2(1-|z|) \geq |\sigma_k| \Longrightarrow |z-\lambda_k| \leq a_1 \min_{\zeta \in \sigma_k} |z-\zeta|.$$

Taking (3.29), (3.30), and (3.31) into account we see that it remains to prove the estimate

$$\sum_{\substack{k \neq n(z), |\sigma_k| \geq 2(1-|z|) \\ \lambda_k \in S_{z/|z|},\beta}} \omega(z,\sigma_k) \log \varepsilon_k^{-1} \leq \text{const.}$$

If $z/|z| \in \sigma_k$, $|\sigma_k| \geq 2(1-|z|)$, then $\omega(z,\sigma_k) \geq \delta_0 > 0$, where δ_0 is an absolute constant. If $k \geq k_0$ this implies [by virtue of (3.13)] that $z \in D(a \log \varepsilon_k^{-1} \sigma_k)$ (here as above k_0 is such that $B/\log \varepsilon_k^{-1} < \delta_0$, $k \geq k_0$), which contradicts $k \neq n(z)$. Let

$$A_z = \{k : k \neq n(z), \lambda_k \in S_{z/|z|,\beta}, |\sigma_k| \geq 2(1-|z|), k \geq k_0\}.$$

Let $A_z^+(A_z^-)$ consist of $k \in A_z$ for which σ_k lies on the right (left) of $z/|z|$ (do not forget to stop somewhere since we are on the circle). It remains to prove

$$(3.32) \qquad \sum_{k \in A_z^-} \omega(z,\sigma_k) \log \varepsilon_k^{-1} \leq \text{const},$$

together with the analogous assertion with A_z^- replaced by A_z^+.

It is clear that for $k \in A_z$, $k \geq k_1 = k_1(\beta)$, the intersection $[0, z/|z|] \cap \partial D(a \log \varepsilon_k^{-1} \sigma_k)$ consists of two points. Let z_k denote the one which is nearest to $z/|z|$. Note that $1-|z_k| > 1-|z|$ if $k \in A_z$, $k \geq k_2$ (k_2 depends only on $\{\varepsilon_k\}$) because otherwise we would have $z \in D(a \log \varepsilon_k^{-1} \sigma_k)$, which contradicts $k \neq n(z)$. Let d_k be the end of σ_k nearest to $z/|z|$. We will use the following two observations:

There is $k_3 \geq \max(k_1, k_2)$ depending only on $\{\varepsilon_k\}$ and β such that

$$k \in A_z, k \geq k_3 \Longrightarrow \omega(z, \sigma_k) \leq a_3 \frac{1 - |z|}{|z/|z| - d_k|},$$

$$k \in A_z, k \geq k \Longrightarrow \omega(z_k, \sigma_k) \geq a_4 \frac{1 - |z_k|}{|z/|z| - d_k|}.$$

These inequalities follow easily from simple geometric considerations. They imply that (as $\log \varepsilon_k^{-1} \omega(z_k, \sigma_k) \leq B$)

$$k \in A_z^-, k \geq k_3 \Longrightarrow \omega(z, \sigma_k) \log \varepsilon_k^{-1} \leq B a_3 a_4^{-1} \frac{1 - |z|}{1 - |z_k|}.$$

Taking this into account we see that to check (3.32) it is sufficient to prove that

$$\sum_{k \in A_z^-, k \geq k_3} (1 - |z|)(1 - |z_k|)^{-1} \leq \text{const.} \tag{3.33}$$

Let $w_1, \ldots, w_N$ be all the points of the set $\{z_k : k \in A_z^-, k \geq k_3\}$ enumerated in such a way that $|w_1| > |w_2| > \cdots > |w_N|$. Note that between $1 - |w_i|$ and $1 - |w_{i+1}|$ there is one and only one point $1 - \lambda_k$, $k \in A_z^-$, $k \geq k_3$. We denote this λ_k by μ_i:

$$1 - |w_i| < 1 - |\mu_i| < 1 - |w_{i+1}|. \tag{3.34}$$

It is clear that this enumeration corresponds to the following enumeration of $\{\sigma_k : k \in A_z^-, k \geq k_3\}$: μ_1 corresponds to the σ_k which is nearest to $z/|z|$, μ_2 corresponds to the σ_k which follows the nearest to $z/|z|$, and so on.

Now we estimate the sum in (3.33) with the help of (3.34):

$$\sum \frac{1 - |z|}{1 - |z_k|} \leq \sum_{1 \leq i \leq N} \frac{1 - |z|}{1 - |w_i|} \leq 2 + \sum_{2 \leq i \leq N} \frac{1 - |\mu_1|}{1 - |\mu_i|} \leq 2 + \sum_{2 \leq i \leq N} \prod_{1 \leq l \leq i-1} \frac{1 - |\mu_l|}{1 - |\mu_{l+1}|}.$$

The last sum is bounded by a constant $C(\beta)$ as the following observation shows:

If σ_1, σ_2 are disjoint arcs of a half-circle T, $T \subset \mathbb{T}$, $|\sigma_2| \geq |\sigma_1|$, $\lambda_i = (1 - |\sigma_i|)\zeta_i$, where ζ_i are the centers of σ_i, $i = 1, 2$, and if $\zeta \notin T$, $\lambda_1, \lambda_2 \in S_{\zeta,\beta}$, then $(1 - |\lambda_1|)(1 - |\lambda_2|)^{-1} \leq q = q(\beta) < 1$.

This finishes the proof of (3.33) and thus the proof of the implication (3.20), (3.13) $\Rightarrow$ (3.17).

The implication (3.21), (3.13) $\Rightarrow$ (3.17) can be proved analogously. ■

B. Slightly Interpolating Sequences

As in the previous section we suppose that intervals σ_n are pairwise disjoint. We say that $\{\sigma_n\}$ is a *slightly interpolating sequence* if $\{(\sigma_n, \varepsilon_n)\}$ is an AFI-sequence with some $\{\varepsilon_n\}$, $\lim_n \varepsilon_n = 0$. It follows from (3.4) that $\{\sigma_n\}$ satisfies this condition if and only if

(3.35) There exists $\{c_k\}$, $\lim_k c_k = \infty$, such that $\inf_n \sum_{k \neq n} c_k \omega(z, \sigma_k) \leq$ const, $z \in \mathbb{D}$.

We say that a sequence of points $\{\lambda_k\}$, $\lambda_k \in \mathbb{D}$, is a *strong interpolating sequence* if

$$\lim_n \prod_{k \neq n} |b_{\lambda_k}(\lambda_n)| = 1. \tag{3.36}$$

Strong interpolating sequences have some remarkable properties, [e.g., see Sundberg and Wolff (1983)]. In what follows $|\lambda_k| = 1 - |\sigma_k|$, $\lambda_k/|\lambda_k|$ is the center of σ_k. We say that $\{\sigma_k\}$ is a *slightly sparse sequence* if

(3.37) there exists $\{R_k\}$, $\lim_k R_k = \infty$, such that $D(R_k\sigma_k) \cap D(R_n\sigma_n) = \varnothing$, $k \neq n$.

Surely, (3.35) implies (3.37).

Theorem 3.38 The following are equivalent.

(3.39) $\{\sigma_k\}$ is a slightly interpolating sequence.
(3.40) $\{\sigma_k\}$ is a slightly sparse sequence and $\{\lambda_k\}$ is a strong interpolating sequence.
(3.41) $\lim_{|z|\to 1}(\sum_{k\geq 1} \omega(z, \sigma_k) - \max_{k\geq 1} \omega(z, \sigma_k)) = 0$.

Proof. The implication (3.39) $\Rightarrow$ (3.40) is an immediate consequence of Theorem 3.16. In fact, (3.39) $\Leftrightarrow$ (3.35) implies that $\prod_{k\neq n} |b_{\lambda_k}(\lambda_n)|^{m_k} \geq \delta > 0$ with some m_k, $m_k \uparrow \infty$. So (3.36) holds. To prove the other implications we need two lemmas.

Lemma 3.42 Let f_k be nonnegative functions on a set $\mathcal{X}$. The following are equivalent.

(3.43) There exist $c_k \geq 0$, $k = 1, 2, \ldots$, such that $\lim_k c_k = \infty$ and $c \equiv \sup_{x\in\mathcal{X}} \inf_n \sum_{k, k\neq n} c_k f_k(x) < \infty$.
(3.44) For every $x \in \mathcal{X}$ there exists a natural number $n(x)$ such that

$$\lim_n \sup_{x\in\mathcal{X}} \sum_{k\geq n, k\neq n(x)} f_k(x) = 0.$$

Proof. (3.43) $\Rightarrow$ (3.44). The assumption implies that $\sum_{k\geq 1} c_k f_k(x) < \infty$ for every $x \in \mathcal{X}$. Let $n(x)$ satisfy $c_{n(x)} f_{n(x)} = \max_k c_k f_k(x)$. Then

$$(\min_{k\geq n} c_k) \sum_{k\geq n, k\neq n(x)} f_k(x) \leq \sum_{k\geq 1, k\neq n(x)} c_k f_k(x) \leq c, \qquad x \in \mathcal{X}.$$

This results in condition (3.44).
(3.44) $\Rightarrow$ (3.43). Let $a_s = \sup_{x\in\mathcal{X}} \sum_{k\geq n_s, k\neq n(x)} f_k(x)$, and let $n_s \uparrow +\infty$ so fast that $\sum_{s\geq 1} a_s < \infty$. Then let $b_s \uparrow +\infty$, $\sum a_s b_s < \infty$, and $c_k = b_s$ for $n_s \leq k < n_{s+1}$. In this case

$$\inf_n \sum_{k,k\neq n} c_k f_k(x) \leq \sum_{k,k\neq n(x)} c_k f_k(x) \leq \sum_{s\geq 1} b_s \sum_{n_s\leq k<n_{s+1}} f_k(x) \leq \sum a_s b_s$$

for every $x \in \mathcal{X}$, and (3.43) follows. ■

Lemma 3.45 Let f_k be nonnegative functions on a locally compact space $\mathcal{X}$, and let the series $\sum_{k\geq 1} f_k$ converge uniformly on every compact subset of $\mathcal{X}$. Then the conjunction of the first two conditions is equivalent to the third one.

(3.46) For every k $\lim_{x\to\infty} \min(f_i(x), f_k(x)) = 0$ uniformly in i, $i \neq k$.
(3.47) There exist $c_k > 0$, $\lim_k c_k = \infty$, such that

$$\sup_{x\in\mathcal{X}} \left(\sum_{k\geq 1} c_k f_k(x) - \max_{k\geq 1} c_k f_k(x) \right) < \infty.$$

(3.48) $\lim_{x\to\infty} (\sum_{k\geq 1} f_k(x) - \max_{k\geq 1} f_k(x)) = 0.$

Proof. (3.46) and (3.47) $\Rightarrow$ (3.48). Let $n(x)$ be a function of $x \in \mathcal{X}$ mentioned in (3.44). Then, by Lemma 3.42, for every $\varepsilon > 0$ there exists $n \in \mathbb{N}$ such that $\sum_{k\geq n, k\neq n(x)} f_k(x) < \varepsilon$ for all $x \in \mathcal{X}$. Let $X_{k,\varepsilon,n} = \{x \in \mathcal{X}: f_k(x) \geq \varepsilon/n\}$, $k = 1, \dots, n$. By virtue of (3.46) there exists a neighborhood $\mathcal{O}$ of ∞ such that

$$|f_m(x)| \leq \frac{\varepsilon}{n}, \qquad \forall x \in X_{k,\varepsilon,n} \cap \mathcal{O}, \forall m \neq k; k = 1, \dots, n.$$

So if $x \in \mathcal{O} \setminus \bigcup_{1\leq k\leq n} X_{k,\varepsilon,n}$, then

$$\sum_{m\geq 1} f_m(x) - \max_{m\geq 1} f_m(x) \leq \sum_{m\neq n(x)} f_m(x) \leq \sum_{1\leq m\leq n} + \sum_{m\geq n, m\neq n(x)} \leq 2\varepsilon.$$

If $x \in X_{k,\varepsilon,n} \cap \mathcal{O}$, $1 \le k \le n$, and $n(x) > n$, then

$$\sum_{m\ge 1} f_m(x) - \max_{m\ge 1} f_m(x) = \sum_{m\ne k} f_m(x) = \sum_{1\le m\le n, m\ne k} + \sum_{m\ge n, m\ne n(x)} + f_{n(x)}(x) \le \frac{\varepsilon}{n}(n-1) + \varepsilon + \frac{\varepsilon}{n} \le 2\varepsilon.$$

And if $n(x) \le n$, then

$$\sum_{m\ge 1} f_m(x) - \max_{m\ge 1} f_m(x) = \sum_{m\ne k} = \sum_{1\le m\le n, m\ne k} + \sum_{m\ge n, m\ne n(x)} < 2\varepsilon.$$

Condition (3.46) follows from (3.48) immediately. At the same time (3.48) and the uniform convergence of the series $\sum f_k$ on compact subsets of $\mathcal{X}$ imply (3.44). By virtue of Lemma 3.42 condition (3.47) holds, too. ∎

We return to the proof of Theorem 3.38.

(3.40) ⇒ (3.39). It is easy to see that (3.36) is equivalent to

$$\lim_{|z|\to 1}\left(\sum_{k\ge 1}\log|b_{\lambda_k}(z)|^{-1} - \max_{k\ge 1}\log|b_{\lambda_k}(z)|^{-1}\right) = 0. \tag{3.49}$$

So $\inf_{\mathbb{D}}\sup_n \prod_{k\ne n}|b_{\lambda_k}(z)|^{m_k} \ge \delta > 0$ for some m_k, $m_k \uparrow \infty$, by Lemma 3.45, and this shows that Blaschke factors $\{b_{\lambda_k}^{m_k}\}$ form an FI-sequence. Choosing ε_n in such a way that $[\log\varepsilon_n^{-1}] = m_n$ and using Theorem 3.16 we see that (3.39) holds.

(3.39) ⇔ (3.41). It follows from Lemma 3.45 that to prove the equivalence (3.41) ⇔ (3.35) [=(3.39)] it is sufficient to prove the implication

(3.50) (3.37) ⇒ for every k $\lim_{|z|\to 1}\min(\omega(z,\sigma_i),\omega(z,\sigma_k)) = 0$ uniformly in i, $i \ne k$.

Suppose the contrary. Let $\{z:\omega(z,\sigma_k) \ge \varepsilon\} \cap \{z:\omega(z,\sigma_{i_n}) \ge \varepsilon\} \cap \{z: 1 - 1/n \le |z| < 1\} \ne \varnothing$. It is clear that $\lim_n i_n = \infty$. The choice of n such that $\varepsilon > BR_{i_n}^{-1/2}$ [B is introduced in (3.14)] shows that $D(B\varepsilon^{-1}\sigma_k) \cap D(R_{i_n}^{1/2}\sigma_{i_n}) \ne \varnothing$. There exists $\varrho = \varrho(\varepsilon, R_k) < 1$ such that for every $z \in D(B\varepsilon^{-1}\sigma_k)$, the disc $D(z,\varrho) \equiv \{\zeta: |\zeta - z|/|1-\bar{\zeta}z| \le \varrho\}$ intersects $D(R_k\sigma_k)$. If R_{i_n} is sufficiently large, $R_{i_n} \ge R(\varrho)$, then $D(R_{i_n}\sigma_{i_n}) \supset \bigcup_{z\in D(R_{i_n}^{1/2}\sigma_{i_n})} D(z,\varrho)$. These observations show that for $n \ge n(\varepsilon)$, $\varnothing \ne D(R_k\sigma_k) \cap D(R_{i_n}\sigma_{i_n})$ which contradicts (3.37). ∎

Surely, there is a direct proof of Theorem 3.38 which does not refer to Theorem 3.16. For example, let us present a direct proof of (3.40) ⇒ (3.39). Let $k(z) = k$, $z \in D(R_k\sigma_k)$ if such a k exists, and $k(z) = \infty$ otherwise. It follows from (3.37) that

$k(z)$ is uniquely determined. By virtue of Lemma 3.42, (3.39) will follow if we prove that

$$\lim_n \sup_{\mathbb{D}} \sum_{k\leq n, k\neq k(z)} \omega(z,\sigma_k) = 0.$$

This will obviously follow from

$$\lim_n \lim_{|z|\to 1} \sum_{k\geq n, k\neq k(z)} \omega(z,\sigma_k) = 0. \tag{3.51}$$

Conditions (3.36) and (3.37) imply that

$$\lim_{|z|\to 1} \prod_{k\neq k(z)} |b_{\lambda_k}(z)| = 0. \tag{3.52}$$

It remains to verify that (3.37) and (3.52) $\Rightarrow$ (3.51).

Let $I_z = \{e^{i\varphi}: |z/|z| - e^{i\varphi}| \leq 1-|z|\}$. Let us split the collection of arcs $\{\sigma_k\}$ into two sets: $\sigma_k \in M$ if $|\sigma_k| \leq 1-|z|$; otherwise $\sigma_k \in Б$. Without loss of generality the sequence $\{R_k\}$ from (3.37) is nondecreasing. Assertion (3.51) follows from

$$\sum_{k\geq N, \sigma_k \in Б, k\neq k(z)} \omega(z,\sigma_k) \leq A/R_N^{1/2} \tag{3.53}$$

$$\lim_{|z|\to 1} \sum_{\sigma_k\in M, k\neq k(z)} \omega(z,\sigma_k) = 0. \tag{3.54}$$

First we prove (3.22). Two cases may occur.

First case: $k(z) = \infty$. In this case it follows from (3.52) that $|\prod_k b_{\lambda_k}(z)| \geq 1 - \varepsilon(|z|)$, where $\lim_{x\to 0}\varepsilon(x) = 0$. So

$$-2\log(1-\varepsilon(|z|)) \geq \log|B(z)|^{-2} = -\sum_{k\geq 1}\log\left(1-\frac{(1-|z|^2)(1-|\lambda_k|^2)}{|1-\bar{\lambda}_k z|^2}\right)$$

$$\geq \sum_{k\geq 1}\frac{(1-|z|^2)(1-|\lambda_k|^2)}{|1-\lambda_k z|^2} \geq \sum_{k\geq 1, \sigma_k\in M}.$$

But $a_2|1-\bar{\zeta}z| \leq |1-\bar{\lambda}_k z| \leq a_1|1-\bar{\zeta}z|$, $\zeta\in\sigma_k$, for every $\sigma_k \in M$. Therefore every member of the last sum can be replaced by $a_1^{-1}\int_{\sigma_k}(1-|z|^2)|1-\bar{\zeta}z|^{-2}\,dm(\zeta) = a_1^{-1}\omega(z,\sigma_k)$ where a_1 is an absolute constant. Thus $\sum_{k,\sigma_k\in M}\omega(z,\sigma_k) \leq 2a\log(1-\varepsilon(|z|))^{-1}$ and (3.54) is proved in the first case.

Second case: $k(z) < \infty$. Replacing $\prod b_{\lambda_k}$ by $\prod_{k\neq k(z)} b_{\lambda_k}$, and $\sum_{\sigma_k\in M}$ by $\sum_{\sigma_k\in M, k\neq k(z)}$ in the above considerations and applying (3.52) once again we prove (3.54) in this case too.

Now we prove (3.53). For an arc $I, I \subset \mathbb{T}$ and $\lambda > 0$, λI denotes as usual the arc with the same center such that $|\lambda I| = \lambda |I|$. We need the following geometric observation:

$$z \in \mathbb{D} \setminus D(R\sigma), 1 - |z| \leq |\sigma| \Longrightarrow a_2 R^{-1/2} I_z \cap \sigma = \varnothing, \tag{3.55}$$

where $a_2, a_2 > 0$, is an absolute constant. See Fig. 2 for the explanation.

It follows from (3.55) that

$$\left(\bigcup_{k\geq N, k\neq k(z), \sigma_k \in Б} \sigma_k \right) \bigcap \left(\bigcap_{k\geq N, k\neq k(z), \sigma_k\in Б} a_2 R_k^{-1/2} I_z \right) = \varnothing. \tag{3.56}$$

The sequence $\{R_k\}$ is assumed nondecreasing, such that

$$a R_N^{-1/2} I_z \subset \bigcap_{\substack{k\geq N, k\neq k(z) \\ \sigma_k \in Б}} a_2 R_k^{-1/2} I_z.$$

This and (3.56) imply (3.53). ■

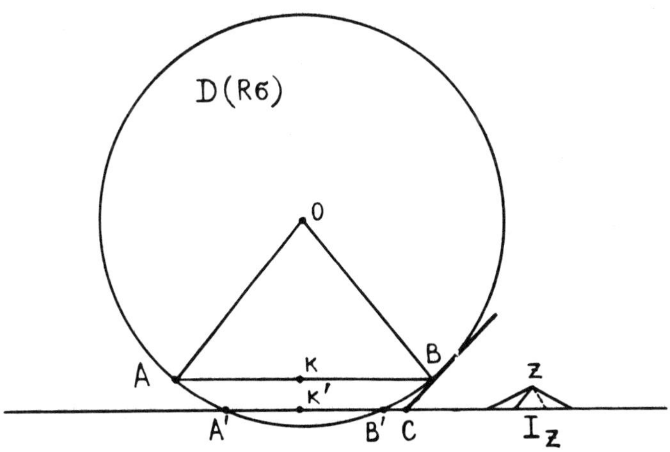

Fig. 2 Here $A'B' = \sigma$, $|A'B'| = 1$, $|K'K| = 1$, $|AB| = 2(R^2 - (R-1)^2)^{1/2} \asymp R^{1/2} \Rightarrow$ angle$AOB \asymp R^{-1/2}$, BC is tangent to $D(R\sigma) \Rightarrow$ angle$ABC \asymp R^{-1/2}$.

C. Concluding Remarks

We conclude the chapter with some remarks.
(3.57) *Remark.* In Corollary 3.15 is a sufficient condition for $\{\sigma_k\}$ to be a slightly interpolating sequence was given:

(3.58) There exists a sequence $\{L_k\}$ such that $\sum L_k^{-1} < \infty$ and $D(L_k\sigma_k)$ are disjoint discs. This condition is not necessary. Let σ_1 be an arbitrary arc. If σ_{k-1} is fixed, let us fix σ_k in such a way that $D(k\sigma_k)$ is tangent to $D[(k-1)\sigma_{k-1}]$. This does not put any restriction on the smallness of σ_k, and it is clear that we can choose the points λ_k ($|\lambda_k| = 1 - |\sigma_k|$, $\lambda_k/|\lambda_k|$ is the center of σ_k) to form a strong interpolating sequence. So $\{\sigma_k\}$ is slightly interpolating. On the other hand, if $D(L_k\sigma_k)$ are disjoint for a sequence $\{L_k\}$ and if $L_j \geq j$, then $L_{j+1} < j+1$. Therefore $\sum_{k, L_k<k} L_k^{-1} \geq \sum_j (2j+1)^{-1} = \infty$. ■

(3.59) *Remark.* One may think that if $\{\sigma_k\}$ is slightly sparse [see (3.37)] and if $\{\lambda_k\}$ satisfies the Carleson condition (C), then $\{\sigma_k\}$ is slightly interpolating. It is not the case, as an easy example shows. So the condition that $\{\lambda_k\}$ is a strong interpolating sequence [see (3.36)] cannot be omitted.

(3.60) *Remark.* If $\{\sigma_k\}$ is a slightly interpolating system of arcs, then $m(\mathbb{T} \setminus \bigcup_{k\geq 1}\sigma_k) > 0$. Indeed, it follows from (3.37) that the set $\sigma = \mathbb{T} \setminus \bigcup_{k\geq 1}\sigma_k$ has a finite set of isolated points. Let ζ be a limit point of σ but not an endpoint of any σ_k. Now put $f_n = 0$ for σ_n lying on the left of ζ and $f_n = 1$ for σ_n lying on the right (do not forget to stop somewhere). Thus there is a function $f, f \in H^\infty$, which (if we assume $m\sigma = 0$) has at ζ different left and right limits, contradiction. In fact, the same reasoning shows that there exists $n \geq 1$ such that $m(I \setminus \bigcup_{k\geq n}\sigma_k) > 0$ for an arbitrary $I, I \not\subset \sigma_k$ $(k = 1, 2, \ldots)$.

(3.61)*One more sufficient condition.* On the contrary, if $\sum_{n\geq 1}|\sigma_n| < 1$, one can distribute intervals (arcs) of lengths $|\sigma_n|$ in such a way that they turn into a slightly interpolating sequence. Moreover, the following elementary proposition provides a useful sufficient condition for $\{(\sigma_n, \varepsilon_n)\}$ to be an FI-sequence.

Lemma 3.62 Let σ_n be intervals (arcs) of $\mathbb{T}$ and ε_n be positive reals $(n = 1, 2, \ldots)$ such that the $\log \varepsilon_n^{-1}$-times enlarged (with respect to their midpoints) intervals $\sigma_n \log \varepsilon_n^{-1}$ are pairwise disjoint. Then $\{(\sigma_n, \varepsilon_n)\}$ is an FI-sequence.

Proof. It is easy to see [for example, from (3.14)] that for any arc $\sigma \subset \mathbb{T}$ and for any $L > 0$ the inequality

$$L\omega(z, \sigma) \leq A\omega(z, L\sigma)$$

holds for $z \in \mathbb{D} \setminus D(L\sigma)$ with an absolute constant A, where $D(L\sigma)$ has the usual meaning (see Section III. A.). (Indeed, this inequality is obvious on the boundary of $\mathbb{D} \setminus D(L\sigma)$ and then on $\mathbb{D} \setminus D(L\sigma)$ itself.) So, for $z \in \mathbb{D} \setminus (\bigcup_{k, k\neq n} D(\log \varepsilon_k^{-1}\sigma_k))$

we have

$$\sum_{k,k\neq n} \omega(z,\sigma_k)\log\varepsilon_k^{-1} \leq A \sum_{k,k\neq n} \omega(z,\sigma_k \log\varepsilon_k^{-1}) \leq A,$$

which proves the lemma. ∎

(3.63) *Remark.* By the way, Lemma 3.62 and Theorem 3.38 imply that if the enlarged arcs $L_k\sigma_k$ are pairwise disjoint for some choice of L_k, $\lim L_k = \infty$, the corresponding points $\lambda_k = \zeta_k(1 - |\sigma_k|)$ (ζ_k is the midpoint of σ_k) satisfy the strong Carleson condition $\lim_k |B_{\lambda_k}(\lambda_k)| = 1$. This means that for any sequence $r_k \uparrow 1$ obeying the Blaschke condition $\sum_{k\geq 1}(1 - r_k) < \infty$, there exist λ_k with $|\lambda_k| = r_k$ ($k = 1, 2, \ldots$) and with the strong interpolating property.

Acknowledgment: We are indebted to S. A. Vinogradov for a useful discussion of properties of strong interpolating sequences.

REFERENCES

Baernstein, A. (1976). Univalence and bounded mean oscillation. *Mich. Math. J. 23*(2): 217–223.

Carleson, L. (1962). Interpolation by bounded analytic functions and the corona problem. *Ann. Math. 76*(2): 547–559.

Carleson, L. (1985). Interpolation with multiplicities. *Ann. Acad. Sci. Fenn., Ser. AI*(10): 235–237.

Garnett, J. (1981). Bounded analytic functions. Academic Press, New York.

Nikolskii, N. K. (1974). Selected problems of weighted approximation and spectral analysis. *Trudy Mat. Inst. Steklov. 120* (English translation in *Proc. Steklov. Inst. Math. 120*, 1976).

Nikolskii, N. K. (1978a) "What problems do spectral theory and complex analysis solve for each other?" Proc. Int. Congress of Mathematicians, Helsinki, Finland, p. 631–638.

Nikolskii, N. K. (1978b). Bases of invariant subspaces and operator interpolation. *Trudy Mat. Inst. Steklov. 130*: 50–123 (English translation in *Proc. Steklov Inst. Math. 4(130)*: 1979).

Nikolskii, N. K. (1986). Treatise on the shift operator (spectral function theory). Springer-Verlag, New York.

Nikolskii, N. K. (1987). Interpolation libre dans l'espace de Hardy. *C. R. Acad. Sci. Paris. Ser. I. 304*(15): 451–454.

Nikolskii, N. K. (1988). A few notes on generalized free interpolation, Uppsala University, Dept. Math. Rep. *3*.

Nikolskii, N. K., Hruščëv, S. V. (1987). Functions models and some problems of spectral function theory. *Trudy Mat. Inst. Steklov. 176* (in Russian).

Sundberg, C., and Wolff, Th. (1983). Interpolating sequences for QA_B. *Trans. Amer. Math. Soc. 276*(2): 551–581.

Vasyunin, V. I. (1976). Unconditionally convergent spectral decompositions and interpolation problems. Thesis, LOMI, Leningrad (in Russian).

Vasyunin, V. I. (1978). Unconditionally convergent spectral decompositions and interpolation problems. *Trudy Mat. Inst. Steklov. 130*: 5–49 (English translation in *Proc. Steklov Math. Inst. 4(130)*(1979): 1–53).

16

The Fock Bundle

Jaak Peetre

0. Introduction.

It was already known to Jacobi (1829) that theta functions satisfy a PDE. Indeed, if we put

$$\vartheta_{e\varepsilon}(t,\xi) = \sum_{n\in\mathbf{Z}} \exp\left(\pi i t(n+\frac{e}{2})^2 + 2\pi i(\xi+\frac{\varepsilon}{2})(n+\frac{e}{2})\right). \tag{1}$$

where ξ, t are complex numbers, $\mathrm{Im}t > 0$, and $e, \varepsilon \in \{0,1\}$, then each function $u = \vartheta_{e,\varepsilon}$ satisfies the equation (Schrödinger or heat equation)

$$4\pi i \frac{\partial u}{\partial t} = \frac{\partial^2 u}{\partial \xi^2}. \tag{2}$$

He also showed that these functions have remarkable invariance properties under two sets of transformation of variables:

$$(t,\xi) \mapsto (\frac{at+b}{ct+d}, \frac{\xi}{ct+d}), \tag{3}$$

$$(t,\xi) \mapsto (t, \xi + tv + \nu). \tag{4}$$

Here a, b, c, d, v, ν $(ad-bc=1)$ are real quantities subject to certain auxiliary arithmetical restrictions (more precisely, v, ν are arbitrary integers, a, b, c, d are integers with additional divisibility properties).

On the other hand, equation (2) too transforms nicely under precisely the same transformations: if u is a solution then each of the functions

$$U_1 = u(\frac{at+b}{ct+d}, \frac{\xi}{ct+d}) \exp(-\pi i \frac{c\xi}{ct+d}) \frac{1}{\sqrt{ct+d}},$$

$$U_2 = u(t, \xi + tv + \nu) \exp(\pi i t e^2 + 2\pi i e \xi)$$

is a solution. Of course, then we need no longer any arithmetical restrictions.

Similar formulas hold in several variables. The theta functions are now defined by the formula

$$\vartheta_{e\varepsilon}(t,\xi) = \sum_{n\in\mathbf{Z}^n} \exp\left(\pi i \langle t(n+\frac{e}{2}), n+\frac{e}{2}\rangle + 2\pi i \langle \xi + \frac{\varepsilon}{2}, n+\frac{e}{2}\rangle\right). \tag{1'}$$

Here $t = (t_{ij})$ is a symmetric $n \times n$ matrix with positive definite imaginary part and $\xi = (\xi_1, \ldots, \xi_n)$ a vector in $\mathbf{C}^n$, e, ε being vectors all of whose entries are zeros or ones. The analogue of equation (2) may be conveniently written as

$$(2') \qquad 4\pi i du = \sum_{i,j}^{n} \frac{\partial^2 u}{\partial \xi_i \partial \xi_j} dt_{ij}.$$

The aim of this paper is to try to understand the invariances mentioned from a higher point of view. In fact, we will introduce a new object called the *Fock bundle*. Technically speaking, it is a Hermitean vector bundle of *infinite rank* (a "Hilbert bundle") over Siegel's generalized upper halfplane (the manifold of all matrices t), whose fibers are Hilbert spaces of entire analytic functions in the variable ξ (Fock spaces). Being an Hermitean vector bundle it carries a natural metric connection. Our main contention is that (1) or - in the case of several variables - (1′) can be viewed as the condition for parallelism for this connection. This has several pleasant features. First of all it becomes obvious that our equation has the desired invariance properties. Furthermore, one gets e.g. a painless construction of the Weil-Shale representation of the metaplectic group, and of the Bargmann-Segal representation of the Heisenberg group as well.

The same ideas can probably also be used in connection with theta function theory but presently we leave theta functions alone, hoping to be able to return this topic on a future occasion.[1]

We remark also that this paper in a way is a continuation of or connected with a series of previous paper (Peetre 1980a, 1980b, Janson, Peetre, and Rochberg 1987, Janson, and Peetre 1989?, Janson, Peetre, and Wallstén 1989?).

In Sec. 1 we set forth the basics of symplectic geometry and give the definition of the Fock bundle. In Sec. 2 we recall the salient facts about connections in Hermitean vector bundles of *finite rank*. We remark that we will in this paper rely on the formal analogy with the finite rank case. Thus most of the following considerations for the Fock bundle (Sec. 3-6) have, strictly speaking, only a heuristic value but, we believe, a strong one. In Sec. 7 we mention a few more issues to which we likewise hope to come back to in the future.

Acknowledgement. My thanks are due to Jan Boman and Björn Gustafsson for patiently listening to me and for other help. Also a casual remark by Dimitry Leĭtes turned out to be helpful.

1. Basic symplectic geometry. Construction of the Fock bundle.

Let V be a symplectic vector space over $\mathbf{R}$ of dimension $2n$ with the symplectic structure σ (= a nondegenerate alternate bilinear form).

Remark. If we instead have a quadratic form on V we speak of an orthogonal space. It is our belief that most of what we do in this paper extends *mutatis mutandis* to this case.

Example 1. $V = \mathbf{R}^n \oplus \mathbf{R}^n$ with points $X = (x, \xi)$,

$$\sigma(X, X') = \langle \xi, x' \rangle - \langle \xi', x \rangle = \sum_{j=1}^{n} \xi_j x'^j - \sum_{j=1}^{n} \xi'_j x^j.$$

[1] Cf. however the closing lines for a few more hints.

Here $\mathbf{R}^n_\xi$ and $\mathbf{R}^n_x$ are two copies of $\mathbf{R}^n$ (one Greek and the other Latin) with the standard duality

$$\langle \xi, x\rangle = \sum_{j=1}^{n} \xi_j x^j,$$

if $\xi = (\xi_1, \ldots, \xi_n), x = (x^1, \ldots, x^n)$. This is the standard example, because if we pick two complementary isotropic subspaces (in the general abstract setting of a symplectic vector space V) then the bilinear form σ induces a nondegenerate pairing between them so that picking further two dual bases we arrive precisely at the identification $V \approx \mathbf{R}^n_\xi \oplus \mathbf{R}^n_x$.

Example 2. V is a complex vector space with an Hermitean structure $(\cdot,\cdot)$. Then we arrive at a symplectic vector space if we forget the complex structure and put

$$\sigma(X, X') = \operatorname{Im}(X, X').$$

In particular, if $V = \mathbf{C}^n$ with coordinates $z = x + i\xi$ and the standard metric

$$(X, X) = \langle z, \bar{z}\rangle = \langle x + i\xi, x - i\xi\rangle$$

we obviously recover ex. 1, as

$$\operatorname{Im}(X, X') = \operatorname{Im}(\langle x, x'\rangle + i\langle \xi, x'\rangle - i\langle x, \xi'\rangle + \langle \xi, \xi'\rangle) = \langle \xi, x'\rangle - \langle \xi', x\rangle.$$

Let $G = Sp(V)$ be the automorphism group of V, that is, the group of all linear selfmaps of V preserving the symplectic form σ. In the case of ex. 1 we get the identification of G with the standard group $Sp(2n, \mathbf{R})$. (In the case of quadratic form we have instead the group $G = O(V) \approx O(2n)$.) We give now an explicit description of this group. Every element g of G is given by a block matrix $\begin{vmatrix} a & b \\ c & d \end{vmatrix}$: if $Y = (\eta, y)$ is the image of (X, ξ) under g, then we may write

$$\begin{aligned} \eta &= a\xi + bx, \\ y &= c\xi + dx; \end{aligned}$$

a, b, c, d are thus $n \times n$ matrices. We have

$$\langle a\xi + bx, c\xi' + dx'\rangle - \langle a\xi' + bx', c\xi' + dx\rangle =$$

$$= \langle c^\top a\xi, \xi'\rangle + \langle c^\top bx, \xi'\rangle + \langle d^\top a\xi, x'\rangle + \langle d^\top bx, x'\rangle -$$
$$-\langle c^\top a\xi', \xi\rangle - \langle c^\top bx', \xi\rangle - \langle d^\top a\xi', x\rangle - \langle d^\top bx', x\rangle,$$

which expression has to be put equal to

$$\langle \xi, x'\rangle - \langle \xi', x\rangle.$$

(Here and in the sequel the superscript $\top$ indicates the transpose of a matrix.) Hence the relations

$$(1) \qquad d^\top a - b^\top c = 1,\ c^\top a = a^\top c,\ d^\top b = b^\top d,$$

which constitute a necessary and sufficient condition for $g = \begin{pmatrix} a & b \\ c & d \end{pmatrix}$ to be in $Sp(2n, \mathbf{R})$. Equivalent formulation:

$$(2) \qquad a^\top d - c^\top b = 1,\ b^\top a = a^\top b,\ d^\top c = c^\top d.$$

Next let V^C be the complexification of V and let $X \mapsto \bar{X}$ be the corresponding conjugation. Properties:

$$\overline{X+Y} = \bar{X} + \bar{Y},$$
$$\overline{kX} = \bar{k}\bar{X} \quad (k \in \mathbf{C}),$$
$$\bar{\bar{X}} = X.$$

Explicit contruction: $V^C = \mathbf{C} \otimes_{\mathbf{R}} V$ with elements $k \otimes X$ (where now $X \in V, k \in \mathbf{C}$), $\overline{k \otimes X} = \bar{k} \otimes X$.

Clearly G acts on V^C and also σ extends to a nondegenerate alternate (skew) form on V^C. A basic observation is now that the form $\sigma(X, \bar{X})$ is skew-Hermitean on V^C and therefore $-i\sigma(X, \bar{X})$ is Hermitean. In fact, it is of signature $+\cdots+-\cdots-$ (same number of pluses and minuses). This leads one to the idea to look at linear subspaces P of V^C of maximal dimension n on which this last skew-Hermitean form is *isotrope* and *positive definite*:

$$1^o \quad \sigma(X, X') = 0 \,\text{for all}\, X, X' \in P,$$
$$2^o \quad -i\sigma(X, \bar{X}) > 0 \,\text{for}\, 0 \neq X \in P.$$

Such a P will be called a *Riemann plane*, because relations of this type were first encountered by Riemann in the theory of algebraic curves (the period matrix of a compact Riemann surface defines a Riemann plane). The set of all Riemann planes $\mathcal{S}$ is called *Siegel's generalized upper halfplane* for the following reason: making the identification in ex. 1 we observe that an element $X = (\xi, x)$ of a Riemann plane P is determined by its second projection x. Hence we can write $\xi = tx$ where t is a uniquely determined complex $n \times n$ matrix. One readily sees that the above properties 1° and 2° also can be written in terms of t as

$$1^o \quad t = t^\top \,(\text{symmetry}),$$
$$2^o \quad \text{Im} t > 0 \,(\text{positive definite}).$$

(It is of course in this form that they appear in Riemann and Siegel.) Thus $\mathcal{S}$ gets parametrized by a set of matrices.

Notice that $\mathcal{S}$ by definition is a subset of a Grassmannian, the set $Gr(V^C)$ of all subspaces of V^C of dimension n:

$$\mathcal{S} \subset Gr(V^C).$$

If we make use of ex. 2 we get a second realization of $\mathcal{S}$ as a bounded symmetric domain (Type II in Élie Cartan's list of classical domains). We write instead $sw = z$; we now use $z = x + i\xi$ and $w = x - i\xi$ as coordinates in V^C. Then

$$sx - is\xi = x + i\xi,$$

$$(s-1)x = i(s+1)\xi,$$

$$\xi = \frac{1}{i}(s+1)^{-1}(s-1)x.$$

Thus s and t are related via the *Cayley transform*

$$t = \frac{1}{i}(s+1)^{-1}(s-1) = \frac{1}{i}(s-1)(s+1)^{-1}.$$

Clearly s is symmetric at the same time as t and, as

$$\text{Im}t = (\bar{s}+1)^{-1}(1-\bar{s}s)(s+1)^{-1} \tag{3}$$

or, equivalently (taking complex conjugates of both sides),

$$\text{Im}t = (s+1)^{-1}(1-s\bar{s})(\bar{s}+1)^{-1}, \tag{3'}$$

we see that $\mathcal{S}$ can be identified with the set of all symmetric (strict) contractions ($\bar{s}s < 1$). This is the Harish-Chandra realization of $\mathcal{S}$.

Proof of (3). We have

$$\bar{t} = -\frac{1}{i}(\bar{s}+1)^{-1}(\bar{s}-1).$$

Hence

$$\text{Im}t = \frac{1}{2i}(t-\bar{t}) = \frac{1}{2}[-(s-1)(s+1)^{-1} - (\bar{s}+1)^{-1}(\bar{s}-1)] =$$

$$= \frac{1}{2}(\bar{s}+1)^{-1}[-(\bar{s}+1)(s-1) - (\bar{s}-1)(s+1)](s+1)^{-1} =$$

$$= \frac{1}{2}(\bar{s}+1)^{-1}[-\bar{s}s + \bar{s} - s + 1 - \bar{s}s - \bar{s} + s + 1](s+1)^{-1} = (\bar{s}+1)^{-1}(1-\bar{s}s)(s+1)^{-1}. \blacksquare$$

Clearly an element g of the group G maps a Riemann plane P onto the subspace gP which also is a Riemann plane. That is, the group acts on $\mathcal{S}$. One can easily show that the action is transitive so $\mathcal{S}$ becomes a homogeneous space. Using the Cayley transform one finds that the isotropy group of the point corresponding to $t = i1$ is $U(n)$. Therefore we have the identification

$$\mathcal{S} \approx Sp(2n, \mathbf{R})/U(n).$$

Let us describe this group action explicitly. Let P be parametrized by the matrix t, that is, $X = (\xi, x) \in P$ iff $\xi = tx$. Consider any $g = \begin{pmatrix} a & b \\ c & d \end{pmatrix}$ in G. Then gP consists of all pairs $(a\xi + bx, c\xi + dx)$ with $(x, \xi) \in P$. Denoting such a pair by (η, y) we may write

$$\eta = (at+b)x,\ y = (ct+d)x,$$

that is,

$$\eta = (at + b)(ct + d)^{-1}y.$$

In other words, gP is parametrized by the matrix $(at+b)(ct+d)^{-1}$. Thus the group action is:

$$\boxed{t \mapsto (at + b)(ct + d)^{-1}}.$$

Now we put into play some vector bundles over $\mathcal{S}$.

First, we have the *tautological bundle* $\mathcal{B} = \cup_{P \in \mathcal{S}} P$, which is such that the fiber over a point P is the vector space P itself. Next, we have the *cotautological bundle* $\mathcal{B}^* = \cup_{P \in \mathcal{S}} P^*$. Now the fiber over P is the dual space $P^* = (V^C)^*/P^\perp$. These bundles have rank n, while the base space $\mathcal{S}$ has dimension $\frac{n(n+1)}{2}$.

The elements of $\mathcal{B}$ are described by pairs (t, ξ) where t is a symmetric square matrix with positive definite imaginary part and ξ is a complex vector. Similarly we may describe the elements of the dual bundle $\mathcal{B}^*$ by pairs (t, λ), as every linear functional, say, Λ on the space P corresponding to the matrix t can be written in the form $\Lambda(X) = \langle \lambda, x \rangle$ where $X = (\xi, x) \in P$.

Each space P has a natural Hermitean structure given by the formula

$$\|X\|^2 = i\sigma(X, \bar{X}) = 2\langle (\mathrm{Im}t)x, \bar{x} \rangle.$$

Thus $\mathcal{B}$ is an Hermitean bundle. We equip P^* with the dual structure, the norm of a linear functional $\Lambda = (t, \lambda)$ thus being given (forgetting a factor $1/2$) by the formula

$$\|\Lambda\|^2 = \langle (\mathrm{Im}t)^{-1}\lambda, \bar{\lambda} \rangle,$$

where as before $\Lambda(X) = \langle \lambda, x \rangle$. Thus $\mathcal{B}^*$ too is an Hermitean bundle. Below we shall make use of the fact that together with the Hermitean structure P^* also comes equipped with a volume element. Explicitly: it is question of the volume element

$$(i/2)^n d\lambda_1 d\overline{\lambda_1} \ldots d\lambda_n d\overline{\lambda_n} / (\det \mathrm{Im}t),$$

which is thus coordinate independent. In addition, we can in each fiber P^* of this bundle define a quadratic form given by

$$Q(\Lambda, \Lambda) = \langle (\mathrm{Im}t)^{-1}\lambda, \lambda \rangle.$$

However, one should bear in mind that, in contrast to all the other notions introduced, Q is *not* intrinsically defined by the symplectic structure on V but depends on the special splitting of V chosen giving rise to the coordinates x, ξ. Or rather, it depends on the conjugation $\lambda \mapsto \bar{\lambda}$ thus obtained upon identifying each fiber P^* with $\mathbf{C}^n$. In particular, we will consider the difference of the two forms available to us, viz. the quantity

$$Q(\lambda, \lambda) - \|\lambda\|^2.$$

In the distinguished coordinates its *real part* thus has the expression:

$$\langle (\mathrm{Im}t)^{-1}\mathrm{Im}\lambda, \mathrm{Im}\lambda \rangle. \tag{4}$$

It is obvious that the group action on $\mathcal{S}$ extends to an action on $\mathcal{B}$ (each element of any fiber P is moved by an element of G to an element of gP) and also, taking adjoints, to $\mathcal{B}^*$. However, also a larger group, viz. the semidirect product $\mathcal{G} = V^* \cdot G$ (V^* is thereby considered as an additive group on which G acts via automorphisms in the obvious way via conjugation) acts on $\mathcal{B}^*$. This group clearly is isomorphic to the group $\mathbf{R}^{2n} \cdot Sp(2n, \mathbf{R})$. Presumably, there is no corresponding group action on $\mathcal{B}$. More precisely, let an element $g \in G$ be given and one l in V^* be given, in other words, we have an element (l, g) in $\mathcal{G}$. Then to each $\Lambda \in P^*$ we assign the linear functional $\tilde{g}\Lambda + l|gP \in (gP)^*$. Here and the sequel we use a tilde˜to denote the adjoint action: $\tilde{g} = (g^{-1})^\top$ and the stroke | stands for taking the restriction to the subspace indicated.

Let us check that the composition of two maps of the above type is again a map of the same type. Thus consider two group elements (l_1, g_1) and (l_2, g_2). Then we have

$$\tilde{g}_1(\tilde{g}_2 + l_2|\tilde{g}_2 P) + l_1|g_1 g_2 P = \tilde{g}_1\tilde{g}_2\Lambda + (\tilde{g}_1 l_2 + l_1)|g_1 g_2 P,$$

wherein we recognize the group law in $\mathcal{G}$: the elements of $\mathcal{G}$ are multiplied as block matrices,

$$\begin{vmatrix} \tilde{g}_1 & l_1 \\ 0 & 1 \end{vmatrix} \cdot \begin{vmatrix} \tilde{g}_2 & l_2 \\ 0 & 1 \end{vmatrix} = \begin{vmatrix} \tilde{g}_1\tilde{g}_2 & \tilde{g}_1 l_2 + l_1 \\ 0 & 1 \end{vmatrix}.$$

Let us write down the above group action explicitly in terms of coordinates. For simplicity, we consider only the separate effect of each factor. That is, we have, on the one hand, pure "homogeneous" transformations:

$$(t, \lambda) \mapsto ((at + b)(ct + d)^{-1}, (ct + d)\tilde{}\lambda) \tag{5}$$

and, on the other hand, pure "translations":

$$(t, \lambda) \mapsto (t, \lambda + te + \varepsilon). \tag{6}$$

Having derived these formulae, it is an easy matter to write down the action of a general group element (l, g) corresponding to the block matrix

$$\begin{vmatrix} \tilde{g} & l \\ 0 & 1 \end{vmatrix} = \begin{vmatrix} 1 & l \\ 0 & 1 \end{vmatrix} \cdot \begin{vmatrix} \tilde{g} & 0 \\ 0 & 1 \end{vmatrix}.$$

Indeed, it is then found that

$$(t, \lambda) \mapsto ((at + b)(ct + d)^{-1}, (ct + d)\tilde{}\lambda + (at + b)(ct + d)^{-1}e + \varepsilon).$$

Derivation of (5): Consider an element $Y = ((at + b)(ct + d)^{-1}y, y)$ in gP. Then $X = g^{-1}Y = (t(ct + d)^{-1}y, (ct + d)^{-1}y)$ is in P. It follows that

$$\tilde{g}\Lambda(Y) = \Lambda(g^{-1}Y) = \Lambda(X) = \langle \lambda, x \rangle = \langle \lambda, (ct + d)^{-1}y \rangle = \langle (ct + d)\tilde{}\lambda, y \rangle. \quad \blacksquare$$

Derivation of (6): Any given linear functional l on V can be written in the form

$$l(X) = \langle e, \xi \rangle + \langle \varepsilon, x \rangle,$$

where $X = (x, \xi)$ denotes an arbitrary element of V. We extend l by linearity to V^C. Let as before $\Lambda(X) = \langle \lambda, x \rangle$ for $X = (tx, x) \in P$. As

$$(l|P)(X) = \langle e, tx \rangle + \langle \varepsilon, x \rangle = \langle te + \varepsilon, x \rangle,$$

it follows that

$$(\Lambda + l|P)(X) = \langle \lambda + te + \varepsilon, x \rangle. \quad \blacksquare$$

Finally, we come to the definition of the *Fock bundle*. It is a vector bundle of *infinite rank* over $\mathcal{S}$, denoted by $\mathcal{F}$, such that the fiber $\mathcal{F}_P$ over any element $P \in \mathcal{S}$ consists of all entire analytic functions on the dual space P^* which are square integrable with respect to the weight function $e^{\pi(\mathrm{Re}Q(X,X) - \|X\|^2)}$. Even better, we will regard the elements not as functions but as *half densities*. That is, we have not to integrate with respect to the Hermitean volume element in P^* introduced above, but rather with respect to its *square root*. Explicitly (see (4)): the norm of an element $u = u(t, \xi)$ is thus given by the formula

$$\|u\|_t^2 = \int_{\mathbf{C}^n} |u(t,\lambda)|^2 \exp\left(-2\pi\langle(\mathrm{Im}t)^{-1}\mathrm{Im}\lambda, \mathrm{Im}\lambda\rangle\right) \times$$

$$\times (i/2)^n d\lambda_1 d\bar{\lambda}_1 \ldots d\lambda_n d\bar{\lambda}_n / (\det \mathrm{Im}t)^{1/2}.$$

Notice that we are here referring to the fixed coordinate system which was used in the definition of the conjugation determining the quadratic form Q.

We require also the Cayley transformed version of the above formula. We put

$$u(t,\lambda) = v(s,w) \cdot \exp(-\frac{\pi}{2}\langle(s+1)^{-1}w, w\rangle) \cdot (\det(1+s))^{-1/2}, \tag{7}$$

where the variables are related by the formulae

$$t = \frac{1}{i}\frac{s-1}{s+1}, \; \lambda = (1+s)^{-1}w$$

(or, equivalently,

$$s = \frac{1+it}{1-it}, \; w = 2(1-it)^{-1}\lambda).$$

In other words, we make a "change of gauge" (cf. Arazy, Fisher, and Peetre (1989?) or Janson, Peetre, and Rochberg (1987)). A simple calculation reveals then that the norm of v is given by

$$\|v\|_s^2 = \int_{\mathbf{C}^n} |v(s,w)|^2 \exp\left(\pi \mathrm{Re}\langle \bar{s}(1-\bar{s}s)^{-1}w, w\rangle - \pi\langle(1-\bar{s}s)^{-1}w, w\rangle\right) \times$$

$$\times (i/2)^n dw_1 d\bar{w}_1 \ldots dw_n d\bar{w}_n / (\det(1-\bar{s}s)^{1/2}.$$

We have only to invoke two more identities which follow easily from (3) or (3′):

$$\langle(\mathrm{Im}t)^{-1}\lambda, \lambda\rangle - \langle(\mathrm{Im}t)^{-1}\lambda, \bar{\lambda}\rangle = \langle\bar{s}(1-s\bar{s})^{-1}w, w\rangle - \langle(1-s\bar{s})^{-1}w, \bar{w}\rangle + \langle(s+1)^{-1}w, w\rangle \tag{8}$$

and

$$\det \mathrm{Im} t = \det(1-\bar{s}s)/|\det(1+s)|^2. \tag{9}$$

Proof of (8): Using (3′) we have

$$\langle (\mathrm{Im}t)^{-1}\lambda, \lambda\rangle - \langle (\mathrm{Im}t)^{-1}\lambda, \bar{\lambda}\rangle =$$

$$= \langle (\bar{s}+1)(1-s\bar{s})^{-1}(s+1)(s+1)^{-1}w, (s+1)^{-1}w\rangle -$$
$$-\langle (\bar{s}+1)(1-s\bar{s})^{-1}(s+1)(s+1)^{-1}w, (\bar{s}+1)^{-1}\bar{w}\rangle =$$
$$= \langle (s+1)^{-1}(\bar{s}+1)(1-s\bar{s})^{-1}w, w\rangle - \langle (1-s\bar{s})^{-1}w, \bar{w}\rangle.$$

The rest follows from the identity

$$(s+1)^{-1}(\bar{s}+1)(1-s\bar{s})^{-1} = (s+1)^{-1} + \bar{s}(1-s\bar{s})^{-1}$$

or, removing the denominators,

$$\bar{s}+1 = 1 - s\bar{s} + (1+s)\bar{s}. \quad \blacksquare$$

Proof of (9): Ovious consequence of the multiplication rule for determinants. ∎

With reference to the norm $\|\cdot\|_t$ and $\|\cdot\|_s$ it will be convenient to write the space as $\mathcal{F}_t$ or $\mathcal{F}^s$ respectively, in place of $\mathcal{F}_P$.

The fact that the definition of the Fock bundle depends of a special splitting of our symplectic vector space V, which at first sight looks unpleasant, will be remedied on later on (see Sec. 4). It turns out that this dependence on the choice of coordinates is compensated by the presence of a certain unitary group action.

2. Connections in Hermitean vector bundles.

To proceed further let us recall some salient features about holomorphic Hermitean vector bundles. For simplicity we consider only the case of finite rank, although the applications we are going to make (to the Fock bundle) will be concerned with Hermitean bundles of infinite rank ("Hilbert bundles").

Let thus $\mathcal{M}$ be any finite dimensional complex manifold and $\pi : \mathcal{M} \to \mathcal{M}$ a holomorphic Hermitean vector bundle of rank r over $\mathcal{M}$, that is, each fiber is a vector space over $\mathbf{C}$ (operations: $u+v$, ku ($k \in \mathbf{C}$)) and carries an Hermitean metric (norm: $\|u\|$, inner product: $(\cdot,\cdot)$). The usual compatibility conditions are assumed to be fulfilled. Of course, we assume also that our bundle is locally trivial: every point in $\mathcal{M}$ has an open neighborhood U such that $\mathcal{H}|U \simeq U \times \mathbf{C}^r$. If we select a trivialization we can thus locally identify elements of $\mathcal{H}$ with vectors $u = (u^1,\ldots,u^r) \in \mathbf{C}^r$ and $\|u\|^2 = \langle hu, u\rangle = \sum_{i,j=1}^r h_{ij}u^i\bar{u}^j$. Here $h = (h_{ij})$ is a positive definite smooth (C^∞) matrix function. (For full details about these notions see e.g. Kobayashi and Nomizu (1963/69).)

A metric connection is a map ∇ from $C^\infty(\mathcal{H})$ into $C^\infty(T^*\mathcal{M}\otimes\mathcal{H})$ such that

1) $\nabla(u+v) = \nabla u + \nabla v$;
2) $\nabla(fu) = f\nabla u + df \otimes u$ (where $f \in C^\infty(\mathcal{M})$);

3) $d\|u\|^2 = (\nabla u, u) + (u, \nabla u) = 2\mathrm{Re}(\nabla u, u)$.
Here $C^\infty(\,)$ stands for the sheaf of smooth germs of the bundle in question, $\mathcal{M}$ being identified with the trivial bundle $\mathcal{M} \otimes \mathbf{C}$ and $T^*\mathcal{M}$ being the cotangent bundle of $\mathcal{M}$. In terms of a trivialization, this means that $\nabla u = du + Au$ or, written out in greater detail, $\nabla u^i = du^i + A^i_j v^j$. Here $A = (A^i_j)$ is a matrix of 1-forms (the connection form). It is a remarkable fact that if we also require that A is a $(1,0)$-form, then the connection is uniquely determined. Then, taking the "holomorphic" and "antiholomorphic" parts of ∇ we can write $\nabla' u = \partial u + Au$ and

4) $\nabla'' u = \bar\partial u$.

The latter condition, apparently, is independent of the trivialization. In other words, we have the

PROPOSITION 1. *There is one and only one map ∇ satisfying the above properties 1)-4).*

Proof. Differentiating we have

$$\langle (dh)u, u\rangle + \langle hdu, u\rangle + \langle hu, du\rangle = \langle hdu + hAu, u\rangle + \langle hu, du + Au\rangle.$$

Taking the holomorphic part it follows that $\partial h = hA$ or, solving out, $A = h^{-1}\partial h$. This establishes the uniqueness. Conversely, we can use this formula to define the connection form for a connection of the desired type. ∎

Thus we can henceforth speak of *the* metric connection[2] on an Hermitean vector bundle. (Notice that this corresponds to the uniqueness of the Levi-Civita connection in Riemannian geometry but is really much easier.)

Remark. 3) can now be written equivalently as 3′) $\partial\|u\|^2 = (\nabla' u, u) + (u, \bar\partial u)$.

Next, let $B : \mathcal{H} \to \mathcal{H}$ be a fiber preserving map, that is, a map which induces a map $b : \mathcal{M} \to \mathcal{M}$ yielding the commutative diagram

$$\begin{array}{ccc} \mathcal{H} & \xrightarrow{B} & \mathcal{H} \\ {\scriptstyle\pi}\downarrow & & {\scriptstyle\pi}\downarrow \\ \mathcal{M} & \xrightarrow{b} & \mathcal{M} \end{array},$$

and such that each restriction $B|\mathcal{H}_t$ ($t \in \mathcal{M}$) is linear and such that, finally, $\|Bu\| = \|u\|$ for all $u \in \mathcal{H}$.

PROPOSITION 2. *The map B preserves the metric connection.*

That is, B "drags" any given section u into a section $\tilde u$ defined by the formula

$$\tilde u(t) = B(b^{-1}(t))u(b^{-1}(t))$$

or

$$\boxed{\tilde u = (b^{-1})^*(Bu)},$$

[2]The authors of Kobayashi, and Nomizu (1963/69) say "Hermitean" connection.

and this section satisfies

$$\boxed{\nabla \tilde{u} = (b^{-1})^*(B\nabla u)}.$$

In particular, if $\tilde{u} = 0$ is parallel ($\nabla u = 0$) then $\tilde{u}$ is parallel ($\nabla \tilde{u} = 0$).

Proof. We get

$$d\tilde{u} = (b^{-1})^* dB \cdot (b^{-1})^* u + (b^{-1})^* B \cdot (b^{-1})^* du. \tag{1}$$

On the other hand, that B preserves the metric means that

$$h(t) = B(t)^* h(b(t)) B(t) \,\text{or}\, h = B^* \cdot b^* h \cdot B.$$

Differentiating this yields

$$dh = dB^* \cdot b^* h \cdot B + B^* \cdot (b^* dh) \cdot B + B^* \cdot b^* h \cdot dB.$$

(Notice that in the above computations the symbol $*$ is used in two different senses: adjoint and pullback!) Writing

$$h^{-1} = B^{-1}(b^* h)^{-1}(B^{-1})^*$$

and multiplying together yields

$$A = B^{-1}(b^* A)B + B^{-1} dB$$

or

$$A = (b^{-1})^* B \cdot (b^{-1})^* A \cdot (b^{-1})^* B^{-1} - (b^{-1})^* dB \cdot (b^{-1})^* B^{-1}.$$

It follows that

$$A\tilde{u} = A \cdot (b^{-1})^* B \cdot (b^{-1})^* u = (b^{-1})^* B \cdot (b^{-1})^* (Au) - (b^{-1})^* (dB) \cdot (b^{-1})^* u. \tag{2}$$

If we now add (1) to (2), we see that

$$d\tilde{u} + Au = (b^{-1})^* B \cdot (b^{-1})^* (du + Au),$$

which is equivalent to the statement. ∎

COROLLARY 1. *If a group acts on the bundle $\mathcal{H}$ via fiber preserving linear isometries then it preserves the metric connection.* ∎

Finally, recall that the equation $\nabla u = 0$ is in general not solvable (even locally). For this, certain integrability conditions are needed. These are conveniently expressed by the vanishing of the *curvature form* $F = dA - \frac{1}{2}[A, A]$. If this the case ($F = 0$ in every trivialization) then we are in the (locally) *flat* situation, i.e., near each point of $\mathcal{M}$ there is a trivialization for which h is the identity matrix. In other words, $\mathcal{H}$ is locally isometric to a product of $\mathcal{M}$ with a factor $\mathbf{C}^r$. (For more details, we refer again to Kobayashi, and Nomizu (1963/69).)

COROLLARY 2. *Let again a group act on the bundle $\mathcal{H}$, now assumed to be globally flat. Then it induces a unitary action on each fiber.*

Proof. Select a "base point" $t_0 \in \mathcal{M}$. Denote by L the vector space of all holomorphic solutions $\nabla u = 0$. Let $a \in \mathcal{H}_{t_0}$ be given. Then there exits a unique $u \in L$ such that $u(t_0) = a$. Let $\tilde{u}$ be the image of u under the action of a group element g. Define $\tilde{u}(t_0) = ga$. Then the correspondence $a \mapsto ga$ has the desired properties. ∎

We now turn to something which perhaps is not so well-known, subbundles of an "orthogonal" bundle. (A continuous analogue is something like a Hilbert space with a reproducing kernel isometrically imbedded in an L^2-space.) Thus, pictorially, we are in the following situation:

$$\begin{array}{ccc} \mathcal{H} & \xrightarrow{\text{id}} & \mathcal{O} \\ {\scriptstyle\pi}\downarrow & & {\scriptstyle\pi}\downarrow \\ \mathcal{M} & \xrightarrow{\text{id}} & \mathcal{M} \end{array},$$

where $\mathcal{O}$ denotes a (not necessarily holomorphic) vector bundle of rank ρ, say, where $\rho \geq r$, admitting ρ orthogonal sections. In other words, the norm of an element u of $\mathcal{H}$ lying over a point $t \in \mathcal{M}$ can be written in the form

$$\|u\|_t^2 = \sum_{i=1}^{\rho} |u_i|^2 h_i(t),$$

where the h_i are certain positive smooth functions on $\mathcal{M}$ and the maps $u \mapsto u_i$ are certain uniquely determined linear functionals defined on the fiber $\mathcal{H}_t$. It follows that there exist uniquely determined section e_i such that for each u in $\mathcal{H}_t$ $(t \in \mathcal{M})$ holds

$$u_i = (u, e_i(t))_t \quad (i = 1, \ldots, \rho).$$

Taking $u = e_k$ $(k = 1, \ldots, \rho)$ it follows that

$$e_{ki}(t) = (e_k(t), e_i(t))_t. \tag{3}$$

(Apparently, the matrix of components (e_{ki}) is the analogue of the reproducing kernel in the continuous case.) Select again a "base point" $t_0 \in \mathcal{M}$. Putting ourselves furthermore in the (globally) flat situation, we can find sections E_i such that for each solution of $\nabla u = 0$ and any $t \in \mathcal{M}$ holds

$$u_i(t) = (u(t_0), E_i(t))_{t_0} \quad (i = 1, \ldots, \rho).$$

The continuous analogue of the following result will play a vital rôle later on (Sec. 6).

PROPOSITION 3. *We have*

$$(E_k(t), E_i(t))_{t_0} = e_{ki}(t) \quad (t \in \mathcal{M},\ i, k = 1, \ldots, \rho). \tag{4}$$

Proof. This is almost trivial. Solving the equation $\nabla u = 0$ with initial data $u(t) = e_i(t)$ it follows, using Prop. 2, that $u(t_0) = E_i(t)$. Using Prop. 2 once more we get from this and formula (3)

$$(E_k(t), E_i(t))_{t_0} = (e_k(t), e_i(t))_t = e_{ki}(t). \quad \blacksquare$$

3. Application to the Fock bundle.
Let us return to the Fock bundle $\mathcal{F}$. Introducing the notation

$$W = e^{-2\pi\langle(\mathrm{Im}t)^{-1}\mathrm{Im}\xi, \mathrm{Im}\xi\rangle} \cdot (\det \mathrm{Im}t)^{-1/2}$$

we can write the norm in $\mathcal{F}$ (see Sec. 1) as

$$\|u\|_t^2 = \int |u|^2 W.$$

(Here and in the sequel we omit the Euclidean volume element $(i/2)^n d\xi_1 d\bar{\xi}_1 \ldots d\xi_n d\bar{\xi}_n$.)

KEY FORMULA. *Let u be any section of $\mathcal{F}$. Then*

$$d\|u\|^2 = 2Re \int (du - \frac{1}{4\pi i} \sum_{i,j=1}^{n} \frac{\partial^2 u}{\partial\xi_i \partial\xi_j} dt_{ij})\bar{u}W$$

or, equivalently,

$$\partial\|u\|^2 = \int (\partial u - \frac{1}{4\pi i} \sum_{i,j=1}^{n} \frac{\partial^2 u}{\partial\xi_i \partial\xi_j} dt_{ij})\bar{u}W + \int u\overline{\bar{\partial}u}W. \quad \blacksquare$$

The proof depends on the following lemma, whose proof again depends on straightforward but tedious matrix manipulations, which we therefore omit.

LEMMA. *We have*

$$4\pi i dW + \sum_{i,j=1}^{n} \frac{\partial^2 W}{\partial\xi_i \partial\xi_j} dt_{ij} + \sum_{i,j=1}^{n} \frac{\partial^2 W}{\partial\bar{\xi}_i \partial\bar{\xi}_j} d\bar{t}_{ij} = 0$$

or, equivalently,

$$4\pi i \partial W + \sum_{i,j=1}^{n} \frac{\partial^2 W}{\partial\xi_i \partial\xi_j} dt_{ij} = 0. \quad \blacksquare$$

Admitting thus the lemma the rest is easy:

$$\partial\|u\|^2 = \int (\partial u \bar{u} + u\overline{\bar{\partial}u})W + \int u\bar{u}\partial W =$$

$$= \int (\partial u\bar{u} + u\overline{\bar{\partial} u})W - \frac{1}{4\pi i}\int u\bar{u}\sum_{i,j=1}^{n}\frac{\partial^2 W}{\partial\xi_i\partial\xi_j}dt_{ij},$$

where by partial integration the last term can be written as

$$-\frac{1}{4\pi i}\int (\sum_{i,j=1}^{n}\frac{\partial^2 u}{\partial\xi_i\partial\xi_j}dt_{ij})\bar{u}W. \quad \blacksquare$$

Comparing with appropriate formulae in Sec. 2 we have now reached the principal aim of this paper: *We can identify the expression*

$$-\frac{1}{4\pi i}\sum_{i,j=1}^{n}\frac{\partial^2 u}{\partial\xi_i\xi_j}dt_{ij}$$

with the connection form for the metric connection in the Fock bundle $\mathcal{F}$. In particular, we see that the equation

(1) $$\nabla u \equiv du + \frac{1}{4\pi i}\sum_{i,j=1}^{n}\frac{\partial^2 u}{\partial\xi_i\partial\xi_j}dt_{ij} = 0 \quad (= (1')\text{ in the Introduction})$$

expresses the condition for a section u of $\mathcal{F}$ to be parallel.

To continue our program we must introduce a group action of $\mathcal{F}$. We shall show that a suitable extension $\tilde{\mathcal{G}}$ of the group $\mathcal{G} = V^* \cdot Sp(V)$ acts on $\mathcal{F}$ via fiber preserving linear isometries. It turns out that $\tilde{\mathcal{G}}$ is essentially the semidirect product of the *Heisenberg group* $Hei(V)$ of V and a double cover of $Sp(V)$, called the *metaplectic group* and denoted $Mp(V)$. From this it will then be possible to conclude that the group operations intertwine with the operator ∇ in (1). This will be done in Sec. 4. It will be convenient to begin with some general considerations.

4. Factor representations.

Consider quite generally a manifold Π and a Lie group Γ. We let Γ act on functions f defined on Π, or on a subset of Π, according to the rule

$$f(p) \mapsto f(gp)M(g,p) \quad (p \in \Pi, g \in \Gamma),$$

where M is a given nonvanishing function defined on the product set $\Gamma \times \Gamma$. In general this action shifts the domain of definition: if f is defined on the subset D then the transformed function is defined on the subset gD. Let the action of the group element g_1 be followed by the action of an element g_2. Then

$$f(p) \mapsto f(g_1g_2p)M(g_1,g_2p)M(g_2,p).$$

In order that this be the same as the effect of letting the product g_1g_2 act on f we must have

(1) $$\boxed{M(g_1g_2,p) = M(g_1,g_2p)\cdot M(g_2,p)}.$$

If this condition is fulfilled we say that we have a *cocycle* or a *factor* (or *multiplier) system.*

Remark. We consider only scalar functions. More generally, one can however also let f be a vector function. Then M has to be a matrix function.

Let us consider some general examples.

Example 1. $M(g,p) \equiv H(p)$. Then (1) reduces to

$$H(g_1 g_2) = H(g_1)H(g_2)$$

so that we have a *homomorphism.*

Example 2. Let A be any (nowhere vanishing) function on Π. Then $M(g,p) = \dfrac{A(gp)}{A(p)}$ satisfies (1). *Check*:

$$\frac{A(g_1(g_2 p))}{A(g_2 p)} \cdot \frac{A(g_2 p)}{A(p)} = \frac{A(g_1 g_2 p)}{A(p)}. \quad \blacksquare$$

We say that we have a *coboundary.* The set of all cocycles which differ by a coboundary from a given cocycle is called the *cohomology class* of the cocycle in question.

Example 3. If $M(g,p)$ is a cocycle so is $M(\gamma g \gamma^{-1}, \gamma p)$ for each $\gamma \in \Gamma$. *Check*:

$$M(\gamma g_1 \gamma^{-1}, \gamma g_2 p) M(\gamma g_2 \gamma^{-1}, \gamma p) = M(\gamma g_1 \gamma^{-1} \cdot \gamma g_2 \gamma^{-1}, \gamma p) = M(\gamma g_1 g_2 \gamma^{-1}, \gamma p). \quad \blacksquare$$

Further, if $M(g,p) = \dfrac{A(gp)}{A(p)}$ then

$$M(\gamma g \gamma^{-1}, \gamma p) = \frac{A(\gamma g \gamma^{-1} \cdot \gamma p)}{A(\gamma p)} = \frac{A(\gamma(gp))}{A(\gamma p)},$$

i.e. it is the coboundary of $A(\gamma p)$.

Example 4. Assume that Π carries a Γ-invariant measure, denoted dp, and that the group leaves invariant the metric

$$\|f\|^2 = \int |f(p)|^2 W(p) dp,$$

where W is a positive weight function, that is,

$$\int |f(gp)|^2 |M(g,p|^2 W(p) dp = \int |f(p)|^2 W(p) dp.$$

Replacing p by gp in the integral to the right we see that

$$|M(g,p)|^2 W(p) = W(gp)$$

or

$$|M(g,p)| = (\frac{W(gp)}{W(p)})^{1/2}. \tag{2}$$

That is, M must be a coboundary "up to a unimodular function". The same conclusion is correct also if dp is not invariant, only "quasi-invariant" in the technical sense: we have only to modify (2) as follows:

$$(2') \qquad |M(g,p)| = (\frac{W(gp)}{W(p)})^{1/2}(\frac{d(gp)}{dp})^{1/2},$$

where the last factor involves the Radon-Nikodym derivative.

Now we begin to specialize. We take $\Pi = \mathcal{S}$, $\Gamma = G = Sp(V)$. Then, choosing as before a splitting of V and identifying $\mathcal{S}$ with a set of matrices, we may take

$$M(g,t) = \det(ct+d) \quad \text{if } g = \begin{vmatrix} a & b \\ c & d \end{vmatrix} \in G, t \in \mathcal{S}.$$

Indeed, we have

$$M(g_1g_2,t)\cdot M(g_2,t) = \det(c_1(a_2t+b_2)(c_2t+d_2)^{-1}+d_1)\cdot\det(c_2t+d_2) =$$

$$= \det(c_1(a_1t+b_2)) + d_1(c_2t+d_2)) = \det((c_1a_2+d_1c_2)t + c_1t_2 + d_1a_2)$$

which is the same as

$$M(g_1g_2,t),$$

taking into account that

$$\begin{vmatrix} a_1 & b_1 \\ c_1 & d_1 \end{vmatrix} \cdot \begin{vmatrix} a_2 & b_2 \\ c_2 & d_2 \end{vmatrix} = \begin{vmatrix} * & * \\ c_1a_2+d_1c_2 & c_1b_2+d_1d_2 \end{vmatrix}.$$

More generally, we may take powers of this particular cocycle. If the exponent is not an integer we must pass to a covering group of $G = Sp(V)$. In particular,

$$M(g,t) = \det(ct+d)^{1/2}$$

is a cocycle for the *metaplectic group* $\tilde{G} = Mp(V)$. The elements of $\tilde{G}$ are pairs of the form

$$(\text{an element } g = |\begin{matrix} a & b \\ c & d \end{matrix} \text{ of } G, \text{a determination of the analytic function } (ct+d)^{1/2}).$$

Next we take $\Pi = \mathcal{B}^*$, $\Gamma = \tilde{\mathcal{G}} = Hei(V)\cdot Mp(V)$. We shall construct a certain factor system in the form

$$(3) \qquad M(g,\Lambda) = \frac{e^{i\pi m(g,\Lambda)}}{\det(ct+d)^{1/2}} \quad \text{where } g \in \tilde{\mathcal{G}},\ \Lambda = (t,\xi) \in \mathcal{B}^*,$$

that is, the function m has to satisfy the additive analogue of formula (1), viz.

$$(1') \qquad m(g_1g_2,p) = m(g_1,g_2p) + m(g_2,p);$$

we will still think of it as a cocycle.

For simplicity we treat each factor separately starting with the second one. Thus we set

$$m(g,\Lambda) = -\langle (ct+d)^{-1}c\xi,\xi\rangle \quad (g \in Sp(V)).$$

The verification of (1′) in this case is tedious but fully possible: We have to show that

$$\langle c_1^\top (c_1(a_2t+b_2)(c_2t+d_2)^{-1} + d_1)^\top (c_2t+d_2)^\top \xi, (c_2t+d_2)^\top \xi\rangle + \langle (c_2t+d_2)^{-1}c_2\xi,\xi\rangle =$$

$$= \langle (ct+d)^{-1}c\xi,\xi\rangle.$$

As before we can contract the two first parantheses in the first term into one:

$$(c_1(a_2t+b_2)(c_2t+d_2)^{-1} + d_1)^\top (c_2t+d_2)^\top = (ct+d)^\top.$$

Thus we have to verify the matrix identity

$$(c_2t+d_2)^{-1}c_1^\top (ct+d)^\top + (c_2t+d_2)^{-1}c_2 = c^\top (ct+d)^\top$$

or

$$c_1^\top + c_2(tc^\top + d^\top) = (c_2t+d_2)c^\top$$

or again

$$c_1^\top + c_2d^\top = d_2c^\top.$$

Now $d = c_1b_2 + d_1d_2$, $c = c_1a_2 + d_1c_2$. Therefore the latter formula is equivalent to

$$c_1^\top + c_2b_2^\top c_1^\top + c_2d_2^\top d_1^\top = d_2a_2^\top c_1^\top + d_2c_2^\top d_1^\top.$$

But, as we are dealing with symplectic matrices then (see Sec. 1)

$$d_2a_2^\top - c_2b_2^\top = 1,\ c_2d_2^\top = d_2c_2^\top.$$

Therefore the sum of the first two terms to the left equals the first term to the right, while the last two of both members are equal. ∎

But it is much simpler to verify directly that m is a coboundary, the coboundary of

$$a(\Lambda) = \frac{1}{2i}\langle (\mathrm{Im}t)^{-1}\xi,\xi\rangle.$$

That is, we have to verify that

$$a(g\Lambda) - a(\Lambda) = m(g,\Lambda).$$

The proof depends on the formula

$$\mathrm{Im}gt = ((ct+d)^{-1})^*\mathrm{Im}t(ct+d)^{-1},$$

which is easily verifiable and which, incidentally, also provides a new proof (the standard one) for the fact that $G = Sp(V)$ operates on $\mathcal{S}$. We have

$$2i \cdot a(\Lambda) = \langle (\mathrm{Im}t)^{-1}(tc^{\top} + d^{\top})(ct + d\widetilde{)}\xi, \xi \rangle,$$

$$2i \cdot a(g\Lambda) = \langle (\mathrm{Im}(gt))^{-1}(ct + d\widetilde{)}\xi, (ct + d\widetilde{)}\xi \rangle =$$

$$= \langle (ct + d)(\mathrm{Im}t)^{-1}(\bar{t}c^{\top} + d^{\top})(ct + d\widetilde{)}\xi, (ct + d\widetilde{)}\xi \rangle = \langle (\mathrm{Im}t)^{-1}(\bar{t}c^{\top} + d^{\top})(ct + d\widetilde{)}\xi, \xi \rangle.$$

Subtracting it follows that

$$a(g\Lambda) - a(\Lambda) = \frac{1}{2i}\langle (\mathrm{Im}t)^{-1}(\bar{t} - t)c^{\top}(ct + d\widetilde{)}\xi, \xi \rangle =$$

$$= -\langle c^{\top}(ct + d\widetilde{)}\xi, \xi \rangle = m(g, \Lambda). \quad \blacksquare$$

Remark. Alternatively, one can base the proof on the formula

(4) $$\langle \partial \log \det \mathrm{Im}t, \xi \otimes \xi \rangle = \langle (\mathrm{Im}t)^{-1}\xi, \xi \rangle.$$

(The brackets to the left refer to the natural duality in matrix space, viz. $\langle a, b \rangle = \mathrm{trace}(ab^{\top})$.) In the Cayley transformed version the analogue of (4), apparently, reads

(4′) $$\langle \partial \log \det(1 - \bar{s}s), \eta \otimes \eta \rangle = \langle \bar{s}(1 - \bar{s}s)^{-1}\eta, \eta \rangle.$$

Notice that (4) reveals an analogue between the special multiplier $m(g, \Lambda)$ and the bracket $\{\,,\,\}_1$ in uniformization theory (see Gunning (1966), Gustafsson, and Peetre (1987)). The former gives thus also a measure for g to differ from an "affine" transformation.

We claim now that the action just constructed induces an action on the Fock bundle $\mathcal{F}$ via fiber preserving linear isometries; this together with the corresponding easier fact for the first factor (the Heisenberg group) is one of the main contentions of this Section.

Only the fact that the operators in question, viz. the transformations

$$f(t, \xi) \mapsto f((at + b)(ct + d)^{-1}, (ct + d\widetilde{)}\xi) \exp\left(-\pi i \langle (ct + d)^{-1}c\xi, \xi \rangle\right) \frac{1}{\det(ct + d)^{1/2}},$$

are unitary has to be verified. But this is easy. Write the norm in $\mathcal{F}_t$ as $\|f\|_t^2 = \int |f|^2 W$ where $W = e^{\pi w}$ with

$$w(t, \Lambda) = 2\langle (\mathrm{Im}t)^{-1}\mathrm{Im}\xi, \mathrm{Im}\xi \rangle = \mathrm{Re}\langle (\mathrm{Im}t)^{-1}\xi, \xi \rangle - \langle (\mathrm{Im}t)^{-1}\xi, \bar{\xi} \rangle.$$

In view of the invariance of the last term in this formula and by our previous computation, we then have

$$w(g\Lambda) - w(\Lambda) = \frac{i}{2}m(g, \Lambda) \quad \mathrm{mod}\ i\mathbf{R};$$

this is the additive version of (2′).

Finally, we come to the corresponding treatment of the first factor.

Recall that the *Heisenberg group* $Hei(V)$ is a central extension of the additive group V. Its element are pairs $g = (X, k)$ with $X \in V$, $k \in \mathbf{C}_1^\times$ and the composition rule is

$$g_1 g_2 = (X_1, k_1) \cdot (X_2, k_2) = (X_1 + X_2, k_1 k_2 e^{i\pi\sigma(X_1, X_2)}).$$

In the following we write $X = (\varepsilon, e)$ with $\varepsilon, e \in \mathbf{R}^n$ corresponding to our selected splitting of V. We define a cocyle in the form

$$M(g, \Lambda) = e^{i\pi m(g,\Lambda)} k$$

where

$$m(g, \Lambda) = \langle te, e\rangle + 2\langle \xi, e\rangle + \langle \varepsilon, e\rangle.$$

Then

$$m(g_1, g_2\Lambda) + m(g_2, \Lambda) = \langle te_1, e_1\rangle + 2\langle \xi + te_2 + \varepsilon_2, e_1\rangle + \langle te_2, e_2\rangle + 2\langle \xi, e_2\rangle + \langle \varepsilon_1, e_1\rangle + \langle \varepsilon_2, e_2\rangle =$$

$$= \langle t(e_1 + e_2), (e_1 + e_2)\rangle + 2\langle \xi, e_1 + e_2\rangle + 2\langle \varepsilon_2, e_1\rangle + \langle \varepsilon_1, e_1\rangle + \langle \varepsilon_2, e_2\rangle,$$

while

$$m(g_1 g_2, \Lambda) = \langle t(e_1 + e_2), e_1 + e_2\rangle + 2\langle \xi, e_1 + e_2\rangle + \langle \varepsilon_1 + \varepsilon_2, e_1 + e_2\rangle.$$

These two expressions differ by the expression

$$\sigma(X_1, X_2) = \langle \varepsilon_1, e_2\rangle - \langle \varepsilon_2, e_1\rangle.$$

This establishes the cocycle relation (1′) in the present case.

Explicitly, we have thus the action

$$f(t, \xi) \mapsto f(t, \xi + te + \varepsilon) \exp\left(i\pi(\langle te, e\rangle + 2\langle \xi, e\rangle + \langle \varepsilon, e\rangle)\right).$$

A simple computation reveals that these transformations preserve the metric. As in this case also each individual fiber is left invariant (in general case, the fibers are permuted by the group operators), we have thus constructed a unitary representation of the Heisenberg group $Hei(V)$ on each fiber $\mathcal{F}_t$ of the Fock bundle $\mathcal{F}$. Taking $t = i1 = i$ we have essentially the *Bargmann-Segal representation* up to a gauge transformation. If we pass to the Harish-Chandra realization of $\mathcal{S}$ we get this representation in its habitual form ($t = i$ corresponds to $s = 0$ and the space $\mathcal{F}_i$ corresponds to $\mathcal{F}^0$). In Sec. 5 we will, similarly, construct the *Shale-Weil representation* of the metaplectic group $\tilde{G} = Mp(V)$ (also known as the "metaplectic representation").

We can thus now make the conclusion, already mentioned at the end of Sec. 3, that the group $\tilde{\mathcal{G}}$ intertwines with the operator ∇. Further application of the presence of the group action will be made in Sec. 6.

We conclude the present Section with an important observation.

Remark. In our previous discussion a serious drawback has been the fact that our definition of the Fock bundle $\mathcal{F}$ depends on an explicit choice of symplectic coordinates in the vector space V. In this remark we outline an *alternative* coordinate free approach.

Recall (see e.g. Kobayashi and Nomizu (1963/69)) that a line bundle $\mathcal{L}$ on a manifold Π is determined by the following data: 1° an open covering $\{U_\alpha\}$ of Π and 2° a family of nowhere vanishing smooth functions $\{\phi_{\alpha\beta}\}$ ("transition functions") on the overlaps $U_\alpha \cap U_\beta \neq \emptyset$ such that on each overlap $U_\alpha \cap U_\beta \cap U_\gamma \neq \emptyset$ holds

$$\phi_{\alpha\beta} \cdot \phi_{\beta\gamma} \cdot \phi_{\gamma\alpha} = 1. \tag{4}$$

Here α, β, γ are supposed to run through some fixed preassigned "index set". Such a system of functions together, with the corresponding covering, will likewise be referred to as a cocycle. (If the manifold is complex and all the transition functions are holomorphic, we have a holomorphic line bundle.) Now we describe the relation to cocycles in the previous sense (factor systems). Let thus a function $M(g,p)$, subject to formula (1), be given. We take the set of indices to be the group Γ acting on the manifold Π, and define

$$\varphi_{gh}(p) = M(gh^{-1}, hp),$$

putting also $U_g = \Pi$ for all g. Then we can of course write

$$\varphi_{hk}(p) = M(hk^{-1}, kp)$$

and so by (1) we obtain

$$\varphi_{gh}(p) \cdot \varphi_{hk}(p) = M(gh^{-1}, hp)M(hk^{-1}, kp) =$$

$$= M(gh^{-1}, hk^{-1}kp)M(hk^{-1}, kp) = M(gh^{-1}hk^{-1}, kp) = M(gk^{-1}, kp) = \varphi_{gk}(p);$$

in other words, (1) is fulfilled. Thus to M there corresponds a line bundle $\mathcal{L}$. Quite generally, the elements of the fiber $\mathcal{L}_p$ at a point p are families $x = \{x_\alpha\}$ of numbers such that $x_\alpha = \varphi_{\alpha\beta}(p)x_\beta$, in the present case thus families $x = \{x_g\}$ such that $x_g = M(gh^{-1}, hp)x_h$. We can now define a group action on $\mathcal{L}$. Let $x = \{x_g\} \in \mathcal{L}_p$ and $g_1 \in \Gamma$ be given. Then we set $g_1 x = \{x_{gg_1}\}$. As we can write

$$x_{gg_1} = M(gh^{-1}, hg_1p)x_{hg_1},$$

it follows that $g_1 x \in \mathcal{L}_{g_1 p}$. Furthermore

$$g_1(g_2 x) = \{x_{gg_1g_2}\} = (g_1g_2)x,$$

so that there is indeed a group action. Thus M determines not only a line bundle $\mathcal{L}$ but also an action of the group Γ on $\mathcal{L}$. Note that $\mathcal{L}$ is topologically trivial (as all U_g coincide with Π). Suppose that, in the general case, we have a family $\{W_\alpha\}$ of positive smooth functions which "transform" according to the rule

$$W_\alpha = |\varphi_{\alpha\beta}|^{-2} W_\beta \text{ on } U_\alpha \cap U_\beta \neq \emptyset.$$

Then we can define the norm of $x = \{x_\alpha\} \in \mathcal{L}_p$ by the formula

$$\|x\|^2 = |x_\alpha|^2 W_\alpha.$$

In the group case the preceding formula is replaced by

$$W_g(p) = |M(gh^{-1}, p)|^{-2} W_h(p).$$

We apply the previous considerations in the case

$$\Gamma = \tilde{\mathcal{G}}, \Pi = \mathcal{B}^*, M(g,p) = \text{ the cocycle in}\,(3).$$

We obtain now a holomorphic line bundle $\mathcal{L}$ over the cotautological bundle $\mathcal{B}^*$ over $\mathcal{S}$ on which the group $\tilde{\mathcal{G}}$ acts. Our previous computations show that $\mathcal{L}$ also carries an invariant metric. We can therefore *redefine* $\mathcal{F}$ as follows. We let the elements of the fiber $\mathcal{F}_P$ over $P \in \mathcal{S}$ be holomorphic sections τ of $\mathcal{L}$ *along* the fiber $\mathcal{B}^*_P$ of $\mathcal{B}^*$ at P such that

$$\int_{\mathcal{B}^*_P} \|\tau\|^2 < \infty,$$

where we integrate with respect to the invariant measure in $\mathcal{B}^*_P$. Thus, compared to Sec. 1, the exponential factor is now incorporated in the norm.

5. The Shale-Weil representation.

This is a simple application of the ideas expressed in Cor. 2 to Prop. 2 in Sec. 2. Indeed, let f be any element of the Fock space $\mathcal{F}_i$ and consider the solution of the equation $\nabla u = 0$, where ∇ is the operator in Sec. 3, with initial datum $u(i) = f$. The application of a transformation g in $Mp(V)$ produces a new solution $\tilde{u}$. We let the new initial datum $\tilde{f} = \tilde{u}(i) \in \mathcal{F}_i$ correspond to f. Clearly the map $f \mapsto \tilde{f}$ is unitary so we obtain a unitary group action (the Shale-Weil or metaplectic representation). Let us make this explicit.

It will be convenient to work with the Harish-Chandra realization of $\mathcal{S}$, i.e. $\mathcal{S}$ is parametrized by symmetric matrices s with $\bar{s}s < 1$, so we are dealing with the standard Fock space $\mathcal{F}^0$, which consists of analytic functions in the variable w. (In this way it becomes easier to make a comparison with the formulae in Peetre (1988?).) Let v be related to u by formula (7) in Sec. 1. Then the analogue of equation (1) in Sec. 3 reads

$$\nabla v \equiv dv + \frac{1}{2\pi} \sum_{i,j=1}^{n} \frac{\partial^2 v}{\partial w_i \partial w_j} ds_{ij} = 0.$$

The basic observation is now the following. Introduce the *"Gauss-Weierstrass functions"*

$$E_{s,w}(z) = \exp(-\frac{\pi}{2}\langle \bar{s}z, z\rangle + \pi\langle \bar{w}, z\rangle)$$

(other names for basically the same objects: *Gabor wavelet, wave packet, coherent state, Gaussian density* etc.). Then we have

PROPOSITION. *The solutions of the differential equation $\nabla v = 0$, where v is a section of $\mathcal{F}$ (that is, $v(s) \in \mathcal{F}^s$, $s \in \mathcal{S}$), with initial datum $v(0) = f \in \mathcal{F}^0$ is given by*

$$(1) \qquad v(s,w) = (f, E_{s,w})_0$$

or, explicitly,

$$v(s,w) = \int_{\mathbf{C}^n} e^{-\frac{\pi}{2}\langle s\bar{z}, \bar{z}\rangle + \pi\langle w, \bar{z}\rangle} e^{-\pi\langle z,\bar{z}\rangle} \times$$
$$\times (i/2)^n dz_1 d\bar{z}_1 \dots dz_n d\bar{z}_n.$$

Proof. It is clear that v is a solution of $\nabla v = 0$. That it satisfies the initial condition $v(0) = f$ follows from the well-known fact (cf. e.g. Janson, Peetre, and Rochberg (1987)) that the function $e_{0,w}$ equals the reproducing kernel of the Hilbert space $\mathcal{F}^0$. ∎

Let g be an element of $Mp(V)$ represented by the block matrix $\begin{vmatrix} a & b \\ c & d \end{vmatrix}$. Then by the preceding discussion

$$\tilde{v}(s,w) = v((as+b)(cs+d)^{-1}, (cs+d\tilde{)}w) \cdot e^{\pi\langle (cs+d)^{-1}cw, w\rangle} \cdot (\det(cs+d))^{-1/2}.$$

Making $s = 0$ in this formula we obtain from (1)

$$\tilde{f}(w) = \int_{\mathbf{C}^n} f(z) e^{-\pi\langle bd^{-1}\bar{z}, \bar{z}\rangle + 2\pi\langle \tilde{d}w, \bar{z}\rangle} e^{-\pi\langle z\bar{z}\rangle} \times$$
$$\times (\text{volume element}) \cdot e^{\pi\langle d^{-1}cw, w\rangle} \cdot (\det d)^{-1/2}.$$

Thus the linear map $f \mapsto \tilde{f}$ is given by the kernel

$$\mathcal{A}_g(w,z) = e^{\pi(\langle Aw, w\rangle + 2\langle B\bar{z}, w\rangle + \langle C\bar{z}, \bar{z}\rangle)} \cdot (\det B)^{1/2},$$

where thus

$$A = d^{-1}c,\ B = d^{-1},\ C = -bd^{-1}.$$

These formulae should be compared with e.g. those in Peetre (1988?), where essentially the same formulae (for $n = 1$) were obtained as a result of a direct computation.

6. Some reproducing formulae.

In this Section we are going to make use of Prop. 3 in Sec. 3 or rather the same idea in the case of infinite rank.

First we write down the reproducing kernel in $\mathcal{F}_t$. It is the function

$$(1) \qquad K_t(\eta,\xi) = K_{t,\xi}(\eta) = e^{-\frac{\pi}{2}\langle (\mathrm{Im}t)^{-1}(\eta-\bar{\xi}), \eta-\bar{\xi}\rangle} / |\det \mathrm{Im}t|^{1/2}.$$

In fact, if Imt is a diagonal matrix this can be read off from Janson, Peetre, and Rochberg (1987) and the general case can be readily brought back to this special case. Again, using

(1) and the transformation properties of reproducing kernels under a gauge transformation (see again Janson, Peetre, and Rochberg (1987)) or by some direct device or other one sees that the reproducing kernel in $\mathcal{F}^s$ is

$$K^s(z,w) = K^{s,w}(z) = \tag{2}$$

$$= e^{\pi(\frac{1}{2}\langle \bar{s}(1-s\bar{s})^{-1}z,z\rangle + \frac{1}{2}\langle s(1-\bar{s}s)^{-1}\bar{w},\bar{w}\rangle + \langle \mathrm{Re}(1-\bar{s}s)^{-1}z,\bar{w}\rangle)} / (\det(1-\bar{s}s))^{1/2}.$$

Proof that (1) implies (2). In the expression for $K_{t\xi}(\eta)$ in (1) we plug $\eta = (s+1)^{-1}z$, $\xi = (s+1)^{-1}w$ and multiply with the necessary factors (see formula (7) in Sec. 1). Then $\eta - \bar{\xi} = (s+1)^{-1}z - (\bar{s}+1)^{-1}\bar{w}$. Thus the exponent becomes (we omit a factor π)

$$\langle(\bar{s}+1)(1-s\bar{s})^{-1}(s+1)(s+1)^{-1}z - (s+1)(1-\bar{s}s)^{-1}(\bar{s}+1)(\bar{s}+1)^{-1}\bar{w}, (s+1)^{-1}z - (\bar{s}+1)^{-1}\bar{w}\rangle -$$

$$-\langle (s+1)^{-1}z,z\rangle - \pi\langle(\bar{s}+1)^{-1}\bar{w},\bar{w}\rangle =$$

$$= \langle (1-s\bar{s})^{-1}z,\bar{w}\rangle + \langle(1-\bar{s}s)^{-1}z,\bar{w}\rangle + \langle \bar{s}(1-s\bar{s})^{-1}sz,z\rangle + \langle s(1-\bar{s}s)^{-1}\bar{w},\bar{w}\rangle.$$

The determinental factors are easy to take care of. The expression for $K^{s,w}(z)$ evolves. ∎

In what follows we prefer to work with $\mathcal{F}^s$ and $K^{s,w}$, rather than with $\mathcal{F}_t$ and $K_{t,\xi}$, this in order to be able to compare with Peetre (1988?).

The analogue of formula (3) in Sec. 2 reads

$$\boxed{(K^{s,w}, K^{s,z})_0 = K^s(z,w)}.$$

Next we invoke the reasoning leading to (4) in Prop. 3 of the same Section. We find

$$\boxed{(E_{s,w}, E_{s,z})_0 = K^s(z,w)}.$$

Explicitly:

$$\int_{\mathbf{C}^n} E_{s,w}(\zeta)\overline{E_{s,z}(\zeta)}e^{-\pi\langle\zeta,\bar{\zeta}\rangle} \times (\text{volume element}) = \tag{3}$$

$$= \exp\left(\frac{\pi}{2}(\langle\bar{s}(1-s\bar{s})^{-1}z,z\rangle + \langle s(1-\bar{s}s)^{-1}\bar{w},\bar{w}\rangle + 2\langle \mathrm{Re}(1-\bar{s}s)^{-1}z,\bar{w}\rangle)\right) / (\det(1-\bar{s}s))^{1/2}.$$

But we can do more! We claim that

$$\int_{\mathbf{C}^n} E_{s,w}(\zeta)\overline{E_{r,z}(\zeta)}e^{-\pi\langle\zeta,\bar{\zeta}\rangle} \times (\text{volume element}) = \tag{4}$$

$$= \exp\left(\frac{\pi}{2}(\langle\bar{s}(1-r\bar{s})^{-1}z,z\rangle + \langle r(1-\bar{s}r)^{-1}\bar{w},\bar{w}\rangle + \right.$$

$$\left. + \langle(1-\bar{s}r)^{-1}z,\bar{w}\rangle) + \langle(1-r\bar{s})^{-1}z,\bar{w}\rangle)\right) / (\det(1-\bar{s}r))^{1/2}.$$

The proof is obvious. Both members of (4) are analytic in r, z and anti-analytic in s, w and if $r = s$, $z = w$ the formula reduces to (3). Thus by a well-known lemma in function theory it must be correct in general. (We are assuming that r, s are symmetric $n \times n$ matrices with $\bar{s}s < 1$, $\bar{r}r < 1$, z, w being arbitrary vectors in $\mathbf{C}^n$.) For the proof see e.g. Janson, Peetre and Rochberg (1987).

If $n = 1$ a formula of the type (4) was proved in Peetre (1988?). We have thus here an alternative "conceptual" proof, essentially not using integral calculus.

7. Comments.

In this section we indicate several topics for future study.

A. At an early stage of this investigation we studied the vector space of L of all solutions of the equation $\nabla u = 0$ (or $\nabla v = 0$) with initial data in a Fock space, which apparently is a Hilbert space,[1] and we looked for a kind of canonical norm for this space (something like an "energy integral"). We do not any longer believe that there exists such a norm. However, the following problem remains: Describe all positive measures μ on $\mathcal{B}^*$, to fix the ideas with $\mathcal{S}$ parametrized by symmetric matrices with positive definite imaginary part, such that

$$\int_{\mathbf{C}^n} |u(t,\xi)|^2 d\mu(t,\xi) \leq C\|u\|_i^2 \text{ for all } u \in L,$$

with some positive constant C. Such measures may be termed *"heat-Carleson" measures* (regarding generalized Carleson measures in general, cf. Peetre (1985), (1986)). Of course, one also has the question of the reversed inequality.

B. The space L is a $\tilde{G}$-invariant Hilbert space with reproducing kernel. It is of note that the reproducing kernel can be identified with the Gauss-Weierstrass functions $E_{s,w}$. As indicated already in Peetre (1988?), it is also of interest to study $\tilde{G}$-invariant Banach spaces (of solutions of $\nabla u = 0$ or $\nabla u = 0$), for instance, the "minimal" space generated by the $E_{s,w}$.

C. Recall (Janson, Peetre, and Rochberg, 1987) that, generally speaking, a *Hankel form* on a Hilbert space of analytic functions is a bilinear form $\Gamma(f, g)$ on this space whose value only depends on the product $f \cdot g$. Returning to our case, we can use the metric connection to move Hankel forms on one fiber of the Fock bundle to another fiber (in other words, by "parallel transport"). It turns out that the new form arising in this way may be viewed as a generalized Hankel form in the sense of Janson, Peetre, and Wallstén (1989?) of a special kind. (Its "Toeplitz part" is again a Gauss-Weierstrass function.) Note that this is practically self-evident on group theoretical grounds but it is instructive to carry out the details.

D. We begun this paper with theta functions. It is therefore appropriate to conclude with a few words about them. Roughly speaking, theta functions arise when passing to a quotient bundle of the cotautological bundle $\mathcal{B}^*$ (a bundle whose fibers are complex tori, thus not any longer a vector bundle). In the definition of the norm we then have to integrate over a fundamental domain. Actually, this can be done in two steps. In the first step one reckons *modulo* transformations $\lambda \mapsto \lambda + te + \varepsilon$ where ε is restricted to be an integer vector ($\varepsilon \in \mathbf{Z}^n$), while at the second step one allows only transformation where both $e, \varepsilon \in \mathbf{Z}^n$. Thus the fibers are "cylinders" respectively tori. It is remarkable that

[1] As all fibers are canonically isomorphic, we may view L as a kind of "master" Fock space.

both type of manifolds also arise in Berezin's theory of quantization of Kähler manifolds (Berezin 1975).

References

Arazy, J., Fisher, S., and Peetre, J. (1989?), Hankel operators on planar domains, *Amer. Math. J.* (to appear).

Berezin, F.A. (1975), General concept of quantization, *Comm. Math. Phys. 40* , 153-174.

Gunning, R.G. (1966), *Lectures on Riemann surfaces*, Princeton University Press, Princeton.

Gustafsson, B., and Peetre, J. (1987), *Notes on projective structures on complex manifolds*, technical report, Stockholm.

Jacobi, C.G.J. (1829), Fundamenta nova theoriae functiorum ellipticum, *Werke I*, pp. 49-239.

Janson, S., and Peetre, J. (1989?), Weak factorization in periodic Fock space, *Math. Nachr.* (to appear).

Janson, S., Peetre, J., and Rochberg, R. (1987), Hankel forms and the Fock space,*Revista Mat. Iberoamer. 3*, 61-138.

Janson, S., Peetre, J., and Wallstén, R. (1989?), A new look on Hankel forms over Fock space, *Studia Math.* (to appear).

Kobayashi, S., and Nomizu, K. (1963/69), *Foundations of differential geometry 1-2*, Interscience, New York.

Peetre, J. (1980a), A class of kernels related to the inequalities of Beckner and Nelson, in: *A tribute to Åke Pleijel*, Uppsala, pp. 171-210.

Peetre, J. (1980b), *On the action of the Mehler transform on some special functions*, technical report, Lund.

Peetre, J. (1985), Paracommutators and minimal spaces, in: S.C. Power (ed.), *Operators and function theory*, Reidel, Dordrecht, pp. 163-224.

Peetre, J. (1986), Invariant function spaces and Hankel operators - a rapid survey, *Expositiones Mathematicae 5*, 3-16.

Peetre, J. (1988?), Some calculations related to Fock space and the Shale-Weil representation, *Integral Equations Operator Theory* (to appear).

Abstract

It was already known to Jacobi that theta functions satisfy a PDE, namely the equation

$$4\pi i\frac{\partial u}{\partial t} = \frac{\partial^2 u}{\partial \xi^2}. \tag{1}$$

He also showed that these functions have remarkable invariance properties under two sets of transformation of variables:

$$(t,\xi) \mapsto (\frac{at+b}{ct+d}, \frac{\xi}{ct+d}),$$

$$(t,\xi) \mapsto (t, \xi + tv + \nu).$$

On the other hand, equation (1) too transforms nicely under precisely the same transformations.

Similar formulas hold in several variables. The analogue of (1) may be written as

$$4\pi i du = \sum_{i,j}^{n} \frac{\partial^2 u}{\partial \xi_i \partial \xi_j} dt_{ij}. \tag{1'}$$

Our aim is to try to understand the invariances mentioned from a higher point of view. In fact, we will introduce a new object called the *Fock bundle*. It is a Hermitean vector bundle of *infinite rank* over Siegel's generalized upper halfplane (the manifold of all symmetric matrices t with positive definite imaginary parts) whose fibers are Hilbert spaces of entire analytic functions in the variable ξ (Fock spaces). Being an Hermitean vector bundle it carries a natural metric connection. Our main contention is that (1) or - in the case of several variables - (1′) can be viewed as the condition for parallelism for this connection. This has several pleasant features. First of all it becomes obvious that our equation has the desired invariance properties. Furthermore, one gets e.g. a painless construction of the Weil-Shale representation of the metaplectic group, and of the Bargmann-Segal representation of the Heisenberg group as well.

Classification: 30C40, 32M15, 47B35.

17

A Local Version of a Theorem of Khinchin

Yoram Sagher and Kecheng Zhou

Abstract

We prove a local Khintchin's inequality for Rademacher functions.

It is well known that for Rademacher functions $\{\phi_k(t)\}_{k=0}^{\infty}$ defined as

$$\phi_0(t) = \begin{cases} 1 & 0 < t < 1/2 \\ -1 & 1/2 < t < 1 \end{cases},$$

$$\phi_0(0) = \phi_0(1/2) = 0, \qquad \phi_0(t+1) = \phi_0(t),$$

and

$$\phi_k(t) = \phi_0(2^k t), \qquad k = 1, 2, \cdots.$$

Khintchin's theorem

$$k_p(\sum |c_k|^2)^{1/2} \le (\int_0^1 |\sum c_k\phi_k(t)|^p\, dt)^{1/p} \le K_p(\sum |c_k|^2)^{1/2}$$

holds, where $0 < p < \infty$, k_p, K_p are constants depending only on p and $\{c_k\}$ is any sequence of complex or real numbers so that $\sum |c_k|^2 < \infty$. In [1] a version of Khintchin inequality on any measurable set $E \subseteq [0,1]$ with $|E| > 0$ and $p = 2$ is given. Here we present a generalization valid for all values of p.

Theorem 1 *For any $0 < p < \infty$, there exist constants A_p, B_p so that for any measurable set $E \subseteq [0,1]$, $0 < |E|$, there exists an $N = N(E)$ so that if $P(t) = \sum_{k=N}^{\infty} c_k\phi_k(t)$, where $\sum |c_k|^2 < \infty$, then*

$$A_p(\sum_{k=N}^{\infty} |c_k|^2)^{1/2} \le (\frac{1}{|E|}\int_E |P(t)|^p\, dt)^{1/p} \le B_p(\sum_{k=N}^{\infty} |c_k|^2)^{1/2}.$$

Proof: Assume that $P(t) = \sum_{k=N}^{n} c_k \phi_k(t)$, where c_k's are real numbers. Consider first the case $p = 2$. We have

$$\int_E | P(t) |^2 \, dt = \int_E P^2(t) dt$$
$$= (\sum_{i=N}^{n} c_i^2) | E | +2 \sum_{N \leq i < j \leq n} \int_E c_i c_j \phi_i(t) \phi_j(t) dt.$$

The second term does not exceed, in absolute value, $2(\sum_{N \leq i<j \leq n} (\int_E \phi_i \phi_j dt)^2)^{1/2} (\sum_{i=N}^{n} c_i^2)$. Given $\epsilon > 0$, we can choose N so that $\sum_{N \leq i<j \leq n} (\int_E \phi_i \phi_j dt)^2 < \epsilon$. This is so because $\int_E \phi_i \phi_j dt$ is the Walsh coefficient of χ_E. We therefore have

$$| \frac{1}{| E |} \int_E P^2(t) dt - \sum_{i=N}^{n} c_i^2 | \leq \epsilon \sum_{i=N}^{n} c_i^2$$

and we have the case $p = 2$.

Consider the case $2 < p < \infty$. The left-hand side inequality is a simple consequence of Holder's inequality and the result for $p = 2$. Again because of Holder's inequality, it is enough to prove the inequality on the right for p even integers. Given $E \subseteq [0, 1]$ with $| E | > 0$. For $k = 0, 1, 2, \cdots; \; i = 1, 2, \cdots, 2^k$, let $I_{k,i}$ be the binary intervals $(i-1)2^{-k} \leq x < i2^{-k}$. Let us show first that it is enough to consider $E = \bigcup_j I_{k_j, i_j}$ with I_{k_j, i_j} disjoint. Given E, we begin with $k = 0$. If $| E \cap I_{0,1} | = | E | \geq \frac{1}{2} | I_{0,1} |$, we choose $I_{0,1}$, and stop. If $| E \cap I_{0,1} | < \frac{1}{2}$, consider $E \cap I_{1,1}$ and $E \cap I_{1,2}$. If $| E \cap I_{1,1} | > \frac{1}{2} | I_{1,1} |$, choose $I_{1,1}$. Note that necessarily $| E \cap I_{1,2} | < \frac{1}{2} | I_{1,2} |$. We proceed to the next level, $k = 2$, disregarding all subsequent divisions of $I_{1,1}$. In this fashion we obtain a sequence of disjoint intervals $\{I_{k_j, i_j}\}$. Let $F = \bigcup_j I_{k_j, i_j}$. Clearly $| F | \leq 2 | E |$. Moreover F contains all points of density of E, so that $\chi_E \leq \chi_F$ almost everywhere. Assume that the local version of

Khintchin's inequality holds for F, then

$$(\frac{1}{|E|}\int_E |\sum c_k\phi_k|^p\, dt)^{1/p} \le (\frac{2}{|F|}\int_F |\sum c_k\phi_k|^p\, dt)^{1/p} \le 2^{1/p}B_p(\sum c_k^2)^{1/2}.$$

Hence we can restrict our attention to sets $F = \bigcup I_{k_j,i_j}$. Let F be such a set. We can write $F = F_1 \bigcup F_2$, where $F_1 \bigcap F_2 = \phi$, F_1 is a union of finitely many binary intervals and $|F_2|^{1/2} \le |F|$. We then have:

$$(\frac{1}{|F|}\int_F |P(t)|^p\, dt)^{1/p} \le (\frac{1}{|F_1|}\int_{F_1} |P(t)|^p\, dt)^{1/p} + (\frac{1}{|F_2|^{1/2}}\int_{F_2} |P(t)|^p\, dt)^{1/p}.$$

For the first term on the right, there exists an N, so that $\phi_k(t)_{k\ge N}$ changes sign at least once on each binary component of F_1. We then have just as in the global version of Khintchin's theorem:

$$(\frac{1}{|F_1|}\int_{F_1} |P(t)|^p\, dt)^{1/p} \le B_p'(\sum_{k=N}^{n} c_k^2)^{1/2},$$

where B_p' depends only on p. For the second term on the right, we have

$$\frac{1}{|F_2|^{1/2}}\int_{F_2} |P(t)|^p\, dt \le (\int_{F_2} |P(t)|^{2p}\, dt)^{1/2}$$
$$\le (\int_0^1 |P(t)|^{2p}\, dt)^{1/2} \le K_{2p}(\sum_{k=N}^{n} c_k^2)^{p/2}.$$

It follows that the local Khintchin's inequality holds in the case $2 < p < \infty$. The extension to complex c_k is trivial.

Consider the case $0 < p < 2$. The right-hand side inequality follows from Holder's inequality and the result for $p = 2$. To prove the left-hand side inequality for $0 < p < 2$, we write $\frac{1}{2} = \frac{1-\theta}{p} + \frac{\theta}{4}$ for some θ, $0 < \theta < 1$, and have

$$(\frac{1}{|E|}\int_E |P(t)|^2\, dt)^{1/2} \le (\frac{1}{|E|}\int_E |P(t)|^p\, dt)^{(1-\theta)/p}(\frac{1}{|E|}\int_E |P(t)|^4\, dt)^{\theta/4}$$
$$\le B_4^\theta(\sum_{k=N}^{n} |c_k|^2)^{\theta/2}(\frac{1}{|E|}\int_E |P(t)|^p\, dt)^{(1-\theta)/p}.$$

It follows immediately that

$$A_p\Big(\sum_{k=N}^{n} \mid c_k \mid^2\Big)^{1/2} \le \Big(\frac{1}{\mid E \mid} \int_E \mid P(t) \mid^p dt\Big)^{1/p}.$$

For infinite sums, we assume $\sum \mid c_k \mid^2 < \infty$. The partial sums $\sum_{k=N}^{n} c_k \phi_k(t)$ converge in L^p because of the right-hand side inequality. We can therefore pass to the limit and get the result for infinite sums.

Corollary: If $\sum c_k \phi_k(t) \in L_p(E)$ for any p and any set $E \subset [0,1]$, $0 < p < \infty$, $\mid E \mid > 0$, then $\sum c_k \phi_k(t) \in L_p([0,1])$ for all $p > 0$.

Reference

1. A. Zygmund, Trigonometric Series, Cambridge University Press, 1959.

18

Extrapolation for Pairs of Related Weights

Carlos Segovia and José L. Torrea

In this chapter we obtain extrapolation theorems for pairs of weights α and β that satisfy the relation $\alpha = \nu^p\beta$, where ν is a given positive function and α and β belong to A_p. The extrapolation theorems obtained are motivated by the result of Steve Bloom [1] on the commutator of the Hilbert transform and are applied to commutators of L^s-Dini singular integral operators; see Theorem (3.7). Theorem (2.4), the extrapolation from $r = \infty$, is a generalization of a result contained in [5], in the same way as Theorem (2.3) generalizes results of [7] and [3]. For notation and the general theory of A_p weights, we indicate [4], for instance. A result related to Theorem (2.3) for $\nu = 1$ was considered by F. J. Ruiz [8].

1. THE CLASS $A_{p,\lambda}^{(\nu)}$

We state a lemma due to Rubio de Francia [7] in García-Cuerva's version [3], which shall be systematically used throughout this paper.

(1.1) Lemma *Let S be a sublinear operator bounded on $L^p(\omega)$, $\omega \geq 0$. Suppose $S(u) \geq 0$ for every $u \in L^p(\omega)$. Then, given $u \in L^p(\omega)$, there exists $U \geq u$ such that*

$$\|U\|_{L^p(\omega)} \leq 2\|u\|_{L^p(\omega)}$$
$$S(U) \leq CU \quad a.e.$$

This research has been partially supported by the Ministry of Education and Sciences of Spain (Programa de Cooperación con Iberoamérica).

Proof. See Lemma 5.1, p. 343 of [4].

Let $\nu \geq 0$ be a measurable function, $1 \leq p < \infty$ and $1 \leq \lambda \leq p$. We shall say that a weight ω belongs to the class $A_{p,\lambda}^{(\nu)}$ if

$$\omega \in A_{p/\lambda} \qquad \text{and} \qquad \nu^p \omega \in A_{p/\lambda}.$$

If $\lambda = 1$ we denote this class by $A_p^{(\nu)}$. Observe that $A_{p,\lambda}^{(\nu)} = A_{p/\lambda}^{(\nu^\lambda)}$.

The following propositions contain basic properties of the class $A_{p,\lambda}^{(\nu)}$. By the observation just made, it is enough to prove them for $\lambda = 1$.

(1.2) Proposition *The class $A_{p,\lambda}^{(\nu)}$ is not empty if and only if $\nu^\lambda \in A_2$.*

Proof. If $\nu \in A_2$, then $\nu = \nu_0 \nu_1^{-1}$, ν_0 and ν_1 belonging to A_1. Letting $\omega = \nu_1 \nu_0^{1-p}$ it follows that $\omega \in A_p$ and $\nu^p \omega = \nu_0 \nu_1^{1-p} \in A_p$. Let us assume now that $\omega \in A_p^{(\nu)}$; then $\omega = \omega_0 \omega_1^{1-p}$ and $\nu^p \omega = \rho_0 \rho_1^{1-p}$, $\omega_0, \omega_1, \rho_0, \rho_1$ belonging to A_1. Then

$$\nu = \rho^{1/p} \omega_1^{1/p'} (\omega_0^{1/p'} \rho_1^{1/p'})^{-1} = \nu_0 \nu_1^{-1},$$

with ν_0 and ν_1 in A_1, as follows by Hölder's inequality.

(1.3) Proposition *Let $\omega \in A_{p,\lambda}^{(\nu)}$. There exists $\varepsilon > 0$ such that if $p < q < p + \varepsilon$, then $\omega \in A_{q,\lambda}^{(\nu)}$.*

Proof. Let $s > 1$ be such that $\nu^p \omega$ and $(\nu^p \omega)^{p'/p}$ satisfy the reverse Hölder condition with exponent s. Let $\varepsilon > 0$ be such that $\varepsilon s' p' \leq 1$. Then, by Hölder's inequality

$$\begin{aligned}
&\left(|Q|^{-1} \int_Q \nu^q \omega\right) \left(|Q|^{-1} \int_Q (\nu^q \omega)^{q'/q}\right)^{q-1} \\
&\quad \leq \left(|Q|^{-1} \int_Q (\nu^p \omega)^s\right)^{1/s} \left(|Q|^{-1} \int_Q \nu^{(q-p)s'}\right)^{1/s'} \\
&\quad \times \left(|Q|^{-1} \int_Q (\nu^p \omega)^{-sq'/q}\right)^{(q-1)/s} \left(|Q|^{-1} \int_Q \nu^{-(q-p)s'q'/q}\right)^{(q-1)/s'} \\
&\quad = I_1 \cdot I_2 \cdot I_3 \cdot I_4.
\end{aligned}$$

Applying Hölder and reverse Hölder, we get

$$I_1 I_3 \leq C \left(|Q|^{-1} \int \nu^p \omega\right) \left(|Q|^{-1} \int_Q (\nu^p \omega)^{-p'/p}\right)^{p-1} \leq C.$$

Recalling that $\nu \in A_2$ implies $\nu = \nu_0\nu_1^{-1}$, ν_0 and ν_1 in A_1, it follows that

$$\begin{aligned} I_2 &\le C\left(|Q|^{-1}\int_Q \nu_0^{(q-p)s'}\right)^{1/s'}\left(|Q|^{-1}\int_Q \nu_1\right)^{-(q-p)} \\ &\le C\left(|Q|^{-1}\int_Q \nu_0\right)^{q-p}\left(|Q|^{-1}\int_Q \nu_1\right)^{-(q-p)} \end{aligned}$$

Analogously,

$$\begin{aligned} I_4 &\le C\left(|Q|^{-1}\int_Q \nu_1^{(q-p)s'p'/p}\right)^{(p-1)/s'}\left(|Q|^{-1}\int_Q \nu_0\right)^{-(q-p)} \\ &\le C\left(|Q|^{-1}\int_Q \nu_1\right)^{q-p}\left(|Q|^{-1}\int_Q \nu_0\right)^{-(q-p)}. \end{aligned}$$

Therefore, $I_2I_4 \le C$. The case $p = 1$ is even simpler.

(1.4) Proposition *If $\omega \in A_{p,\lambda}^{(\nu)}$ and $\lambda < p$, there exists $\varepsilon > 0$ such that $\omega \in A_{q,\lambda}^{\nu}$ for every $p - \varepsilon < q < p$.*

Proof. We have

$$\begin{aligned} &\left(|Q|^{-1}\int_Q \nu^q\omega\right)\left(|Q|^{-1}\int_Q (\nu^q\omega)^{-q'/q}\right)^{q-1} \\ &\quad\le \left(|Q|^{-1}\int_Q \nu^p\omega\right)^{q/p}\left(|Q|^{-1}\int_Q \omega\right)^{1/(q/p)'} \\ &\quad\times \left(|Q|^{-1}\int_Q (\nu^p\omega)^{-q'/q}\right)^{q(q-1)/p}\left(|Q|^{-1}\int_Q \omega^{-q'/q}\right)^{(q-1)/(p/q)'} \le C, \end{aligned}$$

if ε is chosen so that ω and $\nu^p \in A_q$ for every $p - \varepsilon < q < p$.

(1.5) Proposition (Factorization) *The weight ω belongs to $A_{p,\lambda}^{(\nu)}$ if and only if $\omega = \omega_0\omega_1^{1-p/\lambda}$, where $\omega_0 \in A_1^{(\nu^\lambda)}$ and $\omega_1 \in A_1^{(\nu^{-\lambda})}$.*

Proof. If $\omega = \omega_0\omega_1^{1-p}$, then $\omega \in A_p$ and $\nu^p\omega = \nu\omega_0(\nu^{-1}\omega_1)^{1-p} \in A_p$. Conversely, assume $p \ge 2$ and let

$$\begin{aligned} S(u) &= \nu M(\nu^{-1}u) + M(u) + [\omega^{-1}M(\omega u^{p-1})]^{p'-1} \\ &\quad + [(\omega\nu)^{-1}M(\omega\nu u^{p-1})]^{p'-1}, \end{aligned}$$

where M denotes the Hardy–Littlewood maximal operator. If $u \in L^p(\omega)$, it follows that S is positive, sublinear and $\|S(u)\|_{L^p(\omega)} \leq C\|U\|_{L^p(\omega)}$. Then, by Lemma (1.1), there exists $U \geq u$ such that $S(U) \leq cU$ a.e. This implies

$$M(\nu^{-1}U) \leq c\nu^{-1}U, \qquad M(U) \leq cU,$$
$$M(\omega U^{p-1}) \leq c\omega U^{p-1}$$
$$M(\nu\omega U^{p-1}) \leq c\nu\omega U^{p-1}$$

a.e., that is to say, $U \in A_1^{(\nu^{-1})}$ and $\omega U^{p-1} \in A_1^{(\nu)}$. Then, setting $\omega_0 = \omega U^{p-1}$ and $\omega_1 = U$, it follows that $\omega = \omega_0 \omega_1^{1-p}$. If $1 < p < 2$, $\omega^{-p/p'} \in A_{p'}^{(\nu^{-1})}$, $p' > 2$. Then, by the factorization already obtained for $p > 2$, we get the factorization of weights for $1 < p < 2$.

2. EXTRAPOLATION THEOREMS

We begin this section with the following lemma:

(2.1) Lemma *Let $\omega \in A_{p,\lambda}^{(\nu)}$, $1 \leq \lambda \leq r < p$. Then the operator*

$$S(u) = S_1(u) + S_2(u)$$
$$= [\omega^{-1}M(\omega u^{1/t})]^t + [\nu^{-\lambda}\omega M(\nu^\lambda \omega u^{1/t})]^t,$$

where $t = (p - r)/(p - \lambda)$ is sublinear, positive, and satisfies

$$\|S(u)\|_{L^{(p/r')}_{(\omega)}} \leq C\|U\|_{L^{(p/r')}_{(\omega)}}$$

$$(u\omega, S_1(u)\omega) \in A_{r/\lambda} \qquad \textit{and} \qquad (u\nu^r\omega, S_2(u)\nu^r\omega) \in A_{r/\lambda},$$

for every $u \in L^{(p/r)'}_{(\omega)}$, $u \geq 0$.

Proof. Observe that $t(p/r)' = (p/\lambda)'$. Then

$$\int S(u)^{(p/r)'}\omega \leq c \int M(\omega u^{1/t})^{(p/\lambda)'}\omega^{1-(p/\lambda)'}$$
$$+ c \int M(\nu^\lambda \omega \nu^{1/t})^{(p/\lambda)'}\nu^{-\lambda(p/\lambda)'}\omega^{1-(p/\lambda)'}.$$

Since $\omega \in A_{p,\lambda}^{(\nu)}$, we have $\omega^{1-(p/\lambda)'} \in A_{(p/\lambda)'}$ and $\nu^{-\lambda(p/\lambda)'}\omega^{1-(p/\lambda)'} \in A_{(p/\lambda)'}$. Thus

$$\int S(u)^{(p/r)'}\omega \leq c \int u^{(p/r)'}\omega.$$

Now, let ν be as assumed in the hypothesis or $\nu \equiv 1$. We have to prove that

$$\left(|Q|^{-1}\int_Q \nu^r \omega u\right) \times \left(|Q|^{-1}\int_Q \{\nu^r\omega[\nu^{-\lambda}\omega M(\nu^\lambda \omega u^{1/t})]^t\}^{-1/(r/\lambda-1)}\right)^{r/\lambda-1} \leq C.$$

The first factor is bounded (by Hölder's inequality) by

$$\left(|Q|^{-1}\int_Q \nu^p\omega\right)^{1-t}\left(|Q|^{-1}\int_Q \nu^\lambda \omega u^{1/t}\right)^t.$$

The second one, since $M(\nu^\lambda\omega u^{1/t}) \geq |Q|^{-1}\int_Q \nu^\lambda\omega u^{1/t}$, is bounded by

$$\left(|Q|^{-1}\int_Q \nu^{-\lambda(p/\lambda)'}\omega^{1-(p/\lambda)'}\right)^{(r/\lambda-1)}\left(|Q|^{-1}\int_Q \nu^\lambda\omega u^{1/t}\right)^{-t}.$$

Taking into account that $\nu^p\omega \in A_{p/\lambda}$ our claim is proved.

(2.2) Lemma *Let* $1 \leq \lambda < p$, $\lambda \leq r < \infty$, $r \neq p$ *and* $\omega \in A^{(\nu)}_{p,\lambda}$. *If* $s^{-1} = |1 - r/p|$ *and* $0 \leq u \in L^s(\omega)$, *then there exists* $U \in L^s(\omega)$ *such that*

a. $u \leq U$ *a.e.*
b. $\|U\|_{L^s(\omega)} \leq c\|u\|_{L^s(\omega)}$, *and*
c. *if* $r < p$, *then* $U\omega \in A^{(\nu)}_{r,\lambda}$ *and if* $p < r$, *then* $U^{-1}\omega \in A^{(\nu)}_{r,\lambda}$.

Proof. Assume $r < p$. Then $s = (p/r)'$. Applying Lemma (1.1) to the operator $S = S_1 + S_2$ defined in Lemma (2.1), we get a function $U \geq u$ a.e. such that

$$\|U\|_{L^s(\omega)} \leq c\|u\|_{L^s(\omega)} \quad \text{and} \quad S(U) \leq CU \quad a.e.$$

Then, by Lemma (2.1), it follows that

$$U\omega \in A_{r/\lambda} \quad \text{and} \quad \nu^r U\omega \in A_{r/\lambda},$$

that is to say, $U\omega \in A^{(\nu)}_{r,\lambda}$. If $p < r$, let us define $p^* = \lambda(p/\lambda)'$ and $r^* = \lambda(r/\lambda)'$. Then $p^* > r^* > \lambda$. Let $s^{*-1} = 1 - r^*/p^*$. We define $\omega^* = \omega^{1-p^*/\lambda}$ and $U^* = U^{s/s^*}\omega^{p^*/\lambda s^*}$. Then

$$\int (U^*)^{s^*}\omega^* = \int u^s\omega,$$

and $\omega^* \in A_{p^*,\lambda}^{(\nu^{-1})}$. By the first part of this lemma there exists $U^* \geq u^*$ a.e. such that

$$\|U^*\|_{L^{s^*}_{\omega^*}} \leq c\|u^*\|_{L^{s^*}_{\omega^*}},$$

$$U^*\omega^* \in A_{r^*,\omega}^{(\nu^{-1})}.$$

Let U be defined as $U^* = U^{s/s^*}\omega^{p^*/\lambda s^*}$. Then $U \geq u$ a.e. and

$$\int U^s\omega = \int (U^*)^{s^*}\omega^* \leq c\int (u^*)^{s^*}\omega^* = c\int u^s\omega.$$

On the other hand, $(U^*\omega^*)^{1-(r^*/\lambda)'} \in A^{(\nu)}_{\lambda(r^*/\lambda)',\lambda} = A^{(\nu)}_{r,\lambda}$. Since $(U^*\omega^*)^{1-(r^*/\lambda)'} = U^{-1}\omega$, the proof of the case $p < r$ is completed.

Next we state two interpolation theorems which are the main results of this paper.

(2.3) Theorem (Extrapolation from $r < \infty$) *Let T be a sublinear operator defined on C_0^∞ with values in the space of measurable functions. Let $1 \leq \lambda \leq r < \infty$. If the operator satisfies*

$$\|Tf\|_{L^r(\omega)} \leq c_\omega\|f\|_{L^r(\omega)}$$

for every $\omega \in A^{(\nu)}_{r,\lambda}$, then for every p, $\lambda < p < \infty$,

$$\|Tf\|_{L^p(\omega)} \leq c_\omega\|f\|_{L^p(\omega)}$$

holds for every $\omega \in A^{(\nu)}_{p,\lambda}$.

Proof. The theorem follows from Lemma (2.2) as in Theorem (5.19), pp. 448–450 of [4].

(2.4) Theorem (Extrapolation from $r = \infty$) *Let T be a sublinear operator defined on C_0^∞ with values in the space of measurable functions and $1 \leq \lambda < \infty$. If the operator satisfies*

$$\|\omega Tf\|_\infty \leq c_\omega\|\omega f\|_\infty,$$

for every ω such that $\omega^{-\lambda} \in A_1$ and $(\nu\omega)^{-\lambda} \in A_1$, then

$$\|Tf\|_{L^p(\omega)} \leq c_\omega\|f\|_{L^p(\omega)}$$

holds for every $\omega \in A_{p,\lambda}^{(\nu)}$ *and* $p > \lambda$.

Proof. Let $f \in L^p(\omega)$, $\omega \in A_{p,\lambda}^{(\nu)}$, $1 \leq \lambda < p$. We define

$$g = \omega^{1/p(p-1)}|f|\omega^{1/p} \Big/ \left(\int |f|^p\omega\right)^{1/p} \qquad \text{if } |f|\omega^{1/p} \neq 0,$$

and

$$g = \omega^{1/p(p-1)}e^{-\pi|x|^2/p} \qquad \text{if } |f|\omega^{1/p} = 0.$$

Then, $g > 0$ a.e.,

$$\int g^p\omega^{-p'/p} \leq 2 \qquad \text{and} \qquad \||f\omega^{p'/p}g^{-1}\|_\infty = \left(\int |f|^p\omega\right)^{1/p}.$$

Let $h \in L^p(\omega^{1-p'})$. Set

$$S(h) = \omega^{p'-1}\{M[\omega^{1-p'}h)^\lambda]\}^{1/\lambda} + \nu\omega^{p'-1}\{M[(\nu^{-1}\omega^{1-p'}h)^\lambda]\}^{1/\lambda}.$$

This operator S is positive and sublinear; moreover since $\omega \in A_{p,\lambda}^{(\nu)}$,

$$\left(\int S(h)^p\omega^{1-p'}\right)^{1/p} \leq \left(\int M[(\omega^{1-p'}h)^\lambda]^{p/\lambda}\omega\right)^{1/p} + \left(\int M[(\nu^{-1}\omega^{1-p'}h)^\lambda]^{p/\lambda}\nu^p\omega\right)^{1/p} \leq c_\omega\left(\int h^p\omega^{1-p'}\right)^{1/p}.$$

Then, by Lemma (1.1) we obtain a function $G \geq g$ a.e., satisfying

$$\int G^p\omega^{1-p'} \leq c\int g^p\omega^{1-p'} = c\int g^p\omega^{-p'/p} \leq 2c,$$

and $S(G) \leq cG$ a.e. Then

$$M[(\omega^{1-p'}G)^\lambda] \leq c(\omega^{1-p'}G)^\lambda,$$
$$M[(\nu^{-1}\omega^{1-p'}G)^\lambda] \leq c(\nu^{-1}\omega^{1-p'}G)^\lambda$$

showing that $(\omega^{1-p'}G)^\lambda$ and $(\nu^{-1}\omega^{1-p'}G)^\lambda$ belong to A_1. Now since $p'/p = p' - 1$, we have

$$\left(\int |f|^p\omega\right)^{1/p} = \|f\omega^{p'-1}g^{-1}\|_\infty \geq \|f\omega^{p'-1}G^{-1}\|_\infty.$$

Since $(\omega^{p'-1}G^{-1})^{-\lambda}$ and $(\nu\omega^{p'-1}G^{-1})^{-\lambda}$ belong to A_1, we get

$$\begin{aligned}\left(\int |f|^p\omega\right)^{1/p} &\geq c\|\omega^{p'-1}G^{-1}Tf\|_\infty \\ &\geq c'\|\omega^{p'-1}G^{-1}Tf\|_\infty \left(\int G^p\omega^{1-p'}\right))^{1/p} \geq c'\left(\int |Tf|^p\omega\right)^{1/p},\end{aligned}$$

as we wanted to show.

3. COMMUTATOR OF SINGULAR INTEGRALS

Let $k(x) = \Omega(x)|x|^{-n}$ be a kernel defined on $\mathbf{R}^n \sim \{0\}$, such that

i. $\Omega(x)$ is positively homogeneous of degree 0 and

$$\int_{|x|=1} \Omega(x)\,d\sigma_x = 0,$$

ii. for a given s, $1 \leq s \leq \infty$,

$$\omega_s(t) = \sup_{|\rho|<t}\left(\int_{|x|=1} |\Omega(\rho x) - \Omega(x)|^s\,d\sigma_x\right)^{1/s}$$

satisfies the Dini condition

$$\int_0^1 \omega_s(t)\frac{dt}{t} < \infty.$$

(Here ρ is a rotation in $\mathbf{R}^n$ and $|\rho| = \sup_{|x|=1}\|\rho x - x\|$).

A kernel $k(x)$ satisfying these conditions is called an L^s-Dini singular integral kernel. For $s = 1$, see [2], and for $1 < s$, see [6]. It was proved in [6] that these kernels satisfy

$$(3.1)\quad \left(\int_{R<|x|<2R} |k(x-y)-k(x)|^s\,dx\right)^{1/s} \le cR^{-n/s'}\left(|y|/R + \int_{|y|/2R<t<|y|/R} \omega_s(t)\frac{dt}{t}\right),$$

whenever $|y| < R/2$. Moreover, in the same paper [6], the following result was proved:

(3.2) Theorem *If K is a singular integral operator with an L^s-Dini singular integral kernel, then*

i. *if $s' < p < \infty$ and $\omega \in A_{p/s'}$, or*
ii. *if $1 < p < s$ and $\omega^{1-p'} \in A_{p'/s'}$, then*

$$\|Kf\|_{L^p(\omega)} \le c_\omega \|f\|_{L^p(\omega)}.$$

We shall apply the extrapolation theorem (2.4) to obtain weighted L^p estimates for the commutators of these L^s-Dini singular integral operators, i.e., for

$$C_b(f)(x) = \text{p.v.} \int [b(x)-b(y)]k(x-y)f(y)\,dy$$

of the type obtained by S. Bloom [1] for the Hilbert transform. We shall estimate the sharp function of $C_b(f)$, where b is a function in BMO(ν), that is to say,

$$\int_Q |b(x)-m_Q b|\,dx \le c\nu(Q) = c\int_Q \nu,$$

where $\nu \in A_2$.

The following lemmas shall be needed in the estimation of the sharp function of $C_b f$.

(3.3) Lemma *Let Q be a cube and $Q_k = 2^k Q$. Then, if $b \in$ BMO(ν), $\nu \in A_2$, it follows that*

$$|m_Q b - m_{Q_k} b| \le c 2^{nk(1-\eta)}\left(|Q_k|^{-1}\int_{Q_k} \nu\right),$$

where $0 < \eta = \varepsilon/(1+\varepsilon) < 1$ *and the reverse Hölder condition holds for* ν *with exponent* $1+\varepsilon$. *If* $\nu = 1$, *then*

$$|m_Q b - m_{Q_k} b| \leq ck.$$

(3.4) Lemma *If* $\omega^{-\lambda} \in A_1$, *there exists* $\varepsilon > 0$ *such that for every* $1 \leq r \leq \lambda(1+\varepsilon)$, $\omega^{-r} \in A_r$.

(3.5) Lemma *If* $\omega^{-\lambda} \in A_1$, *there exists* $\varepsilon > 0$ *such that* $\omega^{r'} \in A_{r'/\lambda}$ *for every* $1 < r \leq 1+\varepsilon$.

(3.6) Lemma *If* $b \in \mathrm{BMO}(\nu)$ *and* $\nu = \alpha\beta^{-1}$, $\alpha^{-\lambda} \in A_1$, $\beta^{-\lambda} \in A_1$, *then there exists* $\varepsilon > 0$ *such that*

$$\left(|Q|^{-1}\int_Q |b - m_Q b|^r \alpha^{-r}\right)^{1/r} \leq c\beta(x_0)^{-1}$$

holds for $1 < r < \lambda(1+\varepsilon)$ *and* $x_0 \in Q$.

The proof of these lemmas can be found in [1].
Let us define for every cube Q,

$$c_Q = |Q|^{-1}\int_Q \left(\int [m_Q b - b(y)]k(z-y) f_{C4Q}(y)\,dy\right) dz,$$

where $f_{4Q}(x) = f(x)$ if $x \in 4Q$ and zero otherwise. Of course, $f(x) = f_{4Q} + f_{C4Q}$. Then we have

$$\begin{aligned}
|Q|^{-1}\int_Q |C_b f - c_Q| &\leq |Q|^{-1}\int_Q |b - m_Q b||Kf_{4Q}| \\
&+ |Q|^{-1}\int_Q |K[(m_Q b - b)f_{4Q}]| + |Q|^{-1}\int_Q |b - m_Q b| \cdot |Q|^{-1}\int_Q |Kf_{4Q}| \\
&+ |Q|^{-1}\int_Q \left(|b(x) - m_Q b||Q|^{-1}\int_Q\int |k(x-y) - k(z-y)||f_{C4Q}(y)|\,dy\,dz\right)dx \\
&+ |Q|^{-1}\int_Q \left||Q|^{-1}\int_Q \left(\int (m_Q b - b(y))[k(x-y) - k(z-y)]f_{C4Q}(y)\,dy\right)dz\right| dx \\
&+ |Q|^{-1}\int_Q |b - m_Q b| \cdot |Q|^{-1}\int_Q |Kf| \\
&= \sum_{j=1}^{5} \rho_Q^{(j)}(f) + \rho_Q^{(6)}(Kf).
\end{aligned}$$

Let us estimate $\rho_Q^{(j)}$. We shall denote s' by λ.

Estimation of $\rho_Q^{(1)}$

By Hölder's inequality, we have

$$\rho_Q^{(1)} = \left(|Q|^{-1}\int_Q |b - m_Q b|^r \alpha^{-r}\right)^{1/r} \left(|Q|^{-1}\int |Kf_{4Q}|^{r'}\alpha^{r'}\right)^{1/r'}.$$

By Lemma (3.6) and Theorem (3.2), we have

$$\rho_Q^{(1)} \le C\beta^{-1}(x_0)\left(|Q|^{-1}\int_{4Q} |f|^{r'}\alpha^{r'}\right)^{1/r'} \le C\beta^{-1}(x_0)\|f\alpha\|_\infty.$$

Estimation of $\rho_Q^{(2)}$

By Hölder's inequality, we have

$$\rho_Q^{(2)} \le \left(|Q|^{-1}\int_Q |K[(m_Q b - b)f_{4Q}]|^r\,dx\right)^{1/r},$$

and applying Theorem (3.2), we obtain

$$\begin{aligned}\rho_{-1}^{(2)} &\le C\left(|Q|^{-1}\int_{4Q} |m_Q b - b|^r |f|^r\right)^{1/r}\\ &\le C\|f\alpha\|_\infty\left(|Q|^{-1}\int_{4Q} |m_Q b - b|^r \alpha^{-r}\right)^{1/r}.\end{aligned}$$

Then, by Lemma (3.6), we get

$$\rho_Q^{(2)} \le C\beta(x_0)^{-1}\|f\alpha\|_\infty.$$

Estimation of $\rho_Q^{(3)}$

By Hölder's inequality and Theorem (3.2),

$$\begin{aligned}|Q|^{-1}\int_Q |Kf_{4Q}| &\le C\left(|Q|^{-1}\int_{4Q} |f|^r\right)^{1/r}\\ &\le C\|f\alpha\|_\infty\left(|Q|^{-1}\int_{4Q} \alpha^{-r}\right)^{1/r}.\end{aligned}$$

By definition of BMO(ν), we have

$$|Q|^{-1}\int_Q |b - m_Q b| \leq C\left(|Q|^{-1}\int_Q \alpha\beta^{-1}\right)$$
$$\leq C\left(|Q|^{-1}\int_Q \beta^{-r}\right)^{1/r}\left(|Q|^{-1}\int_Q \alpha^{r'}\right)^{1/r'}.$$

Then, by Lemma (3.4), we obtain

$$\rho_Q^{(3)} \leq C\beta(x_0)^{-1}\|f\alpha\|_\infty.$$

Estimation of $\rho_Q^{(4)}$

We have

$$\int |k(x-y) - k(z-y)||f_{C4Q}(y)|\,dy$$
$$\leq \|f\alpha\|_\infty \int_{C4Q} |k(x-y) - k(z-y)|\alpha^{-1}\,dy.$$

If $Q_k = 2^k Q$ and $C_k = Q_k = Q_{k-1}$, then

$$\int_{C4Q} |k(x-y) - k(z-y)|\alpha^{-1}(y)\,dy$$
$$\leq \sum_{k=3}^{\infty}\left(\int_{C_k} |k(x-y) - k(z-y)|^{\lambda'}\,dy\right)^{1/\lambda'}\left(\int_{Q_k} \alpha^{-\lambda}\right)^{1/\lambda}.$$

If 2δ is the length of a side of Q, by (3.1) we get that the last expression is bounded by

$$C\sum\left(2^{-k} + \int_{|x-y|/2^{k+1}\delta < t < |x-y|/2^k\delta} \omega_s(t)\,dt/t\right)\left(|Q|^{-1}\int_{Q_k} \alpha^{-\lambda}\right)^{1/\lambda}.$$

Recalling that $\alpha^{-\lambda} \in A_1$, we obtain

$$\int_{C4Q} |k(x-y) - k(z-y)|\alpha^{-1}(y)\,dy$$
$$\leq C\left(1 + \int_0^1 \omega_s(t)\,dt/t\right)\left(|Q|^{-1}\int_Q \alpha^{-\lambda}\right)^{1/\lambda}.$$

On the other hand, we have

$$|Q|^{-1}\int_Q |b-m_Q b| \le C|Q|^{-1}\int_Q \alpha\beta^{-1}$$
$$\le C\left(|Q|^{-1}\int_Q \alpha^{\lambda'}\right)^{1/\lambda'}\left(|Q|^{-1}\int_Q \beta^{-\lambda}\right)^{1/\lambda}.$$

Thus, $\rho_Q^{(4)} \le C\beta^{-1}(x_0)\|f\alpha\|_\infty$.

Estimation of $\rho_Q^{(5)}$

We have

$$I = \int_{C4Q} |m_Q b - b(y)||k(x-y)-k(z-y)|\alpha^{-1}(y)\,dy$$
$$\le \sum_{k=3}^{\infty}\left(\int_{Q_k}|m_Q b - b|^{\lambda}\alpha^{-\lambda}\right)^{1/\lambda}\left(\int_{C_k}|k(x-y)-k(z-y)|^{\lambda'}\,dy\right)^{1/\lambda'}.$$

Now, since by Lemma (3.3),

$$|m_Q b - m_{Q_k} b| \le C2^{nk(1-\eta)}|Q_k|^{-1}\nu(Q_k) \le$$
$$C2^{nk(1-\eta)}\left(|Q|^{-1}\int_{Q_k}\alpha^{\lambda'}\right)^{1/\lambda'}\left(|Q_k|^{-1}\int_{Q_k}\beta^{-\lambda}\right)^{1/\lambda},$$

by Lemma (3.3), we obtain

$$\left(\int_{Q_k}|m_Q b - b|^{\lambda}\alpha^{-\lambda}\right)^{1/\lambda} \le C2^{nk(1-\eta)}\beta(x_0)^{-1}|Q_k|^{1/\lambda}.$$

We can assume $n(1-\eta) < 1$; then by (3.1) it follows that

$$I \le C\left(1+\sum_k 2^{nk(1-\eta)}\right.$$
$$\left.\times\int_{|x-z|/2^{k+1}\delta<t<|x-z|/2^k\delta}\omega_s(t)\,dt/t\right)\beta(x_0)^{-1}$$
$$\le C\left(1+\int_0^1 \omega_s(t)t^{-n(1-\eta)}dt/t\right)\beta(x_0)^{-1}.$$

Thus, $\rho_Q^{(5)} \le C\beta(x_0)^{-1}\|f\alpha\|_\infty$, if the last integral is finite. Finally, for $\rho_Q^{(6)}(g)$ we have

$$\begin{aligned}\rho_Q^{(6)}(g) &\le C\left(|Q|^{-1}\int_Q \alpha\beta^{-1}\right)\|g\alpha\|_\infty\left(|Q|^{-1}\int_Q \alpha^{-1}\right)\\ &\le C\beta(x_0)^{-1}\|g\alpha\|_\infty.\end{aligned}$$

Let $G_1(f)(x_0) = \sup_{x_0\in Q}\sum_1^5 \rho_Q^{(j)}(f)$ and $G_2 = \sup_{x_0\in Q}\rho_Q^{(6)}(g)$. Then

$$(C_b)f)^{\#}(x_0) = G_1(f) + G_2(Kf),$$

and $\|\beta G_1(f)\|_\infty \le C\|f\alpha\|_\infty$, $\|\beta G_2(g)\|_\infty \le C\|g\alpha\|_\infty$. Substituting $\nu^{-1}f$ and $\nu^{-1}g$ for f and g respectively we get $\|\beta G_1(\nu^{-1}f)\|_\infty \le C\|f\beta\|_\infty$ and $\|\beta G_2(\nu^{-1}g)\|_\infty \le C\|g\beta\|_\infty$, for every β such that $\beta^{-\lambda}$ and $(\nu\beta)^{-\lambda}$ belong to A_1. Then, by Theorem (2.4), we obtain

$$\begin{aligned}\|G_1(\nu^{-1}f)\|_{L^p(\beta)} &\le C_\beta\|f\|_{L^p(\beta)},\\ \|G_2(\nu^{-1}g)\|_{L^p(\beta)} &\le C_\beta\|g\|_{L^p(\beta)},\end{aligned}$$

whenever $\beta \in A_{p,\lambda}^{(\nu)}$. Therefore,

$$\begin{aligned}\|C_bf\|_{L^p(\beta)} \le C\|(C_bf)^{\#}\|_{L^p(\beta)} &\le C(\|G_1(f)\|_{L^p(\beta)} + \|G_2(Kf)\|_{L^p(\beta)}\\ &\le C_\beta(\|f\nu\|_{L^p(\beta)} + \|Kf\nu\|_{L^p(\beta)}) \le C_\beta\|f\|_{L^p(\alpha)},\end{aligned}$$

since $\beta \in A_{p,\lambda}^{(\nu)}$ and consequently $\alpha \in A_{p/\lambda}$.

We have proved the following result:

(3.7) Theorem *Let K be a singular integral operator with an L^s-Dini singular integral kernel. If ν is a weight and $b \in \mathrm{BMO}(\nu)$, then the commutator $C_b(f)$ satisfies*

$$\|C_b(f)\|_{L^p(\beta)} \le C_{\beta,\nu}\|f\|_{L^p(\alpha)},$$

whenever $\alpha = \nu^p\beta$ and $\beta \in A_{p,s'}^{(\nu)}$, under the assumption that

$$\int_0^1 \omega_s(t)t^{-n(1-\eta)}\frac{dt}{t} < \infty,$$

where $\eta = \varepsilon/(1 + \varepsilon)$ *and the reverse Hölder condition holds for* ν *with exponent* $1 + \varepsilon$. *If* $\nu = 1$, *then it is enough to assume that*

$$\int_0^1 \log(1/t)\omega_s(t)\frac{dt}{t} < \infty.$$

REFERENCES

1. S. Bloom, A commutator theorem and weighted BMO. *Trans. Amer. Math. Soc. 292* (1985), 103–122.
2. A. P. Calderón, M. Weiss, and A. Zygmund, On the existence of singular integrals. *Proc. Symp. Pure Math. 10*, Amer. Math. Soc., Providence, R.I. (1967), 56–73.
3. J. García-Cuerva, An extrapolation theorem in the theory of A_p weights. *Proc. Amer. Math. Soc. 83* (1983), 422–426.
4. J. García-Cuerva and J. L. Rubio de Francia, *Weighted Norm Inequalities and Related Topics*. Mathematics Studies, 116, North-Holland, Amsterdam, 1985.
5. E. Harboure, R. A. Macías, and C. Segovia, Extrapolation results for classes of weights. *Amer. J. Math. 110* (1988), 383–397.
6. D. S. Kurtz and R. L. Wheeden, Results on weighted norm inequalities for multipliers. *Trans. Amer. Math. Soc. 255* (1979), 343–362.
7. J. L. Rubio de Francia, Factorization theory and A_p weights. *Amer. J. Math. 106* (1984), 533–547.
8. F. J. Ruiz, *Teoría de Calderón–Zygmund para funciones vectoriales y desigualdades con peso*. Doctoral Thesis, University of Zaragoza, Spain, 1983.

19

A Simple Example of Transference

Fernando Soria, Guido Weiss, and Ales Zaloznik

Many operators in Analysis can be expressed in terms of a convolution kernel acting on a group G and a representation of this group acting on a space of functions, say $L^p(X)$, defined on a general measure space X. Sharp estimates for the norm of such operators on $L^p(X)$ can be expressed in terms of the norm of the convolution operator on $L^p(G)$. Let us be more precise. Suppose G is an amenable group, X is a σ–finite measure space and R is a uniformly bounded representation of G acting on $L^p(X)$. That is, for each $u \in G$, R_u is a linear transformation on $L^p(X)$ satisfying

$$\|R_u F\|_p \leq c\|F\|_p \tag{1}$$

for all such u and $F \in L^p(X)$; moreover, $R_{uv} = R_u R_v$ for all u, v in G. We also assume that R_e is the identity operator if e is the identity element of G and the map $u \longrightarrow R_u$ is strongly continuous. Let $k \in L^1(G)$ have compact support and $N_p(k)$ denote the operator norm of the convolution operator $\varphi(u) \longrightarrow \varphi*k(u) = \int_G \varphi(uv^{-1})k(v)dv$; that is, $N_p(k)$ is the infimum of all $a > 0$ such that

$$\|\varphi*k\|_p \leq a\|\varphi\|_p \tag{2}$$

for all $\varphi \in L^p(X)$. We then introduce the operator H_k acting on functions F on X by letting

---

The second author supported, in part, by NSF Grant MCS 8200884

$$H_k F(x) = \int_G k(u) R_{u^{-1}} F(x)\, du.$$

It is clear that H_k is a bounded operator on $L^p(X)$. In fact, Minkowski's integral inequality gives us

$$\|H_k F\|_{L^p(X)} \le c\|k\|_{L^1(G)} \|F\|_{L^p(X)}. \tag{3}$$

One can establish, however, a much sharper estimate:

Theorem I (The main transference result). H_k, *as an operator on* $L^p(X)$, *has norm not exceeding* $c^2 N_p(k)$.

A proof of this result, as well as many applications and an explanation of why this is much sharper than estimate (3), can be found in [2]. Recently, Berkson and Gillespie [1] have used these ideas to obtain boundedness results for operators that are "transferred" from the circle group **T**. This short note is the result of some conversations the authors had when trying to understand the work of Berkson and Gillespie. We show how to obtain a representation of **T** that acts on $L^p(\mathbb{R})$. We then give an example of how to use this representation and Theorem I to transfer an operator from **T** to $\mathbb{R}$. We would like to emphasize that what we present here is only a special case of what is contained in [1]; moreover, the final result we obtain by this method is a theorem due to Jodeit [4]. On the other hand, our approach is very direct and gives a particularly simple example of an application of the general transference theorem stated above. There are several features in the arguments we present that make our approach different from that found in [1]. It was for these reasons that Berkson encouraged us to present this material. We are also grateful to Berkson for the suggestions, of both mathematical and historical nature, that he made to us.

We begin by describing a uniformly bounded representation R of **T** acting on $L^p(\mathbb{R})$, $1 < p < \infty$. Let $\psi_t(\xi) = e^{2\pi i t[\xi]}$, where $[\xi]$ is the largest integer not exceeding $\xi \in \mathbb{R}$ and $t \in \mathbf{T}$ (we are identifying the circle group with the reals modulo 1). For $F \in \mathcal{S}$ (the space of tempered test functions) we write

$$(R_t F)(x) = (\psi_t(\xi)\hat{F}(\xi))^{\vee}(x), \tag{4}$$

where $\hat{F}$ is the Fourier transform of F and the symbol $^{\vee}$ denotes the inverse Fourier transform of the expression to the left of it. We claim that, if $1 < p < \infty$,

$$\|R_t F\|_{L^p(\mathbb{R})} \leq c\|F\|_{L^p(\mathbb{R})}, \tag{5}$$

where c is independent of $F \in \mathscr{S}$ and t.

We show that this is a consequence of the $L^p(\mathbb{R})$ boundedness of the operator $\mathscr{H}$ defined by

$$(\mathscr{H}F)(x) = \sum_{n \neq 0} \frac{F(x-n)}{n},$$

for $F \in \mathscr{S}$. This, in turn, follows, by a transference argument, from the ℓ^p boundedness of the "original Hilbert transform" operator. The latter is the operator that maps the sequence $a = \{a_n\}$, $n = 0, \pm 1, \pm 2, \ldots$, into the sequence $\tilde{a} = \{\tilde{a}_n\}$, where

$$\tilde{a}_n = \sum_{m \neq 0} \frac{a_{n-m}}{m}.$$

Let us assume, then, that

$$\|\tilde{a}\|_{\ell^p} \leq C\|a\|_{\ell^p}, \tag{6}$$

where C is independent of the sequence a (see [5],[7]). To show

$$\|\mathscr{H}F\|_{L^p(\mathbb{R})} \leq C\|F\|_{L^p(\mathbb{R})} \tag{7}$$

we argue as follows: choose $F \in \mathscr{S}$ and put $F_y(s) \equiv F(y+s)$. Then

$$\|\mathscr{H}F\|_p^p = \int_{-\infty}^{\infty} |\sum_{n\neq 0} \frac{F(x-n)}{n}|^p dx = \sum_{m=-\infty}^{\infty} \int_m^{m+1} |\sum_{n\neq 0} \frac{F(x-n)}{n}|^p dx =$$

$$\int_0^1 \Big\{ \sum_{m=-\infty}^{\infty} \Big| \sum_{n\neq 0} \frac{F_y(m-n)}{n} \Big|^p \Big\} dy \leq C^p \int_0^1 \sum_{m=-\infty}^{\infty} |F_y(m)|^p dy = C^p \|F\|_p^p,$$

the inequality being a consequence of (6).

In order to establish (5) we introduce the function $\varphi_t(\xi) = e^{2\pi it([\xi]-\xi)}$. Thus, $(R_tF)(x) = (e^{2\pi it\xi}\varphi_t(\xi)\hat{F}(\xi))^\vee(x)$: that is, R_t is the composition of the Fourier multiplier operator defined by φ_t and the one involving the function $e^{2\pi it\xi}$; the latter, of course, is the translation operator $F(x) \longrightarrow F(x+t)$, which is an isometry of $L^p(\mathbb{R})$. Hence, it suffices to show that the operator S_t: $F(x) \longrightarrow (\varphi_t(\xi)\hat{F}(\xi))^\vee(x)$ satisfies $\|S_tF\|_{L^p(\mathbb{R})} \leq C\|F\|_{L^p(\mathbb{R})}$ for $F \in \mathscr{S}$, with C independent of F. Observe that φ_t is a 1–periodic function and, thus, a simple calculation of Fourier coefficients and Dini's test gives us the equality,for ξ not an integer,

$$\varphi_t(\xi) = \frac{1-e^{-2\pi it}}{2\pi i} \sum_{n=-\infty}^{\infty} \frac{e^{2\pi in\xi}}{t+n}$$

when $t \in (0,1)$. Since $\lim_{t\to 0+} \frac{1-e^{-2\pi it}}{2\pi i}\frac{1}{t+n} = \begin{cases} 1 \text{ if } n=0 \\ 0 \text{ if } n\neq 0 \end{cases}$, the limiting case $t = 0$ is consistent with the fact that $\varphi_0 \equiv 1$. The behaviour of S_t as t approaches 0 or 1, however, is best seen from the equality

$$(S_t + \frac{1-e^{-2\pi it}}{2\pi i}\mathscr{H})F(x) =$$

$$\frac{1-e^{-2\pi it}}{2\pi i}\sum_{n\neq 0,-1}\frac{-t}{n(n+t)}F(x+n)+\frac{1-e^{-2\pi it}}{2\pi it}F(x)+\frac{1-e^{-2\pi it}}{2\pi i}\frac{t}{t-1}F(x-1).$$

It is clear that the $L^p(\mathbb{R})$ norm of this last expression does not exceed $C\|F\|_{L^p(\mathbb{R})}$, where C is independent of $t \in (0,1)$. It follows from this and (7) that S_t has a uniformly bounded norm as an operator on $L^p(\mathbb{R})$ when $t \in [0,1]$. This establishes (5). Moreover, R_t is a strongly continuous representation of T.

We shall now show how to use this uniformly bounded representation R_t and Theorem I to obtain a theorem that is due to Jodeit [4]. In order to announce this result let us discuss, briefly, L^p multiplier operators on the additive groups $\mathbb{Z}$ (the integers) and $\mathbb{R}$ (the reals). In the first case we are dealing with the Fourier coefficients $c_n = \int_0^1 h(x)e^{-2\pi inx}dx$, $n \in \mathbb{Z}$, of a function $h \in L^p(T)$. In this case, an L^p multiplier is a sequence $m = \{m_n\}$ such that $\{m_n c_n\}$ is the sequence of Fourier coefficients of a function $g \in L^p(T)$ whenever $h \in L^p(T)$; moreover, $\|g\|_p \leq C\|h\|_p$ with C independent of h. It is well known that any such m is a bounded sequence; moreover, the operator T mapping h into g is a convolution operator with kernel

$$k(t) = \sum_{n\in\mathbb{Z}} m_n e^{2\pi int}; \tag{8}$$

That is,

$$g(t) = (Th)(t) = \int_{\mathbf{T}} k(s)h(t-s)ds. \tag{9}$$

Equation (8) defines k as a distribution and we can interpret (9) to be the convolution of the distribution k with, say, a trigonometric polynomial. The boundedness condition $\|Th\|_p \leq C\|h\|_p$ can then be used to extend T to all of $L^p(T)$. In the second case, an L^p multiplier is a bounded function $\mathcal{M}$ such that $\mathcal{M}(\xi)\hat{F}(\xi)$ is the Fourier transform of a function $G \in L^p(\mathbb{R}) \cap L^2(\mathbb{R})$ whenever $F \in L^p(\mathbb{R}) \cap L^2(\mathbb{R})$ and $\|G\|_p \leq C\|F\|_p$ with C independent of F (we have shown ψ_t is the L^p multiplier corresponding to R_t). We shall establish the following result:

Theorem II. *Suppose* $m = \{m_n\}$ *is an* L^p *multiplier on* $\mathbb{Z}$, $1 < p < \infty$. *Let* $\mathcal{M}$ *be defined on* $\mathbb{R}$ *by*

$$\mathcal{M}(\xi) \equiv \sum_{n=-\infty}^{\infty} m_n \chi_{[n,n+1)}(\xi),$$

where χ_E *denotes the characteristic function of* $E \subset \mathbb{R}$. *Then* $\mathcal{M}$ *is an* L^p *multiplier on* $\mathbb{R}$. *Mooreover, the norm of this last operator (on* $L^p(\mathbb{R})$*) does not exceed* c^2 *times the norm of the multiplier operator induced by* m *(on* $L^p(T)$*)), where* c *is the constant in (5).*

Suppose, first that only a finite number of the m_n's are not zero; then, k(t), as defined by equality (8), is a trigonometric polynomial. Thus, if we let G = T, the kernel k satisfies the hypotheses of Theorem I. Let us compute the operator H_k when R is the representation of T acting on $L^p(\mathbb{R})$ that is defined by equality (4). We have

$$(H_kF)^\wedge(\xi) = \int_T k(t)(R_{-t}F)^\wedge(\xi)dt = \int_T k(t)e^{-2\pi it[\xi]}\hat{F}(\xi)dt =$$

$$\hat{F}(\xi)\sum m_n \int_0^1 e^{2\pi int}e^{-2\pi i[\xi]t}dt = \mathcal{M}(\xi)\hat{F}(\xi).$$

Thus, H_k is a convolution operator on $L^p(\mathbb{R})$ whose associated multiplier is $\mathcal{M}$. Theorem I can be applied and we obtain Theorem II in the case k is a trigonometric polynomial. The general case is an easy consequence of this. We can convolve the k in (8) with the Nth Fejér kernel obtaining the trigonometric polynomial

$$k_N(t) = \sum_{n=-N}^{N} m_n\left(1 - \frac{|n|}{N+1}\right)e^{2\pi int}.$$

Applying the result we just obtained to k_N we obtain a convolution operator on $L^p(\mathbb{R})$ whose norm does not exceed $c^2N_p(k)$. Since this estimate is independent of N, we can let this last parameter tend to ∞ and an application of Fatou's lemma gives us Theorem II in the general case.

We end this note with a few comments about this and more general results obtained by this "transference method." First of all, Theorem II is not true when $p = 1$. If it were, H_kF would belong to $L^1(\mathbb{R})$ whenever $F \in L^1(\mathbb{R})$; thus, $\mathscr{M}(\xi)\hat{F}(\xi)$, being the Fourier transform of an integrable function, must always be a continuous function. Since $\mathscr{M}$ is discontinuous at each integer n such that $m_n \neq 0$, this cannot be the case in general. Moreover, the operator $\mathscr{H}$ is not bounded on $L^1(\mathbb{R})$ ($\mathscr{H}F \notin L^1(\mathbb{R})$ if, say, F is $\chi_{[0,1)}$). It is, however, of weak type (1,1). It is only fitting that it was M. Cotlar who led the way to the study of this operator in the "spirit of transference." One of his earliest papers [3] presents an important application of the technique we are employing; in this work, Cotlar studies the Hilbert transform in the Ergodic setting and the operator $\mathscr{H}$ is a special case in his theory. The article of Peterson [5] is a good reference for both Cotlar's results and inequality (6) for the "original Hilbert transform." It was M. Riesz who first pointed out, in 1924, that inequality (6) is valid (see [6]); he proved this estimate in a paper published four years later (see [7], page 241).

References

[1] Berkson, E. R. and Gillespie, **Transference and extension of Fourier multipliers for** $L^p(T)$, preprint.

[2] Coifman, R. R. and Weiss, Guido, **Transference methods in analysis**, CBMS Regional Conf. Series in Math., No. 31, AMS, Providence, R. I. (1973) (reprinted 1986).

[3] Cotlar, M., **A unified theory of Hilbert transforms and ergodic theorems**, Rev. Mat. Cuyana 1 (1955), 105–67.

[4] Jodeit, M. A. Jr, **Restrictions and extensions of Fourier multipliers**, Studia Math., 34 (1970) 215–226.

[5] Peterson, Karl, **Another proof of the existence of the Ergodic Hilbert transform**, Proc. of the American Math. Soc., Vol. 88, Number 1 (1983) 39–43.

[6] Riesz, M., **Les fonctions conjuguées et les séries de Fourier**, Compte Rendues de l'Acad des Sciences, Paris, Tome 178 (1924) 1464–1467.

[7] ———, **Sur les fonctions conjuguées**, Math. Zeitschr. 27 (1928) 218–44.

Universidad Autónoma de Madrid, Washington University in St. Louis, University of Ljubljana

Part II. Functional Analysis and Operator Theory

20

Schur Analysis of a Class of Translation Invariant Forms

Rodrigo Arocena

I. INTRODUCTION: UNITARY EXTENSIONS OF AN ISOMETRY

About ten years ago M. Cotlar observed that the Hilbert transform in a weighted L^2-space is bounded if and only if a related Toeplitz-like kernel is positive definite. That remark was exploited to establish the Cotlar–Sadosky lifting theorem [10], thus starting a unified approach to several problems in analysis, an overview of which can be seen in [11].

In that framework we have proposed the notion of a Toeplitz–Krein–Cotlar (T-K-C) form and obtained by means of it some results in $\mathbb{Z}^2$ concerning extensions of functions of positive type, characterization of stationary processes with positive angle between past and future, and commutant liftings [3]. Those forms can be seen as a way of reducing several problems to the one of describing all the commutative unitary extensions of a certain family of isometries. Thus, in particular, we are led to the study of all the minimal and essentially different unitary extensions of a given isometry, in the sense we now describe.

We say that V is an isometry acting in a Hilbert space H if V is a linear isometric operator from D onto R, which are both closed subspaces of H. The orthogonal complements N, M in H of D, R, respectively, are called the defect subspaces of V. Set $(U,F) \in \mathbf{U}$ if U is a unitary extension of V to a Hilbert space F that contains H and satisfies the minimality condition

$$F = \vee\{U^n H : n \in \mathbb{Z}\};$$

set also $(U,F) \approx (U',F')$ in $\mathbf{U}$ if there exists a unitary operator $\tau \in \mathbf{L}(F,F')$ such that $\tau_{|H}$ is the identity and $\tau U = U'\tau$; then we say that $\mathbf{U}$ is the set of all the minimal unitary extensions of V. This notation will be kept from now on.

A fundamental fact concerning **U** is the following: If $\mathbf{B}(N,M)$ denotes the family of the $\mathbf{L}(N,M)$-valued contractive analytic functions, that is, $\phi \in \mathbf{B}(N,M)$ when ϕ is an analytic function on the unit disk $\mathbb{D}$ such that $\phi(z) \in \mathbf{L}(N,M)$ and $\|\phi(z)\| \leq 1$, $\forall z \in \mathbb{D}$, then there exists a bijection between **U** and $\mathbf{B}(N,M)$. (See [6]). This will be exploited, in particular, in Section VIII and relates the study of **U** with the celebrated Adamjan–Arov–Krein parametrization.

The connections of our study with T–K–C forms are sketched in Section IX, which can be read after this introduction.

The methods we discuss in this paper are applied in [4] to the study of contractive intertwining dilations, that is, of the Sarason–Nagy–Foias commutant lifting theorem.

II. SZEGÖ'S OPERATORS: THE MOST INNOVATIVE EXTENSION

It is natural to ask if **U** contains some naturally distinguished element. Moreover, since a whole family of generalized moment problems can be solved by unitarily extending an isometry, such a salient extension will give in each case a salient solution of the problem. In this section we shall see that a prediction theory approach characterizes an element in **U** that can be considered as the most innovative one. In the next section we shall see that Schur analysis distinguishes the same element. Our approach is related to those of Landau and Constantinescu (see [9], [13]).

For each $(U,F) \in \mathbf{U}$ set $F^+ = \vee\{U^nH\colon n \geq 0\}$, $F^- = \vee\{U^nH\colon n \leq 0\}$, $F_0^+ = \vee\{U^nH\colon n > 0\}$, $F_0^- = \vee\{U^nH\colon n < 0\}$. In terms of prediction theory, F_0^- is the "past" and H the "present." The *Szegö's operator* associated to U is

$$\Delta_+[U] := P_H^F(I - P_{F_0^-}^F)_{|H} \in \mathbf{L}(H), \tag{1}$$

where P_H^F denotes the orthogonal projection of F onto H. If $v \in H$, then $\langle \Delta_+[U]v, v\rangle$ is the distance from v to the past so $\Delta_+[U]$ can be thought of as the prediction-error operator from the past. It is easy to see that

$$0 \leq \Delta_+[U] = P_N^H\Delta_+[U] = \Delta_+[U]P_N^H \leq P_N^H. \tag{2}$$

The kernel $H \cap F_0^-$ of $\Delta_+[U]$ is the "predictable part" of H, while $v \in H$ is "most innovative" when $P_N^H v$ is orthogonal to F_0^- (because $P_D^H v \in F_0^-$). A straightforward verification states that:

(3) Lemma *The following properties are equivalent: (i) $N \perp F_0^-$; (ii) $F^- = F_0^- \oplus N$; (iii) $U^kN \perp H$, $\forall k > 0$; (iv) $\Delta_+[U] = P_N^H$; (v) $F = F^- \oplus UN \oplus \cdots \oplus U^kN \oplus \cdots$.*

The prediction-error operator from the future can be defined in the same way: $\Delta_-[U] := P_H^F(I - P_{F_0^+}^F)_{|H}$. With obvious notation, $\Delta_-[U] = \Delta_+[U^*]$, so $0 \leq \Delta_-[U] = P_M^H\Delta_-[U] = \Delta_-[U]P_M^H \leq P_M^H$ and

(3a) (i) $M \perp F_0^+ \Leftrightarrow$ (ii) $F^+ = F_0^+ \oplus M \Leftrightarrow$ (iii) $U^{-k}M \perp H, \forall k > 0 \Leftrightarrow$

$$\text{(iv) } \Delta_-[U] = P_M^H \Leftrightarrow \text{(v) } F = \cdots \oplus U^{-k}M \oplus \cdots \oplus U^{-1}M \oplus F^+.$$

Now, (3.iii) and (3a.iii) are equivalent. In fact, if $UN \perp H$, then $U^{-1}M \perp (N \oplus U^{-1}R) = H$; assume that $U^{-j}M \perp H$ for $1 \leq j \leq n$: Since $U^{-n}M \perp R = UD$, $U^{-n-1}M \perp D$; if $N \perp U^{-n-1}H \supset U^{-n-1}M$, it follows that $U^{-n-1}M \perp H$, etc. So conditions (3) and (3a) are equivalent; when they hold,

$$F = \left(\bigoplus_{j>0} U^{-j}M\right) \bigoplus H \bigoplus \left(\bigoplus_{k>0} U^{k}N\right)$$

because the second member is U-invariant. Clearly, there exists exactly one $(U,F) \in \mathbf{U}$ with F as above. In that case we have, for $h \in H$ and $n > 0$,

$$P_H^F U^n h = P_H^F U^{n-1}(VP_D^H h + UP_N^H h) = P_H^F U^{n-1}(VP_D^H h) = \cdots = (VP_D^H) nh;$$

since $F = \vee\{U^n H : n \in \mathbb{Z}\}$, U is the minimal unitary dilation [14] of the contraction $(VP_D^H) \in \mathbf{L}(H)$.

(4) Theorem *Given $(U,F) \in \mathbf{U}$ the following conditions are equivalent: (a) $U^k N \perp H$, $\forall k > 0$; (b) $U^{-k}M \perp H$, $\forall k > 0$; (c) $\Delta_+[U] = P_N^H$; (d) $\Delta_-[U] = P_M^H$. There exists only one $(\tilde{U},\tilde{F}) \in \mathbf{U}$ that satisfies them; $\tilde{U}$ is the minimal unitary dilation of $(VP_D^H) \in \mathbf{L}(H)$ and $\tilde{F} = (\bigoplus_{j>0} U^{-j}M) \bigoplus H \bigoplus(\bigoplus_{k>0} U^k N)$.*

We shall say that $(\tilde{U},\tilde{F})$ is the *most innovative minimal unitary extension* of V. Two other interesting characterizations of it can be given.

Each $(U,F) \in \mathbf{U}$ defines a *generalized resolvent* R^U of V given by

$$R^U(z) = P_H^F(I - zU)^{-1}_{|H}, \qquad |z| \neq 1.$$

According to a theorem due to Chumakin [8] the set of functions

$$R(z) = \{I - z[VP_D^H + \phi(z)P_N^H]\}^{-1}, \qquad |z| < 1,$$

with $\phi \in \mathbf{B}(N,M)$ is the same as the set of (the restrictions to $\mathbb{D}$ of) the generalized resolvents of V. From Theorem (4) it follows that

$$(I - zVP_D^H)^{-1} = \sum\{z^n P_H^{\tilde{F}} \tilde{U}^n_{|H} : n \geq 0\} = R^{\tilde{U}}(z),$$

so

(5) The generalized resolvent of V that corresponds to its most innovative unitary extension is the one given by Chumakin's formula with $\phi \equiv 0$.

A Schur-type formula due to Arov and Grossman [6] parametrizes the characteristic functions with scale subspace of the elements of **U**, with parameter in $\mathbf{B}(N,M)$.

If L is a closed subspace of F and $U \in \mathbf{L}(F)$ is a unitary operator, the characteristic function of U with scale subspace L is $S^{U,L} \in \mathbf{B}(L,L)$ given by

$$S^{U,L}(z) = P_L^F(I - zUP_{F-L})^{-1}U_{|L} = [P_L^F(I - zU)_{|L}^{-1}]^{-1}P_L^F(I - zU)^{-1}U_{|L},$$

for $z \in \mathbb{D}$, with $F - L$ denoting the orthogonal complement of L in F. When $(U,F) \in \mathbf{U}$ and $L \subset H$, the following formula

$$(6) \qquad S^{U,L}(z) = S_{12}(z) + S_{11}(z)\theta(z)[I - zS_{21}(z)\theta(z)]^{-1}S_{22}(z), \qquad z \in \mathbb{D},$$

with

$$S_{11}(z) = P_L^H(I - zVP_D^HP_{H-L}^H)_{|M}^{-1},$$
$$S_{12}(z) = [P_L^H(I - zVP_D^H)_{|L}^{-1}]^{-1}P_L^H(I - zVP_D^H)^{-1}VP_{D|L}^H,$$
$$S_{21}(z) = zP_N^H(I - zP_{H-L}^HVP_D^H)^{-1}P_{H-L|M}^H,$$
$$S_{22}(z) = P_N^H(I - zP_{H-L}^HVP_D^H)_{|L}^{-1},$$

gives a correspondence from $\mathbf{B}(N,M)$ onto $\{S^{U,L}: (U,F) \in \mathbf{U}\}$, which is also $1-1$ if $H = D \vee L = R \vee L$. It is worthwhile noting that when (6) holds, θ can be recovered from U by means of

$$\theta(z) = P_M^FU(I - zP_{F-H}U)_{|N}^{-1} \equiv P_M^F(I - zUP_{F-H})^{-1}U_{|M}.$$

Now, $(\tilde{U}P_{\tilde{F}-H}^{\tilde{F}})^n\tilde{U}N = \tilde{U}^{n+1}N \perp M, \qquad \forall n \geq 0$, so

(7) Corollary *The characteristic function with scale subspace $L \subset H$ of the most innovative unitary extension of V is the one given by the Arov–Grossman formula (6) with $\theta \equiv 0$, that is, $S^{\tilde{U},L} = S_{12}$.*

III. SCHUR PARAMETERS OF EACH UNITARY EXTENSION

First Extension of V

Let V_1 be an isometric extension of V with domain $D_1 = H$ and acting in a space $H_1 = H \vee (V_1H)$. Then $H_1 = H \vee (V_1N)$. Set

$$(1) \qquad \gamma_1 = P_M^{H_1}V_{1|N}$$

Then the contraction $\gamma_1 \in \mathbf{L}(N,M)$ is such that $P_H^{H_1}V_1 = VP_D^H + \gamma_1 P_N^H$, so $\langle h + V_1 n, h' + V_1 n'\rangle_{H_1} = \langle h,h'\rangle_H + \langle h,\gamma_1 n'\rangle_H + \langle \gamma_1 n, h'\rangle_H + \langle n,n'\rangle_H$ holds for every $h,h' \in H, n,n' \in N$.

Conversely, let $\gamma_1 \in \mathbf{L}(N,M)$ be any contraction. Set $H_1' = H \times N$ and define β_1: $H_1' \times H_1' \to \mathbb{C}$ by $\beta_1[(h,n),(h',n')] = \langle h,h'\rangle + \langle h,\gamma_1 n'\rangle_H + \langle \gamma_1 n,h'\rangle_H + \langle n,n'\rangle_H$. The vector space H_1' and the positive form β_1 generate a Hilbert space H_1 such that $H \approx H \times \{0\}$ is a closed substance of H_1 and that for any $h \in H$ the vector $(VP_D^H h, P_N^H h) \in H_1'$ defines an element of H_1 which we call $V_1 h$. Clearly, $\|V_1 h\| = \|h\|$, so in this way we get an isometric extension V_1 of V with domain H and acting in $H \vee (V_1 N)$ such that $\gamma_1 = P_M^{H_1} V_{1|N}$.

Let R_1 be the range of such a V_1 and N_1, M_1 its defect subspaces. Since $N_1 = H_1 - D_1 = H_1 - D$ and $H_1 = H \vee (V_1 N)$, $N_1 = \{P_{N_1}^{H_1} V_1 n \colon \in N\}^- = \{(I - P_H^{H_1})V_1 n\colon n \in N\}^-$, so $N_1 = \{(V_1 - \gamma_1)n\colon n \in N\}^-$. For any $h \in H$, $m \in M$ we have $\langle P_{V_1 H}^{H_1} m, V_1 h\rangle = \langle m, V_1 h\rangle = \langle m, V_1 P_N^H h\rangle = \langle m, \gamma_1 P_N^H h\rangle = \langle V_1\gamma_1^* m, V_1 h\rangle$; thus, $(P_{V_1 H}^{H_1})_{|M} = V_1\gamma_1^*$; since $M_1 = H_1 - R_1 = H_1 - V_1 H$ and $H_1 = (V_1 H) \vee H = (V_1 H) \vee M$, it follows that $M_1 = \{P_M^{H_1} m\colon m \in M\}^- = \{(I - P_{V_1 H}^{H_1})m\colon m \in M\}^-$; that is, $M = \{(I - V_1\gamma_1^*)m\colon m \in M\}^-$.

Note that $\theta_1[(V_1 - \gamma_1)n] = D_{\gamma_1} n$ defines a unitary operator from N_1 onto $\mathbf{D}_{\gamma_1}$ such that $V_1 n = \gamma_1 n + \theta_1^{-1} D_{\gamma_1} n$, $\forall n \in N$. (As usual, we set, for any contractive operator $T \in \mathbf{L}(H,K)$, $D_T = (I - T^*T)^{1/2}$ and $\mathbf{D}_T = \{D_T H\}^-$). Thus we may obtain V_1 and H_1 from γ_1 by means of

$$(2) \qquad H_1 = H \oplus \mathbf{D}_{\gamma_1}; \qquad V_1 h = [VP_D^H h + \gamma_1 P_N^H h] \oplus D_{\gamma_1} P_N^H h, \qquad \forall h \in H.$$

Moreover, if V_1' is any isometric extension of V with domain H and acting in $H_1' = H \vee (V_1' H)$ such that $P_M^{H_1'} V_{1|N}' = \gamma_1$, then $(h + V_1 n) \to (h + V_1' n)$ defines a unitary operator $T \in \mathbf{L}(H_1, H_1')$ such that $T_{|H} = I_H$ and $TV_{1|H} = V_1' T_{|H}$.

Summing up:

(3) Proposition *Formulas (1) and (2) give a bijection between the contractions $\gamma_1 \in \mathbf{L}(N,M)$ and the essentially different isometric extensions V_1 of V to H that act in spaces $H_1 = H \vee (V_1 H)$.*

Definition of Schur Parameter

Each $(U,F) \in \mathbf{U}$ is associated to an iteration of the preceding construction. For $m \geq 0$ set $H_m = \vee\{U^n\colon 0 \leq n \leq m\}$ and let V_m be the isometry acting in H_m given by $V_m = U_{|H_{m-1}}$, with $H_{-1} = D$. Thus V_{m+1} is an isometric extension of V_m to H_m that acts in $H_{m+1} = H_m \vee (V_{m+1} H_m)$. If N_m, M_m are the defect subspaces of V_m, we get V_{m+1} from V_m by means of a contraction $\gamma_{m+1} \in \mathbf{L}(N_m, M_m)$ such that

$$(4) \qquad P_{H_m}^{H_{m+1}} V_{m+1} = V_m P_{H_{m-1}}^{H_m} + \gamma_{m+1} P_{N_m}^{H_m},$$

$$\gamma_{m+1} = P_{M_m}^{H_{m+1}} V_{m+1|N_m}.$$

That contraction is obtained from (U,F) by setting $N_m = H_m - H_{m-1}$, $M_m = H_m - UH_{m-1}$, $\gamma_{m+1} = P^F_{M_m} U_{|N_m}$.

We say that $\{\gamma_m : m > 0\}$ is the sequence of *Schur parameters* of $U,F) \in \mathbf{U}$. The name is due to the following fact: in the classical Carathéodory–Fejér problem an isometry V can be defined with the property that there exists a bijection between all the solutions of the problem and $\mathbf{U}$ such that the $\{\gamma_m\}$ associated as above to each $(U,F) \in \mathbf{U}$ are exactly the one dimensional operators given by the numbers that Schur [15] associated to the corresponding solution of the problem [5].

The "degrees of freedom" in the construction of each $U,F) \in \mathbf{U}$ are given by its Schur parameters, because the construction can be carried out as follows:

(i_1) Choose a contraction $\gamma_1 \in \mathbf{L}(N,M)$ and define V_1 as before.

(i_{m+1}) Assuming V_m has been defined, choose a contraction $\gamma_{m+1} \in \mathbf{L}(N_m, M_m)$ and define V_{m+1} in the same way.

(ii) Having defined the sequence $\{V_m : m > 0\}$ that acts in the nested Hilbert spaces $\{H_m : m > 0\}$, let $U_+ \in \mathbf{L}(F_+)$ be the isometry given by $F_+ = \vee\{H_m : m > 0\}$, $U_{+|H_m} = V_{m+1}$;

(iii) Let U be the essentially unique unitary extension of U_+ to $F = \vee\{U^n F_+ : n \in \mathbb{Z}\}$; then $(U,F) \in \mathbf{U}$ and $\{\gamma_m : m > 0\}$ are its Schur parameters.

Note that $H_m = D \oplus N \oplus N_1 \oplus \cdots \oplus N_m$, so $F_+ = H \oplus (\oplus\{N_j : j \geq 1\})$; also $N_{m+1} = \{(U - \gamma_{m+1})h : h \in N_m\}^-$, $M_{m+1} = \{(I - U\gamma^*_{m+1})h : h \in M_m\}^-$. In particular:

(5) Proposition *$(\tilde{U},\tilde{F})$ is the element of* $\mathbf{U}$ *with all its Schur parameters equal to zero.*

IV. CHOICE SEQUENCES

Keeping the same notation as above we have that two unitary operators, $\theta_{m+1} \in \mathbf{L}(N_{m+1}, \mathbf{D}_{\gamma_{m+1}})$ and $\mu_{m+1} \in \mathbf{L}(M_{m+1}, \mathbf{D}_{\gamma^*_{m+1}})$, are defined by $\theta_{m+1}(U - \gamma_{m+1})h = D_{\gamma_{m+1}}h$, if $h \in N_m$, and $\mu_{m+1}(I - U\gamma^*_{m+1})h = D_{\gamma^*_{m+1}}h$, if $h \in M_m$. By means of them we shall associate to each element of $\mathbf{U}$ an *(N,M)-choice sequence*, that is (see [7]), a sequence $\{\Gamma_m : m > 0\}$ of contractions such that $\Gamma_1 \in \mathbf{L}(N,M)$ and $\Gamma_{m+1} \in \mathbf{L}(\mathbf{D}_{\Gamma_m}, D_{\Gamma^*_m})$ for every positive m.

(1) Proposition *Let $(U,F) \in \mathbf{U}$ with Schur parameters $\{\gamma_m : m > 0\}$. Setting $\Gamma_1 = \gamma_1$ and $\Gamma_{m+1} = \mathbf{M}_m \gamma_{m+1} \Theta^*_m$ if $m > 0$, with $\mathbf{M}_1 = \mu_1$, $\mathbf{M}_{k+1} = \mathbf{M}_k \mu_{k+1}$, $\Theta_1 = \theta_1$, $\Theta_{k+1} = \Theta_k \theta_{k+1}$, an (N,M)-choice sequence $\Gamma(U,F) = \{\Gamma_m : m > 0\}$ is defined in such a way that it only depends on (U,F) as an element of* $\mathbf{U}$ *and that the correspondence $(U,F) \to \Gamma(U,F)$ is injective.*

Proof. Clearly, $\Theta_m N_m = \mathbf{D}_{\Gamma_m}$, $\mathbf{M}_m M_m = \mathbf{D}_{\Gamma^*_m}$ and $\Gamma_{m+1} = \mathbf{M}_m \gamma_{m+1} \Theta^*_m$ hold for $m = 1$; assume that the same is true for every $m \leq k$. Then $D_{\Gamma_{k+1}} = \Theta_k D_{\gamma_{k+1}} \Theta^*_k$, $\mathbf{D}_{\Gamma_{k+1}} = \Theta_k(\mathbf{D}_{\gamma_{k+1}}) = \Theta_{k+1}(N_{k+1})$, $D_{\Gamma^*_{k+1}} = \mathbf{M}_k D_{\gamma^*_{k+1}} \mathbf{M}^*_k$ and $\mathbf{D}_{\Gamma^*_{k+1}} = \mathbf{M}_k(\mathbf{D}_{\gamma^*_{k+1}}) = \mathbf{M}_{k+1}(M_{k+1})$, so $\Gamma_{k+2} = \mathbf{M}_{k+1}\gamma_{k+2}\Theta^*_{k+1}$ is a well-defined contraction belonging to $\mathbf{L}(\mathbf{D}_{\Gamma_{k+1}}, \mathbf{D}_{\Gamma^*_{k+1}})$.

If $(U,F) \approx (U',F')$ in $\mathbf{U}$, there exists a unitary operator $\tau \in \mathbf{L}(F,F')$ such that $\tau U^n H = U'^n H$ for every $n \geq 0$, so, with obvious notation, $\gamma'_{m+1}\tau = \tau\gamma_{m+1}$ for every $m \geq 0$, and $\gamma'_1 = \gamma_1$; in this way it can be seen that $\Gamma'_{m+1} \equiv \Gamma_{m+1}$.

Conversely, if $\Gamma(U,F) = \Gamma(U',F')$, then $\gamma_1 = \gamma'_1$ and $\mathbf{M}_k\gamma_{k+1}\Theta^*_{k|\mathbf{D}_{\Gamma_k}} = \mathbf{M}'_k\gamma'_{k+1}\Theta'^*_{k|\mathbf{D}_{\Gamma_k}}, \forall k > 0$, so

(i) $(\mathbf{M}'^*_k\mathbf{M}_k)\gamma_{k+1} = \gamma'_{k+1}(\Theta'^*_k\Theta_k)_{|N_k}, (\Theta'^*_k\Theta_k)D_{\gamma_{k+1}} = D_{\gamma'_{k+1}}(\Theta'^*_k\Theta_k)_{|N_k}, (\mathbf{M}'^*_k\mathbf{M}_k)D_{\gamma^*_{k+1}} = D_{\gamma'^*_{k+1}}(\mathbf{M}'^*_k\mathbf{M}_k)_{|M_k}, \forall k > 0.$

Since $\Theta_k(N_k) = \mathbf{D}_{\Gamma_k} = \Theta'_k(N'_k)$, a unitary operator $\tau \in \mathbf{L}(F_+,F'_+)$ is defined by $\tau_{|H} = I$ and $\tau_{|N_k} = (\Theta'^*_k\Theta_k)_{|N_k}$. It follows that $\tau D_{\gamma_{k+1}} = D_{\gamma'_{k+1}}\tau_{|N_k}$. For h in N_k we have $\tau(U-\gamma_{k+1})h = \theta'^*_{k+1}\Theta'^*_k\Theta_k\theta_{k+1}(U-\gamma_{k+1})h = \theta'^*_{k+1}\tau D_{\gamma_{k+1}}h = \theta'^*_{k+1}D_{\gamma'_{k+1}}\tau h = (U'-\gamma'_{k+1})\tau h$, so

(ii) $\tau(U-\gamma_{k+1})h = (U'-\gamma'_{k+1})\tau h, \forall h \in N_k, k \geq 0.$

Now, $\tau(I-U\gamma^*_1)h = (I-U'\gamma'^*_1)\tau h$ holds for every $h \in M$ so $\tau M_1 = M'_1$ and $\tau_{|M_1} = \mu'^*_1\mu_1$; then (i) shows that $\tau\gamma_2 = \gamma'_2\tau_{|N_1}$. So we assume that $\tau_{|M_k} = \mathbf{M}'^*_k\mathbf{M}_k$: $(I-U_{\gamma^*_k})h \to (I-U'\gamma'^*_k)\tau h, \forall h \in M_{k-1}$, and that

(iii) $\tau\gamma_{k+1}h = \gamma'_{k+1}\tau h, \forall h \in N_k.$

For $h \in M_k$ we have $\tau(I-U\gamma^*_{k+1})h = \tau(h-\gamma_{k+1}\gamma^*_{k+1}h) - \tau(U-\gamma_{k+1})\gamma^*_{k+1}h = \tau h - \gamma'_{k+1}\tau\gamma^*_{k+1}h - (U'-\gamma'_{k+1})\tau\gamma^*_{k+1}h = (I-U'\gamma'^*_{k+1})\tau h$, and $\mathbf{M}'^*_{k+1}\mathbf{M}_{k+1})(I-U\gamma^*_{k+1})h = \mu'^*_{k+1}(\mathbf{M}'^*_k\mathbf{M}_k)\mu_{k+1}(I-U\gamma^*_{k+1})h = \mu'^*_{k+1}(\mathbf{M}'^*_k\mathbf{M}_k)D_{\gamma^*_{k+1}}h = \mu'^*_{k+1}D_{\gamma'^*_{k+1}}(\mathbf{M}'^*_k\mathbf{M}_k)h = (I-U'\gamma'^*_{k+1})\tau h$; consequently, $\tau_{|M_{k+1}} = \mathbf{M}'^*_{k+1}\mathbf{M}_{k+1}$; thus, (i) shows that (iii) holds for every $k \geq 0$. Combining (ii) and (iii) we see that $\tau Uh = U'\tau h, \forall h \in N_k$, so $\tau U_+ = U'_+\tau$ and τ can be extended to a unitary operator from F to F' that leaves invariant every vector in H and intertwines U with U'. Q.E.D.

We shall now see that this association between unitary extensions and choice sequences is a bijection. More precisely:

(2) Theorem *There exists a bijection between the set* $\mathbf{G}$ *of all the* (N,M)*-choice sequences and* $\mathbf{U}$ *that can be obtained as follows. Given* $\Gamma = \{\Gamma_k : k > 0\} \in \mathbf{G}$, *let* $U \in \mathbf{L}(F)$ *be the minimal unitary extension of the isometry* $U_{|F_+}$, *with* $F_+ = H \oplus [\oplus\{\mathbf{D}_{\Gamma_j} : 1 \leq j\}]$, *such that* $U_{|D} = V$, $U_{|N} = \Gamma_1 \oplus D_{\Gamma_1}$, $U_{|\mathbf{D}_{\Gamma_k}} = \eta^*_k\Gamma_{k+1} \oplus D_{\Gamma_{k+1}}$, *where* η_k *is the unitary operator from*

$$M_k = (H \oplus [\oplus\{\mathbf{D}_{\Gamma_j} : 1 \leq j \leq k\}) - U(H \oplus [\oplus\{\mathbf{D}_{\Gamma_j} : 1 \leq j \leq k-1\})$$

onto $\mathbf{D}_{\Gamma^*_k}$, *given by* $\eta_k[I - U(P_{M_{k-1}}U_{|\mathbf{D}_{\Gamma_{k-1}}})^*]h = D_{\Gamma^*_k}\eta_{k-1}h$ *if* $h \in M_{k-1}$, $k \geq 1$, *and* $\eta_0 = 1$. *Then*

(i) $(U,F) \in \mathbf{U}$ *and its Schur parameters are* $\gamma_{k+1} = P_{M_k}U_{|\mathbf{D}_{\Gamma_k}} = \eta^*_k\Gamma_{k+1}$ *for every* $k \geq 0$.

(ii) *The element of* $\mathbf{G}$ *associated to* (U,F) *as in proposition (1) is* Γ.

Proof. Setting $H = H \oplus \mathbf{D}_{\Gamma_1}$ and $V_1 = (VP^H_D + \Gamma_1P^H_N) \oplus D_{\Gamma_1}P^H_N : H \to H_1$, we define an isometric extension V_1 of V with domain H that acts in $H_1 = H \vee (V_1H)$. Also, $N_1 = \mathbf{D}_{\Gamma_1}$ and $\gamma_1 := P^{H_1}_M V_{1|N} = \Gamma_1$. Since $M_1 = \{(I - V_1\Gamma^*_1)m : m \in M\}^-$, it is clear that a unitary operator $\eta_1 \in \mathbf{L}(M_1, \mathbf{D}_{\Gamma_1^*})$ is defined by $\eta_1(I - V_1\Gamma_1^*)m = (D_{\Gamma_1^*})m$, $\forall m \in M$.

By iteration, setting $H_{k+1} = H \oplus D_{\Gamma_1} \oplus \cdots \oplus D_{\Gamma_{k+1}}$ and $V_{k+1} = (V_k P_{H_{k-1}}^{H_k} + \eta_k^*\Gamma_{k+1}P_{N_k}^{H_k}) \oplus D_{\Gamma_{k+1}}P_{N_k}^{H_k}$, we define an isometric extension of V to H_k that acts in H_{k+1} and is such that $\gamma_{k+1} := P_{M_k}^{H_{k+1}}V_{k+1|N_k} = \eta_k^*\Gamma_{k+1}$, while $\eta_{k+1}(I - V_{k+1}\gamma_{k+1}^*)m = (D_{\Gamma_{k+1}^*}\eta_k)m$, $\forall m \in M_k$, gives a unitary operator from M_{k+1} onto $\mathbf{D}_{\Gamma_{k+1}^*}$.

With $F_+ = \vee\{H_k: k > 0\}$ and $U_+ \in \mathbf{L}(F_+)$ such that $U_{+|H_k} = V_{k+1}$, as before, we get $(U, F) \in \mathbf{U}$ as in (i).

Now we apply proposition (4) to this (U, F). If $h \in N_k$, $(U - \gamma_{k+1})h = (V_{k+1} - \gamma_{k+1})h = D_{\Gamma_{k+1}}h$, so θ_{k+1} is the identity in N_{k+1} for every $k \geq 0$. If $h \in M_k$, $\eta_{k+1}(I - U\gamma_{k+1}^*)h = (D_{\Gamma_{k+1}^*}\eta_k)h = (\eta_k D_{\gamma_{k+1}^*})h$, so $\mu_{k+1} = \eta_k^*\eta_{k+1}$ and $\eta_0 = 1$. Thus, Θ_{k+1} equals the identity and $\mathbf{M}_{k+1} = \eta_{k+1}$. The choice sequence associated to (U, F) is given by Γ_1 and $\mathbf{M}_k\gamma_{k+1}\Theta_k = \mathbf{M}_{k+1}\eta_k^*\Gamma_{k+1} = \Gamma_{k+1}$, $\forall k > 0$. Q.E.D.

Remark. If $\Gamma_k \equiv 0$, $\mathbf{D}_{\Gamma_k} \equiv N$ and $D_{\Gamma_k} \equiv I_N$, so the most innovative element $(\tilde{U}, \tilde{F})$ of $\mathbf{U}$ corresponds to the identically zero (N, M)-choice sequence.

V. ON THE RELATION BETWEEN SZEGÖ'S OPERATORS, SCHUR PARAMETERS, AND CHOICE SEQUENCES

For $(U, F) \in \mathbf{U}$, we have $\Delta_-[U] = P_H(I - P_{UF^+})_{|H}$ and $UF^+ = \vee\{UH_{k-1}: k \geq 0\}$, $M_k = (H_k - UH_{k-1}) \perp UH_{k-1}$. Thus $(I - P_{UH_{k-1}})_{|H} = P_{M_k|H} = P_{M_k}(P_{M_{k-1}} + P_{UH_{k-2}})_{|H} = P_{M_k}P_{M_{k-1}|H} = \cdots = P_{M_k}P_{M_{k-1}}\cdots P_{M|H}$. Also, $M_k = \{(I - U_{\gamma_k^*})h: h \in M_{k-1}\}^-$ and $0 = \langle(I - U\gamma_k^*)h, U\gamma_k^*h\rangle$, $\forall h \in M_{k-1}$, so $P_{M_k|M_{k-1}} = (I - U\gamma_k^*)_{|M_{k-1}}$. Consequently:

$$\Delta_-[U] = s\text{-}\lim_{k\to\infty} P_H(I - U\gamma_k^*)\cdots(I - U\gamma_1^*)P_{M|H} \tag{1}$$

and $\|(I - U\gamma_k^*)\cdots(I - U\gamma_1^*)h\|^2$ decreases to $\langle\Delta_-[U]h, h\rangle$, $\forall h \in M$.

If (U, F) is obtained from an (N, M)-choice sequence $\{\Gamma_k: k > 0\}$ as in Theorem IV.2, then $I - U_{\gamma_k^*} = \eta_k^* D_{\Gamma_k^*}\eta_{k-1}$, so

$$\|(I - U\gamma_k^*)\cdots(I - U\gamma_1^*)h\|^2 = \langle D_{\Gamma_1^*}\cdots D_{\Gamma_{k-1}^*}D_{\Gamma_k^*}^2 D_{\Gamma_{k-1}^*}\cdots D_{\Gamma_1^*}h, h\rangle.$$

Thus we obtain

$$\Delta_-[U]_{|M} = s\text{-}\lim_{k\to\infty} D_{\Gamma_1^*}\cdots D_{\Gamma_{k-1}^*}D_{\Gamma_k^*}^2 D_{\Gamma_{k-1}^*}\cdots D_{\Gamma_1^*}, \tag{2}$$

a Verblunsky–Constantinescu-type formula (see [9]), that in the scalar case $(\dim M = 1)$ reduces to $\Delta_-[U]_{|M} = \Pi\{(1 - |\Gamma_k|^2: k \geq 1\}$.

VI. MOMENT PROBLEMS WITH SCALE SUBSPACES

If $L \subset H$ is a closed subspace such that $H = D \vee L = R \vee L$ we shall see that each $(U,F) \in \mathbf{U}$ is determined by the sequence

$$g_n := P_L U^n_{|L} \in \mathbf{L}(L), \qquad n > 0.$$

Now, $H_{k+1} = H_k \vee U^{k+1}L = UH_k \vee L, \forall k > 0$, so $F_+ = D \vee (\vee\{U^kL: k \geq 0\}) = R \vee (\vee\{U^kL: k \geq 0\})$ and consequently $F = D \vee (\vee\{U^kL: k \in \mathbb{Z}\}) = R \vee (\vee\{U^kL: k \in \mathbb{Z}\})$. Let $(U',F') \in \mathbf{U}$; with obvious notation assume that $g'_n \equiv g_n$ and set $\theta(h + Ul_1 + \cdots + U^kl_k) = h + U'l_1 + U'^kl_k$ for any $h \in H, l_1 \cdots l_k \in L$; it is easy to see inductively that θ is an isometry from H_k onto H'_k; it follows that $(U,F) \approx (U',F')$.

We say that $\{g_n: n > 0\} \subset \mathbf{L}(L)$ is the *sequence of moments of (U,F) with scale subspace L*. Thus, if $\Gamma(U,F) = \{\Gamma_n: n > 0\}$ is the corresponding choice sequence, each sequence determines the other one. We now prove that, moreover, $\{g_1,\cdots,g_{k+1}\}$ determines $\{\Gamma_1\cdots\Gamma_{k+1}\}$ and conversely.

If $l,l' \in L$, then

$$\begin{aligned}\langle g_{k+1}l,l'\rangle &= \langle P_{UH_{k-1}}U^{k+1}1 + P_{UH_{k-1}}l' + P_{M_k}l'\rangle \\ &= \langle P_{UH_{k-1}}U^{k+1}l,l'\rangle + \langle P_{UN_k}U^{k+1}l,P_{M_k}l'\rangle,\end{aligned}$$

so we get the following relation between moments and Schur parameters:

$$g_{k+1} = P_L(UP_{H_{k-1}} + \gamma_{k+1}P_{N_k})U^k_{|L}, \qquad \forall k \geq 0. \tag{1}$$

If (U,F) is obtained from the (N,M)-choice sequence $\{\Gamma_n: n > 0\}$ as shown in Theorem IV.2, then

$$g_{k+1} = P_L(UP_{H_{k-1}} \oplus \eta_k^*\Gamma_{k+1}P_{\mathbf{D}_{\Gamma_k}})U^k_{|L}, \qquad \forall k \geq 0, \tag{2}$$

and in particular $g_1 = P_L(VP_D + \Gamma_1P_N)_{|L}$, so $\{\Gamma_1\cdots\Gamma_{k+1}\}$ determines $\{g_1\cdots g_{k+1}\}$.

From $H_k = H_{k-1} \vee U^kL$ it follows that $(P_{N_k}U^kL)^- = N_k$ and from $H_k = L \vee UH_{k-1}$ that $(P_{M_k}L)^- = M_k, \forall k \geq 0$. When $N_k = \mathbf{D}_{\Gamma_k}$, (2) says that

$$\langle \Gamma_{k+1}P_{\mathbf{D}_{\Gamma_k}}U^kl, \eta_kP_{M_k}l'\rangle = \langle g_{k+1}l,l'\rangle - \langle UP_{H_{k-1}}U^kl,l'\rangle, \forall k \geq 0, l,l' \in L. \tag{3}$$

Thus, g_1 determines Γ_1 because $\langle \Gamma_1P_Nl,P_Ml'\rangle = \langle g_1l,l'\rangle - \langle VP_Dl,l'\rangle$; assuming that $\{g_1,\ldots,g_k\}$ determines $\{\Gamma_1,\ldots,\Gamma_k\}, H_k = H \oplus (\oplus\{\mathbf{D}_{\Gamma_j}: 1 \leq j \leq k\})$ and $U_{|H_{k-1}}$ are also determined by them, so (3) shows that $\{g_1,\ldots,g_{k+1}\}$ give Γ_{k+1}.

Different moment problems can be considered as particular cases of the following *moment problems for the isometry V with scale subspace L*:

($\mathbf{P}_k$) Given $\{g_j: 1 \leq j \leq k\} \subset \mathbf{L}(L), \exists(U,F) \in \mathbf{U}$ such that $g_j = P_L U^j_{|L}$, $1 \leq j \leq k$?

($\mathbf{P}$) Given $\{g_j: 1 \leq j\} \subset \mathbf{L}(L), \exists(U,F) \in \mathbf{U}$ such that $g_j = P_L U^j_{|L}, \forall j \geq 1$?

When $\{g_1,\ldots,g_k\}$ are such that ($\mathbf{P}_k$) has a solution, then the same happens for ($\mathbf{P}_{k+1}$) iff g_{k+1} belongs to a "disk," determined by V and $\{g_1\ldots g_k\}$, a well-known result in classical problems. In fact:

(4) Proposition

(i) ($\mathbf{P}_1$) *has solutions for* $g = g_1$ *iff* $g = A_0 + X_0\Gamma Y_0$, *with* $A_0 = P_L V P_{D|L}$, $X_0 = P_{L|M}$, $Y_0 = P_{N|L}$, *and* $\Gamma \in \mathbf{L}(N,M)$ *an arbitrary contraction.*

(ii) Given $\{g_j: 1 \leq j \leq k\} \subset \mathbf{L}(L)$ *such that* ($\mathbf{P}_k$) *has solutions, there exist two subspaces* $\bar{N}_k$, $\bar{M}_k$ *of* H *and three operators* $A_k \in \mathbf{L}(L)$, $X_k \in \mathbf{L}(\bar{M}_k, L)$, $Y_k \in \mathbf{L}(L,\bar{N}_k)$ *such that* ($\mathbf{P}_{k+1}$) *has solutions for* $\{g_1\cdots g_k, g\}$ *iff* $G = A_k + X_k\Gamma Y_k$, *with* $\Gamma \in \mathbf{L}(\bar{N}_k,\bar{M}_k)$ *an arbitrary contraction.*

Proof. (i) follows from (2) with $k = 0$. The first k elements $\{\Gamma_1,\ldots,\Gamma_k\}$ of the choice sequence of any solution (U,F) of ($\mathbf{P}_k$) are determined by $\{g_1,\ldots,g_k\}$, so the same happens with $UP_{H_{k-1}}$, the subspaces of H given by $\bar{N}_k := \mathbf{D}_{\Gamma_k}$, $\bar{M}_k := \mathbf{D}_{\Gamma^*_k}$, and the operators $A_k := P_L U P_{H_{k-1}} U^k_{|L} \in \mathbf{L}(L)$, $X_k := P_L \eta^*_k \in \mathbf{L}(\bar{M}_k, L)$, $\gamma_k := P_{\mathbf{D}_{\Gamma_k}} Y^k_{|L} \in \mathbf{L}(L,\bar{N}_k)$. If ($\mathbf{P}_{k+1}$) can be solved for $\{g_1,\ldots,g_k,g\}$, $g = g_{k+1}$ as in (2) and the equality in (ii) follows. Conversely, if the last is true, any $(U,F) \in \mathbf{U}$ with its choice sequence starting with $\{\Gamma_1\ldots\Gamma_k,\Gamma\}$ solves ($\mathbf{P}_{k+1}$). Q.E.D.

Remark. Thus, g_k always belongs to the "disk" with "center" A_k. If $\Gamma_k \equiv 0$, then g_k is always equal to the center of the corresponding disk and the solution of ($\mathbf{P}$) is $(\tilde{U},\tilde{F})$; in this case, as follows from the definition, the sequence of moments with scale subspace L is $\{P_L(VP^H_D)^n_{|L}: n > 0\}$.

VII. ON A SCHUR TYPE ALGORITHM

The preceding moment problems with scale subspace can be solved by means of the following iterative method. Let $\{g_n: n > 0\} \subset \mathbf{L}(L)$ be given.

Step 0. If N (resp. M) is trivial, ($\mathbf{P}_k$) has a solution iff $g_j = P_L V^j_{|L}$ ($g^*_j = P_L V^{-j}_{|L}$), $1 \leq j \leq k$, and the solution is unique.

Step 1. N and M are assumed to be nontrivial. From (VI.1) it follows that γ_1 exists only if

$$(\mathrm{a}_1) \qquad l \in L \text{ and } P_N l = 0 \Longrightarrow (g_1 - P_L V P_D)l = 0,$$
$$l' \in L \text{ and } P_M l' = 0 \Longrightarrow l' \in [(g_1 - P_L V P_D)L]^\perp$$

when (a_1) holds, since $(P_N L)^- = N$ and $(P_M L)^- = M$, a form $\varepsilon_1: N \times M \to \mathbb{C}$ is defined by

$$(b_1) \qquad \varepsilon_1(P_N l, P_M l') = \langle g_1 l, l' \rangle - \langle V P_D l, l' \rangle$$

Iff $\|\varepsilon\| \leq 1$, $\exists \gamma_1 \in \mathbf{L}(N, M)$, a contraction determined by

$$(c_1) \qquad g_1 = P_L(V P_D + \gamma_1 P_N)_{|L}$$

Then (see Section III) there exists an isometric extension V_1 of V to H that acts in $H_1 = H \vee (V_1 H) = H \vee (V_1 L) = L \vee (V_1 H)$ and satisfies $\gamma_1 = P_M V_{1|N}$. From (c_1) it follows that $\langle V_1 l, l' \rangle = \langle g_1 l, l' \rangle, \forall l, l' \in L$. Thus:

(d_1) $(\mathbf{P}_1)$ has solutions iff (a_1) holds and (b_1) defines a form $\varepsilon_1: N \times M \to \mathbb{C}$ such that $\|\varepsilon_1\| \leq 1$.

In that case, the solutions of $(\mathbf{P}_1)$ are given by the $(U, F) \in \mathbf{U}$ with first Schur parameter equal to γ_1. There exist an infinite number of solutions if $\mathbf{D}_{\gamma_1}, \mathbf{D}_{\gamma_1^*}$ are nontrivial; in the contrary case, there exists only one solution and the algorithm stops here.

Step $(k+1)$ $(k > 0)$. For $1 \leq j \leq k$ we have isometries V_j that extend V_{j-1} to H_{j-1}, act in $H_j = H_{j-1} \vee (V_j H_{j-1}) = H_{j-1} \vee (V_j^j L) = L \vee (V_j H_{j-1})$, and have nontrivial defect subspaces $N_j = (P_{N_j} V_j^j L)^-, M_j = (P_{M_j} L)^-$ If

$$(a_{k+1}) \qquad \begin{aligned} &l \in L \text{ and } P_{N_k} V_k^k l = 0 \Longrightarrow (g_{k+1} - P_L V_k P_{H_{k-1}} V_k^k) l = 0, \\ &l' \in L \text{ and } P_{M_k} l' = 0 \Longrightarrow l' \in [(g_{k+1} - P_L V_k P_{H_{k-1}}) L]^\perp \end{aligned}$$

a form $\varepsilon_{k+1}: N_k \times M_k \to \mathbb{C}$ is defined by

$$(b_{k+1}) \qquad \varepsilon_{k+1}(P_{N_k} V_k^k l, P_{M_k} l') = \langle g_{k+1} l, l' \rangle - \langle V_k P_{H_{k-1}} V_k^k l, l' \rangle$$

Iff $\|\varepsilon_{k+1}\| \leq 1, \exists \gamma_{k+1} \in \mathbf{L}(N_k, M_k)$, a contraction determined by

$$(c_{k+1}) \qquad g_{k+1} = P_L(V_k P_{H_{k-1}} + \gamma_{k+1} P_{N_k}) V_k^k{}_{|L}$$

so there exists an isometry V_{k+1} that extends V_k to H_k, acts in $H_{k+1} = H_k \vee (V_{k+1} H_k) = H_k \vee (V_{k+1}^{k+1} L) = L \vee (V_{k+1} H_k)$, and satisfies $\gamma_{k+1} = P_{M_k} V_{k+1|N_k}$. From (c_{k+1}) it follows that $\langle V_{k+1}^{k+1} l, l' \rangle = \langle g_{k+1} l, l' \rangle, \forall l, l' \in L$. Considering $(U, F) \in \mathbf{U}$ such that $U_{|H_k} = V_{k+1}$ we see that

(d_{k+1}) $(\mathbf{P}_{k+1})$ has solution iff (a_{k+1}) holds and (b_{k+1}) defines a form ε_{k+1}: $N_k \times M_k \to \mathbb{C}$ such that $\|\varepsilon_{k+1}\| \leq 1$.

In that case, the solutions of $(\mathbf{P}_{k+1})$ are given by the $(U,F) \in \mathbf{U}$ with first $(k+1)$ Schur parameters equal to $\gamma_1, \ldots, \gamma_{k+1}$. There exist an infinite number of solutions if $\mathbf{D}_{\gamma_{k+1}}, \mathbf{D}_{\gamma_{k+1}^*}$ are nontrivial; in the contrary case, there exists only one solution and the algorithm stops here.

If the process never comes to an end, that is, if $\gamma_k^* \gamma_k = I$, $\gamma_k \gamma_k^* = I$ never hold, every $(\mathbf{P}_k)$ has an infinite number of solutions and $(\mathbf{P})$ has exactly one solution, the $(U,F) \in \mathbf{U}$ given by the choice sequence $\{\Gamma_k : k > 0\}$ obtained from $\{g_k : k > 0\}$ by means of (VI.3).

VIII. ON A PARAMETRIZATION FORMULA

To determine the sequence $\{g_n : n > 0\}$ of the moments of $(U,F) \in \mathbf{U}$ with scale subspace is the same as determining the function

$$\psi(z) := \sum \{z^n g_{n+1} : n \geq 0\}$$

Since $\psi(z) = P_L(I - zU)^{-1} U_{|L}$, it is related with the characteristic function $S^{(U,L)}$ of (U,F) with scale subspace L (see Section II),

$$S^{U,L}(z) = [P_L(I - zU)^{-1}_{|L}]^{-1} P_L(I - zU)^{-1} U_{|L}, \qquad |z| < 1$$

by

$$S^{U,L}(z) = [I + z\psi(z)]^{-1} \psi(z), \qquad |z| < 1.$$

Since $S^{U,L}$ is a contractive function we get

$$\psi(z) = S^{U,L}(z)[I - zS^{U,L}(z)\}^{-1}, \qquad \forall z \in \mathbb{D}$$

Thus, the Arov–Grossman formula (II.6) gives the following parametrization of all the solutions of the moment problems with scale subspace.

(1) Proposition *These exists a bijection between the set of all the functions*

$$\psi(z) = \sum \{z^n g_{n+1} : n \geq 0\}. \qquad z \in \mathbb{D}.$$

such that $\{g_n : n > 0\}$ *is a solution of the moment problem with scale subspace L for the isometry V and the set* $\mathbf{B}(N,M)$ *of contractive analytic functions. That bijection is given by the formulas*

$$\psi(z) = \alpha(z)[I_L - z\alpha(z)]^{-1},$$
$$\alpha(z) = S_{12}(z) + S_{11}(z)\theta(z)[I_N - zS_{21}(z)\theta(z)]^{-1} S_{22}(z), \theta \in \mathbf{B}(N,M),$$

with $S_{jk}, j,k = 1, 2$, as in (II.6)

IX. APPLICATIONS TO TOEPLITZ–KREIN–COTLAR FORMS

By means of the class of translation invariant forms that we have called Toeplitz–Krein–Cotlar (T-K-C) forms, several moment and interpolation problems can be considered in the general framework that was studied in previous sections. We now sketch that method.

For a given natural number η (equal to 1 or 2 in applications) let $\Omega = \{\omega_j: 1 \leq j \leq \eta\}$ be a family of subsets of $\mathbb{Z}$ and $E = \oplus\{E_j: 1 \leq j \leq \eta\}$ an orthogonal sum of Hilbert spaces. We call A the vector space of functions $h = \{h_j: 1 \leq j \leq \eta\}$ such that each $h_j: \mathbb{Z} \to E_j$ has a finite support supph_j. If S is the shift in A given by $(\mathrm{Sh})(m) \equiv h(m-1)$ and $A_\Omega = \{h \in A: \mathrm{supp} h_j \subset \omega_j, 1 \leq j \leq \eta\}$, we say that $\{B; \mathbb{Z}, \Omega; E\}$ is a T-K-C form if $B: A_\Omega \times A_\Omega \to \mathbb{C}$ is a sesquilinear form such that $B(\mathrm{Sh}, \mathrm{Sh}') = B(h,h')$ if $h, h', Sh, Sh' \in A_\Omega$. Thus, B is invariant under translations whenever that makes sense. When $A_\Omega = A$, $\{B; \mathbb{Z}; E\} \equiv \{B; \mathbb{Z}, \mathbb{Z}^\eta; E\}$ is a classical Toeplitz form.

Now, the data of several moment problems can be given by means of a T-K-C form B in such a way that the problem can be solved iff B is positive, and in that case the solutions can be obtained by extending B to positive classical Toeplitz forms. In general, finding a positive Toeplitz extension of a T-K-C form $\{B; \Gamma; \Omega; E\}$, with Γ an Abelian group, is the same as extending a family of isometries naturally associated with the form to a unitary representation of Γ [3]. In the examples given later, this method can be simplified as follows.

For a given positive T-K-C form $\{B; \mathbb{Z}, \Omega; E\}$, let $\mathbf{T}$ be the set of all the positive Toeplitz forms $B' \equiv \{B'; \mathbb{Z}; E\}$ that extend it. E can be thought of as a subspace of A and $\vee\{S^n E: n \in \mathbb{Z}\} = A$, so each $B' \in \mathbf{T}$ is determined by the numbers $\{B'(S^n v, w): n \geq 0, v, w \in E\}$. Assume also that $E \subset A_\Omega$ (i.e., that $0 \in \omega_j: 1 \leq j \leq \eta$) and that there exists a constant $c > 0$ such that $B(v,v) \leq c\|v\|_E^2$, $\forall v \in E$. Then each B' is determined by the sequence $\{G_n: n > 0\} \subset \mathbf{L}(E)$ such that

$$B'(S^n v, w) \equiv \langle G_n v, w \rangle_E$$

For each $n \in \mathbb{Z}$ set $D'_n = \{h \in A_\Omega: S^n h \in A_\Omega\}$ and $S_n = S^n_{|D'_n}$. If Ω is such that $S(D'_{n+1}) \subset D'_n$ for every $n > 0$, then $S^j(D'_{n+1}) \subset D'_{n+1-j}$ for $1 \leq j \leq n$ and

$$S_{n+1} = (S_{1|D'_{n+1}})^{n+1}, \qquad \forall n > 0$$

The positive form B and the vector space A_Ω generate a Hilbert space H and a natural operator π from A_Ω onto a dense subspace of H, while the linear operators $\{S_n: n \in \mathbb{Z}\}$ generate a family of isometries $\{V_n: n \in \mathbb{Z}\}$ with domains D_n and ranges R_n contained in H, such that $V_0 = I$ and $V_n = (V_{1|D_n})^n = V_{-n}$ if $n > 0$. Set $V = V_1$ and let $\mathbf{U}$ be as before. There exists a bijection between $\mathbf{U}$ and $\mathbf{T}$ that associates each

$(U,F) \in \mathbf{U}$ to $B' \in \mathbf{T}$ given by

$$B'(S^n v, w) \equiv \langle U^n \pi v, \pi w \rangle_F$$

Suppose that $A_\Omega = D_1' + E = SD_1' + E$; then $H = D \vee L = R \vee L$, with $L = (\pi E)^-$. The moments of (U,F) with scale subspace L, $g_n \equiv P_L U^n_{|L}$, are such that $\langle g_n \pi v, \pi w \rangle_L \equiv \langle G_n v, w \rangle_E$; so setting $\mu = \pi_{|E} \in \mathbf{L}(E,L)$ we see that

$$G_n \equiv \mu^* g_n \mu$$

Examples

1. A very simple example of a T-K-C form is furnished by the truncated trigonometric moment problem. Let $\{c_j: -m \leq j \leq m\} \subset \mathbb{C}$ be given. Set $\eta = 1$, $\omega_1 = \{j \in \mathbb{Z}: 0 \leq j \leq m\}$, $E = \mathbb{C}$, and $B(h,h') \equiv \sum c_{j-k} h(j) \bar{h}'(k)$. Then $B \geq 0$ iff there exist positive measures β on $\mathbb{T}$ such that $\hat{\beta}(j) = c_j$, for $-m \leq j \leq m$, and in that case there exists a bijection between the set of all those measures and $\mathbf{T} \approx \mathbf{U}$. In particular, the maximum entropy solution of the problem (see [13]) is given by the most innovative element in $\mathbf{U}$.

2. As a second example we mention the Carathéodory–Fejér interpolation problem of characterizing the sequences $\{c_j: 0 \leq j \leq m\} \subset \mathbb{C}$ such that

$$\{f \in H^\infty(\mathbb{T}): \hat{f}(j) = c_j : 0 \leq j \leq m, \|f\|_\infty \leq 1\}$$

is nonvoid. In this case, setting: $\eta = 2$, $\omega_1 = \{j \in \mathbb{Z}: j \leq m\}$, $\omega_2 = \{j \in \mathbb{Z}: 0 \leq j\}$, $E_1 = E_2 = \mathbb{C}$, $c_\nu \equiv 0$ if $\nu < 0$ and $B(h,h') \equiv \sum\{h_1(u)\bar{h}_1'(u): u \in \omega_1\} + \sum\{c_{u-v} h_1(u) \bar{h}_1'(v) + \bar{c}_{u-v} \bar{h}_1'(u) h_2(v): (u,v) \in \omega_1 \times \omega_2\} + \sum\{h_2(v)\bar{h}_2'(v): v \in \omega_2\}$, classical results can be obtained [5].

3. Consider next the (vectorial) Nehari problem of characterizing the sequences $\{\lambda_m: m \geq 0\} \subset \mathbf{L}(E_1,E_2)$ such that there exist measurable functions $f: \mathbb{T} \to \mathbf{L}(E_1,E_2)$ such that $\|f(t)\| \leq 1$ a.e. and $\hat{f}(-m) = \lambda_m$, $\forall m \geq 0$. This kind of problem was studied [1], [2] by means of "generalized Toeplitz kernels," which can be seen as a particular case of T-K-C forms. In fact, classical results on the Nehari problem can be obtained by considering the form $\{B; \mathbb{Z}, \Omega; E\}$ given by $\eta = 2$, $\omega_1 = \{j \in \mathbb{Z}: j \geq 0\}$, $\omega_2 = \{j \in \mathbb{Z}: 0 \geq j\}$ and

$$\begin{aligned} B(h,h') \equiv & \sum\{\langle h_1(u), h_1'(u)\rangle_{E_1}: u \in \omega_1\} \\ & + \sum\{\langle \lambda_{u-v} h_1(u), h_2'(v)\rangle_{E_2} + \langle h_2(v), \lambda_{u-v} h_1'(u)\rangle_{E_2}: (u,v) \in \omega_1 \times \omega_2\} \\ & + \sum\{\langle h_2(v), h_2'(v)\rangle_{E_2}: v \in \omega_2\}. \end{aligned}$$

In this case, L may be identified with $E = E_1 \oplus E_2$ and the moments $g_n = P_L U^n_{|L}$ with the matrices

$$\begin{bmatrix} 0 & \lambda_n^* \\ \tilde{f}(n) & 0 \end{bmatrix}, \qquad \forall n > 0,$$

for each solution of the problem. Thus, the method we have studied in this paper can be applied in order to obtain some results due to Dym and Gohberg [12]. In particular, from the remark following Proposition (VI.4) it can be seen that the maximum entropy solution is given by the most innovative element of **U**.

REFERENCES

1. Arocena, R., and Cotlar, M. Generalized Toeplitz kernels and Adamjan–Arov–Krein moment problems, *Op. Theory: Adv. Appl. 4* (1982), 37–55.
2. Arocena, R., and Cotlar, M. Dilations of generalized Toeplitz kernels and some vectorial moment and weighted problems, in *Harmonic Analysis*, Springer-Verlag Lect. Notes in Math., Volume 908 (1982), 169–188.
3. Arocena, R. On the extension problem for a class of translation invariant forms, to appear in *J. Operator Theory*.
4. Arocena, R. Unitary extensions of isometries and contractive intertwining dilations, *Op. Theory: Adv. Appl. 41* (1989), 13–23.
5. Arocena, R. On a geometric interpretations of Schur parameters, preprint (1988).
6. Arov, D. Z., and Grossman, L. Z. Scattering matrices in the theory of dilation of isometric operators, *Sov. Math. Doklady 27* (1983), 518–522.
7. Ceausescu, Z., and Foias, C. On intertwining dilations. V, *Acta Sci. Math. (Szeged) 40* (1978), 9–32.
8. Chumakin, M. E. Generalized resolvent of isometric operators, translated from *Sibirskii Mat. Zhurnal 4* (1967), 876–892.
9. Constantinescu, T. A maximum entropy principle for contractive intertwining dilations, *Op. Theory: Adv. Appl. 24* (1987), 69–85.
10. Cotlar, M., and Sadosky, C. On the Helson–Szegö theorem and a related class of modified Toeplitz kernels, *Proc. Symp. Pure Math., Amer. Math. Soc. 25*(1) (1979), 383–407.
11. Cotlar, M., and Sadosky, C. The generalized Bochner theorem in algebraic scattering systems, *Math. Sci. Res. Inst.*, Berkeley, California, MSRI 00621-88 (1988), to appear in *J. Math. Anal. & Appl.*
12. Dym, H., and Gohberg, I. A maximum entropy principle for contractive interpolants, *J. Funct. Anal 65* (1986), 83–125.
13. Landau, H. J. Maximum entropy and the moment problem, *Bull. Amer. Math. Soc. 16*, No. 1 (1987), 47–77.
14. Nagy, B. Sz., and Foias, C. *Harmonic Analysis of Operators on Hilbert Spaces*, Amsterdam-Budapest, 1970.
15. Schur, I. On power series which are bounded in the interior of the unit circle. I, II, first published in German in 1918–1919, English translation in *I. Schur: Methods in Operator Theory and Signal Processing* (I. Gohberg, ed.), *Op. Theory: Adv. Appl. 18* (1986), 31–88.

Mailing Address:
R. Arocena, J. M. Montero 3006, ap. 503
Montevideo, Uruguay

21

Two-Sided Nudelman Interpolation Problems for Rational Matrix Functions

Joseph A. Ball, Israel Gohberg, and Leiba Rodman

In this paper we solve explicitly a very general bitangential Nevanlinna-Pick interpolation problem for rational matrix functions by using exclusively finite dimensional linear algebra techniques. The main tool for converting problems concerning rational matrix functions to linear algebra problems is the idea from systems theory of expressing a given rational matrix function in terms of four matrices A, B, C, D as the transfer function of a linear system. The interpolation conditions themselves can be expressed in terms of these data. In this format interpolation problems involving high order derivatives are handled with the same ease as the simple multiplicity case.

1. Introduction.

Let us first state the interpolation problem we shall call the Nudelman interpolation problem. Let there be given a collection of matrices $(C_-, C_+, A_\pi; A_\zeta, B_+, B_-; \Gamma)$ of appropriate sizes and a subset σ of $\mathbb{C}$. The problem then is to find a rational $M \times N$ matrix function $F(z)$ with no poles in σ and with $\|F(z)\| < 1$ for z in σ which satisfies the interpolation conditions

$$\sum_{z_0 \in \sigma} Res_{z=z_0} F(z) C_-(zI - A_\pi)^{-1} = C_+ \tag{1.1}$$

$$\sum_{z_0 \in \sigma} Res_{z=z_0} (zI - A_\zeta)^{-1} B_+ F(z) = -B_- \tag{1.2}$$

$$\sum_{z_0 \in \sigma} Res_{z=z_0} (zI - A_\zeta)^{-1} B_+ F(z) C_-(zI - A_\pi)^{-1} = \Gamma. \tag{1.3}$$

(Here $Res_{z=z_0} W(z)$ denote the residue of the rational matrix function $W(z)$ at z_0.) This problem we call the Nudelman interpolation problem for the region σ. The one-sided version of the problem (where one considers only a single interpolation condition either of the form (1.1) or (1.2)) was considered by A. A. Nudelman (see [N1, N2]).

In this paper we describe the conditions under which the problem is solvable and give a formula for the set of all solutions for the two cases σ = the right half plane Π^+ and σ = the unit disk

Research of the authors was partially supported by the National Science Foundation.

$\mathcal{D}$. This problem is a natural generalization of bitangential Nevanlinna-Pick interpolation problem and other similar problems.

The method which we use here is based on the approach developed in the paper [BGR1] where a similar problem was solved but without metric constraints. The formulation of the problem leads directly to state space methods for its solution, and encodes in a compact form very general interpolation conditions. This paper may be viewed as an extension to the general case of the results and methods presented in detail in [BGR2] for the simplest one-sided case of the problem. This type of interpolation problem together with solutions has appeared earlier in [F], [H], [RR] for the one-sided case and in [BH1], [BH2], [LA], [K], [D] for various versions of the two-sided problem.

We now state the main results of the paper. Specifically, we state the solution criterion for the case where all three conditions (1.1)-(1.3) are imposed and for the case where one imposes only (1.1) and (1.2), and where σ = the right half plane Π^+. For a general σ, normalization conditions imposed on the data of the problem $\omega = (C_-, C_+, A_\pi; A_\zeta, B_+, B_-; \Gamma)$ are

i) (C_-, A_π) is a *null kernel* pair, i.e. $\bigcap_{j\geq 0} \ker(C_- A_\pi^j) = 0$, *and* $\sigma(A_\pi) \subset \sigma$

ii) (A_ζ, B_+) is a full range pair, i.e. span $\{Im A_\zeta^j B_+ : j \geq 0\} = \mathbb{C}^{\eta_\zeta}$ where $\eta_\zeta \times \eta_\zeta$ is the size of A_ζ, and $\sigma(A_\zeta) \subset \sigma$

iii) Γ is a solution of the Sylvester equation

$$\Gamma A_\pi - A_\zeta \Gamma = B_+ C_+ + B_- C_-. \tag{1.4}$$

It was shown in [BGR1] that rational matrix functions F analytic on σ satisfying the interpolation conditions (1.1)-(1.3) (but possibly not satisfying any metric constraint) always exist if the data set ω satisfies the normalization conditions (i)-(iii). The following result gives the additional requirement needed for existence of a solution with $\|F\|_\infty < 1$ with $\sigma = \Pi^+$.

THEOREM 1.1. *Suppose the data set* $\omega = (C_-, C_+, A_\pi; A_\zeta, B_+ B_-; \Gamma)$ *satisfies the normalization conditions (i) and (ii) and (iii) and* σ *is the right half plane* Π^+. *Let* S_1 *and* S_2 *be the unique solutions of the Lyapunov equations*

$$S_1 A_\pi + A_\pi^* S_1 = C_-^* C_- - C_+^* C_+ \tag{1.5}$$

$$S_2 A_\zeta^* + A_\zeta S_1 = B_+ B_+^* - B_- B_-^* . \tag{1.6}$$

Let $\Lambda = \Lambda(\omega)$ *be the Hermitian matrix*

$$\Lambda = \begin{bmatrix} S_1 & \Gamma^* \\ \Gamma & S_2 \end{bmatrix} . \tag{1.7}$$

Then there exists a rational $M \times N$ *matrix function* F *satisfying the interpolation conditions (1.1)-(1.3) (with* $\sigma = \Pi^+$*) together with the metric constraint* $\sup_{z \in \Pi^+} \|F(z)\| < 1$ *if and only if* Λ *is positive definite. In this case, the set of all solutions* F *is given by*

$$F = (\Theta_{11} G + \Theta_{12})(\Theta_{21} G + \Theta_{22})^{-1} \tag{1.8}$$

where G *is an arbitrary* $M \times N$ *matrix function analytic on* Π^+ *with* $\|G\|_\infty < 1$, *and*

$$\Theta(z) = \begin{bmatrix} \Theta_{11}(z) & \Theta_{12}(z) \\ \Theta_{21}(z) & \Theta_{22}(z) \end{bmatrix}$$

is the matrix function determined from the data set ω *by*

$$\Theta(z) = \begin{bmatrix} I & 0 \\ 0 & I \end{bmatrix} + \begin{bmatrix} C_+ & -B_+^* \\ C_- & B_-^* \end{bmatrix} \begin{bmatrix} (zI - A_\pi)^{-1} & 0 \\ 0 & (zI + A_\zeta^*)^{-1} \end{bmatrix} \Lambda^{-1} \begin{bmatrix} -C_+^* & C_-^* \\ B_+ & B_- \end{bmatrix} . \tag{1.9}$$

THEOREM 1.2. *Suppose the data set* $\omega = (C_-, C_+, A_\pi; A_\zeta, B_+, B_-)$ *satisfies the normalization conditions (i) and (ii) with* $\sigma = \Pi^+$. *Let* S_1 *and* S_2 *be the unique solutions of the Lyapunov equations (1.5) and (1.6). Then there exists a rational* $M \times N$ *matrix function* F *satisfying the interpolation conditions (1.1) and (1.2) (with* $\sigma = \Pi^+$*) and with* $\sup_{z \in \pi^+} \|F(z)\| < 1$ *if and only if there exists a solution* Γ *of the Sylvester equation (1.4) for which the associated matrix* Λ *(1.7) is positive definite. In this case the linear fractional formula (1.8) with* Θ *given by (1.9) describes all interpolants* F *satisfying (1.1) and (1.2) with* $\|F\|_\infty < 1$ *which satisfy the additional interpolation condition (1.3) for a particular choice of* Γ. *One gets all solutions* F *of (1.1) and (1.2) with* $\|F\|_\infty < 1$ *by sweeping through all solutions* Γ *of (1.4) for which the associated matrix* Λ *(1.7) is positive-definite.*

Let us remark that one can consider the Lagrange-Sylvester problem (i.e. interpolation without metric constraints) also with the third interpolation condition (1.3) dropped. In this case, if one assumes that the data set $\omega = (C_-, C_+, A_\pi; A_\zeta, B_+, B_-)$ satisfies the normalization conditions (i) and (ii) then it is shown in [BGR1] that interpolants exist (even for the case of a general set σ) if and only if there exists some solution Γ of the Sylvester equation (1.4). Let us add that when $\sigma(A_\pi) \cap \sigma(A_\zeta)$ is empty, then there exists a unique solution Γ of (1.4) and a matrix function analytic on σ and satisfying (1.1) and (1.2) automatically also satisfies (1.3) with this Γ. In applications the two-sided problem with all three interpolation conditions (if $\sigma(A_\zeta)$ and $\sigma(A_\pi)$ intersect nontrivially) is what usually comes up, for example, for the problem of sensitivity minimization in H^∞-control theory (see [BGR3]).

The paper consists of five sections, the first section being this introduction. In the second section we review what we need from [BGR2] concerning realization and null-pole structure for a completely general regular rational matrix function. Section 3 is a summary of the main results from [BGR1] on bitangential Lagrange-Sylvester interpolation without metric constraints; these results will be used in Section 5 where we solve the problem with metric constraints (the two-sided Nudelman problem). Section 4 develops facts concerning rational matrix functions $\Theta(z)$ with J-unitary values on the imaginary line (resp. unit circle) and J-contractive values in the right half plane Π^+ (resp. unit disk $\mathcal{D}$) where $J = J^* = J^{-1}$ is a symmetry matrix; the special case where $\Theta(z)$ is analytic in Π^+ with only simple zeros in Π^+ was done in [BGR2].

2. Realization and null-pole structure.

Our main tool for the study of interpolation problems for rational matrix functions is the notion of realization from systems theory (see [BGK], [K], [BGR2]). By this we mean the fact that any $M \times N$ rational matrix function $F(z)$ which is analytic at infinity can be written in the form

$$R(z) = D + C(zI - A)^{-1}B \tag{2.1}$$

where A, B, C, D are matrices of sizes $n \times n, n \times N, M \times n$ and $M \times N$ respectively. The form (2.1) for $R(z)$ implies that $R(z)$ is the *transfer function* for the linear system

$$\dot{x}(t) = Ax(t) + Bu(t), x(0) = 0$$

$$y(t) = Cx(t) + Du(t)$$

that is, upon taking Laplace transforms and eliminating the *state* variable x, (2.2) is equivalent to the relation

$$\hat{y}(z) = R(z)\hat{u}(z) \tag{2.3}$$

between the *input* function $u(t)$ and the *output* function $y(t)$; here in general $\hat{\omega}(z)$ denotes the Laplace transform of $\omega(t)$. The realization (2.1) is said to be *minimal* if the size n of the state space is as small as possible among all realization for $R(z)$. As is well known, the realization (2.1) is minimal if and only if (C, A) is a *null-kernel* (also known as *observable*) pair, i.e.

$$\cap_{j\geq 0} Ker CA^j = (0)$$

and (A, B) is a *full-range* (also known as *controllable*) pair, i.e.

$$\sum_{j\geq 0} Im A^j B = \mathbb{C}^n$$

By a standard algorithm one can show that any realization for $F(z)$ can be cut down to a minimal realization for the same $F(z)$.

If two realizations $(A,\ B,\ C,\ D)$ and $(\tilde{A},\ \tilde{B},\ \tilde{C},\ \tilde{D})$ are related according to the formulas

$$\tilde{A} = S^{-1}AS,\ \tilde{B} = S^{-1}B,\ \tilde{C} = CS,\ \tilde{D} = D,$$

where S is any invertible matrix, then, as is easily checked, the associated transfer functions are the same

$$R(z) = D + C(zI - A)^{-1}B = \tilde{D} + \tilde{C}(zI - \tilde{A})^{-1}\tilde{B}.$$

In this case we say that the two realizations $(A,\ B,\ C,\ D)$ and $(\tilde{A},\ \tilde{B},\ \tilde{C},\ \tilde{D})$ are *similar*. According to the *state space isomorphism theorem*, any two *minimal* realizations for the same transfer function $F(z)$ are similar. When $M = N$ and D is invertible, one can interchange the roles of the input $u(t)$ and the output $y(t)$ in (2.2) to get the inverse system

$$\dot{x}(t) = (A - BD^{-1}C)x(t) + BD^{-1}y(t)$$
$$u(t) = -D^{-1}Cx(t) + D^{-1}y(t)$$

The transfer function for this system by (2.3) must be the inverse $R(z)^{-1}$ of the transfer function $R(z)$ for the original system. In this way we arrive at the realization

$$R(z)^{-1} = D^{-1} - D^{-1}C(zI - [A - BD^{-1}C])^{-1}BD^{-1} \tag{2.4}$$

for R^{-1}. Moreover it turns out that (2.4) is a minimal realization for R^{-1} if and only if (2.1) is a minimal realization for $R(z)$.

We say that the rational square $N \times N$ matrix function $R(z)$ is *regular* if det $R(z)$ is not identically zero. We shall need notions of null and pole structure for a regular rational matrix function from [BGR1]. For the case where $R(z)$ is analytic and invertible at infinity, we proceed as follows. Let (2.1) be a minimal realization for $R(z)$. Since we are assuming $D = R(\infty)$ is invertible then (2.4) is a minimal realization for $R(z)^{-1}$. For σ a subset of the complex plane, let $\mathcal{M}$ (respectively $\mathcal{M}^{\times}$) be the spectral invariant subspace for A (resp. for $A^{\times} = A - BD^{-1}C$) corresponding to eigenvalues for A (resp. $A^{\times}$) in σ, and let Q (resp. $Q^{\times}$) be the Riesz projection for A on $\mathcal{M}$ (resp. for $A^{\times}$ on $\mathcal{M}^{\times}$). The collection of linear transformations

$$(C|\mathcal{M}, A|\mathcal{M};\ A^{\times}|\mathcal{M}^{\times},\ Q^{\times}BD^{-1};\ Q^{\times}|\mathcal{M}) \tag{2.5}$$

is called a *null-pole triple* for $R(z)$ with respect to σ. The pair of matrices $(C|\mathcal{M},\ A|\mathcal{M})$ is called a *right pole pair* for R (with respect to σ), the pair $(A^{\times}|\mathcal{M}^{\times},\ Q^{\times}BD^{-1})$ is called a *left null pair* for R (with respect to σ) and $Q^{\times}|\mathcal{M}$ is called the coupling matrix. These notions have been developed in [BGK, BGR2, BR1, BR2, GKLR, GK]. Intuitively the pair $(C|\mathcal{M},\ A|\mathcal{M})$ describes left pole structure and $(A^{\times}|\mathcal{M}^{\times},\ Q^{\times}|BD^{-1})$ describes left null structure of $R(z)$ in σ; $(C|\mathcal{M},\ A|\mathcal{M})$ is called a *right* pole pair since it is a right null pair for $R(z)^{-1}$. These intuitive notions can be made precise in terms of null and pole functions, as is done in [BGR2]; a complete systematic exposition will appear in [BGR4]. Here we will be content with exposition of the basic features of null-pole triples which we shall need in the sequel.

It is convenient to consider, together with (2.5), all triples *similar* to it. Here the similarity of two triples $(C',\ A'_{\pi};\ A'_{\zeta},\ B';\ \Gamma')$ and $(C'',\ A''_{\pi};\ A''_{\zeta};\ B'';\ \Gamma'')$ means that for some invertible linear transformations E and F the equalities

$$C'' = C'E;\ A''_\pi = E^{-1}A'_\pi E;\ A''_\zeta = F^{-1}A'_\zeta F;\ B'' = F^{-1}B_\zeta; \Gamma'' = F^{-1}\Gamma' E$$

hold. Any triple similar to (2.5) will also be called a *null-pole triple* of $R(z)$ with respect to σ, or more simply, *σ-null-pole triple*. Since $A^\times = A - BD^{-1}C$ it is easily verified that

$$Q^\times(A|\mathcal{M}) - A^\times(Q^\times|\mathcal{M}) = (Q^\times BD^{-1})(C|\mathcal{M})$$

where $A,\ B,\ C,\ D,\ Q^\times$ are as in (2.5); thus if $(C_\pi,\ A_\pi;\ A_\zeta,\ B_\zeta;\ \Gamma)$ is any σ-null-pole triple for $R(z)$, then Γ satisfies the Sylvester equation

$$\Gamma A_\pi - A_\zeta \Gamma = B_\zeta C_\pi. \tag{2.6}$$

In this case we refer to Γ as the *null-pole coupling matrix* (or transformation) associated with the right pole pair $(C_\pi,\ A_\pi)$ and left null pair $(A_\zeta,\ B_\zeta)$ for $R(z)$.

We illustrate with a simple example.

Example 2.1 Suppose $R(z)$ is a rational matrix function analytic at infinity with $R(\infty) = I$ and that all zeros and poles of $R(z)$ in the subset $\sigma \subset \mathbb{C}$ are simple. By this latter statement we mean that for each point $z_0 \in \sigma$ the local Smith-McMillan form for $R(z)$ at z_0 has at most one positive partial multiplicity and at most one negative partial multiplicity and any non-zero partial multiplicity is either 1 or -1. Let $z_1, \dots, z_p$ be the zeros and $w_1, \dots, w_q$ be the poles of $R(z)$ in σ. Then (see [BGK]) there exist analytic row vector functions $x^{(i)}(z) = \sum_{j=0}^{\infty} x_j^{(i)}(z - z_i)^j$ such that

i) $x^{(i)}(z_i) = x_0^{(i)} \neq 0$

ii) If $z_i = w_j$ (for some j) is also a pole of $R(z)$, then the product $x(z)R(z)$ has analytic continuation to the point z_i

iii) $x^{(i)}(z)R(z)|_{z=z_i} = 0$.

We refer to the vector $x_0^{(i)}$ as a *left null vector* for $R(z)$ at z_i, and to the pair $(x_0^{(i)}, x_1^{(i)})$ as an *augmented left null chain* for $R(z)$ at z_i. Similarly there exist column vectors $u_1, \dots, u_q$ so that

$$R(z)\varphi(z) = (z - w_j)^{-1}u_j + [\textit{analytic at } w_j]$$

for some column vector function φ_j which is analytic at w_j; we refer to the vector u_j as a *right pole vector* for $R(z)$ at w_j. Define matrices C_π, A_π, A_ζ, B_ζ by

$$C_\pi = [u_1, \ldots, u_q], \; A_\pi = \begin{bmatrix} w_1 & & & 0 \\ & w_2 & & \\ & & \ddots & \\ 0 & & & w_q \end{bmatrix}$$

$$A_\zeta = \begin{bmatrix} z_1 & & & 0 \\ & z_2 & & \\ & & \ddots & \\ 0 & & & z_p \end{bmatrix}, \; B_\zeta = \begin{bmatrix} x_0^{(1)} \\ x_0^{(2)} \\ \vdots \\ x_0^{(p)} \end{bmatrix}.$$

Then from [BGK] it can be shown that (C_π, A_π) is a right σ-pole pair and (A_ζ, B_ζ) is a left σ-null pair for $R(z)$, i.e. there exist matrices $C_{\pi e}$, $A_{\pi e}$, B_π, $B_{\pi e}$, C_ζ, $C_{\zeta e}$, $A_{\zeta e}$, $B_{\zeta e}$ where $A_{\pi e}$ and $A_{\zeta e}$ are square with no spectrum in σ such that $R(z)$ has a minimal realization of the form

$$R(z) = I + [C_\pi, C_{\pi e}] \begin{bmatrix} (zI - A_\pi)^{-1} & 0 \\ 0 & (zI - A_{\pi e})^{-1} \end{bmatrix} \begin{bmatrix} B_\pi \\ B_{\pi e} \end{bmatrix} \tag{2.7}$$

and $R(z)^{-1}$ has a minimal realization of the form

$$R(z)^{-1} = I - [C_\zeta, C_{\zeta e}] \begin{bmatrix} (zI - A_\zeta)^{-1} & 0 \\ 0 & (zI - A_{\zeta e})^{-1} \end{bmatrix} \begin{bmatrix} B_\zeta \\ B_{\zeta e} \end{bmatrix}. \tag{2.8}$$

From (2.8) we deduce another minimal realization for $R(z)$, namely

$$R(z) = I + [C_\zeta, C_{\zeta e}](zI - A^\times)^{-1} \begin{bmatrix} B_\zeta \\ B_{\zeta e} \end{bmatrix} \tag{2.9}$$

where

$$A^\times = \begin{bmatrix} A_\zeta & 0 \\ 0 & A_{\zeta e} \end{bmatrix} - \begin{bmatrix} B_\zeta \\ B_{\zeta e} \end{bmatrix} [C_\zeta \; C_{\zeta e}].$$

As (2.7) and (2.9) are minimal realizations for the same transfer function $R(z)$, these realizations must be similar, i.e., there is an invertible matrix

$$S = \begin{bmatrix} S_{11} & S_{12} \\ S_{21} & S_{22} \end{bmatrix}$$

for which

$$[C_\pi, \; C_{\pi e}] = [C_\zeta, \; C_{\zeta e}]S$$

$$\begin{bmatrix} A_\pi & 0 \\ 0 & A_{\pi e} \end{bmatrix} = S^{-1} A^\times S$$

$$\begin{bmatrix} B_\pi \\ B_{\pi e} \end{bmatrix} = S^{-1} \begin{bmatrix} B_\zeta \\ B_{\zeta e} \end{bmatrix}.$$

Then one can see from the definitions that a σ-null-pole triple for $R(z)$ is given by

$$\tau = (C_\pi,\ A_\pi;\ A_\zeta,\ B_\zeta;\ S_{11}),$$

i.e. S_{11} is the null-pole coupling matrix associated with the right pole pair (C_π, A_π) and left null pair (A_ζ, B_ζ). More explicitly, (see [BR1, BGR1, BGR4]), one can express the coupling matrix $\Gamma = S_{11}$ in terms of augmented null chains and pole vectors introduced above as

$$\Gamma = [\Gamma_{ij}]_{1 \le i \le 1}$$

where

$$\Gamma_{ij} = \begin{cases} (w_j - z_i)^{-1} x_0^{(i)} u_j \ if\ z_i \neq w_j \\ x_1^{(i)} u_j \qquad\qquad if\ z_i = w_j \end{cases}.$$

Note that the entry Γ_{ij} for the case $z_i \neq w_j$ is forced by the condition that Γ satisfy the Sylvester equation (2.6) ∎

We now return to the general situation. Some basic properties of null-pole triples are collected in the following proposition.

PROPOSITION 2.1. *Let $(C,\ A_\pi;\ A_\zeta,\ B;\ \Gamma)$ be a null-pole triple of $R(z)$ with respect to a right half-plane. Then the following properties are valid:*

(i) all the eigenvalues of A_π and of A_ζ are in the open right half-plane;

(ii) the pair (C, A_π) is a null-kernel pair;

(iii) the pair (A_ζ, B) is a full range pair;

(iv) the Sylvester equation

$$\Gamma A_\pi - A_\zeta \Gamma = B_\zeta C_\pi$$

is valid;

(v) the column (resp. row) vector rational function

$$(zI - A_\zeta)^{-1} B R(z) u(z)\ (resp.\ v(z) R(z)^{-1} C (zI - A_\pi)^{-1})$$

is analytic in the right half-plane whenever $u(z)$ (resp. $v(z)$) is a column (resp. row) vector function analytic in the right half plane for which $R(z)u(z)$ (resp. $v(z)R(z)$) is analytic in the right half plane;

(vi) *there exist matrices $\tilde{C}$ and $\tilde{B}$ of suitable sizes such that the representations*

$$R(z)^{-1} = \tilde{C}(zI - A_\zeta)^{-1}B + U(z)$$

and

$$R(z) = C(zI - A_\pi)^{-1}\tilde{B} + V(z)$$

hold, where $U(z)$ and $V(z)$ are rational matrix functions analytic in the right half-plane.

For the proof of various parts of Proposition 3.1 we refer to [BGR2, BGR1, BR2, GK].

Next, we introduce the notion of a null-pole subspace, or singular subspace, of R. By definition, the *null-pole subspace* $\mathcal{S}_\sigma(R)$ consists of all n-dimensional vector rational functions of the form $R(z)x(z)$, where $x(z)$ is an arbitrary n-dimensional vector rational function without poles in σ. This subspace was studied in [BGR5, BR1, BR2, BGR1, BCR]. The description of the null-pole subspace in terms of the null-pole triple is given in the following theorem proved in [BGR1]). It is convenient to use notation $\mathcal{R}_n(\sigma)$ for the set of all n-dimensional rational vector functions without poles in σ. We denote by $Res_{z=z_0}f(z)$ the residue of the rational matrix or vector function f at z_0 (here $z_0 \in \sigma$).

THEOREM 2.2. *Let $(C,\ A_\pi;\ A_\zeta,\ B;\ \Gamma)$ be a null-pole triple of R with respect to σ. Then the null-pole subspace $\mathcal{S}_\sigma(R)$ is given by the formula*

$$\mathcal{S}_\sigma(R) = \{C(zI-A_\pi)^{-1}x+h(z) : x \in \mathbb{C}^{n_\pi},\ h \in \mathcal{R}_n(\sigma) \text{ such that } \sum_{z_0 \in \sigma} Res_{z=z_0}(zI-A_\zeta)^{-1}Bh(z) = \Gamma x\}. \tag{2.7}$$

(Here $n_\pi \times n_\pi$ is the size of A_π.) Conversely, if $\tau = (C,\ A_\pi;\ A_\zeta,\ B;\ \Gamma)$ is a collection of matrices satisfying conditions (i)-(iii) in Proposition 2.1 such that $\mathcal{S}_\sigma(R)$ is characterized by (2.7), then necessarily Γ satisfies the Sylvester equation (2.6) and τ is a null-pole triple for $R(z)$ with respect to σ.

Note that if (C, A_π) is vacuous ($n_\pi = 0$), i.e. $R(z)$ has no poles in σ, then the formula for $\mathcal{S}_\sigma(R)$ simplifies to

$$\mathcal{S}_\sigma(R) = \{h \in \mathcal{R}_n(\sigma) : \sum_{z_0 \in \Pi+} Res_{z=z_0}(zI - A_\zeta)^{-1}Bh(z) = 0\}.$$

On the other hand, if (A_ζ, B) is vacuous (the size n_ζ of A_ζ is equal to 0), i.e. $R(z)$ has no zeros in σ, then $\mathcal{S}_\sigma(R)$ collapses to

$$\mathcal{S}_\sigma(R) = \{C(zI - A_\pi)^{-1}x : x \in \mathbb{C}^{n_\pi}\} + \mathcal{R}_n(\sigma).$$

In the sequel we will use the existence of rational matrix functions with given null-pole triple (with respect to σ). The existence of such functions is ensured provided the conditions (i)-(iv) of Proposition 2.1 are satisfied. More precisely, given matrices C, A_π, A_ζ, B, Γ of suitable sizes which satisfy (i)-(iv) of Proposition 2.1, there exists a rational matrix function R with no poles and zeros on the imaginary axis and at infinity such that $(C,\ A_\pi;\ A_\zeta,\ B;\ \Gamma)$ is a null-pole triple of R with respect to σ. This statement follows from Theorem 2.3 in [GK2] (see also [BR2]). Note that it applies also in the situations when one of the matrices A_π or A_ζ is empty (more precisely, it represents the linear transformation 0 acting on the zero-dimensional vector space). Such situations are interpreted as R having no poles (if A_π is empty) or no zeros (if A_ζ is empty).

We illustrate the characterization of the null-pole subspace in Theorem 2.2 with a simple example.

Example 2.2 Let $R(z) = \begin{bmatrix} 1 & z^{-1} \\ 0 & 1 \end{bmatrix}$ and take $\sigma = \{0\}$. A minimal realization for $R(z)$ is given by

$$R(z) = I + C(zI - A)^{-1}B$$

with

$$C = \begin{bmatrix} 1 \\ 0 \end{bmatrix},\ A = [0],\ B = [0, 1].$$

Thus $A^\times = A - BC$ is given by

$$A^\times = [0].$$

Both A and $A^{\times}$ have spectrum inside σ so the associated Riesz projections for A and $A^{\times}$ are both equal to the identity. We conclude that

$$\tau = (\begin{bmatrix} 1 \\ 0 \end{bmatrix}, [0]; [0], [0,1]; [1])$$

is a σ-null-pole triple for $R(z)$. Thus by Theorem 2.2 the σ-null-pole subspace $\mathcal{S}_\sigma(R) = R \cdot R_2(\sigma)$ for $R(z)$ should be given by

$$\begin{aligned} \mathcal{S}_\sigma(R) = \{ & z^{-1} \begin{bmatrix} 1 \\ 0 \end{bmatrix} x + \begin{bmatrix} h_1(z) \\ h_2(z) \end{bmatrix} : x \in \mathbb{C}, \\ & h_1 \textit{ and } h_2 \textit{ in } \mathcal{R}_1(\sigma) \textit{ such that} \\ & Res_{z=0} z^{-1}[0,1] \begin{bmatrix} h_1(z) \\ h_2(z) \end{bmatrix} = x \}. \end{aligned}$$

More explicitly, the condition on x, h_1, h_2 can be written as

$$h_2(0) = x.$$

Of course for this simple example, it is easy also to check this directly, as follows. A generic element $\begin{bmatrix} f_1 \\ f_2 \end{bmatrix}$ of $\mathcal{S}_\sigma(R)$ has the form

$$\begin{aligned} \begin{bmatrix} f_1(z) \\ f_2(z) \end{bmatrix} &= \begin{bmatrix} 1 & z^{-1} \\ 0 & 1 \end{bmatrix} \begin{bmatrix} g_1(z) \\ g_2(z) \end{bmatrix} \\ &= z^{-1} \begin{bmatrix} g_2(0) \\ 0 \end{bmatrix} + \begin{bmatrix} g_1(z) + z^{-1}(g_2(z) - g_2(0)) \\ g_2(z) \end{bmatrix} \end{aligned}$$

where g_1 and g_2 are any rational functions with no pole at 0. Clearly $\begin{bmatrix} f_1 \\ f_2 \end{bmatrix}$ has the form as required in (2.10) with $x = g_2(0)$, $h_1(z) = g_1(z) + z^{-1}(g_2(z) - g(0))$ and $h_2(z) = g_2(z)$. ∎

If W is a rational matrix function which is analytic and invertible on σ, then certainly $W R_n(\sigma) = R_n(\sigma)$. Hence the σ-null-pole subspace $\mathcal{S}_\sigma(R) = R \cdot \mathcal{R}_n(\sigma)$ for the $R(z)$ and the analogous subspace $\mathcal{S}_\sigma(\tilde{R}) = RW\mathcal{R}_n(\sigma)$ for $\tilde{R} = RW$ are the same if W in analytic and invertible on σ. Therefore by Theorem 2.2, any R and $\tilde{R}$ related in this way have the same σ-null-pole triples. This observation leads to the definition of σ-null-pole triple for any *regular* rational matrix function $R(z)$ (i.e. det $R(z)$ not identically zero), even if $R(z)$ is improper. To do this, let σ_0 be the set of poles and zeros of $R(z)$ in σ. Then (for example by working with the Smith-McMillan form) one can construct a rational matrix function $W(z)$ with no poles or zeros in σ_0 such that $\tilde{R}(z) = R(z) \cdot W(z)$ is analytic with value I at infinity. We then define a σ-null-pole triple for $R(z)$ to be a σ_0-null-pole triple for $\tilde{R}(z)$. By the remarks above, we see that this definition is consistent with the old definition if $R(z)$ is regular at infinity and moreover formula (2.7) describes the σ-null-pole triple even for the case where R is not regular at infinity; for complete details, see [BGR1].

3. Bitangential Lagrange-Sylvester interpolation.

We review here the main results from [BGR1] on the bitangential Lagrange-Sylvester interpolation problem for rational matrix functions. We are given a subset σ of $\mathbb{C}$ and a collection of matrices

$$w = (C_-, C_+, A_\pi; A_\zeta, B_+, B_-; \Gamma)$$

of sizes $N \times n_\pi$, $M \times n_\pi$, $n_\pi \times n_\pi$, $n_\zeta \times n_\zeta$, $n_\zeta \times M$, $n_\zeta \times N$ and $n_\zeta \times n_\pi$ respectively. The collection ω is said to be a *σ-admissible interpolation data set* if ω satisfies the normalization conditions (i)-(iii) given in the Introduction; namely,

i) $(C_-,\ A_\pi)$ is a null-kernel pair (i.e. $\bigcap_{j \geq 0} Ker C_- A_\pi^j = (0)$) and $\sigma(A_\pi) \subset \sigma$

ii) $(B_+,\ A_\zeta)$ is a full-range pair (i.e. $\sum_{j \geq 0} Im A_\zeta^j B_+ = \mathbb{C}^{n_\zeta}$) and $\sigma(A_\zeta) \subset \sigma$

and

iii) Γ satisfies the Sylvester equation

$$\Gamma A_\pi - A_\zeta \Gamma = B_+ C_+ + B_- C_-.$$

The associated interpolation problem (with no associated metric constraints) then is to describe the set of all $M \times N$ rational matrix functions $F(z)$ which satisfy the interpolation constraints (1.1)-(1.3). This formulation enables one to express high multiplicity tangential and bitangential interpolation conditions in a streamlined compact form. We mention here only a few examples to orient the reader; these examples will also be used in later sections to illustrate the theory where metric constraints are imposed.

Example 3.1 Let $w_1, \ldots, w_p$ be p distinct points in σ, let $u_1, \ldots, u_p$ be p nonzero $N \times 1$ column vectors and $v_1, \ldots, v_p$ be p $M \times 1$ column vectors. In addition, let $z_1, \ldots, z_q$ be q distinct points in σ (not necessarily disjoint from $w_1, \ldots, w_p$), let $x_1, \ldots, x_q$ be q nonzero $1 \times M$ row vectors and let $y_1, \ldots, y_q$ be q $1 \times N$ row vectors. If for some pair of indices i and j we have $z_i = w_j$, we set $\xi_{ij} = z_i = w_j$ and also suppose that we are given a complex number ρ_{ij}. Then define a collection of matrices $\omega = (C_-, C_+, A_\pi; A_\zeta, B_+, B_-; \Gamma$ by

$$C_- = [u_1, \ldots, u_p],\ C_+ = [v_1, \ldots, v_p],$$

$$A_\pi = diag.(w_1, \ldots, w_p)$$

$$A_\zeta = diag.(z_1, \ldots, z_q)$$

$$B_+ = col.(x_1, \ldots, x_q)$$

$$B_- = col.(-y_1, \ldots, -y_q)$$

and

$$\Gamma = [\Gamma_{ij}]_{1 \le i \le q; \le j \le \rho}$$

where

$$\Gamma_{ij} = \begin{cases} (w_j - z_i)^{-1}(x_i v_j - y_i u_j) & if\ z_i \neq w_j \\ \rho_{ij} & if\ z_i = w_j. \end{cases}$$

Then one can easily check that ω is σ-admissible and that the interpolation condition (1.1) specializes to

$$F(w_j)u_j = v_j\ for\ 1 \le j \le p \tag{3.1}$$

while (1.2) becomes

$$x_i F(z_i) = y_i\ for\ 1 \le i \le q \tag{3.2}$$

and (1.3) becomes

$$x_i F'(\xi_{ij})u_j = \rho_{ij}\ if\ z_i = w_j. \tag{3.3}$$

This is the simplest *multiplicity-one bitangential interpolation* problem. As particular cases, when one takes $q = 0$ conditions (3.2) and (3.3) are vacuous and one is left with (3.1) alone; this is *multiplicity-one right tangential interpolation.* Similarly, if one takes $p = 0$, conditions (3.1) and

(3.3) are vacuous and one is left with (3.2), i.e. *multiplicity-one left tangential interpolation.* Of course, when the points $w_1, \ldots, w_p, z_1, \ldots, z_q$ in σ are all distinct from one another, the third condition (3.3) is superfluous. ∎

Example 3.2 Let w_o and z_0 be two (not necessarily distinct) points in σ, let $u_1, \ldots, u_p$ be p $N \times 1$ column vectors with $u_1 \neq 0$, let $v_1, \ldots, v_p$ be p $M \times 1$ column vectors. In addition let $x_1, \ldots, x_q$ be q $1 \times M$ row vectors with $x_1 \neq 0$ and let $y_1, \ldots, y_q$ be q $1 \times N$ row vectors. Define matrices C_-, C_+, B_+, B_- as in Example 3.1, We define A_π to be the $p \times p$ Jordan cell with eigenvalue w_0

$$A_\pi = \begin{bmatrix} w_0 & 1 & & 0 \\ & w_0 & \ddots & \\ & & \ddots & 1 \\ 0 & & & w_0 \end{bmatrix}$$

and A_ζ to be the transpose of the $q \times q$ Jordan cell with eigenvalue z_0

$$A_\zeta = \begin{bmatrix} z_0 & & & 0 \\ 1 & z_0 & & \\ & \ddots & \ddots & \\ 0 & & 1 & z_0 \end{bmatrix}$$

In addition, suppose γ_{ij} $(1 \leq i \leq q,\ 1 \leq j \leq p)$ are numbers such that the matrix $\Gamma = [\gamma_{ij}]_{1 \leq i \leq q;\ 1 \leq j \leq p}$ satisfies the Sylvester equation. [Note by a well-known result (see [BGR2] or [LT]) there exists a unique such Γ if $z_0 \neq w_0$; when $z_0 = w_0$ existence of a solution requires some compatibility conditions on the data u_j, v_j, x_i, y_i and when it exists it is not unique.] Then it turns out that $\omega = (C_-,\ C_+,\ A_\pi;\ A_\zeta,\ B_+,\ B_-;\ \Gamma)$ is a σ admissible interpolation data set. Define column vector functions $u(z)$, $v(z)$ and row vector functions $x(z)$, $y(z)$ by

$$u(z) = \sum_{j=1}^{p} u_j (z - w_0)^{j-1}, v(z) = \sum_{j=1}^{p} v_j (z - w_0)^{j-1}$$

$$x(z) = \sum_{j=1}^{q} x_j (z - z_0)^{j-1}, y(z) = \sum_{j=1}^{q} y_j (z - z_0)^{j-1}$$

Then the interpolation condition (1.1) can be shown to specialize to

$$\frac{d^{j-1}}{dz^{j-1}} \{F(z)u(z)\}|_{z=w_0} = \frac{d^{j-1}}{dz^{j-1}} v(z)|_{z=w_0}, \qquad 1 \leq j \leq p \tag{3.4}$$

while (1.2) becomes

$$\frac{d^{i-1}}{dz^{i-1}} \{x(z)F(z)\}|_{z=z_0} = \frac{d^{i-1}}{dz^{i-1}} y(z)|_{z=z_0}, 1 \leq i \leq q \tag{3.5}$$

If $z_0 \neq w_0$, then condition (1.3) is satisfied automatically if (3.4) and (3.5) are satisfied. Otherwise, if $z_0 = w_0$, set ξ_0 equal to the common value $z_0 = w_0$ and define auxiliary row and column vector functions

$$x^{(i)}(z) = \sum_{k=1}^{i-1} x_k (z - \xi_0)^{k-1}, 1 \leq i \leq q$$

and

$$u^{(j)}(z) = \sum_{k=1}^{j-1} u_k (z - \xi_0)^{k-1}, l \leq j \leq p.$$

Then the interpolation condition (1.3) collapses to

$$\frac{1}{(i+j-1)!} \frac{d^{i+j-1}}{dz^{i+j-1}} \{x^{(i)}(z) F(z) u^{(j)}(z)\}|_{z=\xi_0} = \gamma_{ij}, \qquad 1 \leq i \leq q,\ 1 \leq j \leq p. \tag{3.6}$$

That the conditions (3.6) be compatible with (3.4) and (3.5) is equivalent to the matrix $\Gamma = [\gamma_{ij}]$ satisfying the Sylvester equation (1.4); we refer to [BGR1] for details. This example is the general *high-multiplicity bitangential interpolation* problem at the point ξ.

As in Example 3.1, we get simpler one-sided interpolation problems by taking either $p = 0$ *(high-multiplicity right tangential interpolation* at z_0) or $q = 0$ *(high-multiplicity left tangential interpolation* at w_0). Note also that Examples 3.1 and 3.2 coincide when $p = q = 1$. ∎

Example 3.3 Consider a general σ-admissible interpolation data set $w = (C_-, C_+, A_\pi; A_\zeta, B_+, B_-; \Gamma)$ where A_π is assumed to be in Jordan form and A_ζ is assumed to be in transposed Jordan form. Then the interpolation condition (1.1) becomes a collection of conditions of the form (3.4) holding at each eigenvalue w_0 of A_π, condition (1.2) becomes a collection of conditions of the form (3.5) holding at each eigenvalue z_0 of A_ζ, and condition (1.3) specializes to an assortment of conditions of the form (3.6) holding at each common eigenvalue ξ_0 of A_π and A_ζ. Moreover, it is easily seen that for any $n_\pi \times n_\pi$ invertible matrix S and $n_\zeta \times n_\zeta$ invertible matrix T, the transformed data set

$$\tilde{\omega} = (C_- S,\ C_+ S,\ S^{-1} A_\pi S;\ T^{-1} A_\zeta T,\ T^{-1} B_+,\ T^{-1} B_-;\ T^{-1} \Gamma S)$$

is a σ-admissible interpolation data set whenever $\omega = (C_-, C_+, A_\pi; A_\zeta, B_+, B_-; \Gamma)$ is, and moreover a rational matrix function $F(z)$ analytic on σ satisfies the interpolation conditions (1.1)-(1.3) associated with $\tilde{\omega}$ if and only if $F(z)$ satisfies the corresponding conditions associated with ω. In particular, given any σ-admissible interpolation data set ω, we may choose invertible matrices S and T so that $S^{-1}A_\pi S$ is in Jordan form and $T^{-1}A_\zeta T$ is in transposed Jordan form. We conclude that the interpolation problem (1.1)-(1.3) associated with a σ-admissible interpolation data set is equivalent to a collection of interpolation conditions of the form in Example 2.2 holding at various point $z_1, \ldots, z_k$, $w_1, \ldots, w_l$, $\xi_{ij} = z_i = w_j$ in σ. The conditions of the type (3.4)-(3.6) are clumsy and unwieldy to work with (see [BR]); indeed most authors consider only the multiplicity one case with the assumption that $w_1, \ldots, w_p, z_1, \ldots, z_q$ are all distinct so that the third condition can be dropped (see [LA], [K]). As we shall see, when one uses the contour integral form (1.1)-(1.3) it is possible to analyze the full problem in complete generality to a satisfactory conclusion in an elegant fashion.

The main result on Lagrange-Sylvester interpolation from [BGR1] is as follows.

THEOREM 3.1. *Let σ be a subset of $\mathbb{C}$ and let $\omega = (C_-,\ C_+\ A_\pi;\ A_\zeta,\ B_+,\ B_-;\ \Gamma)$ be a σ-admissible interpolation data set. Then there exist rational $M \times N$ matrix functions $R(z)$ analytic on σ which satisfy the associated set of Lagrange-Sylvester interpolation conditions (1.1)-(1.3).*

The set of all solutions can be described as follows. Let φ^{-1} be any rational $N \times N$ matrix function having $\tilde{\tau} = (C_-,\ A_\pi;\ 0,\ 0;\ 0)$ as a σ-null-pole triple and let $\Theta = \begin{bmatrix} \Theta_{11} & \Theta_{12} \\ \Theta_{21} & \Theta_{22} \end{bmatrix}$ be any rational $(M+N) \times (M+N)$ matrix function having $\tau = (\begin{bmatrix} C_+ \\ C_- \end{bmatrix}, A_\pi;\ A_\zeta,\ [B_+,\ B_-];\ \Gamma)$ as a σ-null-pole triple. Then the rational $M \times N$ matrix function R is analytic on σ and satisfies the interpolation conditions (1.1)-(1.3) if and only if R has the form

$$R(z) = (\Theta_{11}(z)G_1(z) + \Theta_{12}(z)G_2(z))(\Theta_{21}(z)G_1(z) + \Theta_{22}(z)G_2(z))^{-1} \tag{3.7}$$

for $M \times N$ and $N \times N$ rational matrix functions $G_1(z)$ and $G_2(z)$ analytic on σ satisfying the additional constraint

$$\varphi(\Theta_{21}G_1 + \Theta_{22}G_2) \text{ is analytic and invertible on } \sigma. \tag{3.8}$$

Remark 3.1. Special cases of Theorem 3.1 arise where A_π is vacuous ($n_\pi = 0$ where $n_\pi \times n_\pi$ is the size of A_π) or A_ζ is vacuous ($n_\zeta = 0$ where $n_\zeta \times n_\zeta$ is the size of A_ζ). When $n_\pi = 0$ one can take $\varphi =$

I_m and the interpolation conditions (1.1)-(1.3) collapse to the single condition (1.2). When $n_\zeta = 0$, (1.1)-(1.3) collapse to (1.1) alone. In general, for a given collection $\tilde{\omega} = (C_-, C_+, A_\pi; A_\zeta, B_+, B_-)$ with (C_-, A_π) a null-kernel pair, (A_ζ, B_+) a full range pair and $\sigma(A_\pi) \cup \sigma(A_\zeta) \subset \sigma$ but with the coupling matrix Γ omitted, a solution $R(z)$ of conditions (1.1) and (1.2) with no poles in σ exists if and only if there exists a solution Γ of the Sylvester equation (1.4). To describe all solutions, one again may use formula (3.7) with G any rational matrix function analytic on σ which satisfies (3.8) but where now one must choose a $\Theta(z) = \Theta_\Gamma(z)$ corresponding to each solution Γ of (1.4). If $\sigma(A_\pi)$ is disjoint from $\sigma(A_\zeta)$, then the solution Γ of (1.4) is unique and (1.3) with this choice of Γ is automatically satisfied if (1.1) and (1.2) are satisfied.

4. J-unitary matrix functions.

In this section we discuss some special connections existing between a rational matrix function Θ and its null-pole triple and null-pole subspace for the special case where Θ is J-unitary either on the imaginary axis or the unit circle. Here J is a signature matrix of the form $J = I_M \oplus -I_N$, and an $(M+N) \times (M+N)$ matrix X is said to be J-unitary if $X^*JX = J$. In the sequel Δ will denote either the open right half plane $\Pi^+ = \{z \in \mathbb{C} = \ Rez > 0\}$ or the unit disk $\mathcal{D} = \{z \in \mathbb{C} : |z| < 1\}$. Then $\partial\Delta$ denotes the boundary of Δ

$$\partial\Delta = \{ix : x \in R\} \cup \{\infty\} \ if \ \Delta = \Pi^+$$

$$\partial\Delta = \{z \in \mathbb{C} : |z| = 1\} \ if \ \Delta = \mathcal{D}.$$

We denote by $\bar{\Delta}$ the closure of Δ in the extended complex plane

$$\bar{\Delta} = \{ix : x \in R\} \cup \{\infty\} \ if \ \Delta = \Pi^+$$

$$\bar{\Delta} = \{z \in \mathbb{C} : |z| = 1\} \ if \ \Delta = \mathcal{D}.$$

In Section 2 we mentioned that any collection of matrices $\tau = (C, A_\pi; A_\zeta, B; \Gamma)$ satisfying conditions (i)-(iv) of Proposition 2.1 is the null pole triple for some rational matrix function $\Theta(z)$. Let us call a collection τ satisfying (i)-(iv) in Proposition 2.1 a *σ-admissible Sylvester data set.* In this section we give the additional condition on a σ-admissible Sylvester data set τ (for the case

$\sigma = \Delta$ is either Π^+ or $\mathcal{D}$) which guarantees that $\Theta(z)$ can be taken to be regular and J-unitary on $\partial\Delta$. This result will be needed for our solution of the two-sided Nudelman interpolation problem in the next section.

THEOREM 4.1. *Let $\tau = (C,\ A_\pi;\ A_\zeta,\ B;\ \Gamma)$ be a collection of matrices of sizes $n \times n_\pi$, $n_\pi \times n_\pi$, $n_\zeta \times n_\zeta$, $n_\zeta \times n$, $n_\zeta \times n_\pi$ respectively which is a σ-admissible Sylvester data set with $\sigma = \Delta$ equal either Π^+ or $\mathcal{D}$, and let $J = J^* = J^{-1}$ be an $n \times n$ signature matrix. Then there exists a rational $n \times n$ matrix function Θ such that*

(i) Θ is regular on $\partial\Delta$ (including infinity if $\Delta = \Pi^+$)

(ii) Θ assumes J-unitary values on $\partial\Delta$

and

(iii) $(C,\ A_\pi;\ A_\zeta,\ B;\ \Gamma)$ is a null-pole triple for Θ with respect to Δ

if and only if the Hermitian matrix

$$\Lambda = \begin{bmatrix} S_1 & \Gamma^* \\ \Gamma & S_2 \end{bmatrix} \tag{4.1}$$

is invertible. Here Λ_1 and Λ_2 are defined to be the unique solutions of the Sylvester equations.

$$S_1 A_\pi + A_\pi^* S_1 = -C^* J C \tag{4.2a}$$

$$S_2 A_\zeta^* + A_\zeta^* S_2 = B J B^* \tag{4.3a}$$

if $\Delta = \Pi^+$, and as the unique solutions of the Stein equations

$$S_1 - A_\pi^* S_1 A_\pi = -C^* J C \tag{4.2b}$$

$$S_2 - A_\zeta S_2 A_\zeta^* = B J B^* \tag{4.3b}$$

if $\Delta = \mathcal{D}$. In this case any such $\Theta(z)$ is given by

$$\Theta(z) = D + [C, -JB^*] \begin{bmatrix} (zI - A_\pi)^{-1} & 0 \\ 0 & (zI + A_\zeta^*)^{-1} \end{bmatrix} \Lambda^{-1} \cdot \begin{bmatrix} C^* J \\ B \end{bmatrix} D \tag{4.4a}$$

if $\Delta = \Pi^+$, and by

$$\Theta(z) = D + (z-1)[C, -JB^*] \cdot$$

$$\cdot \begin{bmatrix} (zI - A_\pi)^{-1} & 0 \\ 0 & (I - zA_\zeta^*)^{-1} \end{bmatrix} \Lambda^{-1} \begin{bmatrix} (I - A_\pi^*)^{-1} C^* J D \\ (A_\zeta - I)^{-1} B D \end{bmatrix} \tag{4.4b}$$

if $\Delta = \mathcal{D}$, where D is any J-unitary $N \times N$ matrix.

The proof of Theorem 4.1 will proceed in several steps. First we introduce the notion of direct sum of admissible Sylvester data sets. For $\nu = 1, 2$ let

$$\tau_\nu = (C^{(\nu)}, A_\pi^{(\nu)}; A_\zeta^{(\nu)}, B^{(\nu)}; \Gamma_{\nu\nu})$$

be a σ-admissible Sylvester data set. Assume that $\sigma_1 \cap \sigma_2$ is empty. We define the *direct sum* of τ_1 and τ_2 to be the collection of matrices

$$\tau_1 \oplus \tau_2 = \left([C^{(1)}, C^{(2)}], \begin{bmatrix} A_\pi^{(1)} & 0 \\ 0 & A_\pi^{(2)} \end{bmatrix}; \begin{bmatrix} A_\zeta^{(1)} & 0 \\ 0 & A_\zeta^{(2)} \end{bmatrix}, \begin{bmatrix} B^{(1)} \\ B^{(2)} \end{bmatrix}; \begin{bmatrix} \Gamma_{11} & \Gamma_{12} \\ \Gamma_{21} & \Gamma_{22} \end{bmatrix}\right)$$

where Γ_{12} and Γ_{21} are the unique solutions of the Sylvester equations.

$$\Gamma_{12} A_\pi^{(2)} - A_\zeta^{(1)} \Gamma_{12} = B^{(1)} C^{(2)} \tag{4.5}$$

$$\Gamma_{21} A_\pi^{(1)} - A_\zeta^{(2)} \Gamma_{21} = B^{(2)} C^{(1)}. \tag{4.6}$$

Note that the assumption that $\sigma_1 \cap \sigma_2$ is empty guarantees that $A_\pi^{(2)}$ and $A_\zeta^{(1)}$ have no common eigenvalues, and hence (4.5) is uniquely solvable for Γ_{12}. A similar remark applies to (4.6). It is not difficult to verify that $\tau_1 \oplus \tau_2$ is a $(\sigma_1 \cup \sigma_2)$-admissible Sylvester data set.

LEMMA 4.2. *For $\nu = 1, 2$ let τ_ν be a σ_ν-admissible Sylvester data set and let Θ be a proper rational matrix function with value I at infinity. Assume that $\sigma_1 \cap \sigma_2$ is empty. Then $\tau_1 \oplus \tau_2$ is a $(\sigma_1 \cup \sigma_2)$-null-pole triple for Θ if and only if for $\nu = 1, 2$ the triple τ_ν is a σ_ν-null-pole triple for Θ.*

PROOF: In what follows $\nu = 1$ or $\nu = 2$. Let $\Theta(z) = I + C(zI - A)^{-1}B$ be a minimal realization for Θ. By Q_ν (resp. $Q_\nu^\times$) we denote the Riesz projection corresponding to the eigenvalues of A (resp. $A^\times = A - BC$) in σ_ν. Put $Q = Q_1 + Q_2$ and $Q_1^\times = Q_2$ and $Q^\times = Q_1^\times + Q_2^\times$ and let $\sigma = \sigma_1 \cup \sigma_2$. Since $\sigma_1 \cap \sigma_2$ is empty, the operator Q is the Riesz projection corresponding to eigenvalues of A in σ. A similar remark holds true for $Q^\times$ with A replaced by $A^\times$.

Let τ_ν be given by (4.4); denote the coupling matrix of $\tau_1 \oplus \tau_2$ by Γ, so

$$\Gamma = \begin{bmatrix} \Gamma_{11} & \Gamma_{12} \\ \Gamma_{21} & \Gamma_{22} \end{bmatrix}$$

where Γ_{11} and Γ_{22} are the coupling matrices of τ_1 and τ_2, respectively, and Γ_{12} and Γ_{21} are defined by (4.5) and (4.6). Let $n_\pi^{(\nu)} \times n_\pi^{(\nu)}$ be the size of $A_\pi^{(\nu)}$ and $n_\zeta^{(\nu)} \times n_\zeta^{(\nu)}$ be the size of $A_\zeta^{(\nu)}$.

Assume that $\tau_1 \oplus \tau_2$ is a σ-null-pole triple for Θ. Then there exist invertible operators

$$T_\pi : \mathbb{C}^{n_\pi^{(1)}} \oplus \mathbb{C}^{n_\pi^{(2)}} \to ImQ$$
$$T_\zeta : \mathbb{C}^{n_\zeta^{(1)}} \oplus \mathbb{C}^{n_\zeta^{(2)}} \to ImQ^\times$$

such that

$$[C^{(1)}\ C^{(2)}]T_\pi^{-1} = C|ImQ \tag{4.7}$$

$$T_\pi \begin{bmatrix} A_\pi^{(1)} & 0 \\ 0 & A_\pi^{(2)} \end{bmatrix} T_\pi^{-1} = A|ImQ \tag{4.8}$$

$$T_\zeta \begin{bmatrix} B^{(1)} \\ B^{(2)} \end{bmatrix} = Q^\times B \tag{4.9}$$

$$T_\zeta \begin{bmatrix} A_\zeta^{(1)} & 0 \\ 0 & A_\zeta^{(2)} \end{bmatrix} T_\zeta^{-1} = A^\times|ImQ^\times \tag{4.10}$$

and

$$\Gamma = T_\zeta^{-1} Q^\times T_\pi \tag{4.11}$$

Since σ_1 and σ_2 are disjoint sets the similarities in (4.8) and (4.10) imply that $T_\pi(\mathbb{C}^{n_\pi^{(1)}} \oplus 0) = ImQ_1, T_\pi(0 \oplus \mathbb{C}^{n_\pi^{(2)}}) = ImQ_2, T_\zeta(\mathbb{C}^{n_\zeta^{(1)}} \oplus 0) = ImQ_1^\times$, $T_\zeta(0 \oplus \mathbb{C}^{n_\zeta^{(2)}}) = ImQ_2^\times$ and thus T_π and T_ζ have the block diagonal form

$$T_\pi = \begin{bmatrix} T_{1,\pi} & 0 \\ 0 & T_{2,\pi} \end{bmatrix} : \mathbb{C}^{n_\pi^{(1)}} \oplus \mathbb{C}^{n_\pi^{(2)}} \to ImQ_1 \oplus ImQ_2$$

and

$$T_\zeta = \begin{bmatrix} T_{1,\zeta} & 0 \\ 0 & T_{2,\zeta} \end{bmatrix} : \mathbb{C}^{n_\zeta^{(1)}} \oplus \mathbb{C}^{n_\zeta^{(2)}} \to ImQ_1^\times \oplus ImQ_2^\times.$$

Note that the operators $T_{\nu,\pi}$ and $T_{\nu,\zeta}$ are invertible. From (4.7)-(4.11) it follows that

$$(C^{(\nu)}T_{\nu,\pi}^{-1}, T_{\nu,\pi}A_\pi^{(\nu)}T_{\nu,\pi}^{-1}) = (C|ImQ_\nu, A|ImQ_\nu) \tag{4.12}$$

$$(T_{\nu,\zeta}A_{\zeta}^{(\nu)}T_{\nu,\zeta}^{-1}, T_{\nu,\zeta}B^{(\nu)}) = (A^{\times}|ImQ_{\nu}^{\times}, Q_{\nu}^{\times}B), \tag{4.13}$$

$$\Gamma_{\nu\nu} = T_{\nu,\zeta}^{-1}Q_{\nu}^{\times}T_{\nu,\pi}. \tag{4.14}$$

To prove the latter identity one uses that (4.13) may be rewritten in the following form:

$$\begin{bmatrix} T_{1,\zeta} & 0 \\ 0 & T_{2,\zeta} \end{bmatrix} \begin{bmatrix} \Gamma_{11} & \Gamma_{12} \\ \Gamma_{21} & \Gamma_{22} \end{bmatrix} \begin{bmatrix} T_{1,\pi}^{-1} & 0 \\ 0 & T_{2,\pi}^{-1} \end{bmatrix} = Q^{\times}|ImQ. \tag{4.15}$$

To prove the converse, assume that for $\nu = 1,2$ the triple τ_ν is a left σ_ν-null-pole triple for Θ. So there exist invertible operators

$$T_{\nu,\pi} : \mathbb{C}^{n_{\pi}^{(\nu)}} \to ImQ_{\nu},\ T_{\nu,\zeta} : \mathbb{C}^{n_{\zeta}^{(\nu)}} \to ImQ_{\nu}^{\times}$$

such that (4.12), (4.13) and (4.14) hold true. We shall prove that $T_{\nu,\pi}$ and $T_{\nu,\zeta}$ satisfy (4.15). In order to do this we have to show that

$$T_{1,\zeta}\Gamma_{11}T_{1,\pi}^{-1} = Q_1^{\times}|ImQ_1,\ T_{2,\zeta}\Gamma_{22}T_{2,\pi}^{-1} = Q_2^{\times}|ImQ_2 \tag{4.16}$$

$$T_{1,\zeta}\Gamma_{12}T_{2,\pi}^{-1} = Q_1^{\times}|ImQ_2,\ T_{2,\zeta}\Gamma_{21}T_{1,\pi}^{-1} = Q_2^{\times}|ImQ_1. \tag{4.17}$$

The identities in (4.16) follow directly from (4.14). From (4.5) and the identities (4.12) and (4.13) we conclude that $Z = T_{1,\zeta}S_{12}T_{2,\pi}^{-1}$ is the unique solution of the Sylvester equation

$$Z(A|ImQ_2) - (A^{\times}|ImQ_1^{\times})Z = Q_1^{\times}B(C|ImQ_2). \tag{4.18}$$

Since $A - A^{\times} = BC$, the solution Z of (4.18) is equal to $Q_1^{\times}|ImQ_2$ which proves the first identity in (4.12). The second is proved in a similar way.

Now define $T_\pi = \begin{bmatrix} T_{1,\pi} & 0 \\ 0 & T_{2,\pi} \end{bmatrix} : \mathbb{C}^{n_{\pi}^{(1)}} \oplus \mathbb{C}^{n_{\pi}^{(2)}} \to ImQ_1 \oplus ImQ_2$ and $T_\zeta = \begin{bmatrix} T_{1,\zeta} & 0 \\ 0 & T_{2\zeta} \end{bmatrix}$: $\mathbb{C}^{n_{\zeta}^{(1)}} \oplus \mathbb{C}^{n_{\zeta}^{(2)}} \to ImQ_1^{x} \oplus ImQ_2^{x}$. Then T_π and T_ζ are invertible and satisfy the identities (4.14) and (4.15) yields (4.9)-(4.12) and the formula (4.17) yields (4.13). Hence $\tau_1 \oplus \tau_2$ is a left σ-null-pole triple for Θ.■

PROOF OF THEOREM 4.1: We consider first the case $\Delta = \Pi^+$. Suppose Λ is invertible. We may then define the rational matrix function $\Theta(z)$ by (4.4a). We note that (4.2a) and (4.3a) combined with (4.1) implies that Λ satisfies the Lyapunov equation

$$\Lambda \begin{bmatrix} A_\pi & 0 \\ 0 & -A_\zeta^* \end{bmatrix} - \begin{bmatrix} -A_\pi^* & 0 \\ 0 & A_\zeta \end{bmatrix} \Lambda = \begin{bmatrix} -C^*J \\ B \end{bmatrix} [C, -JB^*].$$

By inspection we see that Λ is Hermitian. Using these properties it is not difficult to verify that $\Theta(z)$ is J-unitary on the imaginary axis. By inspection we also see that the Riesz spectral projections for $\begin{bmatrix} A_\pi & 0 \\ 0 & -A_\zeta^* \end{bmatrix}$ and $\begin{bmatrix} -A_\pi^* & 0 \\ 0 & A_\zeta \end{bmatrix}$ associated with the right half plane are $\begin{bmatrix} I & 0 \\ 0 & 0 \end{bmatrix}$ and $\begin{bmatrix} 0 & 0 \\ 0 & I \end{bmatrix}$. Therefore the restriction of the null-pole triple

$$([C, -JB^*], \begin{bmatrix} A_\pi & 0 \\ 0 & -A_\zeta^* \end{bmatrix}; \begin{bmatrix} -A_\pi^* & 0 \\ 0 & A_\zeta \end{bmatrix}, \begin{bmatrix} -C^*J \\ B \end{bmatrix}; \Lambda) \tag{4.19}$$

to the right half plane is $(C, A_\pi; A_\zeta, B; \Lambda_{21} = \Gamma)$, and thus $(C, A_\pi; A_\zeta, B; \Gamma)$ is a null-pole triple for $\Theta(z)$ over the right half plane.

Conversely, assume $\Theta(z)$ is J-unitary on the imaginary axis, without poles and zeros on the imaginary axis and at infinity, and $(C, A_\pi; A_\zeta, B; \Gamma)$ is a null-pole triple for $\Theta(z)$ over Π^+. Let $(C', A'_\pi; A'_\zeta, B'; \Gamma')$ be a global null-pole triple for $\Theta(z)$. Then $\Theta(z)$ has a minimal realization

$$\Theta(z) = D + C'(zI - A'_\pi)^{-1}\Gamma'^{-1}B'D$$

with inverse given by

$$\Theta^{-1}(z) = D^{-1} - D^{-1}C'\Gamma'^{-1}(zI - A'_\zeta)^{-1}B'.$$

We compute

$$\Theta(z) = J(\Theta(-\bar{z})^*)^{-1}J = J(D^*)^{-1}J + JB'^*(zI + A'^*_\zeta)^{-1}(\Gamma'^{-1})^*C'^*(D^{-1})^*J$$

and

$$\begin{aligned}\Theta(z)^{-1} = [J(\Theta(-\bar{z})^*)^{-1}J]^{-1} &= J\Theta(-\bar{z})^*J \\ &= JD^*J - JD^*B'^*(\Gamma'^*)^{-1}(zI + A'^*_\pi)^{-1}C'^*J.\end{aligned}$$

We thus conclude that $(-JB'^*, -A'^*_\zeta; -A'^*_\pi, -C'^*J; \Gamma'^*)$ is a global null-pole triple for $\Theta(z)$.

By the definition of a null-pole triple over the left half-plane, it follows that

$$(JB^*, -A_\zeta^*; -A_\pi^*, -C^*J; \Gamma^*)) \tag{4.20}$$

is a null-pole triple for $\Theta(z)$ over the (open) left half-plane.

We have now a null-pole triple $(C, A_\pi; A_\zeta, B; \Gamma)$ for $\Theta(z)$ over Π^+ (which was given in the beginning) and a null-pole triple (4.20) for $\Theta(z)$ over the left half-plane. By Lemma 4.2, the triple (4.19) where $\Lambda = \begin{bmatrix} S_1 & \Gamma^* \\ \Gamma & S_2 \end{bmatrix}$ and S_1 and S_2 satisfy the Lyapunov equations (4.2a) and (4.3a), is a global null-pole triple for $\Theta(z)$. In particular, Λ must be invertible, and then $\Theta(z)$ is given by the formula (4.3a).

The case $\Delta = \mathcal{D}$ is somewhat more complicated since the solution $\Theta(z)$ is not regular at infinity if either A_π or A_ζ is not invertible. Nevertheless this difficulty can be overcome by using the framework of [GK2] (see also [BCR]). Complete details will appear in [BRG4].■

It will also be important to know when the essentially unique rational matrix function $\Theta(z)$ appearing in Theorem 4.1 has J-contractive values (i.e. $J - \Theta(z)^* J \Theta(z) \geq 0$) for z in Δ. To analyze this question it will be convenient to assume that

$$J = \begin{bmatrix} I_M & 0 \\ 0 & -I_N \end{bmatrix}$$

(where $M + N = n$) and to write Θ as a block matrix

$$\Theta(z) = \begin{bmatrix} \Theta_{11}(z) & \Theta_{12}(z) \\ \Theta_{21}(z) & \Theta_{22}(z) \end{bmatrix}$$

conformally with respect to J.

In the initial discussion we concentrate on J-unitary *constant* matrices

$$\Theta = \begin{bmatrix} \Theta_{11} & \Theta_{12} \\ \Theta_{21} & \Theta_{22} \end{bmatrix}.$$

Thus suppose that $\Theta = [\Theta_{ij}]_{1 \leq i,j \leq 2}$ is a J-unitary matrix. The J-unitary property of Θ implies the relations

$$\Theta_{11}\Theta_{11}^* - \Theta_{12}\Theta_{12}^* = I_M \tag{4.21a}$$

$$\Theta_{11}\Theta_{21}^* - \Theta_{12}\Theta_{22}^* = 0 \tag{4.21b}$$

$$\Theta_{21}\Theta_{21}^* - \Theta_{22}\Theta_{22}^* = -I_N. \tag{4.21c}$$

In particular, from (4.21c) $\Theta_{22}\Theta_{22}^* = I + \Theta_{21}\Theta_{21}^*$, so Θ_{22} is invertible.

Now let x_1, x_2, y_1, y_2 be variables ($x_2, y_1 \in \mathbb{C}^M, y_2, x_1 \in \mathbb{C}^N$) which are related according to the relation

$$\begin{bmatrix} y_1 \\ x_1 \end{bmatrix} = \begin{bmatrix} \Theta_{11} & \Theta_{12} \\ \Theta_{21} & \Theta_{22} \end{bmatrix} \begin{bmatrix} x_2 \\ y_2 \end{bmatrix}. \tag{4.22}$$

Define the matrix $U = \begin{bmatrix} U_{11} & U_{12} \\ U_{21} & U_{22} \end{bmatrix}$ by

$$\begin{bmatrix} y_1 \\ y_2 \end{bmatrix} = \begin{bmatrix} U_{11} & U_{12} \\ U_{21} & U_{22} \end{bmatrix} \begin{bmatrix} x_1 \\ x_2 \end{bmatrix}. \tag{4.23}$$

Since Θ_{22} is invertible, U is well-defined and is given by

$$\begin{bmatrix} U_{11} & U_{12} \\ U_{21} & U_{22} \end{bmatrix} = \begin{bmatrix} \Theta_{12}\Theta_{22}^{-1} & \Theta_{11} - \Theta_{12}\Theta_{22}^{-1}\Theta_{21} \\ \Theta_{22}^{-1} & -\Theta_{22}^{-1}\Theta_{21} \end{bmatrix}. \tag{4.24}$$

When U and Θ are related in this way we say that U is the Redheffer transform of Θ and write $U = \mathcal{R}[\Theta]$. Since Θ is J-unitary, we deduce from (4.22) that

$$\|y_1\|^2 - \|x_1\|^2 = \|x_2\|^2 - \|y_2\|^2.$$

Rewrite this equation as

$$\|y_1\|^2 + \|y_2\|^2 = \|x_1\|^2 + \|x_2\|^2.$$

As x_1 and x_2 are arbitrary, we deduce from (4.23) that U is unitary

$$U^*U = I.$$

The converse holds by reversing the argument. We thus have established the following.

LEMMA 4.3. *If $U = \mathcal{R}[\Theta]$ is the Redheffer transform of Θ, then Θ is J-unitary if and only if U is unitary.*

We next consider the case where the matrices Θ and U are related by $U = \mathcal{R}[\Theta]$ and Θ is J-contractive

$$\Theta^* J \Theta \leq J.$$

In terms of the vectors x_2, y_2, y_1, x_1, by (4.22) this means

$$\|y_1\|^2 - \|x_1\|^2 \leq \|x_2\|^2 - \|y_2\|^2.$$

This can be rewritten as

$$\|y_1\|^2 + \|y_2\|^2 \le \|x_1\|^2 + \|x_2\|^2.$$

As x_1 and x_2 are arbitrary, then by (4.23) we conclude that U is a contraction in the usual sense

$$U^*U \le I.$$

We thus have established the following.

LEMMA 4.4. *If $U = \mathcal{R}[\Theta]$ is the Redheffer transform of Θ, then Θ is J-contractive if and only if U is contractive.*

We now return to considering a rational matrix valued function $\Theta(z)$. If

$$\Theta(z) = \begin{bmatrix} \Theta_{11}(z) & \Theta_{12}(z) \\ \Theta_{21}(z) & \Theta_{22}(z) \end{bmatrix}$$

and $\Theta(z)$ is J-unitary on $\partial\Delta$, then from (4.21c) we see that $\Theta_{22}(z)$ is invertible at each regular point of $\partial\Delta$, so in particular det $\Theta_{22}(z)$ does not vanish identically. We may then define the rational matrix function $U(z)$ by $U(z) = \mathcal{R}[\Theta(z)]$. By Lemma 4.3, $\Theta(z)$ being J-unitary on $\partial\Delta$ is equivalent to $U(z)$ being unitary on $\partial\Delta$. By Lemma 4.4 $\Theta(z)$ is J-contractive on Δ if and only if $U(z)$ is contractive on Δ. Thus we have the following result.

PROPOSITION 4.5. *Let $\Theta(z)$ and $U(z)$ be rational matrix functions such that $U(z) = \mathcal{R}[\Theta(z)]$. Then $\Theta(z)$ is J-unitary on $\partial\Delta$ if and only if $U(z)$ is unitary on $\partial\Delta$. Moreover $\Theta(z)$ is J-contractive on Δ if and only if $U(z)$ is contractive on Δ.*

If $U(z)$ is unitary on $\partial\Delta$, then the contractiveness of $U(z)$ in Δ follows just from the analyticity of $U(z)$ in Δ by the maximum modulus theorem. Putting this together with Proposition 4.4 and the formula (4.24) for $U = \mathcal{R}[\Theta]$ in terms of Θ leads to the following results.

COROLLARY 4.6. *Suppose $\Theta(z) = \begin{bmatrix} \Theta_{11}(z) & \Theta_{12}(z) \\ \Theta_{21}(z) & \Theta_{22}(z) \end{bmatrix}$ is a rational matrix function which is J-unitary on $\partial\Delta$. Then Θ is J-contractive on Δ if and only the four functions*

$$\Theta_{12}\Theta_{22}^{-1}, \Theta_{11} - \Theta_{12}\Theta_{22}^{-1}\Theta_{21}, \Theta_{22}^{-1}, \Theta_{22}^{-1}\Theta_{21}$$

have no poles in Δ.

The main result needed in the next section is the following.

THEOREM 4.7. *Let* $\Theta(z) = \begin{bmatrix} \Theta_{11}(z) & \Theta_{12}(z) \\ \Theta_{22}(z) & \Theta_{22}(z) \end{bmatrix}$ *be a rational matrix function which is regular with* J*-unitary values on* $\partial\Delta$, *and let*

$$\tau = (\begin{bmatrix} C_+ \\ C_- \end{bmatrix}, \ A_\pi; \ A_\zeta, \ [B_+, B_-]; \Gamma)$$

be a null-pole triple for $\Theta(z)$ *with respect to* Δ. *Suppose that* τ *enjoys the additional properties*

(a) (C_-, A_π) *is a null kernel pair and*

(b) (A_ζ, B_+) *is a full range pair.*

Then the following are equivalent.

i) Θ *is* J*-contractive on* Δ

ii) The single function $\Theta_{22}^{-1}\Theta_{21}$ *has no poles in* Δ

and

iii) The matrix Λ *given by (4.1)-(4.3) is positive-definite.*

PROOF: The proof of Theorem 4.7 follows in a sequence of steps by pulling together some ideas from [AG] and [BH2]. Here we develop only what is needed for our application; a more thorough self-contained development will appear in [BGR4].

We first need the notion of a positive definite kernel function. An $n \times n$ matrix function $K(z,w)$ defined for z and w in some set E and such that $K(z,w)^* = K(w,z)$ is said to be *positive definite* if for any positive integer r, points $w_1, \ldots, w_r$ in E and vectors $u_1, \ldots, u_r$ in $\mathbb{C}^n$ the $r \times r$ Hermitian matrix with ij entry

$$u_i^* K(w_i, w_j) u_j$$

is positive semidefinite. Of particular interest to us is a kernel of the form

$$K_\Theta(z,w) = \frac{J - \Theta(z) J \Theta(w)^*}{\rho_\Delta(z,w)} \tag{4.25}$$

where

$$\rho_\Delta(z,w) = \begin{cases} z + \bar{w} \ if \ \Delta = \Pi^+ \\ 1 - z\bar{w} \ if \ \Delta = \mathcal{D} \end{cases}.$$

We can now state the following lemma. Throughout the rest of this section we assume that $J = I_M \oplus -I_N$.

LEMMA 4.8. *Suppose Θ is an $(M+N)\times(M+N)$ matrix function which is meromorphic on Δ. Then $\Theta(z)$ is J-contractive on Δ if and only if the associated kernel function $K_\Theta(z,w)$ given by (4.25) and (4.26) is positive-definite.*

PROOF: If $K_\Theta(z,w)$ is a positive-definite kernel function on Δ, then one sees in particular that $K_\Theta(z,z)\geq 0$ on Δ by taking $r=1$ in the definition of positive-definite kernel. This immediately implies that $\Theta(z)$ is J-contractive for z in Δ.

Conversely we must show that each matrix of the form $[u_i^* K_\Theta(w_i,w_j)u_j]_{1\leq i,j\leq r}$ is positive semidefinite from the fact that this is true in general when $r=1$. The result in fact is well-known and a number of proofs are possible. For the case $J=I$, it is a simple consequence of the necessity of the positive-definiteness of the Pick matrix in the theory of Nevanlinna-Pick interpolation. A simple operator-theoretic proof can be found in Proposition 8.1 of [NF]; for a more elementary linear algebra proof for the rational case, combine Theorem 2.4 and 2.12 in [AG]. For the case of a general J, the necessity of the positive-definiteness of the J-Pick matrix for J-contractive meromorphic functions follows from that for the special case $J=I$ by a change of coordinates; this is explained in [BH1]. Alternatively, one can use the operator-theoretic characterization of the kernel function K_Θ in Theorem 3.3 of [BH3] to adapt the proof of Proposition 8.1 in [NF] to the general case. A third approach is to use the Redheffer transform to reduce the general case to the case $J=I$.∎

We next characterize positive-definiteness of the kernel function K_Θ in terms of the state space realization of Θ in Theorem 4.1.

LEMMA 4.9. *Suppose that the rational matrix function $\Theta(z)$ is regular and J-unitary on $\partial\Delta$, has $\tau=(C,A_\pi;A_\zeta,B;\Gamma)$ as a null-pole triple with respect to Δ, and therefore, also has a realization of the form (4.4) as in Theorem 4.1. Then the associated kernel function $K_\Theta(z,w)$ (4.25) is positive definite if and only if Λ (4.1) is positive definite.*

PROOF: For the case $\Delta=\Pi^+$, it is not difficult to identify the matrix Λ as the "associated Hermitian" matrix discussed in [AG] and the result then follows by Theorem 2.4 there. For the case $\Delta=\mathcal{D}$ and $\Theta(z)$ is regular at infinity, the result follows in an analogous fashion from Theorem 3.4

in [AG]. For the general case, one can do an analogous analysis by using the identity

$$K_\Theta(z,w) = X(z)\Lambda^{-1}X(w)^*$$

where

$$X(z) = D[C, -JB^*]\begin{bmatrix}(zI-A_\pi)^{-1} & 0 \\ 0 & (I-zA_\zeta^*)^{-1}\end{bmatrix}$$

if $\Theta(z)$ is given by (4.4b). This identity in turn follows from a tedious but nevertheless straightforward computation.

We next develop the implications of hypotheses (a) and (b) in Theorem 4.7.

LEMMA 4.10. *Let* $\Theta(z) = \begin{bmatrix}\Theta_{11}(z) & \Theta_{12}(z) \\ \Theta_{21}(z) & \Theta_{22}(z)\end{bmatrix}$ *be a rational matrix function which is regular and* J*-unitary valued on* $\partial\Delta$ *and which has*

$$\tau = (\begin{bmatrix}C_+ \\ C_-\end{bmatrix}, A_\pi; A_\zeta, [B_+, B_-]; \Gamma)$$

as a null-pole triple with respect to Δ*. Then:*

i) If (C_-, A_π) *is a null-kernel pair and* $\Theta_{22}^{-1}\Theta_{21}$ *is analytic on* Δ*, then* $\Theta_{11} - \Theta_{12}\Theta_{22}^{-1}\Theta_{21}$ *is also analytic on* Δ*.*

ii) If (A_ζ, B_+) *is a full-range pair and if* $\Theta_{22}^{-1}\Theta_{21}$ *and* $\Theta_{11} - \Theta_{12}\Theta_{22}^{-1}\Theta_{21}$ *are analytic on* Δ*, then* $\Theta_{12}\Theta_{22}^{-1}$ *and* Θ_{22}^{-1} *are also analytic on* Δ*.*

PROOF: Let $\mathcal{S}_\Delta(\Theta)$ be the null-pole subspace for Θ with respect to Δ. By definition we have

$$\mathcal{S}_\Delta(\Theta) = \begin{bmatrix}\Theta_{11} & \Theta_{12} \\ \Theta_{21} & \Theta_{22}\end{bmatrix}\mathcal{R}_{M+N}(\Delta) \tag{4.27}$$

while by Theorem 2.2 we also have

$$\begin{aligned}\mathcal{S}_\Delta(\Theta) = \{&\begin{bmatrix}C_+ \\ C_-\end{bmatrix}(zI-A_\pi)^{-1}x + \begin{bmatrix}h_1(z) \\ h_2(z)\end{bmatrix} : \\ &x \in \mathbb{C}^{n_\pi}, h_1 \in \mathcal{R}_M(\Delta), h_2 \in \mathcal{R}_N(\Delta) \\ &\textit{such that} \\ &\sum_{z_0\in\Delta} Res_{z=z_0}(zI-A_\zeta)^{-1}(B_+h_1(z) + B_-h_2(z)) = \Gamma x\}\end{aligned} \tag{4.28}$$

Now suppose that (C_-, A_π) is a null-kernel pair. Then using the expansion

$$C_-(zI - A_\pi)^{-1}x = \sum_{j=1}^{\infty}(C_-A_\pi^{j-1}x)z^{-j}$$

valid for $|z|$ large and using that (C_-, A_π) is a null-kernel pair, we see that

$$C_-(zI - A_\pi)^{-1}x = 0 \Rightarrow x = 0.$$

From (4.28) we read off that $\mathcal{S}_\Delta(\Theta)$ has the property

$$\begin{pmatrix} f \\ 0 \end{pmatrix} \in \mathcal{S}_\Delta(\Theta) \Rightarrow f \textit{ is analytic on } \Delta. \tag{4.29}$$

We next analyze what property (4.29) means in terms of the representation (4.27) for $\mathcal{S}_\Delta(\Theta)$. Indeed, (4.29) means

$$\begin{aligned} &h_1 \in \mathcal{R}_M(\Delta), h_2 \in \mathcal{R}_N(\Delta) \textit{ such that } \Theta_{21}h_1 + \Theta_{22}h_2 = 0 \\ &\Rightarrow \Theta_{11}h_1 + \Theta_{12}h_2 \textit{ is analytic on } \Delta. \end{aligned} \tag{4.30}$$

Since Θ is J-unitary on $\partial\Delta$, Θ_{22} is invertible on $\partial\Delta$ so in particular det $\Theta_{22} \not\equiv 0$. We can, therefore, solve for h_2 in terms of h_1 in (4.30) ($h_2 = -\Theta_{22}^{-1}\Theta_{21}h_1$) and reexpress (4.30) as

$$h_1 \in \mathcal{R}_M(\Delta),\ \Theta_{22}^{-1}\Theta_{21}h_1 \in \mathcal{R}_N(\Delta) \Rightarrow (\Theta_{11} - \Theta_{12}\Theta_{22}^{-1}\Theta_{21})h_1 \in \mathcal{R}_M(\Delta). \tag{4.31}$$

If we are also given that $\Theta_{22}^{-1}\Theta_{21}$ is analytic on Δ, then the second condition in (4.31) is automatic and we may take h_1 arbitrarily in $\mathcal{R}_M(\Delta)$. It then follows from (4.31) that $\Theta_{11} - \Theta_{12}\Theta_{22}^{-1}\Theta_{21}$ is analytic on Δ. This completes the proof of (i).

Next suppose that (A_ζ, B_+) is a full range pair. Again by working with the Neumann series expansion this time of $(zI - A_\zeta)^{-1}$, one can see from this assumption that any vector y in $\mathbb{C}^{n_\zeta}$ can be expressed as

$$\begin{aligned} y &= \sum_{z_0 \in \Delta} Res_{z=z_0}(zI - A_\zeta)^{-1}B_+h(z) \\ &= \frac{1}{2\pi i}\int_{|z|=R}(zI - A_\zeta)^{-1}B_+h(z)dz \end{aligned}$$

for a vector polynomial (and hence for an element of $\mathcal{R}_M(\Delta)$) $h(z)$.

From (4.28) we read off the consequence

$$\begin{gathered}\textit{For each } h_2 \in \mathcal{R}_N(\Delta) \textit{ there exists } h_1 \in \mathcal{R}_M(\Delta)\\ \textit{so that } \begin{bmatrix} h_1 \\ h_2 \end{bmatrix} \in \mathcal{S}_\Delta(\Theta) \cap \mathcal{R}_{M+N}(\Delta).\end{gathered} \tag{4.32}$$

We next analyze what (4.32) means in terms of the characterization (4.27) for $\mathcal{S}_\Delta(\Theta)$. The result is:

$$\begin{gathered}\textit{For each } h_2 \in \mathcal{R}_N(\Delta) \textit{ there exists}\\ g_1 \in R_M(\Delta) \textit{ and } g_2 \in \mathcal{R}_N(\Delta) \textit{ for which}\\ \Theta_{21}g_1 + \Theta_{22}g_2 = h_2 \textit{ and}\\ \Theta_{11}g_1 + \Theta_{12}g_2 \in \mathcal{R}_M(\Delta).\end{gathered} \tag{4.33}$$

We can solve for g_2 in the first equation of (4.33) ($g_2 = \Theta_{22}^{-1}h_2 - \Theta_{22}^{-1}\Theta_{21}g_1$). If we substitute this in the second equation, we may rewrite (4.33) as

For each $h_2 \in \mathcal{R}_N(\Delta)$ *there exists* $g_1 \in R_M(\Delta)$ *for which*

$$\Theta_{22}^{-1}h_2 - \Theta_{22}^{-1}\Theta_{21}g_1 \in \mathcal{R}_N(\Delta)$$

and

$$(\Theta_{11} - \Theta_{12}\Theta_{22}^{-1}\Theta_{21})g_1 + \Theta_{12}\Theta_{22}^{-1}h_2 \in \mathcal{R}_M(\Delta). \tag{4.34}$$

Now if we assume that $\Theta_{22}^{-1}\Theta_{21}$ and $\Theta_{11} - \Theta_{12}\Theta_{22}^{-1}\Theta_{21}$ are analytic on Δ then the choice of $g_1 \in \mathcal{R}_M(\Delta)$ is irrelevant, and (4.34) simply says that Θ_{22}^{-1} and $\Theta_{12}\Theta_{22}^{-1}$ are analytic on Δ.■

PROOF OF THEOREM 4.7: The equivalence of (i) and (iii) follows from Lemma 4.8 combined with Lemma 4.9 and is even true more generally without hypotheses (a) and (b). The implication (i)⇒(ii) is a consequence of Corollary 4.6 and is also true without hypotheses (a) and (b). It remains to show that (ii)⇒(i) in case (a) and (b) hold. But if $\Theta_{22}^{-1}\Theta_{21}$ is analytic and (a) holds, Lemma 4.10(i) gives that $\Theta_{11} - \Theta_{12}\Theta_{22}^{-1}\Theta_{21}$ is analytic on Δ. If (b) also holds, then (ii) in Lemma 4.10 gives that $\Theta_{22}^{-1}\Theta_{21}$ and $\Theta_{11} - \Theta_{12}\Theta_{22}^{-1}\Theta_{21}$ are analytic on Δ. That Θ is J-contractive on Δ now follows from Corollary 4.6.

5. The two-sided Nudelman interpolation problem.

In this section we prove the theorems stated in the introduction; namely, Theorems 1.1 and 1.2. The introduction considered only the problem for the right half plane. Here we present also the versions for the unit disk $\mathcal{D}$. We first state the disk versions of Theorems 1.1 and 1.2.

THEOREM 5.1. *Suppose that the data set*

$$\omega = (C_-, C_+, A_\pi; A_\zeta, B_+, B_-; \Gamma)$$

satisfies the normalization conditions (i), (ii) and (iii) in the Introduction with $\sigma = \mathcal{D}$. Let S_1 and S_2 be the unique solutions of the Stein equations

$$S_1 - A_\pi^* S_1 A_\pi = C_-^* C_- - C_+^* C_+ \tag{5.1}$$

and

$$S_1 - A_\zeta S_2 A_\zeta^* = B_+ B_+^* - B_- B_-^* \tag{5.2}$$

and let $\Lambda = \Lambda(\omega)$ be the Hermitian matrix

$$\Lambda = \begin{bmatrix} S_1 & \Gamma^* \\ \Gamma & S_2 \end{bmatrix}. \tag{5.3}$$

Then there exists a rational $M \times N$ matrix function F satisfying the interpolation conditions (1.1)-(1.3) (with $\sigma = \mathcal{D}$) together with the metric constraint

$$\sup_{z \in \mathcal{D}} \|F(z)\| < 1$$

if and only if Λ is positive definite.

In this case, the set of all solutions F is given by

$$F = (\Theta_{11} G + \Theta_{12})(\Theta_{21} G + \Theta_{22})^{-1} \tag{5.4}$$

where G is an arbitrary $M \times N$ matrix function analytic on $\mathcal{D}$ with $\sup_{z \in \mathcal{D}} \|G(z)\| < 1$, and $\Theta(z) = \begin{bmatrix} \Theta_{11}(z) & \Theta_{12}(z) \\ \Theta_{21}(z) & \Theta_{22}(z) \end{bmatrix}$ is the matrix function determined by the data set ω by

$$\Theta(z) = \begin{bmatrix} I & 0 \\ 0 & I \end{bmatrix} + (z-1) \begin{bmatrix} C_+ & -B_+^* \\ C_- & B_-^* \end{bmatrix}.$$

$$\begin{bmatrix} S_1 & \Gamma^* \\ \Gamma & S_2 \end{bmatrix}^{-1} \begin{bmatrix} -v_1^* & u_1^* \\ \vdots & \vdots \\ -v_p^* & u_p^* \\ x_1 & -y_1 \\ \vdots & \vdots \\ x_q & -y_q \end{bmatrix}.$$

When $\sigma = \mathcal{D}$, the solutions S_1 and S_2 of the Stein equations (5.1) and (5.2) are given by

$$S_1 = \left[\frac{u_i^* u_j - v_i^* v_j}{1 - \bar{w}_i w_j}\right]_{1 \le i,j \le p}$$

and

$$S_2 = \left[\frac{x_i x_j^* - y_i y_j^*}{1 - z_i \bar{z}_j}\right]_{1 \le i,j \le q}$$

and $\Theta(z)$ is given explicitly as

$$\Theta(z) = \begin{bmatrix} I_M & 0 \\ 0 & I_N \end{bmatrix} + (z-1)\begin{bmatrix} v_1 & \cdots & v_p & -x_1^* & \cdots & -x_q^* \\ u_1 & \cdots & u_p & -y_1^* & \cdots & -y_q^* \end{bmatrix}.$$

$$\begin{bmatrix} (z-w_1)^{-1} & & & & & 0 \\ & \ddots & & & & \\ & & (z-w_p)^{-1} & & & \\ & & & (1-z\bar{z}_1)^{-1} & & \\ & & & & \ddots & \\ 0 & & & & & (1-z\bar{z}_1)^{-1} \end{bmatrix}.$$

$$\cdot \begin{bmatrix} S_1 & \Gamma^* \\ \Gamma & S_2 \end{bmatrix}^{-1} \begin{bmatrix} (1-\bar{w}_1)^{-1} v_1^* & (1-\bar{w}_1)^{-1} u_1^* \\ \vdots & \vdots \\ (1-\bar{w}_p)^{-1} v_p^* & (1-\bar{w}_p)^{-1} u_p^* \\ (z_1 - 1)^{-1} x_1 & -(z_1-1)^{-1} y_1 \\ \vdots & \vdots \\ (z_p-1)^{-1} x_p & -(z_p-1)^{-1} y_p. \end{bmatrix}$$

PROOF OF THEOREMS 1.1 AND 5.1: We break the proof into three pieces.

Step 1 If a solution $F \in \mathcal{R}_{M\times N}(\Delta)$ satisfying (1.1)-(1.3) with $\|F\|_\infty < 1$ exists and if Λ is invertible, then Λ is in fact positive definite and F has the form $F = (\Theta_{11}G + \Theta_{12})(\Theta_{22}G + \Theta_{22})^{-1}$ with $G \in \mathcal{R}_{M\times N}(\Delta)$ with $\|G\|_\infty < 1$.

Step 2 If Λ is positive definite and if $G \in \mathcal{R}_{M\times N}(\Delta)$ has $\|G\|_\infty < 1$, then $F = (\Theta_{11}G + \Theta_{12})(\Theta_{21}G + \Theta_{22})^{-1}$ is a solution.

Step 3 If a solution F exists, then Λ is positive definite (and hence in particular invertible).

Proof of Step 1 Recall from Section 3 that a collection of matrices $\omega = (C_-, C_+, A_\pi; A_\zeta, B_+, B_-; \Gamma)$ associated wit interpolation conditions (1.1)-(1.3) which satisfies the normalization conditions (i)-(iii) in the Introduction we call a *σ-admissible interpolation data set.* Also, as was done in Section 4, we call a collection of matrices $\tau = (C, A_\pi; A_\zeta, B; \Gamma)$ which satisfy conditions (i)-(iv) in Proposition 2.1 a *σ-admissible Sylvester data set.*

Now let $\sigma = \Delta$ where Δ is either the right half plane Π^+ or the unit disk $\mathcal{D}$. Part of the assertion of Theorem 3.1, and a fact easily checked directly, is that

$$\tau = (\begin{bmatrix} C_+ \\ C_- \end{bmatrix}, A_\pi; A_\zeta, [B_+, B_-]; \Gamma) \tag{5.6}$$

is a Δ-admissible Sylvester data set whenever $\omega = (C_-, C_+, A_\pi; A_\zeta, B_+, B_-; \Gamma)$ is a Δ-admissible interpolation data set. Also note that the matrix Λ given by (1.5)-(1.7) if $\Delta = \Pi^+$ and by (5.1)-(5.3) if $\Delta = \mathcal{D}$ is the matrix associated with the Sylvester data set τ (5.6) as in Theorem 4.1 ((4.1)-(4.3)) with $J = I_M \oplus -I_N$. Thus by Theorem 4.1 there exists a rational matrix function $\Theta(z)$ regular and J-unitary valued on $\partial\Delta$ which has τ as a Δ-null-pole triple if and only if Λ is invertible. Moreover, by the formulas (4.3a) and (4.3b) specialized to τ as in (5.6), the unique such Θ with $\Theta(\infty) = I$ if $\Delta = \Pi^+$ (resp. with $\Theta(1) = I$ if $\Delta = \mathcal{D}$) is exactly the $\Theta(z)$ given by (1.9) in Theorem 1.1 (resp. by (5.5) in Theorem 5.1). Moreover by assumption ω is a Δ-admissible interpolation data set, so τ satisfies hypotheses (a) and (b) in Theorem 4.7; thus if Λ is invertible so Θ is defined, then by Theorem 4.7 Λ is positive definite if and only if $\Theta_{22}^{-1}\Theta_{21}$ is analytic on Δ.

We assume a solution $F \in \mathcal{R}_{M\times N}(\Delta)$ of (1.1)-(1.3) with $\sup_{z\in\Delta} \|F(z)\| < 1$ exists and that Λ is invertible. By the discussion above the positive definiteness of Λ follows if we show that $\Theta_{22}^{-1}\Theta_{21}$ is analytic on Δ.

By Theorem 3.1, since F is an element of $\mathcal{R}_{M\times N}(\bar{\Delta})$ satisfying (1.1)-(1.3), F can be represented as

$$F = (\Theta_{11}\tilde{G}_1 + \Theta_{12}\tilde{G}_2)(\Theta_{21}\tilde{G}_1 + \Theta_{22}\tilde{G}_2)^{-1} \tag{5.7}$$

where $\tilde{G}_1 \in \mathcal{R}_{M\times N}(\bar{\Delta})$ and $\tilde{G}_2 \in \mathcal{R}_{N\times N}(\bar{\Delta})$ are such that

$$\varphi(\Theta_{21}\tilde{G}_1 + \Theta_{22}\tilde{G}_2)$$

is analytic and invertible on Δ

where $\varphi^{-1} \in \mathcal{R}_{N\times N}$ has $(C_-, A_\pi; 0, 0; 0)$ as a $\bar{\Delta}$-null-pole triple. Rewrite (5.7) in matrix form as

$$\begin{aligned}\begin{bmatrix} F \\ I \end{bmatrix} \varphi^{-1} &= \begin{bmatrix} \Theta_{12} & \Theta_{12} \\ \Theta_{21} & \Theta_{22} \end{bmatrix} \begin{bmatrix} \tilde{G}_1 \\ \tilde{G}_2 \end{bmatrix} (\Theta_{21}\tilde{G}_1 + \Theta_{22}\tilde{G}_2)^{-1}\varphi^{-1} \\ &= \begin{bmatrix} \Theta_{11} & \Theta_{12} \\ \Theta_{21} & \Theta_{22} \end{bmatrix} \begin{bmatrix} G_1 \\ G_2 \end{bmatrix} \end{aligned} \tag{5.8}$$

where $G_1 = \tilde{G}_1(\Theta_{21}\tilde{G}_1 + \Theta_{22}\tilde{G}_2)^{-1}\varphi^{-1} \in \mathcal{R}_{M\times N}(\Delta)$ and $G_2 = \tilde{G}_2(\Theta_{21}\tilde{G}_1 + \Theta_{22}\tilde{G}_2)^{-1}\varphi^{-1} \in \mathcal{R}_{N\times N}(\Delta)$.

From the bottom component of (5.8)

$$\varphi^{-1} = \Theta_{21}G_1 + \Theta_{22}G_2. \tag{5.9}$$

Now we use that $\|F(z)\| < 1$ for $z \in \partial\Delta$ together with (5.8). Thus, for $z \in \partial\Delta$,

$$\begin{aligned} 0 &> F(z)^*F(z) - I \\ &= [F(z)^* \ I]J \begin{bmatrix} F(z) \\ I \end{bmatrix} \\ &= \varphi(z)^*[G_1(z)^* \ G_2(z)^*]\{\Theta(z)^*J\Theta(z)\} \begin{bmatrix} G_1(z) \\ G_2(z) \end{bmatrix} \varphi(z) \\ &= \varphi(z)^*[G_1(z)^* \ G_2(z)^*]J \begin{bmatrix} G_1(z) \\ G_2(z) \end{bmatrix} \varphi(z) \end{aligned}$$

and thus

$$G_1(z)^*G_1(z) - G_1(z)^*G_2(z) < 0. \tag{5.10}$$

Thus if $G_2(z)x = 0$ for some vector x, then for some $\delta > 0$,

$$-\delta x^*x \geq x^*[G_1(z)^*G_1(z) - G_2(z)^*G_2(z)]x = \|G_1(z)x\|^2$$

which forces $x = 0$. Thus, since $G_2(z)$ has square size $N \times N$, we see that $G_2(z)$ is invertible for $z \in \partial\Delta$. In particular $detG_2(z) \not\equiv 0$ and we may define $G(z) = G_1(z)G_2(z)^{-1} \in \mathcal{R}_{M\times N}$. From (5.10) we see that $G(z)$ is a strict contraction for $z \in \partial\Delta$.

In general we shall write *wno* $w(z)$ for the net change in the argument (or winding number) along $\partial\Delta$ of the scalar continuous function $w(z)$ defined and nonzero-valued on $\partial\Delta$.

Since $\Theta(z)$ is J-unitary on $\partial\Delta, (\Theta_{22}^{-1}\Theta_{21})(z)$ is a strict contraction for $z \in \partial\Delta$; indeed by (4.21) on $\partial\Delta$ we have from

$$\Theta_{22}\Theta_{22}^* = I_N + \Theta_{21}\Theta_{21}^*$$

that Θ_{22} is invertible, and then by

$$\Theta_{22}^{-1}\Theta_{21}\Theta_{21}^*\Theta_{22}^{*-1} = I - \Theta_{22}^{-1}\Theta_{22}^{*-1} < I$$

that $\Theta_{22}^{-1}\Theta_{21}$ is a strict contraction. Since G is also contractive on $\partial\Delta$ we have that $t(\Theta_{22}^{-1}\Theta_{21}G)+I$ is invertible on $\partial\Delta$ for $0 \le t \le 1$, so $wno\ det(t\Theta_{22}^{-1}\Theta_{21}G+I)$ is defined for $0 \le t \le 1$. By a standard homotopy argument

$$wno\ det(\Theta_{22}^{-1}\Theta_{21}G + I) = wno\ detI = 0. \tag{5.11}$$

From (5.9)

$$\begin{aligned} wno\ det\varphi^{-1} &= wno\ det(\Theta_{21}G_1 + \Theta_{22}G_2) \\ &= wno\ det(\Theta_{22}[\Theta_{22}^{-1}G + I]G_2) \\ &= wno\ det\Theta_{22} + wno\ det(\Theta_{22}^{-1}\Theta_{21}G + I) + wno\ detG_2. \end{aligned} \tag{5.12}$$

We have just argued that $wno\ det(\Theta_{22}^{-1}\Theta_{21}G + I) = 0$. Since G_2 is analytic on Δ, by the argument principle

$$wno\ detG_2 \ge 0. \tag{5.13}$$

We claim next that

$$wno\ det\Theta_{22} \ge wno\ det\varphi^{-1}. \tag{5.14}$$

To prove (5.14) we first note from Theorem 2.2 that the subspace $\Theta\mathcal{R}_{M+N}(\Delta)$ is characterized as

$$\begin{aligned} &\Theta\mathcal{R}_{M+N}(\Delta) = \\ &\{\begin{bmatrix} C_+ \\ C_- \end{bmatrix}(zI - A_\pi)^{-1}x + \begin{bmatrix} h_1(z) \\ h_2(z) \end{bmatrix} : \\ &x \in \mathbb{C}^{n_\pi}, h_1 \in \mathcal{R}_M(\Delta), h_2 \in \mathcal{R}_N(\Delta) \text{ such that} \\ &\sum_{z_0 \in \sigma} Res_{z=z_0}(zI - A_\zeta)^{-1}(B_+h_1(z) + B_-h_2(z)) = \Gamma x\}. \end{aligned}$$

Thus

$$
\begin{aligned}
[\Theta_{21}\ \Theta_{22}]\mathcal{R}_{M+N}(\Delta) &= [0\ I]\Theta\mathcal{R}_{M+N}(\Delta) \\
&= \{C_-(zI - A_\pi)^{-1}x + h_2(z) : x \in \mathbb{C}^{n_\pi},\ h_2 \in \mathcal{R}_N(\Delta) \textit{ such that} \\
&\sum_{z_0 \in \sigma} Res_{z=z_0}(zI - A_\zeta)^{-1}B_-h_2(z) - \Gamma x \\
&= -\sum_{z_0 \in \sigma} Res_{z=z_0}(zI - A_\zeta)^{-1}B_+h_1(z) \\
&\qquad \textit{for some } h_1 \in \mathcal{R}_N(\Delta)\}.
\end{aligned} \tag{5.15}
$$

But since by assumption (A_ζ, B_+) is a full range pair, by Lemma 3.2.2(ii) from [BGR1] the map $h \to \sum_{z_0 \in \sigma} Res_{z=z_0}(zI - A_\zeta)^{-1}B_+h(z)$ maps $\mathcal{R}_M(\Delta)$ onto $\mathbb{C}^{n_\zeta}$. Thus given any $h_2 \in \mathcal{R}_N(\Delta)$ one can always solve for an $h_1 \in \mathcal{R}_M(\Delta)$ so that the condition in (5.15) is satisfied. We conclude that

$$[\Theta_{21}\ \Theta_{22}]\mathcal{R}_{M+N}(\Delta) =$$

$$\{C_-(zI - A_\pi)^{-1}x + h(z) : x \in \mathbb{C}^{n_\pi}, h \in \mathcal{R}_N(\Delta)\}.$$

Since by construction φ^{-1} has $(C_-, A_\pi; 0, 0; 0)$ as a Δ-null-pole triple, by Theorem 2.2 again this in turn is equal to

$$[\Theta_{21}\ \Theta_{22}]R_{M+N}(\Delta) = \varphi^{-1}\mathcal{R}_N(\Delta).$$

Hence

$$\Theta_{22}R_N(\Delta) \subset \varphi^{-1}\mathcal{R}_N(\Delta). \tag{5.16}$$

This implies that the poles of $det\Theta_{22}$ in Δ form a subset of the poles of $det\varphi^{-1}$, where we include multiplicities in the count. Since $det\varphi^{-1}$ has no zeros in Δ, (5.14) now follows by the Argument Principle.

When we combine the equality (5.12) with the inequalities (5.11), (5.13) and (5.14) we get the term-wise equalities

$$wno\ det\Theta_{22} = wno\ det\varphi^{-1} \tag{5.17}$$

and

$$wno\ detG_2 = 0. \tag{5.18}$$

From (5.17) we deduce that (5.16) must hold with equality, so

$$\Theta_{22}\mathcal{R}_N(\Delta) = [\Theta_{21}\ \Theta_{22}]\mathcal{R}_{M+N}(\Delta)$$
$$= \Theta_{22}[\Theta_{22}^{-1}\Theta_{21}\ I_N]\mathcal{R}_{M+N}(\Delta).$$

This in turn leads to

$$\Theta_{22}^{-1}\Theta_{21}\mathcal{R}_M(\Delta) \subset \mathcal{R}_N(\Delta)$$

i.e., $\Theta_{22}^{-1}\Theta_{21}$ has no poles in Δ, as desired. Thus by Theorem 4.7 if Λ is invertible then necessarily Λ is positive definite if solutions of the interpolation problem in Theorem 1.1 (if $\Delta = \Pi^+$) or Theorem 5.1 (if $\Delta = \mathcal{D}$) exist.

If we combine (5.18) with (5.8) we can go further. Since G_2 is known to be analytic on Δ, *wno* $detG_2 = 0$ implies that G_2 has no zeros in Δ, so G_2^{-1} and hence also $G = G_1G_2^{-1}$ is analytic on Δ. We already established above that $\|G(z)\| < 1$ for $z \in \partial\Delta$. Hence by the maximum modulus theorem $sup_{z\in\Delta}\|G(z)\| < 1$. Moreover from (5.8)

$$\begin{bmatrix} F \\ I \end{bmatrix} = \begin{bmatrix} \Theta_{11} & \Theta_{12} \\ \Theta_{21} & \Theta_{22} \end{bmatrix} \begin{bmatrix} G \\ I \end{bmatrix} G_2\varphi. \tag{5.19}$$

From the bottom components,

$$I = (\Theta_{21}G + \Theta_{22})G_2\varphi$$

so

$$G_2\varphi = (\Theta_{21}G + \Theta_{22})^{-1}.$$

Then from the top components in (5.19) we get

$$F = (\Theta_{11}G + \Theta_{12})G_2\varphi$$
$$= (\Theta_{11}G + \Theta_{12})(\Theta_{21}G + \Theta_{22})^{-1}.$$

Thus any solution F of the interpolation problem has the prescribed form in Theorem 1.1 (if $\Delta = \Pi^+$) or Theorem 5.1 (if $\Delta = \mathcal{D}$)

Proof of Step 2 Conversely suppose that Λ is positive definite. Then by Theorem 4.7, $\Theta_{22}^{-1}\Theta_{21}$ has no poles in Δ. By reversing an argument in the first part of the proof, we then have

$$\varphi^{-1}R_N(\Delta) = [\Theta_{21}\ \Theta_{22}]\mathcal{R}_{M+N}(\Delta)$$

$$= \Theta_{22}[\Theta_{22}^{-1}\Theta_{21} \ I]\mathcal{R}_{M+N}(\Delta)$$

$$= \Theta_{22}\mathcal{R}_N(\Delta). \tag{5.20}$$

Now let G be an arbitrary element of $\mathcal{R}_{M\times N}(\Delta)$ with $sup_{z\in\Delta}\|G(z)\| < 1$. Then as in the first part of the proof $\|[\Theta_{22}^{-1}\Theta_{21}G](z)\| < 1$ for $z \in \partial\Delta$. As $\Theta_{22}^{-1}\Theta_{21}$ and G are analytic on Δ, by the maximum modulus theorem this continues to hold for z in Δ, so $\Theta_{22}^{-1}\Theta_{21}G + I$ is analytic and invertible on Δ. Thus

$$(\Theta_{21}G + \Theta_{22})\mathcal{R}_N(\Delta) = \Theta_{22}(\Theta_{22}^{-1}\Theta_{21}G + I)\mathcal{R}_N(\Delta)$$
$$= \Theta_{22}\mathcal{R}_N(\Delta) = \varphi^{-1}\mathcal{R}_N(\Delta)$$

where we used (5.20) for the last step. From this we get

$$\varphi(\Theta_{21}G + \Theta_{22})\mathcal{R}_N(\Delta) = \mathcal{R}_N(\Delta)$$

so $\varphi(\Theta_{21}G + \Theta_{22})$ analytic and invertible on Δ. Now by Theorem 3.1 (with $G = G_1$ and $G_2 = I$) it follows that $F = (\Theta_{11}G + \Theta_{12})(\Theta_{21}G + \Theta_{22})^{-1}$ satisfies the interpolation conditions (1.1)-(1.3). Moreover, since $\begin{bmatrix} F \\ I \end{bmatrix} = \Theta \begin{bmatrix} G \\ I \end{bmatrix} W$ (where $W = \Theta_{21}G + \Theta_{22}$) and since Θ is J-unitary on $\partial\Delta$, we have for $z \in \partial\Delta$

$$F(z)^*F(z) - I = [F(z)^* \ I]J\begin{bmatrix} F(z) \\ I \end{bmatrix}$$
$$= W(z)^*[G(z)^* \ I]\Theta(z)^*J\Theta(z)\begin{bmatrix} G(z) \\ I \end{bmatrix} W(z)$$
$$= W(z)^*(G(z)^*G(z) - I)W(z) < 0$$

we see that $\|F(z)\| < 1$ for $z \in \partial\Delta$. As F is analytic on Δ, we conclude by the maximum modulus theorem that $sup_{z\in\Delta}\|F(z)\| < 1$. Thus we have established that if $\Lambda > 0$ then there are many solutions F of the interpolation problem and in fact any F of the indicated linear fractional form in Theorem 1.1 (for $\Delta = \Pi^+$) or Theorem 5.1 (for $\Delta = \mathcal{D}$) is a solution.

Proof of Step 3 It remains only to verify that Λ is positive definite if solutions exist, even if we do not know a priori that Λ is invertible. To do this we first need some operator theoretic preparation. The idea is taken from [BH3], but with all references to Krein spaces suppressed.

For $\Delta = \Pi^+$ and k a natural number we let L_k^2 be the Hilbert space of $\mathbb{C}^k$-vector valued functions norm-square-integrable on the imaginary line $\partial\Delta = \{iy : y \in \mathbb{R}\}$ with respect to the measure $(2\pi)^{-1}dy$, and H_k^2 the Hardy subspace consisting of boundary values of vector functions analytic on Π^+. For $\Delta = \mathcal{D}$, L_k^2 is the analogous space of square integral $\mathbb{C}^k$-valued functions norm-square-integrable on $\partial\Delta = \{e^{i\theta} : 0 \leq \theta \leq 2\pi\}$ with respect to the measure $(2\pi)^{-1}d\theta$, and H_k^2 is the Hardy subspace of boundary values of functions analytic on the unit disk; for details, consult e.g. [RR].

Suppose now that a solution F as in Theorem 1.1 or 5.1 exists. Then the operator $M_F : f \to Ff$ is a strict contraction operator from L_N^2 into L_M^2. We also need some operator interpretations for the block entries of Λ. To do this we introduce observability and controllability operators $\mathcal{C}_+ : H_M^2 \to \mathbb{C}^{n_\zeta}, \mathcal{C}_- : H_N^2 \to \mathbb{C}^{n_\zeta}, \mathcal{O}_+ : \mathbb{C}^{n_\pi} \to H_N^{2\perp}, \mathcal{O}_- : \mathbb{C}^{n_\pi} \to H_M^{2\perp}$ by

$$\begin{aligned}
\mathcal{C}_+ &: h_+ \to \sum_{z_0\in\Delta} Res_{z=z_0}(zI - A_\zeta)^{-1}B_+h_+(z)\\
\mathcal{C}_- &: h_- \to \sum_{z_0\in\Delta} Res_{z=z_0}(zI - A_\zeta)^{-1}B_-h_-(z)\\
\mathcal{O}_+ &: x \to C_+(zI - A_\pi)^{-1}x\\
\mathcal{O}_- &: x \to C_-(zI - A_\pi)^{-1}x.
\end{aligned}$$

Then, if $\Delta = \Pi^+$ the adjoints are given by

$$\begin{aligned}
\mathcal{C}_+^* &: y \to B_+^*(zI + A_\zeta^*)^{-1}y\\
\mathcal{C}_-^* &: y \to B_-^*(zI + A_\zeta^*)^{-1}y\\
\mathcal{O}_+^* &: k_+ \to -\sum_{z_0\in\pi^-} Res_{z=z_0}(zI + A_\pi^*)^{-1}C_+^*k_+(z)\\
\mathcal{O}_-^* &: k_- \to -\sum_{z_0\in\pi^-} Res_{z=z_0}(zI + A_\pi^*)^{-1}C_-^*k_-(z)
\end{aligned}$$

where Π^- is the open left half plane, while, if $\Delta = \mathcal{D}$ then the adjoints are given by

$$\begin{aligned}
\mathcal{C}_+^* &: y \to B_+^*(I - zA_\zeta^*)^{-1}y\\
\mathcal{C}_-^* &: y \to B_-^*(I - zA_\zeta^*)^{-1}y
\end{aligned}$$

$$\mathcal{O}_+^* : k_+ \to -\sum_{z_0 \in \mathcal{D}_e} Res_{z=z_0}(I - zA_\pi^*)^{-1} C_+^* k_+(z)$$

$$\mathcal{O}_-^* : k_- \to -\sum_{z_0 \in \mathcal{D}_e} Res_{z=z_0}(I - zA_\pi^*)^{-1} C_-^* k_-(z)$$

where $\mathcal{D}_e$ is the exterior of the unit disk in the extended complex plane.

We also note that $\mathcal{C}_+$ is surjective since (A_ζ, B_+) by assumption is a full range pair, and that $\mathcal{O}$ is injective since (C_-, A_π) is a null kernel pair (see Lemma 3.2.2 of [BGR1]). Whether $\Delta = \Pi^+$ or $\mathcal{D}$, the interpolation conditions (1.1)-(1.3) can be translated to

$$P_{H_M^{2\perp}} M_F \mathcal{O}_- = \mathcal{O}_+ \tag{5.20}$$

$$\mathcal{C}_+ M_F | H_N^2 = -\mathcal{C}_- \tag{5.21}$$

$$\mathcal{C}_+ P_{H_M^2} M_F \mathcal{O}_- = \Gamma. \tag{5.22}$$

Moreover the operators S_1 and S_2 appearing in Λ can be identified as

$$S_1 = \mathcal{O}_-^* \mathcal{O}_- - \mathcal{O}_+^* \mathcal{O}_+ \tag{5.23}$$

$$S_2 = \mathcal{C}_+ \mathcal{C}_+^* - \mathcal{C}_- \mathcal{C}_-^*. \tag{5.24}$$

To show that $\Lambda = \begin{bmatrix} S_1 & \Gamma^* \\ \Gamma & S_2 \end{bmatrix}$ is positive definite by the Schur complement test we need only show

$$S_2 > 0 \tag{5.25}$$

and that the Schur complement of S_2 is positive definite

$$S_{2s} = S_1 - \Gamma^* S_2^{-1} \Gamma > 0. \tag{5.26}$$

To show (5.25) note that since $||M_F|| < 1$ and $\mathcal{C}_+^*$ is injective (since $\mathcal{C}_+$ is surjective), it follows that

$$||P_{H_N^2}(M_F)^* \mathcal{C}_+^* y||_{H_N^2}^2 < ||\mathcal{C}_+^* y||_{H_M^2} \tag{5.27}$$

for all $y \in \mathbb{C}^{n_\zeta}$. Hence by (5.21) and (5.24),

$$0 < \mathcal{C}_+ \mathcal{C}_+^* - \mathcal{C}_+ M_F P_{H_N^2}(M_F)^* \mathcal{C}_+^*$$

$$= C_+C_+^* - C_-C_-^* = S_2.$$

In particular S_2 is invertible so the Schur complement in (5.26) is defined.

Just as (5.25) was shown to be equivalent to (5.27) we show next that (5.26) is equivalent to

$$\begin{aligned}&\|P_{H_M^{2\perp}\oplus ImC_+^*}M_F[\mathcal{O}_- - C_-^*S_2^{-1}C_+P_{H_M^2}M_F\mathcal{O}_-]x\|^2\\&< \|[\mathcal{O}_- - C_-^*S_2^{-1}C_+P_{H_M^2}M_F\mathcal{O}_-]x\|^2\end{aligned} \tag{5.28}$$

whenever $0 \neq x \in \mathbb{C}^N$. Note that (5.28) is obvious from $\|F\|_\infty < 1$ and the fact that $\mathcal{O}$ is injective and that $\mathcal{O}_-$ and C_-^* have orthogonal image spaces. To verify the equivalence of (5.28) and (5.26) we first check that

$$\begin{aligned}&P_{H_M^{2\perp}\oplus ImC_+^*}M_F[\mathcal{O}_- - C_-^*S_2^{-1}C_+P_{H_M^2}M_F\mathcal{O}_-]x\\&= [\mathcal{O}_+ + C_+^*S_2^{-1}C_+P_{H_M^2}M_F\mathcal{O}_-]x\end{aligned} \tag{5.29}$$

for each $x \in \mathbb{C}^N$. From (5.20) we know $P_{H_M^{2\perp}}M_F\mathcal{O}_-x = \mathcal{O}_+x$, so (5.29) becomes

$$\begin{aligned}&P_{ImC_+^*}(P_{H_M^2}M_F)[\mathcal{O}_- - C_-^*S_2^{-1}C_+P_{H_M^2}M_F\mathcal{O}_-]x\\&= C_+^*S_2^{-1}C_+P_{H_M^2}M_F\mathcal{O}_-x.\end{aligned} \tag{5.30}$$

We have used here that $M_Ff \in H_M^2$ for every $f \in H_M^2$. Since the right side is clearly in $ImC_+^* = (KerC_+)^\perp$, (5.30) is verified if we show

$$\begin{aligned}&C_+P_{H_M^2}M_F[\mathcal{O}_- - C_-^*S_2^{-1}C_+P_{H_M^2}M_F\mathcal{O}_-]\\&- C_+C_+^*S_2^{-1}C_+P_{H_M^2}M_F\mathcal{O}_- = 0.\end{aligned}$$

Use (5.23), (5.20) and (5.24) to get

$$\begin{aligned}&C_+P_{H_M^2}M_F\mathcal{O}_- - C_+M_FC_-^*S_2^{-1}C_+P_{H_M^2}M_F\mathcal{O}_-\\&- C_+C_+^*S_2^{-1}C_+P_{H_M^2}M_F\mathcal{O}_-\\&= \Gamma + C_-C_-^*S_2^{-1}\Gamma - C_+C_+^*S_2^{-1}\Gamma\\&= \Gamma - [C_+C_+^* - C_-C_-^*]S_2^{-1}\Gamma = 0.\end{aligned}$$

This establishes (5.29).

We make use now of the equality (Lemma 6.2.3 in [BGR1])

$$\sum_{z_0 \in \Delta} Res_{z=z_0}(zI - A_\zeta)^{-1} X (zI - A_\pi)^{-1} = 0$$

for any matrix X of suitable size. Consequently, $\mathcal{C}_+\mathcal{O}_+ = 0$ and $\mathcal{C}_-\mathcal{O}_- = 0$. Now by (5.29) and (5.22), the inequality (5.28) amounts to

$$\mathcal{O}_+^*\mathcal{O} + \Gamma^* S_2^{-1}\mathcal{C}_+\mathcal{C}_+^* S_2^{-1}\Gamma < \mathcal{O}_-^*\mathcal{O}_- + \Gamma^* S_2^{-1}\mathcal{C}_-^*\mathcal{C}_- S_2^{-1}\Gamma$$

that is, by (5.24) again,

$$0 < \mathcal{O}_-^*\mathcal{O}_- - \mathcal{O}_+^*\mathcal{O}_+ + \Gamma^* S_2^{-1}[\mathcal{C}_-^*\mathcal{C}_- - \mathcal{C}_+^*\mathcal{C}_+] S_2^{-1}\Gamma = S_1 - \Gamma^* S_2^{-1}\Gamma.$$

This finally gives the required condition (5.26) that S_{2s} be positive definite and completes the proof of positive definiteness of Λ.

It remains only to discuss Theorems 1.2 and 5.2.

PROOF OF THEOREMS 1.2 AND 5.2: Suppose a solution F of (1.1) and (1.2) in $\mathcal{R}_{M\times N}(\Delta)$ with $\|F\|_\infty < 1$ exists. We use (1.3) to define a matrix Γ. Then, as is shown in [BGR1], necessarily Γ satisfies the Sylvester equation (1.4). Then by Theorem 1.1 (if $\Delta = \Pi^+$) or 5.1 (if $\Delta = \mathcal{D}$), the associated matrix Λ is positive definite. The rest also follows now easily from Theorem 1.1 or 5.1.

6. Notes.

The main influences for the approach to interpolation presented in this paper are [H], [BH2], [BR1]. Here we have managed to keep the account elementary completely in the language of linear algebra (the only exception being the proof of Step 3 in Section 5) by suppressing all references to Krein space geometry and Beurling-Lax representation theorems, the key ingredients in [BH2]. The connection between null-pole subspaces and realization and interpolation was first introduced in [BR1, BR2], and further developed in [BGR5, BCR].

Realization formulas for a rational matrix function having prescribed null and pole structure first occurred in [GKLR], with further elaboration in [BR1, BR2, BCR, GK1, GK2]; the case of a

J-unitary matrix function appears in [BR1, BR2, AG]. The latter part of Section 4 concerning the structure of a matrix function J-unitary on $\partial\Delta$ with J-contractive values on Δ comes from [AG] and [BH3]; the proofs are reworked to eliminate the need for Krein space geometry. Finally the winding number arguments in Section 5 are adapted from [H].

REFERENCES

[AG] D. Alpay and I. Gohberg, Unitary rational matrix functions, in *Topics in Interpolation Theory of Rational Matrix Functions* (ed. I. Gohberg), OT 33 Birkhauser (Basel), 1988, pp. 175-222.

[BCR] J. A. Ball, N. Cohen and A. C. M. Ran, Inverse spectral problems for regular improper rational matrix functions, in *Topics in Interpolation Theory of Rational Matrix Functions* (ed. I. Gohberg), OT 33 Birkhauser (Basel), 1988, pp. 123-173.

[BGR1] J. A. Ball, I. Gohberg and L. Rodman, Two-sided Lagrange-Sylvester interpolation problems for rational matrix functions, submitted to Proceedings of the Summer Research Institute in Operator Theory and Operator Algebras, Durham, New Hampshire, 1988.

[BGR2] J. A. Ball, I. Gohberg and L. Rodman, Realization and interpolation of rational matrix functions, in *Topics in Interpolation Theory of Rational Matrix Functions* (ed. I. Gohberg), OT 33 Birkhauser (Basel), 1988, pp. 1-72.

[BGR3] J. A. Ball, I. Gohberg and L. Rodman, Sensitivity minimization and bitangential Nevanlinna-Pick interpolation in contour integral form, submitted to Proceedings of the Signal Processing Workshop, Institute of Mathematics and Applications, Minneapolis, 1988.

[BGR4] J. A. Ball, I. Gohberg and L. Rodman, *Interpolation Problems for Rational Matrix Functions*, monograph in preparation

[BGR5] J. A. Ball, I. Gohberg and L. Rodman, Minimal factorization of meromorphic matrix functions in terms of local data, Integral Equations and Operator Theory 10 (1987), 309-348.

[BH1] J. A. Ball and J. W. Helton, Lie groups over the field of rational functions, signed spectral factorization, signed interpolation, and amplifier design, J. Operator Theory 8 (1982), 19-64.

[BH2] J. A. Ball and J. W. Helton, A Beurling-Lax theorem for the Lie group $U(m,n)$ which contains most classified interpolation, J. Operator Theory 9 (1983), 107-142.

[BH3] J. A. Ball and J. W. Helton, Shift invariant subspaces, passivity, reproducing kernels and H^∞-optimization, in *Contributions to Operatory Theory and its Applications* (ed. I. Gohberg, J. W. Helton, and L. Rodman), Birkhauser (Basel), 1988, pp. 265-310.

[BR1] J. A. Ball and A. C. M. Ran, Local inverse spectral problems for rational matrix functions, Integral Equations and Operator Theory 10 (1987), 349-415.

[BR2] J. A. Ball and A. C. M. Ran, Global inverse spectral problems for rational matrix functions, Linear Algebra and Applications 86 (1987), 237-282.

[BGK] H. Bart, I. Gohberg and M. A. Kaashoek, *Minimal Factorization of Matrix and Operator Functions,* OT 1 Birkhauser (Basel), 1979.

[D] H. Dym, *J Contractive Matrix Functions, Reproducing Kernel Hilbert Spaces and Interpolation*, CBMS volume, Amer. Math. Soc., to appear.

[F] I. I. Fedchina, Tangential Nevanlinna-Pick problem with multiple points, Akad. Nank Armjan. SSR Dokl. 60 (1975), 214-218 [in Russian.]

[GK1] I. Gohberg and M. A. Kaashoek, An inverse spectral problem for rational matrix functions and minimal divisibility, Integral Equations and Operator Theory 10 (1987), 437-465.

[GK2] I. Gohberg and M. A. Kaashoek, Regular rational matrix functions with prescribed pole and zero structure, in *Topics in Interpolation Theory of Rational Matrix Functions* (ed. I. Gohberg) OT 33 Birkhauser (Basel), 1988, pp. 109-122.

[GKLR] I. Gohberg, M. A. Kaashoek, L. Lerer and L. Rodman, Minimal divisors of rational matrix functions with prescribed zero and pole structure, in *Topics in Operator Theory, Systems and Networks* (ed. H. Dym and I. Gohberg), OT 12, Birkhauser (Basel), 1984, pp. 241-275.

[H] J. W. Helton, The distance of a function to H^∞ in the Poincare metric, electral power transfer, J. Functional Analysis 38 (1980), 273-314.

[K] H. Kimura, Conjugation, interpolation and model matching in H^∞, preprint, Osaka University, Osaka, Japan.

[LT] P. Lancaster and M. Tismenetsky, *The Theory of Matrices with Applications*, Academic Press (New York), 1985.

[LA] D. J. N. Limebeer and B. D. O. Anderson, An interpolation theory approach to H_∞ controller degree bounds, Linear Algebra and Applications, to appear.

[NF] B. Sz-Nagy and C. Foias, *Harmonic Analysis of Operators on Hilbert Space,* North Holland (Amsterdam), 1970.

[N1] A. A. Nudelman, On a new problem of moment problem type, Soviet Math. Doklady 18 (1977), 507-510 Doklady Academii Nauk SSSR (1977).

[N2] A. A. Nudelman, A generalization of classical interpolation problems, Soviety Math. Doklady 23 (1981), 125-128; [Doklady Academii Nauk SSSR (1981).

[RR] M. Rosenblum and J. R. Rovnyak, *Hardy Classes and Operator Theory*, Oxford University Press, 1985.

J. A. Ball
Department of Mathematics
Virginia Tech
Blacksburg, Virginia 24061
USA

I. Gohberg
School of Mathematical Sciences
The Raymond and Beverly Sackler Faculty of Exact Sciences
Tel Aviv University
Tel Aviv, Ramat-Aviv 69978
ISRAEL

L. Rodman
Department of Mathematics
College of William and Mary
Williamsburg, Virginia 23185
USA

22

A Metric Property of the Polar Decomposition of Projections

G. Corach, H. Porta, and L. Recht

I. In the space of bounded linear operators in a Hilbert space, the solutions of the equation $p^2 = p$ are called projections and their importance in operator theory can not be overemphasized. In particular, hermitian (and normal) operators can be identified to curves in the set Q of projections, possibly discontinuous curves. In the finite dimensional case, these curves consist of just a finite set of points. A more complicated picture appears when joint spectra are considered, and the theory is far from being complete.

Motivated by spectral theory, the set Q has been studied extensively from the point of view of measure theory, and complete reports are available (for example [Dunford–Schwartz, 1958]).

A different point of view is afforded by geometric interpretations. Since $p^2 = p$ is a quadratic equation with non–zero discriminant, the solution set Q is an algebraic variety without singularities. As such, Q inherits from the ambient space a smooth differential structure. Here the ambient space is the C^*–algebra of all bounded linear operators in the Hilbert space H. The "real part" of Q is the subset $P \subset Q$ of selfadjoint projections. It is a real algebraic smooth manifold and it contains the grassmannians of all dimensions up to the dimension of H. Hence P provides a classifying space for real vector bundles. The larger space Q has a canonical bundle with a connection, and it can be proved that it is a classifying space for bundles-with-connection [Porta, Recht JPAM]. However, it is urgent to remark that P (or Q) does not support a natural Riemannian structure. One reason for this is that the operator norm (= spectral norm) of Hermitian matrices is not differentiable in any dimension higher than 1. The best one can do is consider Finsler structures. This leads to considerable complications and only partial results are known. For example the minimality of geodesics is only known in special cases (see [Port, Recht 1987]).

We study here yet another variational property of the embedding $P \subset Q$, namely the fact that for projections of rank 1, the distance (= geodesic distance) from a fixed element $q \in Q$ to a variable element $p \in P$ is attained in a unique $p = \pi(q)$. This element $\pi(q)$ can be characterized in a variety of ways (see [Corach, Porta, Recht PJM]. Uniqueness fails for projections with rank higher than 1. For details and examples see [Berrizbeitia, Mata, Recht].

II. As indicated above, H will denote a Hilbert space, Q the set of bounded linear $q : H \to H$ with $q^2 = q$ and $P \subset Q$ the self adjoint q. For $q \in Q$ let $2q - 1 = \mu\rho$ be the polar decomposition of $2q - 1$ with $\mu > 0$ and ρ unitary, and write $p = (1 + \rho)/2$. Then p is an orthogonal projection and the map $\pi(q) = p$ is a fibration $\pi : Q \to P$. The fiber $Q_p = \pi^{-1}(p)$ can be described as follows. Let $H = V \oplus N$ be the decomposition $V = \text{image}(p)$, $N = \text{kernel}(p)$. Then all operators in H can be written as 2×2 matrices in terms of $H = V \oplus N$. In this interpretation, the projections q in Q_p have the form

$$q = \begin{bmatrix} a^2 & -ab^* \\ ba & -bb^* \end{bmatrix}$$

where $b : V \to N$ is arbitrary and a is the positive square root $a = (1 + b^*b)^{1/2}$. Therefore Q_p can be parametrized (up to modulus one constants) by the space of bounded linear operators from V to N. These results are given in [Corach, Porta, Recht, Adv M].

III. In the sequel we consider only projections of rank one. Since these projections form a connected component Q_1 of Q, the above constructions carry over and we obtain a fibration $\pi : Q_1 \to P_1$.

In fact, for any dimension, the set $Q_n = \{q;\ q^2 = q,\ \text{trace}(q) = n\}$ is an algebraic smooth manifold which is a component of Q. Taking $P_n = Q_n \cap P$, the corresponding fibrations $\pi : Q_n \to P_n$ result.

All rank one operators $k : H_1 \to H_2$ from a Hilbert space H_1 to a Hilbert space H_2 can be written as $k = x_2 \otimes x_1$ with action $kz = \langle z, x_1 \rangle x_2$. Then $k^* : H_2 \to H_1$ is $k^* = x_1 \otimes x_2$ and $\|k\| = |x_1||x_2|$. Also for $k = x_2 \otimes x_1 : H_1 \to H_2$, $k' = x_3' \otimes x_2' : H_2 \to H_3$ we have $k'k = \langle x_2, x_2' \rangle x_3' \otimes x_1$. Hence $y \otimes x : H \to H$ is a projection if and only if $\langle y, x \rangle = 1$, and it is an orthogonal projection if and only if $x = y$, $|x| = 1$.

Fix $p \in P_1$ and choose $v \in V$ with $|v| = 1$ (the choice of v is unique up to modulus one constant). Then each operator from V to N has the form $b = w \otimes v$ for a unique $w \in N$. Notice that $a = (1 + b^*b)^{1/2}$ is the scalar $a = (1 + |w|^2)^{1/2}$.

Theorem 1. The projections $q \in Q_{1,p}$ have the form $q = y \otimes x$ where $x = (1 + |v|^2)^{1/2}v - w$, $y = (1 + |w|^2)^{1/2}v + w$, for $w \in N$ arbitrary. Further, $\|q\| = |x|^2 = |y|^2 = 1 + 2|w|^2$.

Proof. We use the matrix form

$$q = \begin{bmatrix} a^2 & -ab^* \\ ba & -bb^* \end{bmatrix}$$

to obtain $qv = (1 + |w|^2)v + (1 + |w|^2)^{1/2}w$. Since $\langle v, w \rangle = 0$ we get $|qv|^2 = (1 + |w|^2)(1 + 2|w|^2)$, whence $qv \neq 0$, and qv spans the image of q. Using the matrix form again we see that $sv + n$ (with s a scalar, $n \in N$) is in the kernel of q if and only if

$$\begin{cases} a^2(sv) - ab^*(n) = 0, \\ ba(sv) - bb^*(n) = 0. \end{cases}$$

But this system is equivalent to the single equation $sa(v) = b^*(n)$, or

$$s(1 + |w|^2)^{1/2} = \langle n, w \rangle .$$

Define then $x = (1 + |w|^2)^{1/2}v - w$. We have $\langle x, sv + n\rangle = s(1 + |w|^2)^{1/2} - \langle n, w\rangle$ and therefore $sv + n \in \text{kernel}(q)$ if and only if $\langle x, sv + n\rangle = 0$. This shows that $qv \otimes x$ has the same image and the same kernel as q. Calculate next

$$\begin{aligned}\langle qv, x\rangle &= \langle (1 + |w|^2)v + (1 + |w|^2)^{1/2}w, (1 + |w|^2)^{1/2}v - w\rangle \\ &= (1 + |w|^2)^{3/2} - |w|^2(1 + |w|^2)^{1/2} = (1 + |w|^2)^{1/2}.\end{aligned}$$

Taking $y = qv/(1 + |w|^2)^{1/2}$ we obtain a projection $y \otimes x$ with the same image and kernel as q; thus $q = y \otimes x$. But $qv/(1 + |w|^2)^{1/2} = (1 + |w|^2)^{1/2}v + w$, so the description of q is complete. Clearly $|x|^2 = |y|^2 = 1 + 2|w|^2$, and this finishes the proof.

Just as Theorem 1 describes $Q_{1,p}$ in simple terms using p, we can describe the map $\pi : Q_1 \to P_1$ in simple terms.

Theorem 2. *Let* $q \in Q_1$ *and write* $q = y \otimes x$ *with* $|x| = |y|$. *Then* $\pi(q) = v \otimes v$ *where* $v = \frac{x + y}{2}/(\frac{1 + \|q\|}{2})^{1/2}$.

Proof. Apply Theorem 1: if $\pi(q) = p = v \otimes v$, then q must have the form described in Theorem 1, i.e., $q = y \otimes x$. But from the expressions for x and y we easily get $v = \frac{1}{2}(x + y)/(1 + |w|^2)^{1/2}$ and this is equivalent to the formula in the statement of Theorem 2.

We close this section with two remarks. First, a projection $q = y \otimes x$ can be described by any expression $q = \lambda y \otimes \mu x$ with $\lambda\bar{\mu} = 1$.

With $|x| = |y|$ this implies that also $|\grave{o}| = |\hat{u}|$ and then $x + y$ (hence v) is well defined. Second, Theorem 1 implies that $Q_{1,p}$ can be parametrized by $w \in N$ through the map

$$w \to ((1 + |w|^2)^{1/2}v + w) \otimes ((1 + |w|^2)^{1/2}v - w),$$

and the norm of the right–hand side is equal to $1 + 2|w|^2$.

III. This is the main result:

Theorem 3. For any $q \in Q_1$, the distance $\|q - r\|$, where $r \in P_1$, attains its minimum at $r = \pi(q)$ and only there.

The argument is the same for real or complex scalars, but we show details for the real case only.

Suppose $q \in Q_1$ is fixed, $p = \pi(q)$ and let v, $b = w \otimes v$, $a = (1 + |w|)^{1/2}$ as above. For $r \in P_1$ arbitrary write $r = z \otimes z$ with $|z| = 1$. Decompose $z = pz + (1 - p)z = dv + m$ $(m \in N)$ and assume that z was chosen with $d \geq 0$. We have $d = (1 - |m|^2)^{1/2}$ because $|z| = 1$. Consider now an orthonormal system e_1, e_2, e_3 with $e_1 = v$, $e_2 = w/|w|$ and e_3 chosen such that m is a linear combination of e_2, e_3 : $m = m_2e_2 + m_3e_3$. Under these conditions, $q - r$ leaves the span K of e_1, e_2, e_3 invariant and vanishes on the orthogonal complenent $K^\perp$. Hence the norm of $q - r$ can be calculated by considering its restriction to K. From now on, then, only the restrictions are considered. In terms of the basis e_1, e_2, e_3 we have

$$v = (1,0,0)$$

$$w = (0,c,0) \quad (\text{with } c = |w|)$$

$$x = (a,-c,0)$$

$$y = (a,c,0)$$

$$m = (0,m_2,m_3)$$

$$z = (d,m_2,m_3)$$

and therefore

$$q = y \otimes x = y^{\mathsf{T}}x = \begin{bmatrix} a^2 & -ac & 0 \\ ca & -c^2 & 0 \\ 0 & 0 & 0 \end{bmatrix}$$

$$p = v \otimes v = v^{\mathsf{T}}v = \begin{bmatrix} 1 & 0 & 0 \\ 0 & 0 & 0 \\ 0 & 0 & 0 \end{bmatrix}$$

$$r = z \otimes z = z^{\mathsf{T}}z = \begin{bmatrix} d^2 & dm_2 & dm_3 \\ dm_2 & m_2^2 & m_2m_3 \\ dm_3 & m_2m_3 & m_3^2 \end{bmatrix}.$$

We claim that

$$(\dagger) \qquad \|q - r\| = c^2(1 + 2c^2) + (1 + 2c^2)m_2^2 + (1 + c^2)m_3^2 + 2c(1 + c^2)^{1/2}(c^4 + (1 + 2c_2^2)m_2^2 + c^2m_3^2)^{1/2}.$$

This formula proves that $\|q - r\|$ attains its minimum at $m_2 = m_3 = 0$ and only there. But $m = 0$ gives $r = p$, and the theorem follow. In order to prove (†) we use $\|q - r\|^2 = \|(q - r)(q - r)^{\mathsf{T}}\|$. Now $(q - r)(q - r)^{\mathsf{T}}$ is self adjoint, so its norm coincides with its largest eigenvalue. Observe that the 3×3 matrix $t = (q-r)(q-r)^{\top}$ is singular (because $q-r$ has rank at most 2) and therefore $s = 0$ is a root of the characteristic polynomial $p(s) = \det(t-s)$ of t. Hence $p(s) = -s(s^2 + Bs + C)$, and straightforward calculations give

$$\begin{aligned}
B &= -2c^2(1+2c^2) - 2(1+2c^2)m_2^2 - 2(1+c^2)m_3^2 \\
C &= c^4 - 2c^2(1+2c^2)m_2^2 + 2c^2(1+c^2)m_3^2 \\
&\quad + (1+2c^2)^2 m_2^4 + (1+c^2)^2 m_3^4 + 2(1+c^2)(1+2c^2)m_2^2 m_3^2 \\
(B^2-4C)^{1/2} &= 4c(1+c^2)^{1/2}(c^4 + (1+2c^2)m_2^2 + c^2 m_3^2)^{1/2}
\end{aligned}$$

and therefore the right-hand side of (†) is the largest root $(-B + (B^2 - 4c)^{1/2})/2$ of $p(s) = 0$, as claimed.

IV. A feeling for the structure of Q_1 can be obtained by considering the case where H is the ordinary real plane. For $x^2 + y^2 - z^2 = 1$ define

$$q + q(x,y,z) = \frac{1}{2}\begin{bmatrix} 1+x & y+z \\ y-z & 1-x \end{bmatrix}.$$

Then $q^2 = q$ and since $\operatorname{trace}(q) = 1$ this always gives rank one projections. It is not hard to see that all $q \in Q_1$ can be written in this form, and therefore $(x,y,z) \to q(x,y,z)$, with $x^2 + y^2 - z^2 = 1$, gives a parametrization of Q_1. Observe that $q(x,y,z)^{\top} = q(x,y,-z)$ and $1 - q(x,y,z) = q(-x,-y,-z)$.

Next abbreviate $k = +\sqrt{x^2 + y^2}$ and define

$$p = \pi(q) = q(\tfrac{x}{k}, \tfrac{y}{k}, 0) = \frac{1}{2}\begin{bmatrix} 1 + \frac{x}{k} & \frac{y}{k} \\ \frac{y}{k} & 1 - \frac{x}{k} \end{bmatrix}.$$

Then

$$2q - 1 = \begin{bmatrix} x & y + z \\ y - z & -x \end{bmatrix}$$

$$2p - 1 = \frac{1}{k}\begin{bmatrix} x & y \\ y & -x \end{bmatrix}$$

$$(2q - 1)(2p - 1) = \frac{1}{k}\begin{bmatrix} k^2 + yz & -xz \\ -xz & k^2 - yz \end{bmatrix}.$$

<u>The polar decomposition</u> $2q - 1 = \mu\rho$ <u>is given by</u> $\mu = (2q - 1)(2p - 1)$ <u>and</u> $\rho = 2p - 1$.

<u>Proof</u>. First, $\rho^* = \rho$, $\rho^2 = 1$ are straighforward. Also $k^2 + yz = x^2 + y^2 + yz \geq x^2 + y^2 - (y^2 + z^2)/2 = (1 + x^2)/2 > 0$ and $\det(\mu) = 1$ so that $\mu > 0$.

This describes completely the map $\pi : Q_1 \to P_1$. In particular if p corresponds to $(\cos\theta, \sin\theta, 0)$, then the fiber $Q_{1,p}$ is the branch of hyperbola $x = \operatorname{Cosh}\delta\cos\theta$, $y = \operatorname{Cosh}\delta\sin\theta$, $z = \operatorname{Sinh}\delta$ where $-\infty < \delta < +\infty$. The projections that have a fixed image form an affine manifold. This corresponds to $x^2 + y^2 - z^2 = 1$ being a ruled surface. More precisely: if $q = q(x_0, y_0, z_0)$, then the two lines in $x^2 + y^2 - z^2 = 1$ through (x_0, y_0, z_0) correspond to the sets of projections in Q_1 with the same image or the same kernel as q.

References

Berrizbeitia, P., Mata-Lorenzo, L., and Recht, L., A uniqueness theorem for the unitary part of a reflexion. To appear.

Corach, G., Porta, H., and Recht, L., The geometry of the spaces of projections in a c^*-algebra, to appear in Advances in Mathematics.

Corach, G., Porta, H., and Recht, L., Differential Geometry of systems of projections in Banach spaces, to appear in Pacific Journal of mathematics.

Dunford, N. and Schwartz, J. T., Linear operators, Interscience, 1958.

Kinnunen, S., Linear connections on hypersurfaces of Banach spaces, Ann. Acad. Sci. Fenn., Ser. A, Diss. 65, 1987.

Porta, H. and Recht, L., Minimality of geodesics in Grassmann Manifolds, Proc. Amer. Math. Soc., 100(1987).

Porta, H. and Recht, L., Classification of linear connections, Journal of Pure & Applied Mathematics, 1988.

G. Corach,
Instituto Angentino de Matematica
Buenos Aires

H. Porta
University of Illinois
Urbana, Illinois 61801

L. Recht
Universidad Simon Bolivar
Caracas

23

On Reproducing Kernels and The Continuous Covariance Extension Problem

Harry Dym

CONTENTS

Necessary and sufficient conditions for the existence of a solution to the covariance extension problem in the Wiener algebra of $p \times p$ matrix valued functions on the line (which are adapted from a more general study of extension problems which was developed jointly with I. Gohberg) are presented. A description of the set of all solutions to this problem, when these conditions are met, is then established using reproducing kernel Hilbert space methods. A theorem on reproducing kernel Hilbert spaces with kernels of a special form and its connections with the work of de Branges is also discussed, along with a number of other issues which are intimately related to the problem under study.

The author would like to acknowledge with thanks Renee and Jay Weiss for endowing the chair which supported this research.

1. INTRODUCTION

In this paper we shall discuss the continuous covariance extension problem in the setting of the Wiener algebra of $p \times p$ matrix valued functions on the line. In particular we shall first derive necessary and sufficient conditions for the existence of a solution to this problem and shall then present a description of the set of all solutions to this problem when these conditions are met. The conditions for existence were first obtained by M.G. Krein (1956), where they are stated without proof; the present proof is adapted from Dym and Gohberg (1980), which treats a more general extension problem. A description of the set of all solutions was presented in Dym and Iacob (1984) though the proof given there was flawed because it depended in part on a theorem in the literature on spectral theory of canonical differential equations which turned out to be incorrect (the difficulties are discussed somewhat in Langer (1986) and Krein and Melik Adamyan (1986)). The present proof uses the theory of reproducing kernel Hilbert spaces of vector valued functions which originates with de Branges (1963). It was developed a few years ago (in early 1986) to fit into a monograph Dym (1989b) on applications of reproducing kernel spaces of the type alluded to above to problems of interpolation theory. It was then cut and deferred to an extended version, which is probably still a good few years away, because of lack of space. Accordingly, although it has been rewritten so that it can be read independently, the present article is perhaps best viewed as a supplementary chapter to Dym (1989b). In particular, in terms of the notation used there, it serves as an application of an infinite dimensional $\mathcal{H}(U)$ space.

Since the time the first draft was written, other approaches to the problem under study have appeared: Krein and Melik Adamyan (1986) and Gohberg, Kaashoek and Woerdeman (1988). Nevertheless, the methods used here are quite different and still seem to be of some interest. Also it is well to remark that much of the ensuing development can be generalized to the setting of indefinite inner products and reproducing kernel Krein spaces. The beautiful paper of Krein and Langer (1985), even though it treats only the scalar case, will serve admirably as an introduction to this set of ideas.

The given data for the problem under study is a summable $p \times p$ matrix valued function $h(t)$ which is specified on the interval $-\tau \leq t \leq \tau$. The objective is to find conditions on h which ensure the existence of extensions f of h to the full line which satisfy certain positivity requirements and to furnish a description of these extensions when they exist. To explain the problem more precisely, let $\mathcal{P}_\tau(h)$ denote the set of $p \times p$ matrix valued Fourier transforms

$$\hat{f}(\lambda) = \int_{-\infty}^{\infty} e^{i\lambda t} f(t) dt$$

of $p \times p$ matrix functions f which are summable on the full line $\mathbb{R}$ and are such that

(1) $f(t) = h(t)$ a.e. for $-\tau \leq t \leq \tau$, and

(2) $I_p - \hat{f}(\lambda) > 0$ for every point $\lambda \in \mathbb{R}$.

In this language, the objective is to characterize those h for which the class $\mathcal{P}_\tau(h)$ is nonempty and to provide a description of $\mathcal{P}_\tau(h)$ in this instance.

We shall, as remarked earlier, use a theory of reproducing kernel Hilbert spaces which originates with de Branges (1963) to help solve this problem. The requisite theory of such spaces is developed in Section 2. Actually we begin with a theorem which is more general than needed here, because it has been useful in other applications and at present is not available in published form. Necessary and sufficient conditions for the existence of a solution are then deduced in Section 3. In Section 4 we introduce a space $\mathcal{M}$ of $m \times 1$ vector valued entire functions based on the data of the problem, where $m = 2p$, and identify it as an $\mathcal{H}(U)$ space. This means that $\mathcal{M}$ is a reproducing kernel Hilbert space with a reproducing kernel of the special form (2.1). We then develop a number of identities involving this kernel (not all of which are needed for the problem under study) as well as a number of formulas and identities for continuous analogues of orthogonal matrix polynomials of the first and second kind. Such analogues were first discussed by M.G. Krein (1955), (1956); they play an important role in some of his subsequent studies: Krein and Langer (1985) and Krein and Melik Adamyan (1986). Herein a formula for U (which is uniquely determined by the space $\mathcal{M}$ up to a right constant J unitary factor) is obtained in terms of these continuous analogues of orthogonal matrix polynomials. In Section 5 a representation formula for the set of all solutions to the continuous covariance extension problem in terms of a linear fractional transformation based on this U is obtained, when the requisite conditions for existence are met. In particular, Theorem 5.2 serves to justify Theorem 7.4 of Dym and Iacob (1984) and (2) of Theorem 11.4 of Dym (1989b) which, as noted there, can be used to obtain maximum entropy estimates. These and a number of other ramifications of the analysis developed in the earlier sections are then sketched briefly in Section 6.

Finally some words on notation: The symbols $\mathbb{C}$, $\mathbb{C}^{j\times k}$ and $\mathbb{R}$ will denote the complex numbers, the space of complex $j \times k$ matrices and the real numbers, respectively, whereas $\mathbb{C}^j$ is short for $\mathbb{C}^{j\times 1}$; $\mathbb{C}_+$ [resp. $\mathbb{C}_-$] stands for the open upper [resp. lower] half plane, $\mathbb{T}$ for the unit circle, $\mathbb{D}$ for its interior: $\mathbb{D} = \{\lambda \in \mathbb{C} : |\lambda| < 1\}$ and $\mathbb{E}$ for its exterior with respect to the extended complex plane $\hat{\mathbb{C}} = \mathbb{C} \cup \{\infty\} : \mathbb{E} = \{\lambda \in \hat{\mathbb{C}} : 1 < |\lambda| \le \infty\}$.

If A is a set, then $\overline{A}$ denotes its closure. If A is a matrix or operator, then A^* stands for its adjoint with respect to the standard inner product unless indicated otherwise.

The letter J designates a general $m \times m$ signature matrix, i.e., a constant invertible matrix with $J = J^* = J^{-1}$. Herein a major role will be played by the specific signature matrices

$$J_0 = \begin{bmatrix} I_p & 0 \\ 0 & -I_p \end{bmatrix} \text{ and } J_1 = \begin{bmatrix} 0 & I_p \\ I_p & 0 \end{bmatrix},$$

with $m = 2p$.

The symbol R_α denotes the generalized backwards shift:

$$(R_\alpha f)(\lambda) = \frac{f(\lambda) - f(\alpha)}{\lambda - \alpha} .$$

In Section 2 we shall adopt a flexible notation which permits us to treat problems in $\mathbb{D}$ and problems in $\mathbb{C}_+$ more or less simultaneously by using Δ_+ to denote either $\mathbb{D}$ or $\mathbb{C}_+$ and then invoking the following table:

Δ_+	$\mathbb{D}$	$\mathbb{C}_+$
$\rho_\omega(\lambda)$	$1 - \lambda\omega^*$	$-2\pi i\ (\lambda - \omega^*)$
$b_\omega(\lambda)$	$(\lambda - \omega)/(1 - \lambda\omega^*)$	$(\lambda - \omega)/(\lambda - \omega^*)$
∂	$\mathbb{T}$	$\mathbb{R}$
$< f, g >$	$\frac{1}{2\pi}\int_0^{2\pi} g(e^{i\theta})^* f(e^{i\theta})d\theta$	$\int_{-\infty}^{\infty} g(\lambda)^* f(\lambda)d\lambda$
$\overline{\Delta}_+$	$\overline{\mathbb{D}}$	$\mathbb{C}_+ \cup \mathbb{R}$
Δ_-	$\mathbb{E}$	$\mathbb{C}_-$

Also:

$H^p(\Delta_+)$ denotes the space of scalar valued Hardy functions of class p over Δ_+, $1 \leq p \leq \infty$.

$H^p_{k\times j}(\Delta_+)$ denotes the space of $k \times j$ matrix valued functions each entry of which belongs to $H^p(\Delta_+)$.

$H^p_k(\Delta_+)$ is short for $H^p_{k\times 1}(\Delta_+)$.

The symbol $\underline{\underline{p}}$ is used to denote the orthogonal projection of the space $L_k^2(\partial)$ of square summable $k \times 1$ vector valued functions on ∂ onto $H_k^2(\Delta_+)$ and $\underline{\underline{q}}' = I - \underline{\underline{p}}$.

The analysis in the remaining sections is carried out in $\mathbb{C}_+$. Therein we shall also deal with the following spaces:

$\mathcal{S}_{p\times q}$, the set of $p \times q$ matrix valued functions $S(\lambda)$ which are analytic and contractive (i.e., $S(\lambda)^* S(\lambda) \leq I_q$) in $\mathbb{C}_+$.

$\mathcal{C}_p$, the set of $p \times p$ matrix valued functions $\Phi(\lambda)$ which are analytic with $\Phi(\lambda) + \Phi(\lambda)^* \geq 0$ in $\mathbb{C}_+$.

$\mathcal{W}$, the Wiener algebra of $p \times p$ matrix valued functions of the form $f(\lambda) = c_f + \hat{k}_f(\lambda)$, where $c_f \in \mathbb{C}^{p\times p}$ and $k_f \in L^1_{p\times p}(\mathbb{R})$ (both c_f and k_f are uniquely determined by f).

$$\begin{aligned}
\mathcal{W}_+ &= \{f \in \mathcal{W} : \; k_f(s) = 0 \;\text{ for }\; s < 0\} \\
\mathcal{W}_- &= \{f \in \mathcal{W} : \; k_f(s) = 0 \;\text{ for }\; s > 0\} \\
\mathcal{W}^A &= \{f \in \mathcal{W} : \; c_f = A\} \\
\mathcal{W}^A_\pm &= \mathcal{W}^A \cap \mathcal{W}_\pm .
\end{aligned}$$

2. REPRODUCING KERNEL SPACES

We shall say that a reproducing kernel Hilbert space of $m \times 1$ vector valued functions which are meromorphic in Δ_+ is an $\mathcal{H}(U)$ space if its reproducing kernel $K_\omega(\lambda)$ can be expressed in the form

$$K_\omega(\lambda) = \frac{J - U(\lambda) J U(\omega)^*}{\rho_\omega(\lambda)} \tag{2.1}$$

where U is an $m \times m$ matrix valued function which is J inner over Δ_+. Spaces of this kind were first characterized by de Branges (1963) albeit with an extra technical condition which was later shown to be superfluous by Rovnyak (196?) It turns out that the elements in such a space are really analytic in a region which is symmetric about the boundary of Δ_+; this is what insures that the U in the reproducing kernel is square. In this section we shall treat reproducing kernel spaces with kernels which are essentially of the form (2.1), but with nonsquare U. This setting is more general than needed for the present application. It is presented here because it has proved useful in other applications and is not available elsewhere. We begin with a pair of preliminary lemmas on R_α invariance.

LEMMA 2.1. *If $f \in H^2_m$ and $\alpha \in \Delta_+$, then*

$$\|R_\alpha f\| \leq \frac{\|f\|}{\delta(\alpha)}$$

where

$$\delta(\alpha) = \begin{cases} 1 - |\alpha| & if \quad \Delta_+ = \mathbb{D} \\ (\alpha - \alpha^*)/2i & if \quad \Delta_+ = \mathbb{C}_+ \ , \end{cases}$$

denotes the distance from the point α to the boundary of Δ_+.

PROOF. To begin with, it is readily checked that, for any choice of f and g in H^2_m,

$$\begin{aligned} < R_\alpha f \ , \ g > &= < \frac{f - f(\alpha)}{\lambda - \alpha} \ , \ g > \\ &= < \frac{f}{\lambda - \alpha} \ , \ g > \ , \end{aligned}$$

since $(\lambda - \alpha)^{-1} f(\alpha)$ is orthogonal to H^2_m for $\alpha \in \Delta_+$. But now as $R_\alpha f$ belongs to H^2_m it follows further that

$$\begin{aligned} \|R_\alpha f\|^2 &= < \frac{f}{\lambda - \alpha} \ , \ R_\alpha f > \\ &\leq \|(\lambda - \alpha)^{-1} f\| \ \|R_\alpha f\| \\ &\leq \{\delta(\alpha)\}^{-1} \|f\| \ \|R_\alpha f\| \ . \end{aligned}$$

The rest is plain. ∎

LEMMA 2.2. *Let $\mathcal{M}$ be a closed subspace of H^2_m which is R_α invariant: $R_\alpha \mathcal{M} \subset \mathcal{M}$, for some point $\alpha \in \Delta_+$. Then $\mathcal{M}$ is R_β invariant for every point $\beta \in \Delta_+$.*

PROOF. It is readily checked that the operators R_α , R_β, satisfy the resolvent identity

$$R_\beta = R_\alpha + (\beta - \alpha) R_\alpha R_\beta$$

and hence, by iteration, that

$$R_\beta f - \{R_\alpha f + (\beta - \alpha) R_\alpha^2 f + \ldots + (\beta - \alpha)^{k-1} R_\alpha^k f\} = (\beta - \alpha)^k R_\alpha^k R_\beta f \ .$$

But now the term in curly brackets on the left hand side of the last identity belongs to $\mathcal{M}$ whereas, thanks to Lemma 2.1, the right hand side is subject to the bound

$$\|(\beta - \alpha)^k R_\alpha^k R_\beta f\| \leq \left(\frac{|\beta - \alpha|}{\delta(\alpha)}\right)^k \|R_\beta f\|$$

which tends to zero as $k \uparrow \infty$ when $|\beta - \alpha| < \delta(\alpha)$. Therefore $\mathcal{M}$ is clearly R_β invariant for $|\beta - \alpha| < \delta(\alpha)$ and so too, by an elementary argument, for all $\beta \in \Delta_+$. ∎

THEOREM 2.1. *Let $\mathcal{M}$ be a closed subspace of H_m^2, let J be an $m \times m$ signature matrix which is unitarily equivalent to J_{pq} and let $\mathcal{N} = H_m^2 \ominus J\mathcal{M}$. Suppose further that*

(1) $R_\alpha \mathcal{M} \subset \mathcal{M}$ for some point $\alpha \in \Delta_+$ and

(2) $\mathcal{M} + \mathcal{N} = H_m^2$.

Then $\mathcal{M}$ [resp. $\mathcal{N}$] is a reproducing kernel Krein space, of $m \times 1$ vector valued functions which are analytic in Δ_+, with respect to the indefinite inner product

$$[f, g] = < Jf, g > \tag{2.2}$$

induced by J. Moreover, the reproducing kernel

$$M_\omega(\lambda) = \frac{J - \Theta(\lambda) J' \Theta(\omega)^*}{\rho_\omega(\lambda)} \quad \left[resp. N_\omega(\lambda) = \frac{\Theta(\lambda) J' \Theta(\omega)^*}{\rho_\omega(\lambda)}\right] , \tag{2.3}$$

where

$$J' = \begin{bmatrix} I_{p'} & 0 \\ 0 & I_{q'} \end{bmatrix},$$

$p' \leq p$, $q' \leq q$, $k = p' + q'$ and $\Theta(\lambda)$ is an $m \times k$ matrix valued function such that

$$\Theta(\lambda)^* J \Theta(\lambda) = J' \tag{2.4}$$

for a.e. point $\lambda \in \partial$. Θ is unique up to a right constant J unitary factor and the columns $h_1, \ldots, h_k$ of

$$\frac{\{\rho_\alpha(\alpha)\}^{\frac{1}{2}}}{\rho_\alpha(\lambda)} \Theta(\lambda) = [h_1 \ldots h_k]$$

form a J orthogonal basis for

$$\mathcal{L}_\alpha = \{f \in \mathcal{N} : \ [f, b_\alpha g] = 0 \ \ for\ every\ g \in \mathcal{N}\} \ ,$$

with $[h_i, h_i] = \pm 1$ according as $i \leq p'$ or $i > p'$. Moreover,

(1) if $\mathcal{M}$ is a positive subspace of H_J^2, then $q' = q$;

(2) if $\mathcal{N}$ is full range, then $k = m$ and $J = J'$;

(3) if $\mathcal{M}$ is finite dimensional, then $\mathcal{N}$ is full range.

PROOF. The proof is lengthy and is broken into a number of steps. It is basically a wandering subspace argument which is adapted in part from Ball and Helton (1983). The casual reader may find it easier to focus on the case $\Delta_+ = \mathbb{D}$ with $\alpha = 0$. This contains the main ideas with a minimum of technical detail and, in view of Lemma 2.2, involves no loss of generality for the case $\Delta_+ = \mathbb{D}$. The present more elaborate argument is furnished because, as already noted, of our wish to cover the case $\Delta_+ = \mathbb{C}_+$ also.

STEP 1. *$\mathcal{M}$ and $\mathcal{N}$ are reproducing kernel Krein spaces and $\mathcal{M} \cap \mathcal{N} = 0$.*

PROOF OF STEP 1. By assumption $\mathcal{M}$ is both closed and orthocomplemented in H_J^2. Therefore, by Theorem 3.4 on p.104 of Bognar (1974), $\mathcal{M}$ is a Krein space. The same reasoning applies to $\mathcal{N}$. Now, if $f \in \mathcal{M} \cap \mathcal{N}$, then

$$[f, m] = 0 = [f, n]$$

for every choice of $m \in \mathcal{M}$ and $n \in \mathcal{N}$. Therefore, in view of assumption (2),

$$[f, g] = 0$$

for every $g \in H_J^2$, and this in turn forces $f = 0$ since H_J^2 is a Krein space. In the present setting this is easily seen directly since H_J^2 is a reproducing kernel Krein space with reproducing kernel

$$K_\omega(\lambda) = J/\rho_\omega(\lambda) \ .$$

Thus $\mathcal{M} \cap \mathcal{N} = 0$. Assumption (2) implies further that, for every choice of $\omega \in \Delta_+$ and $\xi \in \mathbb{C}^m$, $K_\omega(\lambda)\xi$ admits a unique decomposition

$$K_\omega(\lambda)\xi = k_{\omega,\xi} + K_\omega\xi - k_{\omega,\xi}$$

with $k_{\omega,\xi} \in \mathcal{M}$ and $K_\omega\xi - k_{\omega,\xi} \in \mathcal{N}$ and that

$$[f, k_{\omega,\xi}] = [f, K_\omega\xi] = \xi^* f(\omega)$$

for every $f \in \mathcal{M}$. Let ξ_j , $j = 1, \ldots, m$, be an orthonormal basis for $\mathbb{C}^m$ and let

$$M_\omega(\lambda) = \sum_{j=1}^{m} k_{\omega,\xi_j}(\lambda)\xi_j^* \ .$$

Then, it follows from the last evaluation, that

$$\begin{aligned} [f, M_\omega\xi] &= \sum_{j=1}^{m} \xi^*\xi_j[f, k_{\omega,\xi_j}] \\ &= \sum_{j=1}^{m} \xi^*\xi_j\xi_j^* f(\omega) \\ &= \xi^* f(\omega) \ . \end{aligned}$$

Therefore $M_\omega(\lambda)$ is a reproducing kernel for $\mathcal{M}$ and, correspondingly,

$$N_\omega(\lambda) = \frac{J}{\rho_\omega(\lambda)} - M_\omega(\lambda)$$

is a reproducing kernel for $\mathcal{N}$.

STEP 2. $b_\alpha \mathcal{N} \subset \mathcal{N}$.

PROOF OF STEP 2. Suppose first that $\Delta_+ = \mathbb{D}$. Then, for every choice of $f \in \mathcal{M}$ and $g \in \mathcal{N}$,

$$\begin{aligned}\left[\frac{\lambda g}{\rho_\alpha} , f\right] &= \left[g , \frac{f}{\lambda - \alpha}\right] \\ &= [g , R_\alpha f] \\ &= 0 .\end{aligned}$$

This proves that $\mathcal{N}$ is invariant under multiplication by $\lambda/\rho_\alpha(\lambda)$ and hence also under multiplication by $b_\alpha(\lambda)$, since

$$b_\alpha(\lambda) = (1 - |\alpha|^2)\frac{\lambda}{\rho_\alpha(\lambda)} - \alpha .$$

Next, if $\Delta_+ = \mathbb{C}_+$, then, for every choice of $f \in \mathcal{M}$ and $g \in \mathcal{N}$,

$$\begin{aligned}-2\pi i[\frac{g}{\rho_\alpha} , f] &= [g , \frac{f}{\lambda - \alpha}] \\ &= [g , R_\alpha f] \\ &= 0 .\end{aligned}$$

This serves to prove that $\mathcal{N}$ is invariant under multiplication by $1/(\lambda - \alpha^*)$ and so too under multiplication by

$$b_\alpha(\lambda) = \frac{\alpha^* - \alpha}{\lambda - \alpha^*} + 1 .$$

STEP 3 *is to verify that* $\mathcal{L}_\alpha$, *the* J *orthogonal complement of* $b_\alpha \mathcal{N}$ *in* $\mathcal{N}$, *is equal to*

$$\mathcal{L}_\alpha = \mathcal{N} \cap (b_\alpha \mathcal{M} + \{\frac{\xi}{\rho_\alpha} : \xi \in \mathbb{C}^m\}) .$$

PROOF. Recall first that

$$H_J^2 = \mathcal{M}[\dot{+}]\mathcal{N} ,$$

where the symbol $[\dot{+}]$ designates that the sum is both direct and orthogonal with respect to the indefinite inner product (2.2). But this in turn clearly implies that

$$b_\alpha H_J^2 = b_\alpha \mathcal{M}[\dot{+}]b_\alpha \mathcal{N}$$

and hence that

$$H_J^2 = \{b_\alpha \mathcal{M}[\dot{+}]b_\alpha \mathcal{N}\}[\dot{+}]\{\frac{\xi}{\rho_\alpha} \ : \ \xi \in \mathbb{C}^m\} \ .$$

Thus, if $f \in \mathcal{N}$, then it can certainly be expressed in the form

$$f = b_\alpha n + b_\alpha m + \frac{\xi}{\rho_\alpha}$$

with $n \in \mathcal{N}$, $m \in \mathcal{M}$ and $\xi \in \mathbb{C}^m$. Therefore, if f is also J orthogonal to $b_\alpha \mathcal{N}$, then

$$\begin{aligned} 0 &= [f \ , \ b_\alpha g] \\ &= [b_\alpha n + b_\alpha m + \frac{\xi}{\rho_\alpha} \ , \ b_\alpha g] \\ &= [n \ , \ g] \end{aligned}$$

for every $g \in \mathcal{N}$ and hence, since $\mathcal{N}$ is a Krein space, $n = 0$ and so

$$\mathcal{L}_\alpha \subset \mathcal{N} \cap \{b_\alpha \mathcal{M} + \frac{\xi}{\rho_\alpha} \ : \ \xi \in \mathbb{C}^m\} \ .$$

On the other hand, if f belongs to the set on the right, then

$$f = b_\alpha m + \frac{\xi}{\rho_\alpha}$$

for some choice of $m \in \mathcal{M}$ and $\xi \in \mathbb{C}^m$ and so

$$\begin{aligned} [b_\alpha g, f] &= [g, m] + \xi^* J b_\alpha(\alpha) g(\alpha) \\ &= 0 \end{aligned}$$

for every $g \in \mathcal{N}$. Thus the opposite inclusion holds also and the proof of this step is complete.

STEP 4. $\mathcal{N} = \mathcal{L}_\alpha [\dot{+}] b_\alpha \mathcal{N}$.

PROOF OF STEP 4. By definition $\mathcal{L}_\alpha$ is clearly J orthogonal to $b_\alpha\mathcal{N}$. Moreover, the sum on the right hand side of the asserted equality is clearly direct because if f belongs to both of the indicated spaces, then there exists a choice of $m \in \mathcal{M}$, $n \in \mathcal{N}$ and $\xi \in \mathbb{C}^m$ such that

$$b_\alpha m + b_\alpha n + \frac{\xi}{\rho_\alpha} = 0 \;.$$

Therefore, for every $\eta \in \mathbb{C}^m$,

$$\begin{aligned} 0 &= [b_\alpha m + b_\alpha n + \frac{\xi}{\rho_\alpha}\,,\,\frac{\eta}{\rho_\alpha}] \\ &= \frac{\eta^* J\xi}{\rho_\alpha(\alpha)} \end{aligned}$$

which implies that $\xi = 0$. Thus

$$m + n = 0$$

which implies further that $m \in \mathcal{M} \cap \mathcal{N} = 0$ and so too that the sum is direct.

Finally, since

$$\mathcal{N} \supset \mathcal{L}_\alpha[\dot{+}]b_\alpha\mathcal{N}\;,$$

thanks to Step 2 and the definition of $\mathcal{L}_\alpha$, it remains only to show that in fact equality prevails in the last inclusion. But, if $f \in \mathcal{N}$, then, as noted in the proof of Step 3, it can be expressed in the form

$$f = b_\alpha m + \frac{\xi}{\rho_\alpha} + b_\alpha n$$

for some choice of $m \in \mathcal{M}$, $\xi \in \mathbb{C}^m$ and $n \in \mathcal{N}$. Thus, by another application of Step 2,

$$f - b_\alpha n = b_\alpha m + \frac{\xi}{\rho_\alpha}$$

belongs to $\mathcal{N}$ and so too to $\mathcal{L}_\alpha$, by Step 3. The rest is plain.

STEP 5. *$\mathcal{L}_\alpha$ is a Krein space.*

PROOF OF STEP 5. This is immediate from Theorem 3.4 on page 104 of Bognar (1974) since, by Step 4, $\mathcal{L}_\alpha$ is a closed subspace of the Krein space $\mathcal{N}$ which is orthocomplemented.

STEP 6. *If f and g belong to $\mathcal{L}_\alpha$, then*

$$g(\lambda)^* J f(\lambda) = \frac{c}{|\rho_\alpha(\lambda)|^2}$$

for a.e. $\lambda \in \partial$, where c is a constant which is independent of λ.

PROOF OF STEP 6. Recall first that the scalar valued functions

$$\varphi_j(\lambda) = \frac{\{\rho_\alpha(\alpha)\}^{\frac{1}{2}}\{b_\alpha(\lambda)\}^j}{\rho_\alpha(\lambda)} ,$$

$j = 0, \pm 1, \ldots$ are orthonormal with respect to the standard inner product and that the set of φ_j with $j \geq 0$ is an orthonormal basis for H^2. Thus any pair of functions f and g belonging to $\mathcal{L} = \mathcal{L}_\alpha$ can be expressed as vector valued Fourier series

$$f = \sum_{j=0}^{\infty} f_j \varphi_j \text{ and } g = \sum_{k=0}^{\infty} g_k \varphi_k$$

with coefficients f_j and g_k in $\mathbb{C}^m$. Moreover, upon taking advantage of the identity

$$\varphi_j(\lambda)\varphi_k(\lambda)^* = \varphi_{j-k}(\lambda)\varphi_0(\lambda)^*$$

which is valid for $\lambda \in \partial$, it follows readily that

$$g(\lambda)^* J f(\lambda) = \sum_{n=-\infty}^{\infty} \left\{ \sum_{k=0}^{\infty} g_k^* J f_{k+n} \right\} \varphi_n(\lambda)\varphi_0(\lambda)^*$$

for a.e. $\lambda \in \partial$. The desired result now drops out by identifying the sum in curly brackets with

$$[f \, , \, (b_\alpha)^n g] = \left[\sum_{j=0}^{\infty} f_j \varphi_j \, , \, \sum_{k=0}^{\infty} g_k \varphi_{k+n} \right]$$

and then invoking Step 4 to conclude that

$$[f \, , \, (b_\alpha)^n g] = 0$$

for $n = \pm 1, \pm 2, \ldots$.

STEP 7. *$\mathcal{L}_\alpha = (\mathcal{L}_\alpha)_+ [\dot{+}] (\mathcal{L}_\alpha)_-$ where $(\mathcal{L}_\alpha)_+$ is a Hilbert space of dimension $p' \leq p$ with respect to the inner product induced by J and $(\mathcal{L}_\alpha)_-$ is a Hilbert space of dimension $q' \leq q$ with respect to the inner product induced by $-J$.*

PROOF OF STEP 7. Let $u_1, \ldots, u_{p'}$ be a J orthonormal basis for $(\mathcal{L}_\alpha)_+$, let $\lambda_0 \in \partial$ be a point at which the u_j have a well defined radial limits and let

$$x_j = u_j(\lambda_0)\rho_\alpha(\lambda_0)\rho_\alpha(\alpha)^{-\frac{1}{2}} .$$

Then

$$x_j^* J x_k \frac{\rho_\alpha(\alpha)}{|\rho_\alpha(\lambda_0)|^2} = u_j(\lambda_0)^* J u_k(\lambda_0) .$$

However, by Step 6, the last identity prevails for a.e. $\lambda \in \partial$ and not just λ_0, and so, upon integrating both sides over ∂, it follows that

$$x_j^* J x_k = \begin{cases} 0 & j \neq k \\ 1 & j = k . \end{cases}$$

Thus the x_j , $j = 1, 2, \ldots, p'$, constitute a J orthonormal basis for a J positive subspace of $\mathbb{C}^m$. Therefore, $p' \leq p$. This completes the proof of the first assertion. The second may be established in much the same way.

STEP 8. *Let h_j , $j = 1, \ldots, k$ denote a J orthogonal basis for $\mathcal{L}_\alpha$ the first p' elements of which are an orthonormal basis for $(\mathcal{L}_\alpha)_+$ and the remaining $q' = k - p'$ elements of which are a basis for $(\mathcal{L}_\alpha)_-$ which is $-J$ orthonormal. Let $\Theta(\lambda)$ denote the $m \times k$ matrix whose ij entry is equal to the i-th component of $\{\rho_\alpha(\alpha)\}^{-\frac{1}{2}}\rho_\alpha(\lambda)h_j(\lambda)$. Then*

$$\frac{J}{\rho_\omega(\lambda)} - M_\omega(\lambda) = N_\omega(\lambda) = \frac{\Theta(\lambda)J'\Theta(\omega)^*}{\rho_\omega(\lambda)} .$$

PROOF OF STEP 8. It follows readily from Step 7 that $\mathcal{L}_\alpha$ is a $k = p' + q'$ dimensional reproducing kernel Krein space of $m \times 1$ vector valued analytic functions with respect to the indefinite inner product (2.2) with reproducing kernel

$$L_\omega(\lambda) = \sum_{j=1}^{p'} h_j(\lambda)h_j(\omega)^* - \sum_{j=p'+1}^{k} h_j(\lambda)h_j(\omega)^* \ .$$

Moreover, it is easily deduced from Step 4, that

$$N_\omega(\lambda) = L_\omega(\lambda) + b_\alpha(\lambda)N_\omega(\lambda)b_\alpha(\omega)^*$$

and hence that

$$N_\omega(\lambda) = \frac{L_\omega(\lambda)}{1 - b_\alpha(\lambda)b_\alpha(\omega)^*} \ .$$

The rest follows from the discussion in the proof of Step 7 upon invoking the identity

$$1 - b_\alpha(\lambda)b_\alpha(\omega)^* = \left\{\frac{\rho_\alpha(\alpha)}{\rho_\alpha(\lambda)\rho_\alpha(\omega)^*}\right\}\rho_\omega(\lambda)$$

and then defining Θ as in the statement of this step.

STEP 9. *The identity (2.4) prevails for a.e. point* $\lambda \in \partial$.

PROOF OF STEP 9. It follows readily from Step 6 and the definition of Θ that

$$\begin{aligned}\eta^*\Theta(\lambda)^*J\Theta(\lambda)\xi\frac{\rho_\alpha(\alpha)}{|\rho_\alpha(\lambda)|^2} &= \left\{\sum_{t=1}^{k}\eta_t h_t(\lambda)\right\}^* J \left\{\sum_{s=1}^{k}\xi_s h_s(\lambda)\right\} \\ &= \left\{\sum_{s,t=1}^{k}\eta_t^* c_{ts}\xi_s\right\}\frac{\rho_\alpha(\alpha)}{|\rho_\alpha(\lambda)|^2}\end{aligned}$$

for a.e. point $\lambda \in \partial$ where c_{ts} is a constant and

$$c_{ts}\frac{\rho_\alpha(\alpha)}{|\rho_\alpha(\lambda)|^2} = h_t(\lambda)^*Jh_s(\lambda) \ .$$

But now, upon integrating the last identity and taking advantage of the properties of h_j, as described in the statement of Step 8, it follows further that

$$\sum_{s,t=1}^{k} \eta_t^* c_{ts}\xi_s = \eta^*J'\xi$$

and hence that (2.4) holds, as asserted.

STEP 10. Θ *is unique up to a right constant J' unitary factor.*

PROOF OF STEP 10. If Ψ is a second such matrix valued function, then, by the uniqueness of the reproducing kernel,

$$\Theta(\lambda)J'\Theta(\omega)^* = \Psi(\lambda)J'\Psi(\omega)^*$$

for every choice of λ and ω in Δ_+. Let β be a point on the boundary of which both Θ and Ψ have nontangential limits. Then,

$$\Theta(\beta)J'\Theta(\beta)^* = \Psi(\beta)J'\Psi(\beta)^* ,$$

whereas, by (2.4),

$$\Theta(\beta)^*J\Theta(\beta) = J' = \Psi(\beta)^*J\Psi(\beta) .$$

Thus,

$$\Theta(\lambda) = \Theta(\lambda)J'\Theta(\beta)^*J\Theta(\beta) = \Psi(\lambda)J'\Psi(\beta)^*J\Theta(\beta) .$$

But this does the trick since

$$\begin{aligned}\{\Psi(\beta)^*J\Theta(\beta)\}^*J'\{\Psi(\beta)^*J\Theta(\beta)\} &= \Theta(\beta)^*J\Psi(\beta)J'\Psi(\beta)^*J\Theta(\beta)\\ &= \Theta(\beta)^*J\Theta(\beta)J'\Theta(\beta)^*J\Theta(\beta)\\ &= J' .\end{aligned}$$

STEP 11. *If $\mathcal{M}$ is a J positive space, then $q' = q$.*

PROOF OF STEP 11. If $\mathcal{M}$ is J positive then

$$\frac{J - \Theta(\omega)J'\Theta(\omega)^*}{\rho_\omega(\omega)} \geq 0 \tag{2.5}$$

for every point $\omega \in \Delta_+$ and hence, upon expressing Θ in the block form

$$\Theta = \begin{bmatrix} A & B \\ C & D \end{bmatrix} , \tag{2.6}$$

where A is $p \times p'$ and D is $q \times q'$, it follows readily that $DD^* \geq I_q + CC^*$. But this in turn implies that the rank of D is equal to q and hence that $q' = q$, as claimed.

STEP 12. *If $\mathcal{N}$ is full range, then $k = m$ and $J = J'$.*

PROOF OF STEP 12. If $\mathcal{N}$ is full range, then there exists a point $\omega \in \Delta_+$ such that

$$\text{span}\{f(\omega) : f \in \mathcal{N}\} = \mathbb{C}^m .$$

Now, if also $k < m$, then there exists a nonzero vector $\xi \in \mathbb{C}^m$ such that

$$\Theta(\omega)^* \xi = 0$$

and hence such that

$$N_\omega(\lambda)\xi = 0$$

for all $\lambda \in \Delta_+$. Thus

$$\xi^* f(\omega) = [f, N_\omega \xi] = 0$$

for every $f \in \mathcal{N}$ which contradicts the assumption that $\mathcal{N}$ is full range. Therefore $k = m$ and $J = J'$, as asserted. ∎

STEP 13. *If $\mathcal{M}$ is finite dimensional, then $\mathcal{N}$ is full range.*

PROOF OF STEP 13. If this were not the case, then for each point $\omega \in \Delta_+$ there would exist a vector $\xi_\omega \in \mathbb{C}^m$ such that

$$\left[f, \frac{J\xi_\omega}{\rho_\omega}\right] = \xi_\omega^* f(\omega) = 0$$

for every $f \in \mathcal{N}$. But this in turn implies that for every point $\omega \in \Delta_+$ there is an element $(\rho_\omega)^{-1} J\xi_\omega$ belonging to $\mathcal{M}$ which plainly contradicts the fact that $\mathcal{M}$ was presumed to be finite dimensional. This completes the proof of both Step 13 and the theorem. ∎

We remark that if the inequality (2.5) extends to Δ_- with Θ as in (2.6), then the upper left hand block of (2.5) implies that

$$AA^* - BB^* \geq I_p$$

on Δ_- and hence that $p = p'$. This, together with Step 11, is the reason that the two sided theory of reproducing kernel Hilbert spaces with kernels $M_\omega(\lambda)$ always involves square Θ with $J = J'$.

The rest of this section is included for the sake of perspective; it does not play a role in the following sections. In particular, the next corollary yields one of the generalizations to indefinite inner product spaces of the Beurling–Lax theorem which is due to Ball and Helton (1983).

COROLLARY 1. *If $\mathcal{M}$ meets the hypotheses of Theorem 2.1, then*

$$\mathcal{N} = closure\left\{\Theta\frac{\eta}{\rho_\omega} : \eta \in \mathbb{C}^k \text{ and } \omega \in \Omega_+\right\} \quad in \quad H_m^2 ,$$

in which Ω_+ designates the set of points in Δ_+ at which Θ has rank k.

PROOF. Clearly

$$\frac{\Theta(\lambda)J'\Theta(\omega)^*\xi}{\rho_\omega(\lambda)} = N_\omega(\lambda)\xi$$

belongs to $\mathcal{N}$ for every point $\omega \in \Delta_+$ and every $\xi \in \mathbb{C}^m$. Therefore,

$$\frac{\Theta(\lambda)\eta}{\rho_\omega(\lambda)} \in \mathcal{N}$$

for every point $\omega \in \Omega_+$ and every $\eta \in \mathbb{C}^k$. Thus $\mathcal{N}$ contains the indicated closure. On the other hand, if $f \in \mathcal{N}$ is orthogonal to $(\rho_\omega)^{-1}\Theta\eta$ for each of the permissible choices of ω and η, then

$$\begin{aligned} 0 &=< f, (\rho_\omega)^{-1}\Theta\eta > \\ &=< f, N_\omega\xi > \\ &= \xi^* J f(\omega) \end{aligned}$$

for every $\xi \in \mathbb{C}^m$ and every $\omega \in \Omega_+$. Therefore $f = 0$ and the asserted equality prevails.

∎

This corollary serves to identify $\mathcal{N}$ as $\mathcal{N} = \Theta H_k^2$ and hence $\mathcal{M} = H_m^2 \ominus J\Theta H_k^2$, providing that the right hand sides are interpreted as appropriate closures in the style of the corollary. Conversely, if $\mathcal{N}$ and $\mathcal{M}$ are known to be of this form and if Θ is known to satisfy (2.4), then it is readily checked by direct calculation that $\mathcal{M}$ and $\mathcal{N}$ are reproducing kernel Krein spaces with reproducing kernels of the form (2.3).

Before proceeding, it is perhaps well to remark that Ball and Helton have established a number of other generalizations of the Beurling–Lax theorem which are not directly deducible from Theorem 2.1. In particular they have treated the case where $\mathcal{M} \cap \mathcal{N} \neq 0$ and $\mathcal{M} + \mathcal{N}$ is closed, but not necessarily equal to H_m^2. On the other hand Theorem 2.1 admits a number of generalizations to reproducing kernel subspaces with respect to indefinite inner product spaces with respect to more general measures than Lebesgue and hence do not fall into the province of generalized Beurling–Lax theorems.

The next theorem, which is basically a corollary of Theorem 2.1 for the special case in which the subspace $\mathcal{M}$ is finite dimensional, has many useful applications; see e.g. Dym (1989b) for a sample and additional references.

THEOREM 2.2. *Let $\mathcal{M} = span\{f_1, \ldots, f_n\}$ be an n dimensional subspace of H_m^2 which is R_α invariant for some point $\alpha \in \Delta_+$, let $\mathbb{P}$ denote the Pick matrix with ij entry*

$$\mathbb{P}_{ij} = [f_j, f_i] \ , \ i, j = 1, \ldots, n \ . \tag{2.7}$$

Then (with respect to the inner product (2.2))

(1) $\mathcal{M}$ is a reproducing kernel Krein space if and only if $\mathbb{P}$ is nonsingular.

(2) $\mathcal{M}$ is a reproducing kernel Hilbert space if and only if $\mathbb{P}$ is positive definite.

Moreover, in both instances, the reproducing kernel $M_\omega(\lambda)$ can be expressed in the form

$$M_\omega(\lambda) = \frac{J - U(\lambda) J U(\omega)^*}{\rho_\omega(\lambda)} \tag{2.8}$$

with

$$U(\lambda) = I_m - \rho_\beta(\lambda) \sum_{i,j=1}^{n} f_i(\lambda)(\mathbb{P}^{-1})_{ij} f_j(\beta)^* J \ , \tag{2.9}$$

where the $f_i(\lambda)$ are rational $m \times 1$ vector valued functions which are analytic in $\overline{\Delta}_+$, $\beta \in \partial$ and Θ is J unitary on ∂.

PROOF. Let $\mathcal{N} = H_m^2 \ominus J\mathcal{M}$. Then it is readily checked that a vector $c \in \mathbb{C}^n$ with components $c_1, \ldots, c_n$, belongs to the kernel of $\mathbb{P}$ if and only if $\Sigma c_j f_j \in \mathcal{M} \cap \mathcal{N}$. Thus

$$\mathcal{M} \cap \mathcal{N} = 0 \text{ if and only if } \mathbb{P} \text{ is nonsingular} \ .$$

But now as $\mathcal{M} \cap \mathcal{N} = 0$ implies that $\mathcal{M} + \mathcal{N}$ is dense in H_m^2 and also $\mathcal{M} + \mathcal{N}$ is closed in H_m^2 since $\mathcal{M}$ is finite dimensional, it follows readily from Theorem 2.1 that $\mathcal{M}$ is a reproducing kernel Krein space if $\mathbb{P}$ is nonsingular. Conversely, if $\mathcal{M}$ is a reproducing kernel Krein space and $f \in \mathcal{M} \cap \mathcal{N}$, then

$$\xi^* f(\omega) = [f, M_\omega \xi] = 0$$

for every point $\omega \in \Delta_+$ and every vector $\xi \in \mathbb{C}^m$. Therefore $\mathcal{M} \cap \mathcal{N} = 0$, if $\mathcal{M}$ is a reproducing kernel Krein space. This completes the proof of the first assertion. The second is immediate from the first and the observation that

$$\left[\sum_{j=1}^{n} c_j f_j, \sum_{k=1}^{n} c_k f_k\right] = \sum_{j,k=1}^{n} c_k^* \mathbb{P}_{kj} c_j \ .$$

Next, by Step 13 of Theorem 2.1, $\mathcal{N}$ is full range.

Finally, it follows readily from the fact that the reproducing kernel for $\mathcal{M}$ can be expressed in the form

$$M_\omega(\lambda) = \sum_{j=1}^{n} \varphi_j(\lambda)\varphi_j(\omega)^*$$

where the φ_j , $j = 1, \ldots, n$, is a J orthonormal basis for $\mathcal{M}$, that

$$M_\omega(\lambda) = F(\lambda)\mathbb{P}^{-1}F(\omega)^* \ ,$$

where

$$F(\lambda) = [f_1(\lambda) \dots f_n(\lambda)] \ .$$

But this in turn implies that

$$U(\lambda)JU(\omega)^* = J - \rho_\omega(\lambda)F(\lambda)\mathbb{P}^{-1}F(\omega)^* \ .$$

The final formula emerges upon presuming $U(\beta) = I_m$ for some point $\beta \in \partial$ and setting $\omega = \beta$. ∎

Theorem 2.2 can be generalized to spaces which do not sit inside H_m^2. The essential features of the H_J^2 inner product are then captured by imposing certain identities on the inner product of the space; see e.g. Theorem 6.12 of Alpay and Dym (1986), the statement and proof of which are adapted in large part from a more general theorem which is due to de Branges (1963). The identities alluded to just above can also be reformulated as additional constraints on the Gram matrix $\mathbb{P}$, as is discussed extensively in Dym (1989a).

3. CONDITIONS FOR THE EXISTENCE OF A SOLUTION

The objective of this section is to characterize these h for which the class $P_\tau(h)$ is nonempty. The second condition in the definition of $P_\tau(h)$ implies that

$$\hat{f}(\lambda) = \hat{f}(\lambda)^* \ , \ \text{for } \lambda \in \mathbb{R} \ ,$$

and hence that

$$f(s) = f(-s)^* \ , \ \text{for a.e. } s \in \mathbb{R} \ ,$$

for every choice of $\hat{f} \in \mathcal{P}_\tau(h)$. Thus, in order for $\mathcal{P}_\tau(h)$ to be nonempty, it is necessary that

$$h(s) = h(-s)^*$$

for a.e. point s in the interval $-\tau \leq s \leq \tau$.

The associated class $\mathcal{Q}_\tau(h)$ of $p \times p$ matrix valued functions

$$\Phi(\lambda) = I_p - 2\int_0^\infty e^{i\lambda s} f(s)ds \tag{3.1}$$

in which f is summable on the half line and

(1) $f(t) = h(t)$ a.e. for $0 \le t \le \tau$, and

(2) $\Phi(\lambda) + \Phi(\lambda)^* > 0$ for every point $\lambda \in \overline{\mathbb{C}}_+$

will prove useful.

THEOREM 3.1. *If $f \in L^1_{p\times p}(\mathbb{R})$, then $\hat{f} \in \mathcal{P}_\tau(h)$ if and only if the matrix valued function Φ which is defined by (3.1) belongs to $\mathcal{Q}_\tau(h)$.*

PROOF. If $\hat{f} \in \mathcal{P}_\tau(h)$, then it follows readily with the help of the identities

$$\frac{\Phi(a+ib) + \Phi(a-ib)^*}{2} = \begin{cases} (b/\pi)\int_{-\infty}^{\infty}\{I_p - \hat{f}(\lambda)\}\{(\lambda-a)^2+b^2\}^{-1}d\lambda & \text{if} \quad b>0 \\ I_p - \hat{f}(a) & \text{if} \quad b=0 \end{cases}$$

that $\Phi \in \mathcal{Q}_\tau(h)$.

The converse is plain. ∎

Let H_τ denote the operator on $L^2_p[0,\tau]$ which is defined in terms of the given data by the rule

$$(H_\tau g)(t) = \int_0^\tau h(t-s)g(s)ds$$

and let $\gamma_\tau(t,s)$, $0 \le t, s \le \tau$, denote the kernel of the integral operator

$$\Gamma_\tau = (I - H_\tau)^{-1} - I$$

on $L^2_p[0,\tau]$, when the indicated inverse exists. For such τ we define the $p \times p$ matrix valued functions

$$E_\tau(\lambda) = I_p + \int_0^\tau e^{i\lambda s}\gamma_\tau(s,0)ds \tag{3.2}$$

and

$$F_\tau(\lambda) = e^{i\lambda\tau}\left\{I_p + \int_0^\tau e^{i\lambda(s-\tau)}\gamma_\tau(s,\tau)ds\right\} . \tag{3.3}$$

They are continuous analogues of the matrix polynomials E_n and F_n which play such an important role in the theory of matrix orthogonal polynomials; see e.g. Dym (1988). Functions of this type were first investigated by M.G. Krein (1955) and (1956). If $I - H_{\tau'}$ is positive definite, then $I - H_\tau$ is positive definite, and hence invertible, for every choice of τ in the interval $0 \le \tau \le \tau'$. Therefore E_τ and F_τ are defined for all such τ. Moreover, if $h(t)$ is continuous for $-\tau' \le t \le \tau'$, then $\gamma_\tau(t,s)$ is continuous in t,s, continuously differentiable in τ for $0 \le t,s \le \tau \le \tau'$ and satisfies the well known identity,

$$\frac{\partial\gamma_\tau(t,s)}{\partial\tau} = \gamma_\tau(t,\tau)\gamma_\tau(\tau,s) . \tag{3.4}$$

The auxiliary identity

$$\frac{\partial\gamma_\tau(\tau-s,\tau-t)}{\partial\tau} = \gamma_\tau(\tau-s,0)\gamma_\tau(0,\tau-t) , \tag{3.5}$$

which may be deduced from (3.4) by identifying $\gamma_\tau(\tau-s,\tau-t)^*$ as the resolvent kernel of an operator of the form H_τ but with $h(t-s)$ replaced by $h(t-s)^*$, will also prove useful.

LEMMA 3.1. *If $I - H_{\tau'}$ is positive definite, then the $p \times p$ matrix valued function $\Lambda_\omega^\tau(\lambda)$ which is defined by the formula*

$$\Lambda_\omega^\tau(\lambda) = \frac{1}{2\pi}\int_0^\tau e^{i\lambda t}\left\{I_p e^{-i\omega^* t} + \int_0^\tau \gamma_\tau(t,s)e^{-i\omega^* s}ds\right\}dt , \tag{3.6}$$

for $0 \le \tau \le \tau'$, can also be expressed in the forms

$$\Lambda_\omega^\tau(\lambda) = \frac{E_\tau(\lambda)E_\tau(\omega)^* - F_\tau(\lambda)F_\tau(\omega)^*}{\rho_\omega(\lambda)} \tag{3.7}$$

and

$$\Lambda_\omega^\tau(\lambda) = \frac{1}{2\pi}\int_0^\tau F_s(\lambda)F_s(\omega)^* ds , \tag{3.8}$$

for every pair of points λ and ω in $\mathbb{C}$.

PROOF. If $h(t)$ is continuous on the interval $|t| \leq \tau'$, then it is readily checked with the aid of (3.4) and (3.5) that

$$\frac{\partial E_\tau(\lambda)}{\partial \tau} = F_\tau(\lambda)\gamma_\tau(\tau, 0) \tag{3.9}$$

and

$$\frac{\partial F_\tau(\lambda)}{\partial \tau} = i\lambda F_\tau(\lambda) + E_\tau(\lambda)\gamma_\tau(0, \tau) \tag{3.10}$$

for $0 \leq \tau \leq \tau'$ and hence that the derivatives of the right hand sides of (3.6), (3.7) and (3.8) with respect to τ are all equal to $F_\tau(\lambda)F_\tau(\lambda)^*/2\pi$. Thus, since the right hand sides themselves vanish for $\tau = 0$, the asserted identities hold. If h is only summable, then the result follows by approximating it in the L^1 sense by a sequence of continuous matrix functions h_n for which the result holds and then passing to the limit (see the discussion in Dym and Gohberg (1980) for help with this if need be). ∎

LEMMA 3.2. *If $I - H_\tau$ is positive definite, then*

$$E_\tau(\lambda)E_\tau^\#(\lambda) = F_\tau(\lambda)F_\tau^\#(\lambda) \tag{3.11}$$

for every point $\lambda \in \mathbb{C}$. Moreover, $E_\tau(\lambda)$ is an invertible matrix for every point $\lambda \in \overline{\mathbb{C}}_+$ and $F_\tau(\lambda)$ is an invertible matrix for every point $\lambda \in \overline{\mathbb{C}}_-$.

PROOF. The identity (3.11) is immediate from (3.7): just cross multiply by $\rho_\omega(\lambda)$ and set $\omega = \lambda^*$. Next, since $I+\Gamma_\tau$ is a positive definite operator on $L_p^2[0,\tau]$, it follows easily from (3.6) that $\Lambda_\omega^\tau(\omega)$ is a positive definite matrix for every point $\omega \in \mathbb{C}$ and therefore, by (3.7), $E_\tau(\omega)$ [resp. $F_\tau(\omega)$] is an invertible matrix for every point $\omega \in \mathbb{C}_+$ [resp. $\omega \in \mathbb{C}_-$]. Finally, if $\omega \in \mathbb{R}$, then

$$\Lambda_\omega^\tau(\omega) = -\frac{1}{2\pi i}\left\{\frac{\partial E_\tau(\omega)}{\partial \lambda}E_\tau(\omega)^* - \frac{\partial F_\tau(\omega)}{\partial \lambda}F_\tau(\omega)^*\right\}.$$

Therefore, if $E_\tau(\omega)^*\xi = 0$ for some point $\omega \in \mathbb{R}$ and vector $\xi \in \mathbb{C}^p$, then, by (3.11), $F_\tau(\omega)^*\xi = 0$ and hence $\Lambda_\omega^\tau(\omega)\xi = 0$ also, by the last identity. Thus $\xi = 0$, since $\Lambda_\omega^\tau(\omega)$ is a positive definite matrix for every point $\omega \in \mathbb{C}$, and hence $E_\tau(\omega)$ is invertible for every point $\omega \in \mathbb{R}$, as is $F_\tau(\omega)$ by the same argument. ∎

LEMMA 3.3. *If $I - H_\tau$ is positive definite, then $(E_\tau)^{\pm 1}$ belongs to the Wiener subalgebra $\mathcal{W}_+^I$, whereas $(e^{-i\lambda\tau}F_\tau)^{\pm 1}$ belongs to the Wiener subalgebra $\mathcal{W}_-^I$.*

PROOF. To begin with E_τ clearly belongs to $\mathcal{W}_+^I$ since $\gamma_\tau(\cdot, 0) \in L^1_{p\times p}[0,\tau]$ as is shown e.g. in Dym and Gohberg (1980). Moreover, Lemma 3.2 guarantees that $E_\tau(\lambda)$ is an invertible $p \times p$ matrix for every point $\lambda \in \overline{\mathbb{C}}_+$, whereas the Riemann-Lebesgue lemma guarantees that $E_\tau(\lambda) \to I_p$ as $\lambda \to \pm\infty$ in $\mathbb{R}$. Therefore, by the matrix version of a well known theorem of Wiener (see e.g. Dym and Gohberg (1980)), $E_\tau^{-1} \in \mathcal{W}_+^I$ also.

The remaining assertions are proved in much the same way. ∎

LEMMA 3.4. *If $I - H_\tau$ is positive definite, then the function $\hat{f}_\tau$ which is defined by the formula*

$$I_p - \hat{f}_\tau(\lambda) = \{E_\tau(\lambda)E_\tau^\#(\lambda)\}^{-1} , \tag{3.12}$$

for $\lambda \in \mathbb{R}$, belongs to $\mathcal{P}_\tau(h)$.

PROOF. It is clear from the definition (3.12) that $I_p - \hat{f}_\tau$ belongs to $\mathcal{W}$ and is a positive definite matrix at every point $\lambda \in \mathbb{R}$. It remains to show that $f_\tau(s) = h(s)$ for $-\tau \le s \le \tau$. It seems to be simplest to do this indirectly by first furnishing what turns out to be an alternate definition of $\hat{f}_\tau$. To this end, it is convenient to introduce the projectors $P_1 = P_{\tau\infty}$, $P_2 = P_{0\tau}$, $P_3 = P_{-\tau 0}$ and $P_4 = P_{-\infty\tau}$ in the Wiener algebra $\mathcal{W}^0$, where

$$P_{ab}\hat{g} = \int_a^b e^{i\lambda s} g(s) ds$$

for $\hat{g} \in \mathcal{W}^0$, and further, to define

$$a_\tau(s) = \begin{cases} \gamma_\tau(s,0) & \text{for } 0 \le s \le \tau \\ 0 & \text{elsewhere} \end{cases}$$

and

$$b_\tau(s) = \begin{cases} \gamma_\tau(0,-s) & \text{for } -\tau \le s \le 0 \\ 0 & \text{elsewhere} \end{cases}$$

so that

$$E_\tau(\lambda) = I_p + \hat{a}_\tau(\lambda) \quad \text{and} \quad E_\tau^{\#}(\lambda) = I_p + \hat{b}_\tau(\lambda) \; .$$

Next, we define the function $\hat{g} = \sum_{j=1}^{4} \hat{g}_j$ with $\hat{g}_j$ in $P_j\mathcal{W}^0$ by the formulas

$$\begin{aligned} \hat{g}_1 &= -\{P_1(\hat{h}\hat{a}_\tau)\}E_\tau^{-1} \\ \hat{g}_2 &= \hat{h}_2 \\ \hat{g}_3 &= \hat{h}_3 \quad \text{and} \\ \hat{g}_4 &= -(E_\tau^{\#})^{-1}P_4(\hat{b}_\tau\hat{h}) \; . \end{aligned}$$

Then clearly,

$$\begin{aligned} P_1\{(I_p - \hat{g})(I_p + \hat{a}_\tau) - I_p\} &= -\hat{g}_1 - P_1\{(\hat{g}_1 + \hat{h} + \hat{g}_4)\hat{a}_\tau\} \\ &= -\hat{g}_1 - \hat{g}_1\hat{a}_\tau - P_1(\hat{h}\hat{a}_\tau) \\ &= 0 \; , \end{aligned}$$

whereas,

$$\begin{aligned} P_2\{(I_p - \hat{g})(I_p + \hat{a}_\tau) - I_p\} &= -\hat{g}_2 + \hat{a}_\tau - P_2\{(\hat{g}_1 + \hat{h} + \hat{g}_4)\hat{a}_\tau\} \\ &= -\hat{h}_2 + \hat{a}_\tau - P_2(\hat{h}\hat{a}_\tau) \\ &= 0 \; , \end{aligned}$$

thanks to the resolvent identity

$$-h(t) + \gamma_\tau(t,0) - \int_0^\tau h(t-s)\gamma_\tau(s,0)ds = 0 \; ,$$

which is valid for a.e. point t in the interval $0 \le t \le \tau$. Therefore

$$(I_p - \hat{g})(I_p + \hat{a}_\tau) = I_p + \chi_-$$

for some choice of $\chi_- \in \mathcal{W}_-^0$. Similar considerations lead to the evaluations

$$P_j\{(I_p + \hat{b}_\tau)(I_p - \hat{g}) - I_p\} = 0$$

for $j = 3$ and $j = 4$ and hence to the conclusion that

$$(I_p + \hat{b}_\tau)(I_p - \hat{g}) = I_p + \chi_+$$

for some choice of $\chi_+ \in \mathcal{W}_+^0$. Thus

$$\begin{aligned}(I_p + \hat{b}_\tau)(I_p - \hat{g})(I_p + \hat{a}_\tau) &= (I_p + \chi_+)(I_p + \hat{a}_\tau) \\ &= (I_p + \hat{b}_\tau)(I_p + \chi_-) ,\end{aligned}$$

which in turn implies that the right hand sides are both equal to I_p. This proves that $\hat{g} = \hat{f}_\tau$ and hence that $f_\tau(s) = h(s)$ for a.e. point s in the interval $-\tau \leq s \leq \tau$, as needed.

∎

For additional insight into the essentially algebraic nature of the preceding proof, the discussion in Section 8 of Dym and Gohberg (1982/3) is suggested. The proofs of both Lemma 3.4 and the next theorem are adapted from Dym and Gohberg (1980).

THEOREM 3.2. *The class $\mathcal{P}_\tau(h)$ is nonempty if and only if the operator $I - H_\tau$ is positive definite.*

PROOF. If $\hat{f} \in \mathcal{P}_\tau$, then

$$\begin{aligned}< (I - H_\tau)g, g > &= \int_0^\tau g(t)^* \{g(t) - \int_0^\tau h(t-s)g(s)ds\} dt \\ &= \frac{1}{2\pi} \int_{-\infty}^{\infty} \hat{g}(\lambda)^* \{I_p - \hat{f}(\lambda)\} \hat{g}(\lambda) d\lambda \\ &\geq 0\end{aligned}$$

for every choice of $g \in L_p^2[0, \tau]$, with equality if and only if $\hat{g}$ and hence g is equivalent to zero. (Actually, since $\hat{g}$ is entire, it will be equal to zero identically in this instance.)

Conversely, if $I - H_\tau$ is positive definite, then $\mathcal{P}_\tau$ is nonempty by Lemma 3.4. ∎

Theorem 3.2 indicates that $I - H_\tau$ is the proper analogue of the Pick matrix for the continuous extension problem. This is reinforced by Theorem 3.1 and the observation that $\mathcal{Q}_\tau(h)$ is the class of solutions to a continuous analogue of the Carathéodory-Féjer interpolation problem. The next step is to search for an analogue of the J_1 inner matrix U which played such a prominent role in theory developed in Dym (1989b).

4. SOME REPRODUCING KERNEL HILBERT SPACES

In this section we shall examine a number of reproducing kernel Hilbert spaces which are related to the extension problem under study and use them to deduce a number of interesting identities.

The notation $H_\tau^\pm$ will be used for the triangular operators which are defined on $L_p^2[0,\tau]$ by the rule

$$(H_\tau^\pm f)(t) = \begin{cases} \int_0^\tau h_\pm(t-s)f(s)ds & \text{for } 0 \le t \le \tau \\ 0 & \text{elsewhere,} \end{cases}$$

where

$$h_+(t) = h(t) \text{ and } h_-(t) = 0 \text{ for } 0 < t \le \tau$$

and

$$h_-(t) = h(t) \text{ and } h_+(t) = 0 \text{ for } -\tau \le t < 0\ .$$

THEOREM 4.1. *If $I - H_\tau$ is positive definite and if $\Phi \in \mathcal{Q}_\tau(h)$, then there exists a J_1 inner matrix U such that the space*

$$\mathcal{M} = \left\{ \begin{bmatrix} \hat{u} \\ \underline{\underline{p}}\Phi^*\hat{u} \end{bmatrix} : u \in L_p^2[0,\tau] \right\}$$

endowed with the indefinite inner product

$$< f_1, f_2 >_{\mathcal{M}} = < J_1 f_1, f_2 >$$

is an $\mathcal{H}(U)$ space. Moreover, $U \in H^{\infty}_{m\times m}(\mathbb{C}_+)$ and its entries are entire functions of exponential type $\leq \tau$. Furthermore, the space $\mathcal{M}$ is independent of the choice of Φ and can in fact be described as the set of elements of the form

$$\left\{ \begin{bmatrix} \hat{u} \\ \{(I - 2H_\tau^-)u\}^\wedge \end{bmatrix} : u \in L_p^2[0,\tau] \right\} .$$

PROOF. It is readily checked that $\mathcal{M}$ is a closed subspace of H_m^2. The rest of the proof, which amounts mostly to verifying that Theorem 2.1 is applicable, is divided into steps.

STEP 1. *If $g \in L^2[0,\tau]$, then*

$$(R_\omega \hat{g})(\lambda) = i \int_0^\tau e^{i\lambda t} \left\{ e^{-i\omega t} \int_t^\tau e^{i\omega s} g(s) ds \right\} dt \tag{4.1}$$

PROOF OF STEP 1. By definition

$$\begin{aligned} (R_\omega \hat{g})(\lambda) &= \int_0^\tau \frac{e^{i\lambda s} - e^{i\omega s}}{\lambda - \omega} g(s) ds \\ &= i \int_0^\tau \left\{ \int_0^s e^{i(\lambda-\omega)t} dt \right\} e^{i\omega s} g(s) ds \end{aligned}$$

and (4.1) drops out by interchanging orders of integration.

STEP 2. *If $u \in L^2[0,\tau]$, then*

$$(\underline{\underline{p}}\Phi^* \hat{u})(\lambda) = \hat{u}(\lambda) - 2 \int_0^\tau e^{i\lambda t} \left\{ \int_t^\tau h(t-s) u(s) ds \right\} dt . \tag{4.2}$$

PROOF OF STEP 2. Since $\Phi \in \mathcal{Q}_\tau(h)$, it can be expressed in the form

$$\Phi(\lambda) = I_p - 2 \int_0^\infty e^{i\lambda s} f(s) ds$$

where $\hat{f} \in \mathcal{P}_\tau(h)$. Thus, since $f(s)^* = f(-s)$ for such f,

$$\begin{aligned}\Phi(\lambda)^* &= I_p - 2\int_{-\infty}^{0} e^{i\lambda s} f(s)ds \\ &= I_p - 2\hat{f}_-(\lambda)\ , \qquad \text{for } \lambda \in \mathbb{R}\ ,\end{aligned}$$

where

$$f_-(s) = \begin{cases} f(s) & \text{for} \quad s < 0 \\ 0 & \text{for} \quad s > 0\ . \end{cases}$$

Thus

$$\underline{\underline{p}}\Phi^* \hat{u} = \hat{u} - 2\int_0^\infty e^{i\lambda t}\left\{\int_0^\tau f_-(t-s)u(s)ds\right\}dt$$

which in turn is readily seen to reduce to (4.2) since $f(t) = h(t)$ for $-\tau \le t \le \tau$.

STEP 3. *$\mathcal{M}$ is R_ω invariant for every point $\omega \in \mathbb{C}$.*

PROOF OF STEP 3. Step 1 guarantees that $\{\hat{u} : u \in L_p^2[0,\tau]\}$ is invariant under R_ω for every choice of $\omega \in \mathbb{C}$. The rest of the argument reduces to establishing the identity

$$R_\omega \underline{\underline{p}}\Phi^* \hat{u} = \underline{\underline{p}}\Phi^* R_\omega \hat{u} \tag{4.3}$$

for $u \in L_p^2[0,\tau]$ and $\omega \in \mathbb{C}$. But now, with the help of Steps 1 and 2 it is readily checked that

$$(R_\omega \underline{\underline{p}}\Phi^* \hat{u})(\lambda) = i\int_0^\tau e^{i\lambda t}\left[e^{-i\omega t}\int_t^\tau e^{i\omega s}\left\{u(s) - 2\int_s^\tau h(s-y)u(y)dy\right\}ds\right]dt\ ,$$

whereas

$$(\underline{\underline{p}}\Phi^* R_\omega \hat{u})(\lambda) =$$
$$i\int_0^\tau e^{i\lambda t}\left[e^{-i\omega t}\int_t^\tau e^{i\omega s}u(s)ds - 2\int_t^\tau h(t-s)e^{-i\omega s}\int_s^\tau e^{i\omega y}u(y)dy\ ds\right]dt\ .$$

Thus, to establish (4.3), it remains only to show that

$$\int_t^\tau e^{i\omega s}\left\{\int_s^\tau h(s-y)u(y)dy\right\}ds = \int_t^\tau h(t-s)e^{i\omega(t-s)}\left\{\int_s^\tau e^{i\omega y}u(y)dy\right\}ds$$

or equivalently, as follows upon interchanging orders of integration on both sides, that

$$\int_t^\tau e^{i\omega y}\left\{\int_t^y e^{i\omega(s-y)}h(s-y)ds\right\}u(y)dy = \int_t^\tau e^{i\omega y}\left\{\int_t^y h(t-s)e^{i\omega(t-s)}ds\right\}u(y)dy\ .$$

But this is easily verified since the terms inside the curly brackets on both sides are readily seen to be the same.

STEP 4. *$\mathcal{M}$ is a Hilbert space with respect to the indefinite inner product of $H^2_{J_1}$.*

PROOF OF STEP 4. If $u \in L^2_p[0,\tau]$ and

$$g = \begin{bmatrix} \hat{u} \\ \underline{\underline{p}}\Phi^* \hat{u} \end{bmatrix} ,$$

then

$$\begin{aligned} < J_1 g, g > &=< \hat{u}, \underline{\underline{p}}\Phi^* \hat{u} > + < \underline{\underline{p}}\Phi^* \hat{u}, \hat{u} > \\ &=< (\Phi + \Phi^*)\hat{u}, \hat{u} > \\ &\geq 0 . \end{aligned}$$

Equality prevails if and only if $\hat{u} = 0$, since $\Phi + \Phi^* > 0$ on $\mathbb{R}$ for $\Phi \in \mathcal{Q}_\tau(h)$. Thus $\mathcal{M}$ is a Hilbert space.

STEP 5. *If $\mathcal{N} = H^2_m \ominus J_1\mathcal{M}$, then*

$$\mathcal{N} = \left\{ \begin{bmatrix} v_1 \\ e^{i\lambda\tau} v_2 - \Phi v_1 \end{bmatrix} : \quad v_1 \in H^2_p \text{ and } v_2 \in H^2_p \right\} .$$

PROOF OF STEP 5. The claim is immediate from the identity

$$\begin{aligned} < J_1 \begin{bmatrix} w \\ \underline{\underline{p}}\Phi^* w \end{bmatrix} , \begin{bmatrix} g_1 \\ g_2 \end{bmatrix} > &=< w, g_2 > + < \underline{\underline{p}}\Phi^* w, g_1 > \\ &=< w, \Phi g_1 + g_2 > \end{aligned}$$

applied to $w \in \{L^2_p[0,\tau]\}^\wedge$ and g_1 and g_2 in H^2_p.

STEP 6. $\mathcal{M} + \mathcal{N} = H^2_m$.

PROOF OF STEP 6. In view of Step 5, it suffices to show that for any choice of g_1 and g_2 in H^2_p, there exists a choice of v_1 and v_2 in H^2_p and $w \in \{L^2_p[0,\tau]\}^\wedge$ such that

$$w + v_1 = g_1$$

and

$$\underline{\underline{p}}\Phi^* w + e^{i\lambda\tau} v_2 - \Phi v_1 = g_2 \ .$$

Let us suppose first that there exists such a solution set. Then, upon taking Φ equal to

$$\Phi_\tau(\lambda) = I_p - 2\int_0^\infty e^{i\lambda s} f_\tau(s)ds \ ,$$

where$\hat{f}_\tau$ is the special element in $\mathcal{P}_\tau(h)$ which is defined by (3.12) and substituting $w = g_1 - v_1$ into the second equation, it is readily seen that

$$\underline{\underline{p}}(\Phi_\tau + \Phi_\tau^*)v_1 = \underline{\underline{p}}\Phi_\tau^* g_1 - g_2 + e^{i\lambda\tau} v_2$$

and hence that v_1 is the solution of the Wiener-Hopf equation

$$\underline{\underline{p}}(I_p - \hat{f}_\tau)v_1 = \varphi_+ \ ,$$

where

$$\varphi_+ = \{p\Phi_\tau^* g_1 - g_2 + e^{i\lambda\tau} v_2\}/2 \ .$$

In fact, upon taking advantage of the factorization formula (3.12), it follows by standard arguments that

$$v_1 = E_\tau \underline{\underline{p}} E_\tau^* \varphi_+ \ . \tag{4.4}$$

Substituting this back into the first equation yields

$$\begin{aligned} w &= g_1 - E_\tau \underline{\underline{p}} E_\tau^* \varphi_+ \\ &= g_1 - E_\tau \underline{\underline{p}} E_\tau^* \{\underline{\underline{p}}\Phi_\tau^* g_1 - g_2\}/2 - E_\tau E_\tau^* e^{i\lambda\tau} v_2/2 \ , \end{aligned}$$

which in turn must belong to $\{L_p^2[0,\tau]\}^\wedge$, since w does. Therefore the orthogonal projection $e^{i\lambda\tau}\underline{\underline{p}}e^{-i\lambda\tau}$ of this quantity onto $\{L_p^2[\tau,\infty)\}^\wedge$ must vanish. But this in turn implies that

$$e^{i\lambda\tau}\underline{\underline{p}}e^{-i\lambda\tau}\{E_\tau E_\tau^* e^{i\lambda\tau} v_2/2\} = e^{i\lambda\tau}\underline{\underline{p}}e^{-i\lambda\tau}[g_1 - E_\tau \underline{\underline{p}} E_\tau^* \{\Phi_\tau^* g_1 - g_2\}/2] \ ,$$

or, equivalently, that v_2 is the unique solution of the Wiener-Hopf equation

$$\underline{\underline{p}}(I_p - \hat{f}_\tau)^{-1} v_2 = \underline{\underline{p}} e^{-i\lambda\tau}[2g_1 - E_\tau \underline{\underline{p}} E_\tau^* \{\Phi_\tau^* g_1 - g_2\}] \, . \tag{4.5}$$

It is now readily checked that if v_2 is defined as the unique solution of the Wiener-Hopf equation (4.5), and v_1 is defined via (4.4), then

$$w = g_1 - v_1 \in \{L_p^2[0,\tau]\}^\wedge$$

(i.e., $\underline{\underline{p}} e^{-i\lambda\tau} w = 0$) and the given system of equations is satisfied.

STEP 7. *$\mathcal{N}$ is full range.*

PROOF OF STEP 7. For each point $\omega \in \mathbb{C}_+$, the functions

$$\begin{bmatrix} \xi/\rho_\omega \\ \{-\Phi(\omega)\xi + e^{i\omega\tau}\eta\}/\rho_\omega \end{bmatrix}$$

belong to $\mathcal{N}$ for every choice of ξ and $\eta \in \mathbb{C}^p$. Thus $\mathcal{N}$ is clearly full range.

STEP 8 *is to complete the proof.*

PROOF OF STEP 8. In view of the preceding steps it is now immediate from Theorem 2.1 that $\mathcal{M}$ is an $\mathcal{H}(U)$ space with reproducing kernel

$$K_\omega(\lambda) = \frac{J_1 - U(\lambda)J_1U(\omega)^*}{\rho_\omega(\lambda)}$$

for every choice of ω and λ in $\mathbb{C}_+$, where U is J_1 inner with respect to $\mathbb{C}_+$. In fact since the elements of $\mathcal{M}$ are vector valued entire functions of exponential type $\leq \tau$, thanks to (4.2), the same holds true for the columns of U. Formula (4.2) also clearly exhibits the fact that the space $\mathcal{M}$ is independent of the choice of $\Phi \in \mathcal{Q}_\tau(h)$ and supplies the second description of $\mathcal{M}$ which is given in the statement of the theorem. ∎

THEOREM 4.2. *If the operator $I - H_\tau$ is positive definite, then the reproducing kernel $K_\omega(\lambda)$ for the space $\mathcal{M}$ defined in Theorem 4.1 is given by the formula*

$$K_\omega(\lambda) = \frac{1}{4\pi}\int_0^\tau \varphi(t,\lambda)\{\varphi(t,\omega)^* + \int_0^\tau \gamma_\tau(t,s)\varphi(s,\omega)^* ds\}dt \, , \tag{4.6}$$

where

$$\varphi(t,\lambda)=\begin{bmatrix}\varphi_1(t,\lambda)\\ \varphi_2(t,\lambda)\end{bmatrix}\ , \tag{4.7}$$

$$\varphi_1(t,\lambda)=e^{i\lambda t}I_p \tag{4.8}$$

and

$$\varphi_2(t,\lambda)=e^{i\lambda t}I_p-2\int_0^t e^{i\lambda s}h(s-t)ds\ , \tag{4.9}$$

for $0\le t\le\tau$.

PROOF. It suffices to show that

(1) $K_\omega\xi\in\mathcal{M}$

and

(2) $<J_1f,K_\omega\xi>=\xi^*f(\omega)$

for every choice of $\omega\in\mathbb{C}_+$, $\xi\in\mathbb{C}^m$ and $f\in\mathcal{M}$. To achieve this it is convenient to write K_ω in the block form

$$K_\omega=\begin{bmatrix}K_\omega^{11} & K_\omega^{12}\\ K_\omega^{21} & K_\omega^{22}\end{bmatrix}$$

and to verify first that the columns of K_ω^{11} and K_ω^{12} belong to $\{L_p^2[0,\tau]\}^\wedge$. Thereafter, it is readily checked with the aid of (4.2) that

$$K_\omega^{2j}\eta=\underline{\underline{p}}\Phi^*K_\omega^{1j}\eta$$

for $j=1,2$, and every choice of $\omega\in\mathbb{C}$, $\Phi\in\mathcal{Q}_\tau(h)$ and $\eta\in\mathbb{C}^p$. Thus (1) holds.

Next, upon expressing $f\in\mathcal{M}$ and $K_\omega\xi$ in the form

$$f=\begin{bmatrix}\hat{u}\\ \underline{\underline{p}}\Phi^*\hat{u}\end{bmatrix}\ ,\quad K_\omega\xi=\begin{bmatrix}\hat{v}\\ \underline{\underline{p}}\Phi^*\hat{v}\end{bmatrix}\quad\text{with}\quad\xi=\begin{bmatrix}\xi_1\\ \xi_2\end{bmatrix}\quad\text{and}\quad\xi_j\in\mathbb{C}^p\ ,$$

it is readily seen that the verification of (2) reduces to checking that

$$<(\Phi+\Phi^*)\hat{u},\hat{v}>=\xi_1^*\hat{u}(\omega)+\xi_2^*(\underline{\underline{p}}\Phi^*\hat{u})(\omega)\ , \tag{4.10}$$

where

$$\hat{v} = K_\omega^{11}\xi_1 + K_\omega^{12}\xi_2 \ .$$

But now as

$$P_2 \frac{\Phi + \Phi^*}{2} \hat{u} = \{(I - H_\tau)u\}^\wedge$$

and

$$K_\omega^{11}\xi_1 = \frac{1}{4\pi}\{(I + \Gamma_\tau)w\}^\wedge \ ,$$

where P_2 is the projection defined in the proof of Lemma 3.4 and

$$w(t) = \varphi_1(t,\omega)^*\xi_1 \ ,$$

it follows readily with the help of the classical Plancherel formula that

$$\begin{aligned} < (\Phi + \Phi^*)\hat{u}, K_\omega^{11}\xi_1 > &= \frac{1}{2\pi} < \{(I - H_\tau)u\}^\wedge, \{(I + \Gamma_\tau)w\}^\wedge > \\ &=< (I - H_\tau)u, (I + \Gamma_\tau)w > \\ &=< u, w > \\ &= \xi_1^* \hat{u}(\omega) \ . \end{aligned} \tag{4.11}$$

This establishes half of the needed evaluation. However, much the same sort of reasoning leads to the supplementary evaluation

$$< (\Phi + \Phi^*)\hat{u}, K_\omega^{12}\xi_2 >=< u, \varphi_2(\cdot,\omega)^*\xi_2 > \tag{4.12}$$

and therefore, in order to complete the proof, it remains only to verify the auxiliary identity

$$\xi_2^* \int_0^\tau \varphi_2(t,\omega)u(t)dt = \xi_2^*(\underline{\underline{p}}\Phi^*\hat{u})(\omega) \ , \tag{4.13}$$

which is an easy consequence of (4.2) and (4.9). ∎

Theorem 4.1 can also be proved another way by invoking Theorem 3.1 of Alpay and Dym (1984). Indeed, in the present setting, the recipe for f_- which is given there in formula (3.3) is equal to $\underline{\underline{p}}\Phi^* f$, for f in the top component of $\mathcal{M}$.

The calculations which appear in the proof of the last theorem have a number of useful implications which we now extract as separate lemmas for future use.

LEMMA 4.1. *If $I - H_\tau$ is positive definite, then the space*

$$\mathcal{M}_0 = \{[I_p \quad 0]f : \ f \in \mathcal{M}\}$$

endowed with the inner product

$$< \hat{u}, \hat{v} >_{\mathcal{M}_0} = < \{(I - H_\tau)u\}^{\wedge}, \hat{v} >$$

is a reproducing kernel Hilbert space of $p \times 1$ vector valued entire functions of exponential type $\leq \tau$ with reproducing kernel

$$2[I_p \quad 0]K_\omega \begin{bmatrix} I_p \\ 0 \end{bmatrix} = \Lambda_\omega(\lambda) \tag{4.14}$$

where $\Lambda_\omega(\lambda) = \Lambda_\omega^\tau(\lambda)$ is given by formula (3.6) (and hence also by (3.7) and (3.8)). Moreover,

$$< \hat{u}, \hat{v} >_{\mathcal{M}_0} = < \frac{\Phi + \Phi^*}{2} \hat{u}, \hat{v} > \tag{4.15}$$

and the mapping

$$\hat{u} \longrightarrow \frac{1}{\sqrt{2}} \begin{bmatrix} \hat{u} \\ \underline{\underline{p}}\Phi^* \hat{u} \end{bmatrix} = \frac{1}{\sqrt{2}} \begin{bmatrix} \hat{u} \\ \{(I - 2H_\tau^-)u\}^{\wedge} \end{bmatrix}$$

is an isometric isomorphism of $\mathcal{M}_0$ onto $\mathcal{M}$ for every choice of $\Phi \in \mathcal{Q}_\tau(h)$.

PROOF. Formula (4.14) is an immediate consequence of matching (3.6) and (4.6). The rest is plain from (4.11) and (4.2). ∎

LEMMA 4.2. *If $I - H_\tau$ is positive definite, then*

$$\mathcal{M}_0^\bullet = \{[0 \quad I_p]f : \ f \in \mathcal{M}\}$$

is a reproducing kernel Hilbert space of $p \times 1$ vector valued entire functions of exponential type $\leq \tau$ with reproducing kernel

$$\begin{aligned} \Lambda_\omega^\bullet(\lambda) &= 2[0 \quad I_p]K_\omega(\lambda) \begin{bmatrix} 0 \\ I_p \end{bmatrix} \\ &= \frac{1}{2\pi} \int_0^\tau \varphi_2(t, \lambda)\{\varphi_2(t, \omega)^* + \int_0^\tau \gamma_\tau(t, s)\varphi_2(s, \omega)^* ds\} dt \end{aligned} \tag{4.16}$$

and inner product

$$< w, v >_{\mathcal{M}_0^\bullet} =< \frac{\Phi^{-1} + \Phi^{*-1}}{2} w, v > \tag{4.17}$$

for every choice of w and v in $\mathcal{M}_0^\bullet$ and $\Phi \in \mathcal{Q}_\tau(h)$. Moreover, the mapping

$$w \longrightarrow \frac{1}{\sqrt{2}} \begin{bmatrix} \underline{\underline{p}}\Phi^{*-1} w \\ w \end{bmatrix}$$

is an isometric isomorphism of $\mathcal{M}_0^\bullet$ onto $\mathcal{M}$.

PROOF. Clearly

$$\Lambda_\omega^\bullet \xi = 2K_\omega^{22}\xi = 2\underline{\underline{p}}\Phi^* K_\omega^{12}\xi \ ,$$

in terms of the notation introduced in the proof of Theorem 4.2. Thus $\Lambda_\omega^\bullet \xi \in \mathcal{M}_0^\bullet$ for every choice of $\omega \in \mathbb{C}_+$ and $\xi \in \mathbb{C}^p$. Moreover, with the aid of (4.12) and (4.13), it is readily verified that

$$\begin{aligned} &< \left(\frac{\Phi^{-1} + \Phi^{-1*}}{2}\right) \underline{\underline{p}}\Phi^* \hat{u},\ 2\underline{\underline{p}}\Phi^* K_\omega^{12}\xi > \\ &=< \Phi^{-1}\underline{\underline{p}}\Phi^*\hat{u},\ \Phi^* K_\omega^{12}\xi > + < \Phi^*\hat{u}, \Phi^{-1}\underline{\underline{p}}\Phi^* K_\omega^{12}\xi > \\ &=< (\Phi + \Phi^*)\hat{u},\ K_\omega^{12}\xi > \\ &= \xi^*(\underline{\underline{p}}\Phi^*\hat{u})(\omega) \ , \end{aligned}$$

for $\hat{u} \in \mathcal{M}_0$ and hence that $\Lambda_\omega^\bullet \xi$ has the requisite reproducing properties. ∎

We remark that the preceding analysis implies that the reproducing kernel K_ω for $\mathcal{M}$ can be expressed in the block form

$$K_\omega = \frac{1}{2} \begin{bmatrix} \Lambda_\omega & \underline{\underline{p}}\Phi^{*-1}\Lambda_\omega^\bullet \\ \underline{\underline{p}}\Phi^*\Lambda_\omega & \Lambda_\omega^\bullet \end{bmatrix} \tag{4.18}$$

where Λ_ω [resp. $\Lambda_\omega^\bullet$] is the reproducing kernel for $\mathcal{M}_0$ [resp. $\mathcal{M}_0^\bullet$] endowed with the inner product (4.15) [resp. (4.17)] and Φ is any element in $\mathcal{Q}_\tau(h)$.

At this point it is convenient to introduce the operators $H_\tau^\bullet$ and $\Gamma_\tau^\bullet$ on $L_p^2[0,\tau]$ via the formulas

$$\begin{aligned} I - H_\tau^\bullet &= (I - 2H_\tau^+)^{-1}(I - H_\tau)(I - 2H_\tau^-)^{-1} \\ &= (I - 2H_\tau^+)^{-1}\{(I - 2H_\tau^+) + (I - 2H_\tau^-)\}(I - 2H_\tau^-)^{-1}/2 \\ &= \{(I - 2H_\tau^-)^{-1} + (I - 2H_\tau^+)^{-1}\}/2 \end{aligned}$$

and

$$\begin{aligned} I + \Gamma_\tau^\bullet &= (I - H_\tau^\bullet)^{-1} \\ &= (I - 2H_\tau^-)(I + \Gamma_\tau)(I - 2H_\tau^+) \ , \end{aligned}$$

when the indicated inverse exists.

LEMMA 4.3. *If $I - H_\tau$ is positive definite, then the reproducing kernel $\Lambda_\omega^\bullet$ for $\mathcal{M}_0^\bullet$ can also be expressed in the form*

$$\Lambda_\omega^\bullet(\lambda) = \frac{1}{2\pi} \int_0^\tau \varphi_1(t,\lambda)\{\varphi_1(t,\omega)^* + \int_0^\tau \dot{\gamma}_\tau(t,s)\varphi_1(s,\omega)^* ds\} dt \tag{4.19}$$

where

$$\begin{aligned} \dot{\gamma}_\tau(t,s) =& \gamma_\tau(t,s) - 2h(t-s) - 2\int_s^\tau \gamma_\tau(t,y)h(y-s)dy - 2\int_t^\tau h(t-y) \times \\ & \{\gamma_\tau(y,s) - 2\int_s^\tau \gamma_\tau(y,x)h(x-s)dx\}dy + 4\int_{max(s,t)}^\tau h(t-y)h(y-s)dy \ , \end{aligned} \tag{4.20}$$

for $0 \le t, s \le \tau$, is the kernel for the integral operator $\Gamma_\tau^\bullet$ introduced above.

PROOF. By definition,

$$\begin{aligned} \varphi_2(t,\omega)^* &= \varphi_1(t,\omega)^* - 2\int_0^t h(t-s)\varphi_1(s,\omega)^* ds \\ &= \{(I - 2H_\tau^+)\varphi_1(\cdot,\omega)^*\}(t) \ . \end{aligned}$$

Therefore, since $(H_\tau^\pm)^* = H_\tau^\mp$ with respect to the standard inner product in $L_p^2[0,\tau]$,

$$\begin{aligned} 2\pi\xi^*\Lambda_\omega^\bullet(\lambda)\eta &= < (I + \Gamma_\tau)\varphi_2(\cdot,\omega)^*\eta, \ \varphi_2(\cdot,\lambda)^*\xi > \\ &= < (I + \Gamma_\tau)(I - 2H_\tau^+)\varphi_1(\cdot,\omega)^*\eta, \ (I - 2H_\tau^+)\varphi_1(\cdot,\lambda)^*\xi > \\ &= < (I + \Gamma_\tau^\bullet)\varphi_1(\cdot,\omega)^*\eta, \ \varphi_1(\cdot,\lambda)^*\xi > \ . \end{aligned}$$

It remains to verify (4.20), but that is straightforward. ∎

Next, we introduce the $p \times p$ matrix valued functions

$$E_\tau^\bullet(\lambda) = I_p + \int_0^\tau e^{i\lambda t}\dot{\gamma}_\tau(t,0)dt \tag{4.21}$$

and

$$F_\tau^\bullet(\lambda) = e^{i\lambda\tau}\left\{I_p + \int_0^\tau e^{i\lambda(t-\tau)}\dot{\gamma}_\tau(t,\tau)dt\right\} , \tag{4.22}$$

which are continuous analogues of the matrix orthogonal polynomials of the second kind.

LEMMA 4.4. *If $I - H_\tau$ is positive definite, then the reproducing kernel $\Lambda_\omega^\bullet$ for $\mathcal{M}_0^\bullet$ which is given by formulas (4.16) and (4.19) can also be expressed as*

$$\Lambda_\omega^\bullet(\lambda) = \frac{E_\tau^\bullet(\lambda)E_\tau^\bullet(\omega)^* - F_\tau^\bullet(\lambda)F_\tau^\bullet(\omega)^*}{\rho_\omega(\lambda)} \tag{4.23}$$

and as

$$\Lambda_\omega^\bullet(\lambda) = \frac{1}{2\pi}\int_0^\tau F_s^\bullet(\lambda)F_s^\bullet(\omega)^*ds . \tag{4.24}$$

PROOF. This is immediate from Lemma 3.1. ∎

LEMMA 4.5. *If $I - H_\tau$ is positive definite and if*

$$\Phi_\tau(\lambda) = I_p - 2\int_0^\infty e^{i\lambda s}f_\tau(s)ds \tag{4.25}$$

where $\hat{f}_\tau$ is the special element in $\mathcal{P}_\tau(h)$ which is defined by (3.12), then

$$\Phi_\tau(\lambda)E_\tau(\lambda) = E_\tau^\bullet(\lambda) \tag{4.26}$$

for every point $\lambda \in \overline{\mathbb{C}}_+$ and

$$\Phi_\tau^\#(\lambda)F_\tau(\lambda) = F_\tau^\bullet(\lambda) \tag{4.27}$$

for every point $\lambda \in \overline{\mathbb{C}}_-$.

PROOF. Let $P_\pm$ denote the natural projection of an element in the Wiener algebra $\mathcal{W}$ onto its component in the subalgebra $\mathcal{W}_\pm$. Then, since $E_\tau^\#$ is invertible in $\mathcal{W}_+^I$ by Lemma 3.3, it follows readily from (3.12) that

$$P_+\{(\Phi_\tau + \Phi_\tau^*)E_\tau\} = 2P_+\{E_\tau^{\#-1}\} = 2I_p$$

and hence that

$$\begin{aligned}\Phi_\tau E_\tau &= 2I_p - P_+(\Phi_\tau^* E_\tau) \\ &= 2I_p - P_+\{(I_p - 2\hat{f}_-)(I_p + \hat{a}_\tau)\} ,\end{aligned}$$

where we have set

$$f_-(s) = \begin{cases} f_\tau(s) & \text{for} \quad s < 0 \\ 0 & \text{for} \quad s > 0 \end{cases}$$

and a_τ is defined as in the proof of Lemma 3.4. But now it is readily seen that

$$\begin{aligned}P_+\{(I_p - 2\hat{f}_-)(I_p + \hat{a}_\tau)\} &= E_\tau - 2P_+\hat{f}_-\hat{a}_\tau \\ &= E_\tau - 2\int_0^\infty e^{i\lambda t}\left\{\int_0^\tau f_-(t-s)a_\tau(s)ds\right\}dt \\ &= E_\tau - 2\int_0^\tau e^{i\lambda t}\left\{\int_t^\tau h(t-s)a_\tau(s)ds\right\}dt\end{aligned}$$

and hence, since

$$\dot{\gamma}_\tau(t,0) = -\gamma_\tau(t,0) + 2\int_t^\tau h(t-y)\gamma_\tau(y,0)dy$$

by (4.20), that (4.26) holds on $\overline{\mathbb{C}}_+$.

Next, set

$$c_\tau(t-\tau) = \begin{cases} \gamma_\tau(t,\tau) & \text{for } 0 \le t \le \tau \\ 0 & \text{elsewhere} . \end{cases}$$

Then

$$F_\tau(\lambda) = e^{i\lambda\tau}\{I_p + \hat{c}_\tau(\lambda)\}$$

and so, by (3.11),

$$\begin{aligned}(\Phi_\tau + \Phi_\tau^*)(I_p + \hat{c}_\tau) &= 2(I_p - \hat{f}_\tau)(I_p + \hat{c}_\tau) \\ &= 2(I_p + \hat{c}_\tau^*)^{-1}\end{aligned}$$

on $\mathbb{R}$. Thus, since $I_p + \hat{c}_\tau^*$ is invertible in $\mathcal{W}_+$ by Lemma 3.3, it follows upon taking the projection P_- of both sides onto the subalgebra $\mathcal{W}_-$, that

$$P_-\{\Phi_\tau(I_p + \hat{c}_\tau)\} + \Phi_\tau^*(I_p + \hat{c}_\tau) = 2I_p .$$

But now, upon setting

$$f_+(s) = \begin{cases} f_\tau(s) & \text{for} \quad s > 0 \\ 0 & \text{for} \quad s < 0 \,, \end{cases}$$

it follows readily that the first term on the left is equal to

$$\begin{aligned} P_-\{(I_p - 2\hat{f}_+)(I_p + \hat{c}_\tau)\} &= I_p + \hat{c}_\tau - 2P_-\hat{f}_+\hat{c}_\tau \\ &= I_p + \hat{c}_\tau - 2\int_{-\infty}^{0} e^{i\lambda t}\left\{\int_{-\tau}^{0} f_+(t-s)c_\tau(s)ds\right\}dt \\ &= I_p + \hat{c}_\tau - 2\int_{0}^{\tau} e^{-i\lambda t}\left\{\int_{t}^{\tau} h(s-t)c_\tau(-s)ds\right\}dt \,. \end{aligned}$$

The desired formula (4.27) drops out readily upon combining terms since

$$\dot{\gamma}_\tau(t,\tau) = -\gamma_\tau(t,\tau) + 2\int_0^t h(t-s)\gamma_\tau(s,\tau)ds$$

by (4.20). ∎

LEMMA 4.6. *If $I - H_\tau$ is positive definite, then*

$$\left\{\underset{=}{p}\frac{\Phi_\tau^* E_\tau \xi}{\rho_\omega}\right\}(\lambda) = \frac{2\{E_\tau(\omega)^*\}^{-1}\xi - E_\tau^\bullet(\lambda)\xi}{\rho_\omega(\lambda)} \tag{4.28}$$

and

$$\left\{\underset{=}{p}\frac{\Phi_\tau^* F_\tau \xi}{\rho_\omega}\right\}(\lambda) = \frac{F_\tau^\bullet(\lambda)\xi}{\rho_\omega(\lambda)} \tag{4.29}$$

for every point $\omega \in \mathbb{C}_+$.

PROOF. Clearly

$$\begin{aligned} \underset{=}{p}\frac{\Phi_\tau^* E_\tau \xi}{\rho_\omega} &= \underset{=}{p}\frac{(\Phi_\tau + \Phi_\tau^*)E_\tau \xi}{\rho_\omega} - \frac{\Phi_\tau E_\tau \xi}{\rho_\omega} \\ &= 2\underset{=}{p}\frac{(I_p - \hat{f}_\tau)E_\tau \xi}{\rho_\omega} - \frac{E_\tau^\bullet \xi}{\rho_\omega} \\ &= 2\underset{=}{p}\frac{(E_\tau^*)^{-1}\xi}{\rho_\omega} - \frac{E_\tau^\bullet \xi}{\rho_\omega} \end{aligned}$$

for $\omega \in \mathbb{C}_+$. But now, for any point $\alpha \in \mathbb{C}_+$ and $\eta \in \mathbb{C}^p$,

$$\eta^* \left\{ \underline{\underline{p}} \frac{(E_\tau^*)^{-1}\xi}{\rho_\omega} \right\} (\alpha) = < \frac{(E_\tau^{*-1})\xi}{\rho_\omega} , \frac{\eta}{\rho_\alpha} >$$
$$= < \frac{\xi}{\rho_\omega} , \frac{E_\tau^{-1}\eta}{\rho_\alpha} >$$
$$= \left\{ \frac{\xi^* E_\tau(\omega)^{-1}\eta}{\rho_\alpha(\omega)} \right\}^*$$
$$= \frac{\eta^* \{E_\tau(\omega)^*\}^{-1}\xi}{\rho_\omega(\alpha)} ,$$

which, when combined with the earlier evaluation, yields (4.28).

Similarly,

$$\underline{\underline{p}} \frac{\Phi_\tau^* F_\tau \xi}{\rho_\omega} = \underline{\underline{p}} \frac{(\Phi_\tau + \Phi_\tau^*) F_\tau \xi}{\rho_\omega} - \frac{\Phi_\tau F_\tau \xi}{\rho_\omega}$$
$$= 2\underline{\underline{p}} \frac{F_\tau^{*-1}\xi}{\rho_\omega} - \frac{\Phi_\tau F_\tau \xi}{\rho_\omega}$$
$$= \{2F_\tau^{*-1} - \Phi_\tau F_\tau\}\xi/\rho_\omega ,$$

since $F_\tau^{*-1} \in e^{i\lambda\tau}\mathcal{W}_+$ by Lemma 3.3. But now, for $\lambda \in \mathbb{R}$, the right hand side in the last sequence of equalities is equal to

$$\left\{ 2\frac{(\Phi_\tau + \Phi_\tau^*)F_\tau}{2} - \Phi_\tau F_\tau \right\} \xi/\rho_\omega = \Phi_\tau^* F_\tau \xi/\rho_\omega$$
$$= F_\tau^\bullet \xi/\rho_\omega ,$$

which serves to complete the proof of (4.29). ∎

LEMMA 4.7. *If $I - H_\tau$ is positive definite, then*

$$\left\{ \underline{\underline{p}} \frac{\Phi_\tau^{*-1} E_\tau^\bullet \xi}{\rho_\omega} \right\} (\lambda) = \frac{2\{E_\tau^\bullet(\omega)^*\}^{-1}\xi - E_\tau(\lambda)\xi}{\rho_\omega(\lambda)} \tag{4.30}$$

and

$$\left\{ \underline{\underline{p}} \frac{\Phi_\tau^{*-1} F_\tau^\bullet \xi}{\rho_\omega} \right\} (\lambda) = \frac{F_\tau(\lambda)\xi}{\rho_\omega(\lambda)} \tag{4.31}$$

for $\omega \in \mathbb{C}_+$.

PROOF. Since $\underline{\underline{p}}\Phi_\tau^{*-1}|_{H_p^2}$ is an invertible operator with inverse $\underline{\underline{p}}\Phi_\tau^*|_{H_p^2}$, formula (4.30) is equivalent to the assertion that

$$\frac{E_\tau^\bullet \xi}{\rho_\omega} = \underline{\underline{p}}\Phi_\tau^* \left\{ \frac{2\{E_\tau^\bullet(\omega)^*\}^{-1}\xi - E_\tau \xi}{\rho_\omega} \right\} .$$

But this is readily verified with the help of formula (4.28) and the evaluation which was used to establish it in the proof of Lemma 4.6. Therefore (4.30) is valid.

Formula (4.31) is selfevident since $F_\tau \xi / \rho_\omega$ belongs to H_p^2 for $\omega \in \mathbb{C}_+$. ∎

THEOREM 4.3. *Suppose that $I - H_\tau$ is positive definite and let*

$$U_\tau = \frac{1}{2}\begin{bmatrix} F_\tau + E_\tau & F_\tau - E_\tau \\ F_\tau^\bullet - E_\tau^\bullet & F_\tau^\bullet + E_\tau^\bullet \end{bmatrix}. \tag{4.32}$$

Then U_τ is J_1 inner with respect to $\mathbb{C}_+$ and $\mathcal{H}(U_\tau)$ is equal to the space $\mathcal{M}$ introduced in Theorem 4.1. Moreover, in addition to the properties exhibited in Lemma 3.3, the entries in U_τ are entire functions of exponential type $\leq \tau$,

$$J_1 - U_\tau(\lambda) J_1 U_\tau^\#(\lambda) = 0 \tag{4.33}$$

for every point $\lambda \in \mathbb{C}$ and

$$\frac{J_1 - U_\tau(\omega) J_1 U_\tau(\omega)^*}{\rho_\omega(\omega)} \geq 0 \tag{4.34}$$

for every point $\omega \notin \mathbb{R}$.

PROOF. In order to verify that $\mathcal{H}(U_\tau) = \mathcal{M}$, it suffices to show that

$$\frac{J_1 - U_\tau(\lambda) J_1 U_\tau(\omega)^*}{\rho_\omega(\lambda)} = K_\omega(\lambda) \tag{4.35}$$

for every choice of λ and ω in $\mathbb{C}_+$, where $K_\omega(\lambda)$ designates the reproducing kernel for $\mathcal{M}$, as given by (4.6). Thus, since

$$\frac{J_1 - U J_1 U^*}{\rho_\omega(\lambda)} = \frac{\begin{bmatrix} EE^* - FF^* & 2I_p - EE^{\bullet *} - FF^{\bullet *} \\ 2I_p - E^\bullet E^* - F^\bullet F^* & E^\bullet E^{\bullet *} - F^\bullet F^{\bullet *} \end{bmatrix}}{2\rho_\omega(\lambda)} \tag{4.36}$$

where the nonstarred functions are evaluated at λ, the starred functions are evaluated at ω and we have temporarily dropped the subscript τ in order to save space, it is readily verified with the help of Lemmas 3.1, 4.2 and 4.4 that the diagonal blocks of the latter match the diagonal blocks of $K_\omega(\lambda)$.

But now in order to verify that the remaining blocks match it suffices to show that

$$\underline{\underline{p}}\Phi_\tau^* \Lambda_\omega^\tau = \frac{2I_p - E_\tau^\bullet(\lambda)E_\tau(\omega)^* - F_\tau^\bullet(\lambda)F_\tau(\omega)^*}{\rho_\omega(\lambda)}$$

and

$$\underline{\underline{p}}\Phi_\tau^{*-1} \Lambda_\omega^\bullet = \frac{2I_p - E_\tau(\lambda)E_\tau^\bullet(\omega)^* - F_\tau(\lambda)F_\tau^\bullet(\omega)^*}{\rho_\omega(\lambda)} ,$$

for ω and λ in $\mathbb{C}_+$. However, this is easily done with the help of formulas (4.28)-(4.30).

Next, since the entries in $K_\omega(\lambda)$ are entire functions in the variable λ, it follows readily from (4.35) that

$$J_1 - U_\tau(\omega^*)J_1U_\tau(\omega)^* = \rho_\omega(\omega^*)K_\omega(\omega) = 0 ,$$

first for $\omega \in \mathbb{C}_+$ and then, by taking the conjugate transpose, for $\omega \in \mathbb{C}_-$, and finally, by taking limits, for $\omega \in \mathbb{R}$ also. This serves to establish (4.33). The inequality (4.34) follows for $\omega \in \mathbb{C}_+$ from the fact that K_ω is a reproducing kernel. However, since U_τ is invertible and

$$U_\tau(\omega)^{-1} = J_1U_\tau^\#(\omega)J_1 ,$$

by (4.33), this also implies that

$$U_\tau(\omega)^{-1}J_1U_\tau(\omega)^{*-1} - J_1 \geq 0$$

for $\omega \in \mathbb{C}_+$ and hence that

$$U_\tau(\omega)^*J_1U_\tau(\omega) - J_1 \geq 0$$

for $\omega \in \mathbb{C}_-$. But since, as is well known (see e.g. Dym (1989b)), the U and U^* can be interchanged without affecting the last inequality and $\rho_\omega(\omega) < 0$ for $\omega \in \mathbb{C}_-$, this clearly serves to establish (4.34) for $\omega \in \mathbb{C}_-$ also. ∎

We remark that it is instructive to verify directly that the right hand side of (3.7) is a reproducing kernel for the space $\mathcal{M}_0$ endowed with the inner product (4.15). The chief tools are the classical Paley-Wiener theorem which serves to help prove that $\Lambda_\omega^\tau \xi$ belongs to $\mathcal{M}_0$ alias $\{L_p^2[0,\tau]\}^\wedge$ and Cauchy's theorem for H_p^2 functions.

5. DESCRIPTION OF SOLUTIONS

In this section we shall derive a description of the set of all solutions to the covariance extension problem when $I - H_\tau$ is positive.

It is convenient to let $\mathcal{G}$ denote the set of $p \times p$ matrix valued functions G which belong to $\mathcal{W}_+^I \cap \mathcal{C}_p$ with $G(\lambda) + G(\lambda)^* > 0$ for every point $\lambda \in \overline{\mathbb{C}}_+$ and to let $\mathcal{S}'$ denote the set of S in $\mathcal{W}_+^0$ for which $I_p - S(\lambda)^* S(\lambda) > 0$ for every point $\lambda \in \overline{\mathbb{C}}_+$.

THEOREM 5.1. *Let $I - H_\tau > 0$ and let $U = U_\tau$ be defined as in (4.32). Then $\Phi \in \mathcal{Q}_\tau(h)$ if and only if it can be expressed in the form*

$$\sigma_\tau[S] = (E_\tau^\bullet - F_\tau^\bullet S)(E_\tau + F_\tau S)^{-1}$$

for some choice of $S \in \mathcal{S}'$.

PROOF. The proof is broken into steps.

STEP 1. *$E_\tau + F_\tau S$ is invertible in $\mathcal{W}_+$ for every $S \in \mathcal{S}'$.*

PROOF OF STEP 1. E_τ is invertible in $\mathcal{W}_+$ by Lemma 3.3 and, by (3.7), $V_\tau = E_\tau^{-1} F_\tau$ is strictly contractive in $\mathbb{C}_+$ and unitary on $\mathbb{R}$. Thus the matrix $I_p + V_\tau(\lambda) S(\lambda)$ is invertible for every point $\lambda \in \overline{\mathbb{C}}_+$ as well as at the point $\lambda = \infty$ for every choice of $S \in \mathcal{S}'$ and hence the desired result drops out from the matrix version of a well-known theorem of Wiener.

STEP 2. *$\Phi = \sigma_\tau[S]$ belongs to $\mathcal{G}$ for every choice of $S \in \mathcal{S}'$.*

PROOF OF STEP 2. To begin with, it is readily checked that

$$\Phi + \Phi^* = -2(E_\tau^* + S^* F_\tau^*)^{-1} \left\{ [S^* \quad I_p] M^* U_\tau^* J_1 U_\tau M \begin{bmatrix} S \\ I_p \end{bmatrix} \right\} (E_\tau + F_\tau S)^{-1} ,$$

where U_τ is given by (4.32) and

$$M = \frac{1}{\sqrt{2}} \begin{bmatrix} I_p & I_p \\ I_p & -I_p \end{bmatrix} .$$

The latter is both unitary and selfadjoint (with respect to the standard inner product) and

$$M^* J_1 M = J_0 .$$

Thus,

$$\begin{aligned} \Phi + \Phi^* &\geq -2(E_\tau^* + S^* F_\tau^*)^{-1} \left\{ [S^* \quad I_p] J_0 \begin{bmatrix} S \\ I_p \end{bmatrix} \right\} (E_\tau + F_\tau S)^{-1} \\ &= 2(E_\tau^* + S^* F_\tau^*)^{-1} (I_p - S^* S)(E_\tau + F_\tau S)^{-1} \\ &> 0 \end{aligned}$$

in $\overline{\mathbb{C}}_+$. The rest is plain.

STEP 3. $\Phi = \sigma_\tau[S]$ *belongs to* $\mathcal{Q}_\tau(h)$ *for every choice of* $S \in \mathcal{S}'$.

PROOF OF STEP 3. By Theorem 3.1 and Lemma 4.5,

$$\sigma_\tau[0] = E_\tau^\bullet E_\tau^{-1} = \Phi_\tau$$

belongs to $\mathcal{Q}_\tau(h)$. Next,

$$\begin{aligned} \sigma_\tau[0] - \sigma_\tau[S] &= \{E_\tau^\bullet E_\tau^{-1}(E_\tau + F_\tau S) - (E_\tau^\bullet - F_\tau^\bullet S)\}(E_\tau + F_\tau S)^{-1} \\ &= (E_\tau^\bullet E_\tau^{-1} F_\tau - F_\tau^\bullet) S (E_\tau + F_\tau S)^{-1} . \end{aligned}$$

But now by (4.26) and (4.27)

$$E_\tau^\bullet E_\tau^{-1} F_\tau + F_\tau^\bullet = (\Phi_\tau + \Phi_\tau^*) F_\tau = 2(F_\tau^*)^{-1}$$

for $\lambda \in \mathbb{R}$ and hence, by Lemma 3.3, it belongs to $e^{i\lambda\tau}\mathcal{W}_+$. Therefore, since

$$(E_\tau + F_\tau S)^{-1} = (I + V_\tau S)^{-1} E_\tau^{-1}$$

belongs to $\mathcal{W}_+$ and $S \in \mathcal{W}_+^0$, it follows that

$$\sigma_\tau[0] - \sigma_\tau[S] \in e^{i\lambda\tau}\mathcal{W}_+^0 .$$

Thus $\sigma_\tau[S] \in \mathcal{Q}_\tau(h)$.

STEP 4. *If* $\Phi \in \mathcal{Q}_\tau(h)$, $Y = [\Phi \quad I_p]$ *and* $X = YU_\tau$, *then*

$$\sum_{i,j=1}^{n} \xi_i^* \frac{X(\omega_i)J_1X(\omega_j)^*}{\rho_{\omega_j}(\omega_i)}\xi_j \geq 0$$

for every choice of points $\omega_1, \ldots, \omega_n$ *in* $\mathbb{C}_+$ *and vectors* $\xi_1, \ldots, \xi_n$ *in* $\mathbb{C}^p$.

PROOF OF STEP 4. Let

$$\mathcal{M}_+ = \{\underline{\underline{p}}J_1Y^*g : \; g \in H_p^2\} .$$

It is a strictly positive space with respect to the J_1 inner product:

$$\begin{aligned} < J_1\underline{\underline{p}}J_1Y^*g, \underline{\underline{p}}J_1Y^*g > &=< (\Phi + \Phi^*)g, g > \\ &> 0 , \end{aligned}$$

unless $g = 0$. Now, if $g_j = \xi_j/\rho_{\omega_j}$ with $\omega_j \in \mathbb{C}_+$, then

$$f_j = \underline{\underline{p}}J_1Y^*g_j = J_1Y(\omega_j)^*\xi_j/\rho_{\omega_j}$$

whereas, if P denotes the orthogonal projection of $\mathcal{M}_+$ onto $\mathcal{M}$, then

$$\begin{aligned} \eta^*(Pf_j)(\alpha) &=< J_1Pf_j, K_\alpha\eta > \\ &=< J_1f_j, K_\alpha\eta > \\ &=< \frac{Y(\omega_j)^*\xi_j}{\rho_{\omega_j}}, K_\alpha\eta > \\ &= \eta^*K_{\omega_j}(\alpha)Y(\omega_j)^*\xi_j , \end{aligned}$$

for every choice of $\eta \in \mathbb{C}^p$ and $\alpha \in \mathbb{C}_+$. Thus upon setting

$$f = \sum_{j=1}^{n} f_j ,$$

it is readily seen that

$$\begin{aligned}\sum_{i,j=1}^{n} \xi_i^* \frac{Y(\omega_i)J_1Y(\omega_j)^*}{\rho_{\omega_j}(\omega_i)} \xi_j &= < J_1 f, f > \\ &\geq < J_1 Pf, Pf > \\ &= \sum_{i,j=1}^{n} < J_1 K_{\omega_j} Y(\omega_j)^* \xi_j, K_{\omega_i} Y(\omega_i)^* \xi_i > \\ &= \sum_{i,j=1}^{n} \xi_i^* Y(\omega_i) K_{\omega_j}(\omega_i) Y(\omega_j)^* \xi_j \\ &= \sum_{i,j=1}^{n} \xi_i^* \frac{\{Y(\omega_i)J_1Y(\omega_j)^* - X(\omega_i)JX(\omega_j)^*\}}{\rho_{\omega_j}(\omega_i)} \xi_j .\end{aligned}$$

The rest is plain.

STEP 5. *If $\Phi \in \mathcal{Q}_\mathcal{T}(h)$, then $\Phi = \sigma_\mathcal{T}[S]$ for some choice of $S \in \mathcal{S}'$.*

PROOF OF STEP 5. Let $XM = [G \;\; H]$, where G and H are both $p \times p$. Then, since $[G(\omega) \;\; H(\omega)]$ has rank p and, by Step 4,

$$G(\omega)G(\omega)^* \geq H(\omega)H(\omega)^*$$

for every point $\omega \in \overline{\mathbb{C}}_+$, G is readily seen to be invertible in $\overline{\mathbb{C}}_+$. Thus

$$XM = G[I_p \quad -S]$$

for some choice of $S \in \mathcal{S}_{p\times p}$. But this in turn implies that

$$\Phi U_{11} + U_{21} = G(I_p - S)/\sqrt{2} ,$$

$$\Phi U_{12} + U_{22} = G(I_p + S)/\sqrt{2}$$

and hence, by (4.26), (4.27) and Lemma 3.3, that

$$\sqrt{2}\, G = \Phi F_\tau + F_\tau^\bullet$$

belongs to $e^{i\lambda\tau}\mathcal{W}_+$ and

$$\sqrt{2}\, GS = -\Phi E_\tau + E_\tau^\bullet$$

belongs to $e^{i\lambda\tau}\mathcal{W}_+^0$. Moreover, the inequality

$$\begin{aligned} G(I_p - SS^*)G^* &= G[I_p \quad -S]J_0[I_p \quad -S]^*G^* \\ &= [\Phi \quad I_p]UMJ_0MU^*[\Phi \quad I_p]^* \\ &= [\Phi \quad I_p]J_1[\Phi \quad I_p]^* \\ &= \Phi + \Phi^* \\ &> 0\ , \end{aligned}$$

which is valid on $\mathbb{R}$, clearly implies that G is invertible on $\mathbb{R}$ and that S is strictly contractive there.

Thus, since $e^{-i\lambda\tau}G \to \sqrt{2}I_p$ as $\lambda \to \pm\infty$, by the Riemann-Lebesgue lemma, it follows from the matrix version of Wiener's theorem that $e^{-i\lambda\tau}G$ is invertible in $\mathcal{W}$ and hence that S belongs to $\mathcal{W}^0$ as well as to $\mathcal{S}_{p\times p}$. Therefore S must in fact belong to $\mathcal{W}_+^0$ and, since the strict contractiveness propagates to all of $\mathbb{C}_+$, it belongs to $\mathcal{S}'$, as needed. ∎

THEOREM 5.2. *Let $I - H_\tau$ be positive definite. Then $\hat{f} \in \mathcal{P}_\tau(h)$ if and only if it can be expressed in the form*

$$I_p - \hat{f} = (E_\tau^*)^{-1}(I_p + S^*V_\tau^*)^{-1}(I_p - S^*S)(I_p + V_\tau S)^{-1}(E_\tau)^{-1} \tag{5.1}$$

on $\mathbb{R}$, or equivalently if and only if it can be expressed in the form

$$I_p - \hat{f} = (F_\tau^*)^{-1}(I_p + SV_\tau)^{-1}(I_p - SS^*)(I_p + V_\tau^* S^*)^{-1}(F_\tau)^{-1} \tag{5.2}$$

on $\mathbb{R}$, where, in both representations,

$$V_\tau = E_\tau^{-1}F_\tau = E_\tau^* F_\tau^{*-1} \tag{5.3}$$

on $\mathbb{R}$ *and* S *is an arbitrary element of* $\mathcal{W}_+^0$ *which is strictly contractive on* $\overline{\mathbb{C}}_+$.

PROOF. Theorems 3.1 and 5.1 imply that $\hat{f} \in \mathcal{P}_\tau(h)$ if and only if $I_p - \hat{f}$ can be expressed in the form

$$I_p - \hat{f} = \{\sigma_\tau[S] + \sigma_\tau[S]^*\}/2$$

on $\mathbb{R}$ for some choice of $S \in \mathcal{S}'$, where U is given by (4.32). However, upon imitating the analysis in the proof of Step 2 of Theorem 5.1, this is readily seen to yield (5.1). Formula (5.2) then follows easily from (5.1) with the aid of the identity

$$(I_p + G^*)^{-1}(I_p - G^*G)(I_p + G)^{-1} = (I_p + G)^{-1}(I_p - GG^*)(I_p + G^*)^{-1} ,$$

which is valid for any strictly contractive $p \times p$ matrix G, upon setting $G = V_\tau S$ and using the second representation for V_τ in (5.3). ∎

6. MAXIMUM ENTROPY, CANONICAL EQUATIONS, CHRISTOFFEL-DARBOUX AND FOURIER ANALYSIS

This section is devoted to a brief sketch of applications of the theory developed in the first section to the topics listed in the title, just above.

To begin with, Theorem 5.2 serves to parametrize $\mathcal{P}_\tau(h)$ in terms of the elements S in $\mathcal{W}_+^0$ which are strictly contractive on $\overline{\mathbb{C}}_+$. The particular choice $S = 0$ corresponds to $\hat{f} = \hat{f}_\tau$. When $I - H_\tau$ is positive definite, as in the present case, then this particular interpolant can also be uniquely characterized as the only element in $\hat{f}$ in $\mathcal{P}_\tau(h)$ with

$$\{(I_p - \hat{f})^{-1} - I_p\}^\vee(t) = 0 \tag{6.1}$$

for $|t| > \tau$. For this reason $\hat{f}_\tau$ was referred to as the band extension in Dym and Gohberg (1980) and (1982/3), which develop a theory of band extensions in a more general setting and contain in particular a proof of the preceding assertion. It is also shown in Dym and Gohberg (1980) that, *under some auxiliary technical constraints on the data h, the limit*

$$\mathcal{E}(\hat{f}) = \lim_{\varepsilon \downarrow 0} \frac{1}{\pi} \int_{-\infty}^{\infty} \frac{\log \det\{I_p - \hat{f}(\lambda)\}}{1+\varepsilon^2\lambda^2} d\lambda \tag{6.2}$$

exists for every $\hat{f} \in \mathcal{P}_\tau(h) \cap L^2_{p\times p}$ *and, moreover, that*

$$\mathcal{E}(\hat{f}_\tau) \geq \mathcal{E}(\hat{f}) \tag{6.3}$$

with equality if and only if $\hat{f}_\tau = \hat{f}$. This is a natural analogue of the maximum entropy characterization of the band extension which occurs in the discrete case. Another analogue is provided by the next theorem which is essentially due to Arov and Krein (1981) and (1983).

THEOREM 6.1. *If* $I - H_\tau$ *is positive definite, then, for each point* $\omega = a + ib$ *in* $\mathbb{C}_+$,

$$\frac{b}{\pi} \int_{-\infty}^{\infty} \frac{\log \det\{I_p - \hat{f}(\lambda)\}}{(\lambda-a)^2+b^2} d\lambda \leq \log \det\{4\pi b \Lambda_\omega^\tau(\omega)\}^{-1} , \tag{6.4}$$

where $\Lambda_\omega^\tau = \Lambda_\omega$ *is given by (3.7). Moreover, strict inequality prevails for every choice of* $\hat{f} \in \mathcal{P}_\tau(h)$ *unless* $F_\tau(\omega) = 0$, *in which case equality is attained by* $\hat{f}_\tau$ *only.*

The proof, which is furnished in a more general setting in Chapter 11 of Dym (1989b), amounts to evaluating and estimating the integral on the left hand side of (6.4) when $I_p - \hat{f}$ is expressed in terms of E_τ, V_τ and an arbitrary strictly contractive $S \in \mathcal{W}_+^0$ as in (5.1). It turns out that (6.4) is valid for the wider class of $\hat{f}$ which is given by (5.1) but with S merely constrained to belong to $\mathcal{W}_+ \cap \mathcal{S}_{p\times p}$ and not being presumed to be either strictly contractive in $\overline{\mathbb{C}}_+$ or to vanish at ∞. Moreover, equality prevails for $\hat{f}$ in this wider class if and only if S is constant with

$$S(\lambda) = -V_\tau(\omega)^*$$

for every choice of $\lambda \in \overline{\mathbb{C}}_+$. But this is incompatible with the requirement that $S \in \mathcal{W}_+^0$, which is necessary in order for $\hat{f}$ to belong to $\mathcal{P}_\tau(h)$, unless $V_\tau(\omega) = 0$.

Let us now suppose that the given data $h(t)$ is continuous for $|t| \leq \tau'$and that $I - H_{\tau'}$ is positive definite. Then, as we have already noted in the proof of Lemma 3.1, E_τ and F_τ are solutions of the system of differential equations (3.9) and (3.10). Moreover, $E_\tau^\bullet$ and $F_\tau^\bullet$ satisfy a similar system but with $\dot{\gamma}_\tau$ in place of γ_τ. Therefore, since

$$\dot{\gamma}_\tau(0,\tau) = -\gamma_\tau(0,\tau) \ ,$$

by (4.20), and

$$\dot{\gamma}_\tau(\tau,0) = -\gamma_\tau(\tau,0) \ ,$$

by taking complex conjugates of the preceding formula, it is now readily checked that

$$\frac{\partial}{\partial\tau} e^{-i\lambda\tau/2} U_\tau(\lambda) = e^{-i\lambda\tau/2} U_\tau(\lambda) \left\{ \frac{i\lambda}{2} J_1 + P(\tau) \right\} \ ,$$

or, equivalently, that

$$\frac{\partial}{\partial\tau} U_\tau(\lambda) = U_\tau(\lambda) \left\{ \frac{i\lambda}{2}(J_1 + I_m) + P(\tau) \right\} \tag{6.5}$$

for $0 \leq \tau \leq \tau'$ where (the "potential")

$$P(\tau) = M \begin{bmatrix} 0 & \gamma_\tau(\tau,0) \\ \gamma_\tau(0,\tau) & 0 \end{bmatrix} M = P(\tau)^* \ .$$

Thus, upon differentiating the formula

$$K_\omega^\tau(\lambda) = \frac{J_1 - U_\tau(\lambda) J_1 U_\tau(\omega)^*}{\rho_\omega(\lambda)} \tag{6.6}$$

for the reproducing kernel for $\mathcal{H}(U_\tau)$ with respect to τ, it follows readily that

$$\begin{aligned} \frac{\partial K_\omega^\tau(\lambda)}{\partial\tau} &= -\frac{1}{\rho_\omega(\lambda)} \left\{ \{\frac{\partial U_\tau(\lambda)}{\partial\tau}\} J_1 U_\tau(\omega)^* + U_\tau(\lambda) J_1 \{\frac{\partial U_\tau(\omega)}{\partial\tau}\}^* \right\} \\ &= -\frac{U_\tau(\lambda)}{\rho_\omega(\lambda)} \left\{ \frac{i(\lambda - \omega^*)}{2}(J_1 + I_m) + P(\tau) J_1 + J_1 P(\tau) \right\} U_\tau(\omega)^* \\ &= \frac{1}{4\pi} U_\tau(\lambda)(J_1 + I_m) U_\tau(\omega)^* \ , \end{aligned}$$

since

$$P(\tau)J_1 + J_1P(\tau) = 0 \ .$$

Consequently,

$$\begin{aligned} K_\omega^\tau(\lambda) &= \int_0^\tau \frac{\partial K_\omega^s(\lambda)}{\partial s} ds \\ &= \frac{1}{4\pi}\int_0^\tau U_s(\lambda)(J_1 + I_m)U_s(\omega)^* ds \\ &= \frac{1}{4\pi}\int_0^\tau U_s(\lambda)M(J_0 + I_m)MU_s(\omega)^* ds \\ &= \frac{1}{2\pi}\int_0^\tau U_s(\lambda)M\begin{bmatrix} I_p & 0 \\ 0 & 0 \end{bmatrix} MU_s(\omega)^* ds \ , \end{aligned}$$

since $K_\omega^0 = 0$. Therefore, upon invoking (4.32) again, it follows that

$$K_\omega^\tau(\lambda) = \frac{1}{4\pi}\int_0^\tau \begin{bmatrix} F_s(\lambda) & 0 \\ F_s^\bullet(\lambda) & 0 \end{bmatrix} \begin{bmatrix} F_s(\omega) & 0 \\ F_s^\bullet(\omega) & 0 \end{bmatrix}^* ds \tag{6.7}$$

which is the proper analogue of the Christoffel-Darboux formula in the present setting and gives yet another way of expressing K_ω^τ, in addition to (4.6) and the formula which is obtained by substituting (4.32) into (4.35) (it is written out in abbreviated form in (4.36)). The better known auxiliary formulas (3.8) and (4.24) emerge as special cases.

Apart from their intrinsic interest as identities, the Christoffel-Darboux formulas serve as a vehicle for developing a Fourier analysis which is appropriate for $\mathcal{H}(U)$ and the corresponding spaces $\mathcal{M}_0$ and $\mathcal{M}_0^\bullet$, respectively. Thus, for example, if (6.7) is introduced in the formula

$$\xi^* g(\omega) =< J_1 g, K_\omega^\tau \xi >$$

for $g \in \mathcal{H}(U_\tau)$ and $\xi \in \mathbb{C}^m$, and if orders of integration are freely interchanged, then one readily obtains the formula

$$g(\omega) = \int_0^\tau \begin{bmatrix} F_s(\omega) & 0 \\ F_s^\bullet(\omega) & 0 \end{bmatrix} \left\{ \frac{1}{4\pi}\int_{-\infty}^{\infty} \begin{bmatrix} F_s(\lambda) & 0 \\ F_s^\bullet(\lambda) & 0 \end{bmatrix}^* J_1 g(\lambda) d\lambda \right\} ds \ , \tag{6.8}$$

which serves to express g as the iteration of a transform and its inverse. Indeed, in the special case that $h = 0$, then

$$F_s(\lambda) = F_s^{\bullet}(\lambda) = e^{i\lambda s} I_p \ , \quad g = \begin{bmatrix} g_1 \\ g_1 \end{bmatrix}$$

with $g_1 \in \{L_p^2[0,\tau]\}^{\wedge}$ and (6.8) is readily seen to reduce to the iteration of a pair of classical Fourier integrals:

$$\begin{bmatrix} g_1(\omega) \\ g_1(\omega) \end{bmatrix} = \int_0^{\tau} \begin{bmatrix} e^{i\omega s} I_p & 0 \\ e^{i\omega s} I_p & 0 \end{bmatrix} \left\{ \frac{1}{4\pi} \int_{-\infty}^{\infty} \begin{bmatrix} 2e^{-i\lambda s} g_1(\lambda) \\ 0 \end{bmatrix} d\lambda \right\} . \tag{6.9}$$

Formula (6.9) clearly exhibits the redundancy which is inherent in the representation (6.8) even when $h \neq 0$. It also illustrates in a very concrete way the fact that, as noted in Lemmas 4.1 and 4.2, $\mathcal{M}_0$ and $\mathcal{M}_0^{\bullet}$ are each isometrically isomorphic to $\mathcal{H}(U_\tau)$. Formula (6.8) also indicates that K_ω is a reproducing kernel for every point $\omega \in \mathbb{C}$ and not just for $\omega \in \mathbb{C}_+$.

7. REFERENCES

Alpay, D., and Dym, H. (1984). Hilbert spaces of analytic functions, inverse scattering, and operator models I, *Integral Equations and Operator Theory*, **7**: 589-641.

Alpay, D., and Dym, H. (1986). On applications of reproducing kernel spaces to the Schur algorithm and rational J unitary factorization, *I. Schur Methods in Operator Theory and Signal Processing* (I. Gohberg, ed.), Birkhäuser Verlag, Basel, pp. 89-159.

Arov, D.Z., and Krein, M.G. (1981). Problems of search of the minimum of entropy in indeterminate extension problems, *Funct. Anal. Appl.*, **15**: 123-126.

Arov, D.Z., and Krein, M.G. (1983). On computations of entropy functionals and their minimums, *Acta Sci. Math.*, **45**: 51-66.

Ball, J.A., and Helton J.W. (1983). A Beurling-Lax theorem for the Lie group $U(m,n)$ which contains most classical interpolation theory, *J. Operator Theory*, **9**: 107-142.

Bognar, J. (1974). *Indefinite Inner Product Spaces*, Springer Verlag, New York.

de Branges, L. (1963). Some Hilbert spaces of analytic functions I, *Trans. Amer. Math. Soc.*, **106**: 445-468.

Dym, H. (1988). Hermitian block Toeplitz matrices, orthogonal polynomials, reproducing kernel Pontryagin spaces, interpolation and extension, *Orthogonal Matrix-valued Polynomials and Applications* (I. Gohberg, ed.), Birkhäuser Verlag, Basel, pp. 79-135.

Dym, H. (1989a). On reproducing kernel spaces, J unitary matrix functions, interpolation and displacement rank, *The Gohberg Anniversary Collection* (H. Dym, M.A. Kaashoek, P. Lancaster and S. Goldberg, eds.), Birkhäuser Verlag, Basel, in press.

Dym, H. (1989b). *J Contractive Matrix Functions, Reproducing Kernel Hilbert Spaces and Interpolation*, American Mathematical Society, Providence, in press.

Dym, H., and Gohberg, I. (1980). On an extension problem, generalized Fourier analysis and an entropy formula, *Integral Equations and Operator Theory*, **3**: 143-215.

Dym, H., and Gohberg, I. (1982/83). Extensions of kernels of Fredholm operators, *Journal d'Analyse Mathématique*, **42**: 51-97.

Dym, H., and Iacob, A. (1984). Positive definite extensions, canonical equations and inverse problems (H. Dym and I. Gohberg, eds.), *Topics in Operator Theory, Systems and Networks*, Birkhäuser Verlag, Basel, pp. 141-240.

Gohberg, I., Kaashoek, M.A., and Woerdeman, H.J. (1988). The band method for positive and contractive extension problems, preprint.

Krein, M.G. (1955). Continuous analogues of theorems on polynomials orthogonal on the unit circle, *Dokl. Akad. Nauk SSSR*, **105**: 637-640.

Krein, M.G. (1956). A contribution to the theory of accelerants and S-matrices of canonical differential systems, *Dokl. Akad. Nauk SSSR*, **111**: 1167-1170.

Krein, M.G., and Langer, H. (1985). On some continuation problems which are closely related to the theory of operators in spaces Π_κ. IV, *J. Operator Theory*, **13**: 299-417.

Krein, M.G., and Melik Adamyan, F.E. (1986). Matrix continuous analogues of the Schur and Caratheodory Toeplitz problems, *Izv. Akad. Nauk Armyan* SSR, **21**: 107-141.

Langer, H. (1986). Review 86g: 34025, *Mathematical Reviews*, American Mathematical Society, Providence, pp. 2960-2961.

Rovnyak, J. (196?). Characterization of spaces $\mathcal{K}(M)$, unpublished manuscript.

Department of Theoretical Mathematics
The Weizmann Institute of Science
Rehovot 76100, Israel

24

On The Singular Values of The Four-Block Operator and Certain Generalized Interpolation Problems

Ciprian Foias and Allen Tannenbaum

Abstract

In a classic paper of Adamjan-Arov-Krein [1], the connection is made between the singular values of the Hankel operator and the problem of interpolation in the disc by L^∞ functions with a bounded number of poles. In this note we extend these results to the four block operator from control theory.

1 Introduction

In the paper [1], Adamjan-Arov-Krein consider the problem of giving an interpolation theoretic interpretation to the singular values of a Hankel operator. More precisely, let $w \in L^\infty$. (All of our Hardy and Lebesgue spaces will be defined on the unit circle in the standard way.) Let $H^\infty(k) := \{q \in L^\infty : q = r + t$ with r rational and having $\leq k$ poles in the unit disc D and $t \in H^\infty\}$. Set

$$\tilde{\sigma_k} := \inf\{\|w - q\|_\infty : q \in H^\infty(k)\}.$$

Moreover, for $P : L^2 \to H^2$ orthogonal projection, define the *Hankel operator* associated to w by

$$H_w := (I - P)w(U)|H^2 : H^2 \to H^{2\perp} =: L^2 \ominus H^2$$

where $w(U) : L^2 \to L^2$ denotes the operator induced by multiplication by w. ($U : L^2 \to L^2$ is the bilateral shift.) Let $s_k(T), k = 0, \ldots, \infty$ denote the kth singular value of a bounded linear operator T, $s_0(T) \geq s_1(T) \geq \cdots \geq s_k(T) \geq s_{k+1}(T) \geq \cdots$. Note by definition $s_0(T) = \|T\|$. Then in [1], it is proven that $s_k(H_w) = \tilde{\sigma_k}$. For $k = 0$, this is the Nehari theorem [10].

Now in control theory from electrical engineering, a certain generalization of the Hankel operator appears. Namely, let $w, f, g, h \in L^\infty$. Consider the *four block operator*

*This work was supported in part by grants from the Research Fund of Indiana University, Department of Energy DE-FG02-86ER25020, National Science Foundation ECS-8704047, DMS-8811084, and by the Air Force Office of Scientific Research AFOSR-88-0020 .

$$A := \begin{bmatrix} (I-P)w(U)|H^2 & (I-P)f(U) \\ g(U)|H^2 & h(U) \end{bmatrix} : H^2 \oplus L^2 \to H^{2\perp} \oplus L^2.$$

Now set $S' := (I-P)U|H^{2\perp}$. Then S'^* is the unilateral shift on $H^{2\perp}$, and clearly A satisfies

$$A(S \oplus U) = (S' \oplus U)A.$$

The four block operator is used in the solution of the most general H^∞ design problems (see [9] for the engineering details). Let

$$\sigma_k := \inf\left\{ \left\| \begin{bmatrix} w-q & f \\ g & h \end{bmatrix} \right\|_\infty : q \in H^\infty(k) \right\},$$

and $s_k(A) =: s_k$. Using the commutant lifting theorem (see [4], [6], [14]), one can prove that $s_0 = \sigma_0$. The point of this note is to prove that in general $s_k = \sigma_k$. (This is done in Theorem 1 below.) Of course when $f = g = h = 0$, we recover the aforementioned results of Adamjan-Arov-Krein [1].

It is important to note that this result is not merely theoretical. Indeed, for a large class of four block operators which appear in engineering applications, one may actually explicitly compute the singular values and vectors and thus the optimal interpolants from Theorem 1 below. This is based on the authors' skew Toeplitz theory [2], [6], [8] which we will briefly discuss in Section 4.

2 Preliminaries

In this section, we will collect some basic facts that we will need throughout the paper. We follow the notation of the Introduction here. Moreover we use the standard definitions and results of operator theory as may be found for example in [13]. By "operator", we will always mean "bounded linear operator" on a separable Hilbert space.

For T an operator, we let $\|T\|_e = s_\infty(T)$ denote the essential norm. We now state the following well-known result whose proof is included for the convenience of the reader:

Lemma 1 *Let H, H_1 be Hilbert spaces equipped with unilateral shifts S, S_1, and let K, K_1 be Hilbert spaces equipped with bilateral shifts U, U_1. Let $T : H \oplus K \to H_1 \oplus K_1$ be an operator such that $(S_1 \oplus U_1)T = T(S \oplus U)$. Then*

$$\|T\|_e \geq \max\{\|P_{K_1}T\|, \|TP_K\|\}$$

where $P_{K_1} : H \oplus K \to K$, and $P_{K_1} : H_1 \oplus K_1 \to K_1$ denote orthogonal projections.

Proof. Let $f \in H \oplus K$, be such that $\|f\| = 1$. Set $f_n := (S^n \oplus U^n)f$, so that $\|f_n\| = 1$, and $f_n \to 0$ weakly as $n \to \infty$. Now

$$\begin{aligned} \|P_{K_1}Tf\| &= \|P_{K_1}(S_1^n \oplus U_1^n)Tf\| \\ &\leq \|(S_1^n \oplus U_1^n)Tf\| \\ &= \|Tf_n\| \end{aligned}$$

which implies that $\|T\|_e \geq \|P_{K_1}T\|$. Similarly $\|TP_K\| = \|P_K T^*\| \leq \|T^*\|_e = \|T\|_e$ from which we get the required result. □

Notice that from the lemma, for the four block operator defined in the Introduction, we immediately have that

$$\|A\|_e \geq \max\left\{\left\|\begin{bmatrix} g & h \end{bmatrix}\right\|_\infty, \left\|\begin{bmatrix} f \\ h \end{bmatrix}\right\|_\infty\right\}.$$

We will use this fact in the next section. In certain important cases from an applied engineering point of view, we can compute $\|A\|_e$ exactly. We will discuss this in Section 4.

3 Main Result

We are now ready to formulate and prove the generalization of the Adamjan-Arov-Krein [1] result discussed in the Introduction. We will follow the above notation and terminlogy in what follows below.

Theorem 1 *Notation as above. Then*

$$\sigma_k := \inf_{q \in H^\infty(k)} \left\|\begin{bmatrix} w-q & f \\ g & h \end{bmatrix}\right\|_\infty = s_k. \tag{1}$$

Proof. From Lemma 1, we have that

$$s_k \geq \|A\|_e \geq \max\left\{\left\|\begin{bmatrix} g & h \end{bmatrix}\right\|_\infty, \left\|\begin{bmatrix} f \\ h \end{bmatrix}\right\|_\infty\right\} =: \mu. \tag{2}$$

We divide the proof into two cases:

Case 1. Assume $s_k > \|A\|_e$. We may suppose without loss of generality that $s_k = 1$, and so $\|A\|_e < 1$. Let (ψ, η) be a Schmidt pair of vectors of A corresponding to s_k, i.e. a non-zero pair of vectors such that

$$A\psi = \eta \qquad A^*\eta = \psi. \tag{3}$$

If we write,

$$\psi = \begin{bmatrix} \psi_1 \\ \psi_2 \end{bmatrix}, \eta = \begin{bmatrix} \eta_1 \\ \eta_2 \end{bmatrix}$$

we see that,

$$g\psi_1 + h\psi_2 = \eta_2, \qquad \overline{f}\eta_1 + \overline{h}\eta_2 = \psi_2. \tag{4}$$

We claim now that given any two Schmidt pairs of vectors corresponding to s_k, (ψ,η) and (ψ',η'), we have that

$$\eta_1\overline{\eta_1'} + \eta_2\overline{\eta_2'} = \psi_1\overline{\psi_1'} + \psi_2\overline{\psi_2'}. \tag{5}$$

Indeed, for $n \geq 0$,

$$\begin{aligned}
&\int_0^{2\pi} e^{in\theta}\eta\cdot\overline{\eta'}\frac{d\theta}{2\pi} = \langle (S'\oplus U)^n A\psi, \eta'\rangle \\
&= \langle A(S\oplus U)^n\psi, \eta'\rangle \\
&= \langle (S\oplus U)^n\psi, A^*A\psi'\rangle \\
&= \langle (S\oplus U)^n\psi, \psi'\rangle \\
&= \int_0^{2\pi} e^{in\theta}\psi\cdot\overline{\psi'}\frac{d\theta}{2\pi}.
\end{aligned}$$

Since the same argument works for (ψ,η) and (ψ',η') interchanged, we immediately get (5).

Claim 1. Let (ψ,η) be any Schmidt pair of A. Thus ψ and η satisfy (4), and from (5) satisfy

$$|\psi_1|^2 + |\psi_2|^2 = |\eta_1|^2 + |\eta_2|^2. \tag{6}$$

Then we claim that there exists $v \in L^\infty$ such that

$$\left\|\begin{bmatrix} v & f \\ g & h\end{bmatrix}\right\| = 1 = s_k, \quad \begin{bmatrix} v & f \\ g & h\end{bmatrix}\psi = \eta, \quad \begin{bmatrix} \overline{v} & \overline{g} \\ \overline{f} & \overline{h}\end{bmatrix}\eta = \psi. \tag{7}$$

Notice that we are considering the equalities of (7) as holding pointwise on the unit circle ∂D.

In order to verify the claim, we first note that since $\mu < 1$, there exists an outer function $\phi \in L^\infty$ such that

$$|\phi|^2 + |h|^2 = 1,$$

and outer functions $f_2, g_2 \in L^\infty$ such that

$$\begin{aligned}
&|f_2|^2 + |f_1|^2 = 1 \\
&|g_2|^2 + |g_1|^2 = 1
\end{aligned}$$

for $f = f_1\phi$ and $g = g_1\phi$. Now from the results of [3], [5] and from (7) above , we need to show that we can write

$$v = -f_1h^*g_1 + f_2g_2\Gamma \tag{8}$$

for $|\Gamma| = 1$, and $h^*(\zeta) = \overline{h(\zeta)}\phi(\zeta)/\overline{\phi(\zeta)}$ ($|\zeta| = 1$), and such that v satisfies

$$\begin{matrix} v\psi_1 + f\psi_2 = \eta_1 \\ \overline{v}\eta_1 + \overline{g}\eta_2 = \psi_1 \end{matrix} . \tag{9}$$

(Again all the inequalities and equalities are regarded as holding pointwise on ∂D.)

Now note that $\psi_1 \neq 0$. Indeed if $\psi_1 = 0$, then from (4) and our hypothesis that $\mu < 1$,

$$|\psi_2|^2 \leq (|f|^2 + |h|^2)(|\eta_1|^2 + |\eta_2|^2) < \|\eta\|^2 = \|\psi\|^2 = |\psi_2|^2$$

a contradiction. Thus without loss of generality we may assume that $\psi_1 = 1$. Using (9), and solving for Γ in (8), we see that we must show

$$\left| \frac{\eta_1 - f\psi_2 + f_1 h^* g_1}{f_2 g_2} \right| = 1. \tag{10}$$

But from (4) (using the fact that $\psi_1 = 1$), we get

$$\psi_2 = \frac{\overline{f}\eta_1 + \overline{h}g}{1 - |h|^2}, \qquad \eta_2 = \frac{g + h\overline{f}\eta_1}{1 - |h|^2}. \tag{11}$$

Thus (10) (after some computation) is equivalent to

$$|\eta_1|^2 = \frac{1 - |h|^2 - |g|^2}{1 - |h|^2 - |f|^2}. \tag{12}$$

But from (6) and (11), we see that

$$1 + \left| \frac{\overline{f}\eta_1 + \overline{h}g}{(1 - |h|^2)} \right|^2 = |\eta_1|^2 + \left| \frac{g + h\overline{f}\eta_1}{1 - |h|^2} \right|^2 \tag{13}$$

which after some simplification can be shown to be equivalent to (12). This completes the proof of the claim.

Notice that if (7) holds for one pair of Schmidt vectors (ψ, η), then it holds for any other pair of Schmidt vectors (ψ', η'). Indeed in order to see this set

$$M := \begin{bmatrix} v & f \\ g & h \end{bmatrix}.$$

Note that $M^* M\psi = \psi$. Let

$$M\psi' = \begin{bmatrix} \eta'_{1\nu} \\ \eta'_2 \end{bmatrix} = \eta'_\nu.$$

We will now show that $\eta'_{1\nu} = \eta'_1$. Using a similar argument as we used to verify (5), we have that for $n \geq 0$,

$$\int_0^{2\pi} e^{in\theta}\eta'_\nu \cdot \overline{\eta}\frac{d\theta}{2\pi} = \langle (S' \oplus U)^n M\psi', \eta\rangle$$

$$= \langle M(S \oplus U)^n \psi', \eta\rangle$$

$$= \langle (S \oplus U)^n \psi', M^* M\psi\rangle$$

$$= \langle (S \oplus U)^n \psi', \psi\rangle$$

$$= \int_0^{2\pi} e^{in\theta}\psi' \cdot \overline{\psi}\frac{d\theta}{2\pi}.$$

Clearly the same argument works for (ψ, η) and (ψ', η'_ν) interchanged, and so using (5), we see that $\eta_1\overline{\eta'_{1\nu}} + \eta_2\overline{\eta'_2} = \eta_1\overline{\eta'_1} + \eta_2\overline{\eta'_2}$. Hence

$$\eta_1(\eta'_{1\nu} - \eta'_1) = 0.$$

If $\eta_1 \neq 0$, then we are clearly done. On the other hand if $\eta_1 = 0$ on a set of positive measure, by (12), we would have that $1 = |g|^2 + |h|^2$ on that set, contradicting our hypothesis that $\mu < 1$.

Set now

$$A' := \begin{bmatrix} (I-P)v(U)|H^2 & (I-P)f(U) \\ g(U)|H^2 & h(U) \end{bmatrix}$$

From our above argument, $(A' - A)\psi = 0$ if $\psi \in \ker(1 - A^*A)$. (Recall we have normalized $s_k = 1$.) Moreover $(A' - A)(\{0\} \oplus L^2) = 0$. Thus $\ker(A' - A)$ is $S \oplus U$-invariant and contains $\ker(1 - A^*A)$ and $\{0\} \oplus L^2$. Hence by the Beurling-Lax theorem, we have that

$$\ker(A' - A) = mH^2 \oplus L^2$$

for some inner $m \in H^\infty$, and mH^2 is the S-invariant subspace such that $\psi_1 \in mH^2$ where

$$\psi = \begin{bmatrix} \psi_1 \\ \psi_2 \end{bmatrix} \in \ker(1 - A^*A).$$

This means that m is a factor of the greatest common inner divisor of all such ψ_1. Recall that S'^* denotes the unilateral shift on $H^{2\perp}$. Then

$$(m(S') \oplus m(U))(A' - A) = (m(S' \oplus U))(A' - A) = (A' - A)(m(S \oplus U)) = 0.$$

Since

$$(A' - A)\begin{bmatrix} \delta \\ \delta_1 \end{bmatrix} = \begin{bmatrix} (I-P)(w-v)\delta \\ 0 \end{bmatrix}, \quad \text{for} \quad \delta \in H^2$$

we have that

$$(I - P)m(w - v)\delta = m(S')(I - P)(w - v)\delta = 0$$

which implies that $m(w - v) \in H^\infty$, and so $w - v \in \frac{1}{m}H^\infty$.

Claim 2. deg $m \leq k$. In order to prove this, we basically follow the proof of Lemma 1.3 from Adamjan-Arov-Krein [1]. Indeed, let d be the degree of m. The operator $Am(S \oplus U) = A'm(S \oplus U)$, and therefore using the identical argument as in [1] pages 37-38, we see that $\|Am(S \oplus U)\| = s_k$, and is a singular value of multiplicity at least $d + \nu$, where ν is the multiplicity of s_k as a singular value of A. Note that since we have assumed that $s_k > \|A\|_e$, we have that $\nu < \infty$.

Next since

$$m(S \oplus U)^* A^* A m(S \oplus U) = A^* m(S' \oplus U)^* m(S' \oplus U) A \leq A^* A,$$

from the minimium-maximum theorem [12] we deduce that

$$s_n(A) \geq s_n(Am(S \oplus U)).$$

Suppose that we have

$$s_{\kappa-1} > s_\kappa = s_{\kappa+1} = \cdots = s_{\kappa+\nu-1} > s_{\kappa+\nu}$$

and that $k = \kappa + r$, $0 \leq r \leq \nu - 1$. Hence we see that $d \leq \kappa \leq k$, since

$$s_{\kappa+\nu}(A) < s_{\kappa+\nu-1}(A) = s_\kappa(A) = s_{d+\nu-1}(Am(S \oplus U)) \leq s_{d+\nu-1}(A)$$

Actually, one can even show that $\kappa = d$. Indeed, since A and A' coincide on $m(S \oplus U)(H^2 \oplus L^2)$, and

$$\dim((H^2 \oplus L^2) \ominus m(S \oplus U)(H^2 \oplus L^2)) = d$$

we can conclude that

$$s_\kappa(A) = \|A'\| = \|A - (A - A')\| \geq s_d(A)$$

so that $\kappa \leq d$. This completes the proof of Claim 2.

We can now finish the proof of Case 1. The above argument shows that $\sigma_k \leq s_k$. The reverse inequality follows from the following argument. Let $\epsilon > 0$ be given. Let $\phi_\epsilon \in H^\infty(k)$ be such that

$$\left\| \begin{bmatrix} w - \phi_\epsilon & f \\ g & h \end{bmatrix} \right\| \leq \sigma_k + \epsilon.$$

Write $\phi_\epsilon = \phi_{1\epsilon}/b_\epsilon$ where $\phi_{1\epsilon}, b_\epsilon \in H^\infty$ and b_ϵ is inner deg $b_\epsilon \leq k$. Then

$$\|b_\epsilon(S' \oplus U)A\| \leq \left\| \begin{bmatrix} b_\epsilon w - \phi_{1\epsilon} & b_\epsilon f \\ b_\epsilon g & b_\epsilon h \end{bmatrix} \right\|_\infty = \left\| \begin{bmatrix} w - \phi_\epsilon & f \\ g & h \end{bmatrix} \right\|_\infty \leq \sigma_k + \epsilon.$$

Thus

$$\|A(P_{b_\epsilon H^2} \oplus I_{L^2})\| = \|A(b_\epsilon(S)b_\epsilon(S)^* \oplus I_{L^2})\| = \|A(b_\epsilon(S) \oplus b_\epsilon(U))(b_\epsilon(S)^* \oplus b_\epsilon(U)^*)\| \leq \sigma_k + \epsilon.$$

Now

$$A = A(P_{H(b_\epsilon)} \oplus \{0\}) + A(P_{b_\epsilon H^2} \oplus I_{L^2})$$

Set $A_\epsilon := A(P_{b_\epsilon H^2} \oplus I_{L^2})$, so that

$$AA^* = A(P_{H(b_\epsilon)} \oplus \{0\})A^* + A_\epsilon A_\epsilon^* \leq A(P_{H(b_\epsilon)} \oplus \{0\})A^* + (\sigma_k + \epsilon)^2 I.$$

But

$$\text{rank } A(P_{H(b_\epsilon)} \oplus \{0\})A^* \leq \dim\ H(b_\epsilon) = \deg\ b_\epsilon \leq k.$$

Hence by virtue of the minimum-maximum theorem [12], we have that AA^* has at most k eigenvalues $> (\sigma_k + \epsilon)^2$, and thus (since ϵ was arbitrary) AA^* (and A^*A) has at most k eigenvalues $> \sigma_k^2$. Therefore $s_k \leq \sigma_k$ as required.

Case 2. Assume $s_k = \|A\|_e$. In this case, the above proof basically goes through word for word except for Claims 1 and 2. The proof of Claim 2 in this case works exactly as in Adamjan-Arov-Krein [1], Corollary 1.2, page 43, to which we refer the reader for the details. So we must show now how to modify the proof of Claim 1 in order to derive the theorem when $s_k = \|A\|_e$. If $\|A\|_e > \mu$, then the above proof of Claim 1 is valid in this setting as well. So suppose $\|A\|_e = \mu = s_k = 1$ (by assumption as above). On the set where $\max\{|f|^2 + |h|^2, |g|^2 + |h|^2\} < 1$, once again the proof of Claim 1 holds. On the complementary set of points (of the unit circle) we may assume without loss of generality that

$$|f|^2 + |h|^2 = 1.$$

As above, we have that

$$\begin{aligned} |\psi_1|^2 + |\psi_2|^2 &= |\eta_1|^2 + |\eta_2|^2 \\ g\psi_1 + h\psi_2 &= \eta_2 \\ \bar{f}\eta_1 + \bar{h}\eta_2 &= \psi_2 \end{aligned} \tag{14}$$

Using precisely the same argument as in Claim 1, we may take

$$\phi = |f|,\ f_1 = \frac{f}{|f|},\ g_1 = \frac{g}{|f|},\ f_2 = 0.$$

Thus from (8)

$$v = \frac{-f\overline{h}g}{|f|^2}$$

and v must satisfy

$$v\psi_1 + f\psi_2 = \eta_1, \qquad \overline{v}\eta_1 + \overline{g}\eta_2 = \psi_1.$$

Hence we are required to verify the validity of

$$\frac{-f\overline{h}g}{|f|^2}\psi_1 + f\psi_2 = \eta_1 \tag{15}$$

from the conditions (14). But solving for η_1 from (14), does indeed give (15) which completes the proof of Case 2, and of the theorem. □

Remarks. The special kind of Parrott type completion given in Claim 1 of the above proof of Theorem 1, leads to a strenghtened version of the Parrott theorem [11], which is formulated and proven in [7]. Basically this gives an extension which is opposite to the classical maximal entropy extension in classical interpolation theory. See [7] for the details.

In the next section we will make several remarks on the computation of the singular values and optimal interpolants of the four block operator.

4 Skew Toeplitz Theory

In this concluding section we would like to briefly consider the computation of the singular vectors and singular values of the four block operator in a very important special case. The methods are too technical to describe in depth here, and are based on the skew Toeplitz theory of [2], [6], [8]. See [6] for all of the details.

Let $m, \tilde{w}, \tilde{f}, \tilde{g}, \tilde{h} \in H^\infty$ with m an arbitrary non-constant inner function, and $\tilde{w}, \tilde{f}, \tilde{g}, \tilde{h}$ rational. In the engineering context, m represents a mathematical model of a physical device to be controlled ("the plant"), and w, f, g, h model certain disturbances [9]. Consider the four block operator

$$A = \begin{bmatrix} (I-P)w(U)|H^2 & (I-P)f(U) \\ g(U)|H^2 & h(U) \end{bmatrix}$$

with $w = \tilde{w}/m, f = \tilde{f}/m, g = \tilde{g}/m, h = \tilde{h}/m$. From Theorem 1 above (also the commutant lifting theorem [6], [13]), one can infer the existence of $q_{opt} \in H^\infty$ such that

$$\left\| \begin{bmatrix} w - q_{opt} & f \\ g & h \end{bmatrix} \right\| = \|A\|.$$

In the H^∞-design engineering problem, q_{opt} corresponds to an optimal compensator used in the control of the plant m in a feedback loop. (The optimality is defined relative to the given disturbance class [9].)

In order to solve such a control problem, one needs of course a method for computing $\|A\|$ and more generally the singular values s_k for this type of four block operator. It turns out that one of the keys in performing this computation, consists in identifying the essential norm of A. We therefore state the following result for whose proof we refer the interested reader to [6]:

Proposition 1 *Notation as above. Let $T : H^2 \ominus mH^2 \to H^2 \ominus mH^2$ denote the compression of the unilateral shift S, that is $T := P_{H^2 \ominus mH^2} S | H^2 \ominus mH^2$, where $P_{H^2 \ominus mH^2} : H^2 \to H^2 \ominus mH^2$ is orthogonal projection. Then*

$$\|A\| = \max\{\alpha, \mu\}$$

where μ is as in (2) above, and

$$\alpha := \sup\left\{ \left\| \begin{bmatrix} \tilde{w}(\zeta) & \tilde{f}(\zeta) \\ \tilde{g}(\zeta) & \tilde{h}(\zeta) \end{bmatrix} \right\| : \zeta \in \sigma_e(T) \right\}.$$

($\sigma_e(T)$ denotes the essential spectrum of T.)

Having found $\|A\|_e$, in [6] we give a determinantal formula for the computation of $s_k^2 \in \sigma(A^*A)$ and the corresponding Schmidt vectors, when $s_k > \|A\|_e$. This is based on the inversion of two Toeplitz operators and the essential inversion of a skew Toeplitz operator. This procedure solves the most general type of H^∞ design problem appearing in control theory.

Let us conclude by indicating how one derives the optimal interpolant (and thus the opimal compensator) from the Schmidt vectors corresponding to a singular value $s_k > \|A\|_e$. Indeed using the notation of the proof of Theorem 1, we have a Schmidt pair ψ, η such that

$$A\psi = s_k \eta.$$

Thus, there exists $q_{opt}^{(k)} \in H^\infty(k)$ with

$$(w - q_{opt}^{(k)})\psi_1 + f\psi_2 = s_k \eta_1$$

$$g\psi_1 + h\psi_2 = s_k \eta_2.$$

From the proof of Case 1 of Theorem 1, we have that $\psi_1 \neq 0$, so that

$$q_{opt}^{(k)} = w - \frac{s_k \eta_1 - f\psi_2}{\psi_1}.$$

From $q_{opt}^{(k)}$, we can derive the corresponding optimal controller in a given systems design problem [6], [9].

In conclusion, for the class of four block operators appearing in control, we have an explicit method based on the skew Toeplitz theory for writing down the interpolants in $H^\infty(k)$ for the generalized Adamjan-Arov-Krein problem.

References

[1] V. M. Adamjan, D. Z. Arov, and M. G. Krein, "Analytic properties of Schmidt pairs for a Hankel operator and the generalized Schur-Takagi problem," *Math. USSR Sbornik* **15** (1971), 31-73.

[2] H. Bercovici, C. Foias, and A. Tannenbaum, "On skew Toeplitz operators, I," *Operator Theory: Advances and Applications* **29** (1988), 21-42.

[3] C. Davis, W. M. Kahan, and H. F. Weinberger, "Norm-preserving dilations and their applications to optimal error bounds," *SIAM J. Numerical Anal.* **19** (1982), 445-469.

[4] A. Feintuch and B. A. Francis, "Uniformly optimal control of linear systems," *Automatica* **21** (1986), 563-574.

[5] C. Foias and A. Frazho, "Redheffer products and lfting contractions on Hilbert space," *J. Operator Theory* **11** (1984), 193-196.

[6] C. Foias and A. Tannenbaum, "On the four block problem, II: the singular system," *Integral Equations and Operator Theory* **11** (1988), 726-767.

[7] C. Foias and A. Tannenbaum, "A strong Parrot theorem," Technical report, October 1988, submitted for publication.

[8] C. Foias, A. Tannenbaum, and G. Zames, "Some explicit formulae for the singular values of certain Hankel operators with factorizable symbol," *SIAM J. Math. Analysis* **19** (1988), 1081-1091.

[9] B. A. Francis, *A Course in H^∞ Control Theory*, Lecture Notes in Control and Information Sciences, vol. 88, Springer-Verlag, 1987.

[10] Z. Nehari, "On bounded bilinear forms," *Annals of Math.* **65** (1957), 153-162.

[11] S. Parrott, "On the quotient norm and the Sz. Nagy-Foias lifting theorem," *J. Functional Analysis* **30** (1978), 311-328.

[12] S. C. Power, *Hankel Operators on Hilbert Space*, Pitman, Boston, 1982.

[13] B. Sz.-Nagy and C. Foias, *Harmonic Analysis of Operators on Hilbert Space*, North Holland, 1970.

[14] N. Young, "The Nevanlinna-Pick problem for matrix-valued functions," *J. Operator Theory* **15** (1986), 239-265.

25

Some Applications of K-Theory and Ext-Theory to Joint Quasitriangularity

Sudhir Gokhale and Norberto Salinas

ABSTRACT

A survey of the problem of joint quasitriangularity for pairs of essentially commuting essentially normal operators is presented. Even in this particular situation, the problem of joint quasitriangularity is far from being solved. A natural general strategy (used already by various authors) for solving this problem is discussed. Several examples where this program has been successful are reviewed. Techniques from K-theory and Ext-theory play a central role in such an approach. This is illustrated in one specially interesting example by giving complete detailed proofs of the corresponding characterization of joint quasitriangularity.

AMS Subject Classification numbers: $47A15, 46L05$

1 Introduction

This is an expository article that deals with the present status of the problem of joint quasitriangularity for pairs of essentially commuting essentially normal operators on Hilbert space.

The notion of quasitriangularity was introduced by P. R. Halmos in [13], and proved to be central in operator theory. An operator T on a (separable, infinite dimensional) Hilbert space $\mathcal{H}$ is called quasitriangular if there exists an increasing sequence $\{P_n\}$ of finite rank projections on $\mathcal{H}$ that tends strongly to the identity operator $1 = 1_{\mathcal{H}}$ and such that $\{\|(1 - P_n)TP_n\|\}$ tends to zero. The sequence $\{P_n\}$ is said to implement the quasitriangularity of T.

The celebrated result of C. Apostol, C. Foias, and D. Voiculescu (see [1]) states that T is quasitriangular if and only if $ind(T - \lambda) \geq 0, \lambda \in C$ whenever $T - \lambda$ is semi-Fredholm. A very interesting (but rather difficult) problem arises when one tries to extend the above spectral characterization to n-tuples of operators acting on $\mathcal{H}$.

The notion of joint quasitriangularity for an n-tuple of operators is straight forward, and it is simply obtained by requiring that the sequence of projections $\{P_n\}$ implements (simultaneously) the quasitriangularity of all the components of the given n-tuple. What is not so clear is the nature of the spectral data needed to extend the above spectral characterization of Apostol, Foias, and Voiculescu to arbitrary n-tuples of operators. In order to state a reasonable conjecture, one is then led to make some special assumptions on the class of n-tuples under consideration. For instance, one can require that the n-tuples be commuting (see [18] for the best results under this hypothesis), or that the n-tuples be essentially commuting and consist of essentially normal operators (i.e. $[T_i, T_j^*]$ be compact, for every i, j). The latter assumptions are specially convenient because one can apply the powerful tools from K-theory and Ext-theory developed by L. Brown, R Douglas, and P Fillmore in [5] and [6] (see also [3] and [8]).

Even for pairs of essentially commuting essentially normal operators, the problem is not settled yet. The first result under these assumptions was obtained by Gayle Kaplan who successfully characterized (in [14]) joint quasitriangularity for pairs of essentially commuting operators when one of them is essentially self-adjoint (or, equivalently, just selfadjoint) and the other is essentially unitary (see below for the precise statement of her result).

For the general problem of joint quasitriangularity, we refer the reader to [16]. In this paper, we restrict ourselves to considering pairs of essentially commuting essentially normal operators. A natural conjecture for the present case was formulated in [16, Section 1], and is still open. This is based on K-theory and Ext-theory information, and goes as follows:

Conjecture: Let (S, T) be a pair of essentially commuting essentially normal operators. Then (S, T) is jointly quasitriangular if and only if every two by two operator matrix whose entries are polynomials in S and T is quasitriangular.

A particular case of the above conjecture was formulated earlier by R. G. Douglas (see [16, Remark 6.8]). This states that if the pair (S,T) satisfies the additional assumption that its joint essential spectrum $\sigma_e(S,T)$ is homeomorphic to a subset Y of $\mathbf{R}^3$, then (S,T) is jointly quasitriangular if and only if every polynomial in S and T is quasitriangular. (Indeed, recall that an invertible 2×2 matrix F of continuous functions on Y is homotopic, within the invertible matrices, to a diagonal matrix whose diagonal entries are 1 and $det(F)$.) Of course, the "only if part" of the above conjecture always holds because of Apostol, Foias and Voiculescu's Theorem stated previously. (Actually, it follows from an earlier result of [9].) Douglas' student, Gayle Kaplan, verified his conjecture for the case that S is essentially self-adjoint and T is essentially unitary. Her final result is very nice and very elegant, and can be phrased as follows: Assume, without loss of generality, that her operator S is a positive contraction and let $R=(1+S)T$. Then, (S,T) is jointly quasitriangular if and only if $0\leq ind(R-\lambda)\leq ind(R-\mu)$ for any pair of complex numbers (λ,μ) such that $|\mu|\leq|\lambda|$, whenever $R-\lambda$ and $R-\mu$ are Fredholm operators (or, equivalently, semi-Fredholm operators).

But, in the equivalent characterization that dealt with Douglas's conjecture, Kaplan only needed linear polynomials in S and T to be quasitriangular. Afterwards, L. Fialkow (in dealing with a problem on elementary operators) asked (see [10]) whether linear polynomials were enough, in general. In [12], we obtained a characterization of joint quasitriangularity for certain solid torus supported pairs (U,E) (i.e. pairs for which the essential spectra $\sigma_e(U,E)$ are certain subsets of a solid torus) that still confirmed Douglas's conjecture (see also [11] for some partial results along these lines), but which yielded an example where Fialkow's question had a negative answer. In fact, given any positive integer n, we produced (in [12]) a pair (U, E) of essentially commuting essentially normal operators such that any polynomial in U and E of degree at most $n-1$ in E is quasitriangular, but (U, E) is not jointly quasitriangular.

In the present note, we develop a slight generalization of the characterization of joint quasitriangularity obtained in [12] (see Theorem 1.1 below). Our techniques are similar to those of [12], but we present them here with a slightly different angle so as to exhibit the general strategy used to understand the phenomenon of joint quasitriangularity (first outlined in [16]). In addition, we improve the example of [12] (mentioned above), by exhibiting, for any pair of positive integers m, n, a pair of essentially commuting essentially normal operators (U, E) such that any polynomial in U and E which has either degree at most $m-1$ in U or degree at most $n-1$ in E is quasitriangular, but (U, E) is not jointly quasitriangular. (In fact, we will exhibit a polynomial in U and E, of degree m in U and degree n in E, that is not quasitriangular, see Corollary 1.3 below.) Although we are convinced that Douglas's conjecture must be true, this example implies that anything stronger than his conjecture cannot be true. On the other hand, if one makes some special assumptions about the K-theoretical nature of $\sigma_e(U, E)$, experience indicates that it should be possible to reduce the test of quasitriangularity to a rather small class of polynomial operators in U and E. For example, this was also confirmed in all the partial results obtained in [7], where torus supported pairs were considered (i.e., the case where both operators are essentially unitary). In order to state our results we need to introduce some more terminology.

Let $\mathbf{D}$ be the closed unit disk in $\mathbf{C}$. In what follows, the pair (U, E) will be assumed to be a solid torus supported pair (see [12]). This means that (U, E) is a pair of essentially commuting essentially normal operators such that the essential spectrum $\sigma_e(U, E)$ (i.e., the joint spectrum of the image of (U,E) in the Calkin algebra) is contained in the solid torus $S^1 \times \mathbf{D}$. This implies, in particular, that U is essentially unitary, and that $\sigma(E) \subseteq \mathbf{D}$. Of course, the cylinder supported pairs of [14] and the cases of torus supported pairs considered in [7] were all solid torus supported pairs.

Let K be a (non-empty) compact, connected subset of $\mathbf{D}$ such that $S^1 \subseteq K$. Of course, it follows that every hole in K (i.e. a bounded component in $\mathbf{C} \setminus K$) is simply connected. We shall further assume that there exists a non-negative rational number α such that K is $2\pi\alpha$ rotation invariant, (i.e. $e^{2\pi\alpha i}K = K$). We say that two holes Ω and Ω' of K are equivalent if $\Omega' = e^{2\pi\alpha ij}\Omega$, for some positive integer j. Let $\Omega_k, k = 1, 2 \ldots$ be an enumeration of the inequivalent holes of K (each one chosen from a different equivalence class). Observe that

$$K = \mathbf{D} \setminus \cup_{j,k} e^{2\pi\alpha ij}\Omega_k$$

Now, rotate the top of the cylinder $[0, 1] \times K$ (with base $\{0\} \times K$) by $2\pi\alpha$ radians and then identify the base and the top of the new twisted cylinder. The resulting compact subset X' of $\mathbf{R}^3$ is homeomorphic to the subset X of the solid torus $S^1 \times \mathbf{D}$ defined by

$$X = \{(e^{2\pi it}, ze^{2\pi\alpha it}) : (t, z) \in [0, 1] \times K\}.$$

Let $\alpha = \frac{m}{n}$ be the irreducible representation of α, where m and n are relatively prime integers, and $n > 0$ (if $\alpha = 0$, we let $n = 1$). If $\alpha \neq 0$, then each hole of K generates a hole of X' that winds around n times "inside X'" before "joining itself".

For each $k = 1, 2 \ldots$, let λ_k be a fixed point of Ω_k, where $\{\Omega_k\}$ is as above. If $0 \notin K$, we let $\lambda_1 = 0$. The following is the main result of this paper, and constitutes the characterization mentioned previously.

Theorem 1.1 *A pair (U, E) of essentially commuting essentially normal operators such that $\sigma_e(U, E) = X$ is jointly quasitriangular if and only if the indices of the Fredholm operators U and $E^n - \lambda_k^n U^m$ are non-negative for $k = 1, 2, \ldots$.*

The above theorem is proved in Section 3. In section 2, we give the necessary background and auxiliary results. If, in Theorem 1.1, we take $\alpha = \frac{1}{n}$, we obtain the main result of [12]. We also derive the following corollaries.

Corollary 1.2 *Let (U, E) be a pair of essentially commuting essentially normal operators such that $\sigma_e(U, E) = S^1 \times K$. Then (U, E) is jointly quasitriangular if and only if U and E are separately quasitriangular.*

Corollary 1.3 *Let m and n be relatively prime positive integers, and let r and s be two numbers in $(0, 1)$ such that $r \leq s$, and the disks*

$$[s + (r\mathbf{D})]e^{2\pi\alpha i j}, j = 1, 2, \ldots, n,$$

are all disjoint subsets of $\mathbf{D}$, *where* $\alpha = \frac{m}{n}$. *Let*

$$K = \mathbf{D} \setminus Int[\cup_{j=1}^{n}[s + (r\mathbf{D})]e^{2\pi\alpha i j}].$$

Further, let X *be defined from* α *and* K, *as above, and let* (U', E') *be a pair of commuting normal operators whose joint spectrum and joint essential spectrum coincide with* X. *Let*

$$U = 1 \oplus U', E = (s + rU_+) \oplus E',$$

where U_+ *is the standard unilateral shift. Then, every polynomial in* (U, E) *that has either degree at most* $m - 1$ *in* U *or degree at most* $n - 1$ *in* E *is quasitriangular, but* $ind(E^n - s^n U^m) = -1$, *so that* (U, E) *is not jointly quasitriangular.*

We would like to thank Professor Kenneth Davidson, from University of Waterloo, for several very helpful discussions.

2 PRELIMINARIES

In this section, we recall some few standard facts of K-theory and Ext-theory that will be needed in the sequel. Our general references are [3,6,8,16,17].

Given a compact subset Y of $\mathbf{C}^2$, every pair of essentially commuting essentially normal operators $\mathfrak{t} = (T_1, T_2)$ on $\mathcal{H}$ such that $\sigma_e(\mathfrak{t}) = Y$ determines an extension $\tau_{\mathfrak{t}}$ of the ideal of compact operators by the C^*-algebra of continuous functions $C(Y)$, i.e. $\tau_{\mathfrak{t}} : C(Y) \mapsto \mathcal{Q}(\mathcal{H})$ is a unital $*$-monomorphism from $C(Y)$ into the calkin algebra $\mathcal{Q}(\mathcal{H})$ over $\mathcal{H}$, such that

$$\tau_{\mathfrak{t}} z_1 = \pi T_1, \tau_{\mathfrak{t}} z_2 = \pi T_2,$$

where $z_j : Y \mapsto \mathbf{C}, j = 1, 2$, are the standard coordinate functions and $\pi : \mathcal{L}(\mathcal{H}) \mapsto \mathcal{Q}(\mathcal{H})$ is the canonical quotient map. The extension $\tau_{\mathfrak{t}}$ is called quasitriangular (see [16]) if $\mathfrak{t}$ is jointly quasitriangular.

As usual, $Ext(Y)$ will denote the (abelian group) of equivalence classes $[\tau]$ of extensions $\tau : C(Y) \mapsto \mathcal{Q}(\mathcal{H})$ (up to unitary equivalence and compact perturbation). The subsemi-group $Ext_{qt}(Y)$ consisting of equivalence classes containing quasitriangular extensions was introduced in [16] and will be central in our discussions. Using standard elementary properties of the index for Fredholm operators, one can define a homomorphism (see [5])

$$\gamma_\infty : Ext(Y) \mapsto Hom(K^1(Y), \mathbf{Z}).$$

Since we are interested in solid torus supported pairs, we shall assume that Y is homeomorphic to a subset of $\mathbf{R}^3$, so that $K^1(Y)$ is isomorphic to the first cohomotopy group of Y and γ_∞ is actually an onto isomorphism (see [4]).

Let $K_p^1(Y)$ be the subsemi-group of $K^1(Y)$ generated by the invertible (analytic) polynomials on Y (see [16, Section 6]). The problem of joint quasitriangularity can now be reduced to studying the validity of the converse to the following immediate consequence of Apostol-Foias-Voiculescu's Theorem (mentioned in the introduction).

Lemma 2.1 *If* $[\tau_{\mathrm{t}}] \in Ext_{qt}(Y)$, *then* $\gamma_\infty[\tau_{\mathrm{t}}]$ *is a function in* $Hom(K^1(Y), \mathbf{Z})$ *which is non-negative on* $K_p^1(Y)$.

In [16], an example of a subset Z of c^2 was given such that $K^1(Z) = \mathbf{Z}$, but $K_p^1(Z) = \{0\}$. However, this did not give a negative result because, in this case, $Ext(Z) = Ext_{qt}(Z)$. In order to prove Theorem 1.1, we shall proceed with the following strategy (see [16]), with $Y = X$. Although this program may not be always useful in producing general positive results, it proved to be quite handy in many cases.

(I) Show that $K_1(Y)$ is generated by $K_p^1(Y)$ by finding a suitable system of independent generators $\{a_k\}$ in $K_p^1(Y)$.

(II) Find standard models of joint quasitriangular pairs s such that $\sigma_e(\mathrm{s}) = Y$, and such that $\gamma_\infty[\tau_{\mathrm{s}}]$ achieves all possible values of non-negative integers on $\{a_k\}$.

(III) Given a solid torus supported pair t such that $\sigma_e(\mathrm{t}) = Y$ and $\gamma_\infty[\tau_{\mathrm{t}}]\{a_k\}$ has non-negative values, we construct (by (II)) a solid torus supported jointly quasitriangular pair s with the same spectral picture as t, i.e. such that

$$\gamma_\infty[\tau_{\mathrm{s}}]\{a_k\} = \gamma_\infty[\tau_{\mathrm{t}}]\{a_k\}.$$

Since $\{a_k\}$ generates $K^1(Y)$ (by (I)), and γ_∞ is an isomorphism (because $Y \subseteq \mathbf{R}^3$), we conclude that t is unitarily equivalent, up to a compact perturbation, to s, and (hence) that t is jointly quasitriangular.

We close this section by recalling the K-theory six term cyclic exact sequence of the mapping torus. Given a unital C^*-algebra $\mathcal{A}$, and a $*$-automorphism ρ, the mapping torus C^*-algebra $\mathcal{T}_\rho(\mathcal{A})$ associated to the pair $(\mathcal{A}, \rho)$ is the C^*-algebra of all continuous functions $\phi : [0,1] \mapsto \mathcal{A}$, such that $\phi(1) = \rho\phi(0)$. Let $\delta : \mathcal{T}_\rho(\mathcal{A}) \mapsto \mathcal{A}$ be the evaluation map at zero, and let $\delta_* : K_*(\mathcal{T}_\rho(\mathcal{A}) \mapsto K_*(\mathcal{A})$ be the map induced in K-theory. Also, let $\Delta : K_0(\mathcal{A}) \mapsto K_1(\mathcal{T}_\rho(\mathcal{A}))$ given by $[\Delta[P](t)] = [e^{2\pi i P t}]$, where P is an idempotent in some matrix algebra over $\mathcal{A}$ (this is essentially the Bott-map). Also, let ID represent the identity homomorphism and $\rho_* : K_*(\mathcal{A}) \mapsto K_*(\mathcal{A})$ denote the homomorphism induced by ρ. Then, the following six term cyclic sequence is exact (see, for instance, [3] or [15]):

$$\begin{array}{ccccc} K_1(\mathcal{A}) & \longrightarrow & K_0(\mathcal{T}_\rho(\mathcal{A})) & \xrightarrow{\delta_*} & K_0(\mathcal{A}) \\ {\scriptstyle \rho_*-ID}\uparrow & & & & \downarrow{\scriptstyle \rho_*-ID} \\ K_1(\mathcal{A}) & \xleftarrow{\delta_*} & K_1(\mathcal{T}_\rho(\mathcal{A})) & \xleftarrow{\Delta} & K_0(\mathcal{A}) \end{array} \tag{1}$$

3 PROOF OF THE RESULTS

As we pointed out in the previous section, in order to complete the proof of Theorem 1.1, we need to carry out steps (I) and (II) of our program for $Y = X$. We start with (I).

Lemma 3.1 *Let $\alpha = \frac{m}{n}$ and K be as in the introduction, and let X be the subset of the solid torus considered in Theorem 1.1. Also, let $z_j : X \mapsto \mathbf{C}, j = 1, 2$, be the coordinate maps.*

(a) If $0 \in K$, then the polynomials z_1 and $z_2^n - \lambda_k^n z_1^m, k = 1, 2, \ldots$ are invertible in $C(X)$ and the classes containing them freely generate $K^1(X)$.

(b) If $0 \notin K$, then $K^1(X)$ is freely generated by the set

$$[z_1], [z_2], [z_2^n - \lambda_k^n z_1^m], k = 2, 3, \ldots.$$

Proof: Let $\mathcal{A} = C(K)$ and let ρ be the $*$-automorphism on $\mathcal{A}$ induced by the rotation on K of angle $2\pi\alpha$ radians. We recall that the set X is the range of the map $\psi : [0,1] \times K \mapsto \mathbf{C}^2$, whose components are given by $\psi_1(t, z) = e^{2\pi i t}$, and $\psi_2(t, z) = z e^{2\pi \alpha i t}$ It readily follows that $C(X)$ is $*$-isomorphic to the C^*-subalgebra of $C([0,1] \times K)$ generated by $\psi_j = z_j \circ \psi, j = 1, 2.$, which is, precisely, the mapping torus $\mathcal{T}_\rho(\mathcal{A})$. Since we shall need to use diagram (1), we first recall that $K^0(K) = K_0(\mathcal{A}) \simeq \mathbf{Z}$ and it is generated by the trivial class [1]. Then, the vertical map $\rho_* - ID$ on the right of (1) is zero. In particular, $Ker(\Delta) = \{0\}$, and, hence, the following sequence is exact:

$$0 \to K_0(\mathcal{A}) \to K_1(\mathcal{T}_\rho(\mathcal{A})) \to Ran(\delta_*) \to 0 \tag{2}$$

We now compute $Ran(\delta_*) = Ker(\rho_* - ID)$. For each $1 \leq j \leq n, k = 1, 2, \ldots$, define the invertible polynomials $a_{j,k} \in \mathcal{A}$ by $a_{j,k}(z) = z - \lambda_k e^{2\pi \alpha i j}$, where $z : K \mapsto \mathbf{C}$ is the coordinate map on K. In order to prove (a), suppose $0 \in K$. Then, $K_1(K) = K_1(\mathcal{A})$ is freely generated by the classes $[a_{j,k}], 1 \leq j \leq n, k = 1, 2, \ldots$. Since $K^1(K)$ is the first cohomotopy group of $\mathcal{A}$, we denote its abelian operation multiplicatively. Thus, it follows that an arbitrary element $[a] \in K_1(\mathcal{A})$ is of the form

$$[a] = \prod_k \prod_{j=1}^n [a_{j,k}]^{q_{j,k}},$$

where, for each j, $\{q_{j,k}\}$ is a sequence in $\mathbf{Z}$ with finitely many non-zero terms. We deduce that $[a] \in Ker(\delta_*)$ if and only if

$$\prod_k \prod_{j=1}^n [\rho a_{j,k}]^{q_{j,k}} [a_{j,k}]^{-q_{j,k}} = [1]. \tag{3}$$

On the other hand,

$$\begin{aligned}(\rho a_{j,k})(z) &= a_{j,k}(ze^{2\pi\alpha i}) = (ze^{2\pi\alpha i} - \lambda_k e^{2\pi\alpha ij}) \\ &= (z - \lambda_k e^{2\pi\alpha i(j-1)})e^{2\pi\alpha i} = a_{j-1,k}(z)e^{2\pi\alpha i}.\end{aligned}$$

Therefore, as elements of $K_1(\mathcal{A})$, we have: $[\rho a_{j,k}] = [a_{j-1,k}], 1 \leq j \leq n, k = 1, 2, \ldots$. In particular, it follows, from (3), that the above element $[a]$ is in $Ker(\rho_* - ID)$ if and only if $q_{1,k} = q_{2,k} = \ldots = q_{n,k}$ for every k. We infer that the sequence $\{[b_k]\}$ defined by $b_k = \prod_{j=1}^n a_{j,k}, k = 1, 2, \ldots$, generates freely $Ran(\delta_*) = Ker(\rho_* - ID)$. Since $\alpha = \frac{m}{n}$, we observe that

$$b_k(z) = \prod_{j=1}^n (z - \lambda_k e^{\frac{2\pi jm}{n}}) = (z^n - \lambda_k^n), k = 1, 2, \ldots.$$

By the exactness of (2), we deduce that $K^1(\mathcal{T}_\rho(\mathcal{A}))$ is freely generated by $\Delta[1] = [e^{2\pi it}] = [\psi_1]$, and the classes containing the invertible elements defined by

$$(\psi_2^n - \lambda_k^n \psi_1^m)(t, z) = z^n e^{2\pi imt} - \lambda_k^n e^{2\pi imt},$$

because the image of these functions under δ, i.e. their values at zero, coincide with $b_k, k = 1, 2, \ldots$. By the observation made at the beginning of this proof, $K^1(X) = K_1(C(X))$ is freely generated by $[z_1]$ and by $[z_2^n - \lambda_k^n z_1^m], k = 1, 2, \ldots$, and (a) follows. For the proof of (b), assume that $0 \notin K$. We recall that in this case we choose $\lambda^1 = 0$, so that $a_{1,j}(z) = z, 1 \leq j \leq n$. Reasoning as above, we see that $Ran(\delta_*)$ is freely generated by $[a_{1,1}]$ and $[\prod_{j=1}^n a_{j,k}], k = 2, 3, \ldots$. The proof of (b) is completed by using the same arguments as in (a). □

We now carry out step (II) of our program.

Lemma 3.2 *Let X be as in Theorem 1.1, and let p and $\{q_k : k = 1, 2, \ldots\}$ be arbitrary non-negative integers.*

(a) Assume that $0 \in K$. Then, there exists a solid torus supported quasitriangular pair $t = (U, E)$ such that

$$\sigma_e(t) = X, ind(U) = p,$$
$$ind(E^n - \lambda_k^n U^m) = q_k, k = 1, 2, \ldots.$$

(b) Assume that $0 \notin K$. Then, there exists a solid torus supported quasitriangular pair t, such that

$$\sigma_e(t) = X, ind(U) = p, ind(E) = q_1,$$
$$ind(E^n - \lambda_k^n U^m) = q_k, k = 2, 3, \ldots.$$

Proof: Assume that $0 \in K$ (the proof of (b) is analogous, as in step (I), and will be omitted). Let $s = (U', E')$ be a pair of commuting normal operators such that $\sigma(s) = \sigma_e(s) = X$. Also, for each (inequivalent) hole Ω_k

of K (see the introduction), let T_k be the adjoint of the Bergman operator (multiplication by the coordinate function) on the Bergman space over $\Omega_k^* = \{z \in \mathbb{C} : \bar{z} \in \Omega_k\}$. Then (see [2]) the self-commutator $[T_k, T_k^*]$ of T_k is trace class, and its trace norm is bounded by $\frac{1}{\pi}\mu(\Omega_k)$, where μ denotes Lebesgue measure. Also, $\sigma_e(T_k) \subseteq \partial\Omega_k$, and $ind(T_k - \lambda_k) = 1$. Now, let p and $\{q_k : k = 1, 2, \ldots\}$ be arbitrary non-negative integers, and let $\mathrm{t} = (U, E)$ be the pair of operators defined as follows:

$$E = 1 \otimes 1_p \oplus \bigoplus_k T_k \otimes 1_{q_k} \oplus E'$$
$$U = U_+^* \otimes 1_p \oplus \bigoplus_k 1 \otimes 1_{q_k} \oplus U',$$

where U_+ is the standard unilateral shift, and 1_r denotes the $r \times r$ identity matrix. It readily follows that t is a quasitriangular pair of commuting essentially nomal operators. Since U' is unitary, $ind(U) = p$. Also, observe that $\{1\} \otimes \cup_k \partial\Omega_k \subseteq X$, and since, by assumption, $S^1 \subseteq K$, it follows that $S^1 \otimes \{1\} \subseteq X$. Thus, we deduce that $\sigma(\mathrm{t}) = \sigma_e(\mathrm{t}) = X$. Finally, for each $k = 1, 2, \ldots$, it is easy to see that

$$\begin{aligned} ind(E^n - \lambda_k^n U^m) &= ind(1 - \lambda_k^n (U_+^*)^{mp}) + q_k ind(T_k^n - \lambda_k^n) + \\ &\quad + ind(\bigoplus_{j \neq k} (T_k^n \otimes 1_{q_j} - \lambda_k^n)) = \\ &= q_k \sum_{j=1}^{n} ind(T_k - \lambda_k e^{2\pi\alpha i j}) = q_k, \end{aligned}$$

as desired.

Remark 3.3

We now complete the proof of the corollaries of Theorem 1.1. Since Corollary 1.2 follows immediately from Theorem 1.1, by taking $\alpha = 0$, we will concern ourselves with Corollary 1.3. First, we observe that the proof that $ind(E^n - s^n U^m) = -1$ follows exactly the same pattern as the final part of the proof of Theorem 1.1 above. Now, we will prove a slightly stronger statement than the first part of Corollary 1.3, namely: if V is any unitary operator that essentially commute with E such that $\sigma_e(V, E) = X$, and $P(V, E)$ is any polynomial in (V, E) that has either degree at most $m - 1$ in V or at most $n - 1$ in E, then $P(V, E)$ is quasitriangular (and, hence, quasidiagonal). Indeed, it suffices to show that if P is an invertible polynomial in $C(X)$ with the required degree properties, then $ind(P(V, E)) = 0$. Let $\mathrm{s} = (V, E)$, and let τ_{s} be the extension determined by s. Since, by Lemma 3.1, $K^1(X)$ is freely generated by $[z_1], [z_2^n - s^n z_1^m]$, and $[P] \in K^1(X)$, we deduce that there exist integers p and q such that

$$\begin{aligned}[P] &= [z_1]^p [z_2^n - s^n z_1^m]^q \\ &= [z_1^p (z_2^n - s^n z_1^m)^q].\end{aligned}$$

Now, assume first that $P(z_1, z_2)$ is of degree at most $n - 1$ in z_2. Since $\{1\} \times K \subseteq X$, the last identity should still be valid for any $(1, z), z \in K$. So, in $C(K)$ we must have $[P(1, z)] = [(z^n - s^n)^q]$. This forces $q = 0$. On the other hand, assume that $P(z_1, z_2)$ is of degree at most $m - 1$ in z_1. By construction, $0 \in K$, so that $S^1 \times \{0\} \subseteq X$, and hence, for every $w \in S^1$, we see that, in $C(S^1)$, we have:

$$[P(w, 0)] = [w^p (-s^n w^m)^q] = [w^{p+mq}].$$

We again conclude that $q = 0$. Therefore, $[P(z_1, z_2)] = [z_1]^p$. Since V is unitary, we apply $\gamma_\infty[\tau_s]$ to both sides of the above equation to obtain:

$$ind(P(V, E)) = p[ind(V)] = 0,$$

as desired. □

References

[1] C. Apostol, C. Foias, and D. Voiculescu, *Some Results on Non-Quasi-triangular Operators IV*, Rev. Roumaine Math. Pures Appl. 18, 1973, 487–513.

[2] C. Berger, and B. Shaw, *Self-commutators of Multi-Cyclic Hyponormal Operators Are Always Trace Class*, Bull. Amer. Math. Soc., 79, 1973, 1139–1199.

[3] B. Blackadar, *K-Theory for Operator Algebras*, M.S.R.I. Publications, Springer Verlag, 5, 1986.

[4] L. Brown, *Universal Coefficient Theorem for Ext and Quasidiagonality*, Proceedings of a Conference on Operator Algebras and Group Representations, Roumania, Pitman, 1981, 60–64.

[5] L. Brown, R. Douglas, and P. Fillmore, *Unitary Equivalence Modulo the Compact Operators and Extensions of c^*-Algebras*, Proceedings of a Conference on Operator Theory, Lectures Notes in Math., 345, Springer Verlag, 1973, 58–128.

[6] L Brown, R. Douglas, and P. Fillmore, *Extensions of C^*-Algebras and K-Homology*, Ann. of Math., 105, 1977, 265–324.

[7] K. Davidson, and N. Salinas, *Joint Quasitriangularity of Torus Supported Pairs*, in preparation.

[8] R. Douglas, *C^*-Algebra Extensions and K-Homology*, Ann. of Math. Studies, Princeton University Press, 95, 1980.

[9] R. Douglas, and C Pearcy, *A Note on Quasitriangular Operators*, Duke Math. J., 37, 1970, 177–188.

[10] L. Fialkow, *Index of an Elementary Operator*, Indiana University Math. J., 35, 1986, 73–102.

[11] S. Gokhale, *Joint quasitriangularity of Certain Pairs of Essentially Normal and Essentially Commuting Operators*, dissertation, University of Kansas, Lawrence, 1983.

[12] S. Gokhale, and N Salinas, *Joint Quasitriangularity of a Solid Torus Supported Pair, and a Question of Fialkow*, J. of Operator Theory, 17, 1987, 201–211.

[13] P. Halmos, *Quasitriangular Operators*, Acta Sci. Math., Szeged, 29, 1968, 283–293.

[14] G. Kaplan, *Joint Quasitriangularity of essentially normal, essentially commuting operators on Infinite Dimensional Hilbert Spaces*, dissertation, S.U.N.Y. at Stony Brook, 1979.

[15] W. Paschke, *On the Mapping Torus of an Automorphism*, Proc., Amer. Soc., 88, 1983, 481–485.

[16] N. Salinas, *Quasitriangular Extensions of C^*-Algebras, and Problems on Joint Quasitriangularity of Operators*, J. of Operator Theory, 10, 1983, 167–205.

[17] J. Taylor, *Banach Algebras and Topology*, in Algebras and Real Analysis, edited by J. Willamson, Academic Press, 1975, 118–187.

[18] D. Voiculescu, *Some Extensions of Quasitriangularity II*, Rev. Roumaine Math. Pures Appl., 18, 1973, 1439–1459.

26

Algebraic Operators and Invariant Linear Manifolds

Domingo A. Herrero

This note provides a characterization of the non-algebraic (bounded linear) operators acting on a complex, separable, infinite dimensional Hilbert space $\mathcal{H}$, which answers a question of R. E. Curto and L. A. Fialkow (personal communication, 1987).

Let $\mathcal{L}(\mathcal{H})$ denote the algebra of all operators acting on the space $\mathcal{H}$.

PROPOSITION 1. *If* $T \in \mathcal{L}(\mathcal{H})$ *is algebraic, and* $\mathcal{M}$ *is a linear manifold invariant under* T, *then* $\dim \mathcal{M}/(T-\lambda)\mathcal{M} < \infty$ *for all* $\lambda \notin \sigma_e(T)$ *(= the essential spectrum of* T*).*

Proof. If $\lambda \notin \sigma(T)$ (= the spectrum of T), then $(T-\lambda)^{-1}$ is a polynomial in T and 1. Therefore,

$$(T-\lambda)\mathcal{M} \subset \mathcal{M} \quad \text{and} \quad (T-\lambda)^{-1}\mathcal{M} \subset \mathcal{M},$$

whence it follows that $(T-\lambda)\mathcal{M} = \mathcal{M}$, and $\dim \mathcal{M}/(T-\lambda)\mathcal{M} = 0$.

If $\lambda \in \sigma(T) \setminus \sigma_e(T)$, and $E(\lambda)$ is the Riesz idempotent

This research was partially supported by a Grant of the National Science Foundation.

associated with the clopen subset $\{\lambda\}$ of $\sigma(T)$, then $E(\lambda)$ is a finite rank operator; $E(\lambda)$ and $(T - \lambda | \ker E(\lambda))^{-1}$ are polynomials in T. If $TM \subset M$, then $E(\lambda)M \subset M$ and $(1 - E(\lambda))M \subset M$, so that M is the algebraic direct sum of $E(\lambda)M$ and $(1-E(\lambda))M$.

It follows (as in the previous case) that

$$(1 - E(\lambda))M \subset (T - \lambda)M \subset M,$$

so that

$$0 \leq \dim M/(T-\lambda)M \leq \dim M/(1-E(\lambda))M = \operatorname{rank} E(\lambda)M \quad \blacksquare.$$

PROPOSITION 2. *If* $T \in L(H)$ *is not algebraic, then for each complex* λ *there is a trace class operator* $K(\lambda)$ *such that* $T \operatorname{ran} K(\lambda) \subset \operatorname{ran} K(\lambda)$ *and the algebraic dimension of the quotient* $\operatorname{ran} K(\lambda)/(T-\lambda) \operatorname{ran} K(\lambda)$ *is the power of the continuum* c.

Proof. H. Salas provided the following example in the paper of R. E. Curto and L. A. Fialkow (1987): let S be the unilateral shift, defined by $Se_n = e_{n+1}$ $(n \geq 0)$ with respect to the orthonormal basis $\{e_n\}_{n=0}^{\infty}$, and let X be the diagonal operator defined by $Xe_n = \alpha_n e_n$ $(n \geq 0)$, where $\alpha_0 = 1$, $\alpha_1 = \frac{1}{2}$, $\alpha_2 = 1$, $\alpha_3 = \frac{1}{2}$, $\alpha_4 = \frac{1}{4}$, $\alpha_5 = 1$, $\alpha_6 = \frac{1}{2}$, $\alpha_7 = \frac{1}{4}$, $\alpha_8 = \frac{1}{8}$, $\alpha_9 = 1, \ldots$; then S ran X   ran X and algebraic dimension ran X/S ran X $= c$.

Observe that if $Ze_n = 2^{-n}e_n$ $(n \geq 0)$, then $ZS = (\frac{1}{2}S)Z$. It is immediate that

$$S \operatorname{ran} ZX = SZ \operatorname{ran} X = ZS \operatorname{ran} X \subset Z \operatorname{ran} X = \operatorname{ran} ZX.$$

Thus, $J = ZX$ is a trace class operator, $N = \operatorname{ran} J$ is invariant under S, and algebraic dimension $N/SN = c$.

Let λ be any point of the complex plane. By replacing (if necessary) $T - \lambda$ by $\alpha(T - \lambda) + \beta$, we can directly assume that $\lambda = 0$ and T is a strict contraction. Since T is not algebraic, it follows from Kaplansky's theorem that there exists y in H such that $\{T^n y\}_{n=0}^{\infty}$ is a linearly independent set of vectors. Let R be the closed linear span of these vectors; clearly, R is an infinite dimensional invariant subspace of T, and $R = T|R$ *is a cyclic strict contraction* [2].

According to Proposition 3 of B. Sz.-Nagy and C. Foiaş article [3], there exists an injective operator $Y:\ell^2 \to R$ with trivial kernel such that $RY = YS$.

Define $M = YN = \text{ran } YJ$. Clearly, YJ is a trace class operator, $M \subset R$, and

$$TM = RM = RYM = YSN \subset YN = M,$$

so that M is invariant under T.

On the other hand, since Y is injective,

$$\dim M/TM = \dim M/RM = \dim YN/RYN = \dim YN/Y(SN) = c. \blacksquare$$

REMARK. Since the intertwinning map Y can be trace class operator, the argument cannot be used to construct $M = \text{ran } K(\lambda)$ (as in Proposition 2) with $K(\lambda)$ a *non-compact operator*.

THEOREM. *The following are equivalent for* T *in* $L(H)$:

(i) T *is not algebraic.*

(ii) *For some* $\lambda \notin \sigma_e(T)$, *there is an invariant linear manifold* M *such that* $M/(T - \lambda)M$ *is infinite dimensional.*

(iii) For each $\lambda \in \mathbb{C}$, *there exists a trace class operator* $K(\lambda)$ *such that* $M = \operatorname{ran} K(\lambda)$ *is invariant under* T, *and* $\dim M/(T-\lambda)M = c$.

REFERENCES

1. Curto, R. E. and Fialkow, L. A. (1987). Similarity, quasisimilarity, and operator factorization (preprint).

2. Kaplansky, I. (1969). Infinite abelian groups, revised edition, The University of Michigan Press, Ann Arbor, p. 40.

3. Sz.-Nagy, B. and Foiaş, C. (1968). Vecteurs cycliques et quasiaffinités, Studia Math., 31: 35-42.

27

A Characterization of Coanalytical Toeplitz Operators in Cowen-Douglas Class

Lin Qing and Norberto Salinas

ABSTRACT

It is shown that the adjoint of a Toeplitz operator whose symbol is a bounded holomorphic function ϕ on the unit disk is in the class $B_m(\Omega)$ of Cowen and Douglas if and only if ϕ is an m-analytic cover of Ω.

AMS subject classification: 47B35, 47B38, 30C25.

In this paper, we prove a general result concerning the canonical models of generalized Bergman kernels that yields, as a very special case, the complete characterization of the analytic Toeplitz operators whose adjoints belong to the class $B_m(\Omega)$ of Cowen and Douglas (see [5]). The same result (see Corollary 3 below) applies to Szego Toeplitz operators (those acting on the Hardy space) and to the Bergman Toeplitz operators (those acting on the Bergman space).

Let Δ be a fixed (bounded) domain in the complex plane $\mathbf{C}$, and let $H^\infty(\Delta)$ denote the Banach algebra of all bounded holomorphic functions on Δ. Given a (bounded) domain $\Omega \subset \mathbf{C}$, we say that a function Φ in $H^\infty(\Delta)$ is an m-analytic cover of Ω whenever $\Omega \subset \Phi(\Delta)$, and for every w in Ω, there exists an invertible function ψ in $H^\infty(\Delta)$ and p points $z_1,\dots,z_p$ in Δ, such that

$$\Phi(z) - w = \psi(z) \prod_{i=1}^{p} (z - z_i)^{k_i},$$

where

$$\sum_{i=1}^{p} k_i = m.$$

It can be proved (see [3] and [8]) that for a given domain $\Omega \subset \mathbf{C}$, a bounded holomorphic function Φ on Δ such that $\Phi^{-1}(\Omega) \subset \Delta$ is an m-analytic cover of Ω if and only if Φ: $\Phi^{-1}(\Omega) \longrightarrow \Omega$ is a proper map of multiplicity m.

As is shown in [6], every operator in the Cowen-Douglas class $B_m(\Omega)$ is unitarily equivalent to the adjoint of the canonical model associated with a generalized Bergman kernel K: $\Omega\times\Omega \longrightarrow \mathcal{L}(\mathbb{C}^m)$. This is the reproducing kernel of a functional Hilbert space $\mathcal{X}$ of $\mathbb{C}^m$ valued holomorphic functions on Ω, with the following properties:

I) $K(\lambda,\mu)^* = K(\mu,\lambda)$, and $K(\lambda,\lambda)$ is a positive invertible operator on $\mathbb{C}^m$, for every $\lambda,\mu \in \Omega$,

II) K is holomorphic in the first variable and antiholomorphic in the second variable (i.e., k is a sesquianalytic function),

III) for every function f in $\mathcal{X}$, for all vectors ξ in $\mathbb{C}^m$ and every $\lambda \in \Omega$, we have: $\langle f(\lambda),\xi\rangle = \langle f,K(.,\lambda)\xi\rangle_{\mathcal{X}}$, and

IV) multiplication by the coordinate function z induces a bounded operator T_z on $\mathcal{X}$ (called the canonical model associated with K) such that $(T_z - \lambda)^*$ is onto, and $\text{Ker}(T_z - \lambda)^* = \{K(.,\lambda)\xi:\ \xi \in \mathbb{C}^m\}$, for every $\lambda \in \Omega$.

The case m = 1 of the above setup is particularly important. Indeed, for m = 1 and under assumptions I) through IV), there exists a norm decreasing injective homomorphism F_K from the commutant $(T_z)'$ of T_z into $H^\infty(\Omega)$, determined by the identity

$$S^* K(.,z) = [F_K(\bar{S})](z)\ K(.,z), \text{ for every S in } (T_z)'$$

and every $z \in \Omega$. In [6], Theorem 5.2 (see also [5]), sufficient conditions are given for the map F_K to be surjective, and hence a topological isomorphism. A necessary condition for this to occur is that the spectrum $\sigma(T_z)$ of the canonical model associated with K coincides with the closure $\bar{\Omega}$ of Ω, so that the essential spectrum $\sigma_e(T_z)$ coincides with $\partial\Omega$.

For the fixed domain $\Delta \subset \mathbb{C}$, we shall assume that F_K is surjective. So, from now on, we let T_ϕ be the operator on $\mathcal{X}$ defined by $T_\phi = (F_K)^{-1}(\phi)$. Thus,

$$(*) \qquad (T_\phi)^*K(.,z) = \bar{\phi}(z)K(.,z), \quad z \in \Delta.$$

In many situations, T_ϕ behaves like an analytic Toeplitz operator; indeed, if Δ is the open unit disk $\mathbb{D}$ and K is either the Szego kernel or the Bergman kernel over $\mathbb{D}$, then T_ϕ is actually an analytic Toeplitz operator.

Proposition 1. Let $\Delta \subset \mathbb{C}$ be a bounded domain and let $K: \Delta\times\Delta \to \mathbb{C}$ be a generalized Bergman kernel such that F_K is surjective, and such that the function $z \to K_z(.) = K(.,z)/\sqrt{K(z,z)}$ satisfies the additional property

V) $K_z \to 0$ weakly, as $\text{dist}(z,\partial\Delta) \to 0$.

Then, for a given $\chi \in H^\infty(\Delta)$, T_χ is Fredholm if and only if χ is bounded away from zero near $\partial\Delta$. Further, for every $\phi \in H^\infty(\Delta)$ we have:

$$(**) \qquad \partial\phi(\Delta) \subset \bigcap_{\varepsilon>0} \phi(\Delta_\varepsilon) = \sigma_e(T_\phi),$$

where $\Delta_\varepsilon = \{z \in \Delta: \text{dist}(z,\partial\Delta) < \varepsilon\}$.

Proof. Assume that χ in $H^\infty(\Delta)$ is bounded away from zero near $\partial\Delta$. Then χ has finitely many zeros $\omega_1,\ldots,\omega_p$ in Δ, and we can write

$$\chi(\zeta) = \psi(\zeta) \prod_{1\le j\le p} (\zeta - \omega_j), \quad \zeta \in \Delta$$

where ψ is in $H^\infty(\Delta)$ and it is bounded away from zero on Δ. This means that ψ is invertible in $H^\infty(\Delta)$, and since F_K is surjective, T_ψ is invertible. Using the fact that K is a generalized Bergman kernel, we deduce that $T_z - \omega_j$ is Fredholm for $1\le j\le p$, and hence T_χ is Fredholm. In order to prove the only if part, assume now that T_χ is Fredholm. By way of contradiction, suppose that there exists a sequence $\{\zeta_j\} \subset \Delta$ such that $\zeta_j \to \zeta_0 \in \partial\Delta$ and $\chi(\zeta_j) \to 0$. But, from (*) and the fact that $\{K_{\zeta_j}\}$ is a sequence of unit vectors, we see that $\|(T_\chi)^*K_{\zeta_j}\| \to 0$, which contradicts (see [10, Section 3]) the fact that $(T_\chi)^*$ is Fredholm and that $\{K_{\zeta_j}\}$ tends weakly to zero, as desired. In order to prove the last assertion of the proposition, assume that $\phi \in H^{\infty}(\Delta)$, and that w is not in $\sigma_e(T_\phi)$. By the first part of the present proof, we deduce that $\phi - w$ is bounded away from zero near $\partial\Delta$. Since ϕ is an open map, we conclude that w is in $\partial\Delta$, as desired.

Theorem 2. Let K be a generalized Bergman kernel on Δ such that the map F_K is surjective and such that condition V) (of Proposition 1) is satisfied. Let ϕ be in $H^{\infty}(\Delta)$ and let Ω be a bounded domain in $\mathbb{C}$. Then the following conditions are equivalent:

a) $(T_\phi)^* \in B_m(\Omega)$.

b) ϕ is an m-analytic cover of Ω.

Corollary 3. Given n in $\mathbb{Z}^+$, let $K_n(z,w) = (1 - z\bar{w})^{-n}$, $z,w \in \mathbb{D} = \{\zeta \in \mathbb{C}: |\zeta| < 1\}$. Also, let $\mathcal{K}_n$ be the space generated by K_n (see [6], Remark 3.3). (Notice that K_1 is the Szego kernel and K_2 is the Bergman kernel.) Then, K_n is a generalized Bergman kernel on $\mathbb{D}$ that satisfies the assumption of Theorem 2. Thus given a function ϕ in $H^{\infty}(\mathbb{D})$ and a bounded domain $\Omega \subset \mathbb{C}$, the following statements are equivalent.

a) ϕ is an m-analytic cover of Ω.

b) The operator $(T_\phi)^*$ on $\mathcal{K}_n$ is in $B_m(\Omega)$.

Corollary 4. Let K_n be the kernel on $\mathbb{D}$ considered in Corollary 3, and let $\phi: \mathbb{D} \to \mathbb{D}$ be a holomorphic map. Then, $(T_\phi)^*$ is in $B_m(\mathbb{D})$ if and only if ϕ is the product of m Blashke factors.

The fact that K_n is a generalized Bergman kernel follows directly from the results of either [1], [4], or [9]. Since the canonical model associated with K_n is a contraction (because the coefficients of K_n are increasing), we see that $\bar{\mathbb{D}}$ is a spectral set and hence F_{K_n} is surjective. Further, the functions $f_m(z) = z^m$, $m = 0,1,\ldots$, form a complete orthogonal system for the space generated by K_n, so condition V) holds, because $K_n(z,z) \to \infty$ as $\mathrm{dist}(z,\partial\mathbb{D}) \to 0$. Hence, the fact that $\partial\Phi(\mathbb{D}) \subset \sigma_e(T_\Phi)$ is always true in this case and therefore Corollary 3 is a consequence of Theorem 2 and Proposition 1. Corollary 4 follows from Corollary 3, and the fact that a holomorphic map $\Phi: \mathbb{D} \to \mathbb{D}$ is proper with multiplicity m if and only if it is the product of m Blashke factors.

In a different direction, one can apply Theorem 2 to prove that the same conclusion as Theorem 2 holds if K is the Bergman kernel on a bounded domain $\Delta \subset \mathbb{C}$ such that $\partial\Delta = \partial\bar{\Delta}$.

<u>Proof of Theorem 2</u>. In order to prove that b) $\Rightarrow$ a), assume that $\Phi: \Phi^{-1}(\Omega) \to \Omega$ is an m-analytic cover of Ω. Then, for each w in Ω, there exist an invertible function ψ in $H^\infty(\Delta)$ and p points $z_1,\ldots,z_p$ in Δ such that for every z in Δ, we have:

$$(***) \qquad \Phi(z) - w = \psi(z) \prod_{i=1}^{p} (z - z_i)^{k_i},$$

where

$$\sum_{i=1}^{p} k_i = m.$$

Since F_K is an isomorphism, we see that T_ψ is invertible in $\mathcal{L}(\mathcal{X})$, and from the fact that

$$(T_\Phi - w)^* = (T_\psi)^* \prod_{i=1}^{p} (T_z - z_i)^{*k_i}$$

it follows that $(T_\Phi - w)^*$ is onto and $\dim[\mathrm{Ker}(T_\Phi - w)^*] = m$ (recall that $(T_z)^*$ is in $B_1(\Delta)$). From (*), we deduce that $K(.,z)$ is in

$$\mathrm{Ker}[T_\Phi - \Phi(z)]^*$$

so that

$$\text{span}\{\text{Ker}(T_\phi - w)^*: w \in \Omega\} \supset \text{span}\{\text{Ker}(T_z - \lambda)^*: \lambda \in \phi^{-1}(\Omega)\}.$$

But, since the closure of the latter set coincides with the closure of $\text{span}\{\text{Ker}(T_z - \lambda)^*: \lambda \in \Delta\}$ (because of analyticity), we conclude that $\text{span}\{\text{Ker}(T_\phi - w)^*: w \in \Omega\}$ is dense in $\mathcal{K}$, and hence a) follows. For the proof that a) $\Longrightarrow$ b), assume that that $(T_\phi)^*$ is in $B_m(\Omega)$. It readily follows, from (*) and the assumption on K, that $\overline{\phi(\Delta)} = \sigma(T_\phi)$. Since $\partial\phi(\Delta) \subset \sigma_e(T_\phi)$, by Proposition 1, and $\phi(\Delta) \subset \sigma(T_\phi)\backslash\sigma_e(T_\phi)$, because $(T_\phi)^*$ is in $B_m(\Omega)$, we see that $\Omega \subset \phi(\Delta)$. Therefore, for every w in Ω there exists z_w in Δ such that $(T_\phi)^*K(.,z_w) = \bar{w}K(.,z_w)$. Since a finite subset of points z in Δ yields a linearly independent set of functions K(.,z) in $\mathcal{K}$ (see [6], Lemma 3.6), and $\dim[\text{Ker}(T_\phi - w)^*] = m$, it follows that each fiber $\phi^{-1}(w)$, $w \in \Omega$, consists of at most m points in Δ. Thus, for each fixed w in Ω, there are points $z_1,\dots,z_p$ in Δ such that (**) is valid for every $z \in \Delta$, where ψ is a nonvanishing function in $H^\infty(\Delta)$ and $p \le m$. We claim that, the function ψ is bounded away from zero. Indeed, since T_ψ is Fredholm, $\sigma(T_\psi) = \overline{\psi(\Delta)}$ (because F_K is surjective) and $0 \notin \psi(\Delta)$, we deduce, from (**), that $0 \notin \overline{\psi(\Delta)}$ and our claim is established. Thus, we proved that, ψ is invertible in $H^\infty(\Delta)$. From (***), it follows that

$$\begin{aligned} m &= \dim[\text{Ker}(T_\phi - w)^*] = \\ &= \dim[\text{Ker}(\prod_{i=1}^{p} (T_z - z_i)^{k_i})^*]. \end{aligned}$$

Since $(T_z - z_i)^*$ is Fredholm and has index 1, $1 \le i \le p$, we see that

$$\sum_{i=1}^{p} k_i = m,$$

and the proof of the theorem is complete.

Corollary 5. Let K be a kernel function as in Proposition 1, and let ϕ be in $H^\infty(\Delta)$. Then,

$$\sigma(T_\phi) \neq \sigma_e(T_\phi)$$

if and only if $(T_\phi)^*$ is in $B_m(\Omega)$, for some domain $\Omega \subset \mathbb{C}$ and some positive integer m.

Proof. The if part is obvious. For the only if part, we notice that, since $\partial\phi(\Delta) \subset \sigma_e(T_\phi)$, it follows that

$$\sigma(T_\phi)\backslash\sigma_e(T_\phi) = \phi(\Delta)\backslash\sigma_e(T_\phi).$$

Let Ω be a connected component of such an open set. Since, $T_\phi - w$ is injective and Fredholm for every w in Ω, we see that $(T_\phi - w)^*$ is onto and has constant nullity for all w in Ω. Further, since $\partial\phi(\Delta) \subset \sigma_e(T_\phi)$ and $\overline{\phi(\Delta)} = \sigma(T_\phi)$, we deduce that $\Omega \subset \phi(\Delta)$. Now, an argument similar to one used in the proof of Theorem 2 shows that span$\{\mathrm{Ker}(T_\phi - w)^*: w \in \Omega\}$ is dense in $\mathcal{H}$, and hence, we conclude that $(T_\phi)^*$ is in $B_m(\Omega)$, for some positive integer m, as desired.

Lin Qing
Institute of Mathematics
Academia Sinica
Beijing, China

Norberto Salinas
Department of Mathematics
University of Kansas
Lawrence, Kansas 66045

REFERENCES

[1] Agler, J., "Hyper-contractions and Subnormality," J. of O. T., vol. 13, 1985, pages 203-217.

[2] Agrawal, O. and Salinas, N., "Sharp Kernels and Canonical Subspaces," Amer. J. of Math., 109, 1987, pages 23-48.

[3] Ahlfors, L. V. and Sario, L., "Riemann Surfaces," Princeton University Press, 1960.

[4] Badri, M., "On Perturbations and Products of Generalized Bergman Kernels," Ph.D Dissertation, University of Kansas, 1985.

[5] Cowen, M. J. and Douglas, R. G., "Complex Geometry and Operator Theory," Acta Mathematica, 141, 1978, pages 187-261.

[6] Curto, R. and Salinas, N., "Generalized Bergman Kernels and the Cowen-Douglas Theory," Amer. J. of Math. 106, 1984, pages 447-488.

[7] Douglas, R. G., "Banach Algebra Techniques in Operator Theory," Academic Press, 1972.

[8] Forester, O., "Lectures on Riemann Surfaces," Springer-Verlag, 1981.

[9] Lin, Q., "Imbedded Operators and Geometric Tensor Products," Ph.D. Dissertation, S.U.N.Y. at Buffalo, 1986.

[10] Salinas, N., "Quasitriangular Extensions of C^*-Algebras and Problems on Joint Quasitriangularity of Operators," J. of O. T., 10, 1983, pages 167-205.

28

Hankel Operators in the Perturbation Theory of Unbounded Self-adjoint Operators

Vladimir V. Peller

§ 1. Introduction

This paper can be considered as a continuation of the paper [13] where Hankel operators were applied in the perturbation theory of unitary and bounded selfadjoint operators. Here we consider the case of unbounded selfadjoint operators and obtain results similar to the results of [13]. Note that the proofs in the case of unbounded operators are considerably more complicated.

We shall consider here the following questions. Let f be a function on $\mathbb{R}$. Is it true that

$$\| f(A) - f(B) \| \leqslant \text{const} \cdot \| A - B \|, \tag{1}$$

whenever A and B are selfadjoint operators (not necessarily bounded) on Hilbert space such that A-B is bounded?

The study of this question was started in [6] where sufficient conditions were found in the case of bounded A and B.

Consider now the situation when $A-B \in \mathbf{S}_1$ (i.e. A-B is of trace class). We are interested in for which f it follows that $f(A)-f(B) \in \mathbf{S}_1$ and the following trace formula holds:

$$\text{trace}\,(f(A)-f(B)) = \int_{-\infty}^{\infty} \xi(t) \cdot f'(t)\,dt, \tag{2}$$

where ξ is the *spectral shift* corresponding to the pair (A,B). The notion of spectral shift was introduced by M. G. Krein (see [9]) for an arbitrary pair (A,B) with $A-B \in \mathbf{S}_1$. He proved that $\xi \in L^1(\mathbb{R})$ and (1) is valid in the case when f' is the Fourier transform of a finite complex measure.

Later in [4] and [5] assumptions on f under which (1) and (2) hold were weakened. In [7] and [8] examples are constructed which show that the condition $f' \in L^\infty$, that seems natural here, does not guarantee (1), (2) (see [13] for further information).

In [13] techniques of Hankel operator were applied to the solution of these problems in the case of unitary and bounded selfadjoint operators. In that paper new sufficient conditions were found and for the first time necessary conditions were obtain that show that the condition $f' \in L^\infty$ is not sufficient. In this paper we obtain similar results in the case of unbounded selfadjoint operators.

The questions under consideration are closely connected with the problem of description of functions f for which the function $\check{f}$ on $\mathbb{R}\times\mathbb{R}$ defined by

$$\check{f}(s,t) \stackrel{\text{def}}{=} \frac{f(s) - f(t)}{s - t}$$

is a multiplier of the class of kernels of trace class operators (see [4], [13]). In this case we say $\check{f} \in \mathfrak{M}$.

In § 3 it will be shown that if f belongs to the Besov class $B^1_{\infty 1}(\mathbb{R})$ (see § 2) then $\check{f} \in \mathfrak{M}$. In § 4 necessary conditions for $\check{f} \in \mathfrak{M}$ will be given. In § 5 the results of §§ 3, 4 will be applied in the perturbation theory of selfadjoint operators. Note that the derivation of trace formula (2) in this case is more complicated than in the case of unitary operators since unlike the Besov space $B^1_{\infty 1}$ of functions on the unit circle, the class $B^1_{\infty 1}(\mathbb{R})$ is not separable and the set of functions whose derivative is the Fourier transform of a finite measure is not dense in $B^1_{\infty 1}(\mathbb{R})$. In § 6 we briefly consider the question of when $A-B \in \mathbf{S}_p$ implies that $f(A)-f(B) \in \mathbf{S}_p$ for $0<p<1$, where $\mathbf{S}_p$ is the Schatten fon Neumann class.

§ 2. Preliminaries

Schur multipliers. Let F, E be Borel spectral measures on separable metric spaces Λ and M and T bounded operator on a Hilbert space $\mathcal{H}$. We consider the double operator integrals

$$\Phi T = \int_\Lambda \int_M \varphi(s,t)\, dF(s)\, T\, dE(t).$$

If $\varphi \in L^\infty(\Lambda \times M)$ then Φ is bounded on the Hilbert – Schmidt class $\mathbf{S}_2$ and its norm equals $\|\varphi\|_{L^\infty}$. If Φ is bounded on $\mathbf{S}_1$, we can define by duality ΦT for any bounded operator T. In this case φ is called a *Schur multiplier of* $\mathbf{S}_1$ (or $\mathcal{B}(\mathcal{H})$, the space of bounded operators on $\mathcal{H}$). The techniques of double operator integrals was developed in [2], [3]. The question of description of the set $\mathfrak{M}(\Lambda \times M)$ of Schur multipliers was considered in [2]. The following characterization of $\mathfrak{M}(\Lambda \times M)$ in the case of discrete measures was obtained in [1]. For the general case see [13] where a more detailed statement of the following characterization and more detailed bibliographical comments are given.

Theorem 1. *Let* $\Phi \in L^\infty$ *and* λ, μ *be positive measures mutually absolutely continuous with* f *and* E *respectively. The following are equivalent:*

1) $\varphi \in \mathfrak{M}(\Lambda \times M)$;

2) *if* k *is a function on* $\Lambda \times M$ *such that the integral operator from* $L^2(\Lambda, \lambda)$ *to* $L^2(M, \mu)$ *with kernel* k *is of trace class, then the integral operator with kernel*

$$(s,t) \mapsto \varphi(s,t)\, k(s,t), \qquad s,\ t \in \Lambda \times M,$$

is also of trace class;

3) *there exists a measure space* $(\mathcal{X}, \sigma)$ *and measurable functions* α *on* $\Lambda \times \mathcal{X}$ *and* β *on* $M \times \mathcal{X}$ *such that*

$$\varphi(s,t) = \int_{\mathcal{X}} \alpha(s,x)\beta(t,x)\, d\sigma(x), \qquad (s,t) \in \Lambda \times M,$$

$$\int_{\mathcal{X}} \|\alpha(\cdot,x)\|_{L^\infty(\Lambda)} \|\beta(\cdot,x)\|_{L^\infty(M)}\, d\sigma(x) < \infty.$$

Note that (2) is equivalent to the fact that the integral operator $\mathcal{J}_\varphi$ with kernel φ from $L^1(\Lambda,\lambda)$ to $L^\infty(M,\mu)$ belongs to the ideal $I\big(L^1(\lambda),L^\infty(\mu)\big)$, i.e. there exists a compactum Q, a finite measure σ on Q, bounded operators $T_1:L^1(\lambda)\to C(Q)$, $T_2: L^1(\sigma)\to L^\infty(\mu)$ such that the following diagram is commutative:

$$\begin{array}{ccc} L^1(\lambda) & \xrightarrow{\mathcal{J}_\mu} & L^\infty(\mu) \\ T_1\downarrow & & \uparrow T_2 \\ C(Q) & \xrightarrow{j} & L^1(\sigma) \end{array},$$

where j is the identical imbedding of $C(Q)$ in $L^1(\sigma)$ (see [13] for details).

Note also that the ideal $I(L^1,L^\infty)$ contains as a closed subspace the ideal $\mathfrak{N}_1(L^1,L^\infty)$ of nuclear operators from L^1 to L^∞

Double operator integrals in perturbation theory. Let A, B be selfadjoint operators on $\mathcal{H}$, E, F their spectral measures, $\varphi\in C^1(\mathbb{R})$. Supposed that B-A is bounded. Then

$$\varphi(B) - \varphi(A) = \int_{\mathbb{R}}\int_{\mathbb{R}} \check{\varphi}(s,t)\, dF(t)\,(B - A)\, dE(s), \tag{3}$$

whenever $\check{\varphi}$ is a Schur multiplier of $\mathcal{B}(\mathcal{H})$ (see [3]). Clearly, (3) implies that $\varphi(B)-\varphi(A)\in \mathbf{S}_1$ whenever $B-A\in \mathbf{S}_1$ and $\check{\varphi}$ is a Schur multiplier. We denote by $\mathfrak{M}_{\mathbb{R}}$ the set of Schur multipliers with respect to arbitrary Borel spectral measures on $\mathbb{R}$ and consider the natural norm on $\mathfrak{M}_{\mathbb{R}}$.

Hankel operators. Let $H^2(\mathbb{C}_+)$ $\big(H^2(\mathbb{C}_-)\big)$ be the Hardy space of functions in the upper (lower) half-plane. They can naturally be identified with subspaces of $L^2(\mathbb{R})$ so that $L^2(\mathbb{R}) = H^2(\mathbb{C}_+)\oplus H^2(\mathbb{C}_-)$. Then for $\varphi\in L^\infty(\mathbb{R})$ the *Hankel operator* $\mathcal{H}_\varphi:H^2(\mathbb{C}_+)\to H^2(\mathbb{C}_-)$ is defined by

$$\mathcal{H}_\varphi f = \mathbb{P}_-\varphi\cdot f.$$

(We denote by $\mathbb{P}_\pm$ the orthogonal projection onto $H^2(\mathbb{C}_\pm)$.)

As mentioned above (see [13] for further discussions) the questions under consideration are closely connected with the study of kernels $\check{\varphi}$. Let T be the integral operator with kernel $\frac{1}{2\pi i}\check{\varphi}$. Then it can be shown that $Tf = \mathbb{P}_-\varphi\cdot f_+ - \mathbb{P}_+\varphi\cdot f_-$, where $f_\pm \stackrel{\text{def}}{=} \mathbb{P}_\pm f$. Note that $\mathbb{P}_-\varphi\cdot f_+ = \mathcal{H}_\varphi f_+$ and the operator $f_- \mapsto \mathbb{P}_+\varphi\cdot f_-$ is analogous to a Hankel operator. Therefore T can be represented as the orthogonal sum of two Hankel operators on $H^2(\mathbb{C}_+)$ and $H^2(\mathbb{C}_-)$.

We need the following criterion of nuclearity of Hankel operators obtained in [11]: $\mathcal{H}_\varphi \in \mathbf{S}_1$ if and only if $\mathbb{P}_-\varphi \in B_1^1(\mathbb{R})$ (see the definition of $B_1^1(\mathbb{R})$ below).

Besov classes (we consider only homogeneous Besov classes). Let V_n, $n\in\mathbb{Z}$, be functions on $\mathbb{R}$ such that the Fourier transforms $\mathcal{F}V_n$ are C^∞-functions supported on $[2^{n-1},2^{n+1}]$, $(\mathcal{F}V_{n+1})(x) = (\mathcal{F}V_n)(2x)$, $n\in\mathbb{Z}$, and

$$\sum_{n\in\mathbb{Z}} (\mathcal{F}V_n)(x) = 1, \qquad x>0.$$

Define V_n^*, $n\in\mathbb{Z}$, by $(\mathcal{F}V_n^*)(x) = (\mathcal{F}V_n)(-x)$.

Then the *Besov class* $B^s_{pq}(\mathbb{R})$ consists of the distributions f on $\mathbb{R}$ such that

$$\left\{ 2^{ns}\left\| f*V_n\right\|_{L^p(\mathbb{R})} \right\}_{n\in\mathbb{Z}} \in \ell^q, \qquad \left\{ 2^{ns}\left\| f*V_n^*\right\|_{L^p(\mathbb{R})} \right\}_{n\in\mathbb{Z}} \in \ell^q,$$

We shall use the notation $B^s_p(\mathbb{R}) = B^s_{pp}(\mathbb{R})$.

The Besov classes admit many other equivalent definitions (see [10]).

Each $f\in B^s_{pq}$ admits a representation

$$f = \sum_{n\in\mathbb{Z}} f*V_n^* + \sum_{n\in\mathbb{Z}} f*V_n + P, \tag{4}$$

where P is a polynomial. We say that the Fourier transform $\mathcal{F}f$ is supported on $(0,+\infty)$ (on $(-\infty,0)$) if $f = \sum_{n\in\mathbb{Z}} f*V_n$ ($f = \sum_{n\in\mathbb{Z}} f*V_n^*$).

It is not always convenient to consider Besov classes modulo polynomials. In this paper we shall deal with the spaces $B^1_{\infty1}(\mathbb{R})$ and $B_p^{1/p}(\mathbb{R})$. In the first case we assume that $\deg P\leqslant 1$ in (4) while in the second case P is constant.

§ 3. Sufficient conditions

In this section we show that $\check{\varphi} \in \mathfrak{M}_{\mathbb{R}}$ if $\varphi \in B^1_{\infty 1}(\mathbb{R})$. In § 4 we shall see that this condition is fairly close to be necessary. But it is not necessary as seen from the remark given at the end of this section.

Theorem 2. *Let* $\varphi \in B^1_{\infty 1}(\mathbb{R})$. *Then* $\check{\varphi} \in \mathfrak{M}_{\mathbb{R}}$.

Proof. Let $\varphi = \varphi_- + \varphi_0 + \varphi_1$, where $\mathcal{F}\varphi_-$ and $\mathcal{F}\varphi_+$ are supported on $(-\infty, 0)$ and $(0, +\infty)$, and φ_0 is a linear function. Clearly $\check{\varphi}_0 \in \mathfrak{M}_{\mathbb{R}}$. The facts that $\check{\varphi}_- \in \mathfrak{M}_{\mathbb{R}}$ and $\check{\varphi}_+ \in \mathfrak{M}_{\mathbb{R}}$ have the same proof. So we assume that $\mathcal{F}\varphi$ is supported on $(0, +\infty)$.

We have $\varphi = \sum_{n \in \mathbb{Z}} \varphi * V_n$ and $\sum_{n \in \mathbb{Z}} 2^n \|\varphi * V_n\|_{L^\infty} < \infty$. Clearly, it is sufficient to show that $\check{\varphi} \in \mathfrak{M}_{\mathbb{R}}$ and $\|\check{\varphi}\|_{\mathfrak{M}_{\mathbb{R}}} \leqslant \text{const} \cdot 2^n \|\varphi\|_{L^\infty}$, whenever $\varphi \in L^\infty(\mathbb{R})$ and $\operatorname{supp} \varphi \subset [2^{n-1}, 2^{n+1}]$.

Let q be a C^∞-function on $\mathbb{R}$ such that $q \geqslant 0$, $q(x) = 0$ for $x \leqslant 1/2$ and $q(x) + q(1/x)$ for $x > 0$. It is easy to construct such a function. Put $r(x) = q(x-1)$, $x \in \mathbb{R}$.

Denote by Q_t and R_t distributions on $\mathbb{R}$ such that $(\mathcal{F}Q_t)(x) = q(x/t)$ and $(\mathcal{F}R_t)(x) = r(x/t)$.

Let us show that

$$\check{\varphi}(x,y) = \int_0^\infty \big((S_t^* \varphi) * Q_t\big)(x)\, e^{ity}\, dt + \int_0^\infty \big((S_t^* \varphi) * Q_t\big)(y)\, e^{itx}\, dt, \tag{5}$$

where $(S_t^* f)(t) = e^{-itx} f(x)$.

Before we prove (5), let us show that

$$\int_0^\infty \big\|\big((S_t^* \varphi) * Q_t\big)\big\|_{L^\infty} dt \leqslant \text{const} \cdot 2^n \|\varphi\|_{L^\infty}. \tag{6}$$

Then by Theorem 1 it would follow from (5) that $\check{\varphi} \in \mathfrak{M}_{\mathbb{R}}$ and $\|\check{\varphi}\|_{\mathfrak{M}_{\mathbb{R}}} \leqslant \text{const} \cdot 2^n$.

We have

$$\big\|\big((S_t^* \varphi) * Q_t\big)\big\|_{L^\infty} = \|\varphi * R_t\|_{L^\infty}.$$

Note that $\varphi * R_t = \mathbf{0}$ for $t > 2^{n+2}/3$. For $t \leqslant 2^{n-1}/3$ we have $\varphi * R_t = \varphi$, so $\|\varphi * R_t\|_{L^\infty} = \|\varphi\|_{L^\infty}$. For $2^{n-1}/3 \leqslant t \leqslant 2^{n+2}/3$ we define R_t^* by

$$(\mathscr{F} R_t^*) = \begin{cases} (\mathscr{F} R_t)(x), & x \leqslant 12t \\ (\mathscr{F} R_t)(12t - x), & x \geqslant 12t \ . \end{cases}$$

It is easy to see that for $2^{n-1}/3 \leqslant t \leqslant 2^{n+2}$ we have $\varphi * R_t = \varphi * R_t^*$. It is also clear that $R_t^* \in L^1(\mathbb{R})$ and $\|R_t^*\|_{L^1}$ does not depend on t. Hence

$$\| \varphi * R_t \|_{L^\infty} \leqslant \text{const} \cdot \|\varphi\|_{L^\infty} .$$

Therefore

$$\int_0^\infty \|\varphi * R_t\|_{L^\infty} \leqslant \text{const} \cdot 2^n \|\varphi\|_{L^\infty} .$$

Let us now prove (5). Assume first that $\mathscr{F}\varphi \in L^1(\mathbb{R})$. We have

$$\int_0^\infty \big((S_t^*\varphi) * Q_t\big)(x)\, e^{ity}\, dt = \int_0^\infty \int_0^\infty (\mathscr{F}\varphi)(s+t)\, r(s/t)\, e^{isx}\, e^{ity}\, ds\, dt .$$

Similarly

$$\int_0^\infty \big((S_t^*\varphi) * Q_t\big)(y)\, e^{itx} dt = \int_0^\infty \int_0^\infty (\mathscr{F}\varphi)(s+t)\, r(t/s)\, e^{isx}\, e^{ity}\, ds\, dt .$$

Since $r(t/s) + r(s/t) = 1$, for $s, t > 0$, it follows that to prove (5), it is sufficient to show that

$$\frac{\varphi(x) - \varphi(y)}{x - y} = \int_0^\infty \int_0^\infty (\mathscr{F}\varphi)(s+t)\, r(t/s)\, e^{isx}\, e^{ity}\, ds\, dt . \tag{7}$$

But this equality can easily be verified by making the substitution $u = s+t$, $v = t$.

It remains to consider the general case. Let $\varphi \in L^\infty(\mathbb{R})$ and $\operatorname{supp} \mathscr{F}\varphi \subset [2^{n-1}, 2^{n+1}]$. Consider a smooth function h on $\mathbb{R}$ such that $h \geqslant 0$, $\operatorname{supp} h \subset [-1,1]$, $\int_{-1}^{1} h(x)dx = 1$. Put $h_\varepsilon(x) = (1/\varepsilon) \cdot h(x/\varepsilon)$, $\varepsilon > 0$. Consider now the function φ_ε defined by $\mathcal{F}\varphi_\varepsilon=(\mathcal{F}\varphi)*h_\varepsilon$. Clearly, $\mathcal{F}\varphi_\varepsilon\in L^1(\mathbb{R})$, $\lim\limits_{\varepsilon\to 0}\|\varphi_\varepsilon\|_{L^\infty}=\|\varphi\|_{L^\infty}$, and $\lim\limits_{\varepsilon\to 0}\varphi_\varepsilon(x)=\varphi(x)$ for $x\in\mathbb{R}$.

We have already proved that (5) is valid for φ_ε with $\varepsilon>0$. Now the result follows by making $\varepsilon\to 0$. ∎

Note that we have proved that the integral operator $\mathcal{J}_{\check{\varphi}}$ with kernel $\check{\varphi}$ belongs to the ideal $I(L^1(\lambda),L^\infty(\mu))$ for any Borel measures λ and μ (see § 2).

Corollary. *Let* $\varphi\in B^1_{\infty 1}(\mathbb{R})$. *Then the Hankel operator* $\mathcal{H}_\varphi\mathbb{P}_+$, *acting from* L^1 *to* L^∞ *belongs to the ideal* $I(L^1,L^\infty)$.

This can easily be derived from (5) and (6) (see [13] for details).

Remark. As in the case of the unit circle or a finite interval it can be shown that the condition $\varphi\in B^1_{\infty 1}(\mathbb{R})$ is not necessary for $\check{\varphi}\in\mathfrak{M}_{\mathbb{R}}$. Indeed, it can be seen from [13] that there exists a function φ with compact support such that $\varphi\notin B^1_{\infty 1}(\mathbb{R})$ and $\check{\varphi}\in\mathfrak{M}_I$ for any finite interval I. But since $\operatorname{supp}\varphi$ is compact, it follows that $\check{\varphi}\in\mathfrak{M}_{\mathbb{R}}$.

§ 4. Necessary conditions.

In this section we obtain necessary conditions for $\check{\varphi}\in\mathfrak{M}_{\mathbb{R}}$. It is shown in [13] that if $\varphi\in\mathfrak{M}_I$ for any finite interval I, then $\varphi|I\in B^1_1(I)$ for any finite interval I. It follows that for $\check{\varphi}\in\mathfrak{M}_{\mathbb{R}}$ it is necessary that $f|I\in B^1_1(I)$ for any finite interval I.

We obtain here a stronger necessary condition that is fairly close to the sufficient condition $\varphi\in B^1_{\infty 1}(\mathbb{R})$ obtained in § 3.

Definition. A function φ on $\mathbb{R}$ is said to belong to the class $\mathcal{L}(\mathbb{R})$ if the Hankel operators $\mathcal{H}_\varphi$ and $\mathcal{H}_{\overline{\varphi}}$ are bounded operators from the Hardy class $H^1(\mathbb{C}_+)$ to $B^1_1(\mathbb{R})$.

This means that the Hankel operators $\mathcal{H}_\varphi$ and $\mathcal{H}_{\overline{\varphi}}$ defined on a dense subset of $H^1(\mathbb{C}_+)$ extend by continuity to bounded operators. Clearly, $\varphi\in\mathcal{L}(\mathbb{R})$ if and only if the integral operator with kernel $\check{\varphi}$ is a bounded operator from $H^1(\mathbb{C}_+)\oplus H^1(\mathbb{C}_-)$ to $B_1^1(\mathbb{R})$.

Theorem 3. *Let φ be a function on $\mathbb{R}$ such that $\check{\varphi}\in\mathfrak{M}_{\mathbb{R}}$. Then $\varphi\in\mathcal{L}(\mathbb{R})$.*

Before we prove Theorem 3, we compare the condition $\varphi\in\mathcal{L}(\mathbb{R})$ with the sufficient condition $\varphi\in B^1_{\infty 1}(\mathbb{R})$ obtained in § 3. Let J be the closure of the set of smooth rapidly decreasing functions in the norm

$$\|h\|_J = \sup_{z\in\mathbb{C}_+} |h(z)\operatorname{Im} z|.$$

Let now N be the space of functions f analytic in $\mathbb{C}_+$ that admit a representation

$$f = \sum_{n\geq 0} g_n h_n, \qquad \sum_{n\geq 0} \|g_n\|_{H^1} \|h_n\|_J < \infty . \tag{8}$$

As in [13] where the case of the unit disc is treated, it is easy to see that $\mathcal{H}_{\overline{\varphi}}$ maps $H^1(\mathbb{C}_+)$ into $B_1^1(\mathbb{R})$ if and only if φ determines a continuous linear functional on N defined on a dense subset of N by

$$(\varphi,f) = \int_{\mathbb{R}} \varphi(x)\, f(-x)\, dx.$$

Clearly, each function in N belongs to the space M consisting of functions f satisfying

$$\lim_{y\to 0,+\infty} y \int_{\mathbb{R}} |f(x+iy)|\, dx = 0.$$

It is well known that the conjugate space M^* coincides with the space of functions of class $B^1_{\infty 1}(\mathbb{R})$ analytic in $\mathbb{C}_+$. If each function in M admitted a representation (8), we would have $\mathcal{L}(\mathbb{R})=B^1_{\infty 1}(\mathbb{R})$. But this is not the case since, as we have mentioned in § 3, the condition $\varphi\in B^1_{\infty 1}(\mathbb{R})$ is not sufficient for $\check{\varphi}\in\mathfrak{M}_{\mathbb{R}}$.

Thus the condition $\varphi\in\mathcal{L}(\mathbb{R})$ is slightly weaker than $\varphi\in B^1_{\infty 1}(\mathbb{R})$ and the difference between them is the same as the difference between N and M.

We proceed to the proof of Theorem 3. It is analogous to the proof of Theorem 5 of [13] but it is technically more complicated.

Proof of Theorem 3. Let us prove that $\mathcal{H}_{\bar{\varphi}}$ is a bounded operator from $H^1(\mathbb{C}_+)$ to $B_1^1(\mathbb{R})$. The proof of the corresponding assertion for $\mathcal{H}_\varphi$ is the same. For $k \in L^2(\mathbb{R}\times\mathbb{R})$ denote by $\mathcal{J}_k$ the integral operator on $L^2(\mathbb{R})$ with kernel k.

We consider integral operators $T=\mathcal{J}_k$ on L^2 with kernel k of the form

$$k(x,y) = \int_0^\infty \int_0^\infty \alpha(u,v)\, e^{-ixu} e^{-iyv}\, du\, dv, \tag{9}$$

where $\alpha \in L^1(\mathbb{R}_+\times\mathbb{R}_+)$. Clearly, in this case $TL^2(\mathbb{R}) \subset H^2(\mathbb{C}_-)$ and $T|H^2(\mathbb{C}_-)=\mathbf{0}$. Let us estimate the norm of $\mathcal{J}_{\check{\varphi}k}$ in $\mathbf{S}_1$. Assume first that

$$\int_0^\infty \left|(\mathcal{F}\varphi)(w)\right| w\, dw < \infty . \tag{10}$$

Then we estimate the norm of $\mathcal{H}_{\bar{\varphi}}$ as an operator from $H^1(\mathbb{C}_+)$ to $B_1^1(\mathbb{R})$ in terms of $\|\check{\varphi}\|_{\mathfrak{M}_{\mathbb{R}}}$.

Let us show that $\mathbb{P}_+\mathcal{J}_{\check{\varphi}k}\mathbb{P}_-$ is a Hankel operator. Since $\check{\varphi} \in \mathfrak{M}_{\mathbb{R}}$, it would follow that $\mathbb{P}_+\mathcal{J}_{\check{\varphi}k}\mathbb{P}_- \in \mathbf{S}_1$, so the symbol of $\mathbb{P}_+\mathcal{J}_{\check{\varphi}k}\mathbb{P}_-$ must belong to $B_1^1(\mathbb{R})$ (see § 2).

We apply formula (7). It is easy to see that $\mathbb{P}_+\mathcal{J}_{\check{\varphi}k}\mathbb{P}_-$ is an integral operator whose kernel can be represented as

$$\int_0^\infty\int_0^\infty\int_0^\infty\int_0^\infty \alpha(u,v)\,(\mathcal{F}\varphi)(s+t+u+v)\, e^{isx} e^{ity}\, du\, dv\, ds\, dt =$$

$$= \int_0^\infty\int_0^\infty\int_0^\infty \beta(w)\,(\mathcal{F}\varphi)(s+t+w)\, e^{isx} e^{ity}\, dw\, ds\, dt ,$$

where $\beta(w) = \int_0^w \alpha(r,w-r)\,dr$. It follows that $\mathbb{P}_+\mathcal{J}_{\check{\varphi}k}\mathbb{P}_-$ is a Hankel operator whose symbol has Fourier transform

$$\int_0^\infty \beta(w)\,(\mathcal{F}\varphi)(u+w)\,dw$$

Note that since k ranges over the class of kernels (9) such that $\|\mathcal{J}_k\|_{\mathbf{S}_1} \le 1$, the inverse Fourier transform of β ranges over a dense subset of the unit ball of H^1 (this follows from the fact that any function from the unit ball of H^1 can be represented as a product of two functions from the unit ball of H^2).

Therefore for $f \in H^1(\mathbb{C}_+)$ the function g, whose Fourier transform is defined by

$$(\mathcal{F}g)(u) = \int_0^\infty (\mathcal{F}\varphi)(w)\,(\mathcal{F}f)(u+v)\,dw\,,$$

belongs to $B_1^1(\mathbb{R})$. But this just means that the Hankel operator $\mathcal{H}_{\bar{\varphi}}$ is a bounded operator from $H^1(\mathbb{C}_+)$ to $B_1^1(\mathbb{R})$. Moreover the above reasoning shows that

$$\|\mathcal{H}_{\bar{\varphi}}\|_{H^1 \to B_1^1} \leqslant \text{const} \cdot \|\varphi\|_{\mathfrak{M}_{\mathbb{R}}}\,. \tag{11}$$

To get rid of (10), we approximate a function φ in $\mathfrak{M}_{\mathbb{R}}$ by a sequence $\{\varphi_n\}$ so that φ_n satisfies (10), $\|\varphi_n\|_{\mathfrak{M}_{\mathbb{R}}} \leqslant$ const, and $\lim_{n\to\infty} \varphi_n(x) = f(x)$ for $x \in \mathbb{R}$. Then (11) would imply that $\mathcal{H}_{\bar{\varphi}}$ is a bounded operator from $H^1(\mathbb{C}_+)$ to $B_1^1(\mathbb{R})$.

It is not difficult to construct such a sequence $\{\varphi_n\}$. First, using the convolution with Fejér kernels, one can approximate φ by functions whoes Fourier transforms have compact support. Then we can use the same kind of approximation that has been used in the proof of Theorem 2. ∎

§ 5. Applications to selfadjoint operators.

In this section the results obtained in §§ 3,4 are applied in the perturbation theory of (not necessarily bounded) selfadjoint operators on Hilbert space.

Theorem 4. *Let* $\varphi \in B_{\infty 1}^1(\mathbb{R})$. *The following assertions hold*:

1) *if* A *and* B *are selfadjoint operators then*

$$\|\varphi(A) - \varphi(B)\| \leqslant \text{const} \cdot \|\varphi\|_{B_{\infty 1}^1(\mathbb{R})} \|A - B\|;$$

2) *if* A *and* B *are selfadjoint operators,* $A - B \in \mathbf{S}_1$ then $\varphi(A) - \varphi(B) \in \mathbf{S}_1$ *and trace formula* (2) *holds.*

Proof. As already noticed in the introduction, 1) follows directly from (3) and Theorem 2.

Let us now prove 2). Again it follows from (3) and Theorem 2 that $\varphi(A)-\varphi(B)\in \mathbf{S}_1$. Let us establish formula (2).

As in the proof of Theorem 2 it is sufficient to consider the case $\operatorname{supp}\varphi\subset[2^{n-1},2^{n+1}]$. In the case $\mathscr{F}\varphi\in L^1(\mathbb{R})$ formula (2) was proved by M. G. Krein [9]. In the general case as in the proof of Theorem 2 we approximate φ by the functions φ_ε. We have

$$\operatorname{trace}\big(\varphi_\varepsilon(A)-\varphi_\varepsilon(B)\big) = \int_{-\infty}^{\infty}\xi(\lambda)\,\varphi_\varepsilon'(\lambda)\,d\lambda,$$

where ξ is the spectral shift.

Clearly,

$$\lim_{\varepsilon\to 0}\int_{-\infty}^{\infty}\xi(\lambda)\,\varphi_\varepsilon'(\lambda)\,d\lambda = \int_{-\infty}^{\infty}\xi(\lambda)\,\varphi'(\lambda)\,d\lambda.$$

The proof will be completed if we show that

$$\lim_{\varepsilon\to 0}\operatorname{trace}\big(\varphi_\varepsilon(A)-\varphi_\varepsilon(B)\big) = \big(\varphi(A)-\varphi(B)\big).$$

To prove this equality we make use of formula (3). We have

$$\varphi_\varepsilon(A)-\varphi_\varepsilon(B) = \int_{\mathbb{R}}\int_{\mathbb{R}}\check{\varphi}_\varepsilon(x,y)\,dF(y)\,(A-B)\,dE(x),$$

where E, F are spectral measures of A and B respectively. We have by (5)

$$\varphi_\varepsilon(A)-\varphi_\varepsilon(B) = \int_0^{\infty}\int_{\mathbb{R}}\int_{\mathbb{R}}\big((S_t^*\varphi_\varepsilon)*Q_t\big)(x)\,e^{ity}\,dF(y)\,(A-B)\,dE(x) +$$

$$+\int_0^{\infty}\int_{\mathbb{R}}\int_{\mathbb{R}}\big((S_t^*\varphi_\varepsilon)*Q_t\big)(y)\,e^{itx}\,dF(y)\,(A-B)\,dE(x) =$$

$$= \int_0^{\infty} e^{itB}(A-B)\big((S_t^*\varphi_\varepsilon)*Q_t\big)(A)\,dt + \int_0^{\infty}\big((S_t^*\varphi_\varepsilon)*Q_t\big)(B)\,(A-B)\,e^{itA}\,dt.$$

Since $(S_t^*\varphi_\varepsilon)*Q_t=\mathbf{0}$ for $t>2^{n+2}/3$, we can consider integration from 0 to $2^{n+2}/3$. It is shown in the proof of Theorem 2 that

$$\big\|(S_t^*\varphi_\varepsilon)*Q_t\big\|_{L^\infty} \leqslant \text{const}\cdot\|\varphi_\varepsilon\|_{L^\infty} \leqslant \text{const}\cdot\|\varphi\|_{L^\infty}.$$

It is easy to see that

$$\lim_{\varepsilon\to 0}\Big((S_t^*\varphi_\varepsilon)*Q_t\Big)(x) = \Big((S_t^*\varphi)*Q_t\Big)(x), \qquad x\in\mathbb{R}.$$

It follows that

$$\lim_{\varepsilon\to 0}\Big((S_t^*\varphi_\varepsilon)*Q_t\Big)(A) = \Big((S_t^*\varphi)*Q_t\Big)(A),$$
$$\lim_{\varepsilon\to 0}\Big((S_t^*\varphi_\varepsilon)*Q_t\Big)(A) = \Big((S_t^*\varphi)*Q_t\Big)(A)$$

in the strong operator topology.

Let us show that

$$\lim_{\varepsilon\to 0}\ \text{trace}\ e^{itB}(A-B)\Big((S_t^*\varphi_\varepsilon)*Q_t\Big)(A) = \text{trace}\ e^{itB}(A-B)\Big((S_t^*\varphi)*Q_t\Big)(A),$$

$$\lim_{\varepsilon\to 0}\ \text{trace}\ \Big((S_t^*\varphi_\varepsilon)*Q_t\Big)(B)(A-B)e^{itA} = \text{trace}\ \Big((S_t^*\varphi_\varepsilon)*Q_t\Big)(B)(A-B)e^{itA}.$$

This follows from the following easily verifiable fact: if $T\in\mathbf{S}_1$, $\{R_n\}$ is a sequence of bounded operators strongly converging to R, then

$$\lim_{n\to\infty}\text{trace}\ TR_n = \text{trace}\ TR.$$

Therefore

$$\lim_{\varepsilon\to 0}\ \text{trace}\Big(\varphi_\varepsilon(A) - \varphi_\varepsilon(B)\Big) =$$

$$= \text{trace}\int_0^\infty e^{itB}(A-B)\Big((S_t^*\varphi_\varepsilon)*Q_t\Big)(A)\,dt\ +$$

$$+\ \text{trace}\int_0^\infty \Big((S_t^*\varphi_\varepsilon)*Q_t\Big)(B)(A-B)e^{itA}\,dt =$$

$$= \text{trace}\int_0^\infty\int_{\mathbb{R}}\int_{\mathbb{R}}\Big((S_t^*\varphi)*Q_t\Big)(x)\,e^{ity}\,dF(y)(A-B)\,dE(x)\ +$$

$$+\ \text{trace}\int_0^\infty\int_{\mathbb{R}}\int_{\mathbb{R}}\Big((S_t^*\varphi)*Q_t\Big)(y)\,e^{itx}\,dF(y)(A-B)\,dE(x) =$$

$$= \text{trace}\Big(\varphi(A) - \varphi(B)\Big)$$

by (3) and (5).■

Theorem 5. *Let $\varphi\in C^1(\mathbb{R})$ and $\varphi\notin\mathcal{L}(\mathbb{R})$. The following assertions hold:*

1) there exist selfadjoint operators A, $\{B_n\}_{n\geq 1}$ such that

$$\lim_{n\to\infty} \|A - B\| = 0, \qquad \lim_{n\to\infty} \frac{\|\varphi(A) - \varphi(B_n)\|}{\|A - B_n\|} = \infty ;$$

2) *there exist selfadjoint operators* A *and* B *such that* $A-B \in \mathbf{S}_1$ *but* $\varphi(A)-\varphi(B) \notin \mathbf{S}_1$.

The proof of this theorem is similar to that of Theorem 7 of [13].

§ 6. Perturbations of class $\mathbf{S}_p$ with $p<1$.

In [14] the following question was considered for $0 < p < 1$: for which functions f on the unit circle $f(U) - f(V) \in \mathbf{S}_p$, whenever U, V are unitary operators on Hilbert space such that $U - V \in \mathbf{S}_p$? A sufficient condition ($f \in B^{1/p}_{p\infty}$) and a necessary condition (the Hankel operators H_f and $H_{\bar{f}}$ map H^1 into $B^{1/p}_p$) were found there.

The same problem can be considered for selfadjoint operators. It can be reduced to the problem to describe functions φ on $\mathbb{R}$ for which the function $\check{\varphi}$ is a Schur multiplier of the class $\mathbf{S}_p$.

As in the case of unitary operators we can obtain the following necessary condition.

Theorem 6. *Let* $0 < p < 1$ *and* $\varphi \in C^1(\mathbb{R})$. *Suppose that* $\varphi(A) - \varphi(B) \in \mathbf{S}_p$, *whenever* A *and* B *are selfadjoint operators such that* $A-B \in \mathbf{S}_p$. *Then the Hankel operators* $\mathcal{H}_\varphi$ *and* $\mathcal{H}_{\bar{\varphi}}$ *are bounded operators from* $H^1(\mathbb{C}_+)$ *to* $B^{1/p}_p(\mathbb{R})$.

The proof of this theorem is similar to that of Theorem 3 and is based on the following description of the Hankel operators of class $\mathbf{S}_p$: $\mathcal{H}_\varphi \in \mathbf{S}_p$ if and only if $\mathbb{P}_-\varphi \in B^{1/p}_p(\mathbb{R})$ (see [11], [12], [15]).

I conjecture that the condition $\varphi \in B^{1/p}_{\infty p}(\mathbb{R})$ is sufficient.

References

[1] **G. Benett,** *Schur multipliers,* Duke Math. J., **44** (1977), 603-639.

[2] **M. S. Birman** and **M. Z. Solomyak,** *Double operator Stieltjes integrals* (Russian), Problems of Math. Phys., Leningrad Univ., **1** (1966), 33-67.

[3] **M. S. Birman** and **M. Z. Solomyak,** *Double operator Stieltjes integrals.* II (Russian), Problems of Math. Phys., Leningrad Univ., **2** (1967), 26-60.

[4] **M. S. Birman** and **M. Z. Solomyak,** *Double operator Stieltjes integrals.* III. *Limit passage in the integrand* (Russian), Problems of Math. Phys., Leningrad Univ., **6** (1973), 27-53.

[5] **M. S. Birman** and **M. Z. Solomyak,** *Remarks on the spectral shift* (Russian), Zapiski Nauchn. Semin. LOMI, **27** (1972), 33-46.

[6] **Yu. L. Daletskii** and **S. G. Krein,** *Integration and differentiation of Hermitian operators and an application to perturbation theory* (Russian), Trudy Seminar. Funktsional. Analiz, Voronezh, **1** (1956), 81-106.

[7] **Yu. B. Farforovskaya,** *An example of a Lipschitz function of a selfadjoint operator giving a non-nuclear increment under a nuclear perturbation* (Russian), Zapiski Nauchn. Semin. LOMI, **30** (1972), 146-153.

[8] **Yu. B. Farforovskaya,** *An estimate of the norm* $\|f(A)-f(B)\|$ *for selfadjoint operators* A *and* B (Russian), Zapiski Nauchn. Semin. LOMI, **56** (1976), 143-172.

[9] **M. G. Krein,** *On some new investigations in the perturbation theory of selfadjoint operators* (Russian), in: The First Summer Math. School, Kiev, 1964, 103-187.

[10] **J. Peetre,** *New thoughts on Besov spaces,* Duke Univ. Press, Durham, N.C., 1976.

[11] **V. V. Peller,** *Hankel operators of class* $\mathfrak{S}_p$ *and their applications (rational approximation, Gaussian processes, the problem of majorizing operators)* (Russian), Mat. Sbornik, **113** (1980), 538-581.
Translation: Math. USSR Sbornik, **41** (1982), 443-479.

[12] **V. V. Peller,** *A description of Hankel operators of class* $\mathfrak{S}_p$ *for* $p>0$, *an investigation of the rate of rational approximation, and other applications* (Russian), Mat. Sbornik, **122** (1983), 481-510.
Translation: Math. USSR Sbornik, **50** (1985), 465-494.

[13] **V.V.Peller,** *Hankel operators in the perturbation theory of unitary and selfadjoint operators* (Russian), Funktsional. Anal. i Prilozhen., **19** (1985), 37-51.

[14] **V.V.Peller,** *For which* f *does* $A-B\in \mathbf{S}_p$ *imply that* $f(A)-f(B)\in \mathbf{S}_p$*?*, Operator Theory: Advances and Appl., Birkhäuser Verlag, **24** (1987), 289-294.

[15] **S.Semmes,** *Trace ideal criteria for Hankel operators and applications to Besov spaces,* Integral Equat. Operator Theory, **7** (1984), 241-281.

29

Composition Operators as Integral Operators

Donald Sarason

I. INTRODUCTION

Composition operators induced by holomorphic self-maps of the unit disk, D, have received increasing attention of late. A survey by C. Cowen (1989) describes recent progress. These operators act on a variety of spaces of holomorphic functions in D. The theme of the present paper is that they act also on spaces of functions on the unit circle, ∂D, and in that role can often be advantageously viewed as integral operators. Basically, the action on a function on ∂D is via composition with the harmonic extension of the function.

Let b be a holomorphic map of D into D satisfying $b(0) = 0$. (The last restriction is not a serious one for the questions to be addressed here, namely, boundedness and compactness.) Let L^p $(1 \leqq p \leqq \infty)$ denote the L^p space of normalized Lebesgue measure on ∂D. The functions in these spaces will always be regarded as extended harmonically into D by means of Poisson's formula. For h in L^p we let $C_b h$ denote the function on ∂D whose value at $e^{i\theta}$ is $h(b(e^{i\theta}))$.

In Section II it is shown that C_b acts as an operator of

norm 1 on each L^p, and a necessary and sufficient condition for C_b to be compact on L^1 is derived. An interesting question, not resolved here, concerns the relation between the last condition and a known necessary and sufficient condition for C_b to be compact on the Hardy spaces H^p. The ensuing sections develop further the compactness criterion for C_b on L^1. In Section III it is shown that C_b acts in a natural way on M, the space of finite complex Borel measures on ∂D (the dual of C, the space of continuous functions on ∂D). It is shown that C_b is compact on L^1 if and only if C_b maps M into L^1. Section IV contains a sufficient condition, involving H^1, for the compactness of C_b on L^1. In the concluding Section V, the adjoint, $C_b{}^*$, of C_b is introduced, and it is shown that C_b is compact on L^1 if and only if $C_b{}^*$ maps L^∞ into C. It is shown also that the weak compactness of C_b on L^1 implies its compactness.

The following simple version of the Schur boundedness test (J. Schur, 1911) will be used.

LEMMA 1. *Let K be a measurable function on the Cartesian product of ∂D with itself. If there are positive numbers A and B such that*

$$\frac{1}{2\pi}\int_{-\pi}^{\pi} |K(e^{i\theta},e^{it})|\, d\theta \leqq A \quad \text{for almost all } t,$$

$$\frac{1}{2\pi}\int_{-\pi}^{\pi} |K(e^{i\theta},e^{it})|\, dt \leqq B \quad \text{for almost all } \theta,$$

then the integral operator T with kernel K, defined by

$$(Th)(e^{i\theta}) = \frac{1}{2\pi}\int_{-\pi}^{\pi} h(e^{it})K(e^{i\theta},e^{it})\,dt,$$

is a bounded operator on L^p $(1 < p < \infty)$ *of norm at most* $A^{1/p}B^{1/p'}$ $(p' = p/(p-1))$. *If the first of the two conditions holds, then the integral operator* T *on* M *with kernel* K, *defined by*

$$(Tm)(e^{i\theta}) = \int K(e^{i\theta},e^{it})\,dm(e^{it}),$$

is a bounded operator of M *into* L^1, *with norm at most* A.

The simple proof is included for completeness. To prove the first part, let h be in L^p $(1 < p < \infty)$ and let $g = Th$. By Hölder's inequality,

$$|g(e^{i\theta})|^p \leq \frac{B^{p/p'}}{2\pi}\int_{-\pi}^{\pi} |h(e^{it})|^p|K(e^{i\theta},e^{it})|\,dt,$$

which, when integrated with respect to θ, gives, after an application of Fubini's theorem, the desired inequality, namely $\|g\|_p \leq A^{1/p}B^{1/p'}\|h\|_p$. To prove the second part, let m be a measure in M and let $g = Tm$. Then obviously

$$|g(e^{i\theta})| \leq \int |K(e^{i\theta},e^{it})|\,d|m|(e^{it}).$$

Again, integration with respect to θ and an application of Fubini's theorem yields the desired inequality.

The following known consequence of the boundary unicity

theorem for H^∞ will be needed.

LEMMA 2. *If* S *is a closed subset of* ∂D *of measure* 0, *then* $b^{-1}(S)$ *has measure* 0.

In fact, under the hypotheses on S there is a nonzero function h in the disk algebra that vanishes on S (Garnett, 1981, p. 125). Then $C_b h$ is a nonzero function in H^∞ with radial limit 0 at each point of $b^{-1}(S)$, implying that the latter set has measure 0.

A final bit of notation: $\mathcal{L}(X)$ will denote the space of bounded linear operators on the Banach space X.

II. BOUNDEDNESS AND COMPACTNESS ON L^p

If $|b| < 1$ almost everywhere on ∂D, then, in view of Poisson's formula, C_b is the integral operator whose kernel is the function

$$K_b(e^{i\theta}, e^{it}) = \frac{1 - |b(e^{i\theta})|^2}{|e^{it} - b(e^{i\theta})|^2},$$

in other words,

$$(C_b h)(e^{i\theta}) = \frac{1}{2\pi}\int_{-\pi}^{\pi} K_b(e^{i\theta}, e^{it}) h(e^{it})\, dt$$

for h in L^p. We define K_b as above even in case $|b| = 1$ on a set of positive measure.

LEMMA 3. <u>If</u> $|b(e^{i\theta})| < 1$ <u>then</u> $\|K_b(e^{i\theta}, \cdot)\|_1 = 1$. <u>Also</u> $\|K_b(\cdot, e^{it})\|_1 \leqq 1$ <u>for all</u> t. <u>The last inequality is an equality for all</u> t <u>if</u> $\|b\|_\infty < 1$. <u>It is an equality for almost all</u> t <u>if and only if</u> $|b| < 1$ <u>almost everywhere on</u> ∂D.

Indeed, the first statement follows because, if $|b(e^{i\theta})| < 1$, then $K_b(e^{i\theta}, \cdot)$ is the Poisson kernel for the point $b(e^{i\theta})$, a function of unit norm in L^1. That $\|K_b(\cdot, e^{it})\|_1 = 1$ for all t if $\|b\|_\infty < 1$ follows from the equality

$$K_b(e^{i\theta}, e^{it}) = 1 + 2 \operatorname{Re} \sum_{n=1}^{\infty} e^{-int} b(e^{i\theta})^n.$$

If $\|b\|_\infty < 1$, the series on the right converges uniformly on ∂D, and each term integrates to 0 against $d\theta$ because of the condition $b(0) = 0$. Thus, in the general case, if $0 < r < 1$, then $\|K_{rb}(\cdot, e^{it})\|_1 = 1$ for all t, and, letting $r \to 1$ and applying Fatou's lemma, we can conclude that $\|K_b(\cdot, e^{it})\|_1 \leqq 1$ for all t. To prove the last statement in the lemma one need only integrate K_b with respect to both variables and use Fubini's theorem; the total integral is 1 in case $|b| < 1$ almost everywhere on ∂D and less than 1 otherwise.

PROPOSITION 1. C_b <u>is a bounded operator on</u> L^p <u>of norm</u> 1 $(1 \leqq p < \infty)$. <u>Moreover</u> $C_{rb} \to C_b$ <u>in the strong operator topology of</u> $\mathfrak{L}(L^p)$ <u>as</u> r <u>tends to</u> 1.

In fact, for the case where $|b| < 1$ almost everywhere on ∂D the first statement follows immediately from Lemma 3 and Schur's test (Lemma 1). (Note that the norm of C_b on L^p is at least 1 because C_b fixes the constant functions.) The first statement for the general case follows from this special case in conjunction with the second statement, so it only remains to establish the second statement.

It will first be shown that $\lim_{r \to 1} C_{rb}$ exists in the strong operator topology of $\mathcal{L}(L^p)$. Because the operators C_{rb} are uniformly norm bounded it will suffice, in order to establish this, to show that $\lim_{r \to 1} C_{rb}h$ exists in L^p norm for a dense set of functions h in L^p. As the existence of the last limit is obvious when h is a trigonometric polynomial, the existence of the strong operator limit follows.

To complete the proof it suffices to remark that $C_{rb}h \to C_b h$ almost everywhere for each h in L^p. (The point is that h has a radial limit almost everywhere on ∂D, and the set of θ such that $b(e^{i\theta})$ lies in the exceptional set is of measure 0, as one can deduce from Lemma 2.)

That C_b has norm 1 in $\mathcal{L}(L^p)$ is basically Littlewood's subordination principle (Duren, 1970), which thus emerges in the present treatment as a corollary of Schur's test.

An obvious necessary condition for C_b to be compact on any L^p, first pointed out by H. J. Schwartz (1969), is that $|b| < 1$ almost everywhere on ∂D. (For, if the condition fails, then the sequence $(b^n)_1^\infty$ does not tend to 0 in L^p norm, although it is the image under C_b of a sequence that tends weakly to 0.) An obvious sufficient condition that C_b be compact on every L^p is $\|b\|_\infty < 1$.

PROPOSITION 2. _If_

(c) $$\|K_b(\cdot, e^{it})\|_1 = 1 \quad \underline{\text{for all}} \quad t,$$

then C_b _is compact on_ L^p $(1 \leqq p < \infty)$.

In fact, assume (c) holds. Since C_{rb} is compact on L^p for $r < 1$, it will suffice to show that $C_{rb} \to C_b$ in $\mathcal{L}(L^p)$ norm as $r \to 1$. To do that we consider the map

$$re^{it} \to K_{rb}(\cdot, e^{it})$$

from $\bar{D}$ to L^1. It is a consequence of (c) that this map is continuous relative to the L^1 norm, as one can see on the basis of the known result that a sequence in L^1 converges in norm if it converges almost everywhere and the corresponding sequence of norms converges to the norm of the limit function. Consequently

$$\|K_b(\cdot, e^{it}) - K_{rb}(\cdot, e^{it})\|_1 = o(1)$$

uniformly in t as $r \to 1$. Since also obviously

$$\|K_b(e^{i\theta}, \cdot) - K_{rb}(e^{i\theta}, \cdot)\|_1 \leqq 2$$

for all r and θ, the norm convergence of C_{rb} to C_b in $\mathcal{L}(L^p)$ $(1 \leq p < \infty)$ follows immediately from Schur's test.

One can obtain an alternative proof of Proposition 2 by use of a general compactness criterion for operators on L^1 spaces (Dunford and Schwartz, 1958, p. 507) to handle the case $p = 1$ and then interpolation between L^1 and L^∞ to settle the remaining cases.

The converse of Proposition 2 holds for $p = 1$.

PROPOSITION 3. <u>If C_b is compact on L^1 then (c) holds</u>.

In fact, assume C_b is compact on L^1, and fix a point e^{it} in ∂D. For $0 < r < 1$, let h_r be the Poisson kernel for the point re^{it}:

$$h_r(z) = \frac{1 - |rz|^2}{|e^{it} - rz|^2}.$$

Each h_r has unit norm in L^1, so there is a sequence $(r_n)_1^\infty$ tending to 1 such that $(C_b h_{r_n})_1^\infty$ converges in L^1 norm. But $C_b h_r = K_{rb}(\,\cdot\,, e^{it})$, which for $r < 1$ has unit norm in L^1, and $K_{rb}(\,\cdot\,, e^{it}) \to K_b(\,\cdot\,, e^{it})$ almost everywhere as $r \to 1$. It follows that $K_b(\,\cdot\,, e^{it})$ has unit norm in L^1, as desired.

There remains a puzzle concerning the relation between the compactness of C_b on L^1 and its compactness on the other L^p spaces. By the Riesz projection theorem, if $1 < p < \infty$, then C_b is compact on L^p if and only if it is compact on H^p. J. H. Shapiro and P. D. Taylor (1973) have shown that if C_b is compact on one of the spaces H^p then it is compact on all of them $(0 < p < \infty)$. Thus, for example, the compactness of C_b on L^1 implies its compactness on H^2. The reverse implication would be surprising, but I have been unable to produce a counterexample.

Shapiro (1987) has obtained a necessary and sufficient condition, in terms of the Nevanlinna counting function for b, for the compactness of C_b on H^2. That condition is therefore implied by condition (c), an implication that does not seem directly apparent. A purely function-theoretic proof of the implication would be of interest.

Shapiro and Taylor show in their paper that C_b will fail to be compact on H^2 (and hence also on L^1) if b has an angular derivative in the sense of Carathéodory at any point of ∂D. This provides a simple way to construct examples of functions b for which (c) fails. Nontrivial examples for which (c) holds can be obtained from exposed points of the unit ball of H^1. Suppose f is an outer function of unit norm in H^1. Let F_f be the Herglotz integral of $|f|$:

$$F_f(z) = \frac{1}{2\pi}\int_{-\pi}^{\pi} \frac{e^{i\theta}+z}{e^{i\theta}-z}\,|f(e^{i\theta})|\,d\theta.$$

The function $b_f = (F_f-1)/(F_f+1)$ then maps D into D and vanishes at the origin. I have proved in an earlier paper (Sarason, 1989) that b_f satisfies (c) if f is an exposed point of the unit ball of H^1, and I have conjectured the reverse implication. A proof of the conjecture would go a long way toward elucidating the structure of the exposed points of the unit ball of H^1. Although that structure is not now well understood, there is nevertheless an abundance of examples; for instance, the functions of unit norm in H^1 whose reciprocals are in H^1 are exposed. In the paper just cited it is shown, on the basis of the connection with exposed points, that (c) holds for the product of the function $(1-z)/2$ with a Blaschke product vanishing at the origin whose zeros lie on the real axis and tend to the point -1.

It should be mentioned that the function b_f is not the general holomorphic self-map of the unit disk vanishing at the origin; it is rather, to within a constant multiple of unit modulus, the general nonextreme point of the unit ball of H^∞ that vanishes at the origin. I do not have an example of an extreme point that satisfies (c).

III. COMPACTNESS ON M

If h is in L^1, then the Poisson integral of $C_b h$ equals the composition of the Poisson integral of h with b. This is evident in case $\|b\|_\infty < 1$, whence the general case follows by a limit argument, because $C_{rb}h \to C_b h$ in norm as r tends to 1 (by Proposition 1).

If m is a measure in M, then the L^1 norm on any circle centered at the origin of the Poisson integral of m is at most $\|m\|$. Because C_b has norm 1 as an operator on L^1, the same estimate holds for the composition of the Poisson integral of m with b, implying that that composition is the Poisson integral of a measure. We define $C_b m$ to be the latter measure. In view of the preceding remarks, this definition for absolutely continuous measures is compatible with the earlier one for L^1 functions.

Notice that if m is a probability measure then so is $C_b m$, since its Poisson integral is positive and takes the value 1 at the origin (due to the condition $b(0) = 0$).

PROPOSITION 4. The following conditions are equivalent.

(i) C_b is compact on L^1.

(ii) C_b is compact on M.

(iii) C_b maps M into L^1.

Of course, in (iii), L^1 is being identified with the space of absolutely continuous measures in M. Because L^1 becomes via this identification a subspace of M, the implication (ii) => (i) is trivial.

To prove that (i) implies (ii) and (iii), assume that (i), and hence also (c), holds. If $0 < r < 1$ then C_{rb} is obviously compact on M and maps M into L^1, so it will suffice to show that $C_{rb} \to C_b$ in $\mathcal{L}(M)$ norm as r tends to 1. If m is in M, then clearly the Poisson integral of $C_{rb}m$ converges pointwise to the Poisson integral of $C_b m$ as r tends to 1. Thus, to prove that $C_{rb} \to C_b$ in $\mathcal{L}(M)$ norm, it will suffice to prove that $\lim_{r\to 1} C_{rb}$ exists in $\mathcal{L}(M)$ norm. That can be accomplished by the same argument as was used in the proof of Proposition 2. Namely, as a consequence of condition (c) we have

$$\|K_{rb}(\ \cdot\ ,e^{it}) - K_{sb}(\ \cdot\ ,e^{it})\|_1 \ = \ o(1)$$

uniformly in t as r and s tend to 1. The last part of Lemma 1 thus implies that $\|C_{rb} - C_{sb}\|_{\mathcal{L}(M)} \to 0$ as r and s tend to 1, which yields the desired convergence.

A proof that (iii) implies (i) will complete the proof of Proposition 4. Assume (iii) holds. Fix a point e^{it} in ∂D, and let δ_t be the Dirac measure at that point. The Poisson integral of $C_b\delta_t$ is then the function $(1-|b|^2)/|e^{it}-b|^2$, whose boundary function is $K_b(\ \cdot\ ,e^{it})$. Since $C_b\delta_t$ is by assumption absolutely continuous, it therefore coincides with the measure $\frac{1}{2\pi} K_b(e^{i\theta},e^{it})\, d\theta$. Since $C_b\delta_t$ is a probability measure, we can conclude that $\|K_b(\ \cdot\ ,e^{it})\|_1 = 1$, i.e., condition (c) holds, and the proof is complete.

IV. A SUFFICIENT CONDITION FOR COMPACTNESS ON L^1

Shapiro and Taylor prove in their paper that C_b is a Hilbert-Schmidt operator on H^2 if and only if $(1-|b|)^{-1}$ is in L^1. They use this result to show, by a simple argument, that C_b is a nuclear operator on H^1 if it is a Hilbert-Schmidt operator on H^2. Their argument shows that C_b is even nuclear on L^1 if it is a Hilbert-Schmidt operator on H^2.

Let H_1 be the real-variable version of H^1, that is, the space of functions in L^1 whose conjugates are also in L^1, with the norm

$$\|h\|_{H_1} = \|h\|_1 + \|\tilde{h}\|_1$$

($\tilde{h}$ denoting the conjugate function of h). The argument of Shapiro and Taylor shows that C_b maps L^1 into H_1 if C_b is a Hilbert-Schmidt operator on H^2.

The next proposition is suggested by the foregoing results.

PROPOSITION 5. <u>If</u> C_b <u>maps</u> L^1 <u>into</u> H_1, <u>then</u> C_b <u>is compact on</u> L^1.

In fact, assuming C_b maps L^1 into H_1, we can conclude, by the closed graph theorem, that C_b is a bounded map of L^1 into H_1. We shall show that C_b is compact on L^1 by verifying that condition (c) is satisfied. For w a point of D, let P_w denote the Poisson kernel for w, namely, $P_w(z) = (1-|wz|^2)/|1-\bar{w}z|^2$. Since $\|P_w\|_1 = 1$ we have $\|C_bP_w\|_{H_1} \leqq$ const. for all w in D. The Fourier projection ($h \to \frac{1}{2}(h+i\tilde{h}) + \frac{1}{2}h(0)$) is bounded on H_1, and the Fourier projection of C_bP_w is $(1-\bar{w}b)^{-1}$, as one sees from the equality

$$C_bP_w = \frac{1}{1-\bar{w}b} + \frac{w\bar{b}}{1-w\bar{b}}.$$

Hence $(1-\bar{w}b)^{-1}$ is in H^1, with $\|(1-\bar{w}b)^{-1}\|_1 \leqq$ const., for all w in D. If e^{it} is any point of ∂D then, by letting

w tend to e^{it} from D and using Fatou's lemma, we can conclude that $(e^{it}-b)^{-1}$ is in H^1. Hence also $(e^{it}+b)/(e^{it}-b)$ is in H^1. The real part of the last function is therefore the Poisson integral of its boundary function. Beacuse the function equals 1 at the origin and the boundary function of the real part if $K_b(\cdot, e^{it})$, condition (c) follows.

Incidentally, Proposition 5 is not an instance of a general fact about operators from L^1 to H_1, for it is simple to construct a bounded operator of L^1 into H_1 that is not compact on L^1 (one is given in Section 5).

The compactness condition in Proposition 5 is only a sufficient one, not a necessary one, as the following example shows. Let ρ be a positive function of unit norm in L^1 that is bounded away from 0 but with $\rho \log^+ \rho$ nonintegrable. Let f be the outer function with modulus ρ, and form the function b_f as in Section II. Since f^{-1} is in H^1 (in fact, in H^∞), the function f is an exposed point of the unit ball of H^1. This implies that C_{b_f} is compact on L^1, as explained in Section II. If C_{b_f} mapped L^1 into H_1, then the function $(1+b_f)/(1-b_f)$ would be in H^1, by the proof of Proposition 5. But the real part of $(1+b_f)/(1-b_f)$ on ∂D is ρ, whose conjugate function is nonintegrable by a well-known theorem of A. Zygmund (Duren, 1970, p. 80). Hence C_{b_f} does not map L^1 into H_1.

V. THE ADJOINT OF C_b

For $1 \leqq p < \infty$ the space $L^{p'}$, where $p' = p/(p-1)$, is the dual of L^p under the pairing

$$\langle h,g\rangle = \frac{1}{2\pi}\int_{-\pi}^{\pi} h\bar{g}\, d\theta \qquad (h \in L^p, \quad g \in L^{p'}).$$

If $q < p$ then $L^{q'} \subset L^{p'}$, so the adjoint of C_b as an operator on L^p is defined also on $L^{q'}$, and it is easily seen to coincide there with the adjoint of C_b as an operator on L^q. We may thus speak without ambiguity of a single operator $C_b{}^*$ on $\cup_{p>1} L^p$.

If g is a trigonometric polynomial of degree n, then $C_b{}^*g$ is a trigonometric polynomial of degree at most n. This is so because C_b maps the annihilator (in L^2, say) of the trigonometric polynomials of degree n into itself (due to the condition $b(0) = 0$). Since the trigonometric polynomials are dense in C and $C_b{}^*$ is bounded on L^∞, it follows that $C_b{}^*$ maps C into C.

PROPOSITION 6. C_b *is compact on* L^1 *if and only if* $C_b{}^*$ *maps* L^∞ *into* C.

To prove this, assume first that C_b is compact on L^1. Then $|b| < 1$ almost everywhere on ∂D, and C_b is the integral operator with kernel K_b. Thus $C_b{}^*$ is the integral operator with the adjoint kernel, in other words,

$$(C_b^* g)(e^{it}) = \frac{1}{2\pi} \int_{-\pi}^{\pi} g(e^{i\theta}) K_b(e^{i\theta}, e^{it}) \, d\theta.$$

Moreover, condition (c) holds, which implies, as noted in the proof of Proposition 2, that the map $e^{it} \to K_b(\cdot, e^{it})$ of ∂D into L^1 is norm continuous. The continuity of $C_b^* g$ for g in L^∞ is now obvious.

For the other direction, assume $C_b^* L^\infty \subset C$. It will be shown first, by contradiction, that $|b| < 1$ almost everywhere on ∂D. Suppose $|b| = 1$ on a subset of ∂D of positive measure. Let S be a closed nowhere dense subset of ∂D such that the set $R = b^{-1}(S)$ is contained in ∂D and has positive measure. (Standard reasoning produces such a set S.) The set S also has positive measure, by Lemma 2. If h is a function in $L^1(\partial D \setminus S)$ (i.e., a function in L^1 that is carried by $\partial D \setminus S$), then $C_b h$ is in $L^1(\partial D \setminus R)$. By duality, C_b^* maps $L^\infty(R)$ into $L^\infty(S)$. In particular, then, the function $C_b^* \chi_R$ is carried by S. The function $C_b^* \chi_R$ is not the zero function, as one can see by integrating it against the constant function 1. Since S is closed and nowhere dense, it is impossible for $C_b^* \chi_R$ to be continuous, contrary to hypothesis. Hence $|b| < 1$ almost everywhere on ∂D. It follows that C_b^* is an integral operator, as in the first part of the proof.

A lemma is now needed. It will be stated for the unit circle, although it holds in general measure spaces (Dunford and Schwartz, 1958, p. 295).

LEMMA 4. *If the sequence* $(h_n)_1^\infty$ *in* L^1 *converges weakly to the function* h *and also converges to* h *almost everywhere, then* $h_n \to h$ *in norm.*

The condition that $C_b^*L^\infty \subset C$ tells us that the map $e^{it} \to K_b(\cdot, e^{it})$ from ∂D into L^1 is weakly continuous. Using Lemma 4 we conclude that the map is continuous in L^1 norm. By Lemma 3, $K_b(\cdot, e^{it})$ has unit norm for almost all t, so, by continuity, it has unit norm for all t, in other words, condition (c) holds. Therefore C_b is compact on L^1, as desired, and the proof of Proposition 6 is complete.

The operator C_b on M can be regarded as the adjoint of the operator C_b^* on C. To verify this it suffices to check that, for all m in M, the equality $\int h\, d(C_b m) = \int C_b^* h\, dm$ holds for a set of functions h that spans C. The equality is easily established when $h = P_w$, the Poisson kernel for the point w of D. In fact, a simple argument shows that $C_b^* P_w = P_{b(w)}$, and the equality $\int P_w\, d(C_b m) = \int P_{b(w)}\, dm$ follows immediately from the definition of $C_b m$. As the functions P_w span C, the statement at the beginning of the paragraph has been justified. It will be used in the proof of the following result.

PROPOSITION 7. *If* C_b *is weakly compact on* L^1 *then* C_b *is compact on* L^1.

In fact, assume C_b is weakly compact on L^1. By Proposition 4, it will suffice to show that C_b then maps M into L^1, which will be deduced from the known theorem (Dunford and Schwartz, 1958, p. 482) that an operator T on a Banach space X is weakly compact if and only if T^{**} maps X^{**} into X. Let C_b^{**} denote the adjoint of C_b^{*} as an operator on L^∞. Then, by the theorem just mentioned, C_b^{**} maps $(L^\infty)^*$ into L^1. Let m be any measure in M. By the Hahn-Banach theorem, m can be extended from a bounded linear functional on C to a bounded linear functional, μ, on L^∞. To complete the proof it only remains to show that $C_b m = C_b^{**}\mu$, in other words, that $\int h\, d(C_b m) = \int h\, d(C_b^{**}\mu)$ for all functions h in C. The last equality follows immediately from the remarks preceding the statement of the proposition.

In connection with Proposition 7 it is worth exhibiting an operator on L^1 that is weakly compact but not compact. One can be produced very simply as follows. Let $(I_n)_1^\infty$ be a disjoint sequence of subarcs of ∂D, and define the operator T on L^1 by

$$(Th)(e^{i\theta}) = \sum_1^\infty \left(\int_{I_n} h(e^{it})\, dt\right) e^{i\theta}.$$

That T is weakly compact follows from a general criterion of Dunford and Pettis (Dunford and Schwartz, 1958, p. 507); it follows also from the observation that T is absolutely 1-summing (Pisier, 1986). Since the functions

$$(\text{length } I_n)^{-1} \chi_{I_n}$$

form a bounded sequence in L^1 whose image sequence under T has no norm-convergent subsequence, the operator T is not compact. (The noncompactness of T follows also from the Dunford-Pettis criterion.) This is also an example of a non-compact operator on L^1 that maps L^1 into H^1.

The question of whether there is an analogue of Proposition 7 with H^1 in place of L^1 seems interesting. Let BMO and VMO denote the usual spaces of functions of bounded mean oscillation and vanishing mean oscillation on ∂D, and let BMOA and VMOA denote their intersections with H^1. As is well known, H^1 is the dual of VMOA and BMOA is the dual of H^1 under the pairing introduced at the beginning of this section, insofar as it is defined. (For each function in one of the spaces, it is defined for a dense set of functions in the paired space.)

The space VMO consists of those functions on ∂D that can be written as sums of continuous functions and conjugates of continuous functions. Since C_b^* preserves C and commutes with the conjugation operator, it preserves VMO and hence also VMOA. Similarly, as BMO consists of those functions writable as sums of functions in L^∞ and conjugates of functions in L^∞, and as C_b^* preserves L^∞, it follows that C_b^* preserves BMO and hence also BMOA. One easily sees that C_b on H^1 is the adjoint of C_b^* on VMOA, and C_b^* on BMOA is the adjoint of C_b on H^1. By the criterion mentioned during the proof of Proposition 7, together with the equivalence of the weak compactness of an operator and the weak compactness of its adjoint (Dunford and Schwartz, 1958, p. 485), C_b is weakly compact on H^1 if and only if C_b^* maps BMOA into VMOA. The question of whether the analogue of Proposition 7 for H^1 holds is thus equivalent to the question of whether the inclusion $C_b^*(\text{BMOA}) \subset \text{VMOA}$ is sufficient for the compactness of C_b on H^1.

REFERENCES

Cowen, C. C. (1989). Composition operators on Hilbert spaces of analytic functions: A status report, *Proceedings of Symposia in Pure Mathematics*, Amer. Math. Soc., Providence (forthcoming).

Dunford, N., and Schwartz, J. T. (1958). *Linear Operators, Part I*, Interscience, New York.

Duren, P. L. (1970). *Theory of H^p Spaces*, Academic Press, New York.

Garnett, J. B. (1981). *Bounded Analytic Functions*, Academic Press, New York.

Pisier, G. (1986). *Factorization of Linear Operators and Geometry of Banach Spaces*, CBMS Regional Conference Series in Mathematics, No. 60, Amer. Math. Soc., Providence.

Sarason, D. (1989). Exposed points in H^1, I, *Operator Theory: Advances and Applications* (forthcoming).

Shapiro, J. H. (1987). The essential norm of a composition operator, Ann. of Math. 125, pp. 375-404.

Shapiro, J. H., and Taylor, P. D. (1973). Compact, nuclear, and Hilbert-Schmidt composition operators on H^2, Indiana Univ. Math. J. 23, pp. 471-496.

Schur, J. (1911). Bemerkung zur Theorie der beschränkten Bilinearformen mit unendlich vielen Veränderlichen, J. Reine Angew. Math. 140, pp. 1-28.

Schwartz, H. J. (1969). Composition Operators on H^p, Thesis, University of Toledo.

Part III. Partial Differential Equations

30

Uniform Estimates for Regularization of Free Boundary Problems

H. Berestycki, L. A. Caffarelli, L. Nirenberg

Introduction:

Free boundary problems are often approximated by regularizing ones. To obtain information about the solution to the original problem one tries to establish results for the approximating one which carry over in the limit. A useful reference is Friedman's book [F].

In this paper we study such a regularizing problem in a domain Ω with smooth boundary $\partial\Omega$, of the form: u is a nonnegative bounded solution of

$$Lu = a_{ij}(x)u_{ij} + b_i(x)u_i + c(x)u = \beta_\epsilon(u); \tag{1.1}$$

here L is a linear elliptic operator. The nonlinear term $\beta_\epsilon(u)$ is a nonnegative L^∞ function satisfying

$$\beta_\epsilon \text{ has its support in } [0, \epsilon] \text{ and } \beta_\epsilon \leq B/\epsilon. \tag{1.2}$$

In addition we sometimes assume

$$\beta_\epsilon \geq \frac{\kappa}{\epsilon} \text{ for some } \kappa > 0, \quad \text{on } [a\epsilon, b\epsilon], \text{ for some positive } a < b < 1. \tag{1.3}$$

In particular, in sections 5 and 7 we consider

$$\beta_\epsilon(u) = \frac{1}{\epsilon}\beta\left(\frac{u}{\epsilon}\right) \tag{1.4}$$

where β is a continuous function with support on [0,1], and positive on (0,1) and with

$$\int_0 \beta(s)ds = M > 0. \tag{1.5}$$

A principal feature of our work is to obtain estimates up to the boundary, and when studying behaviour near the boundary we will assume that u satisfies on $\partial\Omega$, or on a portion of $\partial\Omega$,

$$\mu(x) \cdot \nabla u = 0, \tag{1.6}$$

where $\mu(x)$ is a smooth vector on $\partial\Omega$ which always points outside Ω. We stress however that for estimates in compact subsets of Ω, no boundary condition is required.

As we will see, this problem is an approximation (as $e \searrow 0$) for a free boundary problem in which a nonnegative function v satisfies

$$Lv = 0 \quad \text{in } \Omega^+ = \{x \in \Omega \big| v(x) > 0\}$$

and a condition on the free boundary $\partial\Omega^+ \cap \Omega$—in case $a_{ij} \in C^1$. Namely, if a part σ of the free boundary is a smooth hypersurface, with $v > 0$ on one side and $v = 0$ on the other, near σ, then v satisfies the boundary condition:

$$a_{ij}\nu_i\nu_j|\nabla v|^2 = 2M \quad \text{on } \sigma. \tag{1.7}$$

Here ν is the unit normal to Ω^+.

In §3 we will establish uniform Lipschitz continuity of the u_ϵ on compact subsets of $\overline{\Omega}$—independent of ϵ. In order to advance our understanding of the limit situation, we think of the set

$$\{x \in \Omega \big| a\epsilon < u_\epsilon(x) < b\epsilon\}, \tag{1.8}$$

here $a < b$ are fixed numbers between zero and 1, as a band which approximates the free boundary of the limit problem. Under suitable hypotheses we will show in section 4 that for any compact subset J of $\overline{\Omega}$ there are positive constants $c_0 = c_0(J)$, ρ, independent of ϵ such that

$$c_0^{-1}d(x,\Omega_\epsilon) \geq u_\epsilon \geq c_0 d(x,\Omega_\epsilon) \tag{1.9}$$

provided $d(x, \Omega_\epsilon) \le \rho$ and ϵ is sufficiently small. Here

$$\Omega_\epsilon = \{x \in \Omega \,\big|\, u_\epsilon(x) \le \epsilon\}. \tag{1.10}$$

In addition, in section 6, we will show that with $C = C(a)$, independent of ϵ,

$$\text{Vol}\,[\{a\epsilon < u < b\epsilon\} \cap B_r] \le C\epsilon r^{n-1} \tag{1.11}$$

for any ball B_r of radius r; here $0 < a < b < 1$.

So our point of view is to study the free boundary and the limit solution by establishing uniform estimates near the set where u_ϵ is close to ϵ.

In this same spirit, Redondo [R], has proved regularity of the free boundary for the obstacle problem by establishing uniform regularity for the approximating level surfaces of a suitably regularized problem.

Our interest in these questions arose in connection with earlier work by two of the authors concerned with properties of solutions of a particular second order semilinear elliptic equation in an infinite cylinder in R^n. This is a model in flame combustion describing the propagation of a curved premixed flame. The problem is the following: T is a positive function < 1 in the infinite cylinder $S = R \times \omega$, where ω is a bounded domain in R^{n-1} with smooth boundary. We use coordinates $x = (x_1, y)$, $y = (y_1, \ldots, y_{n-1}) \in \omega$; $T_1 = T_{x_1}$ etc. The function T which represents the temperature in a travelling front satisfies

$$(1.12)\quad \begin{cases} \Delta T - c\alpha(y)T_1 + f(T) = 0 \quad \text{in } S, \\ \text{and Neumann boundary condition} \\ T_\nu = 0 \quad \text{on } \partial S. \end{cases}$$

In addition

$$\lim_{x_1 \to -\infty} T(x_1, y) = 0, \qquad \lim_{x_1 \to +\infty} T(x_1, y) = 1 \tag{1.13}$$

uniformly in y. The function T and the positive constant c are the unknowns. The function α is a given continuous positive function on $\bar{\omega}$, and f is a given Lipschitz continuous function with support on $[\theta, 1]$, $0 < \theta < 1$; f is positive on $(\theta, 1)$. Some further regularity is assumed near 1, and $f'(1) < 0$.

The problem is described in detail in [BL] where existence of (c, T) was proved. In [BN2] it was proved that the solution (c, T) is unique (up to x_1–translation of T) and that $T_1 > 0$. See also [BLN] for a description of some of the results and earlier results for $n = 1$.

It is of physical interest to consider a family f_ϵ of functions f, with $\theta \to 1$, such that

$$\int_\theta^1 f_\epsilon(s)ds \to \text{ some positive constant } N.$$

It is expected that the solutions (c_ϵ, T_ϵ), after suitable x_1–translation of T_ϵ tend to (c, T) where $T \equiv 1$ for x_1 large, and that the region $\{T < 1\}$ and $\{T \equiv 1\}$ touch at a free boundary surface γ, the flame front. In the region where $T < 1$ it satisfies $\Delta T - c\alpha T_1 = 0$. Several years ago Berestycki and Nirenberg proved that this was the case of $n = 2$; they also proved that the "free boundary curve" γ lies in a bounded region. They did not prove regularity of γ but showed that at regular points of γ the function T satisfies the free boundary condition (1.7):

$$T_\nu = \sqrt{2N} \tag{1.14}$$

where ν represents the exterior normal on γ to the region where $T < 1$.

If we set $u = \frac{1-T}{1-\theta}$, then, with an appropriate function β_ϵ we obtain:

$$\begin{cases} \Delta u_\epsilon - c_\epsilon \alpha(y)\partial_1 u_\epsilon &= \beta_\epsilon(u_\epsilon) \quad \text{in } S \\ \partial_\nu u_\epsilon &= 0 \quad \text{on } \partial S. \end{cases} \tag{1.15}$$

Also $\partial_1 u_\epsilon < 0$ (by [BN2]).

Using the results of section 3 we will prove in section 7 that for $n \geq 1$, the u_ϵ are uniformly Lipschitz continuous independent of ϵ; that for a suitable subsequence, $\epsilon = \epsilon_i \to 0$, $c_\epsilon \to c > 0$, and $u_\epsilon \to v$. For v we will establish the properties described above. The free boundary condition (1.14) for $T = 1 - (1-\theta)v$ at regular points of the free boundary surface γ is a consequence of Theorem 5.1.

Returning to (1.1), let us formulate the conditions on L. We always assume uniform ellipticity:

$$a_{ij}\xi_i\xi_j \geq m|\xi|^2, \qquad m > 0, \tag{1.16}$$

and

$$\begin{cases} \text{the coefficients are measurable, and bounded in absolute value} \\ \text{by a constant } C_1. \end{cases} \tag{1.17}$$

We shall usually also assume

$$a_{ij}(x) \text{ are continuous in } \overline{\Omega} \tag{1.18}$$

$$c(x) \leq 0, \tag{1.19}$$

or even

$$c \equiv 0. \tag{1.20}$$

Whenever (1.18) is assumed, our estimates will depend on the modulus of continuity of the a_{ij}. This is always to be understood and will not be mentioned again.

Our solutions u are assumed to be in $W^{2,n}_{\text{loc}}(\overline{\Omega})$ and to be strong solutions (see [GT] chapter 9) i.e. the equation holds almost everywhere.

A basic result which is used several times is the Harnack inequality of Krylov and Safonov for positive solutions of $Lu = 0$. We use the form in Cor. 9.25 of [GT]. In addition we need such a result up to the boundary in case (1.6) holds. So in section 2 we present an extension. Conditions (1.18), (1.19) are not required there but they will be assumed in the remaining sections.

An important property for equations of the form

$$Lu = f \tag{1.21}$$

(under conditions (1.16)–(1.18)) in the intersection G of a ball $B_r(x_0)$ with Ω, for solutions satisfying (1.6) on $\partial\Omega \cap B_r$, is the interior estimate

$$|\nabla u(x_0)| \leq K\left(\frac{\sup|u|}{r} + \sup|f|\right). \tag{1.22}$$

Here K depends on the smoothness of $\partial\Omega \cap B_r$ (i.e., on bounds for derivatives of functions representing $\partial\Omega$ in B_r) but does not depend on the distance $d(x_0, \partial\Omega)$ of x_0 to $\partial\Omega$. K also depends of course on C_1 and m, and on the smoothness of μ. This follows from standard elliptic theory, for example, $W^{2,p}$ estimates for $p > n$ (see [GT]).

If we also assume (1.19) and that

$$\begin{cases} \text{for some } x_1 \in \partial B_r(x_0), \text{ and some } t > 0, \\ \overline{B_{tr}(x_1) \cap B_r(x_0)} \subset \Omega, \end{cases} \tag{1.23}$$

then there exists a unique solution of (1.21) in G satisfying (1.6) on $\partial\Omega \cap B_r$ and, say, $u = 0$ on the remainder of ∂G, and we have

$$|u(x)| \leq K \sup|f| \quad \text{in } G \tag{1.24}$$

with K depending as before on G and on C_1, m, and μ and also on t. The solution is in $W^{2,p}$ in any compact subset of $\overline{G} \setminus (\partial\Omega \cap \partial B_r)$.

The paper is organized as follows:

§**2.** Harnack inequality up to the boundary.

§3. Uniform Lipschitz continuity in compact subsets of $\overline{\Omega}$.

§4. Nondegeneracy of "minimal solutions," i.e. the proof of (1.9).

§5. Study of the limit v of u_ϵ as $\epsilon \to 0$ and the proof of the boundary condition (1.7) on a smooth portion of the free boundary $\gamma = \partial\{v > 0\}$.

§6. Here we show that in any compact subset, the $(n-1)$ dimensional Hausdorff measure $H^{n-1}(\gamma \cap B_r) \le Cr^{n-1}$. This is a first step in the study of regularity of γ. In this paper we do not carry out such a study but we plan to do so subsequently.

§7. Application of our results to study the free boundary flame propagation problem (1.12)–(1.15) when $\epsilon \to 0$.

We often use the expression: some constant is under control. That means it depends just on C_1, m and other things occurring in the hypotheses; it does not depend on u_ϵ or on their limit v.

2. A boundary form of Harnack's inequality:

The Harnack inequality is a very important tool in the study of positive solutions of linear elliptic equations. It has different forms; see [GT], [K], and [BP] for recent formulations.

In this section we give a simple extension up to the boundary of the Krylov–Safonov form of it for positive solutions satisfying boundary conditions of Neumann or mixed types on a smooth portion of the boundary.

Let Ω be a bounded domain in R^n. In Ω consider a nonnegative strong solution $u \in W^{2,n}(\Omega)$ of

$$Lu = a_{ij}(x)u_{ij} + b_i(x)u_i + cu = 0 \tag{2.1}$$

for which (1.16) and (1.17) hold.

THEOREM 2.1. *Let u be as above and assume that for an open connected and smooth subset Γ of $\partial\Omega$, u is of class C^1 on $\Omega \cup \Gamma$, and satisfies*

$$\mu(x) \cdot \nabla u + \gamma(x) u = 0 \quad \text{on } \Gamma, \tag{2.2}$$

where γ and the vector μ are smooth on Γ, and μ is nowhere tangential to Γ. Then for any compact subset $\tilde{\Omega}$ of $\Omega \cup \Gamma$ there is a constant M such that

$$\max_{\tilde{\Omega}} u \leq M \min_{\tilde{\Omega}} u. \tag{2.3}$$

Here M depends only on Ω, $\tilde{\Omega}$, C_1, m, and bounds on the C^2 norms on Γ of μ and γ.

PROOF: With the aid of the Krylov–Safonov inequality, see Cor. 9.25 in [GT], we see that it suffices to prove such an inequality locally near Γ. Near a point on Γ we may find a local C^2 transformation of x variables which straightens out the boundary so that it becomes $x_n = 0$ and, nearby, Ω lies in $x_n < 0$. Another C^2 change of variables renders $\mu = (0, \ldots, 0, 1)$. Finally, setting

$$v = e^{x_n \gamma(x_1, \ldots, x_{n-1})} u$$

we see that v satisfies a similar equation and the boundary condition $v_n = 0$ on $x_n = 0$. Clearly $\tilde{\Omega} \cap \Gamma$ can be covered by a finite number of neighbourhoods in each of which such transformations can be found with uniformly bounded C^2 norms.

Thus it suffices to prove the inequality (2.3) for v in a half ball

$$\Omega = \{x_n < 0\} \cap \{|x| < R\} = B_R^-,$$

with

$$\tilde{\Omega} = \{x \big| x_n \leq 0, |x| \leq R/2\}.$$

v satisfies an equation of the form (2.1) but with modified constants C and m of (1.16), (1.17); also $v_n = 0$ on $x_n = 0$. But now inequality (2.3) for v is an immediate consequence of Cor. 9.25 in [GT]. Namely extend v and the equation to $B_R = \{|x| < R\}$

by making v, a^{nn}, $a^{\alpha\beta}$, b^{α}, c even in x_n and extending $a^{n\alpha}$, b^n as odd; $\alpha < n$. Because $v_n = 0$ on $x_n = 0$, the extended function continues to be a strong solution of the extended equation, and it belongs to $W^{2,n}(B_R) \cap C^1(B_R)$. Now just apply Cor. 9.25 in [GT] to v.

Here is an application of Theorem 2.1 which we will use later.

THEOREM 2.2. *Let G be an open convex set lying in $B_1 = \{x < 1\}$ in R^n, with $0 \in \overline{G}$. Assume that the boundary of G consists of a smooth compact hypersurface γ (with smooth boundary lying on ∂B_1) and of a portion of ∂B_1. Let $u \in W^{2,n}(G)$ be a nonnegative strong solution of (2.1); assume that the coefficients satisfy the preceding conditions and (1.19): $c \leq 0$. Assume also that $u \in C(\overline{G}) \cap C^1(G \cup \Gamma)$. On Γ we suppose that u satisfies the Neumann boundary condition*

$$u_\nu = 0.$$

where ν is the exterior normal on Γ. Assume finally that at some point $\bar{x}$ on ∂B,

$$u(\bar{x}) = 0 \quad \text{and} \quad \overline{\lim_{t\to 1}} \frac{u(t\bar{x})}{1-t} \leq 1. \tag{2.4}$$

Then

$$u(0) \leq N \tag{2.5}$$

where N is a constant depending on Γ (i.e. bounds on derivatives of functions expressing it in terms of local coordinates) and on C_1, m.

The theorem is of interest even if $G = B_1$. Clearly it may be extended to more general (nonconvex) domains under suitable conditions. Note that the constants do not depend on the distance from 0 to Γ; it may be zero.

PROOF: Let $u(0) = k$. By the preceding theorem we have $u \geq kM^{-1}$ in $B_{\frac{1}{2}} \cap G$, for a constant M depending on Γ (as above) and on C, m. We now follow a standard argument (used in the classical proof of the Hopf lemma). In $D = G \setminus B_{\frac{1}{2}}$, consider the comparison function

$$h = e^{-\lambda|x|^2} - e^{-\lambda};$$

$h > 0$ in D and $h = 0$ on ∂B_1. A direct calculation shows that $Lh \geq 0$ if λ is fixed sufficiently large. Furthermore, because of the convexity of G we see that $h_\nu \leq 0$ on Γ. Consequently in D, the function

$$z = u - \frac{kM^{-1}h}{e^{-\lambda/4} - e^{-\lambda}}$$

satisfies

$$Lz \leq 0.$$

On $|x| = \frac{1}{2}$ and $|x| = 1$, we have $z \geq 0$, while on Γ $z_\nu \geq 0$, because G is starshaped about 0. Using the maximum principle and the Hopf lemma we see that in $\overline{D}$, z cannot have a negative minimum at any point on $\Gamma \cap B_1$. Hence $z \geq 0$ in $\overline{D}$ and its minimum occurs at $\bar{x}$.

But then

$$\overline{\lim_{t\to 1}} \frac{z(t\bar{x})}{1-t} \geq 0.$$

This implies that

$$1 \geq \frac{2\lambda M^{-1}k}{e^{-\lambda/4} - e^{-\lambda}}$$

and we obtain (2.5) with

$$N = \frac{M}{2\lambda}(e^{-\lambda/4} - e^{-\lambda}).$$

3. Uniform Lipschitz Continuity

Consider a solution u_ϵ, $0 < u_\epsilon < 1$ of (1.1) in Ω, $\partial\Omega$ is smooth. We do not require Ω to be bounded (we have in mind the infinite cylinder S) but we require that $\partial\Omega \cap B_R$ for any ball B_R, has the following:

PROPERTY P: there is a number $r = r(R)$ and a constant $\overline{C} = \overline{C}(R)$ (depending also on Ω) such that for any point $x_0 \in \partial\Omega \cap B_R$, a region in $\overline{\Omega}$ containing $\overline{\Omega} \cap \{|x - x_0| \leq r\}$ can be mapped one–to–one by a C^2 mapping ϕ onto $\{y \,\big|\, y_n \leq 0, |y| \leq r\}$ with $\phi(x_0) = 0$, and such that

$$||\phi||_{C_2}, ||\phi^{-1}||_{C^2} \leq \overline{C}.$$

In addition, at points x on $\partial\Omega$, ϕ maps the direction μ into the normal direction to $y_n = 0$.

We assume that β_ϵ in L^∞, has its support in $[0, \epsilon]$ and satisfies

$$|\beta_\epsilon(u)| \leq \frac{B}{\epsilon} \tag{3.1}$$

with B independent of ϵ. Concerning the operator L we assume (1.16)–(1.20). Our strong solution u_ϵ in $W^{2,n}_{\text{loc}}(\overline{\Omega})$ actually belongs to $W^{2,p}_{\text{loc}}(\Omega)$ for any $p > 1$—by standard elliptic estimates, and so is in $C^1(\Omega)$. If in addition we assume (1.6) then $u_\epsilon \in W^{2,p}_{\text{loc}}(\overline{\Omega})$ for $p > 1$, and so in $C^1(\overline{\Omega})$.

We prove:

THEOREM 3.1. *(a) In any compact subset J of Ω the function u_ϵ satisfies a Lipschitz condition with Lipschitz constant k depending only on C_1, B, m and distance of J to $\partial\Omega$.*
(b) Assume also that u_ϵ satisfies (1.6), then for any $R > 0$, in $\overline{\Omega} \cap B_R$, the function u_ϵ satisfies a Lipschitz condition with Lipschitz constant k depending only on Ω, R, C_1, B, m, and $\overline{C}(R+1)$ and bounds on derivatives of μ.

In section 7 we will apply Theorem 3.1 (b) to obtain uniform Lipschitz continuity for our solutions T_ϵ of (1.11), (1.12) with f_ϵ in place of f.

PROOF OF THEOREM 3: We wish to establish for any x_0 in J or in $B_R \cap \Omega$,

$$|\nabla u(x_0)| \leq k, \tag{3.2}$$

with k as described. We will use the property P for $B_{R+1} \cap \Omega$.

For convenience we will simply write u_ϵ as u and we will use the following notation:

$$\Omega_\delta = \{x \in \overline{\Omega} \big| 0 \leq u(x) \leq \delta\} \quad \text{for } \delta > 0.$$

By a slight abuse of notation we will continue to use this notation even after we have made the local changes of variable ϕ in the balls B_r.

We will concentrate on part (b) of the Theorem. The proof of (a) is similar, but simpler, it does not require the local transformations of variables near $\partial\Omega$. In fact we may just ignore $\partial\Omega$.

In working with the maps ϕ centred at any point in $B_{R+1} \cap \partial\Omega$, let $s \geq 1$ be such that distances by the maps ϕ and ϕ^{-1} are stretched by at most s. We may take $s = \overline{C}(R+1)$.

In deriving (3.2) we assume ϵ small, we will distinguish several cases. Here $r = r\,(R+1)$ of Property P.

Case 1. The distance of x_0 from Ω_ϵ is at least $\frac{r}{6s}$. Then in $B_{\frac{r}{6s}}(x_0) \cap \Omega$, u satisfies the linear equations $Lu = 0$. (3.2) then follows from (1.22).

In treating the next case we will rely on the following simple

LEMMA 3.1. *Let $G = \Omega \cap B_1$ where B_1 is a unit ball with centre, say the origin, in $\overline{\Omega}$, and assume (1.23), with $x_0 = 0$ and some t. Suppose $0 \leq u \in W^{2,p}(G)$, $p > n$, that, Lu is bounded, and (1.6):*

$$\mu \cdot \nabla u = 0 \quad \text{on } \partial\Omega \cap \overline{G}.$$

Then

$$|\nabla u(0)| \leq C(u(0) + ||Lu||_{L^\infty(G)}) \tag{3.3}$$

where C depends only on μ and Ω (i.e. the smoothness as usual), C_1, m, t.

PROOF: Denote the L^∞ norm of Lu by κ. Set $u = u_1 + u_2$ where u_1 is the solution of

$$\begin{aligned} Lu_1 &= Lu \quad \text{in } G \\ \mu(x) \cdot \nabla u_1 &= 0 \quad \text{on } \partial\Omega \cap \overline{G} \\ u_1 &= 0 \quad \text{on the remainder of } \partial G, \end{aligned}$$

with $u_1 \equiv 0$ if $G \subset \Omega$. Then, by (1.24) and (1.22),

$$|u_1(0)| + |\nabla u_1(0)| \leq C\kappa. \tag{3.4}$$

Now $Lu_2 = 0$ and $\mu \cdot \nabla u_2 = 0$ on $\partial\Omega \cap \overline{G}$, while $u_2 \geq 0$ on the remainder of ∂G. By the maximum principle (see [BN1]) it follows that $u_2 \geq 0$ in G. Furthermore

$$u_2(0) \leq u(0) + C\kappa =: \kappa_1.$$

Applying Theorem 1.1 we find

$$u_2 \leq M\kappa_1 \quad \text{in } B_{\frac{1}{2}} \cap \Omega.$$

But then by (1.22) it follows that

$$|\nabla u_2(0)| \leq C\kappa_1.$$

Combining this with (3.4) we obtain (3.3). Q.E.D.

Observe that if G lies entirely in Ω then $u_1 = 0$ and in place of Theorem 1.1 we just use the standard form, Cor. 9.25 in [GT], and standard elliptic estimates.

Turn to the next case for (3.2).

Case 2. $x_0 \in \Omega_\epsilon$. Consider a ball B_ϵ of radius ϵ around x_0, which we take to be the origin. The case that $B_\epsilon \subset \Omega$ is the simpler one so we will suppose this does not hold and we set $G_\epsilon = B_\epsilon \cap \Omega$. We now scale things to achieve that $\epsilon = 1$; namely we set

$$v(x) = \frac{1}{\epsilon} u(\epsilon x).$$

So v is defined in the scaled $\tilde{G} = \{x \,\big|\, \epsilon x \in G_\epsilon\} = B_1 \cap$ (scaled Ω). Because of (3.1) we see that in $\tilde{G}$, v satisfies

$$|a_{ij}(\epsilon x) v_{ij} + \epsilon b_i(\epsilon x) v_i + \epsilon^2 c(\epsilon x) v| \le B.$$

Also $v(0) \le 1$ and $\mu(\epsilon x) \cdot \nabla v = 0$ on the scaled part of $\partial\Omega$.

We may therefore apply Lemma 3.1. Note in fact that for ϵ small, (1.23) holds with $t = 1/4$, that the scaled $\partial\Omega$ is even smoother than the original $\partial\Omega$ and that $\mu(\epsilon x)$ is smoother than $\mu(x)$. Thus we find $|\nabla v(0)| \le C$. So $|\nabla u(0)| = |\nabla v(0)| \le C$, and (3.2) is established in this case. Note that we did not require $c = 0$.

Finally we have to consider the case that $0 < d(x_0, \Omega_\epsilon) < r/6s$. This will be handled with aid of the following

LEMMA 3.2. *There exists a constant A depending only on (as usual) the smoothness of $\partial\Omega$, μ, R, C_1, B, m, $\overline{C}(R+1)$ such that if*

$$x \in B_R \cap \Omega \quad \text{and} \quad d(x, \Omega_\epsilon) < r/3s,$$

then

$$u(x) - \epsilon \le A d(x, \Omega_\epsilon). \tag{3.5}$$

Postponing the proof of the lemma let us first complete that of Theorem 3.1. We have to verify (3.2) in the last case that $0 < d = d(x_0, \Omega_\epsilon) < r/6s$. In the region

$\{|x_0 - x| < d\} \cap \Omega$, which lies within a distance $2d < r/3s$ of Ω_ϵ, the function $(u - \epsilon)$ satisfies $L(u - \epsilon) = 0$. Here we use (1.20). By Lemma 3.2 we know that $u - \epsilon < 2Ad$ there, so applying the estimate (1.22) we see that

$$|\nabla u(x_0)| \leq \frac{K}{d} \cdot 2Ad = 2KA.$$

Finally, the

PROOF OF LEMMA 3.2: We break it up into two cases; the proofs for both cases are essentially the same but the first is simpler and is the only one needed for Theorem 3.1(a).

Case (i) $d(x, \partial\Omega) > r/3s$. We denote again $d = d(x, \Omega_\epsilon)$. We may suppose the point x under consideration is the origin. In the ball $|x| < d$ we have $u - \epsilon > 0$, and $u - \epsilon = 0$ at some point on the boundary. By case 2 of the theorem, which we have already established, we know that at that point,

$$|\nabla u| \leq k.$$

In the unit ball B_1 consider the scaled function

$$v(x) = \frac{u(dx) - \epsilon}{d};$$

v is a positive solution of the scaled equation

$$a_{ij}(dx)v_{ij} + db_i(dx)v_i = 0.$$

Only here do we use the fact that $c = 0$. Furthermore there is a point on ∂B_1 where $v = 0$ and $|\nabla v| \leq k$. We may therefore apply Theorem 2.2 in the whole ball B_1 and infer that

$$v(0) \leq kN.$$

Consequently we have

$$u(0) - \epsilon \leq kNd$$

i.e. (3.5) with $A = kN$.

Case (ii) $0 < d(x, \Omega_\epsilon) < r/3s$ and $d(x, \partial\Omega) \leq r/3s$. In this case let x_0 be the closest point on $\partial\Omega$ to x. According to hypothesis P there is a C^2 mapping ϕ of a region containing $B_r(x_0) \cap \Omega$ onto the half ball $\{y_n < 0, |y| < r\} = B_r^-$, with x_0 mapping to the origin. By the definition of s the point x maps to a point y_0 in the half ball with $|y_0| < r/3$. In addition, the distance d from y_0 to the set of points in B_r^- where the transformed $u \leq \epsilon$, is less than $r/3$. Thus the region $G = \{y \,\big|\, |y - y_0| \leq d, y_n < 0\}$ lies entirely in B_r^-; it is of course convex.

In G, the function u satisfies an equation $\tilde{L}u = 0$ with $\tilde{L}$ as in the theorem (though the constants C and m have changed). In addition by property P, u satisfies $u_{y_n} = 0$ on $y_n = 0$; $u - \epsilon > 0$ in G and vanishes at some point on its curved boundary. As before, by case 2 of the theorem, $|\nabla u| \leq ks$ at that point.

We now take the point y_0 as origin and perform the same scaling as in the previous case: the function

$$v(y) = \frac{u(dy) - \epsilon}{d}$$

is defined in the scaled region $\tilde{G}$. It is positive and vanishes at a point on the curved boundary, and there $|\nabla v| \leq ks$. As before we may apply Theorem 2.2 in its general form and conclude that $v(0) \leq ks^2N$. Back in the original domain Ω, we therefore have

$$u(x) - \epsilon \leq ks^2Nd(x, \Omega_\epsilon)$$

for our original point x.

The proofs of Lemma 3.2 and of Theorem 3.1 are complete.

4. Nondegeneracy of "minimal solutions"

We continue our study of (1.1) by proving nondegeneracy of the solution near the set where it is $\leq \epsilon$ i.e., the solution grows at some fixed rate as we go away from that set.

Let $u \in W^{2,p}_{\text{loc}}(\overline{\Omega})$ be a nonnegative strong solution of (1.1), (1.6), i.e.,

$$Lu = \beta_\epsilon(u) \quad \text{a.e. in } \Omega \tag{4.1}$$

$$\mu(x) \cdot \nabla u = 0 \quad \text{on } \partial\Omega. \tag{4.2}$$

with μ pointing out of Ω. In addition we assume (1.3). (We repeat that for purely interior estimates in compact subsets of Ω, we do not require (4.2). Concerning L we require the conditions (1.16) and (1.17).

The function u is assumed, furthermore, to be a *minimal solution.* This means: whenever we have a strong supersolution $h \in W^{2,p}(G)$ in a bounded subdomain G of Ω, i.e.,

$$Lh \leq \beta_\epsilon(h) \quad \text{in } G, \tag{4.3}$$

which satisfies, in addition,

$$h \geq u \quad \text{on } \partial G \cap \Omega \tag{4.4}$$

$$\mu(x) \cdot \nabla h \geq 0 \quad \text{on } \partial\Omega \cap \overline{G}, \tag{4.5}$$

then

$$h \geq u \quad \text{in } G. \tag{4.6}$$

This condition may seem awkward to verify but we will show in section 7, that it holds in our flame propagation problem of section 1.

We assume that Ω satisfies Property P of section 3. Recalling (1.10), set

$$d_\epsilon(x) = d(x, \Omega_\epsilon).$$

THEOREM 4.1. *Assume the conditions above. With* $B_R = \{|x| < R\}$, *consider the compact region* $J_R = \overline{B}_R \cap \overline{\Omega}$. *There are positive numbers* ϵ_0, ρ *and* c_0, *depending only on* C_1, m, B, a, b, κ, Ω *and the* C^2 *norm of* μ *on* $B_{R+1} \cap \partial\Omega$, *such that for* $x_0 \in J_R$, $\epsilon \le \epsilon_0$,

$$u(x_0) \ge c_0 d_\epsilon(x_0) \quad \text{if } d_\epsilon(x_0) \le \rho. \tag{4.7}$$

PROOF: The proof involves careful construction of suitable supersolutions. As in the proof of Theorem 3.1 we will work with $r = r(R+1)$ of Property P, and with the number s described shortly after (3.2). We distinguish several cases. We will always require $d_\epsilon < 1$ and $d_\epsilon < r/3s$.

Case 1. The distance of x_0 from $\partial\Omega$ is at least $r/3s$. We choose x_0 as origin and set

$$d_\epsilon(0) = 2\delta.$$

In $B_{2\delta}$, centred at the new origin, u satisfies $Lu = 0$. By the Harnack inequality,

$$u \le Cu(0) \quad \text{in } B_\delta$$

for some $C \ge 1$, under control. If δ is small we will construct in B_δ a C^1 supersolution $h \in W^{2,\infty}$ satisfying (4.3), (4.4). $h = h(r)$ will depend only on $r = |x|$ and will satisfy

$$h(0) = a\epsilon < u(0)$$

(the number a enters in (1.3), and also $h(\delta) \ge \delta D^{-1}$ for some $D > 0$ under control. By our hypothesis that u is a "minimal solution" it follows that we cannot have $h \ge u$ everywhere on ∂B_δ. Therefore

$$\frac{\delta}{D} \le h(\delta) \le Cu(0) \tag{4.8}$$

and this has the desired form (4.7).

We take for $h(r)$ the following function

$$(4.9)\qquad \begin{cases} h &= \epsilon a & \text{for } 0 \le r \le r_0 \\ h &= \epsilon a + \frac{k}{2}(r - r_0)^2 & \text{for } r_0 \le r \le \lambda \\ h &= H - \frac{A}{2}(r - \delta)^2 & \text{for } \lambda \le r \le \delta. \end{cases}$$

The numbers $r_0(\epsilon)$, λ, k, H and A are all to be chosen. First we require that h be in C^1 at λ and that

$$h(\lambda) = \epsilon b,$$

see (1.3). This means

$$(4.10)\qquad k = \frac{2\epsilon(b - a)}{(\lambda - r_0)^2},$$

$$(4.11)\qquad H = \frac{A}{2}(\delta - \lambda)^2 + \epsilon b.$$

and $k(\lambda - r_0) = A(\delta - \lambda)$, so we require

$$(4.12)\qquad A(\delta - \lambda) = \frac{2\epsilon(b - a)}{\lambda - r_0}.$$

In order to verify (4.3) it suffices to establish

$$(4.13)\qquad Lh \le \frac{\kappa}{\epsilon} \quad \text{for } r_0 < r < \lambda$$

since there $a\epsilon \le h \le b\epsilon$ and so (1.3) applies, and

$$(4.14)\qquad Lh \le 0 \quad \text{for } \lambda < r < \delta.$$

Consider first (4.13) on (r_0, λ). We have

$$\begin{aligned} \frac{1}{2}\partial_i(r - r_0)^2 &= x_i\left(1 - \frac{r_0}{r}\right) \\ \frac{1}{2}\partial_{ij}(r - r_0)^2 &= \delta_{ij}\left(1 - \frac{r_0}{r}\right) + \frac{r_0}{r^3}x_i x_j. \end{aligned}$$

Thus

$$
(4.15)\qquad \begin{aligned} Lh &= k\left[\left(1-\frac{r_0}{r}\right)\Sigma a_{ii} + \frac{r_0}{r^3}a_{ij}x_ix_j + \left(1-\frac{r_0}{r}\right)b_ix_i\right] + ch \\ &\le C(k+\epsilon b) \end{aligned}
$$

with C under control. Hence (4.13) will hold if, by (4.10),

$$
C\left(\frac{2\epsilon^2}{(\lambda-r_0)^2}(b-a)+\epsilon^2 b\right) \le \kappa.
$$

It suffices to require

$$
(4.16)\qquad \epsilon^2 \le \frac{\kappa}{2b}
$$

and

$$
(4.17)\qquad \frac{\epsilon^2}{(\lambda-r_0)^2} = \frac{\kappa}{2(b-a)C}.
$$

We then use (4.17) to fix $\lambda - r_0$.

Finally, to verify (4.14) we see as in (4.15), that in (λ, δ),

$$
Lh = -A\left[\left(1-\frac{\delta}{r}\right)\Sigma a_{ii} + \frac{\delta}{r^3}a_{ij}x_ix_j + \left(1-\frac{\delta}{r}\right)b_ix_i\right] + ch.
$$

Using (1.16) and (1.17) we find

$$
(4.18)\qquad \begin{aligned} Lh &\le nC_1A\left(\frac{\delta}{r}-1\right) - A\frac{m\delta}{r} + nAC_1(\delta-\lambda) + C_1H \\ &\le A\left(-nC_1 + (nC_1-m)\frac{\delta}{\lambda} + nC_1(\delta-\lambda)\right) + C_1H. \end{aligned}
$$

Now fix

$$
(4.19)\qquad \lambda = \frac{nC_1 - m}{nC_1 - m/2}\delta,
$$

then $r_0(\epsilon)$ is determined from (4.17). We have

$$
Lh \le -A\left(\frac{m}{2} - nC_1\delta\right) + C_1H
$$

so if we assume

$$\delta \leq \frac{m}{4nC_1} \tag{4.20}$$

then we find, using also (4.11),

$$\begin{aligned} Lh &\leq -A\frac{m}{4} + C_1\frac{A}{2}(\delta-\lambda)^2 + C_1\epsilon b \\ &= -\frac{m}{4}\frac{2\epsilon(b-a)}{(\lambda-r_0)(\delta-\lambda)} + C_1\frac{\epsilon(b-a)}{(\lambda-r_0)}(\delta-\lambda) + C_1\epsilon b \\ &\leq \sqrt{\frac{\kappa(b-a)}{2C}}\left(-\frac{m}{2\delta} + C_1\delta\right) + C_1\epsilon \qquad \text{by (4.12),} \\ &\leq 0 \end{aligned}$$

provided

$$\delta \leq \frac{1}{2}\left(\frac{m}{C_1}\right)^{\frac{1}{2}} \tag{4.21}$$

and

$$C_1\epsilon \leq \frac{1}{2}\sqrt{\frac{\kappa(b-a)}{2C}}(mC_1)^{\frac{1}{2}}. \tag{4.22}$$

Thus h is a supersolution and has all the properties stated earlier—we have only to check that

$$h(\delta) \geq \delta/D$$

for some D under control. In fact from (4.11) we see that

$$\begin{aligned} h(\delta) = H &\geq \frac{A}{2}(\delta-\lambda)^2 \\ &= \frac{\epsilon(b-a)}{\lambda-r_0}(\delta-\lambda) \\ &= \sqrt{\frac{\kappa(b-a)}{2C}}\left(1-\frac{\lambda}{\delta}\right)\delta \qquad \text{by (4.12)} \\ &= \delta/D. \end{aligned}$$

with

$$D = \sqrt{\frac{2C}{\kappa(b-a)}} \frac{m}{2nC_1 - m}$$

by (4.19), and so under control.

We have established (4.7) provided ϵ satisfies (4.22), and $\delta = \frac{1}{2}d_\epsilon$ satisfies (4.20) and (4.21).

Case 2. The distance of x_0 from $\partial\Omega$ is not greater that $r/(3s)$. Set $d_\epsilon = d$. First we proceed as in the proof of Lemma 3.2, case (ii). Carrying out that mapping we find that the transposed function u is defined in the region (recall x_0 has been mapped to a point y_0 which we take to be the origin)

$$\{y \,\big|\, |y| < sd, \quad y_n \le \sigma\}$$

with some constant $\sigma \ge 0$. There $u > \epsilon$ so it satisfies an equation of the form (4.1), with the transformed operator L. On the flat boundary $y_n = \sigma$, u satisfies the Neumann condition

$$u_n = 0 \quad \text{on } y_n = \sigma.$$

This transposed function is still a minimal solution of the (transformed) equation $Lu = \beta_\epsilon$.

We now argue just as in case (i). With $\delta = sd/2$, we see by our Harnack inequality, Theorem 2.1, that $u \le Cu(0)$ in

$$G_\delta = \{y \,\big|\, |y| < \delta, \quad y_n \le \sigma\}.$$

We use the same supersolution $h(r)$ given by (4.9). Since h is nondecreasing in r we see that the transformed form of (4.5) holds:

$$h_{y_n} \ge 0 \quad \text{on } y_n = \sigma.$$

The rest of the argument of Case 1 applies. Of course the constants C_1 and m have changed but we may still conclude that

$$u(0) \ge c\delta$$

—assuming (4.22), (4.20) and (4.21)—with some $c > 0$ under control. Back in the original x–variables this means

$$u(x_0) \geq c_0 d_\epsilon.$$

Q.E.D.

5. Preliminary results as $\epsilon \to 0$.

In this section we derive some facts concerning the limit function obtained as $\epsilon \to 0$. In particular, under suitable conditions, we will prove (1.7).

Consider our solution $u_\epsilon \in W^{2,p}_{\text{loc}}(\overline{\Omega})$ of (1.1), $p > n$, $0 \leq u \leq 1$, satisfying (1.6), under the conditions in section 3 i.e. we assume (1.16)–(1.20), so $c = 0$, and (1.2):

$$\beta_\epsilon \leq \frac{B}{\epsilon}.$$

We also assume condition P. We note, as before, that in studying compact subsets of Ω there is no need to assume (1.6) or property P.

By Theorem 3.1, in any compact subset of $\overline{\Omega}$ the u_ϵ are uniformly Lipschitz. Thus if let $\epsilon \to 0$ through a suitable sequence we have $u_\epsilon \to v$ in $\overline{\Omega}$ uniformly on compact subsets. Clearly $0 \leq v \leq 1$, and v is locally Lipschitz in $\overline{\Omega}$, so $\{v > 0\}$ is relatively open in $\overline{\Omega}$.

If $v(x_0) > 0$ then for some $r > 0$, $v \geq \delta > 0$ in $G_r = \overline{B}_r(x_0) \cap \overline{\Omega}$. Hence for ϵ sufficiently small, $u_\epsilon \geq \delta/2$ and satisfies $Lu_\epsilon = 0$ in G_r. Using standard elliptic estimates we infer that for any fixed $p > n$,

$$||u_\epsilon||_{W^{2,p}(G_r/2)} \leq C \quad \text{independent of } \epsilon.$$

It follows that $v \in W^{2,p}$ in any compact set J in $\overline{\Omega}$ in which $v > 0$. In particular v satisfies (1.6) at any point of $\partial\Omega$ where $v > 0$ and $Lv = 0$ in $\{v > 0\}$. Furthermore, in J the functions $u_\epsilon \to v$ in C^1.

In the rest of this section we will assume that the leading coefficients

$$a_{ij} \in C^1(\overline{\Omega}). \tag{5.1}$$

We will show that $u_\epsilon \to v$ in H^1 strongly on any compact subset of $\overline{\Omega}$. We carry this out for a ball; it will be seen to be merely a consequence of the fact that $Lu \geq 0$.

LEMMA 5.1. *Assume the conditions above. Let $G = \Omega \cap B$ where B is a ball. Then $u_\epsilon \to v$ in $H^1(G)$.*

PROOF: The u_ϵ converge weakly in $H^1(G)$ to v so it suffices to prove that (we write u for u_ϵ):

$$\ell := \limsup_{\epsilon\to 0} \iint_G a_{ij} u_i u_j = \iint_G a_{ij} v_i v_j. \tag{5.2}$$

Clearly $\ell \geq$ the right hand side of (5.2). To prove (5.2), multiply the equation (1.1) by u and integrate over G. Using Green's theorem we find (double integral sign means volume integration; single means boundary integration), since $u\beta_\epsilon \geq 0$

$$\begin{aligned} 0 &\leq \iint_G u(a_{ij}u_{ij} + b_i u_i) \\ &= \int_{\partial G} u a_{ij} u_i \nu_j dS + \iint_G -a_{ij}u_i u_j - u a_{ij,j} u_i + u b_i u_i. \end{aligned}$$

So

$$\iint_G a_{ij} u_i u_j \leq \int_{\partial G} u a_{ij} u_i \nu_j dS + \iint_G -u a_{ij,j} u_i + u b_i u_i.$$

Here $a_{ij,j}$ is $\partial_j a_{ij}$. Let $\epsilon \to 0$. We claim that the right hand side converges to

$$I := \int_{\partial G\cap\{v>0\}} v a_{ij} v_i \nu_j + \iint_{G\cap\{v>0\}} - v a_{ij,j} v_i + v b_i v_i. \tag{5.3}$$

This is easy to see, for $|\nabla u| \leq C$, $u \to v$ uniformly and $u_i \to v_i$ weakly in $L^2(G)$, and $u_i \to v_i$ uniformly on any set $\{x \in G \,\big|\, v \geq \tau > 0\}$ $\forall \tau > 0$. Hence $\ell \leq I$.

For any fixed $\delta > 0$ let $v^\delta = \max\{v - \delta, 0\}$. Using Green's theorem and the fact that $Lv = 0$ where $v > 0$, we see that (recall $|\nabla u|$ is bounded)

$$0 = \iint_G v^\delta Lv = \int_{\partial G} v^\delta a_{ij} v_i \nu_j - \iint_{G \cap \{v > \delta\}} a_{ij} v_i v_j + \iint_G v^\delta (b_i - a_{ij,j}) v_i.$$

Letting $\delta \to 0$ we find

$$\iint_{G \cap \{v>0\}} a_{ij} v_i v_j = I. \tag{5.3$'$}$$

Hence (5.2) is proved.

Before proceeding, it's worth describing some simple examples even in one dimension. Consider the equation

$$u_{xx} = \frac{1}{\epsilon} \chi_\epsilon(u)$$

where χ_ϵ is the characteristic function of the interval $(0, \epsilon)$. The function

$$u_1 = \begin{cases} \dfrac{\epsilon}{2} + \dfrac{x^2}{2\epsilon} & \text{for } |x| \le \epsilon \\ |x| & \text{for } |x| \ge \epsilon \end{cases}$$

is a C^1 solution. So are

$$u_2 = \begin{cases} \dfrac{x^2}{2\epsilon} & \text{for } |x| \le \sqrt{2}\epsilon \\ \sqrt{2}|x| - \epsilon & \text{for } |x| \ge \sqrt{2}\epsilon, \end{cases}$$

$$u_3 = \begin{cases} \epsilon - \sqrt{2}x & \text{for } x \le 0 \\ \epsilon - \sqrt{2}x + \dfrac{x^2}{2\epsilon} & \text{for } 0 \le x \le \sqrt{2}\epsilon, \\ 0 & \text{for } x \ge \sqrt{2}\epsilon, \end{cases}$$

Note that on the interval $[-1, 1]$, $u_2 > u_1$ at the end points, for ϵ small, but $u_2(0) < u_1(0)$. Thus u_1 is not a "minimal solution" in the sense of section 4. It is not difficult to see that u_3 is "minimal". As $\epsilon \to 0$ these functions tend respectively to $\tilde{u}_1 = |x|$, $\tilde{u}_2 = \sqrt{2}|x|$ and $\tilde{u}_3 = \sqrt{2}x^-$. The left derivative of $\tilde{u}_3$ at the origin is $-\sqrt{2}$ so it satisfies (1.7) there.

We turn now to the proof of condition (1.7) for our $v = \lim u_\epsilon$. Here we work only in Ω not at $\partial\Omega$ so we don't require (1.6) or regularity of $\partial\Omega$. The solution $\tilde{u}_1$ above shows that the set where the limit function v vanishes need not have interior.

Assume the conditions above, including (5.1). Let $B_r(0)$ be a closed ball in Ω and let $\sigma \subset \Omega$ be a smooth (connected) hypersurface (containing 0) on which v vanishes and such that in B_r, v is positive on one side of σ and zero on the other. Set $G = B_r \cap \{v > 0\}$. By elliptic regularity theory it follows that $v \in W^{2,p}(\overline{G})$ for any $p > 1$. We will sometimes write B for B_r.

THEOREM 5.1. *Assume the conditions above and, in addition (1.4) and (1.5), more precisely:*

$$(5.4) \qquad \begin{cases} \beta_\epsilon(u) = \dfrac{1}{\epsilon}\beta\left(\dfrac{u}{\epsilon}\right) & \text{where } \beta \text{ is a continuous function which is positive} \\ \text{on } (0,\, 1) \text{ and } 0 \text{ otherwise.} & \displaystyle\int_0^1 \beta = M. \end{cases}$$

Then (1.7) holds:

$$(5.5) \qquad a_{ij}(x)\nu_i\nu_j|\nabla v|^2 = 2M \quad \text{on } \sigma.$$

Here ν is the exterior normal to G and $|\nabla v|$ is the gradient in $\overline{G}$ of v.

PROOF OF THEOREM 5.1: Set

$$\mathcal{B}_\epsilon(u) = \int_0^u \beta_\epsilon(s)ds = \int_0^{u/\epsilon} \beta(s)ds.$$

By (5.4),

$$\mathcal{B}_\epsilon(u) = M \quad \text{for } u \geq \epsilon. \tag{5.6}$$

With $\zeta \in C_0^\infty(B_r)$, multiply (1.1) by ζu_k and apply Green's theorem:

$$\begin{aligned} 0 &= \iint \zeta u_k(a_{ij}u_{ij} + b_i u_i - \beta_\epsilon) \\ &= \iint -\zeta a_{ij}u_i u_{jk} - \zeta_j a_{ij}u_j u_k - \zeta a_{ij,j}u_i u_k + \zeta u_k b_i u_i + \mathcal{B}_\epsilon(u)\zeta_k \\ &= \iint \frac{1}{2}\zeta_k a_{ij}u_i u_j + \frac{1}{2}\zeta a_{ij,k}u_i u_j - \zeta_j a_{ij}u_j u_k - \zeta a_{ij,j}u_i u_k \\ &\qquad + \zeta u_k b_i u_i + \mathcal{B}_\epsilon \zeta_k. \end{aligned}$$

As $\epsilon \to 0$ through our sequence, we may suppose $\mathcal{B}_\epsilon$ converges weakly in $L^2(B)$ to ρ, $0 \leq \rho \leq M$. From (5.6) we infer that $\rho = M$ in G.

Now let $\epsilon \to 0$ and use Lemma 5.1: $u_\epsilon \to v$ strongly in $H^1(B)$. Therefore

$$\begin{aligned} -\iint_B \rho\zeta_k &= \iint\limits_{B\cap\{v>0\}} \Big[\frac{1}{2}\zeta_k a_{ij}v_i v_j + \frac{1}{2}\zeta a_{ij,k}v_i v_j - \zeta_j a_{ij}v_j v_k - \zeta a_{ij,j}v_i v_k \\ &\qquad + \zeta v_k b_i v_i\Big]. \end{aligned} \tag{5.7}$$

Using Green's theorem once more we find

$$\begin{aligned} 0 &= \iint\limits_{B\cap\{v>0\}} \zeta v_k(a_{ij}v_{ij} + b_i v_i) \\ &= \text{R.H.S. of (5.7)} + \int\limits_{\sigma\cap B} \left(\zeta v_k a_{ij}v_i\nu_j - \frac{1}{2}\zeta a_{ij}v_i v_j \nu_k\right) dS. \end{aligned} \tag{5.8}$$

Now on σ, where $v = 0$ we have, since $v > 0$ in G,

$$v_i = -\nu_i|\nabla v|.$$

Thus (5.7) and (5.8) yield

$$\frac{1}{2}\int\limits_{\sigma\cap B} \zeta a_{ij}\nu_i\nu_j|\nabla v|^2\nu_k dX = \iint_B \rho\zeta_k. \tag{5.9}$$

In (5.9), if we take ζ to be supported in the interior of $B_r \setminus G$, then

$$\iint_{B_R \setminus G} \rho\zeta_k = 0 \quad \text{for every such } \zeta.$$

It follows that $\rho \equiv M_1$, a nonnegative constant in $B \setminus G$. Consequently for any $\zeta \in C_0^\infty(B)$ we find, again via Green's Theorem, that for $M - M_1 = \overline{M}$,

$$\iint \rho\zeta_k = \int_{\sigma\cap B} \overline{M}\zeta\nu_k dS$$

where ν is the exterior unit normal to G. It follows with the aid of (5.9) that $\forall\, \zeta \in C_0^\infty(B)$,

$$\int_{\sigma\cap B} \zeta\nu_k(a_{ij}\nu_i\nu_j|\nabla v|^2 - 2\overline{M})dS = 0$$

which yields (5.5) with $\overline{M}$ in place of M. To complete the proof we will show that $M_1 = 0$. Note that $\overline{M}$ cannot be zero, for if it were, our v would be in $W^{2,p}(B)$ and would vanish on $B \setminus G$. By unique continuation, necessarily $v \equiv 0$ on B—contradiction.

So suppose $0 < M_1 < M$. By our condition (5.4), for some positive $a < b < 1$ we have, $\forall\, \epsilon > 0$,

$$\mathcal{B}_\epsilon(\epsilon a) < M_1 < \mathcal{B}_\epsilon(\epsilon b).$$

Let D be a closed disc in $B_r \setminus G$ with centre at x_0 and radius τ. Since the $a_{ij} \in C^1$, the operator

$$\tilde{L} = a_{ij}\partial_{ij}$$

has a Green's function in D. It follows that if $w \in W^{2,p}(D)$, $p > n$, then for some $t > 0$,

$$w(x_0) \le \max_{\partial D} w + \frac{1}{t}\max(\tilde{L}w)^- - t\iint_{|x-x_0|<\tau/2} (\tilde{L}w)^+.$$

Apply this to u_{ϵ_i} which we call u^i. Since $|\nabla u^i| \le k$ we see that

$$u^i(x_0) \le 1 + \frac{C}{t} - t\iint_{|x-x_0|<\tau/2} \beta_{\epsilon_i}(u^i(x)) \tag{5.10}$$

with C independent of i.

At every point x where $\mathcal{B}_{\epsilon_i}(u^i(x))$ converges to M_1, for sufficiently large i,

$$a\epsilon < u^i(x) < b\epsilon.$$

Consequently, since $\beta > 0$ on (0,1),

$$\epsilon_i \beta_{\epsilon_i}(u^i(x)) \geq \kappa > 0$$

for some fixed $\kappa > 0$, so

$$B \geq \epsilon_i \beta_{\epsilon_i}(u^i(x)) \geq \kappa.$$

Set $g_i(x) = \epsilon_i \beta_{\epsilon_i}(u^i(x))$. Then g_i converges weakly in L^2, and almost everywhere, in D to a function g satisfying

$$g \geq \kappa.$$

Multiplying (5.10) by ϵ_i we find

$$\epsilon_i u^i(x_0) \leq \epsilon_i + \frac{C}{t}\epsilon_i - t \iint_{|x-x_0|<\tau/2} g_i.$$

Letting $i \to \infty$ we find

$$0 \leq -t\kappa \text{ Vol } \{|x - x_0| < \tau/2\}.$$

Contradiction. Thus $M_1 = 0$.

Theorem 5.1 is proved.

REMARK: Our derivation of the condition (1.7) on the free boundary differs from the one usually used in deriving such conditions in other free boundary problems. Usually one multiplies (1.1) by a test function ζ and integrates, applies Green's theorem etc. We multiplied instead by ζu_k.

6. The free boundary:

In sections 3 and 4 we established some estimates for "minimal solutions" u, $0 \leq u \leq 1$, of

$$Lu = a_{ij}(x)u_{ij} + b_i u_i = \beta_\epsilon(u) \quad \text{in } \Omega \tag{6.1}$$

and (1.6)—assuming property P, and that the coefficients of L satisfy (1.16)–(1.18) and also that the β_ϵ satisfy (1.2), (1.3). We proved that in $B_R(0) \cap \overline{\Omega}$, $R > 0$,

$$|\nabla u_\epsilon| \leq k(R), \tag{6.2}$$

and that for $\epsilon \leq \epsilon_0(R)$, $d_\epsilon = d(x, \{u \leq \epsilon\}) \leq \rho(R)$, and some $c_0(R) > 0$,

$$u(x) \geq c_0 d_\epsilon(x).$$

We had $\rho(R) \leq r(R+1)/(3\overline{C}(R+1))$, see Property P.

In this section we investigate the limit $v = \lim u_\epsilon$ as $\epsilon \to 0$, obtained in section 5, and derive some information about the smoothness of the free boundary. Except for Lemma 6.1 we no longer refer to u_ϵ in this section; we just consider a nonnegative function $v \leq 1$ defined in a region $B_R \cap \overline{\Omega}$ where Ω has smooth boundary and satisfies Property P. We assume

$$|\nabla v| \leq k \tag{6.3}$$

$$v(x) \geq c_0 d(x) \quad \text{if } d(x) = d(x, \{v = 0\}) \leq \rho. \tag{6.4}$$

We will assume $\rho \leq r(R+1) \leq 1$, see Property P. v is a solution of (1.1) and (1.6) wherever $v > 0$, and $v \in W^{2,p}(\overline{\Omega} \cap \{v \geq \delta\})$ for any $\delta > 0$. As before, we assume (1.16), (1.17), that μ (in (1.6)) is smooth and points outside Ω. We will also assume condition (5.2):

$$a_{ij} \in C^1(\overline{\Omega}).$$

The main result of the section concerns the closed set $\gamma = \partial\{v > 0\}$ in Ω:

THEOREM 6.1. *Under the conditions above, let B_r, $0 < r \leq 1$ be an open ball in B_{R-1} with centre in $\overline{\Omega}$. Then the $(n-1)$–dimensional Hausdorff measure of $(\gamma \cap B_r)$ satisfies*

$$H^{n-1}(\gamma \cap B_r) \leq Cr^{n-1}. \tag{6.5}$$

Here C is under control; it is independent of v and of r.

The results in this section are based on ideas in [AC] and in [C]; compare Theorem 6.1 with Representation theorem 4.5 in [AC].

We begin by showing the ϵ–approximation analogue (1.11) of (6.5). Namely—here we use (1.3)—

LEMMA 6.1. *For our function u_ϵ,*

$$Vol\,[B_r \cap \{a\epsilon < u_\epsilon < b\epsilon\}] \leq C\epsilon r^{n-1}. \tag{6.6}$$

PROOF: Multiplying (6.1) by $u^\epsilon = \min(u_\epsilon, \epsilon)$, and integrating over $G = B_r \cap \Omega$ we find by Green's theorem, here $G_\epsilon = G \cap \{u < \epsilon\}$, (we write $u_\epsilon = u$),

$$\begin{aligned}\iint_{G_\epsilon} a_{ij}u_iu_j + \iint_G u\beta_\epsilon &= \int_{\partial G} u^\epsilon a_{ij}u_i\nu_j + \iint_G -u^\epsilon a_{ij,j}u_i + u^\epsilon b_i u_i \\ &\leq C\epsilon r^{n-1}.\end{aligned}$$

by (6.2) and estimates for the coefficients and for $|\nabla a_{ij}|$ in B_r. According to (1.3), $\beta_\epsilon \geq \kappa/\epsilon$ where $a\epsilon \leq u \leq b\epsilon$. Hence the left hand side of (6.6) is

$$\leq \frac{C}{\kappa a}\epsilon r^{n-1}. \qquad \text{Q.E.D.}$$

In the proof of Theorem 6.1 we will need an analogue of this for v. In the following lemma the constant C is under control.

LEMMA 6.2. *Under the conditions of Theorem 6.1* $\exists C$ *independent of* $\tau > 0$ *and* v, *such that*

$$\iint_{B\cap\{0<v<\tau\}} |\nabla v|^2 \leq C\tau r^{n-1}. \tag{6.7}$$

PROOF: We combine the proofs of Lemmas 6.1 and 5.1. Set $G = B_r \cap \Omega$. With $0 < \delta < \tau$ set $v^\delta = \max\{v - \delta, 0\}$, $v^\delta_\tau = \min\{v^\delta, \tau\}$, $v_\tau = \min\{v, \tau\}$. Set $G = B_r \cap \Omega$, $G_\tau = G \cap \{0 < v < \tau\}$. If we proceed as in the proof of Lemma 5.1, but with v^δ_τ in place of v^δ, we find, see (5.3), (5.3)$'$,

$$\iint_{G_\tau} a_{ij} v_i v_j = \int_{\partial G\cap\{v>0\}} v_\tau a_{ij} v_i \nu_j + \iint_{G\cap\{v>0\}} v_\tau (b_i - a_{ij,j}) v_i.$$

Since $|v_i| \leq k$ we conclude that (6.7) holds.

We will make use of a form up to the boundary of Lemma 7 of [C]. Here v is as above.

LEMMA 6.3. *There exists a positive constant* $C \leq 1$, *with* C^{-1} *under control, and a constant* $\rho_1 \leq \rho$, ρ_1^{-1} *under control, such that if* B_δ *is a ball in* B_R *with centre on* γ *and radius* $\delta \leq \rho_1$, *then*

$$\sup_{B_\delta} v \geq C\delta. \tag{6.8}$$

By the maximum principle and (1.6), we see that this maximum occurs on ∂B_δ. In connection with this we will also use

LEMMA 6.4. *Under the conditions of Lemma 6.3 with* v *satisfying (6.8),* $\exists$ *a constant* $K > 0$ *under control, independent of* δ, *such that*

$$\int_{B_\delta\cap\Omega} |\nabla v|^2 \geq K^{-1} \text{ Vol } B_\delta. \tag{6.9}$$

This is a calculus lemma, and follows from standard interpolation inequalities, but here is a simple

PROOF: The more difficult case is that B_δ is not contained in Ω so we will just treat this. We may suppose $\rho_1 \le r(R)/(3\overline{C}(R))$. Let x_0 be the closest point on $\partial\Omega$ to the centre of B_δ. By Property P we may map a region containing $\{|x - x_0| \le r\} \cap \overline{\Omega}$ one–to–one by ϕ onto $\overline{B}_r^- = \{|y| \le r\} \cap \{y_n \le 0\}$ with $\phi(x_0) = 0$. The centre of B_0 maps to a point y' with $|y'| \le r/3$. By (6.3), in B_r^-, v satisfies $|\nabla v| \le k\overline{C}$, $v(y') = 0$ and $v \ge C\delta$ at some point $\bar{y}$, $|\bar{y} - y'| \ge \dfrac{\delta}{\overline{\overline{C}}}$.

Hence for λ small, but with λ^{-1} under control, we may infer that

$$v < \frac{C\delta}{4} \quad \text{in } \tilde{B} = \{y \,\big|\, |y - y'| < \lambda, y_n \le 0\}$$

and

$$v > \frac{3}{4}C\delta \quad \text{in } \tilde{\tilde{B}} = \{|y - \bar{y}| < \lambda\} \cap B_r^-.$$

It follows that the integral of $|\nabla v|$ on a straight segment going from $\tilde{B}$ to $\tilde{B}$ is at least $C\delta/2$. We may construct a right circular cylinder Σ in G with base of radius $\lambda/10$ and lying in $\tilde{B}$, with the other flat end lying in $\tilde{\tilde{B}}$, and with length (along the generator) $\ge \delta/2$. The integral of $|\nabla v|$ on any segment parallel to the generator, and ending on the flat ends is at least $C\delta/2$. Thus

$$\iint_\Sigma |\nabla v| \ge \frac{C\delta}{2} \cdot \text{ area of base.}$$

By Schwarz inequality,

$$\iint_\Sigma |\nabla v|^2 \ge \left(\iint_\Sigma \nabla v\right)^2 (\mathrm{Vol}\ \Sigma)^{-1}$$

$$\ge \delta^n / K_1$$

with K_1 under control (depending on C, k, $\overline{C}$). Going back to the original coordinates we obtain (6.9).

Postponing the proof of the lemma 6.3 we now give the

PROOF OF THEOREM 6.1: Cover $\gamma \cap B_r$ by balls B_j centred on γ of radius $\tau < \rho$ such that at most $C(n)$ of them overlap. τ is arbitrary and so the theorem will follow if we can establish (here A = surface area)

$$\sum A(\delta B_j) \leq C r^{n-1},$$

or, equivalently,

$$\sum \text{Vol } B_j \leq C\tau r^{n-1}, \tag{6.10}$$

with C independent of τ and r.

Applying Lemmas 6.3 and 6.4 we see that

$$\int_{B_j \cap \Omega} |\nabla v|^2 \geq \tau^n / K.$$

Hence

$$\begin{aligned}\sum \text{Vol } B_j &\leq K \sum_j \int_{B_j \cap \Omega} |\nabla v|^2 \\ &\leq KC(n) \int_{\Omega \cap B_j} |\nabla v|^2.\end{aligned} \tag{6.11}$$

By (6.3), $v \leq k\tau$ on B_j. Also each B_j is in the ball B_{2r} with the same centre as B_r; $B_{2r} \subset B_R$. We may therefore apply Lemma 6.2 in the ball B_{2r}, with τ replaced by $k\tau$, and infer that the right hand side of (6.11) is bounded by $C\tau r^{n-1}$ with C under control. Inequality (6.10) is established.

We turn to Lemma 6.3; we will follow the proof of Lemma 7 in [C], with minor modifications. The lemma is nontrivial, in fact its conclusion shows that γ cannot be too wild. For instance, in the plane, a cusp is excluded or even a very narrow angle:

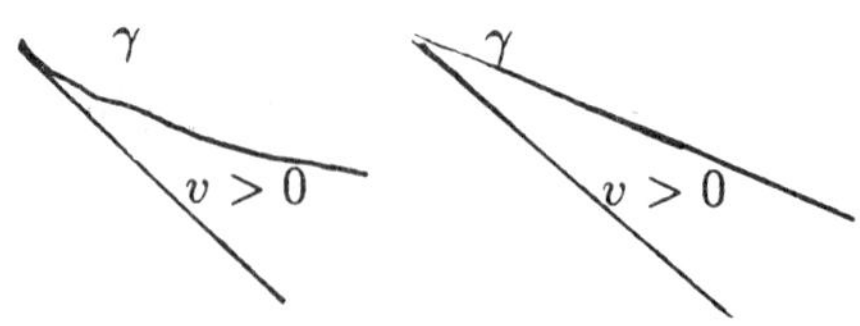

To prove the lemma we will construct a polygonal path with initial point on γ, on the vertices of which v grows linearly.

This will be done with the aid of the following result concerning our function v satisfying (6.3), (6.4) etc.

LEMMA 6.5. *Suppose $v(x_0) > 0$ and $a = d(x_0) \leq \rho_1$. On $\partial B_a(x_0)$ there is a point x_1 with*

$$v(x_1) \geq (1+\lambda)v(x_0). \tag{6.12}$$

Here $\lambda^{-1} > 0$ is under control; it is independent of x_0 and of v.

PROOF: We have to consider the cases that $B_a(x_0)$ lies entirely in Ω or does not. We will just treat the latter. We will make a local change of variable which flattens out $\partial\Omega$. Property P is not useful here. Instead we make a local transformation of variable ψ of $\overline{B}_a \cap \overline{\Omega}$ onto $G = \overline{B}_a \cap \{x_n \leq \sigma\}$, $0 \leq \sigma \leq a$, which maps x_0 to the origin, and $\partial\Omega$ onto $x_n = \sigma$. For ρ_1 sufficiently small we can do this keeping the C^2 norms of ψ and ψ^{-1} bounded by some constant (which we still call $\overline{C}$) under control. However we do not assume that ψ maps the vector field μ onto the direction normal to $x_n = \sigma$. It just maps μ onto some other C^2 vector field $\tilde{\mu}$ on $x_n = \sigma$, so there, the transposed v satisfies

$$\tilde{\mu} \cdot \nabla v = 0.$$

By (6.4), $v(0) \geq c_0 a$. If we consider the function

$$w(x) = \frac{1}{a} v(ax)$$

it corresponds to setting $a = 1$, for w satisfies

$$a_{ij}(ax)w_{ij} + ab_i(ax)w_i = 0,$$

so (1.6) and (1.7) still hold. We will therefore suppose $a = 1$. σ should be replaced by σ/a, but since there is no restriction other than nonnegativity we will continue to call it σ; now $\sigma \leq 1$.

There is a point on ∂B_1 where $v = 0$. So by (6.3), for $c_1 = (6k\overline{C})^{-1}$ there is a ball

$$\tilde{B} = \{|x - z| < d\} \text{ lying in } x_n \leq \sigma - c_1, d = c_0 c_1, \tag{6.13}$$

with $|z| = 1$ such that

$$v(x) \leq \frac{1}{2}c_0 \leq \frac{1}{2}v(0) \quad \text{in } \tilde{B} \cap G =: \Gamma. \tag{6.14}$$

We may suppose $v(0) = 1$. By the maximum principle and the Hopf lemma we see that in $\overline{G}$, v assumes its maximum on ∂B_1.

We will give an indirect argument proving the existence of such a λ in (6.12). Suppose there is no such λ. Then there is a sequence of elliptic operators L_j satisfying (1.16), (1.17) and with a fixed modulus of continuity for the leading coefficients, a sequence σ_j, $0 \leq \sigma_j \leq 1$; a sequence of balls $\tilde{B}_j$ as in (6.13) with fixed radius d, having the properties described above relative to the Γ_j defined in (6.14); a sequence of vector fields $\tilde{\mu}_j$, of unit length and uniformly bounded C^2 norm, and with last component $\geq \tau > 0$; and also a sequence of solutions v_j of $L_j v_j = 0$ in $G_j = B_1 \cap \{x_n < \sigma_j\}$, with $v_j(0) = 1$, $v_j \leq 1/2$ in Γ_j and $\sup_{G_j} v_j \to 1$. For a suitable subsequence, $\sigma_j \to \sigma$, so G_j goes to some G, $\Gamma_j \to \Gamma$, the leading coefficients of the L_j will converge uniformly to some continuous a_{ik} in $\overline{G}$, with the same modulus of continuity, and the coefficients of the first order terms in L_j converge weakly in every L_p, $p < \infty$ to functions b_k. So (1.16)–(1.18) hold for the limit L. Also the $\tilde{\mu}_j$ converge in $C^{1,\delta}$, $\delta > 0$, to $\tilde{\mu}$ etc. Furthermore in any compact set J of $G \cup (\overline{G} \cap \{x_n = \sigma\})$ the v_j have uniformly bounded $W^{2,p}$ norms and so converge weakly in $W^{2,p}(J)$ and strongly in $C^1(J)$ to a function $\tilde{v}$ which is a solution of $L\tilde{v} = 0$. $\tilde{\mu} \cdot \nabla \tilde{v} = 0$ on $x_n = \sigma$, even if $\sigma = \lim \sigma_j$ is 0 or 1. In case $\sigma = 1$, the condition $\tilde{\mu} \cdot \nabla \tilde{v} = 0$ on $x_n = \sigma$ is of course lost; no matter. The limit function $\tilde{v}$ satisfies $\tilde{v}(0) = 1$, and $\sup \tilde{v} = 1$. By the maximum principle $\tilde{v} \equiv 1$, contradicting the fact that $\tilde{v} \leq 1/2$ on Γ. (In case $\sigma < 1$, the Hopf lemma shows that the maximum of $\tilde{v}$ cannot occur at any point of $\{x_n = \sigma\} \cap B_1$.)

At last, the

PROOF OF LEMMA 6.3: We may suppose B_δ centred at the origin. Let x_0 be a point in $\overline{\Omega}$ where $v > 0$, with $|x_0| < \delta/4$. So $r_0 = d(x_0) < \delta/4$ and, by (6.4), $v(x_0) \geq c_0 d(x_0)$. By Lemma 6.5 there is a point x_1 in $\overline{\Omega}$, with $|x_0 - x_1| = r_0$ such that

$$v(x_1) \geq (1 + \lambda) v(x_0).$$

Applying Lemma 6.5 recursively we find points x_k in $B_\delta \cap \overline{\Omega}$ with

$$|x_k - x_{k-1}| = r_{k-1} = d(x_{k-1})$$

and such that

$$v(x_k) \geq (1 + \lambda) v(x_{k-1}).$$

We can continue to do this provided $x_1, \ldots, x_{k-1}$ lie in B_δ.

By (6.3) and (6.4) we have

$$c_0 r_{k-1} \leq v(x_{k-1}),$$

so

$$v(x_k) - v(x_{k-1}) \geq \lambda v(x_{k-1}) \geq c_0 \lambda r_{k-1}$$

and

$$v(x_k) \geq c_0 r_0 + c_0 \lambda (r_0 + \ldots + r_{k-1}). \tag{6.15}$$

Since $1 \geq v(x_k) \geq (1 + \lambda)^k v(x_0)$, there must be a last point x_k, of this sequence, in B_δ, i.e. $x_{k+1} \notin B_\delta$. Since $|x_k - x_{k-1}| = r_{k-1}$ we have $r_k \leq 2r_{k-1}$.

Suppose $|x_k| < \delta/2$, then $|x_{k+1} - x_k| = r_k < \delta/2$ and so $x_{k+1} \in B_\delta$. Impossible. So $|x_k| \geq \delta/2$. From the triangle inequality it follows that

$$\sum_0^{k-1} r_i \geq \frac{\delta}{2}.$$

By (6.15), $v(x_k) \geq c_0\lambda\delta/2$. We have proved (6.8) with $C = c_0\lambda\delta/2$.

Q.E.D.

7. The combustion problem:

In this section we will apply the results of section 3, and the free boundary condition (1.7) derived in section 5, to the combustion problem (1.12). We wish to study what happens to (T_ϵ, c_ϵ) as $\epsilon \to 0$. Here f is a given Lipschitz function which is positive on $(\theta, 1)$, $\theta > 0$ and zero outside. f is assumed to be of class $C^{1,\delta}$ for some $\delta > 0$ on some interval $[1-\eta, 1]$ and $f'(1) < 0$; $\int_0^1 f(T)dT = N$, and

$$f_\epsilon(T) = \frac{1}{\epsilon} f\left(1 - \frac{1-T}{\epsilon}\right). \tag{7.1}$$

From [BL] we know that for each $\epsilon < 1$ there is a unique solution (T_ϵ, c_ϵ), $c_\epsilon > 0$, $0 < T_\epsilon < 1$ satisfying (1.12):

$$\begin{cases} \Delta T_\epsilon - c_\epsilon \alpha(y)\partial_1 T_\epsilon + f_\epsilon(T_\epsilon) & = 0 \quad \text{in } S \\ T_{\epsilon\nu} & = 0 \quad \text{on } \partial S \end{cases} \tag{7.2}$$

and the condition (1.13) at infinity:

$$\lim_{x_1 \to -\infty} T_\epsilon(x_1, y) = 0, \qquad \lim_{x_1 \to +\infty} T_\epsilon(x_1, y) = 1$$

uniformly in y. The function α is a continuous positive function on $\bar{\omega}$.

In studying what happens as $\epsilon \to 0$ we will make use of results in [BN2], in particular that $\partial_1 T_\epsilon > 0$ in $\overline{S}$. For convenience we will usually denote T_ϵ, c_ϵ by T, c, and we set $c\alpha(y) = b(y)$.

A crucial part of the work in [BN2] was concerned with the asymptotic behaviour of T as $x_1 \to \pm\infty$. It was shown that

$$(7.3)\qquad \begin{cases} T(x_1,y) - \gamma_- e^{\lambda x_1}\phi(y) &= o(e^{\lambda x_1}) \quad \text{as } x_1 \to -\infty, \\ 1 - T(x_1,y) - \gamma_+ e^{\tau x_1}\psi(y) &= o(e^{\tau x_1}) \quad \text{as } x_1 \to +\infty. \end{cases}$$

Here γ_-, γ_+ are positive number, λ is the (principal) positive eigenvalue (there is only one) of

$$(7.4)\qquad \begin{cases} (\Delta y - \lambda b(y) + \lambda^2)\phi &= 0 \quad \text{in } \omega \\ \phi_\nu &= 0 \quad \text{on } \partial\omega \end{cases}$$

with positive eigenfunction ϕ on $\bar{\omega}$. τ is the (principal) negative eigenvalue (also unique) with eigenfunction $\psi > 0$ on $\bar{\omega}$ of

$$(7.4)'\qquad \begin{cases} (\Delta_y + f'(1) - \tau b(y) + \tau^2)\psi &= 0 \quad \text{in } \omega \\ \psi_\nu &= 0 \quad \text{on } \partial\omega. \end{cases}$$

Using the "method of sliding domains" the uniqueness of (T, c) was proved in [BN2]. Since we will use it again we refer the reader to that paper. An additional useful fact from [BN2] is that λ and τ are both increasing in b, i.e.

$$(7.5)\qquad \begin{cases} \text{if } b' \gneqq b \text{ then the corresponding principal positive and negative} \\ \text{eigenvalues } \lambda', \tau' \text{ satisfy } \lambda' > \lambda > 0 > \tau' > \tau. \end{cases}$$

Set

$$(7.6)\qquad u = \frac{1-T}{1-\theta} \text{ and } \beta(u) = \frac{f(T)}{1-\theta} = \frac{f(1-(1-\theta)u)}{1-\theta},$$

so $0 < u < (1-\theta)^{-1}$. Then

$$(7.7)\qquad \int_0^1 \beta(u)du = M = \frac{N}{(1-\theta)^2}.$$

and

$$\begin{aligned}\beta_\epsilon(u) := \frac{f_\epsilon(T)}{1-\theta} &= \frac{1}{\epsilon(1-\theta)} f\left(1 - \frac{1-T}{\epsilon}\right) \\ &= \frac{1}{\epsilon(1-\theta)} f\left(1 - (1-\theta)\frac{u}{\epsilon}\right) \\ &= \frac{1}{\epsilon}\beta\left(\frac{u}{\epsilon}\right).\end{aligned}$$

β_ϵ has its support on $[0, \epsilon]$ and is positive in $(0, \epsilon)$. By (7.2), u satisfies

$$\text{(7.8)} \qquad \begin{cases} \Delta u - b(y)u_1 = \beta_\epsilon(u) & \text{in } S \\ u_\nu = 0 & \text{on } \partial S \end{cases}$$

and $\lim_{x_1 \to -\infty} u(x_1, y) = (1-\theta)^{-1}$, $\lim_{x_1 \to +\infty} u(x_1, y) = 0$ uniformly in y. Also $u_1 < 0$.

In preparation for letting $\epsilon \to 0$ we first prove

THEOREM 7.1. *(i) u_ϵ is a "minimal solution" of (7.8) in the sense of section 4.* *(ii) There exist positive constants $\underline{c}$, $\bar{c}$ such that*

$$\text{(7.9)} \qquad \underline{c} \le c_\epsilon \le \bar{c} \quad \forall \epsilon.$$

(iii) On S the functions T_ϵ satisfy a Lipschitz condition with constant k independent of ϵ.

PROOF: (i) Recalling the definition of minimal solution, let $h \in W^{2,p}(G)$, $p > n$, be a strong supersolution of (7.8) in a bounded subdomain G of S, i.e. (denoting (T_ϵ, c_ϵ) by (T, c)),

$$\begin{aligned} \Delta h - bh_1 &\le \beta_\epsilon(h) \quad \text{in } G \\ h_\nu &\ge 0 \quad \text{on } \partial S \cap \overline{G} \end{aligned}$$

and satisfying

$$h \ge u \quad \text{on } \partial G \cap S.$$

We have to prove that $h \geq u$ in G.

To this end we will use the method of sliding domains, or sliding method. Since $\beta_\epsilon(h) = 0$ for $h < 0$, we observe first that $h > 0$ on $\overline{G}$. This holds on $\partial G \cap S$ and so holds everywhere as we see with the aid of the maximum principle and the Hopf lemma.

Now slide u to the left a distance t. That is, in G, consider the function $u^t(x_1, y) = u(x_1 + t, y)$. It satisfies (7.8) and as $t \to \infty$, $u^t \to 0$ in G. So for $t = t_1$ large we have $u^t < h$ on $\overline{G}$. Next, start decreasing t from t_1. For every $t > 0$, $u^t < h$ on $\partial G \cap S$. This follows from the fact that $u \leq h$ there and $u_1 < 0$. Decrease t in $[0, t_1]$ until, perhaps, we reach a first value $t_0 > 0$ such that $u^{t_0} \leq h$ on $\overline{G}$ but with equality at some point. If no such $t_0 > 0$ exists, then $u \leq h$ and we are through.

So suppose there is such a $t_0 > 0$. Since β_ϵ is a Lipschitz function, the function $w = h - u$ satisfies (by the Theorem of the mean)

$$(\Delta - b\partial_1 w) = \tilde{c}w \quad \text{in } G$$

with $\tilde{c}(x_1, y)$ bounded measurable. $w \geq 0$, and equals zero at some point X, and $w > 0$ on $\partial G \cap S$. By the minimum principle (see [BN1]) X cannot be in G, and by the Hopf lemma it cannot be on ∂S either, for there $w_\nu \leq 0$. Contradiction; (i) is proved.

REMARK: The same proof shows that T is also a "minimal solution" of (7.2), i.e., that u is also a "maximal solution" of (7.8) in an obvious sense. In particular if H is a solution of (7.2) in G with $H = T$ on $\partial G \cap S$ and $H_\nu = 0$ on $\partial G \cap \partial S$, then $H \equiv T$. This does not mean that every solution of (7.8) in a bounded region G with given positive boundary values is unique in G. As we have seen, it *is* true for our global solution T which is monotone in x_1.

(ii) We will prove the claim with $\underline{c}$, $\bar{c}$ depending only on α and on $N = \int_0^1 f_\epsilon dT$, so we will drop the ϵ and just consider f.

Let α_0, $\alpha_1 > 0$ be the minimum and maximum of $\alpha(y)$, and let (c_0, T_0), (c_1, T_1) be the corresponding solutions of (7.2), for $n = 1$, when α is replaced by α_0 and α_1 respectively. By uniqueness these are the solutions in any dimension n. It was shown in [BN2] that

$$c_1 \leq c \leq c_0.$$

To prove (ii) it suffices therefore to establish the estimate

$$\underline{c} \leq c_1, \quad c_0 \leq \bar{c} \tag{7.10}$$

with suitable $\underline{c}$, $\bar{c}$ depending only on α_1, α_0 and N.

We have

$$\ddot{T} - c_1\alpha_1\dot{T}_1 + f(T_1) = 0 \qquad -\infty < x_1 < \infty$$

$$T_1(-\infty) = 0, \ T_1(+\infty) = 1.$$

Multiplying by $\dot{T}_1$ and integrating over R we obtain

$$-c_1\alpha_1 \int_{-\infty}^{\infty} \dot{T}_1^2 + F(1) = 0. \tag{7.11}$$

Here $F(T) = \int_0^T f(s)ds$, so $F(1) = N$. There are no other terms because $\dot{T}_1$ decays exponentially at $\pm\infty$. On the other hand if we multiply the equation by $(1 - T_1)$ and integrate, we find, after integrating by parts,

$$\int_{-\infty}^{\infty} \dot{T}_1^2 - \frac{1}{2}c_1\alpha_1 + \int_{-\infty}^{\infty} (1 - T_1)f(T_1)dx_1 = 0. \tag{7.12}$$

Hence

$$\int_{-\infty}^{\infty} \dot{T}_1^2 \leq \frac{1}{2}c_1\alpha_1$$

and so

$$N = F(1) \leq \frac{1}{2}c_1^2\alpha_1^2.$$

This yields the first inequality of (7.10) with $\underline{c} = \alpha_1^{-1}\sqrt{2N}$.

To prove the second part of (7.10) we use the fact that every f_ϵ vanishes for $T \le \theta$. After shifting T_0 we may suppose $T_0(0) = \theta$; then

$$\ddot{T}_0 - c_0\alpha_0\dot{T}_0 = 0 \quad \text{for } x_1 \le 0.$$

Consequently, with $b = c_a\alpha_0$, we have

$$T = \theta e^{bx_1} \quad \text{for } x_1 \le 0,$$

so

$$\int_{-\infty}^{0} \dot{T}_0^2 = \frac{b}{2}\theta^2.$$

Hence

$$\int_{-\infty}^{\infty} \dot{T}_0^2 \ge \frac{b}{2}\theta^2.$$

But (7.11) also holds for T_0:

$$b\int_{-\infty}^{\infty} \dot{T}_0^2 = N.$$

Combining this with the preceding inequality we obtain the second inequality in (7.10) with $\bar{c} = \sqrt{2N}(\theta\alpha_0)^{-1}$. (ii) is proved.

(iii) This follows directly from Theorem 3.1: The function u_ϵ satisfies (7.8), and $\beta_\epsilon \le B/\epsilon$ where $B = (1-\theta)^{-1}\max f$. By Theorem 3.1, on $[-1,1]\times\bar{\omega}$, u_ϵ satisfies a Lipschitz condition with Lipschitz constant independent of ϵ. Since this holds for any x_1–translation of u, and $0 \le u \le 1$, we infer that u_ϵ is Lipschitz continuous on $\overline{S}$ with constant independent of ϵ. So the same is true of T.

The proof of Theorem 7.1 is complete, and we now consider the limit problem as $\epsilon \to 0$—leading to a free boundary.

We have shown that the solutions T_ϵ of (7.2) satisfy a uniform Lipschitz condition and that the c_ϵ satisfy (7.9). We wish to choose a sequence $\epsilon_i \to 0$ for which the c_{ϵ_i} and the T_{ϵ_i} converge.

In order to ensure that the limit of the T_{ϵ_i} is not identically 0 or 1, we first translate each T_ϵ so that

$$\max_{y \in \bar{\omega}} T_\epsilon(0, y) = \theta.$$

Then we choose a sequence $\epsilon_i \to 0$ such that c_{ϵ_i} tend to some limit c in $\underline{c} \leq c \leq \bar{c}$, and the T_{ϵ_i} converge uniformly on every compact subset to a function $V \leq 1$. The functions $u_{\epsilon_i} = (1-\theta)(1-T_{\epsilon_i})$ of (7.6) converge to a function $v = (1-\theta)(1-V)$ uniformly on compact subsets of $\overline{\Omega}$. As we showed in section 5 the function v, and hence also V have bounded $W^{2,p}$ norms, for fixed $p > n$, on every compact subset of $\{V < 1\}$. Clearly $V_1 \geq 0$. Also

$$\text{(7.13)} \qquad \begin{cases} \max V(0, y) = \theta \text{ and } V \text{ satisfies a Lipschitz condition} \\ \text{with some constant } k. \end{cases}$$

We will make use of an extension of (7.12) for solutions (c, T) of (7.2) when $n > 1$. This is obtained in the same way. Multiply the equation by $(1-T)$ and integrate over S. With the aid of Green's theorem—recall that $|\nabla T|$ decays exponentially at $\pm\infty$—we find

$$\iint_S |\nabla T|^2 - \frac{c}{2}\int_\omega \alpha + \iint (1-T) f_\epsilon(T) = 0,$$

so

$$\iint |\nabla T_\epsilon|^2 \leq \frac{\bar{c}}{2}\int_\omega \alpha.$$

It follows that

$$\iint |\nabla V|^2 \leq \frac{\bar{c}}{2}\int \alpha.$$

Next we observe that since $T = T_\epsilon \leq \theta$ for $x_1 \leq 0$ it satisfies

$$\Delta T - c_\epsilon \alpha T_1 = 0$$

there. Using (7.9) we see that for small $\epsilon > 0$ the function $h = \theta e^{\epsilon x_1}$ is a supersolution; i.e.

$$\Delta h - c_\epsilon \alpha h_1 < 0. \tag{7.14}$$

From the maximum principle (and the Hopf lemma applied on ∂S) we see that

$$T_\epsilon \leq h \quad \text{for } x_1 \leq 0.$$

It follows that

$$V \leq \theta e^{\epsilon x_1} \quad \text{for } x_1 \leq 0. \tag{7.15}$$

Before presenting our main results for V, some lemmas.

PROPOSITION 7.1. $\lim_{x_1 \to -\infty} V(x_1, y) = 0$, $\lim_{x_1 \to +\infty} V(x_1, y) = 1$, *uniformly for* $y \in \bar{\omega}$.

PROOF: (i) We have just proved the first statement. Since V is nondecreasing in x_1 and $V \leq 1$ we see that

$$\xi(y) = \lim_{x_1 \to +\infty} V(x_1, y)$$

exists for every y and $|\nabla \xi| \leq k$. We claim that ξ is constant. Suppose not. Then, by Lipschitz continuity, there is a ball B in ω with centre $\bar{y}$, and a point $y_0 \in \omega$ such that for suitable constants $a < b$,

$$\xi \leq a \text{ in } B \text{ and } \xi(y_0) > b.$$

Since V is nondecreasing in x_1 it follows that for some large R

$$V(x_1, y) \leq a \quad \text{for } x_1 \geq R, y \in B$$

$$V(x_1, y_0) > b \quad \text{for } x_1 \geq R.$$

By Lipschitz continuity,

$$V(x_1, y) > \frac{a+b}{2} \quad \text{for } x_1 \geq R,\ |y - y_0| < \delta$$

for some small δ. By shrinking δ, or the ball B we may suppose B has radius δ.

If P is a smooth curve in ω joining $\bar{y}$ to y_0 then the integral of $|\nabla y V|$ along P, keeping $x_1 \geq R$ fixed is at least $\frac{b-a}{2}$. By forming a tube Σ made up of rigid shifts of P so as to join points $\bar{y}+\eta$ in B to $y = y_0 + \eta$ in $\{|y - y_0| < \delta\}$ we see easily that

$$\int_{\Sigma} |\nabla V| dy \geq \rho_0 > 0.$$

Here ρ_0 is a positive constant. Hence

$$\int |\nabla V|^2 dy \geq \rho_1 > 0$$

where ρ_1 is a positive constant. If we now integrate this with respect to x_1 from R to ∞ we conclude that

$$\iint_S |\nabla V|^2 = \infty.$$

Contradiction.

(ii) We now claim that

$$\lim_{x_1 \to +\infty} V(x_1, y) = \xi \quad \text{uniformly in } y.$$

To prove this it suffices to show that given any $\epsilon > 0$, $\exists R$ such that

$$V(x_1, y) \geq \xi - \epsilon \quad \text{for } x_1 \geq R.$$

This is easily done. Cover $\bar{\omega}$ by closed balls $B_1, \ldots, B_j$ with centres $y_1, \ldots, y_j$ and radius $\epsilon/(2k)$, see (7.13). For some $R > 0$ we have

$$V(x_1, y_i) \geq \xi - \epsilon/2 \quad \text{for } x_1 \geq R, i = 1, \ldots, j.$$

Hence by (7.13), for any y in $\bar{\omega}$,

$$V(x_1, y) \geq \xi - \epsilon \quad \text{for } x_1 \geq R.$$

(iii) Finally we show that $\xi = 1$. Suppose $\xi < 1$, then V satisfies

$$\Delta V - c\alpha V_1 = 0 \quad \text{in } S.$$

According to (7.15), $V \leq \theta e^{\epsilon x_1}$ for $x_1 \leq 0$. It follows that for A large,

$$V \leq A e^{\epsilon x_1/2} \quad \text{in } S.$$

Now decrease A until equality first occurs. It must occur for $A \geq \theta$ and at a point in $\overline{S}$. But $e^{\epsilon x_1/2}$ is also a supersolution, i.e. it satisfies (7.14). Then such contact is impossible. The proposition is proved.

Finally, our principal result about V—recall that it satisfies (7.13), is nondecreasing in x_1, and satisfies $\Delta V - c\alpha\partial_1 V = 0$, and hence $V_1 > 0$, where $\{V < 1\}$. Set $\gamma = \partial\{V < 1\}$.

THEOREM 7.2. *(a) There is a constant R such that*

$$V \equiv 1 \text{ for } x_1 \geq R.$$

Hence γ lies in $0 < x_1 \leq R$ and is compact.
(b) There is a positive number c_0 such that if $V(x_1, y) < 1$ and $d(x_1, y) =$ distance of (x_1, y) to γ is less than or equal to 1, then

$$V(x_1, y) \leq 1 - c_0 d(x_1, y).$$

(c) On regular parts of γ in S,

$$T_\nu = \sqrt{2N}$$

where ν is the exterior normal to $\{V < 1\}$ on γ.

$$\text{Recall that } \int_0^1 f(s)ds = N.$$

PROOF: We have only to prove (a). (b) then follows from Theorem 3.1 applied to u_ϵ, and its limit v—for in Theorem 7.1 we proved u_ϵ is a "minimal solution". (c) follows from Theorem 5.1 applied to v.

Our proof of (a) involves several steps.

(i) We observe first that given $\eta > 0 \ \exists\ \epsilon_0 > 0$ and $r > 0$ such that

$$T_\epsilon(x_1, y) \geq 1 - \eta \quad \text{for } x_1 \geq r, \quad y \in \bar{\omega}, \quad \epsilon \leq \epsilon_0.$$

Indeed for any $\eta > 0 \ \exists\ r$ such that

$$V(x_1, y) \geq 1 - \eta/2 \quad \text{for } x_1 \geq r.$$

This follows from Proposition 7.1. But we also have

$$|T_\epsilon(r, y) - V(r, y)| \leq \eta/2 \quad \text{for } y \in \bar{\omega}, \epsilon \leq \epsilon_0 \text{ small.}$$

Hence for $x_1 \geq r$,

$$\begin{aligned} T_\epsilon(x_1, y) > T_\epsilon(r, y) \ &\geq V(r, y) - \eta/2 \\ &\geq 1 - \eta. \end{aligned}$$

(ii) We will make use of the one dimensional problem

$$\ddot{\zeta} - b\dot{\zeta} + f_\epsilon(\zeta) = 0 \quad \text{on } R. \tag{7.17}$$

$$\zeta(-\infty) = 1 - \delta, \quad \zeta(+\infty) = 1. \tag{7.18}$$

Here $\delta < 1$ is a small positive number to be chosen, and ϵ is then taken so small that $\epsilon(1 - \theta) < \delta$. This looks different from the problems we have discussed because of the condition $\zeta(-\infty) = 1 - \delta$. However it means that the function

$$\xi = \frac{\zeta - 1}{\delta} + 1$$

is a solution of

$$\ddot{\xi} - b\dot{\xi} + \frac{1}{\delta} f_\epsilon(1 - \delta(1 - \xi))$$

$$\xi(-\infty) = 0, \quad \xi(+\infty) = 1.$$

which is of the form we have been discussing. It therefore has a unique solution (ξ, b), up to translation of ξ. Consequently so does (7.17), (7.18).

We will derive a lower bound on $b = b_\epsilon$ by the method used in the proof of Theorem 7.1 (ii). Multiplying equation (7.17) by $\dot{\zeta}$ and integrating we find as before

$$-b\int_{-\infty}^{\infty} \dot{\zeta}^2 + N = 0.$$

If we multiply (7.17) by $1 - \zeta$ and integrate, we obtain, after integrating by parts,

$$\int_{-\infty}^{\infty} \dot{\zeta}^2 - \frac{b}{2}\delta^2 + \int (1-\zeta) f_\epsilon = 0$$

Thus

$$\frac{N}{b} = \int \dot{\zeta}^2 \leq \frac{b}{2}\delta^2$$

or

$$b \geq \frac{\sqrt{2N}}{\delta^2}.$$

We now fix δ so that

$$\frac{\sqrt{2N}}{\delta^2} > \bar{c}\alpha_1,$$

recall that $c_\epsilon \leq \bar{c}$ and $\alpha_1 = \max \alpha$. With δ so fixed we require, as stated, $\epsilon(1-\theta) < \delta$. In fact we now fix θ' with $1 - \delta < \theta' < 1$ and consider only smaller ϵ:

$$\epsilon(1-\theta) < 1 - \theta'. \tag{7.19}$$

Then we have our corresponding solution $(b_\epsilon, \zeta_\epsilon)$ of (7.17), (7.18).

Let us go back to (7.16) and fix $\eta = 1 - \theta'$. Then, for some r, ϵ_0,

$$T_\epsilon(r, y) \geq \theta' \quad \text{for } \epsilon \leq \epsilon_0.$$

By a suitable shift we now require for $\epsilon \leq \epsilon_0$ and satisfying (7.19),

$$\zeta_\epsilon(r) = \theta'.$$

Recall also that

$$b_\epsilon \geq \frac{\sqrt{2N}}{\delta^2} > \bar{c}\alpha_1 \geq c_\epsilon \alpha. \tag{7.20}$$

(iii) We claim that

$$T_\epsilon \geq \zeta_\epsilon \quad \text{for } x_1 \geq r. \tag{7.21}$$

To prove this we compare their behaviours at $+\infty$. The behaviour of T_ϵ is described by the second equation in (7.3), and there is a corresponding formula for ζ_ϵ:

$$1 - \zeta_\epsilon(x_1) - \gamma' e^{\tau' x_1} = o(e^{\tau' x_1}) \text{ as } x_1 \to +\infty.$$

In virtue (7.20) and (7.5), $0 > \tau' > \tau$, and this implies that $T_\epsilon > \zeta_\epsilon$ near $+\infty$.

We now employ the sliding arguments once more—in the region $x_1 \geq r$. If we slide T_ϵ to the left it increases and in fact it is eventually bigger than ζ_ϵ in $x_1 \geq r$. Now start shifting back to the right. During the shift it remains greater than ζ_ϵ for $x_1 = r$. But it cannot touch ζ_ϵ at some other point for $\zeta = \zeta_\epsilon$ is a subsolution, namely

$$\begin{aligned} \Delta\zeta \quad &-c_\epsilon \alpha(y)\zeta_{x_1} + f_\epsilon(\zeta) \\ &\geq \Delta\zeta - b_\epsilon \zeta_{x_1} + f_\epsilon(\zeta) = 0 \end{aligned}$$

because if (7.20) and the fact that $\zeta_{x_1} > 0$. Thus on moving back to the right, T_ϵ can first touch ζ_ϵ only in its original position, proving (7.21).

(iv) Consider $(\zeta_\epsilon, b_\epsilon)$ and let $\epsilon \to 0$. As in the proof of (7.9) we have

$$\sqrt{2N} \leq b_\epsilon \leq \sqrt{2N}\theta^{-1}.$$

We find as before when considering u_ϵ, for a subsequence of our previous ϵ_i, that ζ_{ϵ_i} tends to a function $\tilde{\zeta}$, uniformly on compact intervals. It follows from (7.21) that $V \geq \tilde{\zeta}$. Thus to complete the proof of part (a) of the the theorem we have only to show that $\tilde{\zeta}(R) = 1$ for some finite R.

Suppose this is not the case. Then $1-\delta < \tilde{\zeta} < 1$ everywhere, and $\tilde{\zeta}(r) = \theta' > 1-\delta$. But $\tilde{\zeta}$ satisfies

$$\ddot{\tilde{\zeta}} - b\dot{\tilde{\zeta}} = 0,$$

where $b = \lim b_{\epsilon_i}$. This has no nonconstant bounded solution. Contradiction.

Theorem 7.2 is proved.

We recently learned that Theorem 2.1 is closely related to results obtained by A.I. Nazarov: On estimates of Hölder constants for solutions of initial boundary problems with inclined derivatives for parabolic equations, Zapiski Nauchnykh Sem. LOMI 183, 19(1987), p. 130–.

Acknowledgement: The first author was partly supported by ARO–DAAL 03–88–K–0047 while visiting the Courant Institute. The others were partly supported by NSF grants DMS 8804567, DMS 8504033 and INT 8714668.

Univ. Paris VI
Lab. Analyse Numérique
4 Pl. Jussieu
55–65, 5me ét.
75252 Paris Cedex 05
France

Institute for Advanced Study
Princeton N.J. 08540
U.S.A.

Courant Institute
251 Mercer St.
New York, NY 10012
U.S.A.

References

[AC] H.W. Alt, L.A. Caffarelli. *Existence and regularity for a minimum problem with free boundary*, J. Reine Angew. Math. 325, (1981) 105–144.

[BL] H. Berestycki and B. Larrouturou. *A semilinear elliptic equation in a strip arising in a two-dimensional flame propagation model*, J. Reine Angew. Math., to appear.

[BLN] H. Berestycki, B. Larrouturou, and L. Nirenberg. *A nonlinear elliptic problem describing the propagation of a curved premixed flame*, NATO Advanced Research Workshop on Math'l. Modelling in Combustion and Related Topics.

[BN1] H. Berestycki and L. Nirenberg. *Monotonicity, symmetry and antisymmetry of solutions of semilinear elliptic equations*, Journal of Geometry and Physics, to appear.

[BN2] H. Berestycki and L. Nirenberg. *Some qualitative properties of solutions of semilinear elliptic equations in cylindrical domains*, to appear.

[BP] M. Bardi and B. Perthame. *An estimate of the Harnack constant for elliptic equations*, preprint.

[C] L.A. Caffarelli. *A Harnack inequality approach to the regularity of free boundaries. Part III: Existence, compactness and dependence on X*, to appear.

[F] A. Friedman. *Variational Principles and Free Boundary Problems*, Wiley–Interscience, New York, 1982.

[GT] D. Gilbarg and N.S. Trudinger. *Elliptic Partial Differential Equations of Second Order*, Springer–Verlag, 2nd ed., 1983.

[K] N.V. Krylov. *Nonlinear elliptic and parabolic equations of the second order*, "Nauka", Moscow, 1985; Engl. trans. Reidel, 1987.

[R] J.C. Redondo. *Uniform regularity for level sets of solutions of the penalized obstacle problem*, Ph.D. Thesis, Courant Inst., 1989.

31

L^p Estimates for The Three-dimensional Systems of Ellastostatisc on Lipschitz Domains

B. E. J. Dahlberg and C. E. Kenig

The purpose of this work is to study the solvability of the Dirichlet problem, with L^p and Hölder continuous data, and the traction boundary value problem with L^p data for the Lamé systems of linearized elastostatics on an arbitrary bounded Lipschitz domain Ω in $\mathbf{R}^3$. We obtain the optimal estimates for the sharp ranges of L^p spaces. The corresponding results for $p = 2$ in $\mathbf{R}^n$, $n \geq 3$, were obtained in (Dahlberg, *et al.*, 1988). The extension of the results in the present paper to the case of $\mathbf{R}^n$, $n \geq 4$ remains an open problem. The corresponding theorems for the Dirichlet and Neumann problem for Laplace's equation in a Lipschitz domain are by now well understood in all dimensions, thanks to the results in (Dahlberg, 1979), (Jerison-Kenig, 1981), (Verchota, 1984), and (Dahlberg-Kenig, 1987).

Our main results are as follows: Let Ω be a bounded Lipschitz domain in $\mathbf{R}^3$, with connected boundary. Let $\vec{u} = (u_1, u_2, u_3)$ be a solution to the Lamé system

$$\mu\Delta\vec{u} + (\lambda + \mu)\nabla \operatorname{div} \vec{u} = 0 \quad \text{in } \Omega, \tag{0.1}$$

Supported in part by the NSF

where $\mu > 0, \lambda > -2\mu/3$ are constants (Lamé moduli). We consider the Dirichlet problem for (0.1),

$$\text{(D)} \qquad \vec{u}|_{\partial\Omega} = \vec{g} \quad \text{a.e. on } \partial\Omega$$

in the sense of non-tangential convergence.

We show that there exists $\varepsilon = \varepsilon(\Omega) > 0$ such that, given $\vec{g} \in L^p(\partial\Omega, d\sigma)$, $2 - \varepsilon < p \leq \infty$, then there exists a unique solution $\vec{u}$ of (D) with $\|(\vec{u})^*\|_p < \infty$, where $(\ )^*$ denotes the non-tangential maximal function (Theorems 1.3 and 1.4). Note that the case $p = \infty$ is also included (Theorem 1.4). If, in addition, $\vec{g} \in \Lambda^\alpha(\partial\Omega)$ (the space of Hölder continuous functions of order α), and $\alpha < \alpha_0 = \alpha_0(\Omega)$, then $\vec{u} \in \Lambda^\alpha(\overline{\Omega})$ (Theorem 1.4). If $\vec{g} \in L^q_1(\partial\Omega)$, i.e., it has first derivatives in $L^q(\partial\Omega)$, and $1 < q < 2 + \varepsilon$, then there is a unique solution of (D) with $(\nabla\vec{u})^* \in L^q(\partial\Omega, d\sigma)$ (Theorem 1.3). The limitations $2 - \varepsilon < p \leq \infty$, $0 < \alpha < \alpha_0$, and $1 < q < 2 + \varepsilon$ are all sharp.

We also consider the traction boundary value problem for (0.1).

$$\text{(T)} \qquad \lambda(\operatorname{div}\vec{u})N + \mu[\nabla\vec{u} + \nabla\vec{u}^t]N\Big|_{\partial\Omega} = \vec{g}$$

a.e. on $\partial\Omega$ in the sense of non-tangential convergence. Here ^{t} denotes the transpose of a matrix, and N denotes the outward unit normal to $\partial\Omega$. Then, if $\vec{g} \in L^q(\partial\Omega)$, $1 < q < 2 + \varepsilon$, and verifies 6 (necessary) linear conditions, there exists a unique (modulo a 6 dimensional linear space) $\vec{u}$ satisfying (T), with $(\nabla\vec{u})^* \in L^q(\partial\Omega, d\sigma)$ (Theorem 1.5). Moreover, the range $1 < q < 2 + \varepsilon$ is sharp.

Our approach to obtain Theorems 1.3, 1.4, and 1.5, follows our earlier work (Dahlberg-Kenig, 1987), to reduce these results to the corresponding 'atomic estimate' (Lemma 1.6). It is in our proof of the 'atomic estimate' that the restriction to 3 dimensions appears. Finally, we follow the elegant arguments of (Pipher-Verchota, in preparation), to show that the 'atomic estimate' also yields L^∞ bounds and Hölder continuity for (D).

It is important to note the connection of our work with the recent work of (Pipher-Verchota, preprint) and (Pipher-Verchota, in preparation), on the Dirichlet problem for the biharmonic equation

$$
(0.2) \qquad \begin{cases} \Delta^2 u = 0 \text{in } \Omega \\ u|_{\partial\Omega} = g \in L^p_1(\partial\Omega) \\ \frac{\partial u}{\partial N}\Big|_{\partial\Omega} = f \in L^p(\partial\Omega) \end{cases}
$$

with the estimate $\|u^*\|_{L^p(\partial\Omega)} < \infty$. The L^2 theory for this problem was carried out in [**3**], in all dimensions. (Pipher-Verchota, preprint) extended it to the range $2 - \varepsilon < p < \infty$, and $p = \infty$ and Λ^α estimates in (Pipher-Verchota, in preparation), when $\Omega \subset \mathbf{R}^3$. Moreover, they gave counterexamples for $n \geq 4$, to the estimates in the same range of p's. Their counterexamples were carried out in the exterior of 'slender ' cones, by separation of variables. A similar approach for the Lamé systems (0.1) fails to give any counterexample. Thus, unlike in the case of the biharmonic equation, the situation for the Lamé systems (0.1) remains wide open in $\mathbf{R}^n$, $n \geq 4$.

Finally, we refer the reader to (Dahlberg, *et al.*, 1988) and (Fabes,*et al.*, 1988) for the notations, results and definitions that will be used throughout this paper.

In closing, it is our pleasure to acknowledge many helpful conversations with J. Pipher and G. Verchota. We are particularly indebted to them for explaining to us their unpublished work , in which the argument for the passage from the 'atomic estimate' to L^∞ and Hölder estimates was discovered.

We will start out by recalling the main results in (Dahlberg, *et al.*, 1988).

THEOREM 1.1. (Dahlberg, *et al.*, 1988): Let Ω be a bounded Lipschitz domain with connected boundary in $\mathbf{R}^n$, $n \geq 3$. Let $\mu > 0$, $\lambda > -2\mu/n$. Then, given an $\vec{g} \in L^2(\partial\Omega)$ there exists a unique $\vec{u}$ in Ω, solving (0.1), and verifying (D), such that $\vec{u}^* \in L^2(\partial\Omega)$. In fact, $\|\vec{u}^*\|_{L^2(\partial\Omega)} \leq C\|\vec{g}\|_{L^2(\partial\Omega)}$ where C depends only on the Lipschitz character of $\partial\Omega$. If in addition $\vec{g} \in L^2_1(\partial\Omega)$, $\|\vec{u}^*\|_{L^2(\partial\Omega)} + \|\nabla\vec{u}^*\|_{L^2(\partial\Omega)} \leq C\|\vec{g}\|_{L^2_1(\partial\Omega)}$. Moreover, there exists $\varepsilon = \varepsilon(\Omega) > 0$ (which depends only on the Lipschitz character of $\partial\Omega$) so that the above results extend to $L^p(\partial\Omega)$, $L^p_1(\partial\Omega)$, $2 - \varepsilon < p < 2 + \varepsilon$.

To explain the results on the traction problem (T), we need to introduce some more notation.

Let Ψ denote the space of vector-valued functions on $\mathbf{R}^n$ satisfying the $n(n+1)/2$ equations

$$D_i\psi^j + D_j\psi^i = 0 \qquad 1 \leq i, j \leq n.$$

Ψ is spanned by $n(n+1)/2$ linear functions, all of which are solutions of (0.1). In fact, it is easy to see that $\vec{\psi} = A\vec{X} + \vec{C} \in \Psi$ if and only if $A + A^t = 0$.

Define $L^2_\Psi(\partial\Omega) = \{\vec{f} \in L^2(\partial\Omega) : \int_{\partial\Omega} \vec{f} \cdot \vec{\psi}\, dQ = 0 \text{ for all } \vec{\psi} \in \Psi\}$. It is a subspace of codimension $n(n+1)/2$ of $L^2(\partial\Omega)$.

THEOREM 1.2. *Let Ω be a bounded Lipschitz domain with connected boundary in $\mathbf{R}^n$, $n \geq 3$. Let $\mu > 0$, $\lambda > -2\mu/n$. If $\vec{g} \in L^2_\Psi(\partial\Omega)$, there exists, up to elements of Ψ, a unique $\vec{u}$ verifying (0.1) in Ω and (T), with $\|\nabla\vec{u}^*\|_{L^2(\partial\Omega)} < \infty$. In fact, $\vec{u}$ will satisfy $\|\vec{u}^*\|_{L^2(\partial\Omega)} + \|\nabla\vec{u}^*\|_{L^2(\partial\Omega)} \leq C\|\vec{g}\|_{L^2(\partial\Omega)}$, where C depends only on the Lipschitz character of Ω. Moreover, there exists $\varepsilon = \varepsilon(\Omega) > 0$ (which depends only on the Lipschitz character of Ω) so that the above results extend to $L^p_\Psi(\partial\Omega)$, $2 - \varepsilon < p < 2 + \varepsilon$.*

The above theorems also have analogues for the case when $\Omega = \{(x, y) : y > \varphi(x)\}$ is the domain above a Lipschitz graph. (Here $\varphi : \mathbf{R}^{n-1} \to \mathbf{R}$, $|\varphi(x) - \varphi(x')| \leq M|x - x'|$.) In the case of Theorem 1.1, in this case we have the estimate $\|\vec{u}^*\|_{L^p(\partial\Omega)} \leq C\|\vec{g}\|_{L^p(\partial\Omega)}$ and $\|\nabla\vec{u}^*\|_{L^p(\partial\Omega)} \leq C\|\nabla_T\vec{g}\|_{L^p(\partial\Omega)}$ where $C = C(M)$, $2 - \varepsilon < p < 2 + \varepsilon$, $\varepsilon = \varepsilon(M)$, and $\nabla_T\vec{g}$ denotes the tangential gradient of $\vec{g}$. In the case of Theorem 1.2, for the graph case, the subspace Ψ plays no role (since its elements do not belong to $L^p(\partial\Omega)$), and the estimate becomes $\|\nabla\vec{u}^*\|_{L^p(\partial\Omega)} \leq C\|\vec{g}\|_{L^p(\partial\Omega)}$, $C = C(M)$, $2 - \varepsilon < p < 2 + \varepsilon$, $\varepsilon = \varepsilon(M)$.

We are now in a position to state our main results.

THEOREM 1.3. *Let Ω be a bounded Lipschitz domain with connected boundary in $\mathbf{R}^3$. Let $\mu > 0$, $\lambda > -2\mu/3$. Then, there exits $\varepsilon = \varepsilon(\Omega) > 0$ (which depends only on the Lipschitz character of Ω), so that, given any $\vec{g} \in L^p(\partial\Omega)$, $2 - \varepsilon < p < \infty$, there exits a unique $\vec{u}$ in Ω, solving (0.1), and verifying (D), such that $u^* \in L^p(\partial\Omega)$. In fact, $\|\vec{u}^*\|_{L^p(\partial\Omega)} \leq C\|\vec{g}\|_{L^p(\partial\Omega)}$*

where C depends only on the Lipschitz character of $\partial\Omega$.

If, on the other hand, $\vec{g} \in L_1^p(\partial\Omega)$, $1 < p < 2+\varepsilon$, there exits a unique $\vec{u}$ in Ω solving (0.1), and verifying (D), such that $\vec{u}^*$ and $\nabla\vec{u}^* \in L^p(\partial\Omega)$. In fact

$$\|\vec{u}^*\|_{L^p(\partial\Omega)} + \|\nabla\vec{u}^*\|_{L^p(\partial\Omega)} \leq C\|\vec{g}\|_{L_1^p(\partial\Omega)},$$

where C depends only on the Lipschitz character of $\partial\Omega$.

THEOREM 1.4. *Let Ω, μ, λ be as in Theorem 1.3. Then, given any $\vec{g} \in L^\infty(\partial\Omega)$, there exists a unique $\vec{u}$ in Ω, solving (0.1), and verifying (D), such that $\vec{u} \in L^\infty(\Omega)$. In fact, $\|\vec{u}\|_{L^\infty(\Omega)} \leq C\|\vec{g}\|_{L^\infty(\partial\Omega)}$, where C depends only on the Lipschitz character of $\partial\Omega$. Moreover, there exists $\alpha_0 = \alpha_0(\Omega)$ (which depends only on the Lipschitz character of Ω) such that, if $\vec{g} \in \Lambda^\alpha(\partial\Omega)$, $0 < \alpha < \alpha_0$, then $\vec{u} \in \Lambda^\alpha(\overline{\Omega})$. In fact, we have*

$$\|\vec{u}\|_{\Lambda^\alpha(\overline{\Omega})} + \sup_{X\in\Omega} \delta(X)^{\alpha-1}|\nabla\vec{u}(X)| \leq C\|\vec{g}\|_{\Lambda^\alpha(\partial\Omega)},$$

where C depends only on the Lipschitz character of Ω, and $\delta(X) = dist(X, \partial\Omega)$.

THEOREM 1.5. *Let Ω, μ, λ be as in Theorem 1.3. Then, there exists $\varepsilon = \varepsilon(\Omega) > 0$ (which depends only on the Lipschitz character of Ω), so that, given any $\vec{g} \in L_\Psi^p(\partial\Omega)$, $1 < p < 2+\varepsilon$, there exists, up to elements of Ψ, a unique $\vec{u}$ verifying (0.1) in Ω and (T), with $\|\nabla\vec{u}^*\|_{L^p(\partial\Omega)} < \infty$. In fact, $\vec{u}$ will satisfy*

$$\|\vec{u}\|_{L^p(\partial\Omega)} + \|\nabla\vec{u}^*\|_{L^p(\partial\Omega)} \leq C\|\vec{g}\|_{L^p(\partial\Omega)},$$

where C depends only on the Lipschitz character of $\partial\Omega$.

We will now sketch the proof of Theorems 1.3, 1.4 and 1.5. We will give the complete proof of the existence part of the above theorems, in the case when Ω is a Lipschitz graph, i.e., $\Omega = \{(x,y) : y > \varphi(x)\}$, where $\varphi : \mathbf{R}^2 \to \mathbf{R}$, $|\varphi(x) - \varphi(x')| \leq M|x - x'|$. The general case of existence, as well as the uniqueness assertions follow from well known techniques as in (Dahlberg-Kenig, 1987), (Dahlberg, *et al.*, 1988), (Fabes, *et al.*, 1988), (Pipher-Verchota, preprint), and (Pipher-Verchota, in preparation). As in (Dahlberg-Kenig, 1987), in the case of a Lipschitz graph, the existence results in Theorem 1.3 and Theorem 1.5, reduce to an 'atomic estimate'. As was mentioned before, this is the place where the restriction to $n = 3$ appears.

LEMMA 1.6. *Suppose that $\vec{u}$ solves (0.1) in Ω, and that, either* $\operatorname{supp}\vec{u}|_{\partial\Omega} \subset \{(x,\varphi(x)) : |x| < 1\}$, *and* $\|\nabla_T\vec{u}\|_{L^\infty(\partial\Omega)} \leq 1$ *or* $\operatorname{supp} T\vec{u}|_{\partial\Omega} = \vec{a} \subset \{(x,\varphi(x) : |x| < 1\}$ *with* $\|\vec{a}\|_{L^\infty(\partial\Omega)} \leq 1$ *and* $\int_{\partial\Omega}\vec{a} = 0$. *Then, there exists* $\alpha_0 = \alpha_0(M)$ *so that, if* $0 < \alpha < \alpha_0$, *we have*

$$\int_{\partial\Omega}(\nabla\vec{u})^*(|Q|+1)^\alpha \, d\sigma(Q) \leq C, \tag{1.7}$$

where $C = C(M)$.

In the proof of Lemma 1.6, we can assume, without loss of generality, that $\varphi(0) = 0$. We will give the detailed proof in the second case, the first one being slightly simpler. We need to introduce some notation, as in Lemma 2.7 of (Dahlberg-Kenig, 1987). Let $\Lambda(R) = \{(x,\varphi(x)) : R < |x| < 2R\}$, for $Q \in \partial\Omega$, $\gamma(Q) = \{X \in \Omega : |X - Q| < 2\operatorname{dist}(X,\partial\Omega)\}$ and for $Q \in \Lambda(R)$, $\gamma_1(Q) = \{X \in \gamma(Q) : |X - Q| < R\}$. Now, for $\tau \in I = [\frac{1}{8},\frac{1}{2}]$, set $\Omega_\tau = \{(x,y) : \varphi(x) < y < \varphi(x) + \theta R, \tau R < |x| < \tau^{-1}R\}$, where $R > 1$, and θ is chosen so that, for $Q \in \Lambda(R)$, $\overline{\gamma_1(Q)} \subset\subset \Omega_\tau$. Note that each Ω_τ is a bounded Lipschitz domain. We will first need a version of Cacciopoli's inequality in our setting.

LEMMA 1.8. *Suppose that $\vec{v}$ verifies (0.1) in $\Omega_{1/8}$, and that either* $T\vec{v}|_{\partial\Omega_{1/8}\cap\partial\Omega} \equiv 0$ *or* $\vec{v}|_{\partial\Omega_{1/8}\cap\partial\Omega} \equiv 0$. *Then,*

$$\Big(\int_{\Omega_{1/4}} |\nabla\vec{v} + \nabla\vec{v}^t|^2\Big)^{1/2} \leq C\Big(\frac{1}{R^2}\int_{\Omega_{1/8}} |\vec{v}|^2\Big)^{1/2},$$

where $C = C(M)$.

PROOF: As in (Dahlberg, *et al.*, 1988), write (0.1) in the form $\frac{\partial}{\partial x_i} a_{ij}^{rs} \frac{\partial}{\partial x_j} v^s = 0$, where $a_{ij}^{rs} = a_{ji}^{sr}$ are constants, such that $n_i a_{ij}^{rs} \frac{\partial v^s}{\partial x_j}\Big|_{\partial\Omega} = T\vec{v}|_{\partial\Omega}$, where $N = (n_1, n_2, n_3)$ is the outward unit normal. Note that, if we introduce the quadratic form $a(\xi,\zeta) = a_{ij}^{rs}\xi_i^r\zeta_j^s$ on 3×3 matrices, then $a(\xi,\zeta) = a(\zeta,\xi)$, and $a(\xi,\xi)\ge 0$. (In fact, $a(\xi,\xi)\simeq|\xi+\xi^t|^2$.) Since a is symmetric, and non-negative,

$$2a(\xi,\zeta)\le a(\xi,\xi)+a(\zeta,\zeta). \tag{1.9}$$

Choose now $\psi\in C_0^\infty(\mathbf{R}^n)$, $\psi\equiv 1$ on $\Omega_{1/4}$, $\operatorname{supp}\psi\cap\overline{\Omega}\subset\subset\overline{\Omega}_{1/8}$, with $|\nabla\psi|\le C/R$. $a(\nabla\vec{v},\nabla(\vec{v}\psi^2)) = \psi^2 a(\nabla\vec{v},\nabla\vec{v}) + 2\psi a(\nabla\vec{v},\zeta)$, where $\zeta_i^r = v^r\frac{\partial\psi}{\partial x_i}$

$$\begin{aligned}\int_\Omega a(\nabla\vec{v},\nabla(\vec{v}\psi^2)) &= \int_\Omega a_{ij}^{rs}\frac{\partial v^s}{\partial x_j}\frac{\partial}{\partial x_i}(v^r\psi^2)\\ &= -\int_\Omega\frac{\partial}{\partial x_i}(a_{ij}^{rs}\frac{\partial v^s}{\partial x_j})v^r\psi^2+\int_{\partial\Omega}n_i a_{ij}^{rs}\frac{\partial v^s}{\partial x_j}v^r\psi^2,\end{aligned}$$

by the divergence theorem. Both terms vanish by our hypothesis, and so

$$\int_\Omega\psi^2 a(\nabla\vec{v},\nabla\vec{v}) = -2\int_\Omega a(\nabla\vec{v},\zeta).$$

Our desired inequality now follows from (1.9).

REMARK: Note that the above proof in fact shows that if (0.1) holds in Ω_τ, $T\vec{v}|_{\partial\Omega_\tau\cap\partial\Omega}\equiv 0$, and $\rho<\tau$, $1/8\le\rho<\tau\le 1/4$, we have

$$\Big(\int_{\Omega_\rho}|\nabla\vec{v}+\nabla\vec{v}^t|^2\Big)^{1/2}\le\frac{C}{(\tau-\rho)^N}\Big(\frac{1}{R^2}\iint_{\Omega_\tau}|\vec{v}|^2\Big)^{1/2}, \tag{1.10}$$

for a suitable N.

COROLLARY 1.11. *Under the hypothesis of Lemma 1.8, there exists $C>0$ so that*

$$\Big(\frac{1}{R^n}\int_{\Omega_{1/4}}|\vec{v}|^2\Big)^{1/2}\le C\Big\{\frac{1}{R^n}\int_{\Omega_{1/8}}|\vec{v}|+\frac{R}{R^n}\int_{\Omega_{1/8}}|\nabla\vec{v}|\Big\}. \tag{1.12}$$

PROOF: We will first use Korn's inequality (Lemma 1.18 in Dahlberg, *et al.*, 1988). We can then choose an antisymmetric matrix B_ρ, (in fact, $B_\rho = ⨍_{\Omega_\rho}\big[\frac{\nabla\vec{u}-\nabla v^t}{2}\big]$, where $⨍$ denotes mean value), so that $\vec{w}_1(X) = \vec{v}(X) - B_\rho X$ verifies

$$\int_{\Omega_\rho} |\nabla \vec{w}_1|^2 \le C \int_{\Omega_\rho} |\nabla \vec{w}_1 + \nabla w_1^t|^2. \tag{1.13}$$

We can also choose a constant vector $\vec{c}_\rho$, (in fact $\vec{c}_\rho = \fint_{\Omega_\rho} \vec{w}_1$) so that , for $\vec{w} = \vec{w}_1 - \vec{c}_\rho$, we have

$$\Big(\frac{1}{R^n} |\vec{w}|^{\overline{p}}\Big)^{\frac{1}{\overline{p}}} \le C\Big(\frac{R^2}{R^n} \int_{\Omega_\rho} |\nabla \vec{w}|^2\Big)^{1/2} \tag{1.14}$$

where $\overline{p} = \frac{2n}{n-2} > 2$ (Poincaré's inequality). Notice that (1.13) also holds for $\vec{w}$, since $\nabla \vec{w}_1 = \nabla \vec{w}$, and that $\nabla \vec{w} + \nabla \vec{w}^t = \nabla \vec{v} + \nabla \vec{v}^t$,

$$\begin{aligned}
\Big(\frac{1}{R^n} \int_{\Omega_\rho} |\vec{v}|^{\overline{p}}\Big)^{1/\overline{p}} &\le \Big(\frac{1}{R^n} \int_{\Omega_\rho} |\vec{w}|^{\overline{p}}\Big)^{1/\overline{p}} + |\vec{c}_\rho| + CR|B_\rho| \\
&\le C\Big(\frac{R^2}{R^n} \int_{\Omega_\rho} |\nabla v + \nabla \vec{v}^t|^2\Big)^{1/2} + |\vec{c}_\rho| + CR|B_\rho| \\
&\le \frac{C}{(\tau-\rho)^N}\Big(\frac{1}{R^n} \int_{\Omega_\tau} |\vec{v}|^2\Big)^{1/2} + |\vec{c}_\rho| + CR|B_\rho|,
\end{aligned}$$

by 1.10. All in all, we have,

$$\Big(\frac{1}{R^n} \int_{\Omega_\rho} |\vec{v}|^{\overline{p}}\Big)^{1/\overline{p}} \le \frac{C}{(\tau-\rho)^N}\Big(\frac{1}{R^n} \int_{\Omega_\tau} |\vec{v}|^2\Big)^{1/2} + \frac{C}{R^n} \int_{\Omega_{1/8}} |\vec{v}| + \frac{CR}{R^n} \int_{\Omega_{1/8}} |\nabla \vec{v}|,$$

where $\overline{p} > 2$. (1.12) is now a simple consequence of this estimate. To see this we use the following fact (see for instance Fabes-Stroock, 1984 for a proof). If

$$\Big(\frac{1}{R^n} \int_{\Omega_\rho} f^{\overline{p}}\Big)^{1/\overline{p}} \le \frac{C}{(\tau-\rho)^N}\Big(\frac{1}{R^n} \int_{\Omega_\tau} f^2\Big)^{1/2},$$

where $f \ge 0, \overline{p} > 2$, then

$$\Big(\frac{1}{R^n} \int_{\Omega_{1/4}} f^2\Big)^{1/2} \le C\Big(\frac{1}{R^n} \int_{\Omega_{1/8}} f\Big).$$

In order to apply this fact, let $I = \frac{1}{R^n}\int_{\Omega_{1/8}}|\vec{v}| + \frac{CR}{R^n}\int_{\Omega_{1/8}}|\nabla\vec{v}|$, $f = (\frac{|\vec{v}|}{I}+1)$. The corollarly follows.

Proof of Lemma 1.6: Let $I = \int_{\partial\Omega}(\nabla\vec{u})^*(|Q|+1)^\alpha\, d\sigma(Q)$. Then

$$I \leq C\int\limits_{|x|<2}(\nabla\vec{u})^*\, d\sigma(Q) + C\sum_{j=1}^{\infty}2^{j\alpha}\int\limits_{\Lambda(2^j)}(\nabla\vec{u})^*\, d\sigma(Q) \leq C + C\sum_{j=1}^{\infty}2^{j\alpha}\int\limits_{\Lambda(2^j)}(\nabla\vec{u})^*,$$

by the L^2 theory (Theorem 1.2 in the graph case). For $R \geq 2$, let $a(R) = \int_{\Lambda(R)}(\nabla\vec{u})^*$, and let $\gamma_2(Q) = \gamma(Q)\backslash\gamma_1(Q)$, $m_i(Q) = \sup\{|\nabla\vec{u}(X)| : X \in \gamma_i(Q)\}$, $i = 1, 2$. Observe first that if $X \in \gamma_2(Q)$, $\vec{u}$ verifies (0.1) in $B = \{Y : |Y - X| \leq \delta R\}$, where $\delta = \delta(M)$ is small enough. Interior Schauder estimates then show that

$$|\nabla\vec{u}(X)| \leq \frac{C}{R^{n+1}}\int_B|\vec{u}(Y)|\, dY \leq \frac{C}{R^n}\int_{\widetilde{\Lambda}(R)}(\vec{u})^*(Q)\, d\sigma(Q),$$

where $\widetilde{\Lambda}(R) = \Lambda(R/2) \cup \Lambda(R) \cup \Lambda(2R)$. Thus, for $Q \in \Lambda(R)$, $m_2(Q) \leq \frac{C}{R^n}\int_{\widetilde{\Lambda}(R)}(\vec{u})^*(Q)\, d\sigma(Q)$, and so, with $\varepsilon_0 = \varepsilon/2 - \varepsilon$, where ε is as in the graph version of Theorem 1.3, we have:

$$\begin{aligned}\int_{\Lambda(R)} m_2(Q)\, d\sigma(Q) &\leq \frac{C}{R}\int_{\widetilde{\Lambda}(R)}(\vec{u})^*(Q)\, d\sigma(Q)\\ &\leq CR\Big(\frac{1}{R^2}\int_{\widetilde{\Lambda}(R)}(\vec{u})^{*2-\varepsilon}\Big)^{1/2-\varepsilon} \leq C\frac{R}{R^{2/(2-\varepsilon)}} = C/R^{\varepsilon_0},\end{aligned}$$

where we have used the fact that $\sigma(\widetilde{\Lambda}(R)) \simeq R^2$, and Theorem 1.3 in the graph case for $p = 2-\varepsilon$. In fact, by Stokes theorem, we have $\vec{u}(X) = \mathcal{D}(\vec{u})(X) + S(T\vec{u})(X)$, where $S(\vec{f})(X) = \int_{\partial\Omega}\Gamma(X-Q)\vec{f}(Q)d\sigma(Q)$, and $\mathcal{D}(\vec{f})(X) = \int_{\partial\Omega}T_Q\Gamma(X-Q)\vec{f}(Q)d\sigma(Q)$; moreover, $\mathcal{D}(\vec{f})\big|_{\partial\Omega} = (\frac{1}{2}I + K^*)\vec{f}$, where, by the L^p version of Theorem 2.6 in (Dahlberg *et al* (1988)), $\frac{1}{2}I + K^*$ and $-\frac{1}{2}I + K^*$ are invertible operators on $L^p(\partial\Omega)$. Hence, $\|\vec{u}\|_{L^p(\partial\Omega)} \leq C\|S(\vec{a})\|_{L^p(\partial\Omega)} \leq C$ since $\int_{\partial\Omega}\vec{a} = 0$ and thus $S(\vec{a})(X) = O(|X|^{1-n})$ for large X. (We are indebted to G. Verchota for this last argument.)

In order to estimate $\int_{\Lambda(R)} m_1(Q)\, d\sigma(Q)$, we will use the L^2 estimate in Theorem 1.2, for the bounded Lipschitz domains Ω_r. Let $\vec{v}$ be the solution given in Theorem

1.2, in Ω_τ, with $T\vec{v}|_{\partial\Omega_\tau} = T\vec{u}|_{\partial\Omega_\tau}$. By our uniqueness result, $\vec{u} = \vec{v} + B_\tau X + \vec{C}_\tau$, where B_τ is the antisymmetric matrix $B_\tau = \fint_{\Omega_\tau}\{\frac{\nabla(\vec{u}-\vec{v})-\nabla(\vec{u}-\vec{v})^t}{2}\}$. Also, let $\tilde{B}_\tau = \fint_{\Omega_\tau}\{\frac{\nabla\vec{u}-\nabla\vec{u}^t}{2}\}$. By the estimate in Theorem 1.2, we have:

$$\int_{\partial\Omega_\tau}(\nabla\vec{v})^{*2} \le C\int_{\partial\Omega_\tau}|T\vec{u}|^2 \le C\int_{\partial\Omega_\tau\cap\Omega}|\nabla\vec{u}+\nabla\vec{u}^t|^2,$$

since $T\vec{u}|_{\partial\Omega_\tau\cap\partial\Omega} \equiv 0$. Thus

$$\begin{aligned}\int_{\Lambda(R)} m_1^2 &\le \int_{\partial\Omega_\tau}(\nabla\vec{v})^{*2} + |B_\tau|^2 \le C\int_{\partial\Omega_\tau}(\nabla\vec{v})^{*2} + |\tilde{B}_\tau|^2 \\ &\le C\int_{\partial\Omega_\tau\cap\Omega}|\nabla\vec{u}+\nabla\vec{u}^t|^2 + \int_{\partial\Omega_\tau}|\tilde{B}_\tau|^2,\end{aligned}$$

where the next to last inequality follows because

$$\left|\fint_{\Omega_\tau}\{\frac{\nabla\vec{v}-\nabla\vec{v}^t}{2}\}\right| \le C\fint_{\partial\Omega_\tau}(\nabla\vec{v})^*.$$

For $h > 0$, but small, write $\Omega_\tau = \Omega^1_{\tau,h}\cup\Omega^2_{\tau,h}$, where $\Omega^1_{\tau,h} = \{(x,y)\in\Omega_\tau : \varphi(x) < y < \varphi(x)+hR\}$. Write also $\tilde{B}_\tau = \tilde{B}_{1,\tau}+\tilde{B}_{2,\tau}$, where $\tilde{B}_{i,\tau} = \frac{1}{|\Omega_\tau|}\int_{\Omega^i_{\tau,h}}\{\frac{\nabla\vec{u}-\nabla\vec{u}^t}{2}\}$. Once more, if $Y\in\Omega^2_{\tau,h}$, $\vec{u}$ verifies (0.1) in $B = \{X : |X-Y| < \delta R\}$, $\delta = \delta(M,h)$. Using again interior Schauder estimates for $Y\in\Omega^2_{\tau,h}$, $|\nabla\vec{u}(Y)| \le \frac{C_h}{R^n}\int_{\tilde{\Lambda}(R)}(\vec{u})^*$, and thus, $|\tilde{B}_{2,\tau}| \le \frac{C_h}{R^n}\int_{\tilde{\Lambda}(R)}(\vec{u})^*$. On the other hand, $|\tilde{B}_{1,\tau}| \le Ch\int_{\tilde{\Lambda}(R)}(\nabla\vec{u})^*$. All in all,

$$\int_{\Lambda(R)} m_1^2 \le Ch^2\Big(\int_{\tilde{\Lambda}(R)}(\nabla\vec{u})^*\Big)^2 R^{n-1} + C_hR^{n-1}\Big(\frac{1}{R}\int_{\tilde{\Lambda}(R)}(\vec{u})^*\Big)^2 + C\int_{\partial\Omega_\tau\cap\Omega}|\nabla\vec{u}+\nabla\vec{u}^t|^2$$

Integrating the above inequality in $\tau\in[1/4,1/2]$, we obtain

$$\Big(\fint_{\Lambda(R)} m_1^2\Big)^{1/2} \le Ch\fint_{\tilde{\Lambda}(R)}(\nabla\vec{u})^* + \frac{C_h}{R}\fint_{\tilde{\Lambda}(R)}(\vec{u})^* + \Big(\frac{C}{R^n}\iint_{\Omega_{1/4}\setminus\Omega_{1/8}}|\nabla\vec{u}+\nabla\vec{u}^t|^2\Big)^{1/2}$$

If we now apply Lemma 1.8, and (1.12), together with the argument used above, we see that,

$$\Big(\fint_{\Lambda(R)} m_1^2\Big)^{1/2} \le Ch\fint_{\tilde{\Lambda}(R)}(\nabla\vec{u})^* + \frac{C_h}{R}\fint_{\tilde{\Lambda}(R)}(\vec{u})^*,$$

and so

$$\int_{\Lambda(R)} m_1 \leq Ch \int_{\widetilde{\Lambda}(R)} (\nabla\vec{u})^* + \frac{C_h}{R} \int_{\widetilde{\Lambda}(R)} (\vec{u})^* \leq Ch \int_{\widetilde{\Lambda}(R)} (\nabla\vec{u})^* + C_h/R^{\varepsilon_0} R^{(n-1)},$$

as in the bound for m_2. In all, $a(R) \leq Ch \int_{\widetilde{\Lambda}(R)} (\nabla\vec{u})^* + C_h/R^{\varepsilon_0}$, where h is yet to be chosen. Thus,

$$I \leq C + Ch \sum_{j=1}^{\infty} 2^{j\alpha} \int_{\widetilde{\Lambda}(2^j)} (\nabla\vec{u})^* + C_h \sum_{j=1}^{\infty} 2^{j\alpha} 2^{-j\varepsilon_0}.$$

Choose now $\alpha_0 < \varepsilon_0$, we then have

$$I \leq C + ChI + C_h,$$

and the lemma follows by choosing h so small that $Ch < 1/2$.

Note that, strictly speaking, we must *a priori* know that $I < \infty$ for the above arugment to apply. To reduce matters to that case, one can approximate our φ by φ with compact support. In that case, it is not hard to see that $|\nabla\vec{u}(X)| = O(|X|^{-n})$ at ∞. We then use the previous argument to show $I \leq C$, C independent of the size of the support of φ, and then pass to the limit.

REMARK: . We can substitute the L^∞ estimates $\|\vec{a}\|_{L^\infty(\partial\Omega)} \leq 1$, $\|\nabla_T\vec{u}\|_{L^\infty(\partial\Omega)} \leq 1$, by the corresponding L^p estimates, for any $2 \leq p < \infty$, and still obtain the same result in Lemma 1.6.

Theorems 1.3 and 1.5 follow from Lemma 1.6 in the same way as in (Dahlberg-Kenig, 1987).

We now turn to the proof of Theorem 1.4. We follow the corresponding arguments in (Pipher-Verchota, in preparation)

LEMMA 1.15. *Let Ω, α_0 be as in Lemma 1.6. Let $G(X,Y)$ be the matrix Green's function for (0.1) constructed in Theorem 3.6 of (Dahlberg, et al., 1988). Then, if $X = (0,1)$, we have*

$$\int_{\partial\Omega} |\nabla_Q G(X,Q)|(1+|Q|)^\alpha \, d\sigma(Q) \leq C$$

for $0 \leq \alpha < \alpha_0$, *where* $C = C(M)$.

PROOF: Let $\Gamma(X,Y)$ be the Kelvin matrix fundamental solution of (0.1). (See Dahlberg, *et al.*, 1988, (0.1)). Then $|\Gamma(X,Y)| \simeq |X-Y|^{2-n}$, $|\nabla\Gamma(X,Y)| \simeq |X-Y|^{1-n}$.

Moreover, $G(X,Y) = \Gamma(X,Y) - v^X(Y)$, where v^X is the matrix valued solution to the Dirichlet problem for (0.1), with $v^X\big|_{\partial\Omega} = \Gamma(X,-)|_{\partial\Omega}$. By the above estimate on $\nabla\Gamma$, the fact that $\operatorname{dist}(X,\partial\Omega) \cong 1$, and Theorem 1.3 in the graph case, there exist $p = p(M) > 2$ so that $\int_{\partial\Omega}[(\nabla v^X)^*]^p \leq C$, $C = C(M)$. Let us now find a Lipschitz function ψ, with $\varphi \leq \psi$, $\varphi(x) = \psi(x)$ for $|x| \geq 4$, $\|\nabla\psi\|_\infty \leq 10M$, $X \in {}^c\tilde{\Omega}$, where $\tilde{\Omega} = \{(x,y) : y > \psi(x)\}$, and $\operatorname{dist}(X,\partial\tilde{\Omega}) \simeq 1$. We are now ready to prove our estimate:

$$\int_{\partial\Omega\cap\{(x,\varphi(x)):|x|<4\}} |\nabla_Q G(X,Q)|(1+|Q|)^\alpha \, d\sigma(Q) \leq C,$$

by the pointwise bound on $|\nabla\Gamma|$ and the above bound on $(\nabla v^X)^*$. For the rest, let $u(Y) = G(X,Y)$, $Y \in \tilde{\Omega}$. We know that

$$\int_{\partial\tilde{\Omega}\cap\{(x,\psi(x)):|x|<4\}} |\nabla_T u|^p \leq C,$$

where p is as above, by the pointwise bound for $|\nabla\Gamma|$, and the above bound on $(\nabla v^X)^*$. Moreover, $\operatorname{supp} u|_{\partial\tilde{\Omega}} \subset \{(x,\psi(x)) : |x| < 4\}$, since $G(X,-)|_{\partial\Omega} \equiv 0$. Lemma 1.6 (see the Remark after its proof) gives the desired estimate.

PROOF OF THEOREM 1.4: We start with the case $\vec{g} \in L^\infty(\partial\Omega)$. Since the class of Lipschitz graphs is dilation and translation invariant, it is enough to see that, if Ω, X are as in Lemma 1.15, then

$$|\vec{u}(X)| \leq C\|\vec{g}\|_{L^\infty(\partial\Omega)}, \quad C = C(M)$$

But (see Dahlberg, *et al.*, 1988)

$$\vec{u}(X) = \int_{\partial\Omega} \frac{\partial G}{\partial\rho_Q}(X,Q)\vec{g}(Q)\, d\sigma(Q),$$

where $|\frac{\partial G}{\partial \rho_Q}(X,Q)| \le C|\nabla_Q G(X,Q)|$. Thus, the case $\alpha = 0$ of Lemma 1.15 implies the desired bound. Let us now turn to the case when $\vec{g} \in \Lambda^\alpha(\partial\Omega)$, where $0 < \alpha < \alpha_0$, α_0 as in Lemma 1.15. Recall that

$$\|\vec{g}\|_{\Lambda^\alpha(\partial\Omega)} = \sup_{Q,Q' \in \partial\Omega} \frac{|\vec{g}(Q) - \vec{g}(Q')|}{|Q - Q'|^\alpha}$$

As is well known, it suffices to show that

$$\sup_{X \in \Omega} \delta(X)^{\alpha-1} |\nabla \vec{u}(X)| \le C \|\vec{g}\|_{\Lambda^\alpha(\partial\Omega)}$$

Fix $X \in \Omega$, and let $R = \delta(X)$. Using translations and dilations once more, we see that it is enough to see that if Ω, X are as in Lemma 1.15, then

$$|\nabla \vec{u}(X)| \le C\|\vec{g}\|_{\Lambda^\alpha(\partial\Omega)}, \quad C = C(M)$$

In order to show this, note that

$$\nabla \vec{u}(X) = \int_{\partial\Omega} \nabla_X \frac{\partial}{\partial \rho_Q} G(X,Q)[\vec{g}(Q) - \vec{g}(0)]\, d\sigma(Q).$$

If we now use interior Schauder estimates in $B = \{Y : |X - Y| < \delta\}$, $\delta = \delta(M)$, together with the estimate $|\vec{g}(Q) - \vec{g}(0)| \le C|Q|^\alpha \|\vec{g}\|_{\Lambda^\alpha(\partial\Omega)}$ and Lemma 1.15, our estimate follows.

REFERENCES

B. E. J. Dahlberg, (1979). On the Poisson integral for Lipschitz and C^1 domains, *Studia Math.*, 13–24.

B. E. J. Dahlberg and C. E. Kenig (1987). Hardy spaces and the Neumann problem in L^p for Laplace's equation in Lipschitz domains, *Annals of Math.*, **125**, 437–465.

B. E. J. Dahlberg, C. E. Kenig and G. C.Verchota (1986). The Dirichlet problem for the biharmonic equation in a Lipschitz domain, *Annales de l'Institut Fourier*, Grenoble, **36**, 109–135.

B. E. J. Dahlberg, C. E. Kenig and G. C. Verchota (1988). Boundary value problems for the systems of elastostatics in Lipschitz domains, *Duke Math. Jour.,* **57**, 795–818.

E.B. Fabes, C. E. Kenig and G. C. Verchota (1988). The Dirichlet problem for the Stokes system on Lipschitz domains, *Duke Math. Jour.*, **57**, 769–793.

E. B. Fabes and D. W. Stroock (1984). The L^p integrability of Green's function and fundamental solutions for elliptic and parabolic equations, *Duke Math. J.*, **51**, 997–1016.

D. S. Jerison and C. E. Kenig (1981). The Neumann problem on Lipschitz domains, *Bull. AMS*, **4**, 203-207.

J. Pipher and G. C. Verchota The Dirichlet problem in L^p for biharmonic functions on Lipschitz domains, preprint.

J. Pipher and G. C. Verchota The maximum principle for biharmonic functions, in preparation.

G. C. Verchota (1984). Layer potentials and regularity for the dirichlet problem for Laplace's equation in Lipschitz domains, *J. of Funct. Anal.*, **59**, 572–611.

32

Growth and Geometry of Eigenfunctions of the Laplacian

H. Donnelly[1] and C. Fefferman[2]

Let M be a smooth, connected, compact Riemannian manifold without boundary. A principal theme of our paper [DF1] is that an eigenfunction of the Laplacian satisfying $-\Delta u=\lambda u$ on M behaves like a polynomial of degree $\sqrt{\lambda}$. This is very natural in view of elementary examples such as spherical harmonics. Already the simplest polynomial $u(t)=t^{\sqrt{\lambda}}$ on $[-1,1]$ leads us to guess the basic growth properties of eigenfunctions. The prototype of such growth properties is the estimate

$$\mathrm{MAX}_{B(x,r)}|u| \leq \left(\frac{Cr}{r'}\right)^{C\sqrt{\lambda}} \cdot \mathrm{MAX}_{B(x,r')}|u| \quad \text{for } o<r'<r, \tag{0}$$

which we proved in [DF1]. Our purpose here is to prove more refined estimates, all motivated by the properties of $t^{\sqrt{\lambda}}$. We shall apply our estimates to the geometry of nodal domains. Recall that the nodal domains for a real-valued eigenfunction u are defined as the connected components of the set where $u\neq 0$. Our results are as follows.

Theorem 1 (Growth Properties): Let M and u be as above. Then for $r<r_0$ we have

---

Both authors partially supported by NSF.
[1]Dept. of Mathematics, Purdue University
[2]Dept. of Mathematics, Princeton University

(A) $\int_{B\left(x,(1+\frac{1}{\sqrt{\lambda}})r\right)} |u|^2 \le C\int_{B(x,r)} |u|^2$

(B) (L^2 - Bernstein's inequality) $\left(\int_{B(x,r)} |\nabla u|^2\right)^{1/2} \le$

$\frac{C\sqrt{\lambda}}{r} \left(\int_{B(x,r)} |u|^2\right)^{1/2}$

(C) (Weak L^∞-Bernstein inequality) $\text{MAX}_{B(x,r)} |\nabla u| \le \frac{C\lambda^K}{r} \text{MAX}_{B(x,r)} |u|$

with $K = \frac{n+2}{4}$, n=dim M.

(D) (BMO estimate for $\log|u|$) Let E be a measurable subset of a ball B. Then $\text{MAX}_B|u| \le \left(\frac{C \text{ vol}B}{\text{vol}E}\right)^{C\lambda^K} \text{MAX}_E|u|$, with $K = \frac{n(n+2)}{4}$.

The constant r_0 depends only on the manifold M.

In (C) and (D) we expect $K=1/2$, but we do not know how to prove such sharp results. It would be interesting to obtain L^p or even L^∞ versions of (A) and (B). S. Chanillo has pointed out to us that our deduction of (D) from (C) is close to the previous work of Garafolo and Lin [GF], which may lead to a better constant K.

Theorem 2 (Geometry of Nodal Domains): Let M, u, λ be as above, with u real-valued. Pick any ball $B \subset M$, and let $\Omega \subset B$ be any of the connected components of $\{x \in B | u(x) \neq 0\}$. If Ω intersects the middle half of B, then $\text{vol}\Omega \ge c\lambda^{-K} \text{vol}B$ with $K = \frac{n+(n+2)n^2}{2}$.

(Again, the value of K is clearly not best possible.)

To understand the meaning of Theorem 2, let $M=\mathbb{R}^2/Z^2$ be a flat torus, and let B be a disc of radius $1/2$. Then $u(x,y)=e^{Sx}\cos(Ty)$, defined in B, is a local eigenfunction of the Laplacian, with eigenvalue $\lambda = T^2-S^2$. Suppose we fix λ and let S and T tend to infinity. The components of $\{z \in B | u(z) \neq 0\}$ are horizontal strips of thickness $\sim T^{-1}$, which can be arbitrarily small. Theorem 2 shows that the restriction of a global eigenfunction to B cannot behave like this. Our proof of Theorem 2 is based on the rapid growth of u in very thin components $\Omega \subset B$, which is clear in the above example.

Note that Courant's nodal domain theorem is of no use here (though one is strongly tempted to apply it), since a nodal domain in M can split into many components when intersected with B. Theorem 2 may be viewed as a local analogue of Courant's nodal domain theorem.

Our results may be extended to manifolds with boundary using the ideas in [DF2], but we do not pursue this. Also, we make no attempt to minimize our assumptions on the smoothness of the manifold M. This awaits results with sharp dependence upon λ. Thus, we suppose that M is covered by coordinate charts in which the metric has bounded geometry. Recall that a metric $ds^2 = \sum_{ij} g_{ij}\, dx^i dx^j$ on the unit ball $\bar{B} = \{|x| \le 1\} \subset \mathbb{R}^n$ has bounded geometry if

$$\|g_{ij}\|_{C^{m^*}(\bar{B})},\ \|g^{ij}\|_{C^{m^*}(\bar{B})} \le C^* \tag{BG}$$

for large constants m^* and C^*.

The rest of this paper is devoted to the proofs of Theorems 1 and 2.

We begin with the proof of Theorem 1(A). The basic estimate which we need is as follows.

Lemma A (Carleman inequality): Let u be a smooth function of compact support on a small ball $B(p,h) \subset M$, and define $r(x) = \mathrm{dist}(x,p)$, $\bar{r}(x)=\int_0^{r(x)} \exp(-\nu s^2)ds = r(x) + O\left(r^3(x)\right)$ for a large constant ν. If $\alpha > C(1+\sqrt{\lambda})$, and if u vanishes in a ball $B(p,\delta) \subset B(p,h)$, then we have the estimate

$$\int_{B(p,h)} \left(\bar{r}(x)\right)^{-2\alpha} (\Delta u+\lambda u)^2 \, dvol(x) \geq c\alpha^4 \int_{B(p,\delta+\frac{c\delta}{\alpha})} \bar{r}(x)^{-2\alpha-4} |u(x)|^2 dvol(x) .$$

Proof: We make technical improvements in the proof of Prop. 2.10 in [DF1], which in turn comes from Aronszajn [A]. Let us recall the set-up from [DF1]. In $B(p,h)$ we introduce geodesic polar coordinates $(r,t_1 \ldots t_{n-1})$, where $r = r(x)$ and $(t_1 \ldots t_{n-1})$ are coordinates on the standard unit sphere S^{n-1}.

Our metric and its volume element take the form $ds^2 = dr^2 + r^2\gamma_{ij}dt_i dt_j$, $dvol = r^{n-1}\sqrt{\gamma}drdt$, with $\gamma = \det(\gamma_{ij})$. We introduce a local conformal change in the metric and volume element. For a large positive constant ν, define $\bar{g}_{ij} = \exp(-2\nu r^2)g_{ij}$. The geodesic lines starting from p coincide for the two metrics, and one has $\bar{r}= \int_0^r e^{-\nu s^2}ds = r+O(r^3)$ near the origin. The modified volume element $\overline{dvol}$ is obtained by multiplying the volume element of $(\bar{g}_{ij})$ by the factor $\psi = \exp\left(\frac{2}{3}\nu(n-2)\bar{r}^2\right)$. In geodesic polar coordinates one has

$$\overline{ds^2} = \overline{dr^2} + \bar{r}^2\bar{\gamma}_{ij}\, dt_i dt_j \quad , \quad \overline{dvol} = \bar{r}^{n-1}\psi\sqrt{\bar{\gamma}}\; \overline{dr}dt .$$

Let $\bar{\Delta}$ denote the Laplacian for $(\bar{g}_{ij})$, and let $\bar{\lambda} = \lambda \exp(2\nu r^2)$. Suppose $u \in C_0^\infty\big(B(p,h)\big)$, and assume $u=0$ near p. For $\alpha > C(1+\sqrt{\lambda})$, we want to derive a lower bound for the integral

$$I = \iint \bar{r}^{-2\alpha} \left|(\bar{\Delta}+\bar{\lambda})u\right|^2 \bar{r}^{n-1} \sqrt{\gamma}\ \psi\ d\bar{r}dt \quad . \tag{1}$$

Writing $\bar{r} = e^{-p}$, $\beta = \alpha - \frac{n}{2} + 2$, $u = e^{-\beta p} w$, $\theta = \frac{\partial}{\partial p}(\log\sqrt{\gamma})$, and writing w', w'' for $\frac{\partial w}{\partial p}$, $\frac{\partial^2 w}{\partial p^2}$, we may express (1) in polar coordinates to obtain the formula:

$$I = \iint \left|w'' - (n-2+2\beta-\theta)w' + \beta(\beta+n-2-\theta)w + \Delta_p w + \bar{\lambda} e^{-2p} w\right|^2 \sqrt{\gamma}\ \psi\ dpdt.$$

Here $\Delta_p w = \frac{1}{\sqrt{\gamma}} \frac{\partial}{\partial t_i}\left(\sqrt{\gamma}\ \gamma^{ij} \frac{\partial w}{\partial t_i}\right)$.

Removing the θ terms from I gives

$$I_o = \iint \left|w'' - (n-2+2\beta)w' + \beta(\beta+n-2)w + \Delta_p w + \bar{\lambda} e^{-2p} w\right|^2 \sqrt{\gamma}\ \psi\ dpdt.$$

Clearly, by the triangle inequality,

$$I \geq \frac{1}{2} I_o - I_1 \text{ with } I_1 = \iint \theta^2 \left|w' - \beta w\right|^2 \sqrt{\gamma}\ \psi\ dpdt.$$

The discussion of Prop. 2.10 in [DF1] shows that $I_1 < \frac{1}{4} I_o$ for small h. Therefore $I \geq \frac{1}{4} I_o$. Now we write (for a function f to be picked later)

$$I_o = I_2 + I_3 + I_4 + I_5 \qquad \text{with}$$

$$I_2 = \iint \left|w'' + \beta(\beta+n-2)w + \Delta_p w + \bar{\lambda}\ e^{-2p}\ w\right|^2 \sqrt{\gamma}\ \psi\ dpdt$$

$$I_3 = (2\beta+n-2)^2 \iint \left|w' + fw\right|^2 \sqrt{\gamma}\ \psi\ dpdt$$

$$I_4 = -2(2\beta+n-2) \iint w' \left[w''+\beta(\beta+n-2)w+\Delta_p w+\bar{\lambda}e^{-2p}w+(2\beta+n-2)fw\right] \sqrt{\bar{\gamma}}\, \psi \, dpdt$$

$$I_5 = -(2\beta+n-2)^2 \iint f^2 w^2 \sqrt{\bar{\gamma}}\, \psi \, dpdt.$$

This elementary formula and our eventual choice of f are different from their analogues in [DF1], and are responsible for the improvement in our results.

Integration by parts in t gives

$$I_4 = -2(2\beta+n-2) \iint \Big[w'w''+\beta(\beta+n-2)w'w - \bar{\gamma}^{ij} \frac{\partial w'}{\partial t_i} \frac{\partial w}{\partial t_j} + \bar{\lambda}e^{-2p}w'w + (2\beta+n-2)fw'w\Big]\sqrt{\bar{\gamma}}\, \psi \, dpdt.$$

Integration by parts in p then shows that

$$I_4 = +(2\beta+n-2)\iint\Big[(w')^2(\psi\sqrt{\bar{\gamma}})'+\beta(\beta+n-2)w^2(\psi\sqrt{\bar{\gamma}})'-(\psi\sqrt{\bar{\gamma}}\,\bar{\gamma}^{ij})' \frac{\partial w}{\partial t_i} \frac{\partial w}{\partial t_j} + (\bar{\lambda}e^{-2p}\psi\sqrt{\bar{\gamma}})'w^2 + (2\beta+n-2)w^2 (f\psi\sqrt{\bar{\gamma}})\Big] dtdp.$$

For $\nu >> 1$ one has $(\psi\sqrt{\bar{\gamma}})'>0$, and $-(\psi\sqrt{\bar{\gamma}}\,\bar{\gamma}^{ij})'>0$ as a matrix. (See [DF1] and Aronszajn [A] for these local calculations.)

Hence $I_4 \geq (2\beta+n-2) \iint \left[(2\beta+n-2)\ (f\psi\sqrt{\bar{\gamma}})' + (\bar{\lambda}e^{-2p}\, \psi\, \sqrt{\bar{\gamma}})'\right]w^2\, dtdp$

$$= (2\beta+n-2) \iint \left[(2\beta+n-2)f' + (2\beta+n-2)\mu f + \frac{(\bar{\lambda}e^{-2p}\psi\sqrt{\bar{\gamma}})'}{\psi\sqrt{\bar{\gamma}}} \right]w^2\psi\sqrt{\bar{\gamma}}\, dtdp$$

with $\mu = \frac{\partial}{\partial p} \ln(\psi\sqrt{\bar{\gamma}})$.

Since I_2 and I_3 are obviously positive, we conclude that

$$I_0 = I_2 + I_3 + I_4 + I_5 \geq$$

$$(2\beta+n-2)^2 \iint \left[f'+ \mu f + \frac{(\bar{\lambda}e^{-2p}\psi\sqrt{\bar{\gamma}})'}{(2\beta+n-2)\psi\sqrt{\bar{\gamma}}} - f^2\right]w^2\, \psi\, \sqrt{\bar{\gamma}}\, dtdp \qquad (2).$$

Suppose $\beta > C_{\#}(1+\sqrt{\lambda})$ for a large constant $C_{\#}$. Then

$$|(2\beta+n-2)^2 \iint \frac{(\bar{\lambda}e^{-2p}\psi\sqrt{\bar{\gamma}})'}{(2\beta+n-2)} w^2\psi\sqrt{\bar{\gamma}}\, dtdp| \le c_{\#}\beta^3 \iint e^{-2p}w^2\psi\sqrt{\bar{\gamma}}\, dtdp, \quad (2a)$$

with $c_{\#}$ small for $C_{\#}$ large. Now we make use of the discussion of I_0 given in [DF1]. By looking at the term called I_2 in [DF1] we see that

$$I_0 \ge \iint 2(\beta(\beta+n-2)+\bar{\lambda}e^{-2p})w\cdot(2\beta+n-2)f^{\#}\, w\cdot\psi\sqrt{\bar{\gamma}}\, dtdp$$

with $f^{\#} = \frac{1}{2}\frac{\partial}{\partial p}\ln(1+\frac{\bar{\lambda}}{\beta(\beta+n-2)}e^{-2p}) + \frac{1}{2}\mu$ and $\mu \ge 100\, e^{-2p}$. Hence, after possibly taking $C_{\#}$ larger, we get $f^{\#} \ge 10\, e^{-2p}$, and therefore $I_0 \ge 10\,\beta^3 \iint e^{-2p}\, w^2\psi\sqrt{\bar{\gamma}}\, dtdp$. Comparing this with (2a) we get $|(2\beta+n-2)^2 \iint \frac{(\bar{\lambda}e^{-2p}\psi\sqrt{\bar{\gamma}})'}{(2\beta+n-2)} w^2\psi\sqrt{\bar{\gamma}}\, dtdp| \le \frac{1}{10} I_0$. Therefore, (2) yields

$$2I_0 \ge (2\beta+n-2)^2 \iint \left[f'+\mu f-f^2\right] w^2\psi\sqrt{\bar{\gamma}}\, dtdp. \quad (2b)$$

We shall pick f to make the expression in square brackets in (2b) equal to a desired weight function. Specifically, let $\phi(t)$ be a smooth function on $\mathbb{R}^1$, with $\phi(t)=0$ for $t\le 0$ and $\phi(t)>0$ for $t>0$. For an arbitrary large number p_*, we attempt to solve

$$f' + \mu f - f^2 = \beta^2\phi\big(\beta(p-p_*)\big) \quad (3)$$

with $f = 0$ for $p \le p_*$.

Making the change of scale $f = \beta F$, $z = \beta(p-p_*)$ transforms (3) into an equation of the form

$$\frac{dF}{dz} = H(z) + G(z)F + F^2 , \quad F(o) = 0$$

with H and G smooth functions, uniformly in C^2. Consequently, the standard existence theorem for ordinary differential equations gives a solution to (3) for $0 \le \beta(p-p_*) \le c_1$, with c_1 a fixed, small positive constant. That is, (3) can be solved for $0 \le p \le p_* + \frac{c_1}{\beta}$.

Therefore, if $\operatorname{supp} W \subset \{0 \le p \le p_* + \frac{c_1}{\beta}\}$, then (2b) shows that

$$2I_o \ge (2\beta+n-2)^2 \beta^2 \iint \phi\Big(\beta(p-p_*)\Big) \, |w|^2 \psi\sqrt{\overline{\gamma}} \, dtdp.$$

For $0 < c_2 < c_3 < c_1$ we have $\phi(z) > c_4$ for $c_2 \le z \le c_3$ and $c_4 = c_4(c_2,c_3)$. Hence, the previous estimate yields

$$I_o \ge c_4 \, \beta^4 \iint_{p_*+\frac{c_2}{\beta} \le p \le p_* + \frac{c_3}{\beta}} |w|^2 \, \psi\sqrt{\overline{\gamma}} \, dtdp.$$

Setting $r_*=e^{-p_*} << 1$, we recall the definitions of w and p to rewrite the above in the form:

$$I_o \ge c_5\beta^4 \iint_{r_*(1-\frac{c_6}{\beta}) < \overline{r} < r_*(1-\frac{c_7}{\beta})} \overline{r}^{-2\alpha-4} \, u^2 \, \overline{r}^{n-1}\sqrt{\overline{\gamma}} \, \psi \, d\overline{r}dt \qquad (c_6>c_7>0 \text{ and } c_5>o).$$

Recalling $I \ge \frac{1}{4}I_o$ and the definition of I, we get:

$$\iint \overline{r}^{-2\alpha}\left|(\overline{\Delta}+\overline{\lambda})u\right|^2 \overline{r}^{n-1}\sqrt{\overline{\gamma}} \, \psi \, d\overline{r}dt \ge \frac{c_5}{4} \beta^4 \iint_A \overline{r}^{-2\alpha-4}|u|^2 \, \overline{r}^{n-1}\sqrt{\overline{\gamma}} \, d\overline{r}dt$$

$$A = \left\{r_*(1-\frac{c_6}{\beta})<\overline{r}<r_*(1-\frac{c_7}{\beta})\right\} \tag{4}$$

We proved this assuming $u \in C_0^\infty\left(\left\{r_*(1-\frac{c_6}{\beta}) < \bar{r} < h\right\}\right)$ with $h \ll 1$. Passing from $(\bar{\Delta}+\bar{\lambda})u$ to $(\Delta+\lambda)u$ as in [DF1], and picking p_* so that $r_* = e^{-p_*} = \bar{\delta}/(1-\frac{c_6}{\beta})$, we obtain

$$\int_{B(p,h)} \bar{r}^{-2\alpha} \left|\Delta u+\lambda u\right|^2 \text{dvol} \geq c_8\alpha^4 \int_{\bar{\delta} < \bar{r} < \bar{\delta} + \frac{c_9\delta}{\alpha}} \bar{r}^{-2\alpha-4} |u|^2 \text{ dvol} .$$

This is equivalent to the conclusion of the Lemma, since for $\bar{\delta}=\delta+O(\delta^3)$ we have $B(p,\delta)=\{\bar{r} < \bar{\delta}\}$, $B(p, \delta + \frac{c_{10}\delta}{\alpha}) \backslash B(p,\delta) \subset \left\{\bar{\delta}<\bar{r}<\bar{\delta}+\frac{c_9\bar{\delta}}{\alpha}\right\}$. The proof of the Lemma is complete. □

The proof of inequality (A) rests on the above Lemma, and on the cruder Carleman inequality

$$\int_{B(p,h)} \bar{r}^{-2\alpha} (\Delta u+\lambda u)^2 \text{ dvol} \geq c \int_{B(p,h)} \bar{r}^{-2\alpha}u^2 \text{ dvol},$$ which

we proved in [DF1], in the setting of Lemma (A). Combining our two Carleman inequalities, we have

$$\int_{B(p,h)} \bar{r}^{-2\alpha}(\Delta u+\lambda u)^2 \text{ dvol} \geq c\int_{B(p,h)} \bar{r}^{-2\alpha}u^2 \text{ dvol} + c\alpha^4 \int_{B(p,\delta+\frac{c^*\delta}{\alpha})} \bar{r}^{-2\alpha-4} |u|^2 \text{ dvol} \tag{5}$$

for $u \in C_0^\infty\left(B(p,h)\backslash B(p,\delta)\right)$ and $\alpha > C(1+\sqrt{\lambda})$.

Now suppose F is a global eigenfunction, satisfying $-\Delta F = \lambda F$ on M. We introduce a cutoff function $\chi \in C_0^\infty\left(B(p,h)\backslash B(p,\delta)\right)$, and then apply the basic estimate (1) to $u = \chi F$. Note that $\Delta u+\lambda u = \chi(\Delta F+\lambda F) + 2\nabla\chi\cdot\nabla F + F\Delta\chi = 2\nabla\chi\cdot\nabla F + F\Delta\chi$. We may take χ to satisfy the following conditions:

$$\chi = 1 \text{ in } B(p,\tfrac{h}{2})\backslash B\left(p,\delta + \frac{c^{\#}\delta}{10\alpha}\right);$$

$$|\nabla\chi| \le \frac{C\alpha}{\delta} \text{ in } B\left(p,\delta + \frac{c^{\#}\delta}{10\alpha}\right);$$

$$|\nabla\chi| \le C \text{ in } B(p,h)\backslash B(p,{}^{h}/_{2}).$$

Thus, (5) applied to $u = \chi F$ yields the following:

$$\int_{B(p,h)\backslash B(p,{}^{h}/_{2})} \bar{r}^{-2\alpha}(|\nabla F|^2 + |F|^2)\, dvol + \int_{B\left(p,\delta+\frac{c^{\#}\delta}{10\alpha}\right)\backslash B(p,\delta)} \bar{r}^{-2\alpha}\left(\frac{\alpha^2}{\delta^2}|\nabla F|^2 + \frac{\alpha^4}{\delta^4}|F|^2\right) dvol \ge$$

$$c\int_{B(p,{}^{h}/_{2})\backslash B\left(p,\delta+\frac{c^{\#}\delta}{10\alpha}\right)} \bar{r}^{-2\alpha}|F|^2\, dvol + c\alpha^4 \int_{B\left(p,\delta+\frac{c^{\#}\delta}{10\alpha}\right)\backslash B\left(p,\delta+\frac{c^{\#}\delta}{10\alpha}\right)} \bar{r}^{-2\alpha-4}|F|^2\, dvol \qquad (6)$$

Now $\displaystyle\int_{B(p,h)\backslash B\left(p,\frac{h}{2}\right)} \bar{r}^{-2\alpha}|\nabla F|^2\, dvol \le \left(\max_{B(p,h)\backslash B\left(p,\frac{h}{2}\right)} \bar{r}^{-2\alpha}\right)\int_M |\nabla F|^2\, dvol$

$$= \left(\max_{B(p,h)\backslash B\left(p,\frac{h}{2}\right)} \bar{r}^{-2\alpha}\right)\lambda\|F\|^2_{L^2(M)} \le C\lambda\left(\max_{B(p,h)\backslash B\left(p,\frac{h}{2}\right)} \bar{r}^{-2\alpha}\right)\max_M|F|^2 \qquad (7)$$

while for a suitable ball $B' \subset B\left(p,\frac{h}{10}\right)\backslash B\left(p,\frac{h}{20}\right)$ we have

$\int_{B'} |F|^2\, dvol \ge c(h)\cdot e^{-C\sqrt{\lambda}}\max_M|F|^2$ by the results of [DF1], and therefore

$$\int_{B\left(p,\frac{h}{2}\right)\backslash B\left(p,\delta+\frac{c^{\#}\delta}{10\alpha}\right)} \bar{r}^{-2\alpha}|F|^2\, dvol \ge c(h)\left(\text{MIN}_{B'}\, \bar{r}^{-2\alpha}\right)e^{-C\sqrt{\lambda}}\max_M|F|^2 \qquad (8).$$

Comparing (7) and (8), we see that

$$\int_{B\left(p,\frac{h}{2}\right)\setminus B\left(p,\delta+\frac{c^{\#}\delta}{10\alpha}\right)} \bar{r}^{-2\alpha} |F|^2 \, dvol \geq \int_{B(p,h)\setminus B\left(p,\frac{h}{2}\right)} \bar{r}^{-2\alpha}\left(|\nabla F|^2 + |F|^2\right) dvol \tag{9}$$

provided we take $\alpha \geq C'\sqrt{\lambda}$. From (6) and (9), we now obtain

$$\frac{C\alpha^4}{\delta^4} \int_{B\left(p,\delta+\frac{c^{\#}\delta}{10\alpha}\right)\setminus B(p,\delta)} \bar{r}^{-2\alpha-4}|F|^2 dvol + \frac{C\alpha^2}{\delta^2} \int_{B\left(p,\delta+\frac{c^{\#}\delta}{10\alpha}\right)\setminus B(p,\delta)} \bar{r}^{-2\alpha} |\nabla F|^2 \, dvol \geq c\alpha^4 \int_{B\left(p,\delta+\frac{c^{\#}\delta}{\alpha}\right)\setminus B\left(p,\delta+\frac{c^{\#}\delta}{10\alpha}\right)} \bar{r}^{-2\alpha-4} |F|^2 \, dvol \tag{10}$$

for $\alpha \geq C'\sqrt{\lambda}$. Fixing $\alpha = C'\sqrt{\lambda}$ and observing that $\bar{r}^{-2\alpha}$ has constant order of magnitude in $B\left(p,\delta + \frac{c^{\#}\delta}{\alpha}\right)\setminus B(p,\delta)$, we can simplify (10) to the form:

$$\frac{\alpha^2}{\delta^2} \int_{B\left(p,\delta+\frac{c^{\#}\delta}{10\alpha}\right)\setminus B(p,\delta)} |F|^2 dvol + \int_{B\left(p,\delta+\frac{c^{\#}\delta}{10\alpha}\right)\setminus B(p,\delta)} |\nabla F|^2 \, dvol \geq c\,\frac{\alpha^2}{\delta^2} \int_{B\left(p,\delta+\frac{c^{\#}\delta}{\alpha}\right)\setminus B\left(p,\delta + \frac{c^{\#}\delta}{10\alpha}\right)} |F|^2 \, dvol \tag{11}$$

On the other hand, elementary elliptic theory shows that

$$|\nabla F(x)|^2 \leq C\left(\frac{\alpha}{\delta}\right)^{n+2} \int_{y\in B\left(x,\ \frac{c^{\#}\delta}{10\alpha}\right)} |F(y)|^2 \, dvol\ (y) \tag{12}$$

since $-\Delta F = \lambda F$ and $\frac{c^{\#}\delta}{10\alpha} < \lambda^{-1/2}$. To verify (12), just rescale to the unit ball and note that the rescaled eigenvalue is $O(1)$.

Integrating (12) over $x \in B\left(p,\delta + \frac{c^{\#}\delta}{10\alpha}\right)\setminus B(p,\delta)$ and changing the order of integration, we obtain

$$\int_{B\left(p,\delta+\frac{c^{\#}\delta}{10\alpha}\right)\setminus B(p,\delta)} |\nabla F(x)|^2 \, dvol(x) \le C\left(\frac{\alpha}{\delta}\right)^{n+2} \int_{\substack{x\in B\left(p,\delta+\frac{c^{\#}\delta}{10\alpha}\right)\setminus B(p,\delta) \\ y\in B\left(x,\frac{c^{\#}\delta}{10\alpha}\right)}} |F(y)|^2 \, dvol(y)\, dvol(x)$$

$$\le C\frac{\alpha^2}{\delta^2} \int_{y\in B\left(p,\delta+\frac{c^{\#}\delta}{5\alpha}\right)\setminus B\left(p,\delta-\frac{c^{\#}\delta}{10\alpha}\right)} |F(y)|^2 \, dvol(y) .$$

Putting this into (11), we get

$$C\int_{B\left(p,\delta+\frac{c_{\#}\delta}{5\alpha}\right)\setminus B\left(p,\delta-\frac{c_{\#}\delta}{10\alpha}\right)} |F(y)|^2 \, dvol(y) \ge \int_{B\left(p,\delta+\frac{c_{\#}\delta}{\alpha}\right)\setminus B\left(p,\delta+\frac{c^{\#}\delta}{10\alpha}\right)} |F(x)|^2 \, dvol(x) .$$

Adding $\int_{B\left(p,\delta+\frac{c^{\#}\delta}{10\alpha}\right)} |F|^2 \, dvol$ to both sides of the inequality, we find that

$C \int_{B\left(p,\delta+\frac{c_{\#}\delta}{5\alpha}\right)} |F|^2 \, dvol \ge \int_{B\left(p,\delta+\frac{c_{\#}\delta}{\alpha}\right)} |F|^2 \, dvol$. Applying this with $\delta/\left(1+\frac{c_{\#}}{5\alpha}\right)$ in place of δ, we get

$$C\int_{B(p,\delta)} |F|^2 \, dvol \ge \int_{B\left(p,\delta+\frac{c\delta}{\alpha}\right)} |F|^2 \, dvol . \tag{13}$$

Recalling that $\alpha = C'\sqrt{\lambda}$ and iterating (13) finitely many times yields

$C'' \int_{B(p,\delta)} |F|^2 \, dvol \ge \int_{B\left(p,\delta+\frac{\delta}{\sqrt{\lambda}}\right)} |F|^2 \, dvol$, which is the desired inequality (A). □

Next we prove the Bernstein inequalities (B) and (C). From elementary elliptic theory, we recall the following result.

Lemma: Suppose ds^2 has bounded geometry on the unit ball $B \subset \mathbb{R}^n$, and suppose $-\Delta_g u=\lambda u$ in B, with $|\lambda| \leq 1$. Then $|\nabla u(o)|^2 \leq C\int_B |u(y)|^2\, dy$.

Rescaling to a small ball, we deduce the following consequence.

Cor: Suppose ds^2 has bounded geometry, and suppose $-\Delta_g u=\lambda u$ in $B(x,r)$ with $|\lambda| r^2 \leq 1$. Then

$$|\nabla u(x)|^2 \leq \frac{C}{r^{2+n}} \int_{B(x,r)} |u(y)|^2\, dy\ . \tag{14}$$

We already used this inequality in the proof of Theorem 1(A).

To derive inequality (B), we apply (14) with $r = \frac{\delta}{\sqrt{\lambda}}$ and then integrate over all $x \in B(x_0,\delta)$. The result is

$$\int_{B(x_0,\delta)} \left|\nabla u(x)\right|^2 dx \leq \frac{C}{r^{n+2}} \int_{\substack{x\in B(x_0,\delta)\\ y\in B(x,r)}} |u(y)|^2 dy\,dx \leq \frac{C'}{r^2} \int_{B(x_0,\delta+r)} |u(y)|^2 dy\ .$$

Inequality (A) now yields

$$\int_{B(x_0,\delta)} |\nabla u(x)|^2 dx \leq \frac{C''}{r^2} \int_{B(x_0,\delta)} |u(y)|^2\, dy\ ,\ \text{i.e.}$$

$$\left(\int_{B(x_0,\delta)} |\nabla u|^2\right)^{1/2} \leq \frac{C''\sqrt{\lambda}}{\delta} \left(\int_{B(x_0,\delta)} |u|^2\right)^{1/2},$$ which is inequality (B).

The proof of (C) is even easier. With $x\in B(x_0,\delta)$ and $r=\frac{\delta}{\sqrt{\lambda}}$, we invoke (A) and (14) to write

$$|\nabla u(x)|^2 \le \frac{C}{r^{n+2}} \int_{B(x,r)} |u|^2 \le \frac{C}{r^{n+2}} \int_{B(x_0,\delta+r)} |u|^2 \le \frac{C'}{r^{n+2}} \int_{B(x_0,\delta)} |u|^2$$

$\le \frac{C''}{r^{n+2}} \cdot \mathrm{MAX}_{B(x_0,\delta)} |u|^2 \cdot \delta^n$. That is, $|\nabla u(x)| \le \frac{C''}{\delta} \lambda^{\frac{n+2}{4}} \mathrm{MAX}_{B(x_0,\delta)} |u|$

for $x \in B(x_0,\delta)$, which is inequality (C).

The proof of (D) is as follows. Fix a ball $B_0 \subset M$. For $\alpha < \sup_{B_0} |u|$,

define $E_\alpha = \{x \in B_0 \,\big|\, |u(x)| < \alpha\}$,

$\mathscr{B}_\alpha = \{\text{open balls } B \subset B_0 \,\big|\, |u(x)| < \alpha \text{ for all } x \in B\}$

and $\mathscr{B}^*_\alpha = \{$ maximal $B \in \mathscr{B}_\alpha$ under inclusion $\}$.

Every $x \in E_\alpha$ belongs to some $B \in \mathscr{B}_\alpha$, hence also (by Zorn's lemma, say) to some $B \in \mathscr{B}^*_\alpha$. Vitali's covering lemma therefore gives a collection $\{B_\nu\}$ of pairwise disjoint balls $B_\nu \in \mathscr{B}^*_\alpha$, with

$$\sum_\nu \mathrm{vol} B_\nu \ge c \ \mathrm{vol}\, E_\alpha. \tag{15}$$

Now let us think about the behavior of u on a ball $B(x,r) \in \mathscr{B}^*_\alpha$. Since $B(x,r) \in \mathscr{B}_\alpha$, we have $\sup_{B(x,r)} |u| \le \alpha$. Since $B(x,r)$ is maximal in $\mathscr{B}_\alpha$, we must have $\sup_{B(x,r)} |u| = \alpha$. Thus, let $y \in \overline{B(x,r)}$ satisfy $|u(y)| = \alpha$. For $z \in B(x,r) \cap B(y,s)$ with $s > o$ to be picked in a moment, we have $\alpha > |u(z)| \ge \alpha - s \cdot \mathrm{MAX}_{\overline{B(x,r)}} |\nabla u| \ge \alpha - s \cdot \frac{C\lambda}{r}^{\frac{n+2}{4}} \alpha$, by virtue of (C). Taking $s = \frac{1}{2} C^{-1} \lambda^{-(\frac{n+2}{4})} r$, we get $\alpha > |u(z)| \ge \frac{\alpha}{2}$ in $B(x,r) \cap B(y,s)$. In particular

$$\mathrm{vol}\{z \in B(x,r) \,\big|\, \alpha > |u(z)| \ge \tfrac{\alpha}{2}\} \ge c' \lambda^{\frac{-n(n+2)}{4}} \mathrm{vol} B(x,r) \text{ for } B(x,r) \in \mathscr{B}^*_\alpha .$$

Applying this to the balls $B_\nu \in \mathscr{B}_\alpha^*$ and summing on ν, we find that

$$\mathrm{vol}\{z\in B_o \mid \alpha>|u(z)| \geq \tfrac{\alpha}{2}\} \geq \sum_\nu \mathrm{vol}\{z\in B_\nu \mid \alpha>|u(z)| \geq \tfrac{\alpha}{2}\} \geq c'\lambda^{\frac{-n(n+2)}{4}} \sum_\nu \mathrm{vol}\, B_\nu$$

$$\geq c''\lambda^{\frac{-n(n+2)}{4}} \mathrm{vol}\, E_\alpha, \text{ by (15).}$$

In other words, $\mathrm{vol}E_{\alpha/2} \leq (1-c''\lambda^{-n(n+2)/4})\mathrm{vol}E_\alpha$ for $\alpha < \sup_{B_o}|u|$.

Using this for $\alpha = 2^{-k} \sup_{B_o}|u|$, $k=1,2,3\cdots$, we conclude that $\mathrm{vol}\{x\in B_o \mid |u(x)|< \eta \sup_{B_o}|u|\} \leq C\eta^{(c''\lambda^{-n(n+2)/4})}\mathrm{vol}\, B_o$ for $o < \eta < 1$, which is equivalent to (D).

Now we come to the proof of Theorem 2.

We break up the argument into a series of lemmas. The point of the first lemma and its corollaries is to establish rapid growth of eigenfunctions on nodal sets of small volume.

Lemma 1: Let $ds^2=\sum_{ij} g_{ij}\, dx^i dx^j$ be a metric of bounded geometry on the unit ball $B \subset \mathbb{R}^n$, and let Δ_g be the corresponding Laplacian. Suppose u is a real-valued solution of $-\Delta_g u=\lambda u$ in B, with $0 < \lambda \leq 1$.

Let Ω be a union of some of the components of $\{x\in B \mid u(x)>0\}$.

Then

$\left|u\chi_\Omega(o)\right| \leq C^\# \int_B \frac{|u\chi_\Omega(y)|}{|y|^{n-2}}\, dy$, with $C^\#$ depending only on m^*, C^* in the bounded geometry estimate (BG).

Proof: Fix $r \in \left[\frac{1}{2}, \frac{3}{4}\right]$, and take $B_r = \{|x|<r\} \subset \mathbb{R}^n$. Let $G_r(y)$ and $P_r(y)$ denote the Green's function and Poisson kernel for $-\Delta_g$ on B_r, so that

$$F(o) = \int_{B_r} G_r(y)\left(-\Delta_g F(y)\right) dy + \int_{\partial B_r} P_r(y)F(y)\, darea(y),\ F \in C^\infty(B) \tag{16}$$

We have $0 \le G_r(y) \le \dfrac{C^\#}{|y|^{n-2}}$ and $0 \le P_r(y) \le C^\#$, since the metric has bounded geometry. (In the two-dimensional case, our Lemma and its proof require trivial modifications which we omit.)

We shall apply (16) to a smooth variant of $F = u\chi_\Omega$. Specifically, let $\phi_\epsilon(t) = \epsilon\phi(t/\epsilon)$, where $\phi \in C^\infty(\mathbb{R}^1)$ satisfies $\phi(t)=0$ for $t \le 1$, $\phi(t)=t-1$ for $t \ge 2$, and $\phi''(t) \ge 0$ everywhere. Then define $F_\epsilon = \phi_\epsilon(u)\cdot\chi_\Omega$. Thus, $F_\epsilon \in C^\infty(B)$, since $\phi_\epsilon(u)=0$ near $\partial\Omega$. Moreover,

$$-\Delta_g F_\epsilon = \chi_\Omega\cdot\left[-\phi'_\epsilon(u)\Delta_g u - \phi''_\epsilon(u)|\nabla_g u|^2\right] \le \chi_\Omega\cdot\left[-\phi'_\epsilon(u)\Delta_g u\right] = \lambda u\phi'_\epsilon(u)\chi_\Omega .$$

Hence (16) applied to F_ϵ yields

$$\phi_\epsilon(u)\chi_\Omega(o) \le C^\# \int_{B_r} \frac{\lambda u\phi'_\epsilon(u)\chi_\Omega(y)}{|y|^{n-2}}\, dy + C^\# \int_{\partial B_r} \phi_\epsilon(u)\chi_\Omega(y)\, darea(y) .$$

Now we can take the limit as $\epsilon \to 0$ to conclude that

$$u\chi_\Omega(o) \le C^\#\lambda \int_{B_r} \frac{u\chi_\Omega(y)}{|y|^{n-2}}\, dy + C^\# \int_{\partial B_r} u\chi_\Omega(y)\, darea(y),\ r \in \left[\frac{1}{2}, \frac{3}{4}\right] .$$

Integrating from $r = \frac{1}{2}$ to $r = \frac{3}{4}$ and recalling that $0 < \lambda \le 1$, we obtain the conclusion of the Lemma. □

Cor 1: In addition to the hypotheses of Lemma 1, suppose $\text{vol}\Omega < c_1$, where the small constant c_1 depends on m^*, C^* in the bounded geometry estimate (BG). Then

$$|u\chi_\Omega(o)| \le \frac{1}{2} \sup_B |u\chi_\Omega| \ .$$

Proof: Just write

$$|u\chi_\Omega(o)| \le C^{\#} \int_{|y|<1} |u\chi_\Omega(y)| \frac{dy}{|y|^{n-2}} \le C^{\#} \sup_B |u\chi_\Omega| \cdot \int_\Omega \frac{dy}{|y|^{n-2}}.$$

If $\text{vol}\Omega < c_1 << 1$ then $\int_\Omega \frac{dy}{|y|^{n-2}} < \frac{1}{2C^{\#}}$, proving the corollary. □

An obvious rescaling of Cor. 1 to a small geodesic ball yields:

Cor 2: Suppose $\Sigma g_{ij}\, dx^i dx^j$ has bounded geometry in the unit Ball B_1 and let u be a real-valued solution of $-\Delta u=\lambda u$ in a ball $B=B(x_o,\delta)\subset B_1$ with $o<\delta<\lambda^{-1/2}$. Let Ω be a union of some of the connected components of $\{x\in B \,|\, u>o\}$. If $\lambda>1$ and if $\frac{\text{vol}\Omega}{\text{vol}B} < c_1$, then $|u\chi_\Omega(x_o)| \le \frac{1}{2} \sup_B |u\chi_\Omega|$. The constant c_1 depends only on m^* and C^* in the bounded geometry estimate.

<u>Cor 3</u>: Let Δ_g denote the Laplacian in a metric of bounded geometry, and suppose $-\Delta_g u=\lambda u$ in $B(x_o,\delta)$ with u real, $\lambda>0$, and $0 < \delta\le\lambda^{-1/2}$. Let Ω be a union of some of the components of $\{x\in B(x_o,\delta) \,|\, u(x)>0\}$. If $\lambda>1$ and if $\frac{\text{vol}\Omega}{\text{vol}B(x_o,\delta)} < \eta^n < 1$, then

$$\sup \left\{ u(y) \;|\; y\in B(x_o,\delta\cdot(1-C_3\eta)) \cap \Omega \right\} \le \frac{1}{2} \sup\left\{u(y) \,|\, y\in B(x_o,\delta)\cap\Omega \right\} .$$

The constant C_3 depends only on m^*, C^* in the bounded geometry estimate (BG).

Proof: For $y \in B(x_0, \delta\cdot(1-C_3\eta))$, apply Cor. 2 to $B=B(y,C_3\eta\delta) \subset B(x_0,\delta)$. We have

$$\mathrm{vol}(\Omega \cap B) \le \mathrm{vol}\Omega \le \eta^n\, \mathrm{vol}B(x_0,\delta) = \eta^n\, (C_3\eta)^{-n}\, \mathrm{vol}B < c_1\, \mathrm{vol}B$$

if we take $C_3 >> 1$ depending on c_1. Hence Cor. 2 shows that

$$u\chi_\Omega(y) \le \frac{1}{2} \sup_B u\chi_\Omega \le \frac{1}{2} \sup_{B(x_0,\delta)} u\chi_\Omega$$

for $y \in B(x_0, \delta\cdot(1-C_3\eta)$. □

Lemma 2: Suppose $-\Delta_g u = \lambda u$ in $B(x_0,\delta)$, with u real, $\lambda>1$, $0<\delta\le\lambda^{-1/2}$ and g having bounded geometry. Let Ω be a union of some of the components of $\{x\in B(x_0,\delta)\,|\,u(x)>0\}$, and assume $x_0\in\Omega$. Suppose $\frac{\mathrm{vol}\Omega}{\mathrm{vol}B(x_0,\delta)} < \eta^n < \frac{1}{2}\eta_0^n \le \frac{1}{2}$. Then there is a positive number r_0 satisfying:

(a) $0 < r_0 \le \frac{\eta}{\eta_0}\delta$

(b) $\frac{\mathrm{vol}\left(\Omega\cap B(x_0,r_0)\right)}{\mathrm{vol}B(x_0,r_0)} \ge \eta_0^n$

(c) $\sup_{\Omega\cap B(x_0,r_0)} |u| \le \left(\frac{r_0}{\delta}\right)^{c_4/\eta_0} \cdot \sup_{B(x_0,\delta)} |u|.$

The constant c_4 depends only on m^*, C^* in the bounded geometry estimate.

Proof: Just take $r_0<\delta$ as large as possible subject to (b). Since $x_0\in\Omega$ and $\mathrm{vol}\Omega \le \eta^n \mathrm{vol}\, B(x_0,\delta)$, (a) is obvious. Moreover, $\mathrm{vol}\big(\Omega\cap B(x_0,r)\big)/\mathrm{vol}B(x_0,r) < \eta_0^n$ for $r_0<r<\delta$, so Cor. 3 above yields

$$F\big(r\cdot(1-C_3\eta)\big)\le \frac{1}{2}F(r) \quad \text{for } r_0 < r < \delta, \text{ with}$$

$$F(r) = \sup_{\Omega\cap B(x_0,r)} |u|. \quad \text{Hence,}$$

$$F(r_0) \le \left(\frac{r_0}{\delta}\right)^{C_4/\eta} F(\delta), \text{ i.e.}$$

$$\sup_{\Omega\cap B(x_0,r_0)}|u| \le \left(\frac{r_0}{\delta}\right)^{C_4/\eta} \sup_{\Omega\cap B(x_0,\delta)}|u|,$$

which implies (c). □

Now let u be the restriction of a global eigenfunction to $B(x_0,\delta)$, satisfying the hypotheses of Lemma 2, with η_0,η to be picked. We have

$$\sup_{B(x_0,r_0)\cap\Omega}|u| \le \left(\frac{r_0}{\delta}\right)^{C_4/\eta_0}\cdot \sup_{B(x_0,\delta)}|u| \quad \text{, by Lemma 2 (c).}$$

Also

$$\sup_{B(x_0,r_0)}|u| \le \left(\frac{C\ \mathrm{vol}\ B(x_0,r_0)}{\mathrm{vol}\ \Omega\cap B(x_0,r_0)}\right)^{C\lambda^K}\cdot \sup_{B(x_0,r_0)\cap\Omega}|u| \text{ with } K = \frac{n(n+2)}{4}$$

by Theorem 1 (D).

In view of Lemma 2 (b), this implies

$$\sup_{B(x_0,r_0)}|u| \le \left(C\ \eta_0^{-n}\right)^{C\lambda^K}\cdot \sup_{B(x_0,r_0)\cap\Omega}|u| .$$

Estimate (0) for global eigenfunctions tells us that

$$\sup_{B(x_0,\delta)}|u| \le \left(\frac{C\delta}{r_0}\right)^{C\sqrt{\lambda}} \sup_{B(x_0,r_0)}|u| \quad .$$ Putting these

estimates together we get

$$\sup_{B(x_o,\delta)}|u| \le \left(\frac{C\delta}{r_o}\right)^{C\sqrt{\lambda}} \sup_{B(x_o,r_o)}|u| \le \left(\frac{C\delta}{r_o}\right)^{C\sqrt{\lambda}}\left(C\eta_o^{-n}\right)^{C\lambda^K} \sup_{B(x_o,r_o)\cap\Omega}|u|$$

$$\le \left(\frac{C\delta}{r_o}\right)^{C\sqrt{\lambda}}\left(C\eta_o^{-n}\right)^{C\lambda^K}\cdot\left(\frac{r_o}{\delta}\right)^{c_4/\eta}\cdot \sup_{B(x_o,\delta)}$$

Hence $\left(\frac{Cr_o}{\delta}\right)^{(c_4/\eta)-C\sqrt{\lambda}}\cdot\left(C\eta_o^{-n}\right)^{C\lambda^K} \ge 1.$

Since $\frac{Cr_o}{\delta} < \frac{C\eta}{\eta_o}$ by Lemma 2 (b), we see that

$$\left(\frac{C\eta}{\eta_o}\right)^{\frac{c_4}{\eta} - C\sqrt{\lambda}}\cdot\left(C\eta_o^{-n}\right)^{C\lambda^K} \ge 1, \text{ provided } \frac{c_4}{\eta} - C\sqrt{\lambda} \ge 0. \tag{17}$$

Taking $\eta_o = \hat{c}\lambda^{-K}$ ($0<\hat{c} << 1$ depending on C, c_4 in (17)), and setting $\eta = \eta_o^2$, we get $\frac{c_4}{\eta} - C\sqrt{\lambda} \ge 0$, so that (17) yields $\left(C\eta_o\right)^{\frac{c_4}{\eta} - C\sqrt{\lambda} - nC\lambda^K} \ge \left(C^{-1}\right)^{(n+1)C\lambda^K}$, which is clearly false, for suitable c. Thus we have derived a contradiction from the hypotheses of Lemma 2 for global eigenfunctions, with $\eta_o = \hat{c}\lambda^{-K}$ and $\eta = \eta_o^2$. In other other words, we have proved the following result.

Lemma 3: Suppose M has bounded geometry, and suppose $-\Delta_g u = \lambda u$ on M with u real and $\lambda>1$. Let $B(x_o,\delta) \subset M$ be a ball with radius $\delta < \lambda^{-1/2}$. Let Ω be a union of components of $\{x\in B(x_o,\delta) \mid u(x)>0\}$, and assume $x_o\in\Omega$. Then

$$\frac{\operatorname{vol}\Omega}{\operatorname{vol}B(x_o,\delta)} \ge c\,\lambda^{\frac{-n^2(n+2)}{2}}$$

Lemma 3 trivially implies Theorem 2. If B is any ball in M, then let $\Omega \subset B$ be a component of $\{x\in B \mid u(x)>0\}$. Assume $x_o\in\Omega$ belongs to the middle half of B. Taking $\delta = \min\left(\frac{1}{2}\lambda^{-1/2}, \frac{1}{2}\text{ radius }(B)\right)$, we simply apply Lemma 2 to $\Omega \cap B(x_0, \delta)$ in the ball $B(x_0, \delta)$. The proof of Theorem 2 is complete. □

REFERENCES

[A] N. Aronszajn, A unique continuation theorem for solutions of elliptic partial differential equations of second order, J. Math. Pures Appl. 36 (1957), 235-249.

[DF1] H. Donnelly and C. Fefferman, Nodal Sets of eigenfunctions on Riemannian manifolds, Inv. Math. 93 (1988), 161-183.

[DF2] H. Donnelly and C. Fefferman, Nodal Sets of eigenfunctions: Riemannian manifolds with boundary, preprint.

[GL] N. Garafalo and F. H. Lin, Unique continuation for elliptic operators: A geometric-variational approach, Comm. Pure Appl. Math 40 (1987), 347-366.

33

Trying to Beat Heisenberg

F. Alberto Grünbaum

I. Introduction

The Heisenberg uncertainty principle establishes a lower limit to the simultaneous resolution of position and momentum for a particle. In the hands of W. Pauli and H. Weyl [1] the "principle" was converted into an inequality featuring an arbitrary function and its Fourier transform. The "principle" has then been found to rear its head in many applications far removed from quantum mechanics: it basically says that you cannot ***simultaneously*** concentrate a signal in the time and frequency domains.

In many areas of application this is a curse and not a blessing. Furthermore, in certain areas, like Nuclear Magnetic Resonance (NMR), the relation between input and output is fairly complicated and only approximately given by the Fourier transform. In spite of this, it is fairly common to assert that such an injunction against "double concentration" should hold. See for instance [2]. The problem of "double concentration" in NMR has become an active area of research recently in connection with medical imaging applications of NMR.

Partially supported by National Science Foundation grant DMS87-20007.

The purpose of this note is to look into a special case of this problem. We have chosen the case of the "scattering transform" since it serves as a model for several "natural" nonlinear extensions of the Fourier transform. We will deal here with a case of the Schrödinger equation. In terms of our initial motivation, namely, Nuclear Magnetic Resonance, one should deal with a 2-component Zakharov-Shabat system, as explained in [3]. The results in that case are similar to those described here, so we stick to this simpler situation.

In closing this introduction it is good to recall that Heisenberg never made any claims in the ***nonlinear case.*** This paper is thus not an effort to impugn Heisenberg but rather to disassociate his name from some unproven "extensions" of his beautiful result.

For simplicity we consider the simplest possible case: a one-dimensional discrete model of the Schrödinger equation. The same problem should be looked into in other setups.

II. The scattering transform and the Born approximation

Given a potential $q(x)$, $x \in R$, one considers solutions $y(x,k)$ of the Schrödinger equation

$$-y'' + q(x)y = k^2 y \quad .$$

Under appropriate decay conditions on $q(x)$ one can consider, for real k, solutions f_i, $i = 1,2$, satisfying

$$f_1(x,k) \sim e^{ikx}, \qquad x \longrightarrow +\infty$$
$$f_2(x,k) \sim e^{-ikx}, \qquad x \longrightarrow -\infty \quad .$$

In terms of these solutions it is then customary to define the "scattering data" $a(k)$, $b(k)$ by means of the expressions

$$f_2(x,k) = b(k)f_1(x,k) + a(k)f_1(x,-k)$$

It is furthermore possible to show under enough assumptions on q that

$$2ikb(k) = \int_{-\infty}^{\infty} e^{-iky} q(y)f_2(y,k)dy \quad .$$

This expression is the basis of our analysis.

It is possible to use the expression above to obtain an infinite series expansion of the quantity

$$2ikb(k)$$

in terms of the potential $q(y)$. Neglecting all terms except the linear one, you obtain the "Born approximation"

$$2ikb(k) \cong \int_{-\infty}^{\infty} e^{-iky} q(y)e^{-iky} \equiv \hat{q}(2k)$$

Here $\hat{q}$ denotes the Fourier transform of q.

This result is well known and can be made quite rigorous: it states that the gradient at $q \equiv 0$ of the "scattering transform"

$$q(y) \longrightarrow 2ikb(k)$$

is given by (essentially) the Fourier transform.

The difficulties is computing $b(k)$ explicitly are many. Only a handful of examples are available. Needless to say the 3-dimensional case is even harder: this has made the "Born approximation" a very common tool even in cases where the basic assumptions are no longer valid.

III. The Heisenberg inequality in R and T

If $f(x)$ is defined on the real line, $\hat{f}(k)$ denotes its Fourier transform, and all the terms below are finite, one has

$$\frac{\int x^2 |f(x)|^2 dx}{\int |f(x)|^2 dx} \quad \frac{\int k^2 |\hat{f}(k)|^2 dk}{\int |\hat{f}(k)|^2 dk} \geq \frac{1}{4} \quad .$$

We will not give here the proof of this inequality, rather we will prove a similar statement for a function $f(\theta)$ defined on the unit circle T, $0 \leq \theta \leq 2\pi$. Our motivation for doing this is twofold: first, it is exactly this version that we will need; second, we have not seen this in print. The proof given here is a trivial modification of the one given in [1]. For simplicity, assume that $f(\theta)$ is real and even, $f(\theta) = f(-\theta)$. One consequence of this is that in the expansion

$$f(\theta) = \sum c_n e^{in\theta}$$

the coefficients c_n are real valued. These two trivial assumptions shorten the proof a bit. It is also harmless to assume that $f(\theta)$ is smooth in $-\pi < \theta < \pi$.

By integration by parts,

$$\begin{aligned}
\int_{-\pi}^{\pi} f(\theta) f'(\theta)\theta d\theta &\equiv \tfrac{1}{2} \int_{-\pi}^{\pi} (f^2(\theta))'\theta d\theta \\
&= \tfrac{1}{2} |\theta f^2(\theta)|_{-\pi}^{\pi} - \tfrac{1}{2} \int_{-\pi}^{\pi} f^2(\theta) d\theta \\
&= \tfrac{1}{2}\pi(f^2(\pi) + f^2(-\pi)) - \tfrac{1}{2} \int_{-\pi}^{\pi} f^2(\theta) d\theta \quad .
\end{aligned}$$

If one assumes that $f(\theta)$ is continuous at $\pm\pi$, this can be expressed as

$$\int_{-\pi}^{\pi} f(\theta) f'(\theta)\theta d\theta = \tfrac{1}{2} \int_{-\pi}^{\pi} (f^2(\pi) - f^2(\theta)) d\theta \quad .$$

Now by Schwarz's inequality we get

$$\left(\int f(\theta)\theta f'(\theta)d\theta\right)^2 \leq \int f^2(\theta)\theta^2 d\theta \int f'(\theta)^2 d\theta \quad .$$

Finally from

$$f(\theta) = \sum c_n e^{in\theta}$$

we get

$$f'(\theta) = \sum inc_n e^{in\theta}$$

and

$$\int_{-\pi}^{\pi} f(\theta)^2 d\theta = 2\pi \sum c_n^2 ; \qquad \int_{-\pi}^{\pi} f'(\theta)^2 d\theta = 2\pi \sum n^2 c_n^2 \quad .$$

Therefore we get

$$2\pi \int f^2(\theta)\theta^2 d\theta \cdot \sum n^2 c_n^2 \geq \tfrac{1}{4} \left(\int_{-\pi}^{\pi} (f^2(\pi) - f^2(\theta))d\theta \right)^2 .$$

If we are now willing to make the extra assumption that

$$f(\pi) = f(-\pi) = 0$$

we get a more standard looking form of the Heisenberg inequality on the circle, namely,

$$\frac{\int_{-\pi}^{\pi} \theta^2 f^2(\theta)d\theta}{\int_{-\pi}^{\pi} f^2(\theta)d\theta} \cdot \frac{\sum_{-\infty}^{\infty} n^2 c_n^2}{\sum_{-\infty}^{\infty} c_n^2} \geq \frac{1}{4} \quad .$$

The importance of the assumption $f(\pm\pi) = 0$ is obvious. This standard form of the inequality is violated by $f(\theta) \equiv 1$.

IV. The discrete Schrödinger equation and its "scattering transform"

Replace the real line by the integers, denoted here by Z. We take as our model of the Schrödinger equation the doubly infinite difference system of equations for $y_i(z)$

$$y_{i-1} - b_i y_i + y_{i+1} = \left(z + \frac{1}{z}\right) y_i \quad , \qquad -\infty < i < \infty .$$

The appropriate range for z is the unit circle, i.e. $z = e^{i\theta}$ and the role of the potential is taken by vector b_i, $-\infty < i < \infty$.

If we had $b_i \equiv 0$ then two linearly independent solutions of the equations above are z^i and z^{-i}. If the potential b_i vanishes fast enough, and $y_i \approx z^{-i}$ as $i \longrightarrow -\infty$, then the solution will look as $i \longrightarrow +\infty$ as

$$y_i \cong \alpha(z) z^{-i} + \beta(z) z^{i} \quad .$$

This gives rise to relations between the "scattering data" $\alpha(z)$, $\beta(z)$ and the potential along the same lines that we saw in the continuous case considered earlier. The pertinent relation for us is

$$\left(z - \frac{1}{z}\right)\beta(z) = \sum b_j z^{-2j} + \text{nonlinear terms in } b \ .$$

The first term on the right-hand side is exactly the Fourier tranform

$$\sum b_j e^{-ij\theta}$$

of the sequence b_j evaluated at the argument 2θ instead of θ.

This leads us to consider the following ***main problem***: Given a potential b_j, $-\infty < j < \infty$, which decays at $\pm\infty$ fast enough and such that

$$\left(z - \frac{1}{z}\right)\beta(z)$$

vanishes at $\theta = \pm\pi$, is it true that the pair

$$\{b_j\}, \qquad \left(z - \frac{1}{z}\right)\beta(z) = 2i \sin\theta \ \beta(\theta)$$

satisfies an inequality of the form

$$\frac{\int_{-\pi}^{\pi} \theta^2 |\sin\theta\ \beta(\theta)|^2 d\theta}{\int_{-\pi}^{\pi} |\sin\theta\ \beta(\theta)|^2 d\theta} \cdot \frac{\sum j^2 |b_j|^2}{\sum |b_j|^2} \geq C > 0 \tag{1}$$

with C a universal constant.

Notice, first of all, that since we are dealing with a finite integration interval, the first ratio on the left-hand side is bounded above by π^2 by the mean value theorem.

The question above is thus equivalent to asking if one can keep the ratio

$$\frac{\sum j^2 |b_j|^2}{\sum |b_j|^2}$$

bounded away from zero by the requirement that the expression

$$\left(z - \frac{1}{z}\right)\beta(z) = 2i \sin\theta\ \beta(\theta)$$

should vanish at $\theta = \pi$.

The answer ***turns out to be*** NO, as we show in the next section.

V. A class of "concentrated" potentials

Given any positive integer $p \geq 1$ consider the potential given by

$$b_j = \begin{cases} b & \text{if} \quad j = \pm p \\ b_0 & \text{if} \quad j = 0 \\ 0 & \text{otherwise} \end{cases} .$$

As we will see below, the choice

$$b_0 = \frac{2b}{pb-1} \tag{2}$$

guarantees that the "scattering transform" $\left(z - \frac{1}{z}\right)\beta(z)$ will vanish at $z = -1$. Since this condition is met, we can now consider the value of the second factor in (1). We readily get

$$\frac{\sum j^2 |b_j|^2}{\sum |b_j|^2} = \frac{2p^2b^2(1 - pb)^2}{4b^2 + 2(1 - pb)^2b^2}$$

and it is clear that this quantity can be made as close to zero as desired by choosing

$$b \approx \frac{1}{p} .$$

The conclusion is thus that the contemplated "extension" of Heisenberg's inequality to this "nonlinear situation" is not valid. The explicit expression for $\beta(z)$ is given below. We get, after a bit of simplification:

$$\left(z - \frac{1}{z}\right)\beta(z) = b_0 + b(z^{2p} + z^{-2p}) + b(b + b_0)\frac{z^{4p} - 1}{z^{2p-1}(z^2-1)} + b_0b^2\frac{(z^{2p} - 1)^2}{z^{2p-2}(z^2-1)^2} .$$

When $z = -1$ we get

$$b_0 + 2b - b(b + b_0)2p + b_0b^2p^2$$

and setting this equal to zero gives

$$b_0 = \frac{2b}{pb - 1}$$

as mentioned in (2) above.

It is clear that there are many valid ways of measuring the "concentration" of a potential or of its "scattering data". The one chosen above is a convenient one, but others may be more significant.

References

[1] H. Weyl (1928), *The Theory of Groups and Quantum Mechanics* (Dover Publications, New York).

[2] D. Hoult (1977), "Zeugmatography: a criticism of the concept of a selective pulse in the presence of a field gradient," *Journal of Magnetic Resonance* 26, p.165.

[3] F. A. Grünbaum (1989), "Soliton mathematics in signal processing," Minneapolis Workshop in Signal Processing, 1988; publication of the Inst. of Mathematics and Applications, Univ. of Minnesota, T. Kailath, editor.

34

Smoothing Properties of Solutions to The Benjamin-Ono Equation

Gustavo Ponce

§. **Introduction.** In this paper we consider the initial value problem (IVP) for the Benjamin-Ono equation

$$(1.1) \qquad \begin{cases} \partial_t u - H\partial_x^2 u + u\partial_x u = 0 & x \in \mathbb{R}, t > 0 \\ u(x,0) = u_0(x) \end{cases}$$

where $u = u(x,t), \partial_t = \partial/\partial t, \partial_x = \partial/\partial x$, and H is the Hilbert transform, i.e.,

$$(1.2) \qquad (Hf)(x) = \text{p.v.}\,\frac{1}{\pi}\int \frac{f(y)}{x-y}\,dy = F^{-1}(-i\ \text{sgn}(\xi)\widehat{f}(\xi))$$

where $\widehat{\ }$ and F^{-1} denote the Fourier transform and its inverse, respectively.

The equation (1.1) arises in the study of internal waves in deep stratified fluids (see T. B. Benjamin [**2**], and H. Ono [**19**]), and presents properties strikingly similar to the Korteweg-de Vries (KdV) equation

$$(1.3) \qquad \partial_t u + \partial_x^3 u + u\partial_x u = 0 \qquad x \in \mathbb{R}, t > 0.$$

Some of the properties are: (1.1) defines a hamiltonian system, satisfies inifinitely many conservation laws (see [**4**], [**5**]), can be solved by an analogue of the inverse scattering method (see [**1**]), and possesses (multi)-soliton solutions (see [**5**], [**6**]). Moreover, in [**13**] R. J. Iório has established results concerning the existence and uniqueness of solutions to the IVP (1.1) similar to those previously known for the KdV equation. However, he proved a remarkable difference in the persistence properties of solutions (i.e., the solution $u(t)$ at the time t belongs to the same function space X as does the initial data and describe a continuous curve in X) of these equations in weighted Sobolev spaces $H^s \cap L^2(|x|^s\,dx)$ (see notation below) when $s \geq 4$, (Theorem 5.4 in [**13**], and Theorem 8.3 in [**15**]).

On the other hand, T. Kato [**15**], and S. N. Kruzhkov, and A. V. Faminskii [**18**] have shown that solutions of the KdV equation possess a local smoothing effect. This smoothing effect was used in [**15**] and [**18**] to derive existence results for rough initial data and to improve the regularity of known solutions. Both proofs rely in a crucial way on the algebraic property of the symbol of the (linearized) KdV equation. In [**10**], P. Constantin and J. C. Saut have shown that this local smoothing phenomenon is inherent in the dispersive nature of the equation. They gave a proof of this property for general linear dispersive equations. Using Duhamel's Principle, they extended their results to some semilinear equations. It is clear that this method does not provide strong estimates for equations with nonlinear terms involving derivatives. In particular, they do not recover the results in [**15**], [**18**] for the KdV equation. Also we may remark that simultaneously with [**10**], P. Sjölin [**21**] and L. Vega [**25**] have proved this local smoothing property in solutions of the linear Schrödinger equation, (for previous global smoothing L^p results for this equation, see R. S. Strichartz [**24**]). Since these equations (dispersive) are time reversible, global smoothing effects in Sobolev spaces H^s have to be excluded whenever persistence properties in these spaces have been established.

None of the above methods of proof seem to be appropriate to obtain the corresponding results for the BO equation. The difficulties appear in the calculus of inequalities involving fractional derivatives (roughly speaking, the expected improvement in smoothness is 1/2), and the combination of the nonlinear term with the nonlocal character of the associated linear equation. In particular, the explicit Leibniz rule for the Hilbert transform (i.e.,$H(f \cdot g) = f \cdot Hg + Hf \cdot g + H(Hf \cdot Hg)$) does not appear to be useful.

In the present paper after proving the necessary calculus of inequalities we obtain the desired result for the BO equation when the uniqueness of the solution is known. Also we show a local smoothing effect for a H^1-weak solution of the equation in (1.1).

Our proof relies on several commutator estimates involving Bessel potentials of fractional order and multiplication operators. Some of them were deduced and used by T. Kato and G. Ponce in [**17**] to show strong well-posedness in Lebesgue spaces L^p_s of the Euler and Navier-Stokes equations (uniformly in the viscosity). Although our estimates (section 2) are quite appropriate for the problems considered here we remark that related results which also provide pointwise information will be given by R. R. Coifman and Y. Meyer in a forthcoming paper ([**9**]).

The plan of this paper is as follows. In section 2 we prepare the calculus of inequalities necessary for the proofs in the later sections. In section 3, we prove the appropriate local smoothing effect for solutions of the IVP (1.1) with data $u_0 \in H^s$ with $s > 3/2$. In section 4, we establish the improvement of 1/4-local derivatives a.e. in time for a weak solution of the IVP (1.1) with data $u_0 \in H^1$.

Notation.

- The norm in $L^p(\mathbb{R}^n), 1 \leq p \leq \infty$ will be denoted by $\|\cdot\|_p$.
- $J^s = (1-\Delta)^{s/2}$ and $I^s = (-\Delta)^{s/2}$ denote the Bessel and Riesz potential of order $-s$ respectively. In particular, for $n = 1$ and $s = 1$, $I = H \cdot \partial_x$ where H is the Hilbert transform. We will use that $H \cdot H = -1$.
- $L^p_s \equiv J^{-s}L^p$ whose norm will be denoted by $\|\cdot\|_{s,p} = \|J^s \cdot\|_p$. When $p = 2$ we will write H^s instead of L^2_s.
- $[A; B] \equiv A \cdot B - B \cdot A$ where A, B are operators. In particular, $[J^s; f]g \equiv J^s(f \cdot g) - f \cdot J^s g$ in which f is regarded as a multiplication operator.
- $S(\mathbb{R}^n)$ denotes the Schwartz class.
- $H^s_{\text{loc}}(\Omega)$ where Ω is an open set in $\mathbb{R}^n$ and $s > 0$ denotes the functions $u \in L^2_{\text{loc}}(\Omega)$ which satisfy that for every $\phi \in C_0^\infty(\Omega)$, $\phi \cdot u \in H^s(\mathbb{R}^n)$.

§2. Calculus of Inequalities.

Here we will deduce several estimates involving fractional derivation (i.e., Bessel Potential of fractional order) which will be used in following sections.

LEMMA 2.1. *For each $s \in \mathbb{R}$ there is a constant $c > 0$ such that for all $\varphi \in (\mathbb{R}^n)$ and $f \in H^{s-1}(\mathbb{R}^n)$*

$$\|[J^s; \varphi]f\|_2 \leq c\|\varphi\|_{\ell,\infty}\|J^{s-1}f\|_2 \tag{2.1}$$

where $\ell = |s| + n + 2$.

Remark: By using the techniques given in the proofs below one can show stronger and more general results than that in (2.1). However, we shall not pursue the problem here since the estimate (2.1) is general enough for our purposes.

PROOF: See [**12**] Lemma 6.14 p. 249.

In the following lemmas f, g denote functions in $S(\mathbb{R}^n)$. The extension of these results to functions for which the right-hand side of (2.2)–(2.4) make sense is standard.

LEMMA 2.2. *If $s > 0$ and $1 < p < \infty$, then $L_s^p \cap L^\infty$ is an algebra. Moreover,*

$$\|fg\|_{s,p} \leq c(\|f\|_\infty \|g\|_{s,p} + \|f\|_{s,p}\|g\|_\infty). \tag{2.2}$$

PROOF: When s is a positive integer, the above result is well known. The case $0 < s < 1$ was proved by R. S. Strichartz ([**23**], Theorem 2.1, Chapter 2). For the proof of the general case we refer to T. Runst ([**20**], p. 332), and T. Kato and G. Ponce ([**17**] Appendix, Lemma X4). In [**20**] the spaces $F_s^{p,q}$ used there coincide with the spaces L_s^p when $q = p$. I would like to thank Prof. R. S. Strichartz for pointing out to me this reference.

LEMMA 2.3. *If $s > 0$ and $1 < p < \infty$, then*

$$\|[J^s; f]g\|_p \leq c(\|\partial f\|_{p_1}\|J^{s-1}g\|_{p_2} + \|J^s f\|_{p_3}\|g\|_{p_4})$$

where $p_1, p_4 \in (1, \infty]$ such that

$$\frac{1}{p} = \frac{1}{p_1} + \frac{1}{p_2} = \frac{1}{p_3} + \frac{1}{p_4}.$$

PROOF: The proof of (2.3) when $p_1 = p_4 = \infty$ was given in ([**17**] Appendix, Lemma X1). The necessary modifications for the general case are contained in the proof of the following lemma.

LEMMA 2.4. *If $s > 0$ and $1 < p < \infty$, then*

$$\begin{aligned} &\|J^s(fg) - f(J^s g) - (J^s f)g\|_p \\ &\qquad \leq c\Big(\|\partial f\|_{p_1}\|J^{s-1}g\|_{p_2} + \|J^{s-1}f\|_{p_2}\|\partial g\|_{p_1} + \|J^{s-\theta}f\|_{p_3}\|J^\theta g\|_{p_4}\Big) \end{aligned} \tag{2.4}$$

for any $\theta \in [0, s]$ and any $p_1 \in (1, \infty]$, $p_2, p_3, p_4 \in (1, \infty)$ such that

$$\frac{1}{p} = \frac{1}{p_1} + \frac{1}{p_2} = \frac{1}{p_3} + \frac{1}{p_4}.$$

The proof of this lemma (and the previous two lemmas) is based on the following result due to R. R. Coifman and Y. Meyer ([**7**] p. 179; [**8**] p. 154).

THEOREM 2.5. *Let $\sigma \in C^\infty(\mathbf{R}^n \times \mathbf{R}^n - (0,0))$ satisfy*

$$|\partial_\xi^\alpha \partial_\eta^\beta \sigma(\xi,\eta)| \leq c_{\alpha,\beta}(|\xi| + |\eta|)^{-|\alpha|-|\beta|} \tag{2.5}$$

for $(\xi,\eta) \neq (0,0)$ and any multi-indices $\alpha, \beta \in (\mathbf{Z}^+)^n$. If $\sigma(D)$ denotes the bilinear operator

$$\sigma(D)(f,g)(x) = \iint e^{i\langle x,\xi+\eta\rangle}\sigma(\xi,\eta)\widehat{f}(\xi)\widehat{g}(\eta)\,d\xi d\eta$$

then

$$\|\sigma(D)(f,g)\|_p \leq c\|f\|_{p_1}\|g\|_{p_2} \tag{2.6}$$

for any $p \in (1,\infty)$, and any $p_1 \in (1,\infty]$ such that $\frac{1}{p} = \frac{1}{p_1} + \frac{1}{p_2}$.

REMARK: The case $p = p_2 \neq 2$ was proved by G. David and J. L. Journé ([**11**] p.382).

For the proof of this theorem it is only necessary to assume that the estimate (2.5) holds for $|\alpha|, |\beta| \leq k$, with k depending on n and p.

PROOF OF LEMMA 2.4: The proof is similar to that provided in detail in [**16**] for the Lemma 2.3, so that a sketch will suffice. We have that

$$\begin{aligned}
&\Big(J^s(fg) - f\cdot J^s g - J^s f\cdot g\Big)(x)\\
&\quad = c\iint e^{i\langle x,\xi+\eta\rangle}\Big\{(1+|\xi+\eta|^2)^{s/2} - (1+|\xi|^2)^{s/2} - (1+|\eta|^2)^{s/2}\Big\}\widehat{f}(\xi)\widehat{g}(\eta)\,d\xi\,d\eta\\
&\quad = c\sum_{j=1}^{3}\sigma_j(D)(f,g)(x),
\end{aligned}$$

with

$$\sigma_j(\xi,\eta) = \{(1+|\xi+\eta|^2)^{s/2} - (1+|\xi|^2)^{s/2} - (1+|\eta|^2)^{s/2}\}\phi_j\left(\frac{|\xi|}{|\eta|}\right) \tag{2.7}$$

and where the ϕ_j are C^∞-functions on $\mathbf{R}$ satisfying

$$\begin{aligned}
&0 \leq \phi_j \leq 1,\ j = 1,2,3;\\
&\phi_1 + \phi_2 + \phi_3 \equiv 1 \text{ on } \mathbf{R}^+;\\
&\text{supp } \phi_1 \subseteq (-\infty, 1/3], \text{supp } \phi_2 \subseteq [1/4, 4], \text{supp } \phi_3 \subseteq [3,\infty)\\
&\qquad \phi_1(y) = \phi_3(1/y) \text{ for any } y \leq 0.
\end{aligned}$$

Since $\sigma_1(\xi,\eta)=\sigma_3(\eta,\xi)$ it is enough to estimate $\sigma_1(\cdot,\cdot)$ and $\sigma_2(\cdot,\cdot)$.

First we consider $\sigma_1(D)(f,g)$, whose symbol $\sigma_1(\xi,\eta)$ can be written as

$$\begin{aligned}\sigma_1(\xi,\eta)&=\Big\{\big((1+|\xi+\eta|^{\,2})^{s/2}-(1+|\eta|)^{s/2}\big)+\big(1-(1+|\xi|^2)^{s/2}\big)-1\Big\}\phi_1\Big(\frac{|\xi|}{|\eta|}\Big)\\&\equiv\sigma_{1,1}(\xi,\eta)+\sigma_{1,2}(\xi,\eta)+\sigma_{1,3}(\xi,\eta).\end{aligned}$$

By combining the argument used in ([**17**] Appendix, proof of Lemma X1, step 1) and Theorem 2.5 we obtain that

$$\|\sigma_{1,1}(D)(f,g)\|_p\le c\|\partial f\|_{p_1}\|J^{s-1}g\|_{p_2}. \tag{2.8}$$

The term $\sigma_{1,2}(D)(f,g)$ equals that denoted by $\sigma_{3,2}$ in ([**17**] Appendix, proof of Lemma X1, step 2). Thus by using Theorem 2.5 it follows that

$$\|\sigma_{1,2}(D)(f,g)\|_p\le c\|\partial f\|_{p_1}\|J^{s-1}g\|_{p_2} \tag{2.9}$$

It is easy to see that $\sigma_{1,3}$ satisfies (2.5), hence

$$\|\sigma_{1,3}(D)(f,g)\|_p\le c\|f\|_{p_3}\|g\|_{p_4}. \tag{2.10}$$

Next, we consider $\sigma_2(D)(f,g)$ whose symbol $\sigma_2(\xi,\eta)$ is given by the expression

$$\sigma_2(\xi,\eta)=\Big\{(1+|\xi+\eta|^2)^{s/2}-((1+|\xi|^2)^{s/2}+(1+|\eta|^2)^{s/2})\Big\}\,\phi_2\Big(\frac{|\xi|}{|\eta|}\Big)=\sigma_{2,1}(\xi,\eta)+\sigma_{2,2}(\xi,\eta).$$

The second term above, $\sigma_{2,2}$, can be easily handled since

$$\sigma_{2,2}(\xi,\eta)\widehat{f}(\xi)\widehat{g}(\eta)=-\frac{(1+|\xi|^2)^{s/2}+(1+|\eta|^2)^{s/2}}{(1+|\xi|^2)^{s'/2}\cdot(1+|\eta|^2)^{s''/2}}(J^{s'}f)^\wedge(\xi)(J^{s''}g)^\wedge(\eta)\cdot\phi_2\Big(\frac{|\xi|}{|\eta|}\Big)$$

($0\le s',s''$ and $s'+s''=s$) satisfies the hypothesis of Theorem 2.5. Then we have that

$$\|\sigma_{2,2}(D)(f,g)\|_p\le c\|J^{s-\theta}f\|_{p_3}\|J^{\theta}g\|_{p_4}. \tag{2.11}$$

Finally, we estimate the term $\sigma_{2,1}(D)(f,g)$ by using complex (multi-linear) interpolation. First we extend $\sigma_{2,1}(\xi,\eta) \equiv \sigma_{2,1}^{\ell}(\xi,\eta)$ to complex values of $\ell = \tau + it$ with $0 \le \tau \le k$ (see the remark of Theorem 2.5).

On the one hand, using Mihlin's Theorem (see for example [**3**]) and Theorem 2.5 we get that

$$\|\sigma_{2,1}^{it}(D)(f,g)\|_p \le c(t)\|f\|_{p_3}\|g\|_{p_4}.$$

On the other hand if $\ell = k + it$, Theorem 2.5 leads to an estimate of

$$\sigma_{2,1}^{k+it}(D)(f,g) = \frac{(1+|\xi+\eta|^2)^{(k+it)/2}}{(1+|\xi|^2)^{k_1/2}\cdot(1+|\eta|^2)^{k_2/2}}\phi_2(\div)(J^{k_1}f)^\wedge(\xi)(J^{k_2}g)^\wedge(\eta)$$

(where $0 \le k_1, k_2,\ k_1 + k_2 = k$) of the form

$$\|\sigma_{2,1}^{k+it}(D)(f,g)\|_p \le c(t)\|J^{k_1}f\|_{p_3}\|J^{k_2}g\|_{p_4}$$

Since $c(t)$ is of polynomial order, by complex interpolation (see [**3**], [**22**]) between the previous cases we find that

$$\|\sigma_{2,1}(D)(f,g)\|_p \le c\|J^{s-\theta}f\|_{p_3}\|J^{\theta}g\|_{p_3}. \tag{2.12}$$

Collecting the estimates (2.8)–(2.12) we complete the proof of Lemma 2.4.

§3. The Local Smoothing Effect for Solutions in $H^s, s > 3/2$. We start by stating the following result concerning the well-posedness of the problem (1.1).

THEOREM 3.1. *If $s > 3/2$, then for any $u_0 \in H^s(\mathbb{R})$ there exists $T = T(\|u_0\|_s) > 0$, and a unique solution u to (1.1) such that*

$$u \in C([0,T]; H^s). \tag{3.1}$$

The map $u_0 \to u$ from H^s to the space in (3.1) is continuous. Also, when $s \ge 2$, T in (3.1) can be taken arbitrarily large (global solution).

For the proof of the existence part we refer to [**13**]. The continuous dependence on the data can be proved by the same arguments used in [**14**] or in [**16**]. The gap between 3/2 and 2 is due to the form and the number of conservation laws necessary to establish the global existence.

Our main result in this section is

THEOREM 3.2. *Let $s > 3/2$, $0 < T < \infty$. If $u \in C([0,T] : H^s)$ is the solution of the IVP(1.1) for $u_0 \in H^s$, then*

$$u \in L^2([0,T] : H^{s+1/2}_{\text{loc}}(\mathbb{R})). \tag{3.3}$$

REMARK: The above result implies that $u(t) \in H^{s+1/2}_{\text{loc}}$ for a.e $t \in [0,T]$. Thus $u(t) \in C^{s-\varepsilon}$ for a.e $t \in [0,T]$, and any $\varepsilon > 0$. Moreover, for $s \in (3/2, 2)$ we have that the equation is realized in $L^2_{\text{loc}}([0,T] \times \mathbb{R})$. This is a significant improvement of the result in Theorem 3.1 from which the equation holds in $C([0,T] : H^{s-2})$.

PROOF: Let $\phi : \mathbb{R} \to \mathbb{R}$ be any increasing function such that $\phi' \in C_0^\infty(\mathbb{R})$. Since $u(t)$ is a solution of the equation

$$\partial_t u - H\partial_x^2 u + u\partial_x u = 0 \tag{3.4}$$

it follows that

$$\partial_t Hu + \partial_x^2 u + H(u\partial_x u) = 0, \tag{3.5}$$

and formally

$$\begin{aligned}\int \Big\{ \partial_t J^s u J^s u + \partial_t J^s Hu J^s Hu - J^s H\partial_x^2 u J^s u + J^s \partial_x^2 u J^s Hu \\ + J^s(u\partial_x u) J^s u + J^s H(u\partial_x u) J^s Hu \Big\} \phi(x)\, dx = 0.\end{aligned} \tag{3.7}$$

The first and second term in (3.7) can be written as

$$\frac{1}{2}\frac{d}{dt}\int (J^s u J^s u + J^s Hu J^s Hu)\phi. \tag{3.8}$$

The third and fourth terms in (3.7) are equal to

$$\begin{aligned}&\int J^s H\partial_x u J^s \partial_x u \phi + \int J^s H\partial_x u J^s u\phi' - \int J^s \partial_x u J^s H\partial_x u\phi - \int J^s \partial_x u J^s Hu\phi' \\ &= \int J^s H\partial_x u J^s u\phi' + \int J^s HH\partial_x u J^s Hu\phi' \\ &= \int J^s Iu J^s u\phi' + \int J^s IHu J^s Hu\phi'.\end{aligned} \tag{3.9}$$

Since the above terms are similar, it suffices to consider only one of them. Thus

$$\int J^s I u J^s u \phi' = \int J^{s+1} u J^s u \phi' + \int J^s (I - J) u J^s u \phi', \tag{3.10}$$

using that $f(x) = |x| - (1 + x^2)^{1/2} \in L^\infty(\mathbb{R})$ the last term in the right-hand side (rhs) of (3.10) is bounded by

$$c\|J^s u\|_2^2 \tag{3.11.}$$

By Plancherel's formula the first term in the rhs of (3.10) can be written as

$$\int J^{s+1/2} u J^{s+1/2} u \phi' + \int J^{s+1/2} u [J^{1/2}; \phi'] J^s u = \alpha_1 + \alpha_2,$$

we keep α_1 and express α_2 in the following form

$$\alpha_2 = \int J^s u J^{1/2} (J^{1/2} (\phi' J^s u) - \phi' J^{s+1/2} u) = \int J^s u ([J; \phi'] J^s u - [J^{1/2}; \phi'] J^{s+1/2} u)$$

Thus Lemma 2.1 leads to the estimate

$$|\alpha_2| \leq c\|J^s u\|_2^2 \tag{3.12}$$

Next we consider the fifth term in (3.7) which is equal to

(3.12)

$$\int u J^s \partial_x u J^s u \phi + \int [J^s; u] \partial_x u J^s u \phi = -\frac{1}{2} \int \partial_x u J^s u J^s u \phi - \frac{1}{2} \int u J^s u J^s u \phi' + \int [J^s; u] \partial_x u \cdot J^s u \phi.$$

Lemma 2.3 implies that the above terms can be bounded by

$$c\|u\|_{1,\infty} \|J^s u\|_2^2. \tag{3.13}$$

Finally, we have the sixth term in (3.7)

$$\int J^s H(u \partial_x u) J^s H u \phi = \int (J^s H(u \partial_x u) - u J^s H \partial_x u) J^s H u \phi + \int u J^s H \partial_x u J H u \phi.$$

Using integration by parts, the last term above is bounded by (3.13). Then it suffices to estimate the L^2-norm of the expression inside parentheses in the first term of the rhs above. Thus

(3.14)
$$J^sH(u\partial_x u) - uJ^sH\partial_x u = \frac{1}{2}J^s(I-J)(u\cdot u) + \frac{1}{2}J^{s+1}(u\cdot u) - uJ^{s+1}u - uJ^s(I-J)u.$$

By Lemma 2.2 and Mihlin's theorem the first and fourth terms in the rhs of (3.14) when evaluated in L^2 are bounded by

$$\|u\|_\infty \|J^s u\|_2. \tag{3.15}$$

For the second and third terms in (3.14) we use Lemma 2.4 (with $p = p_2 = 2$, $p_3 = 4$, and $\theta = (s+1)/2$), and the Sobolev embedding theorem to conclude that

$$\|J^{s+1}(u\cdot u) - 2uJ^{s+1}u\|_2 \le c\|J^s u\|_2^2 \tag{3.16}$$

when $s > 3/2$.

Since $\|J^s u(t)\|_2$ is *a priori* bounded, by collecting the results in (3.8)–(3.16) and integrating in time we obtain that

$$\int_0^T \int \left\{(J^{s+1/2}u)^2 + (J^{s+1/2}Hu)^2\right\}\phi'(x)\,dx dt \le M$$

where M depends only on $\|u_0\|_s, T$, and ϕ. From Lemma 2.1 it follows that for any $\varphi \in C_0^\infty(\mathbb{R})$

$$J^{s+1/2}(u\varphi) \in L^2([0,T] : L^2(\mathbb{R})) = L^2([0,T]\times \mathbb{R}) \tag{3.17}$$

which completes the proof. The above formal derivation of (3.17) can be justified by using the continuous dependence of the solution with respect to the initial data, (see Theorem 3.1).

§4. Local Smoothing Effects in H^1-Solutions of the Benjamin-Ono Equation. Using the conservation laws for the BO equation (see [5]) and the results in section 2, we obtain the following existence result in H^1

THEOREM 4.1. *Let $u_0 \in H^1(\mathbb{R})$. Then there is a solution u to the IVP (1.1) such that*

$$u \in C_w([0,\infty) : H^1) \cap C([0,\infty); L^2) \cap L^2([0,T] : H^{5/4}_{\text{loc}}) \tag{4.1}$$

for any $T > 0$.

PROOF: First we approximate u_0 with a sequence $\{u_{0j}\} \subseteq H^2$ such that $\|u_0 - u_{0j}\|_{1,2} \to 0$ and $\|u_{0j}\|_{1,2} = \|u_0\|_{1,2}$. Let $\{u_j\} \subseteq C([0,\infty) : H^2)$ be the solutions of (1.1) with data $\{u_{0j}\}$. By the conservation laws it follows that $\{u_j\}$ is bounded in $C([0,\infty) : H^1)$ by a constant depending only on $\|u_0\|_{1,2}$.

For the solutions $\{u_j\}$ we repeat the argument in the proof of Theorem 3.2 with $s = 3/4$. Hence it suffices to modify only the estimates (3.13) and (3.16) in that proof. Omitting the index j we have that

$$\left|\int \partial_x u J^s u J^s u\phi\right| \le c\|\partial_x u\|_2 \|J^s u\|_4^2 \le c\|u(t)\|_{1,2}^3 \le \tilde{c}_1(\|u_0\|_{1,2})$$

and by Lemma 2.3

$$\|[J^s; u]\partial_x u\|_{4/3} \le c\|u\|_{1,2}\|J^s u\|_4 \le \tilde{c}_2(\|u_0\|_{1,2}).$$

These handle the estimate in (3.13) for the present case. The necessary changes in (3.16) follows directly from Lemma 2.4 (with $p = 4/3, p_1 = 2, p_2 = 4, p_3 = p_4 = 8/3$, and $\theta = \frac{(s+1)}{2} = \frac{7}{8}$) and the Sobolev embedding theorem. Therefore, using Lemma 2.3 we can conclude that for any $\varphi \in C_0^\infty(\mathbb{R})$,

$$\int_0^T \int |J^{3/4}(u_j(x,t)\varphi(x))|^2\, dxdt + \int_0^T \int |J^{3/4} u_j(x,t)|^2 \varphi^2(x)\, dxdt \le M$$

where M depends only on $\|u_0\|_{1,2}, T$, and φ.

From the equation we have that the maps $t \mapsto u_j(t) \in H^1$ are weakly equicontinuous in $[0,T]$ for any $T > 0$. Then $\{u_j\}$ contains a subsequence (denoted again by $\{u_j\}$) which converges weakly in H^1 uniformly in t. By a compactness argument and the diagonal process we may assume that

$$u_j \to u \quad \text{in} \quad L^2([0,T] \times (-a,a))$$

for any $T, a > 0$. Thus it is easy to show that u is a solution of (1.3), and that $u \in L^2([0,T]; H^{5/4}_{\text{loc}})$ for any $T > 0$. To complete the proof it suffices to note that $\|u_j(t)\|_2 = \|u_{0j}\|_2 \to \|u_0\|_2$. Since $(u_j(t))$ converges weakly (uniformly in t) to $u(t)$ it follows that $u \in C([0,\infty) : L^2(\mathbb{R}))$, with $\|u(t)\|_2 = \|u_0\|_2$.

ACKNOWLEDGEMENTS

The author would like to thank Jerry L. Bona for bringing this problem to his attention. Also he is grateful to R. R. Coifman, T. Kato, and C. Kenig for several valuable conversations. This work was supported in part by a NSF grant.

References

[1] Ablowitz, M. J. and Fokas, A. S., *The inverse scattering transform for the Benjamin-Ono equation—a pivot to multidimensional problems*, Stud. Appl. Math. **68**(1983),1–10.

[2] Benjamin, T. B., *Internal waves of permanent form in fluids of great depth*, J. Fluid Mech. **29** (1967), 559–592.

[3] Bergh, J. and Löfström, "Interpolation Spaces", Springer, 1970.

[4] Bock, T. L. and Kruskal, M. D., *A two parameter Miura transformation of the Benjamin-Ono equation*, Phys. Letters **74** (1979), 173–176.

[5] Case, K. M., *Benjamin-Ono related equations and their solutions*, Proc. Natl. Acad. Sci. U.S.A. **76** (1979), 1–3.

[6] Chen, H. H., Lee Y. C. and Pereira, N. R., *Algebraic internal wave solitons and the integrable Calogero-Moser-Sutherland N-body problem*, Phys. Fluid **22** (1979), 187–188.

[7] Coifman, R. R. and Meyer, Y., *Commutateurs D'Inégrales Singulières et Opérateurs Multilinéaires*, Ann. Inst. Fourier, Grenoble **28** (1978), 177-202.

[8] Coifman, R. R. and Meyer, Y., *Au delá des opérateurs pseudodifférentieles*, Astérisque 57, Société Mathématique de France ,1978.

[9] Coifman, R. R., (private communication)

[10] Constantin, P. and Saut, J. C., *Local smoothing properties of dispersive equations*, J. AMS **1** ((1988), 413–439.

[11] David, G. and Journé, J. L., *A boundedness criterion for generalized Calderón-Zygmund operators*, Ann. Math. **120** (1984), 371–397.

[12] Folland, G., "Introduction to Partial Differential Equations", Princeton University Press, 1976.

[13] Iório, R. J., *On the Cauchy problem for the Benjamin-Ono equation*, Comm. P.D.E. **11** (1986), 1031–1081.

[14] Kato, T., *Quasilinear equations of evolution with applications to partial differential equations*, Lecture Notes in Math. **448**, Springer, 1975.

[15] Kato, T., *On the Cauchy problem for the (generalized) Korteweg-de Vries equation*, Advances in Mathematics, Supplementary Studies, Studies in Applied Mathematics, **8** (1983), 93–128.

[16] Kato, T. and Ponce, G., *On nonstationary flows of viscous and ideal fluids in* $L^p_s(\mathbb{R}^2)$, Duke Math. J. **55** (1987), 487–499.

[17] Kato, T. and Ponce, G., *Commutator Estimates and the Euler and Navier-Stokes equations*, Comm. Pure Applied Math. **41** (1988), 891–907.

[18] Kruzhkov, S. N. and Faminskii, A. V., *Generalized solutions to the Cauchy problem for the Korteweg-de Vries equation*, Math. U.S.S.R. Sbornik **48** (1984), 93–138.

[19] Ono, H., *Algebraic solitary waves in stratified fluids*, J. Phys. Soc. Japan **39** (1975), 1082–1091.

[20] Runst, T., *Mapping properties of nonlinear operators in Triebel-Lizovkin and Besov type*, Analysis Mathematica **12** (1986), 313–346.

[21] Sjölin, P., *Regularity of solutions to the Schrödinger equation*, Duke Math. J. **55** (1987), 699–715.

[22] Stein, E. and Weiss, G., " Introduction to Fourier Analysis on Euclidean Spaces," Princeton University Press, 1971.

[23] Strichartz, R. S., *Multipliers in fractional Sobolev spaces*, J. Math. Mech. **16** (1967), 1031–1060.

[24] Strichartz, R. S., *Restrictions of Fourier Transform to quadratic surfaces and decay solutions of wave equations*, Duke Math. J. **44** (1977), 705–714.

[25] Vega, L., Doctoral Thesis, Universidad Autonoma, Madrid, 1987.

Part IV. The Mathematical Work of Mischa Cotlar

Mathematical Publications of Mischa Cotlar

1. Aritmética abstracta (Abstract arithmetic). Boletín de la Facultad de Ingeniería, Montevideo, Uruguay (1937).

2. Sobre las derivadas areolares simétricas de funciones de una variable compleja dual (On symmetric areolar derivatives of functions of one dual complex variable), Jointly with Juan Carlos Vignaux. An. Soc. Ci. Argentina 121 (1936), 128–133. Zbl. *14*.168.

3. Teoría de anágenos (Theory of anagenes). An. Soc. Ci. Argentina *127* (1939).

4. Familias Normales de Funciones no Analíticas (Normal Families of Non-Analytic Functions). An. Soc. Ci. Argentina *129* (1940), 3–25. MR *2*, 189.

5. Sobre Conjuntos No Medibles y Generalización de la Integral de Lebesgue (On Non-Measurable Sets and a Generalization of the Lebesgue Integral). Publ. Inst. Mat. Univ. Nac. Litoral *2* (1940), 149–176. MR *3*, 227.

6. Funciones Univalentes Sobre un Conjunto de Puntos del Contorno de un Dominio de Holomorfismo (Functions Which are Univalent on a Subset of the Boundary of a Domain of Regularity). Publ. Inst. Mat. Univ. Nac. Litoral *4* (1942), 47–96. MR *4*, 155; Zbl. *61*.153.

7. Los Orígenes de la Teoría del Algoritmo Wronskiano Dentro de la Doctrina Pitagórica. (The Origins of the Theory of the Wronskian Algorithm within the Pythagorian Doctrine), Jointly with Beppo Levi and Pedro Capelli. Math. Notae *3* (1943), 74–100. MR *5*, 57; Zbl. *60*.005.

8. Consideraciones Acerca de una Proposición de W. H. Young (Considerations Concerning a Proposition of W. H. Young), Jointly with Beppo Levi. Math. Notae *4* (1944), 145–155. MR *6*, 120.

9. Algunas Preguntas vinculadas a una Posible Extensión del Principio de Conservación de Dominios (Some Questions Related to a Possible Extension of the Principle of Conservation of Domains), Jointly with Pedro F. Capelli. Publ. Inst. Mat. Univ. Nac. Litoral *6* (1944), 63–96. MR *8*, 22; Zbl. *61*.153.

10. Las Familias Totalmente Normales de Funciones Holomorfas (Totally Normal Families of Holomorphic Functions), Jointly with Juan Carlos Vignaux. Publ. Inst. Mat. Univ. Nac. Litoral *7* (1944), 1–151. MR *6*, 261; Zbl. *60*.201.

11. Integrales Asimptóticas de Laplace–Stieltjes (Asymptotic Laplace–Stieltjes integrals), Jointly with Juan Carlos Vignaux. Univ. Nac. La Plata. Publ. Fac. Ci. Fisicomat. No. 180, Vol. 3, núm. 4. Serie segunda, 14, Contribuciones (1944), 345–400. MR *6*, 268; Zbl. *60*.250.

12. Un Método de Construcción de Estructuras y sus Aplicaciones a Espacios Topológicos y Aritmética Abstracta (A method of Construction of Structures and its Application to Topological Spaces and Abstract Arithmetic). Univ. Nac. Tucumán. Revista A. *4* (1944), 105–157. MR *7*, 225; Zbl. *60*.059.

13. Una Generalización de los Factoriales (A Generalization of Factorials). Math. Notae *5* (1945), 89–107. MR *7*, 242; Zbl. *60*.024.

14. Ejercicios Sobre la Función Coseno (Exercises on the Cosine Function), Jointly with Beppo Levi. Math. Notae *5* (1945), 193–214. MR *7*, 506.

15. Sobre Funciones de Conjunto no-Aditivas y una Generalización de la Descomposición de Vallée–Poussin (On Nonadditive set Functions and Generalization of the Decomposition of Vallée Poussin), Jointly with Yanny Frenkel. Univ. Nac. Tucumán. Revista A. 5 (1946), 227–253; MR *8*, 572.

16. Estudio de una Clase de Polinomios de Bernoulli (Study of a Class of Bernoulli Polynomials). Math. Notae *6* (1946), 69–95. MR *9*, 30; Zbl. *61*.140.

17. Un Método para Obtener Congruencias de Números de Bernoulli (A Method for Obtaining Congruences of Bernoulli Numbers). Math. Notae *7* (1947), 1–29. MR *9*, 175; Zbl. *30*.014.

18. Una Teoría General de la Integral Basada en una Extensión del Concepto de Límite (A General Theory of Integral Based on an Extension of the Concept of Limit), Jointly with Yanny Frenkel. Univ. Nac. Tucumán. Revista A. 6 (1947), 113–159. MR *10*, 107; Zbl. *32*.196.

19. Una Extensión del Teorema de Rolle Para Transformaciones Continuas del Plano (An Extension of Rolle's Theorem for Continuous Transformations of the Plane). Math. Notae *8* (1948), 79–84. MR *10*, 557; Zbl. *32*.199.

20. Grupos Semiordenados y l-Ideales de Riesz–Birkhoff (Semiordered Groups and Riesz–Birkhoff l-Ideals), Jointly with Eduardo Zarantonello. Publ. Inst. Mat. Univ. Nac. Litoral *8* (1948), 105–192. MR *10*, 99; Zbl. *32*.166.

21. Sobre la Variación de Funciones Discontínuas y Multivaluados de Variable Real (On the Variation of Discontinuous and Multivalued Functions of a Real Variable), Jointly with Emilio O. Roxin. Revista Unión Mat. Argentina *14* (1949), 38–46. MR *11*, 336; Zbl. . 34.325.

22. Medidas Invariantes en Espacios Topológicos Compactos (Invariant Measures in Compact Topological Spaces), Jointly with Rodolfo A. Ricabarra. Math. Notae *9* (1949), 52–77. MR *16*, 228.

23. Sobre un Teorema de E. Hopf (On a Theorem of E. Hopf), Jointly with Rodolfo A. Ricabarra. Revista Unión Mat. Argentina *14* (1949), 49–63. MR *12*, 85; Zbl. *37*.076.

24. Mayorantes y Minorantes No-Aditivas en la Teoría de la Integral de Perron–Denjoy (Non-Additive Majorants and Minorants in the Theory of the Perron–Denjoy Integral), Jointly with Yanny Frenkel. Revista Acad. Ci. Madrid *44* (1950), 411–426. MR *13*, 121; Zbl. *45*.025.

25. Sobre la Integral de Kolmogoroff (On Kolmogoroff's integral), Jointly with Yanny Frenkel. Univ. Buenos Aires. Contrib. Ci. Ser. A. *1* (1950), 46–63. MR *13*, 634; Zbl. *41*.380.

26. La Integral de Carathéodory (The Carathéodory Integral), Jointly with Rodolfo A. Ricabarra. Memorias Real Acad. Ci. Exact. Fis. Nat. Madrid *4* (1950), 1–47. MR *16*, 228.

27. Sobre Transformaciones de Conjuntos y Operadores de Koopman (On Transformations of Sets and Koopman's Operators), Jointly with Rodolfo A. Ricabarra. Revista Unión Mat. Argentina *14* (1950), 232–254. MR *12*, 719; Zbl. *45*.389.

28. Sobre los Fundamentos de la Teoría Ergódica (On the Foundations of Ergodic Theory). *Sympo. Latinoamer., Diciembre 1951.* Centro de Coop. Científ. UNESCO (1952), 71–84. MR *14*, 566; Zbl. *48*.92.

29. On the Existence of Characters in Topological Groups, Jointly with Rodolfo A. Ricabarra. Amer. J. Math. 76 (1954), 375–388. MR *16*, 111; Zbl. *55*.257.

30. El Problema de los Momentos y la Teoría de Operadores Hermitianos (The Problem of Moments and the Theory of Hermitian Operators). *Segundo Symp. Matem. Latinoamer., Villavicencio, Argentina, Julio, 1954*, Centro de Coop. Científ. UNESCO (1954), 71–85. MR *16*, 1082; Zbl. *58*.335.

31. On a Theorem of Beurling and Kaplansky. Pacific J. Math. *4* (1954), 459–465. MR *16*, 146; Zbl. *55*.266.

32. Sobre los Teoremas Ergódicos (On Ergodic Theorems). Math. Notae *14* (1955), 85–119. MR *19*, 843.

33. Teoría Algebráica de la Media y Teorema de Hahn–Banach (On the Algebraic Theory of the Mean and the Hahn–Banach Theorem). Rev. Un. Mat. Argentina *17* (1955), 9–24. MR *18*, 383.

34. A Combinatorial Inequality and its Applications to L^2-Spaces. Rev. Mat. Cuyana *1* (1955), 41–55. MR *18*, 219.

35. A General Interpolation Theorem for Linear Operations. Rev. Mat. Cuyana *1* (1955), 57–84. MR *19*, 563.

36. Some Generalizations of the Hardy–Littlewood Maximal Theorem. Rev. Mat. Cuyana *1* (1955), 85–104. MR *19*, 564.

37. A Unified Theory of Hilbert Transforms and Ergodic Theorems. Rev. Mat. Cuyana *1* (1955), 105–167. MR *18*, 893.

38. On the Integral of Carathéodory, Jointly with Rodolfo Ricabarra. Univ. Nac. La Plata, Publ. Fac. Ci. Fisicomat. Serie Segunda, Rev. *5* (1956), 153–161. MR *19*, 844; Zbl. *80*.271.

39. On the Convexity Theorem of Riesz–Thorin and Marcinkiewicz, Jointly with María Luisa Bruschi. Univ. Nac. La Plata. Publ. Fac. Ci. Fisicomat. Serie Segunda, Rev. *5* (1956), 162–172. MR *20*, 1218.

40. On Almost Orthogonal Operators in L^p-Spaces, Jointly with Rafael Panzone. Acta Sci. Math. Szeged 19 (1958), 165–171. MR *21*, #809.

41. "Condiciones de Continuidad de Operadores Potenciales y de Hilbert" (Continuity Conditions for Potential and Hilbert Operators). Cursos y Seminarios de Matem., Fasc. 2, Facultad de Ciencias Exactas y Naturales, Univ. de Buenos Aires, Buenos Aires, 1959. MR *22*, #6985; Zbl. *99*.377.

42. Operadores Potenciales Generalizados y Sumas Ortogonales (Generalized Potential Operators and Orthogonal Sums). Rev. Un. Mat. Argentina *19*, 112–134 (1960). MR *24*, #A986.

43. Generalized Potential Operators, Jointly with Rafael Panzone. Rev. Un. Mat. Argentina *19*, 3–41 (1960). MR *24*, #A842.

44. On Some Inequalities of Potential Operators, Jointly with Eduardo L. Ortiz. Univ. Nac. La Plata, Publ. Fac. Ci. Fisicomat. Serie Segunda, Rev. *8*, 16–34 (1962). MR *26*, #1699.

45. Semigrupos Arquimedianos y Álgebras de Operadores (Archimedian Semigroups and Operator Algebras). *Proc. Third Brazilian Math. Colloq. 1961*, Summa Brasil. Math. *5* (1962), 5–46. MR *35*, #2805.

46. "Introducción a la Teoría de la Representación de Grupos" (Introduction to the Theory of Group Representations). Cursos y Seminarios de Matem., Fasc. 11, Facultad de Ciencias Exactas y Naturales, Univ. de Buenos Aires, Buenos Aires, 1963. MR *35*, #1711.

47. "Introducción al Álgebra. Nociones de Álgebra Lineal" (Introduction to Algebra. Concepts of Linear Algebra), Jointly with Cora Ratto de Sadosky. Fourth edition, Ed. Univ. de Buenos Aires, Buenos Aires, 1966. MR *35*, #4061.

48. Some Remarks on the Marcinkiewicz Convexity Theorem in the Upper Triangle, Jointly with C. A. Berenstein, N. Kerzman & P. Krée. Studia Math. *29* (1967), 79–95. MR *37*, #6780.

49. "Nociones de Espacios Normados" (Concepts in the Theory of Normed Spaces), Jointly with Roberto Cignoli. Preliminary edition. Ed. Univ. de Buenos Aires, Buenos Aires, 1967. MR *40*, #669.

50. On Quasi-Homogeneous Bessel Potential Operators, Jointly with Cora Sadosky. *Singular Integrals*, Proc. Symp. Pure Math. Amer. Math. Soc. *10* (1967), 275–287. MR *52*, #4037; Zbl. *58*.1.

51. "Equipación con Espacios de Hilbert" (Equipping with Hilbert spaces). Cursos y Seminarios de Matem., Fasc. 15, Facultad de Ciencias Exactas y Naturales, Univ. de Buenos Aires, Buenos Aires, 1968. MR *41*, #5922.

52. "Convolution Operators and Factorization." McGill Analysis Seminar, McGill University, Montreal, 1972.

53. "An Introduction to Functional Analysis," Jointly with Roberto Cignoli. Translated from the Spanish by A. Torchinsky and A. González Villalobos. Texts in Advanced Mathematics, North-Holland Publ. Co., Amsterdam-London, 1974. MR *53*, #8845.

54. A Moment Theory Approach to the Riesz Theorem on the Conjugate Function with General Measures, Jointly with Cora Sadosky. Studia Math. *53* (1975), 75–101. MR *53*, #1142; Zbl. *303*.42024.

55. Transformée de Hilbert, Théorème de Bochner et le Problème des Moments, I, Jointly with Cora Sadosky. C. R. Acad. Sci. Paris Ser. A-B *285* (1977), 433–436. MR *57*, #10344.

56. Transformée de Hilbert, Théorème de Bochner et le Problème des Moments, II, Jointly with Cora Sadosky. C. R. Acad. Sci. Paris Ser. A-B *285* (1977), MR *57*, #7012; Zbl. *388*.44008.

57. On the Helson–Szegö Theorem and a Related Class of Modified Toeplitz Kernels, Jointly with Cora Sadosky. *Harmonic Analysis in Euclidean Spaces*, Proc. Symp. Pure Math. Amer. Math. Soc. *35* (1979), 383–407. MR *81j*: 42022.

58. Characterization of Two Measures Satisfying Riesz Inequality for the Hilbert Transform in L^2, Jointly with Cora Sadosky. Acta Cient. Venez. *30* (1980), 346–348. MR *81d*: 44004.

59. Continuous Generalized Toeplitz Kernels in $\mathbb{R}$, Jointly with Rodrigo Arocena. *Special issue in honor of Antonio Monteiro*, Portugal. Math. *39* (1980), 419–439. MR *86i*:42007.

60. Weighted Inequalities in L^2 and Lifting Properties, Jointly with Rodrigo Arocena & Cora Sadosky. *Mathematical Analysis and Applications*, Adv. Math. Suppl. Stud. *7A* (1981), 95–128. MR *83e*: 42016.

61. A Generalized Herglotz–Bochner Theorem and L^2-Weighted Inequalities with Finite Measures, Jointly with Rodrigo Arocena. *Harmonic Analysis Conf. in honor of A. Zygmund*, Wadsworth Intl. Math. Series (1982), 258–269. MR *85f*: 42022.

62. On Some L^p Versions of the Helson–Szegö Theorem, Jointly with Cora Sadosky, in *Harmonic Analysis Conference in honor of A. Zygmund*, Wadsworth Intl. Math. Series (1982), 306–317. MR *85i*: 42015.

63. Majorized Toeplitz Forms and Weighted Inequalities with General Norms, Jointly with Cora Sadosky, in *Harmonic Analysis, Minneapolis 1981*, Lecture Notes in Math. *908*, Springer-Verlag (1982), 139–168. MR *84h*: 42019.

64. Dilation of Generalized Toeplitz Forms and Weighted Inequalities with General Norms, Jointly with Rodrigo Arocena, in *Harmonic Analysis, Minneapolis 1981*, Lecture Notes in Math. *908*, Springer-Verlag (1982), 169–188. MR *84j*: 47006.

65. Generalized Toeplitz Kernels and Adamjan–Arov–Krein Moment Problems, Jointly with Rodrigo Arocena. *Toeplitz Centennial, Tel Aviv, 1981*, Operator Theory: Adv. Appl., *4* (1982), 37–55. MR *84a*: 42021.

66. On a Lifting Theorem and Its Relation to Some Approximation Problems, Jointly with Rodrigo Arocena. *Functional Analysis, Holomorphy and Approximation Theory, Rio de Janeiro, 1980*, North-Holland Math. Stud., *71* (1982), 1-25. MR *85c*: 41057.

67. Generalized Toeplitz Kernels, Hankel Forms and Sarason's Commutation Theorem, Jointly with Rodrigo Arocena. Acta Cient. Venezolana *33* (1982), 89–98. MR *85c*: 47927.

68. Moment Theory and Continuity of the Hilbert and Poisson Transform in L^2 Spaces. *Functional Analysis, Holomorphy, and Approximation Theory, Río de Janeiro, 1979*, Lecture Notes in Pure and Appl. Math., *83*, M. Dekker (1983), 33–57. MR *84d*: 44008.

69. Vector-Valued Inequalities of Marcinkiewicz–Zygmund and Grothendieck Type for Generalized Toeplitz Kernels, Jointly with Cora Sadosky, in *Harmonic Analysis, Cortona 1982*, Lecture Notes in Math. *992* Springer-Verlag (1983), 278–308. MR *85i*: 42017.

70. Inegalités à Poids pour les Coefficients Lacunaires de Certaines Fonctions Analytiques, Jointly with Cora Sadosky. C. R. Acad. Sci. Paris *A 299* (1984), 591–594. MR *86c*: 30067.

71. Generalized Toeplitz Kernels, Stationarity and Harmonizability, Jointly with Cora Sadosky. J. Analyse Math. *44* (1984/85), 117–133. MR *87g*: 60039.

72. A Lifting Theorem for Subordinated Invariant Kernels, Jointly with Cora Sadosky. J. Functional Anal. *67* (1986), 345–359. MR *87g*: 47045.

73. Lifting Properties, Nehari Theorem and Paley Lacunary Inequality, Jointly with Cora Sadosky. Rev. Matem. Iberoamericana *2* (1986), 55–71. MR *88d*: 47007.

74. Toeplitz Kernels, Scattering Structures and Covariant Systems, Jointly with R. Arocena & José Rafael León. *Aspects of Mathematics and Its Applications*, North-Holland Math. Library *34* (1986), 77–95. MR *87j*: 47037.

75. Rodolfo A. Ricabarra (August 1925–November 1984) (obituary), Jointly with L. Recht. Rev. Un. Mat. Argentina *32* (1985) 159–161. MR *88e*: 01047.

76. Some Aspects of the Trigonometric Moment Problem. *First Conference on Analysis, Mérida, 1986*, Univ. de los Andes, Mérida, Venezuela (1986), 65–82, MR *88i*: 42020.

77. Prolongements des Formes de Hankel Generalisées en Formes de Toeplitz, Jointly with Cora Sadosky. C. R. Acad. Sci. Paris *I 305* (1987), 167–170.

78. Toeplitz Liftings of Hankel Generalized Forms, Jointly with Cora Sadosky, *Function Spaces and Applications Lund 1986*, Lecture Notes in Math. *1302*, Springer-Verlag (1988), 22–43.

79. Integral Representations of Bounded Hankel Forms Defined in Scattering Systems with a Multiparametric Evolution Group, Jointly with Cora Sadosky. Operator Theory: Adv. & Appl. *35* (1988), 357–375.

80. Eigenfunction Expansions of Covariance Kernels, Jointly with Jose León & Cristina Pereyra. Acta Cientif. Venez., (in press).

81. The Generalized Bochner Theorem in Algebraic Scattering Systems, Jointly with Cora Sadosky. MSRI #00621–88. *Analysis at Urbana II, Proc. of Special Year at the University of Illinois, 1987*, London Math. Soc. Lecture Notes Ser. *138*, Cambridge University Press (1989), 144–169.

82. Two-parameter Lifting Theorems and Double Hilbert Transforms in Commutative and Non-Commutative Settings, Jointly with Cora Sadosky. MSRI #03821–88. J. Math. Anal. & Appl., (in press).

83. Nonlinear Lifting Theorems, Integral Representations and Stationary Processes in Algebraic Scattering Systems, Jointly with Cora Sadosky, in *Proc. International Conference on Operator Theory in Honor of Israel Gohberg*. Operator Theory: Adv. & Appl. *41* (1989), 97–123.

84. The Helson–Szegö Theorem in L^p of the Bidimensional Torus, Jointly with Cora Sadosky, *Proc. Conference on PDE & Harmonic Analysis at Florida Atlantic University*, 1988, Contemporary Math., (in press).

85. Non-Unitary Versions of the Nehari and the Sz. Nagy–Foias Theorems, Jointly with Cora Sadosky. *Proc. of Workshop on Matrix and Operator Theory, Rotterdam, 1989*, Operator Theory: Adv. & Appl. (submitted).

The Early Papers of Mischa Cotlar (1936-1955)

John Horváth

All the early works of Mischa Cotlar were written in Spanish and published almost exclusively in South American journals which are difficult to find. For this reason, some of his results were later rediscovered by others and named after them. I will discuss the contents of these early papers, subdividing them roughly according to their subject matter as follows:

1) Complex variables,

2) Real variables, measure theory and integration,

3) Ordered groups and vector spaces,

4) Harmonic analysis.

The numbers in square brackets refer to the works of Mischa Cotlar listed as his Mathematical Publications, the letters in curly brackets to the works of other authors listed at the end of the article by Cora Sadosky.

1. Complex Variables [2, 4, 6, 9, 10, 11, 13, 14, 16, 17]. Some among Mischa Cotlar's first publications were inspired by Juan Carlos Vignaux and Pedro Capelli, and were in part written with their collaboration. They are related to the theory of normal families due to Paul Montel, with whom Cotlar had a fruitful correspondence (see e.g. the footnote on p. 6 of [4]).

In paper [4] Cotlar starts out with a formula which can be found in the first section of Riemann's doctoral dissertation {Rie}, and which more succinctly can be written as

$$dw = \frac{\partial w}{\partial z}dz + \frac{\partial w}{\partial \bar{z}}d\bar{z},$$

where

$$Sw = \frac{\partial w}{\partial z} = \frac{1}{2}\left(\frac{\partial w}{\partial x} - i\frac{\partial w}{\partial y}\right) \quad \text{and} \quad Dw = \frac{\partial w}{\partial \bar{z}} = \frac{1}{2}\left(\frac{\partial w}{\partial x} + i\frac{\partial w}{\partial y}\right).$$

A function f defined in a domain Δ of the complex plane $\mathbb{C}$ is holomorphic if $Df = 0$, or equivalently if $Sf(z) = f'(z)$ for $z \in \Delta$. Then $S^n f$ is holomorphic for every $n \in \mathbb{N}$ and so one has trivially the inequalities

$$(1) \qquad |D^m S^{n-m} f(z)| \le \frac{1}{2^n}|S^n f(z)| \quad \text{for} \quad z \in \Delta \quad \text{and} \quad 1 \le m \le n.$$

Furthermore a function f which is bounded and holomorphic in Δ satisfies for every $n \in \mathbb{N}$ the Cauchy inequality

$$(2) \qquad |S^n f(z)| \le n!\frac{M}{R^n},$$

where $M = \sup_{z\in\Delta}|f(z)|$ and R is the distance of z from the boundary of Δ. Cotlar takes these two properties as the basis of a definition and says that a function f defined in Δ belongs to the class (N) if its derivatives $D^m S^n f$ of all orders exist and conditions (1) and (2) are satisfied.

Cotlar's main concern is, however, with a more general class. A function

f defined in Δ is (NG) if the derivatives $D^m S^n f$ all exist and if at any $z \in \Delta$ conditions (1) and (2) are satisfied for $n \le 2$ and for all n such that $Sf(z) = S^2 f(z) = \ldots = S^{n-1} f(z) = 0$. He then proceeds to prove for (NG) functions many results which for holomorphic functions are classical:

- if f is a non-constant (NG) function, then every zero of f has finite order;
- the zeros of a non-constant (NG) function are isolated points;
- the principle of the argument about the number of zeros of a function in a domain, and Rouché's theorem;
- if the maximum of $|f(z)|$ is attained at an interior point of Δ, then f is a constant;
- the theorems of Bloch and Schottky;
- if a family of (NG) functions omits two values, then it is normal;
- the Landau-Carathéodory theorem: if (f_n) is a sequence belonging to an additive group of (N) functions such that the f_n omit two values, and if $(f_n(z))$ converges at infinitely many points z of Δ, then (f_n) converges uniformly on every compact subset of Δ.

As it is now universally known, a holomorphic function f defined in the unit disk $D = \{z \in \mathbb{C} : |z| < 1\}$ is said to be <u>univalent</u> (schlicht) if $z_1 \neq z_2$ implies $f(z_1) \neq f(z_2)$. Functions which are univalent in D have properties that are not shared by general holomorphic functions. In his paper [6] Cotlar emphasizes one such property discovered by Leopold Fejér. The latter showed that even if f is holomorphic in D and continuous in the closure $\bar{D}$ of D, its power series $\sum_{n=0}^{\infty} c_n z^n$ can diverge on a dense subset of the boundary $\Gamma = \{z \in \mathbb{C} : |z| = 1\}$ of D {Lan}. If, however, f is furthermore univalent in D, then $\sum_{n=0}^{\infty} c_n e^{in\theta}$ converges uniformly on Γ {Lan, §13, p. 65}.

In [6] Cotlar generalizes this and other properties of univalent functions. Let f be a holomorphic function defined in D satisfying $f(0) = 0$. If $\xi \in \Gamma$ is such that for every sequence $z_n \in D$ which tends to ξ the sequence $f(z_n)$ converges to the same limit $f(\xi)$, then f is said to be defined at ξ and to have value $f(\xi)$. Let γ be an arc of Γ such that f is defined at every point of γ and that f is a continuous function on γ; Cotlar says that γ is a <u>proper arc</u> for the function f if every point of $f(\gamma)$ can be joined with the point at infinity by a continuous curve which does not meet $f(D)$. He proves the following generalization of Fejér's theorem: If f is a bounded holomorphic function in D, continuous on $\bar{D}$, then $\sum_{n=0}^{\infty} c_n e^{in\theta}$ converges to $f(\xi)$ at every point $\xi = e^{i\theta}$ of a proper arc with the possible exception of a non-dense subset.

In his proofs Cotlar applies the theory of subordination due to J.E. Littlewood and W.W. Rogosinski: f is <u>subordinated</u> to the univalent function F if there exists a holomorphic function σ defined in D such that $\sigma(0) = 0$, $|\sigma(z)| \le 1$ and $f(z) = F(\sigma(z))$ {Hi, Sec. 18.5}. Cotlar uses the observation that if γ is a proper arc for f, then σ is holomorphic on γ. It is remarkable that in 1941 Cotlar was familiar with subordination, which even now is not widely known and whose first detailed treatment appeared only in 1944 in Littlewood's idiosyncratic book {Li}.

Let us quote two more of the many results of the paper. If f is univalent in D, then the Koebe-Bieberbach distortion theorem {Hi, Th. 17.4.7} states that

$$|f(z)| \le |f'(0)|\frac{|z|}{(1-|z|)^2}.$$

Assume now that, instead of f being univalent, there exists an arc of length $\ell > 0$ which is proper for f and on which f is univalent. If ν is the integral part of $2\pi/\ell$, then

$$|f(z)| \le \left[\sum_{j=1}^{\nu} C(\ell, j)|f^{(j)}(0)|\right]\frac{|z|}{(1-|z|)^2},$$

where the $C(\ell, j)$ are positive constants whose values are given explicitly in the paper. Furthermore the number of zeros of f in the disk $|z| < r$ is at most equal to

$$\nu(\ell, r) = \frac{\pi}{2 \text{ arc } \tan\left[\frac{1-r}{1+r}\tan\frac{\ell}{4}\right]}.$$

Let us also mention a result according to which a family $\{f\}$ of functions which has a common proper arc of length ℓ, on which all the f are univalent, is quasi-normal {Va} of order $\nu(\ell, r)$ in the disk $|z| < r$. In all these result the arc of length ℓ can be replaced by a set of measure ℓ.

In the papers [9] and [10], written jointly with P. Capelli and J.C. Vignaux, respectively, Cotlar combines the theory of holomorphic functions with what is called the theory of functions of a real variable. A holomorphic function f maps inner points of its domain Δ into inner points of $f(\Delta)$. The authors of [9] ask whether the extension of f to the boundary of Δ maps points of density in the sense of Lebesgue into points of density. They do not answer the question but show why the study of points of density of subsets of the boundary should be of interest. Let E be a subset of the unit circle Γ and ξ a point of density of E. The authors introduce the ε-<u>thickness</u> <u>of</u> <u>densification</u> $d(\varepsilon)$ of ξ with respect to E as the largest number d such that for any arc $I \subset \Gamma$ containing ξ with length $|I| \le d$ one has $|I \cap E|/|I| \ge 1 - \varepsilon$. Assume that E has measure $\ell > 0$ and that f is a bounded holomorphic function in the unit disk D such that $|f(e^{i\theta})| \le M$ and $|f(e^{i\theta})| \le m$ on E. A typical result of [9], reminiscent of

Nevanlinna's two-constant theorem {Nev, Chap. III, §2, No. 36, p. 42}, is the following inequality:

$$|f(z)| \le m + \frac{1-|z|^2}{d(\varepsilon)^2}(2\pi-\ell) + \varepsilon$$

valid for any $\varepsilon > 0$ and z belonging to the ray 0ξ. An analogous inequality holds for harmonic functions and an application to $\log|f(z)|$ yields the F. and M. Riesz uniqueness theorem. The authors also generalize the inequality to the case when f is only integrable on Γ in the sense of Lebesgue or even of Denjoy. A recent paper of Cotlar's student José de Abreu G. {Ab} improves [9] in two directions: instead of the L^{∞}-norm of $f(e^{i\theta})$ he considers the BMO-norm, and besides points of density he considers Marcinkiewicz points.

Montel's theory of normal families concerns sequentially compact sets of holomorphic functions for the topology of uniform convergence on compact subsets of D. Also the Ostrowski-Hinčin theorem {Bi, vol. II, Chap. III, §8, no. 3., p. 157} deduces from the convergence of (f_n) on a subset of positive measure of Γ the uniform convergence of (f_n) on closed sets completely interior to D. If, however, (f_n) converges uniformly on an arc $AB \subset \Gamma$, then according to a theorem of Montel $(f_n(z))$ converges uniformly in a set which comes up to Γ and is shaded in Figure 1. In the book-long paper [10] the authors find common generalizations of the Ostrowski-Hinčin and the Montel theorems, and also investigate

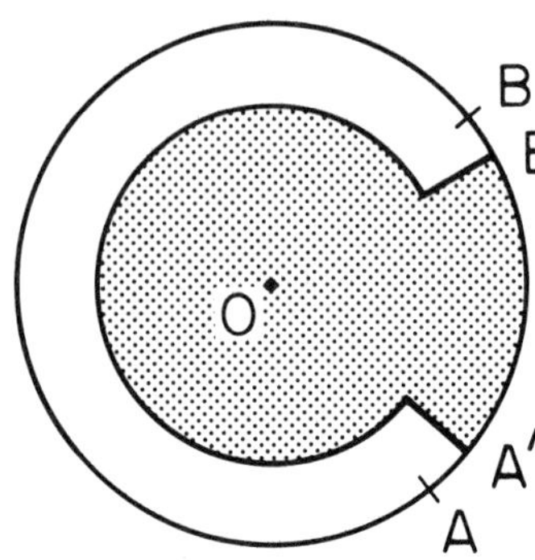

Figure 1

sequentially compact sets for other convergence structures.

Remaining first in the framework of real variables, they consider families of not necessarily continuous functions, and besides uniform convergence they also consider pointwise convergence, convergence in the mean, weak L^p- convergence, etc. Correspondingly, a family $\{F\}$ of complex-valued functions defined on an open interval $I \subset \mathbb{R}$ is said to be (α)-normal if every sequence (F_n) of elements of $\{F\}$ contains a subsequence (F_{n_i}) which converges uniformly on every compact subinterval I_1 of I; it is (α_1)-normal if (F_{n_i}) converges uniformly almost everywhere in I_1; it is (β)-normal if (F_{n_i}) converges pointwise on I. Among the other concepts of normalcy introduced by the authors, let us mention that $\{F\}$ is (δ)-normal if (F_{n_i}) converges in measure and (ε)-normal if (F_{n_i}) converges in the mean. The classical Ascoli-Arzelà theorem refers to (α)-normal families of continuous functions. To broaden their scope the authors consider, besides the usual concept of continuity, also some generalizations. Thus e.g., f is $(C,1)$-continuous at t_0 if

$$\lim_{\delta\to 0} \frac{1}{\delta} \int_\alpha^\beta [f(t) - f(t_0)]dt = 0,$$

where (α,β) is an interval containing t_0 and $\delta = \beta - \alpha$. If in the last integral the integrand is replaced by its absolute value, we obtain the notion of C^*-continuity. The usual Lipschitz conditions figure among the continuity concepts, and they say that f is <u>upper</u>-(M_∞)-<u>semicontinuous</u> at t_0 if

$$f(t_0) = \lim_{p\to 0} \lim_{\delta\to 0} \left\{ \frac{1}{\delta} \int_\alpha^\beta [f(t)]^p dt \right\}^{1/p} .$$

They prove more than fifty criteria for their various notions of normalcy of which we quote only a few to give an idea of their flavor.

• If $\{F\}$ consists of positive functions, is uniformly bounded, and equally and uniformly upper-(M_∞)-semicontinuous, then it is (α)-normal.

• Let $\{F\}$ consist of odd $(C,1)$-continuous functions on $\mathbb{R}$ having period 2π which are integrable on $(0,2\pi)$ and whose Fourier-coefficients decrease in absolute value. If the F are equally continuous in the mean, and at one point their Fourier-series converge absolutely and uniformly with respect to the F, then $\{F\}$ is (α)-normal.

Several criteria are related to Mandelbrojt's quasi-analyticity of lacunary trigonometric series {Ma, Chap VIII, no. 53, p. 127}.

• If $\{F\}$ is a uniformly bounded family of functions which is either uniformly $(C,1)$-continuous at each $t \in I$ or uniformly upper-(M_∞)-semicontinuous at each $t \in I$, then $\{F\}$ is (β)-normal.

Among the many generalizations of the Ostrowski-Hinčin and the Montel theorems, let us quote the following:

Let (f_n) be a bounded sequence of functions holomorphic in D and continuous on the arc $AB \subset \Gamma$. Assume that the f_n satisfy the following conditions:

1) on AB we have $\int_{AB} \log^+ |f_n(e^{i\theta})| d\theta < N$ for all n,

2) a Picard-type condition: there are two values which the functions f_n omit in D,

3) in the sector OAB one has $|f_n(z)| \le M(r)$ for $|z| \le r$, where $M(r)$ satisfies

$$\log M(r) = O\left(\log \frac{1}{1-r}\,\delta^{\frac{1}{2}\log\log\frac{1}{1-r}}\right), \quad \frac{1}{2} \le \delta < 1.$$

If (f_n) converges on a set $E \subset AB$ of positive measure, then it converges uniformly on every radius OP, where P is a point of density of E.

If E is a subset of Γ, the <u>irradiation</u> of E is the union of all the radii OP as P varies in E. A subset $E \subset \Gamma$ has <u>equal degree of density</u> if

$$\lim_{|I|\to 0} \frac{|I|-|I\cap E|}{|I|^2}$$

is the same strictly positive number for every point $\xi \in E$, where, as above, I is an arc containing ξ.

Let $\{f\}$ be a family of functions holomorphic in D and defined on $E \subset \Gamma$. Cotlar and Vignaux say that $\{f\}$ is <u>totally</u> (α)-<u>normal</u> in $D \cup E$ if any sequence of elements of $\{f\}$ contains a subsequence which converges uniformly on every compact subset of D and also on the irradiation of E', where E' is a subset of E having equal degree of density. Similar definitions hold for the other types of normalcy. The authors establish properties of totally normal families, some of which are the analogues of results concerning Montel's normal families. For instance, if $\{f\}$ is totally (α)-normal in $D \cup E$ and if $(f_n) \subset \{f\}$ converges on an infinite sequence of points of D which has a limit point in D, then (f_n) converges uniformly on $D \cup E$.

In the last chapter, which contains 28 numbered theorems and 60 corollaries, the authors present criteria for total normalcy. The criteria for total-(α)-normalcy are divided into two groups: those with hypotheses in D and those with hypotheses on Γ. Let us state the first of the criteria:

• For any $z \in D$ let $C(z)$ be the disk $\{\zeta \in D : |\zeta - z| \le 1 - |z|\}$. If there exists an $\alpha > 0$ and an increasing sequence of numbers $0 < r_n < 1$ converging to 1 such that for all n, for all $z \in D$ with $|z| = r_n$, and for all f one has

$$\iint_{C(z)} |f'(z)|^{\alpha} dxdy \le M(1-r_n)^2,$$

and the f' are continuous on Γ, then $\{f\}$ is totally (α)-normal in $D \cup \Gamma$.

Ever since he started studying mathematics as a teenager, Cotlar was fascinated by the theory of numbers. In the articles [13], [16], and [17] he considers sequences of polynomials and of numbers defined by formal power series, and obtains a great quantity of results pertaining to number theory and to combinatorics. Paper [13] starts with a very clear expository account of the general interpolatory polynomials of Paul Appell {Bou, Chap. VI, §1} and René Lagrange {Lag}.

The classical Bernoulli polynomials $B_n(x)$ are defined by the formal power series expansion

$$\frac{te^{xt}}{e^t-1} = \sum_{n=0}^{\infty} B_n(x)\frac{t^n}{n!}.$$

They were introduced by Jakob Bernoulli towards the end of the 17th century to study the sums of nth powers of consecutive integers. In fact

$$\Delta B_n(x) = B_n(x+1) - B_n(x) = nx^{n-1} \tag{3}$$

and thus

$$\frac{1}{n+1}[B_{n+1}(k) - B_n(0)] = 1^n + 2^n + \ldots + (k-1)^n$$

for every positive integer k. The Bernoulli polynomials satisfy the relations

$$\frac{d}{dx}B_n(x) = nB_{n-1}(x),$$

and are given more explicitly by

$$B_n(x) = \sum_{j=0}^{n} \binom{n}{j} B_{n-j} x^j,$$

where the B_n, defined by the power series

$$\frac{t}{e^t-1} = \sum_{n=0}^{\infty} B_n \frac{t^n}{n!},$$

are the celebrated Bernoulli numbers which figure in the title of [17], in the power series expansion of $\tan x$ {Bou, Chap. VI, §2}, in the formula giving the value of the Riemann zeta function at positive even integers, and in various criteria affirming the truth of Fermat's last theorem {Rib, Lecture VI}.

Having recalled these concepts, Cotlar considers the operator $\underset{\omega}{\Delta}$ defined by

$$\underset{\omega}{\Delta} f(x) = \frac{f(x+\omega)-f(x)}{\omega}.$$

For $\omega = 1$ this is the difference operator Δ we met in (3), and for $\omega = 0$ it becomes the differentiation operator $D = \frac{d}{dx}$. Appell says that a sequence $(P_n(x))$, where $P_n(x)$ is a polynomial of degree n, is $\underset{\omega}{\Delta}$-harmonic if $\underset{\omega}{\Delta} P_n(x) = nP_{n-1}(x)$. If we define the generalized factorials by

$$\sigma_n(x;\omega) = x(x-\omega)\dots(x-(n-1)\omega),$$

then for an arbitrary sequence $(a) = (a_n)$ the polynomials

$$\sigma_n(x;\omega|(a)) = \sum_{j=0}^{n} \binom{n}{j} a_{n-j}\sigma_j(x;\omega) \tag{4}$$

form a $\underset{\omega}{\Delta}$-harmonic sequence and, conversely, every $\underset{\omega}{\Delta}$-harmonic sequence is of this form; the Bernoulli polynomials correspond to $\omega = 0$, $a_n = B_n$. The generalized factorials have the generating function

$$(1+\omega t)^{x/\omega} = \sum_{n=0}^{\infty} \sigma_n(x;\omega) \frac{t^n}{n!}$$

(with $\exp(xt)$ for $\omega = 0$), and therefore the generating function of the polynomials (4) is $(1+\omega t)^{x/\omega}A(t)$, where

$$A(t) = \sum_{n=0}^{\infty} a_n \frac{t^n}{n!}.$$

More generally, Cotlar introduces with R. Lagrange the operators

$$\underset{\omega(b)}{\Delta} f(x) = \sum_{n=0}^{\infty} \frac{b_n}{n!} \underset{\omega}{\overset{n+1}{\Delta}} f(x),$$

where (b_n) is an arbitrary sequence and $\underset{\omega}{\overset{n}{\Delta}}$ is the n-fold iterate of $\underset{\omega}{\Delta}$. A sequence $(P_n(x))$ of polynomials is $\underset{\omega(b)}{\Delta}$-harmonic if $P_n(x)$ has degree n and $\underset{\omega(b)}{\Delta} P_n(x) = nP_{n-1}(x)$ for $n \geq 1$.

The (generalized) <u>Bernoulli polynomials</u> of R. Lagrange are defined by

$$L_n(x) = \sigma_n(x;\omega;(a_n)|(c_n)),$$

where (a_n) and (c_n) are arbitrary sequences,

$$\sigma_n(x;\omega;(a_n)|(c_n)) = \sum_{j=0}^{n} \binom{n}{j} c_{n-j}\sigma_j(x;\omega;(a_n)),$$

and the $\sigma_j(x;\omega;(a_n))$ have generating function $(1+\omega tA(t))^{x/\omega}$. If the sequence $(\bar{a}_n)$ is defined by

$$\sum_{n=0}^{\infty} \frac{\bar{a}_n}{n!}(tA(t))^{n+1} = t,$$

then the sequence $(L_n(x))$ is $\underset{\omega(\bar{a})}{\Delta}$-harmonic, and every $\underset{\omega(\bar{a})}{\Delta}$-harmonic sequence has this form. Two sequences of Bernoulli polynomials are <u>equivalent</u> if the corresponding sequences (c_n) are equal and the sequences (a_n) (i.e., the power series $A(t)$) differ by a factor $\delta \neq 0$.

After this preparatory material, Cotlar introduces the main topic of his investigations: the three sequences of polynomials $(\varphi_n(x))$, $(g_n(x))$ and $(M_n(x))$. A sequence of $(P_n(x))$ of polynomials is <u>permutable</u> if $P_n(k) = P_k(n)$ for every pair n, k of positive integers. There exists exactly one equivalence class of permutable sequences of Bernoulli polynomials. Its elements depend on a parameter δ, are denoted $(\varphi_n(x;\delta))$ and have the generating function $(1+\frac{t}{\delta})^x e^t$. Cotlar writes $\varphi_n(x) = \varphi_n(x;1)$.

The <u>conjugate</u> $\bar{P}_n(x)$ of a polynomial $P_n(x)$ is defined by

$$\bar{P}_n(x) = (-1)^n P_n(-x-1),$$

and the $g_n(x)$ are defined as the conjugates of the $\varphi_n(x)$, i.e.,

$$g_n(x) = (-1)^n \varphi_n(-x-1).$$

They have the generating function $(1-t)^{-x-1}e^{-t}$ and satisfy

$$(-1)^k g_{k-1}(-n) = (-1)^n g_{n-1}(-k)$$

for all integers $n, k \geq 0$. In particular $f_n^k = g_n(k)$ is the number of permutations of $n+k$ letters $a_1, \ldots, a_k, b_1, \ldots, b_n$ such that none of the letters b_j is left fixed. In [13] Cotlar establishes relations between the f_n^k and studies their generating functions.

Finally, the generating function of $(M_n(x))$ is the product of the generating functions of $(\varphi_n(x))$ and $(g_n(x))$, i.e.

$$\frac{1}{1-t}\left(\frac{1+t}{1-t}\right)^x.$$

The polynomials $M_n(x)$ are given explicitly by

$$M_n(x) = \sum_{j=0}^{n} \binom{n}{j} \frac{2^j}{j!} \sigma_j(x),$$

where $\sigma_n(x) = \sigma_n(x;1)$. They form the only permutable sequence in which each term is its own conjugate. They are connected with the Bernoulli numbers in the following way: if we introduce the differential operator

$$D_{(M)} = \sum_{n=2}^{\infty} \frac{2B_n(2^n-1)}{n!} D^{n-1},$$

then

$$D_{(M)}M_n(x) = M_{n-1}(x). \tag{5}$$

For the coefficients Cotlar adopts the notation

$$M_n(x) = \sum_{j=0}^{n} M_n^j x^{n-j},$$

and similarly for $\varphi_n(x)$ and $g_n(x)$. He proves an enormous amount of relations for the polynomials, of which we quote only a few:

$$M_n'(x) = 2\{M_{n-1}(x) - \tfrac{1}{3}M_{n-3}(x) + \tfrac{1}{5}M_{n-5}(x) -+ \dots\},$$

$$\varphi_n^k = \varphi_{n-1}^k - (n-2)\varphi_{n-1}^{k-1} + (n-1)\varphi_{n-2}^{k-2},$$

$$g_n^k = g_{n-1}^k + (n-1)(g_{n-1}^{k-1} + g_{n-2}^{k-2}),$$

$$M_n^k = \frac{1}{n}M_{n-1}^{k-1} + \frac{2}{n}M_{n-1}^{k} + \frac{n-1}{n}M_{n-2}^{k-2}.$$

If p is an odd prime number, the preceding formulas imply a number of congruences modulo p, as for instance

$$M_{p-1}^{2r} \equiv M_p^{2r+1} \equiv 0 \pmod p \quad \text{for} \quad 3 \le 2r+1 \le p-2,$$

$$M_{p-1}^{2k-1} \equiv 2M_{p-1}^{2k}, \; M_{p-3}^{2k-1} \equiv M_{p-1}^{2k-1} + 2M_{p-1}^{2k-3} \pmod p.$$

If a is an integer not divisible by p, let

$$\kappa(a) = \frac{a^{p-1}-1}{p}$$

be its Fermat quotient. Further, let

$$W_p = \frac{(p-1)!+1}{p}$$

be the Wilson quotient, and for a real number x, let $E(x) = [x]$ be its integral part. With these notations, Cotlar proves several general congruences for the M_p^k. The following is a special case of one of them (formula (48) of [14]): for integers r,s such that $0 \le r < p-1$, $1 \le s \le p-1$ one has

$$\sum_{a=1}^{p-1} a^r + 2\sum_{a=1}^{p-1} a^r E\left(\frac{sa}{p}\right) \equiv -2\sum_{a=1}^{p-1} a^{r+1}\kappa(a)$$

$$- 2s\kappa(s)\sum_{a=1}^{p-1} a^{r+1} - 2sW_p\sum_{a=1}^{p-1} a^{r+1} - \delta_r s^{-r} M_p^{r+1} \pmod p,$$

where $\delta_r = 0$ if $r = p-2$ or $p-1$, and $\delta_r = 0$ otherwise.

From (5) one obtains for $2 \le 2k \le p-1$ the congruence

$$\frac{2(2^{2k}-1)}{2k}B_{2k} \equiv M_{p-1}^{2k-1} \pmod p$$

between Bernoulli numbers and coefficients of $M_{p-1}(x)$. The general congruences referred to above give for special values of the parameters many classical congruences, as the one due to Euler:

$$-\frac{2(2^{2k}-1)}{2k}B_{2k} \equiv \sum_{a=1}^{p-1} (-1)^a a^{2k-1} \pmod p,$$

and the one to Voronoï {Rib, (5.2), p. 107}:

$$\frac{s^{2k}-1}{s^{2k-1}}\frac{B_{2k}}{2k} \equiv \sum_{a=1}^{p-1} E\left(\frac{sa}{p}\right)a^{2k-1} \pmod p$$

for $2 \le 2k < p-1$, $1 \le s \le p-1$. But Cotlar also proves new congruences, as for instance,

$$\frac{2(2^{2k}-1)}{s^{2k-1}}\frac{B_{2k}}{2k} \equiv -\sum_{a=1}^{p-1}(-1)^{r(as)}a^{2k-1} \pmod{p},$$

where $1 \le s \le p-1$ and $r(x) = x - pE(\frac{x}{p})$ is the remainder of x modulo p. He emphasizes that in his congruence the factor $E(\frac{sa}{p})$ of Voronoï's congruence is replaced by the simpler factor $(-1)^{r(as)}$.

The paper [14], written jointly with Beppo Levi, also has arithmetic connotations. In it the authors prove that for almost every real number x the sequence $(\cos 2\pi(n_k x+\alpha_k))$, or equivalently the sequence of fractional parts of $n_k x + \alpha_k$, is everywhere dense in $(0,1)$ provided that the sequence of integers (n_k) is strictly increasing. They also discuss conditions under which $|\cos 2\pi n_k x - \cos 2\pi n_\ell x|$ remains above a fixed positive number.

A function f defined in the right half-plane of $\mathbb{C}$ is <u>represented asymptotically</u> by the formal Laplace integral $\int_0^\infty e^{-zt}\varphi(t)dt$ if for each $a > 0$

$$f(z) - \int_0^a e^{-zt}\varphi(t)dt = o(e^{-az})$$

as $\operatorname{Re} z \longrightarrow \infty$. J.C. Vignaux studied this concept in two articles {Vi 1,2}, and in paper [11], written by Cotlar and Vignaux, the authors extend the investigations to Laplace-Stieltjes integrals. They generalize on the one hand results concerning convergent Laplace-Stieltjes integrals (e.g., inversion formulas), and on the other hand results about asymptotic power series (e.g., the existence of a function represented asymptotically by an arbitrarily given power series).

2. Real Variables, Measure Theory and Integration [5, 8, 15, 18, 19, 21, 24, 25, 26, 38]. Another field in which Cotlar acquired an expertise early in his career is the theory of functions of a real variable, in particular the theory of the integral.

Beppo Levi, with whom he corresponded about its content, wrote a prologue to Cotlar's first article [5] on the subject. Lebesgue's general measure problem asks whether it is possible to associate with every subset E of $\mathbb{R}^p$ a positive number $\mu(E)$ such that

1) if E_1 and E_2 are congruent sets, then $\mu(E_1) = \mu(E_2)$,

2) if (E_n) is a sequence of pairwise disjoint subsets of $\mathbb{R}^p$, then $\mu(\bigcup_n E_n) = \sum_n \mu(E_n)$,

3) if E is the p-dimensional cube with vertices $(0,\dots,0)$ and $(1,\dots,1)$, then $\mu(E) = 1$.

Banach, Kuratowski and others proved that the problem has no solution. The purpose of Cotlar's paper is to show that making some slight changes in the hypotheses the question does have a positive answer.

Denote by m the Lebesgue measure on $\mathbb{R}^p$, and by m_i and m_e the inner and the outer Lebesgue measure, respectively. Let us write $E \le F$ if $m(E \cap \complement F) = 0$. If both $E \le F$ and $F \le E$, then Cotlar identifies E and F, i.e., he considers sets only modulo sets of measure zero.

According to a classical definition {Sak, Chap. IV, §10, p. 128}, a point $x \in \mathbb{R}^p$ is a point of dispersion of a set $E \subset \mathbb{R}^p$ if

$$\lim_{m(Q)\to 0} \frac{m(E \cap Q)}{m(Q)} = 0,$$

where Q is a cube with center x. Lebesgue's density theorem {Sak, Chap. IV, (1.2), p. 129} states that if E is a measurable subset of $\mathbb{R}^p$ then almost every point of E is a point of dispersion of $\complement E$. Cotlar completes the theorem by proving that if E is an arbitrary subset of $\mathbb{R}^p$, then the points of E that are points of dispersion of $\complement E$ form a measurable subset of E. This motivates the following definition:

A non-empty subset N of $\mathbb{R}^p$ is a <u>nucleus</u> if (with the exception of a set of measure zero) no point of N is a point of dispersion of $\complement N$.

A necessary and sufficient condition for a set N to be a nucleus is that $m_i(N) = 0$ and $m_e(N) > 0$; as Cotlar points out at the end of his paper, this allows to extend most of his theory to abstract measure spaces.

The result quoted above proves that every non-measurable subset E of $\mathbb{R}^p$ can be decomposed in the form $E = E_0 \cup N$, where E_0 is measurable and N is a nucleus. Cotlar shows that this decomposition is unique (up to sets of measure zero).

If N is a nucleus, then $\complement N$ is not measurable, hence the union of a measurable set and of a nucleus $\bar{N}$ called the conjugate of N. The set $N \cup \bar{N}$ is obviously measurable and this enables Cotlar to define his <u>pseudo-measure</u> μ as follows:

- if E is measurable, then $\mu(E) = m(E)$,
- if E is non-measurable and has decomposition $E = E_0 \cup N$, then $\mu(E) = m(E_0) + \frac{1}{2}m(N \cup \bar{N})$.

Let E and F be two subsets of $\mathbb{R}^p$. Cotlar says that $E = E_0 \cup N_1$ and $F = F_0 \cup N_2$ are <u>non-rampant</u> if $m(E \cap F) = 0$, $m(N_1 \cap N_2) = 0$ and $m(\bar{N}_1 \cap \bar{N}_2) = 0$. The anlgue of postulate 2) in Lebesgue's problem is then the following theorem: If (E_n) is a sequence of pairwise non-rampant subsets of $\mathbb{R}^p$, then $\mu(\bigcup_n E_n) = \sum_n \mu(E_n)$.

Cotlar defines pseudo-measurable functions and their pseudo-integrals over arbitrary sets. One can extend to them the main properties of the Lebesgue integral which they generalize.

Cotlar's next four works on integration [15, 18, 24, 25] were written in collaboration with Yanny Frenkel, his wife. Their total length is over one hundred pages, and they are related mostly to the general concept of integral studied by A.N. Kolmogorov {Kol}.

Kolmogorov considers a collection $\mathcal{M}$ of sets, stable under intersections, and a set function F on $\mathcal{M}$ which is not assumed to be additive or even single-valued. He introduces two kinds of integrals: for the first one he considers countable partitions $E = \bigcup_n E_n$ of a set $E \in \mathcal{M}$ by pairwise disjoint sets $E_n \in \mathcal{M}$, for the second one only finite partitions. The integrals $\int_E^* F(dE)$ and $\int_E F(dE)$ are then defined as the Moore-Smith limits of the sums $\sum_n F(E_n)$ with respect to the directed sets of partitions of E.

In [18] Cotlar and Frenkel consider a non-empty collection $\mathcal{P}$ of subsets of a set S such that if $P \in \mathcal{P}$ and $P \subset Q$, then $Q \in \mathcal{P}$. They define cluster values of a possibly multi-valued real function f on S with respect to $\mathcal{P}$. The largest and the smallest cluster value are then $\overline{\lim}\, f(s)$ and $\underline{\lim}\, f(s)$, respectively, and if the two coincide, their common value is the limit of f with respect to $\mathcal{P}$. Next, they apply these concepts to the case when S is some set of sequences $s = (E_n)$ of disjoint subsets of a real interval I (or of an abstract space), where it is not required that $\bigcup E_n = I$. The integral $\int_I F(dE)$ is the limit of $f(s) = \sum_n F(E_n)$ with respect to a class $\mathcal{P}$ of subsets of S satisfying appropriate conditions; it generalizes those of Kolmogorov and Burkill. The upper and lower integrals are defined as the corresponding upper and lower limits, and the authors establish relations between them which generalize theorems concerning Riemann, Lebesgue and Stieltjes integrals.

Kolmogorov gave a sufficient condition in terms of the additivity and continuity of F for the existence of his integral. In [25] the authors introduce the concept of a <u>differentially</u> σ-<u>continuous</u> function F, and with its help they give a necessary and sufficient condition in order that if one of the upper and lower integrals exists, then the other exists also and the two are equal.

In [15] Cotlar and Frenkel take up not only Kolmogorov's ideas but also those of Rosalind Cecily Young {You 1,2}. They introduce two kinds of semi-order relations on the set of real-valued functions on a set, which are later in [20] extended to an abstract setting. For each of these semi-orders they show that there exists a largest additive minorant of a given set function of bounded variation, and prove generalizations of the Lebesgue and de la Vallée Poussin decompositions.

Article [24] deals with the integrals of Perron and Denjoy. The authors introduce generalized Perron and Denjoy majorants f and minorants g of a function p on an interval. f and g are continuous but no necessarily additive set functions, satisfying $\underline{D}f(x) \geq p(x) \geq \bar{D}f(x)$ on an appropriate set, whose Burkill integrals are continuous functions of an interval. In terms of f and g they generalize a theorem of Marcinkiewicz {Sak, (3.13), p. 253} about the Perron and the Denjoy integrability of p.

In two of the six papers written jointly with Rodolfo A. Ricabarra [26, 38], the authors introduce a concept of integral which includes that of Daniell-Stone-Bourbaki and also Carathéodory's extension of a measure, and which unifies the Lebesgue and the Riemann integral. They point out that their theory was conceived before Bourbaki's and Stone's treatments were published.

They start out with a subadditive and positively homogeneous function p defined on a sublattice L of the vector space $\mathcal{F}(T)$ of all real-valued functions on a set T. They introduce the following subspaces of L:

$A(L)$ = the maximal subspace of L on which p is additive,

$B(L)$ = the intersection of all maximal sublattices of $A(L)$,

$C(L)$ = the set of all $x \in A(L)$ such that x and $-x$ satisfy the condition

$$p(\sup(x,y)) + p(\inf(x,y)) = p(x) + p(y) \quad \text{for all} \quad y \in L.$$

They prove that either $C(L)$ is void or $C(L) = B(L) = A(L)$, and the latter happens if and only if $p(z^+) + p(z^-) = p(z)$ for $z \in L$. The elements of $C(L)$ are said to be <u>integrable in the sense of Carathéodory</u> with respect to p.

A <u>Riemann integral</u> μ is a positive linear form defined on a sublattice of $\mathcal{F}(T)$. An <u>outer Riemann-Carathéodory integral</u> is a subadditive, positively homogeneous, monotone, extended real-valued function m on $\mathcal{F}(T)$ which satisfies $m(x) = m(x^+) + m(x^-)$ whenever $m(|x|) < \infty$. A function x on T is <u>integrable</u> in the sense of Riemann-Carathéodory with respect to m, if it is integrable in the above sense, choosing $p = m$ and for L the lattice of those x for which $m(|x|) < \infty$. The authors introduce a concept of measurability to characterize integrable functions. They also define m-separated subsets of T and show that integrable functions are uniform limits of linear combinations of characteristic functions of such sets, and therefore the Riemann integral is the limit of Riemann sums.

A <u>Lebesgue integral</u> μ is a Riemann integral such that if (x_n) is an increasing sequence in $\mathcal{F}(T)$ such that $\sup \mu(x_n) < \infty$, and if $\lim_n x_n(t) = x(t)$, then $\lim_n \mu(x_n) = \mu(x)$. An <u>outer Lebesgue-Carathéodory integral</u> is an outer Riemann-Carathéodory integral m such that $m\left(\sum_1^\infty x_n\right) \le \sum_1^\infty m(x_n)$, whenever (x_n) is a sequence of positive elements of $\mathcal{F}(T)$. A second notion of measurability is defined, and integrable functions are characterized with its help.

The note [8] written with Beppo Levi considers the following result of W.H. Young {Ho, §136, p. 183}: If (G_n) is a sequence of subsets of a closed set of finite measure in $\mathbb{R}^p$, and if there exists a number $C > 0$ such that the inner measure of each G_n is $\geq C$, then there exists a set E with inner measure $\geq C$ such that each $x \in E$ belongs to infinitely many sets G_n. Cotlar and Levi point out that the same infinite collection of sets G_n cannot be chosen for all x.

Rolle's theorem is extended in [19] to a differentiable map $f : (x,y) \longmapsto (u,v)$ defined in the closure of a plane domain D. If $u(x,y)$ is constant on the boundary of D, then the Jacobian of f vanishes on an irreducible set whose closure is D.

The joint paper [21] with Emilio O. Roxin introduces the concept of total variation $V_f(I)$ of a not necessarily single-valued function f defined on an interval I. If $\omega_f(x)$ is the oscillation of f at x and for $y \in \mathbb{R}$ we denote by $N_f(y,c(I))$ the number of those $x \in I$ for which $\omega_f(x) = 0$ and $f(x) = y$, then

$$V_f(I) = \int_{-\infty}^{\infty} N_f(y,c(I))dy + \sum_x \omega_f(x).$$

This generalizes a formula due to S. Banach {He-St, (17.34), p. 270}.

3. <u>Ordered Groups and Vector Spaces</u> [1, 3, 12, 20, 33, 45]. Mischa Cotlar's very first publications [1, 3] were on the abstract theory of anagenes, to which he returned in [12]. Let there be given a family Γ of subsets of a set E, and a binary relation $x\rho C$ between elements $x \in E$ and $C \in \Gamma$. If

C and D belong to Γ and $x\rho D$ for every $x \in C$, then one writes $C\rho D$. Cotlar assumes that $x\rho C$ and $C\rho D$ imply $x\rho D$. An anagene is a set $C \in \Gamma$ such that x belongs to C if and only if x belongs to every D such that $D\rho C$. This abstraction includes many concepts as special cases, such as closed sets, measurable sets and closed ideals in partially ordered sets. In particular, it provides a method to complete partially ordered sets.

Cotlar developed an early interest in ordered algebraic structures, later reaffirmed while in contact with Antonio Monteiro. The joint paper with Eduardo H. Zarantonello [20] deals with partially ordered (or preordered) semigroups and vector semigroups. The authors consider structures with a number of properties, such as lattice-ordered, Dedekind-complete, regular and archmedian semigroups and establish relations between them. They introduce (ℓ)-ideals and present a decomposition theorem which is the abstract analogue of the Lebesgue decomposition into absolutely continuous and singular parts. The authors prove an isomorphism between lattice ordered vector groups with unit and the space of Borel measurable functions on a totally disconnected compact Haüsdorff space, modulo sets of first category. This representation theorem was later rediscovered by other mathematicians. In a final chapter, the authors present an abstract analogue of the Kolmogorov integral. They consider a complete lattice-ordered semigroup of set-functions f defined on a σ-field K. Such an f is said to be of bounded variation if there exists a totally additive set function F such that $f(E) \le F(E)$ for all $E \in K$. The smallest such F is then the Kolmogorov upper integral Wf of f.

In [33] Cotlar generalizes the extension theorem concerning linear forms, named after Hahn and Banach, to a semigroup setting. This was immediately taken up by G. Aumann {Au}, who writes: "The theorems of Cotlar . . . can be generalized in a natural way without changing their method of proof and thus they obtain a remarkably rounded form." These results were further generalized by H. Dinges {Di} and B. Fuchsteiner {Fu}.

In 1961 Cotlar presented a lecture at the Third Brazilian Mathematical Colloquium, in which he gave an expository account [45] of his two papers [20] and [33], adding some new results, in particular about a new class of partially ordered semigroups he names after von Neumann.

4. Harmonic Analysis [29, 31]. Let G be a commutative Hausdorff topological group with identity element e. A <u>character</u> of G is a continuous map χ from G into the group Γ of complex numbers with absolute value 1 such that $\chi(st) = \chi(s)\chi(t)$ and $\chi(s^{-1}) = \chi(s)$ for all $s, t \in G$. The group G is said to have sufficiently many characters if for any $s \in G$ such that $s \neq e$ there exists a character χ with $\chi(s) \neq \chi(e) = 1$. A theorem due to D.A. Raikov states that a locally compact group always has sufficiently many characters.

In their joint paper [29] Cotlar and Ricabarra investigate the existence of a character on G which separates e from an element $s \in G$ different from e. They call a subset B of G <u>big</u> if G can be covered by a finite number of translates $sB = \{sb : b \in B\}$ of B. Their result is the following: Given $s \neq e$ there exists a character χ such that $\chi(s) \neq 1$ if and only if there exists a big neighborhood V of e such that $s \notin V^6$. They also prove that in contrast there always exists a big neighborhood W of e such that $s \notin W^2$. In their proof the authors use the Krein-Milman theorem of which they had already given a detailed expository account in their paper [22], where they prove some of the results of Kryloff and Bogoliuboff using functional analysis, and in particular use the theory to show the existence of a measure invariant under a flow. They use also the existence due to S. Banach and J. von Neumann of a finitely additive, translation-invariant measure on G, and furthermore a generalization of a theorem of Godement they prove in their paper, concerning the spectral synthesis of a single function on G. For references to results similar to that of [29] consult the Notes to §23 in {He-Ro}.

Assume now that G is locally compact. Denote by $\hat{G}$ the set of all characters of G, which is a group under pointwise multiplication. If we equip $\hat{G}$ with the topology of uniform convergence on compact subsets of G, then $\hat{G}$ becomes a locally compact group: the <u>dual group</u> of G. The value $\chi(s)$ of $\chi \in \hat{G}$ at $s \in G$ will be denoted by $\langle s, \chi\rangle$ and is considered as a function on $G\times\hat{G}$.

The vector space $L^1(G)$ of integrable functions on G with respect to Haar measure μ becomes a Banach algebra if the convolution $(f*g)(s) = \int f(st^{-1})g(t)d\mu(t)$ is taken for product. The Fourier transform $\hat{f}$ of $f \in L^1(G)$ is the bounded continuous function on $\hat{G}$ defined by $\hat{f}(\chi) = \int f(s)\overline{\langle s, \chi\rangle}d\mu(s)$. Wiener's classical Tauberian theorem can be reformulated as follows: if $\hat{f}(\chi)$ is never zero, then the smallest closed subspace of $L^1(G)$ containing all translates $f_t : s \longmapsto f(st^{-1})$ of f (or, equivalently, the smallest closed ideal containing f) is all of $L^1(G)$.

Wiener's theorem motivated A. Beurling to pose in 1947 the problem of spectral synthesis. If I is a closed ideal in $L^1(G)$, denote by $S(I)$ the set of all $\chi \in \hat{G}$ such that $\hat{g}(\chi) = 0$ for all $g \in I$. Similarly, for $f \in L^1(G)$ let $S(f) = \{\chi \in \hat{G} : \hat{f}(\xi) = 0\}$. The question is whether $S(I) \subset S(f)$ implies $f \in I$.

We now know through the work of Laurent Schwartz and P. Malliavin that the answer to the question is in general negative. If, however, $S(I)$

consists of a single point, then the answer is positive; this was proved by Beurling for the real line and by Kaplansky for an arbitrary G. Henry Helson {He} gave a "direct and simple proof" of the theorem, and gave also the following generalization inspired by a 1939 paper of Ditkin: If the boundary of $S(I)$ intersects the boundary of $S(f)$ in a reducible set (i.e., a set which contains no non-empty perfect subset), then $S(I) \subset S(f)$ implies $f \in I$ (see also {Lo, 37C, p. 151}).

Cotlar shows in [31] that proofs of both theorems can be based on lemmas which for $G = \mathbb{R}$ figure already in Wiener's original proof of his theorem.

The Mathematical Contributions of Mischa Cotlar Since 1955

Cora Sadosky

Ever since 1944 Cotlar was interested in general methods of decomposition of operators, forms and other structures into "simple" components and, from this point of view, he first focused on the theory of the integral (since 1940), and then on the theory of ideals in vector lattices (see [20], [33]. [45]) and on problems in ergodic theory ([22], [23], [27], [28]). When starting his doctoral dissertation under Zygmund (1952), the latter suggested the treatment of some open problems in the theory of singular integrals, that he had developed with Calderón. Cotlar dealt with singular integrals also through the method of elementary decompositions. Since the classical spectral decomposition of forms assume properties of symmetry (commutation) or invariance with respect to certain transformations, and since the Hilbert transform commutes with translations and dilations, these properties led Cotlar to a decomposition of the Hilbert operator into sums that are similar to the spectral decomposition of a compact operator.

The numbers in square brackets refer to the works of Mischa Cotlar listed as his Mathematical Publications, the letters in curly brackets to the works of other authors listed at the end of this article.

This, in turn, led to a general lemma of "almost orthogonal" sums, nowadays known as Cotlar's Lemma, that gives the L^2 boundedness of the singular integrals, as well as a general theory of decomposition of operators (detailed in [41] and [52]) and a "local interpretation" leading to the method of the maximal operator (in particular for the singular integrals).

After developing, through this focus on decomposition theory, the study of singular integrals ([34]-[37], [41]) and of potential operators ([41]-[44]), he centered his interest in moment theory, the decomposition theory _par excellence_.

As a preparation he wrote an expository survey of the basic results of moment theory, from the first works by Sturm, Sylvester and Brockard to the then recent developments of M. G. Krein and Achiezer [30].

With a similar purpose Cotlar faced the first notions of group representations from the viewpoint of representation or decomposition of functionals ([46], especially the Appendix on von Neumann semigroups), and then he wrote, in 1965, another expository survey [51], this time on the basic results of Grothendieck's theory of nuclear spaces, on Gelfand's triples, and on the Krein-Berezanskii theory of kernels, that was finally published several years later, in 1968, due to the military intervention of the Argentinian universities in 1966.

In this last monograph [51], as in a previous summary on potential operators and orthogonal sums [42], Cotlar discussed a possible connection between Krein's ideas in moment theory and those on the Hilbert transform. The study of such connection became the main object of his last work to date, done in collaboration with Cora Sadosky.

THE FOUR SEMINAL PAPERS IN "Revista Matemática Cuyana"

After Cotlar obtained the Ph.D. degree in Chicago, he and Yanny returned to Argentina in 1953. Academic work was out of the question in Buenos Aires and La Plata, where university authorities required proof of political affiliation to the ruling party as precondition to employment. Meanwhile, a haven for mathematics was being created in Mendoza, the Instituto de Matemática of the Department of Scientific Research (DIC), inspired by the exiled Portuguese mathematician A. Monteiro and under the auspices of the Rector of the Universidad de Cuyo. Cotlar was named Director of the Instituto and was joined there by a distinguished group of young mathematicians, that included A. Monteiro, R. Ricabarra, Oscar Varsavsky, O. Villamayor, and E. Zarantonello.

One of their first preocupations was to create a vehicle for their research, and so appeared the Revista Matemática Cuyana. Only three issues of the Revista were ever published, since the following year the DIC was disbanded after the overthrow of the central government. The entire second issue consisted of four articles of Cotlar ([34], [35], [36] and [37]), that expand on the ideas of his doctoral dissertation and contain the bulk of his unifying theory of operators through decompositions.

In the first few lines of the first article it is stated that the purpose of the foursome is "to unify the theory of Hilbert transforms and ergodic theorems." This is done by introducing, in [34], new methods to work in L^2 without relying on the Fourier transform, in [35], new interpolation results to work in L^p, in [36], maximal singular integral operators to prove pointwise existence and, finally, by using in [37] all those results together to reach the desired unification.

Unified treatment of the Hilbert transform and ergodic theory

The Hilbert transform

$$H f(x) = \int_{-\infty}^{\infty} f(x-t)\, dt/t = \int_{-\infty}^{\infty} f(\sigma_t x)\, dt/t$$

has formal sense if f is defined in an abstract measure space Σ, in which a group of measure-preserving transformations $\sigma_t: \Sigma \to \Sigma$, i.e., an ergodic system, acts. Cotlar set out to extend to such ergodic Hilbert transforms the main properties of the classical ones, such as convergence of the principal value integral, pointwise and in the mean, and continuity of the operator H in L^p, $1 < p < \infty$. (See also [32].)

L^2 continuity: the method of "almost orthogonality"

The L^2 continuity of the classical Hilbert transform follows directly by using the Fourier transform isomorphism, which is not available in the ergodic setting. What is shared by both the classical and ergodic H are the basic properties of commuting with translations and (positive) dilations, which lead to the decomposition of H into an "almost orthogonal" sum $\sum T_n$, in such a way that $T_i T_j$ is very small when $|i-j|$ is large, more precisely,

$$\|T_i T_j\| \leq c\ 2^{-|i-j|} \qquad \text{and} \qquad \|T_n\| \leq C, \forall n. \tag{1}$$

Such sums are similar to the orthogonal sums in the spectral decomposition of a compact selfadjoint operator $A = \sum \lambda_n P_n$, where $P_i P_j = 0$ for $i \neq j$ and $\sup |\lambda_n| = \|A\|$.

This analogy suggested the following result on "almost orthogonal" operators, that was proved through a combinatorial inequality.

LEMMA: Given a collection (T_n) of selfadjoint operators T_n that commute with each other and satisfy condition (1), the series $\sum T_n$ is strongly convergent to a limit T such that $\|T\| \leq 5\ C$.

From this Lemma and from the preceding decomposition of the operator H follows the continuity of H in L^2, both for the Calderón-Zygmund singular integrals and for the Hilbert transforms in ergodic systems. The method, though, did not yield conditions for the continuity of H in other $L^2(\mu)$. That study was done many years later, in collaboration with C. Sadosky, on the basis of the relation of H with the shift operators. In spite of the fact that H commutes with translations $x \to x+a$, it does not commute with the shifts $S_t f(x) = e^{itx} f(x)$, and neither does the analytic projector P associated to H, $P = 1/2\ (-iI + H)$. That is, the range of P does not reduce S_t, but it has the weaker property that the range is invariant under S_t. This weaker property provides the framework for a theory of invariant forms in scattering systems, leading to weighted L^2-continuity results for H.

Although published in a quite unaccesible place, the Lemma, thereafter known as "Cotlar's Lemma," immediately called the attention of B. Sz.-Nagy and of J. Peetre. Sz.-Nagy praised the result and provided another proof of it (SzN). Peetre applied the Lemma in (Pee), comparing it with the approach to the Hilbert transform given by Hörmander in (Hor). It was subsequently extended for operators from L^p to $L^{p'}$ in joint work with Panzone [40], [43] (see also [42] and, especially [41] and [52]).

In 1968 E. M. Stein asked Cotlar, then at Rutgers, about the possible extension of the result to noncommuting operators, and in a few days each of them independently provided suitable modifications of the original proof. Stein used the noncommutative version of the Lemma in his celebrated joint paper with A.W. Knapp (Kn-St), which, widely referred, became the usual quotation source for the result. The extension to the noncommutative case proved crucial in the applications, in particular in the work of Calderon and Vaillancourt (Cal-Va 1), (Cal-Va 2), of C. Fefferman and Beals in the theory of pseudodifferential operators (Fe 2), and in Charles Fefferman's proof of the Carleson-Hunt theorem for Fourier series (Fe 1).

Yet it is in "Au delà des opérateurs pseudodifférentiels," by R. Coifman and Yves Meyer (Co-Me), where the results of the <u>Revista Cuyana</u>, as the Lemma, the maximal operator inequality for singular integrals and the concept of localization of operators, were given appropiate relevance. In fact, Coifman and Meyer developed the methods further, and later their students David and Journé used the Lemma in a crucial way in their celebrated "T 1" theorem (Da-Jou). Recently A. Carbery and A. Seeger have obtained an L^p substitute of the Lemma through the Littlewood- Paley theory (Car-See).

Interpolation results

"A general interpolation theorem for linear operations" [35] starts with several covering lemmas of Vitali type, including a sharpening of the so-called Wiener Covering Lemma, later used to establish in [36] the Hardy-Littlewood Maximal Theorem and its generalizations. The definition of "condition (p)" (a condition proved equivalent to what is now called "weak type") is given through the simultaneous satisfaction of two "half weak type" conditions. Cotlar proved and used the Marcinkiewicz interpolation theorem in the diagonal (without bounds for the norms of the operators) in his doctoral dissertation, before learning about the result through Professor Zygmund (Zygmund published the full version of the Marcinkiewicz theorem in (Zy), which appeared only in 1956, but he lectured about it in his Seminar, while Cotlar was his student in Chicago). In [35] there are a number of localized and "Kolmogorov type" versions of the Riesz and Marcinkiewicz interpolation theorems, and they are applied to obtain the L^p continuity of the Hilbert transform in one and two variables.

Once the L^2 result was established in [34], to obtain the L^p result through interpolation and duality, was indeed to follow the pattern of the Zygmund school. What is really striking is that in [37] the L^p continuity of the Hilbert transform is obtained without using the weak type (1,1) of the operator. It is done by **extrapolation**, through an argument that, unknwon to Cotlar, had already been used by M. Riesz and by Bochner. First it was checked that the operator H satisfies the identity $(Hf(x))^2 = (f(x))^2 + 2\ H(f.\ Hf)(x)$ for all simple functions f. This implies that H, being continuous in L^2, is continuous in L^4, which in turn entails its continuity in L^8, and so on. The continuity in L^p, for other p, follow again by interpolation ($p > 2$) and duality ($p < 2$). This approach was later used in more general situations by Gohberg and M. G. Krein (Go-Kr), Lindestrauss and Tzafriri (Li-Tz), and, again, to deal with the Hilbert transforms, by Rubio de Francia (RdF) and by Cotlar and Sadosky [62].

The theory of the maximal operator

Several variants of the maximal method of estimating an operator by estimating the corresponding maximal one, that have become since part of the general folklore, were introduced in [36]. The paper deals mainly with maximal operators that are to a given operator T what the Hardy-Littlewood maximal operator is to the identity operator I, and with other forms of "localization" of operators. It also gives maximal theorems in product spaces, pioneering much later work on the subject.

In particular, it gives what Coifman and Meyer call "Cotlar's inequality":

$$T_* f(x) \leqslant C_n((Tf)^*(x) + \|T\| f^*(x)) \qquad (2)$$

where T_* is the maximal operator for T and g^* is the Hardy-Littlewood maximal function of g. This inequality allowed Cotlar to prove the weak type (1,1) of the maximal operator for the singular integrals, and, through this result, their pointwise convergence theorems in L^p.

Inequality (2) has the peculiarity that T appears in both sides. It is to be noted that $T_* f$ cannot be bounded by f^*, since then the singular integrals would be bounded from L^∞ to L^∞. Coifman and Meyer (Co-Me, p. 102) compare this inequality with the Fefferman-Stein inequality involving the sharp function and the Hardy-Littlewood maximal function (Fe-St).

The general nature of the results of [34]. [35] and [36] allow their use in the ergodic setting mentioned at the beginning of this section. Thus in [37] the basic results on mean and pointwise convergence of singular integrals are extended to ergodic systems, and unified with the ergodic theorems of Birkhoff, von Neumann and Wiener.

THE THREE VOLUMES IN THE SERIES "Cursos y Seminarios"

In 1957, by the initiative and under the care of Cora Ratto de Sadosky, a Lecture Notes Series was started at the Facultad de Ciencias Exactas of the University of Buenos Aires.

It was <u>Cursos y Seminarios</u> and it became a major link of Buenos Aires with the outside mathematical world. This was due, in great part, to the fact that the first three volumes in the series contained important original results and were written by Laurent Schwartz, Mischa Cotlar and Alberto Calderón, in that order.

Cotlar published only a small number of papers during the golden period (1957-1966) of mathematics in Buenos Aires, and most of his contributions were reported in <u>Cursos y Seminarios</u>, of which he published three volumes, numbered 2 [41], 11 [46] and 15 [51].

Volume 2

"Condiciones de continuidad de operadores potenciales y de Hilbert" [41], product of a course and a seminar held in 1958-59, presents an introduction to the Calderón-Zygmund theory of singular integrals, through the viewpoint introduced by Cotlar in the four papers of the <u>Revista Matematica Cuyana</u>.

An operator $T: f \to Tf$, acting on functions, is called <u>local</u> if supp $Tf \subset$ supp f, so that

$$f_1 f_2 = 0 \quad \text{implies} \quad f_1(Tf_2) = (Tf_1)(Tf_2) = 0.$$

In the case of a local operator T, if $R^n = Q_1 \cup Q_2 \cup \ldots$ is a partition of the space in disjoint cubes, and $\varphi_j = 1_{Q_j}$, then

$$T = \sum T\varphi_j \quad \text{with} \quad (T\varphi_i)(T\varphi_j) = 0 \quad \text{for } i \neq j. \tag{3}$$

On the other hand, the spectral theorem says that every compact selfadjoint operator T, with $\|T\| \leq 1$, can be written as

$$T = \sum T_j \quad \text{with} \quad T_i T_j = 0 \quad \text{for } i \neq j, \quad \|T_j\| \leq 1, \tag{4}$$

where $T_j = \lambda_j P_j$, for P_j projector and $|\lambda_j| \leq 1$. This property, as well as the fact that the differential operators are characterized by (3), suggest to consider operators of the form $\sum T\varphi_j$ or $\sum T_j$, where the $T\varphi_j$, or the T_j, are orthogonal in a sense more general that the one in (3) or (4).

Cotlar considered three different generalizations of (3) and (4). The first consisted in replacing (4) by the "almost orthogonality" condition

$$\|T_i T_j\| \leq (1/2)^{|i-j|}, \quad \|T_j\| \leq 1 \quad \text{for all } i,j, \tag{4a}$$

that was already mentioned when refering to [34]. The second, to replace (3) by $(T\varphi_i)(T\varphi_j)$ = compact, and, more specifically, since $(T\varphi_i)(T\varphi_j) = [\varphi_i, T]\,\varphi_j$, by

$$[\varphi_j, T] \in \mathcal{V}_p = \{\text{compact operators in } L^p\}. \tag{3a}$$

Since condition (3) implies

$$Tf(x) = \lim_Q |Q|^{-1}\int_Q T(\varphi_Q f)(s)\,ds = \gamma(x) \sup_{Q\ni x} \int_Q T(\varphi_Q f)\,ds \tag{3b}$$

formula (3b) leads to a third type of generalization of the decomposition into orthogonal elements.

In many problems of analysis there appear operators of the kind $T: f \to Tf = \sum f * k_j$, where the kernels k_j satisfy

$$\|k_j\| \leq 1, \quad k_j(x) = 0 \text{ if } |x| > 2^j, \quad \int k_j = 0$$

and

$$\|k_j(x+h) - k_j(x)\| \leq (1/2)^j\,|h|. \tag{5}$$

Operators of type (5) admit representations $T = \sum T_j$ or $T = \sum T\varphi_j$ satisfying the generalized orthogonality conditions (4a), (3a) and (3b), and both the Hilbert transform and the Calderón-Zygmund singular integral operators are of type (5). Cotlar proved the following: (i) Operators satisfying (4a) are bounded,

$\|T\| \leq 5$ (Almost Orthogonality Lemma), and this provides a proof of the L^2 boundedness of the Calderón-Zygmund singular integrals.

(ii) Condition (3b) entails an inequality for the maximal operator T associated with the singular integral operator T, that, in turn, yield the weak type (1,1) of T and the pointwise convergence of $Tf = \sum f * k_j$.

(iii) Condition (3a) is key to establishing the Fredholm-type properties of the Calderón-Zygmund operators.

A detailed study of the operators of type (5) appears in [41], with emphasis on the orthogonality conditions (3b) and (4a) and their applications to singular integrals and potential operators. The volume also includes an exposition of the M. Riesz-Thorin and the Marcinkiewicz interpolation theorems, the Sobolev immersion theorems and various variants and applications ot these results.

Since [41] appeared several years earlier than the classic books by Stein (St) and Stein and Weiss (St-We), many young Argentine mathematicians were introduced to the Calderón-Zygmund theory through its reading.

In 1960 Bela Sz.-Nagy suggested to Cotlar the possibility of printing a version of [41] as a book in the Springer-Verlag series. Several circumstances prevented this project, and [41], --written in Spanish and rapidly out of print--, became one more of the unaccesible sources of Cotlar's work.

Volume 11

"Representación de grupos" [46] is an introduction to spectral theory, group representations and harmonic analysis in locally compact groups. It was originally written as an introduction to a course on group representation given in Buenos Aires by J. Dieudonné, and recorded in the same series of Cursos y Seminarios as volume 14.

In the first part, the spectral theory of bounded operators and the notions of primary representations and factors are presented through the method of extension of positive functionals in archimedean semigroups and ordered semirings. In this part, Cotlar followed the ideas of M. H. Stone and G. Mackey, as well as his own on extension of positive invariant functionals [33] and on ideals in ordered vector spaces, that appeared in joint work with Zarantonello [20]. In the last chapter of this first part, these ideas help formulate an abstract representation theory, anticipating some later works in mathematical logic.

The second part deals with the basic notions of symmetric and normed algebras, and the representation and decomposition of symmetric and C^*-algebras. The last chapter consists of an introduction to harmonic analysis in commutative groups and unitary representations of the compact and locally compact groups.

Volume 15

"Equipación en espacios de Hilbert" [51] corresponds to

three courses given by Cotlar between 1962 and 1964 at Buenos Aires and Darmouth College. It was written in 1964, before the publishing of the books by A. Pietsh (Pie), K. Maurin (Mau), and Iu. Berezanskii (Ber) on closely related subjects. The publication of the volume was interrupted in 1966, during the military intervention of the University of Buenos Aires, and it appeared only in 1968, after being rescued by A. González Villalobos.

The volume was originally intended as an introduction to the course by L. Schwartz already refered as volume 1 in the series <u>Cursos y Seminarios.</u>. It contains an introduction to the theory of nuclear spaces and summarizes the principal ideas and results of the classic works of L. Schwartz (Sch), A. Grothendieck (Gro) and Gelfand-Vilenkin (Ge-Vi). The results are explained in detail and with much motivation, unifying the treatments of those different authors, and relating them with the M. G. Krein-Berezanskii theory of decomposition of positive definite kernels.

Section 1 consists of a summary of the notions of reproducing kernels, Gelfand triples and the real and complex interpolation methods of Lions and Calderón. Sections 2 to 5 deal with the basic properties of nuclear spaces, Schwartz's theorem of the kernel and some abstract extensions of it, Grothendieck operators, nuclear (F) spaces, Kothe spaces, tensor products of topological vector spaces, L^1 and C , von Neumann-Schatten theory and unconditional summability. Section 6 deals with the relation with approximation theory and ε-entropy, and Section 7 with the eigenfunction expansions of Gelfand-Kostuchenko and Krein-Berezanskii, the concept of linear stochastic processes, measures in linear processes and invariant measures in nuclear spaces. In the last Section there is a list of open problems, that were dealt with in later work in collaboration with C. Sadosky.

McGill Lecture Notes

"Convolution operators and factorization" [52], published by the McGill Analysis Seminar in 1972, constitute an attempt to relate Cotlar's method of almost-orthogonal decompositions with the Krein-Gohberg factorization theory. It contains a Preface, seven chapters and an Epilogue.

In the first part, two aspects of the method of almost-orthogonal decompositions are given, together with applications to the theory of Fredholm operators.

The second part is a presentation of the factorization theory for functions f defined in the circle and matrix- (or operator-) valued, in terms of $f = f_+ \Delta f_-$, where f_+ is analytic in the interior of the disc, f_- is analytic in the exterior of the disc and Δ is a diagonal matrix. This theory is then related to the Grothendieck-Birkhoff theorem on decomposition of holomorphic bundles, and to Bott's periodicity theorem. In this part Cotlar followed ideas of Atiyah, Schubin, Gohberg, Bunghart and Orth.

The last part presents an abstract version of Wiener-Hopf

theory, a la Gohberg-Krupnik. Some of the questions raised in this notes were subsequently explored in joint work with R. Arocena and C. Sadosky.

THE TEXTBOOKS

Cotlar wrote, in collaboration with R. Cignoli, a textbook on functional analysis, which was published in Spanish as a two volume set by EUDEBA, the University of Buenos Aires Press [49].

The fist volume contains a detailed discussion on convex sets in vector spaces, on ordered vector spaces, and on extension of positive functionals with applications to Banach-Mazur limits and different moment problems. It deals with Banach and Hilbert spaces, C(K) spaces, basic notions of approximation theory by polynomials, positive and continuous functionals in C(K) and in $C_o(R)$, Lebesgue spaces L^p, and representation of continuous functionals. The second volume contains a treatment of the spectral theory for bounded and unbounded operators in Hilbert space, and the theory of Fredholm operators.

An English edition of the book appeared as [53] in North-Holland. It is an expanded version, containing chapters on the abstract theory of integration, from the viewpoint of [20], through ideals in ordered spaces and vector lattices, on Banach algebras, and on C^*-algebras representation. It also contains notions of interpolation theory for spaces and operators, notions of Fourier integrals, basic properties of

L^p-boundedness for the Hilbert transform (through the method of almost-orthogonal decomposition), and Gohberg's method for the Noether equation.

The other textbook, "Introducción al algebra" [47], written in collaboration with Cora Ratto de Sadosky as a result of their teaching the first year undergraduate course in linear algebra, shared with his first mathematical reader, the Russian book by Professor Grave, the quality of being both deep and unassuming. It treats the basic concepts of abstract and linear algebra with only high-school prerrequisites, and introduce the student to graduate-level techniques. It contains an introduction on number systems, and deals with algebraic structures (relations and operations, propositional algebra, groups, rings, fields and polynomials), and linear algebra (vector spaces, matrices and linear transformations, linear and p-linear forms, determinants, systems of linear equations, inner products and eigenvalues). The strong presentation points include the construction of the splitting field of a polynomial, and the introduction of determinants as p-linear alternating forms.

This algebra textbook has many editions, attesting to its pedagogical success. Unfortunately, Cotlar's excellent lecture notes on measure theory, real analysis and the theory of topological vector spaces and distributions were never published in book form, though they circulated among students for many years.

THE WORK SINCE 1973

In 1973 Cotlar and C. Sadosky started a program to study the relation between moment problems and the Hilbert transform, that led to the development of a lifting theory in so-called algebraic scattering systems. The first results of this program appeared in [54], and then lifting and integral representation theorems in [55], [56] and [57]. A more detailed exposition of the L^2 theory connecting moment problems and Hilbert transforms in the onedimensional case, as well as unifiying a number of applications to prediction theory and the H^1-BMO duality theory appeared in [60] in collaboration with R. Arocena. In more recent work ([72], [77], [82], [84]), Cotlar and Sadosky have given a concrete shape to the ideas from which the program stems.

Cotlar's Lemma showed that the Hilbert transform basic property on boundedness in L^2 is related to a more general property of operators admitting certain type of decomposition. Since the Lemma was not suited to characterize the positive measures μ such that H is bounded in $L^2(\mu)$, it was natural to think that such characterization could also be related to a general property of certain operators. Such measures μ being of a special class, that we shall call here R_2, we have the following moment problem: to characterize the Fourier transforms $\hat{\mu}$ of such μ, i.e., those sequences $s: Z \to C$ such that there exists a measure $\mu \in R_2$ with $\hat{\mu}(n) = s(n)$, for all n Z. A first answer, that nevertheless does not allow anything concrete to be said about the measure, is that the kernel $K: Z\times Z \to C$, defined by

$$K(m,n) = (M-1)\, s(m-n) \qquad \text{for sign } m = \text{sign } n$$

and

$$K(m,n) = (M+1)\, s(m-n) \qquad \text{for sign } m \neq \text{sign } n$$

for some $M \geq 1$, has to be positive definite. This kernel K is not translation invariant, or Toeplitz, but its restriction to each quadrant $Z_i \times Z_j$, i, j = 1,2, where $Z_1 = \{0, 1, 2, \ldots\}$, $Z_2 = \{-1, -2, \ldots\}$, coincides with a Toeplitz kernel $K_{ij}(m,n) = k_{ij}(m-n)$. Kernels of this type were called GTKs ("generalized Toeplitz kernels"), and were later studied by several other authors.

Since, by the Bochner theorem, every positive definite Toeplitz kernel is the Fourier transform of a positive measure, Cotlar and Sadosky set to extend this result to to the GTKs, and, more generally they started a systematic study of these kernels, their representations and their applications to Hilbert transform theory.

In [54] some basic properties of the classical moment theory were extended to the moment problem mentioned above, and the corresponding reduced moment problem was first considered. In [55] and [57] the analog of the theorem of Bochner was established, first for GTKs acting in R and then in Z, which in the discrete case reads: a GTK, $K \sim (K_{ij})$, is positive definite if and only if there exist four finite measures μ_{ij}

on the circle, such that for $i, j = 1, 2$, $K_{ij}(m,n) = \hat{\mu}_{ij}(m-n)$ for $(m,n) \in Z_i x Z_j$, and such that the matrix $(\mu_{ij}(E))$ is positive definite for every Borel set E.

From this Generalized Bochner Theorem (GBT) it follows that if a positive measure $\mu \in R_2$, then there exists another (complex) measure μ', such that $\hat{\mu}'(n) = \hat{\mu}(n)$ for all n negative, and $|\mu'(E)| \leq \mu(E)$ for all Borel sets E. This led to a variant and a refinement of the theorem of Helson and Szegö (He-Sze) in prediction theory, that characterizes the mesures in R_2. Moreover the GBT provided a solution to the open problem of characterizing pairs of measures μ and ν, such that H acts continuously from $L^2(\mu)$ to $L^2(\nu)$ [58].

Another proof of the GBT, based on the spectral theorem, as well as a unified viewpoint to obtain as applications from it the different pieces of the H^1-BMO duality theory in the circle and variants and refinements of prediction theory results of Helson, Sarason, Koosis, Ibragimov and Rozanov were given in [60], a joint paper with Arocena and Sadosky dealing with all aspects of the L^2 theory.

In joint papers of Cotlar with Arocena, that followed [60], other applications of the GBT were explored, and the study of the GTKs was further developed by both together ([61], [64], [65]. [66], [67] and [74], in collaboration with J. Leon), and in later work by Arocena, followed by papers of the students of Cotlar and Arocena at the Universidad Central de Venezuela: P. Alegría, R. Bruzual, M. Domínguez, S. Marcantognini and M. D. Morán.

In the joint paper with Arocena [65] a simplified proof of the GBT is given in the case of the circle, along with two unicity theorems, and its link with the classical theorems of Nehari ⟨Ne⟩ and Adamjan, Arov and Krein ⟨Ad-Ar-Kr⟩ is indicated. In [64] vector-valued versions of the GBT are given as well as dilations results, and the possible connection with the Sz.-Nagy -Foias theory is again considered. In [66] the GBT is approached from the viewpoint of extension of functionals, and a nonconstructive parametrization formula is sketched (a parametrization theorem generalizing that of Adamjan, Arov and Krein appeared later in ⟨Aro 1⟩). The joint paper [61] is devoted to the GBT in R, and it is shown there that the conditions of unicity for this theorem yield a simplified proof of Koosis theorem ⟨Koo⟩ for T and R. In [67] Arocena and Cotlar characterized the complex measures μ such that the matrix (m_{ij}), $m_{11} = m_{22} = \mu$, $m_{12} = m_{21} = |\mu|$ is weakly positive (i.e., gives rise to a positive definite GTK) and deduce as a corollary the theorems of Nehari and of Devinatz and Widom on the invertibility of the Toeplitz operators. This is a survey paper that includes a summary of the work done by Arocena, Cotlar and Sadosky up to 1981.

In [62], [63], [69], Cotlar and Sadosky extended the GBT and its applications, especially the theorems of Helson-Szego and Nehari, to L^p spaces, $1 < p < \infty$, obtaining new results for weighted inequalities for the Poisson transform, and establishing connections between the GBT and Grothendieck's theorem. In [70] and [73] they developed the methods used in the previous papers and dealt with generalizations of the Paley inequality for lacunary series, and in [71] they studied stationary and harmonizable stochastic processes having GTKs as covariance kernels.

Every GTK K is given by a quadruple of Toeplitz kernels (K_{ij}), and K is positive definite if $K_{12} = \overline{K_{21}}$ and

$$\left|\sum K_{12}(m,n)\, a_m b_n\right| \leq \left(\sum K_{11}(m,n)\, a_m \overline{a_n}\right)^{1/2} \left(\sum K_{22}(m,n)\, b_m \overline{b_n}\right)^{1/2}$$

for all pairs of finite sequences (a_n), (b_n) such that $a_n = 0$ for $n < 0$ and $b_n = 0$ for $n \geqslant 0$, which we shall write as

$$K_{12} \leq (K_{11},\ K_{22}) \quad \text{in} \quad Z_1 x Z_2.$$

The GBT implies, in particular, the following <u>lifting property</u>: if $K \sim (K_{ij})$ is a positive definite GTK there exists a Toeplitz kernel K' such that $K' = K_{12}$ in $Z_1 x Z_2$ and

$$K' \leq (K_{11},\ K_{22}) \quad \text{in all} \quad ZxZ.$$

In [72] Cotlar and Sadosky proved that this lifting property still holds if Z is replaced by an abstract set X in which acts a bijection $S: X \rightarrow X$, Z_1 and Z_2 are replaced by two subsets, X_1 and X_2, such that

$$S X_1 \subset X_1 \quad \text{and} \quad S^{-1} X_2 \subset X_2$$

and K (K_{ij}) is "Toeplitz with respect to S", i.e.:

$$K_{ij}(Sx, Sy) = K_{ij}(x, y) \quad \text{for} \quad i,\ j = 1,\ 2.$$

Moreover, this abstract lifting property also holds for operator-valued kernels.

In [67] the GTKs were presented in the context of a generalization of Sarason commutant theorem. Later on Arocena proved in (Aro 2) that the GBT contains the commutant theorems of Sarason and of Sz.-Nagy-Foias, and the inverse implication was proved by Cotlar and Sadosky in [72], thus establishing the logical equivalence between the two lifting theorems.

The relation of the GTKs with generalized Lax-Phillips scattering structures, with extensions to covariant systems and applications to Levy-Khinchine type results was considered in joint work with Arocena and Leon [74]. These subjects were later developed in detail by Arocena in (Aro 1) and, in a different direction, by Cotlar and Sadosky in [72], [78], [79], [81], [82] and [83].

In another direction and in joint work with J. León and Cristina Pereyra [80], Cotlar studied the connection of the GTKs with white noise and the Langevin equation. The theory of the GTKs is nowadays being developed not only by some researchers working directly with Cotlar, but by researchers in Israel, Japan, the German Democratic Republic and the Soviet Union as well (see (Dy), (Dy-Go), (Fr-Ki), (Kat), (Na), (Ya) and their references).

In [77] Cotlar and Sadosky extended the result of [72] to a more general lifting theorem for sesquilinear forms $B_{ij}: V \times V \to C$, where V is a vector space with a fixed linear isomorphism $S: V \to V$ (or a group S_t of isomorphisms), with two subspaces W_1 and W_2, satisfying

$$S W_1 \subset W_1 \qquad \text{and} \qquad S^{-1} W_2 \subset W_2.$$

Assuming that, for $i, j = 1, 2$, the B_{ij} are S-Toeplitz and such that $B_{12} \leq (B_{11}, B_{22})$ in $W_1 \times W_2$, the lifting theorem asserts the existence of another S-Toeplitz sesquilinear form B such that $B = B_{12}$ in $W_1 \times W_2$ and B (B_{11}, B_{22}) in all $V \times V$.

The systems $[V; W_1, W_2; S]$, described above, are called algebraic scattering systems (a.s.s.). In terms of liftings in a.s.s., it became clear that **also the boundedness property of the Hilbert transform in weighted L^2 spaces is related to a more general theory of decomposition of quadruples of Toeplitz forms that satisfy a boundedness condition in an a.s.s.** In other words, the lifting property and the GBT are a natural substitute of the method of almost-orthogonal operators to deal with weighted --instead of unweighted-- L^2-norm inequalities.

In their more recent joint work ([79], [82], [83], [84], [85]), Cotlar and Sadosky extended the lifting theorem to a.s.s. with two isomorphisms, or with two evolution groups, and gave two-dimensional versions of the GBT, as well as of the theorems of Helson-Szego and Nehari, and also quantized versions of those results in non-commutative settings like the algebra of Hilbert-Schmidt operators, the Heisenberg group or the Fock space. The theory of forms acting in a.s.s. is only in its beginnings and there are a number of problems in consideration, among them the deeper links with the Krein and the Grothendieck theories.

Howard University

Washington, DC 20059

References

{Ab} José de Abreu G.: Acotaciones radiales de funciones H^1 y refinamientos de teoremas de Montel y Ostrowski-Khintchine. Acta Científica Venezolana 39 (1988), 1-8.

{Ad-Ar-Kr} V.M. Adamjan, D.Z. Arov & M.G. Krein: Infinite Hankel matrices and generalized Carathéodory-Fejér and I. Schur problems. Funct. Anal. Appl. 2 (1968), 1-17 (Russian).

{Aro1} R. Arocena: Scattering functions, linear systems, Fourier transforms of measures and unitary dilations to Krein spaces: a unified approach. Publ. Math. d'Orsay, 85-02 (1985), 1-57.

{Aro2} R. Arocena: Toeplitz kernels and dilations of intertwining operators. Int. Eqs. & Op. Theory 6 (1983), 759-778.

{Au} Georg Aumann: Über die Erweiterung von additiven monotonen Funktionen auf regulär geordneted Halbgruppen. Arch. Math. 8 (1957), 422-427.

{Ber} I.U. Berezanskii: Expansions in eigenfunctions of selfadjoint operators. Transl. Math. Monographs, AMS, Providence, RI, 1968.

{Bi} Ludwig Bieberbach: Lehrbuch der Funktionentheorie, vol. II, 4th ed. Teubner, 1934.

{Bou} Nicolas Bourbaki: Eléments de mathématique, Livre IV, Fonctions d'une variable réelle (théorie élémentaire), Chaps. IV-VII. Hermann, 1951.

{Cal-Va1} A.P. Calderón & R. Vaillancourt: On the boundedness of pseudo-differential operators. J. Math. Soc. Japan 23 (1971), 374-378.

{Cal-Va2} A.P. Calderón & R. Vaillancourt: A class of bounded pseudo-differential operators. P.N.A.S., USA 69 (1972), 1185-1187.

{Cal-Zy} A.P. Calderón & Antoni Zygmund: On singular integrals. Acta Math. 88 (1952), 85-139.

{Car-See} A. Carbery & A. Seeger: Conditionally convergent series of linear operators on L^p spaces and L^p estimates for pseudo-differential operators. Proc. London Math. Soc. 57 (1989).

{Co-Me} R.R. Coifman & Y. Meyer: Au-delà des opérateurs pseudo-différentiels. Societé Mathématique de France, Astérisque 57, 1978.

{Da-Jou} G. David & J.-L. Journé: A boundedness criterion for generalized Calderón-Zygmund operators. Annals of Math. 120 (1984), 371-397.

{Di} Hermann Dinges: Decomposition in ordered semigroups. J. Functional Analysis 5 (1970), 436-483.

{Dy} Harry Dym: On a Szegö formula for a class of generalized Toeplitz kernels. Integral Eqs. and Operatory Thy. 8 (1985), 427-431.

{Dy-Go} H. Dym & I. Gohberg: A maximum entropy principle for contractive interpolants. J. Functional Anal. 65 (1986), 83-125.

{Fe1} C. Fefferman: Pointwise convergence of Fourier series. Ann. Math. 98 (1973), 551-572.

{Fe2} C. Fefferman: L^p bounds for pseudo-differential operators. Israel J. Math. 14 (1973), 413-417.

{Fe-St} C. Fefferman & E.M. Stein: H^p spaces in several variables. Acta Math. 129 (1972), 137-193.

{Fr-Ki} B. Fritzche & B. Kirstein: On generalized Nehari problems. Math. Nachr. 138 (1988), 217-237.

{Fu} Benno Fuchssteiner: Sandwich theorems and lattice semigroups. J. Functional Analysis 16 (1974), 1-14.

{Ge-Vi} I.M. Gelfand & N.J. Vilenkin: Generalized functions, vol. 4. Fizmatgiz, Moscow, 1961 (Russian). English transl.: Academic Press, New York, 1964.

{Go-Kr} I. Gohberg & M.G. Krein: The theory of Volterra operators in Hilbert spaces and applications. Nauka Ed., Moscow, 1967 (Russian). English transl.: Transl. Math. Monographs, AMS, vol. 24, Providence, RI, 1970.

{Gro} A. Grothendieck: Résumé de la théorie métrique des produits tensoriels topologiques. Bol. Soc. Mat. Sao Paulo 8 (1956), 1-79.

{He} Henry Helson: Spectral synthesis of bounded functions. Ark. Mat. 1 (1952), 497-502.

{He-Sze} H. Helson & G. Szegö: A problem in prediction theory. Ann. Mat. Pura Appl. 51 (1960), 107-138.

{He-Ro} Edwin Hewitt & Kenneth. A. Ross: Abstract harmonic analysis, vol. I. Grundlehren der Mathematischen Wissenschaften, vol. 115, Springer, 1963.

{He-St} Edwin Hewitt & Karl Stromberg: Real and abstract analysis. Springer, 1965.

{Hi} Einar Hille: Analytic function theory, vol. II. Ginn & Co. 1962.

{Ho} E.W. Hobson: The theory of functions of a real variable and the theory of Fourier's series, vol. I, 3rd ed. Cambridge University Press, 1927; Dover, 1957.

{Hör} Lars Hörmander: Estimates for translation invariant operators on L^p spaces. Acta Math. 104 (1960), 93-139.

{Kat} V.E. Katznelson: Integral representation of hermitian positive kernels of mixed type and generalized Nehari problem. Funk. Anal. i Priloz. 43 (1985), 54-70 (Russian).

{Kn-St} A.W. Knapp & E.M. Stein: Intertwining Operators on Semi-Simple Lie Groups. Annals of Math. 93 (1971), 489-578.

{Kol} Andrei N. Kolmogorov: Untersuchungen über den Integralbegriff. Math. Annalen 103 (1930), 654-696.

{Koo} P. Koosis: Moyennes quadratiques pondérées de fonctions périodiques et leur conjuguées harmoniques. C.R. Acad. Sci. Paris A 291 (1980), 255-257.

{Lag} René Lagrange: Mémoire sur les suites de polynômes. Acta Math. 51 (1928), 201-309.

{Lan} Edmund Landau: Darstellung and Begründung einiger neuerer Ergebnisse der Funktionentheorie, 2nd ed. Springer, 1929.

{La-Ph} P. Lax & R.S. Phillips: Scattering Theory. Academic Press, New York, 1967.

{Li-Tz} J. Lindenstrauss & L. Tzafriri: Classical Banach Spaces, II. Ergeb. Math. vol. 92, Springer-Verlag, Berlin-Heidelberg-New York, 1977.

{Li} J.E. Littlewood: Lectures on the theory of functions. Oxford University Press, 1944.

{Lo} Lynn H. Loomis: An introduction to Abstract Harmonic Analysis. D. Van Nostrand, Toronto-New York-London, 1953.

{Ma} Szolem Mandelbrojt: Séries de Fourier et classes quasi-analytiques de fonctions. Gauthier-Villars, 1935.

{Mau} K. Maurin: Generalized eigenfunction expansions and unitary representations of topological groups. Monogr. Matem. 48, PWN, Warsaw, 1968.

{Nak} T. Nakazi: Bounded Hankel forms with weighted norms. Pacific J. Math. (in press).

{Ne} Z. Nehari: On bilinear forms. Ann. Math. 68 (1957), 153-162.

{Nev} Rolf Nevanlinna: Eindeutige analytische Funktionen, 2nd. ed. Grundlehren der Mathematischen Wissenschaften, vol. 46, Springer, 1953.

{Pe1} Jaak Peetre: Une charactérisation abstraite des opérateurs différentiels. Math. Scand. 7 (1959), 211-218.

{Pe2} Jaak Peetre: Espaces d'interpolation et théorème de Soboleff. Ann. Inst. Fourier (Grenoble), 16, fasc. 1 (1966), 279-317.

{Pie} A. Pietsch: Nuclear locally convex spaces. Ergeb. Math., vol. 66, Springer-Verlag, Berlin-Heidelberg-New York, 1972.

{Rib} Paulo Ribenboim: 13 lectures on Fermat's last theorem. Springer, 1979.

{Rie} Bernhard Riemann: Grundlagen für eine allgemeine Theorie der Functionen einer veränderlichen complexen Grösse. Collected Works, 3-45.

{RdF} J.L. Rubio de Francia: Linear operators in Banach space lattices and weighted L^2-inequalities. Math. Nachr. 133 (1987), 197-209.

{Sak} Stanislaw Saks: Theory of the integral. Monografie Matematyczne, Tome 7, 1937.

{Sch} Laurent Schwartz: Produits tensoriels topologiques d'espaces vectoriels topologiques. Espaces vectoriels topologiques nucléaires. Applications. Séminaire Schwartz, Paris, 1954.

{St} E.M. Stein: Singular integrals and differentiability properties of functions. Princeton Univ. Press, Princeton, 1970.

{St-We} E.M. Stein & G. Weiss: Introduction to Fourier analysis on Euclidean spaces. Princeton Univ. Press, Princeton, 1971.

{SzN} Béla Sz.-Nagy: Note on sums of almost orthogonal operators. Acta Sci. Math. Szeged 18 (1957), 189-191.

{SzN-Fo} B. Sz.-Nagy & C. Foias: Analyse Harmonique des opérateurs de l'espace de Hilbert. Masson-Akad. Kiadó, Paris-Budapest, 1967.

{Va} Georges Valiron: Familles normales et quasi-normales de fonctions méromorphes. Mémorial des Sciences Mathématiques, fasc. 38. Gauthier-Villars, 1929.

{Vi1} J.C. Vignaux: On the asymptotic representation of functions by integrals. An. Soc. Ci. Argentina 138 (1944), 27-39, 97-119, 249-260.

{Vi2} J.C. Vignaux: Asymptotic series and integrals. Univ. Nac. La Plata. Publ. Fac. Ci. Fisicomat. No. 180, Vol. 3, núm 4. serie segunda, 14, Contribuciones, 401-412 (1944).

{Ya} T. Yamamoto: On the generalization of the theorem of Helson and Szegö. Hokkaido Math. J. 14 (1985), 1-11.

{You1} Rosalind Cecily Young: Functions of Σ defined by addition or functions of intervals in n-dimensional formulation. Math. Zeitschrift 29 (1928-29), 171-216.

{You2} Rosalind Cecily Young: On Riemann integration with respect to a continuous increment. Math. Zeitschr. 29 (1928-29), 217-233.

{Zy} A. Zygmund: On a theorem of Marcinkiewicz concerning interpolation of operations. J. Math. Pures Appl. 35 (1956), 223-248.